GENOMES

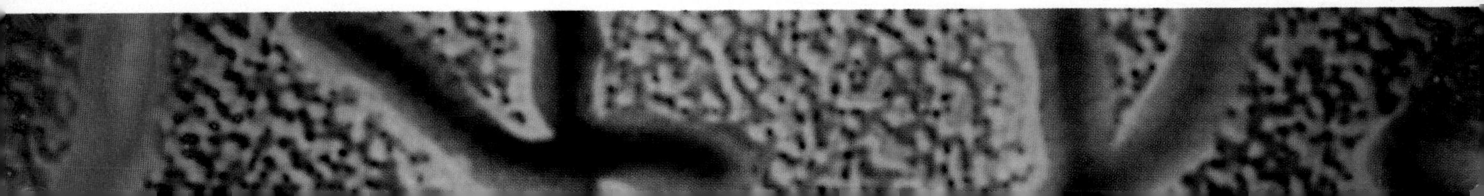

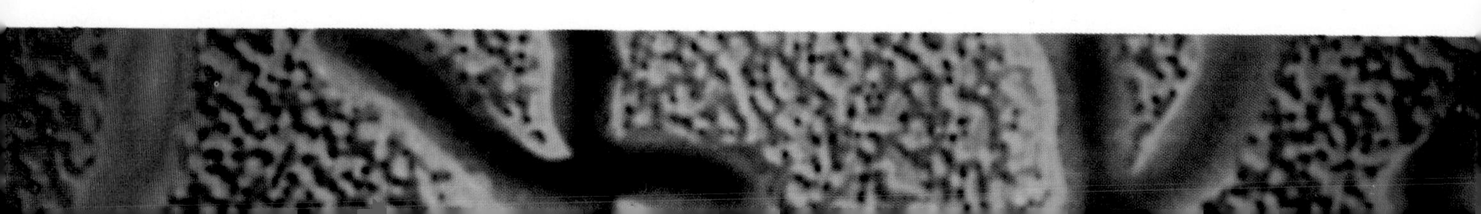

GENOMES

T.A. BROWN

Department of Biomolecular Sciences, UMIST, Manchester, M60 1QD, UK

BIOS SCIENTIFIC PUBLISHERS

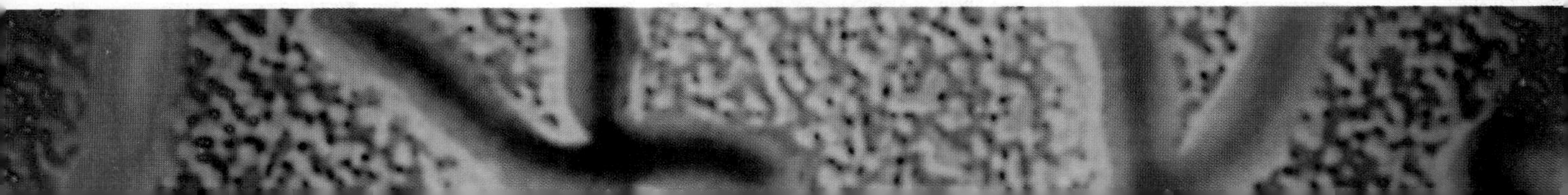

First published 1999

A CIP catalogue record for this book is available from the British Library.

ISBN 1 85996 201 7

BIOS Scientific Publishers Ltd
9 Newtec Place, Magdalen Road, Oxford OX4 1RE, UK
Tel. +44 (0) 1865 726286. Fax +44 (0) 1865 246823
World Wide Web home page: http://www.bios.co.uk/

Published in the United States of America, its dependent territories and Canada by John Wiley & Sons, Inc., by arrangement with BIOS Scientific Publishers Ltd.

Copies may be obtained from:

USA
John Wiley & Sons Inc.,
605 Third Avenue, New York,
NY 10158–0012, USA

Canada
John Wiley & Sons (Canada) Ltd,
22 Worcester Road, Rexdale,
Ontario M9W 1L1, Canada

Production Editor: Fran Kingston
Typeset by J&L Composition Ltd, Filey, North Yorkshire, UK
Illustrations drawn by J&L Composition Ltd, Filey, North Yorkshire, UK
Printed by The Bath Press Ltd, Bath, UK

Chapters and Concepts

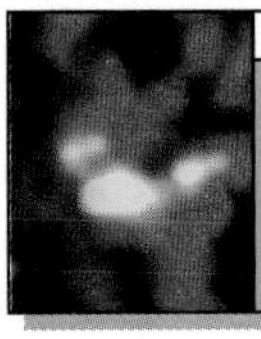

PART 1

How Genomes are Studied

Chapter 1: What is a Genome? 1

- *The genome is the entire DNA content of a cell, including all of the genes and all of the intergenic regions*
- *The human genome contains approximately 80 000 genes but the coding regions of these genes take up only 3% of the genome*
- *The yeast genome contains 6000 genes and has a more compact organization*
- *The genomes of some plants are dominated by repetitive DNA sequences*
- *Prokaryotic genomes are small with very little space between genes*
- *Understanding the information contained in genome sequences will be the major challenge of the early 21st century*

Chapter 2: Mapping Genomes by Genetic Techniques 13

- *Genetic mapping uses genetic techniques such as crossbreeding and pedigree analysis to construct genome maps*
- *Genes were originally used as markers but these are now supplemented with DNA markers such as RFLPs, SSLPs and SNPs*
- *The basis of genetic mapping is the calculation of recombination frequency by linkage analysis*
- *If planned breeding experiments are possible then two-point or multipoint test crosses are used*
- *Genetic mapping with humans is complicated by the limitations of pedigree analysis*
- *Genetic analysis of bacteria requires gene transfer from one cell to another*

Chapter 3: Mapping Genomes by Physical Techniques 37

- *Physical mapping uses molecular biology techniques such as hybridization analysis and PCR to examine DNA molecules for the positions of distinctive sequence features*
- *Restriction mapping is a useful method for physical mapping of small genomes*
- *FISH is a technique that enables the positions of DNA markers in a chromosome to be visualized by fluorescent labeling*
- *STS mapping with radiation hybrid panels and clone libraries is the most powerful type of physical mapping*
- *Considerable progress in mapping the human genome was made in the mid-1990s, culminating in a combined genetic and physical map with a marker density approaching one per 100 kb*

Chapter 4: Sequencing Genomes 59

- *Two methods exist for rapid DNA sequencing but the chain termination procedure is the method of choice for genome sequencing*
- *A single sequencing experiment produces only a few hundred bp of sequence*
- *A contiguous DNA sequence for a small genome can be assembled by random shotgun sequencing*
- *For large genomes, overlapping clone contigs are built up with fragments cloned in high capacity vectors such as BACs*
- *A shotgun sequencing method that uses map data to aid sequence assembly may provide a quicker method for sequencing large genomes*
- *Rapid sequence acquisition from large genomes is possible by EST sequencing and sequence skimming*

Chapter 5: Understanding a Genome Sequence 85

- *Understanding the information contained in genome sequences will be the major challenge of the early 21st century*
- *Genes can be located in DNA sequences by searching for ORFs, but the procedure is less successful with eukaryotic genomes*
- *Genes can be located experimentally by searching for expressed sequences*
- *Homology analysis is a powerful means of assessing the function of unknown genes*
- *A function can sometimes be assigned to a gene by determining the effect that its inactivation has on the phenotype of the organism*
- *Information on the genetic content of a large genome can be obtained by studying a smaller, related genome*
- *Studies of the transcriptome and proteome hold the key to understanding genome activity*

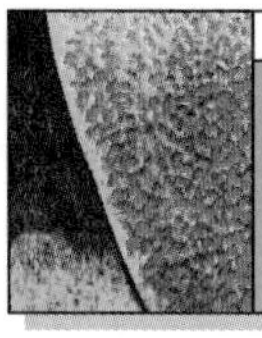

PART 2

How Genomes Function

Chapter 6: Genome Anatomies 115

- *Eukaryotic nuclear genomes vary in length from less than 10 Mb to over 100 000 Mb*
- *DNA is packaged in the nucleus by association with histone proteins in chromatin*
- *Many nuclear genes are members of multigene families*
- *The mitochondria and chloroplasts of eukaryotes have their own genomes, but they do not contain many genes*
- *Most, but not all, bacterial genomes are circular DNA molecules*
- *Gene families called operons are common in bacterial genomes*
- *Eukaryotic genomes contain clustered and dispersed repetitive DNA sequences*
- *Many dispersed repeats, in eukaryotes and bacteria, are active or ancient transposable elements*

Chapter 7: The Role of DNA-binding Proteins 143

- *DNA and proteins are linear, unbranched polymers*
- *The DNA double helix can adopt a variety of conformations*

- *The functional, three-dimensional structure of a protein is determined by its amino acid sequence*
- *Various techniques can be used to locate protein binding sites on DNA molecules*
- *X-ray crystallography and NMR are powerful methods for determining protein structures*
- *Binding proteins are thought to locate their attachment sites mainly by direct sequence readout of the double helix*
- *DNA-binding proteins have special structural motifs that enable them to form a close attachment with the double helix*

Chapter 8: Initiation of Transcription: the First Step in Gene Expression 171

- *Initiation of transcription is the primary control point for gene regulation*
- *Chromatin structure and nucleosome positioning influence patterns of gene expression*
- *DNA-dependent RNA polymerases synthesize RNA on a DNA template*
- *The transcription initiation complex is assembled at the promoter sequence near the start of a gene*
- *Eukaryotic initiation requires general transcription factors and other accessory proteins*
- *In bacteria, initiation rates are controlled by promoter structure and by repressor and activator proteins*
- *In eukaryotes, initiation is controlled by transcription factors that activate or, less frequently, repress the initiation complex*

Chapter 9: Synthesis and Processing of RNA 195

- *As well as mRNAs that code for protein, cells also contain various noncoding RNAs*
- *Most RNAs are made as precursors that must be processed to produce the mature molecules*
- *RNA polymerases transcribe genes until a termination signal is reached*
- *Eukaryotic mRNAs are processed by capping, polyadenylation and intron splicing*
- *All pre-rRNAs and pre-tRNAs are processed by cutting events and by chemical modification*
- *Some eukaryotic mRNAs are chemically modified in such a way that their coding properties are changed*
- *mRNA degradation is a specific process that removes coding RNAs whose protein products are no longer required*

Chapter 10: Synthesis and Processing of the Proteome 231

- *tRNAs form the link between the mRNA and the polypeptide being synthesized*
- *The genetic code specifies the amino acid sequence of the polypeptide*
- *Aminoacyl-tRNA synthetases attach amino acids to tRNAs which then base-pair to the appropriate codons*
- *Ribosomes coordinate and catalyze protein synthesis*
- *Bacteria and eukaryotes use different mechanisms for initiation of translation, but later stages of the process are similar in all organisms*
- *In order to become functional, all proteins must be folded, some must be cut into segments and chemically modified, and a few must have their inteins removed*
- *Protein turnover results in degradation of proteins whose functions are no longer needed*

Chapter 11: Regulation of Genome Activity 263

- *Transient changes in genome activity enable the cell to respond to external stimuli*
- *Transient changes occur in response to external signaling compounds that enter the cell or act via cell surface receptors*
- *Permanent changes in genome activity underlie differentiation*
- *Permanent changes in genome activity can be brought about by DNA rearrangements, changes in chromatin structure, and feedback loops*
- *During developmental processes, coordination between genomes in different cells can be achieved by intercellular signaling*
- *Gradients of signaling compounds can convey positional information to cells*
- *Genes that control the body plan of the fruit fly are similar to genes that carry out equivalent functions in vertebrates including humans*

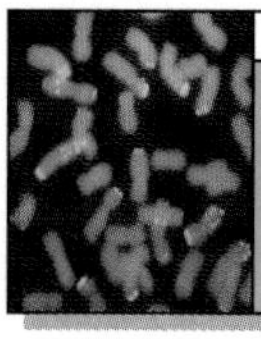

PART 3 How Genomes Replicate and Evolve

Chapter 12: Genome Replication 299

- *Genome replication occurs every time the cell divides*
- *The double helix structure indicated a possible mode of DNA replication, which was subsequently confirmed by the Meselson–Stahl experiment*
- *The topological problem is solved by DNA topoisomerases which relieve torsional stress ahead of the replication fork*
- *Origins of replication are well defined in bacteria and yeast but less so in eukaryotes*
- *During replication, the leading strand is copied continuously but the lagging strand is copied discontinuously*
- *Strand synthesis requires a primer*
- *An array of enzymes and proteins are involved in the events at the replication fork*
- *Termination of replication is poorly understood*
- *The ends of linear DNA molecules will shorten unless special measures are taken to repair them: in eukaryotic chromosomes this problem is solved by telomere synthesis*
- *Cell cycle checkpoints ensure that DNA replication is coordinated with cell division*

Chapter 13: The Molecular Basis of Genome Evolution 329

- *Mutation and recombination are the key processes responsible for genome evolution*
- *A mutation is a change in the nucleotide sequence of the genome, caused by a replication error or by a mutagen*
- *Replication errors are infrequent but can lead to point mutations; replication slippage can occur when short repetitive sequences are copied*
- *A variety of chemical and physical mutagens cause damage to DNA molecules*
- *Mutations have various effects on genome function*
- *Some cells can modify their repair processes to induce hypermutation and bacteria may be able to carry out programmed mutations*

- *Most mutations can be corrected by DNA repair processes which include excision procedures for removing damaged nucleotides and mismatch repair systems for correcting replication errors*
- *Recombination results in restructuring of the genome*
- *Models for recombination involve the formation of heteroduplexes containing one or more Holliday structures*
- *Site-specific recombination is responsible for insertion of the λ genome into E. coli DNA*
- *Transposition of DNA transposons and retroelements involves recombination events*

Chapter 14: Patterns of Genome Evolution 367

- *The first biochemical systems were based on RNA*
- *The first DNA genomes appeared after RNA molecules had evolved protein-coding properties*
- *Genomes acquire new genes by genome and gene duplications*
- *After duplication, a new gene copy might degrade into a pseudogene or might evolve a new function*
- *The sequences of some gene copies stay the same because of concerted evolution*
- *Genome evolution also involves the rearrangement of domains in existing genes and, in some cases, domain shuffling*
- *Some organisms have probably acquired genes from other species*
- *Transposable elements have had an important influence on genome evolution, by acting as sites for homologous recombination and by moving to new locations in the genome*
- *Introns may have evolved early or they may have evolved late*
- *The human genome is 97% identical to that of the chimpanzee*

Chapter 15: Molecular Phylogenetics 391

- *Molecular phylogenetics enables the evolutionary relationships between DNA sequences to be deduced*
- *Gene trees are not the same as species trees*
- *The reconstruction of phylogenetic trees involves alignment of homologous DNA sequences followed by any one of several approaches to tree building*
- *Molecular clocks can be used to estimate the time of divergence of ancestral sequences, but there is no universal molecular clock so calibration is important*
- *Molecular phylogenetics has been applied to questions such as the relationships between humans and other primates, the time of the earliest life on Earth, the origins of AIDS and the relationships between prions of different species*
- *Molecular phylogenetics and related analyses are being used to study the origin of modern humans and the patterns of past human migrations*

Contents

Abbreviations **xxi**

Preface **xxiii**

Acknowledgements **xxv**

An Introduction to GENOMES **xxvii**

PART 1

How Genomes are Studied **xxx**

Chapter 1 What is a Genome? 1

1.1 The Human Genome *3*
1.1.1 The physical structure of the human genome 3
1.1.2 The genetic content of the human genome 6
Box 1.1 Some key facts about genes 7
1.2 Genomes of Other Organisms *7*
1.2.1 Genomes of eukaryotes 8
1.2.2 Genomes of prokaryotes 9
1.3 Why are Genome Projects Important? *11*

Chapter 2 Mapping Genomes by Genetic Techniques 13

2.1 Genetic and Physical Maps *15*
2.2 Markers for Genetic Maps *15*
2.2.1 Genes were the first markers to be used 15
Box 2.1 A quick guide to Mendelian genetics 17
2.2.2 DNA markers for genetic mapping 18
Restriction fragment length polymorphisms (RFLPs) 18
Simple sequence length polymorphisms (SSLPs) 18
Single nucleotide polymorphisms (SNPs) 21
2.3 Approaches to Genetic Mapping *22*
2.3.1 Linkage analysis is the basis of genetic mapping 22
The behavior of chromosomes during meiosis explains why genes display partial and not complete linkage 22
From partial linkage to genetic mapping 25
2.3.2 Carrying out linkage analysis with different types of organism 26
Linkage analysis when planned breeding experiments are possible 27
Gene mapping by human pedigree analysis 29
Box 2.2 Multipoint crosses 30
Genetic mapping in bacteria 32

Chapter 3 Mapping Genomes by Physical Techniques 37

3.1 Restriction Mapping *39*
3.1.1 The basic methodology for restriction mapping 39
3.1.2 The scale of restriction mapping is limited by the sizes of the restriction fragments 41
Restriction mapping with rare cutters 41
Monitoring the results of restriction with a rare cutter 44

3.2 *FISH – Fluorescent* In Situ *Hybridization* 45
3.2.1 *In situ* hybridization with radioactive or fluorescent probes 45
3.2.2 FISH in action 47
Low-resolution FISH with metaphase chromosomes 47
High-resolution FISH with extended chromosomes and DNA fibers 47
3.3 *Sequence Tagged Site (STS) Mapping* 48
3.3.1 Any unique DNA sequence can be used as an STS 49
3.3.2 Fragments of DNA for STS mapping 49
Radiation hybrids 50
A clone library can also be used as the mapping reagent for STS analysis 51
3.4 *Progress Towards the Human Genome Map* 54

Chapter 4 Sequencing Genomes **59**

4.1 *The Methodology for DNA Sequencing* 60
4.1.1 Chain termination DNA sequencing 60
Chain termination sequencing in outline 61
Box 4.1 The chemical degradation sequencing method 61
Chain termination sequencing requires a single-stranded DNA template 64
Box 4.2 DNA polymerases for chain termination sequencing 66
The primer determines the region of the template DNA that will be sequenced 67
Thermal cycle sequencing offers an alternative to the traditional methodology 67
Fluorescent primers are the basis of automated sequence reading 68
4.1.2 Radical departures from conventional DNA sequencing 69
4.2 *Assembly of a Contiguous DNA Sequence* 70
4.2.1 Sequence assembly by the shotgun approach 70
The potential of the shotgun approach was proven by the *Haemophilus influenzae* sequence 70
The strengths and limitations of the shotgun approach 73
4.2.2 Sequence assembly by the clone contig approach 73
Cloning vectors for large pieces of DNA 73
Clone contigs can be built up by chromosome walking, but the method is laborious 75
Newer, more rapid methods for clone contig assembly 78
4.2.3 The directed shotgun approach 79
4.3 *The Sequencing Phase of the Human Genome Project* 80
4.3.1 The Human Genome Project is now largely based on BAC libraries 81
4.3.2 Rapid sequence acquisition can be directed at interesting regions of the genome 81
EST sequencing is a rapid way of accessing genes 81
Sequence skimming is equally rapid 81
4.3.3 The sequencing phase of the Human Genome Project raises ethical questions 82

Chapter 5 Understanding a Genome Sequence **85**

5.1 *Locating Genes in DNA Sequences* 86
5.1.1 Gene location by sequence inspection 86
Genes are open reading frames 86
Simple ORF scans are less effective with higher eukaryotic DNA 87
Homology searches give an extra dimension to sequence inspection 89
5.1.2 Experimental techniques for gene location 89
Hybridization tests can determine if a fragment contains expressed sequences 89
cDNA sequencing enables genes to be mapped within DNA fragments 90
Methods are available for precise mapping of the ends of transcripts 91
Exon–intron boundaries can also be located with precision 92
5.2 *Determining the Function of a Gene* 92
5.2.1 Computer analysis of gene function 93
Homology reflects evolutionary relationships 93

Homology analysis can provide information on the function of an entire gene or of segments within it 94
Homology analysis in the yeast genome project 95
5.2.2 Assigning gene function by experimental analysis 96
Gene inactivation is the key to functional analysis 96
Methods for inactivating specific genes depend on homologous recombination 97
The phenotypic effect of gene inactivation is sometimes difficult to discern 98
Gene overexpression can also be used to assess function 98
5.2.3 More detailed studies of the activity of a protein coded by an unknown gene 100
Directed mutagenesis can be used to probe gene function in detail 100
Reporter genes and immunocytochemistry can be used to locate where and when genes are expressed 102
Protein–protein interactions can be examined by phage display and the yeast two-hybrid system 103
5.3 Comparative Genomics 103
5.3.1 Comparative genomics as an aid to gene mapping 103
5.3.2 Comparative genomics in the study of human disease genes 106
5.4 From Genome to Cell 106
5.4.1 Patterns of gene expression 107
Gene expression is assayed by measuring levels of RNA transcripts 107
The proteome 109

PART 2 **How Genomes Function** **112**

Chapter 6 Genome Anatomies **115**

6.1 The Anatomy of the Eukaryotic Genome 116
6.1.1 Eukaryotic nuclear genomes 117
Packaging of DNA into chromosomes 117
The special features of metaphase chromosomes 118
Where are the genes in a genome? 119
Box 6.1 Unusual chromosome types 120
Families of genes 120
Box 6.2 How many genes in the human genome? 122
Box 6.3 Two examples of unusual gene organization 124
Pseudogenes and other gene relics 124
6.1.2 Eukaryotic organelle genomes 125
Physical features of organelle genomes 125
The genetic content of organelle genomes 125
The origins of organelle genomes 126
6.2 The Anatomy of the Prokaryotic Genome 128
6.2.1 The physical structure of the prokaryotic genome 128
The traditional view of the bacterial 'chromosome' 129
Complications on the *E. coli* theme 129
6.2.2 The genetic organization of the prokaryotic genome 132
Operons are characteristic features of prokaryotic genomes 133
Speculations regarding the content of the minimal genome and the identity of distinctiveness genes 134
6.3 The Repetitive DNA Content of Genomes 134
Box 6.4 The organization of the human genome 135
6.3.1 Tandemly repeated DNA 136
Satellite DNA is found at centromeres and elsewhere in eukaryotic chromosomes 136
Mini- and microsatellites 136
6.3.2 Interspersed genome-wide repeats 137

	Transposition via an RNA intermediate	137
	DNA transposons	139
6.3.3	Repetitive DNA and eukaryotic chromosome architecture	140
Chapter 7	**The Role of DNA-binding Proteins**	**143**
7.1	*The Structure of DNA*	*144*
7.1.1	Nucleotides and polynucleotides	145
	The chemical structures of the four nucleotides	145
	Polynucleotides	145
	RNA	145
7.1.2	The double helix	145
	The key features of the double helix	147
	The double helix has structural flexibility	147
Box 7.1	**Base-pairing in RNA**	**147**
7.2	*Proteins*	*150*
7.2.1	The four levels of protein structure	150
	Amino acid diversity underlies protein diversity	153
Box 7.2	**Noncovalent bonds in proteins**	**153**
	Primary structure dictates protein function	154
7.3	*Methods for Studying DNA-binding Proteins*	*156*
7.3.1	Locating a protein-binding site on a DNA molecule	156
	Gel retardation identifies DNA fragments that bind to proteins	156
	Protection assays pinpoint binding sites with greater accuracy	157
	Modification interference identifies nucleotides central to protein binding	157
7.3.2	Purification of a DNA-binding protein	157
7.3.3	Techniques for studying protein and DNA–protein structures	160
	X-ray crystallography has broad applications in structure determination	160
	NMR gives detailed structural information for small proteins	160
7.4	*Interactions Between DNA and DNA-binding Proteins*	*162*
7.4.1	The DNA partner in the interaction	163
	Direct readout of the nucleotide sequence	163
	The nucleotide sequence has a number of indirect effects on helix structure	163
7.4.2	The protein partner	164
	The helix-turn-helix motif is present in prokaryotic and eukaryotic proteins	164
Box 7.3	**RNA-binding motifs**	**165**
	Zinc fingers are common in eukaryotic proteins	166
	Other DNA-binding motifs	167
7.4.3	DNA and protein in partnership	167
	Contacts between DNA and proteins	167
	Can sequence specificity be predicted from the structure of a recognition helix?	168
Chapter 8	**Initiation of Transcription: the First Step in Gene Expression**	**171**
8.1	*Accessing the Genome*	*173*
8.1.1	Effects of chromatin packaging on eukaryotic gene expression	173
	Heterochromatin, euchromatin and chromatin loops	173
	Structural and functional domains	174
	Nucleosome positioning is important in modulating gene expression	175
Box 8.1	**DNA methylation and gene expression**	**177**
8.2	*Assembly of the Transcription Initiation Complexes of Prokaryotes and Eukaryotes*	*178*
8.2.1	RNA polymerases	178
Box 8.2	**Mitochondrial and chloroplast RNA polymerases**	**178**
8.2.2	Recognition sequences for transcription initiation	179
	Bacterial RNA polymerases bind to promoter sequences	179
	Eukaryotic promoters are more complex	179
8.2.3	Assembly of the transcription initiation complex	181

	Transcription initiation in *E. coli*	181
Box 8.3	**Initiation of transcription in the archaea**	**181**
	Transcription initiation with RNA polymerase II	182
	Transcription initiation with RNA polymerases I and III	182
8.3	*Regulation of Transcription Initiation*	*185*
8.3.1	Strategies for controlling transcription initiation in bacteria	185
	Promoter structure determines the basal level of transcription initiation	185
	Regulatory control over bacterial transcription initiation	186
Box 8.4	***Cis* and *trans***	**188**
8.3.2	Control of transcription initiation in eukaryotes	188
	Activators of eukaryotic transcription initiation	189
Box 8.5	**Regulation of RNA polymerase I initiation**	**189**
	Contacts between transcription factors and the preinitiation complex	190
	A few repressors of eukaryotic transcription initiation have been identified	190
	Control over the activity of transcription factors	191
Chapter 9	**Synthesis and Processing of RNA**	**195**
9.1	*The RNA Content of a Cell*	*196*
9.1.1	Coding and noncoding RNA	196
9.1.2	Precursor RNA	197
9.1.3	RNA in time and space	198
	The population of RNAs in a cell changes over time	198
	In eukaryotes, RNAs must be transported from nucleus to cytoplasm, and possibly back again	199
9.2	*Synthesis and Processing of mRNAs*	*201*
9.2.1	Synthesis of bacterial mRNAs	201
	The elongation phase of bacterial transcription	201
	Termination of bacterial transcription	203
	Control over the choice between elongation and termination	203
Box 9.1	**Antitermination during the infection cycle of bacteriophage λ**	**207**
9.2.2	Synthesis of eukaryotic mRNAs by RNA polymerase II	207
	Capping of RNA polymerase II transcripts occurs immediately after initiation	207
	Elongation of eukaryotic mRNAs	209
	Termination of mRNA synthesis is combined with polyadenylation	210
9.2.3	Intron splicing	212
	Conserved sequence motifs indicate the key sites in GU–AG introns	213
Box 9.2	**Processed pseudogenes**	**214**
	Outline of the splicing pathway for GU–AG introns	214
	snRNAs and their associated proteins are the central components of the splicing apparatus	215
	SR proteins are mediators in splice site selection	216
	AU–AC introns are similar to GU–AG introns but require a different splicing apparatus	216
9.3	*Synthesis and Processing of Noncoding RNAs*	*219*
9.3.1	Transcript elongation and termination by RNA polymerases I and III	219
9.3.2	Cutting events involved in processing of bacterial and eukaryotic pre-rRNA and pre-tRNA	219
9.3.3	Introns in eukaryotic pre-rRNA and pre-tRNA	221
	Eukaryotic pre-rRNA introns are self-splicing	221
	Eukaryotic tRNA introns are variable but all are spliced by the same mechanism	222
9.4	*Processing of Pre-RNA by Chemical Modification*	*222*
Box 9.3	**Other types of intron**	**225**
9.4.1	Chemical modification of pre-rRNAs	225
Box 9.4	**Chemical modification without snoRNAs**	**226**
9.4.2	RNA editing	226

9.5 *Turnover of mRNAs* 227
Box 9.5 More complex forms of RNA editing 228

Chapter 10 Synthesis and Processing of the Proteome **231**
10.1 *The Role of tRNA in Protein Synthesis* 232
10.1.1 Aminoacylation: the attachment of amino acids to tRNAs 232
All tRNAs have a similar structure 233
Aminoacyl-tRNA synthetases attach amino acids to tRNAs 234
10.1.2 Codon–anticodon interactions: the attachment of tRNAs to mRNA 235
The genetic code specifies how an mRNA sequence is translated into a polypeptide 235
Codon–anticodon recognition is made flexible by wobble 238
Box 10.1 The origin and evolution of the genetic code 240
10.2 *The Role of the Ribosome in Protein Synthesis* 240
Box 10.2 The compactness of vertebrate mitochondrial genomes 242
10.2.1 Ribosome structure 242
Ultracentrifugation was used to measure the sizes of ribosomes and their components 242
Probing the fine structure of the ribosome 242
10.2.2 Translation in detail 243
Initiation in bacteria requires an internal ribosome binding site 243
Initiation in eukaryotes is mediated by the cap structure and poly(A) tail 245
Elongation is similar in bacteria and eukaryotes 246
Box 10.3 Initiation of eukaryotic translation without scanning 247
Box 10.4 Regulation of translation initiation 248
Termination requires release factors 248
10.3 *Post-translational Processing of Proteins* 248
Box 10.5 Frameshifting 250
10.3.1 Protein folding 250
Not all proteins fold spontaneously in the test tube 250
Box 10.6 Translation in the archaea 251
In cells, folding is aided by molecular chaperones 252
10.3.2 Processing by proteolytic cleavage 253
Cleavage of the ends of polypeptides 253
Proteolytic processing of polyproteins 254
10.3.3 Processing by chemical modification 255
10.3.4 Inteins 256
10.4 *Protein Turnover* 258
10.4.1 Degradation of ubiquitin-tagged proteins in the proteasome 259

Chapter 11 Regulation of Genome Activity **263**
11.1 *Transient Changes in Genome Activity* 266
11.1.1 Signal transmission by import of the extracellular signaling compound 267
Lactoferrin is an extracellular signaling compound which acts as a transcription factor 267
Some imported signaling compounds directly influence the activity of pre-existing transcription factors 268
Some imported signaling compounds influence genome activity indirectly 268
11.1.2 Signal transmission mediated by cell surface receptors 270
The simplest signal transduction pathways involve direct activation of transcription factors 270
More complex signal transduction pathways have many steps between receptor and genome 273
Signal transduction via second messengers 275
11.2 *Permanent and Semipermanent Changes in Genome Activity* 277
11.2.1 Genome rearrangements 278

Yeast mating types are determined by gene conversion events 278
Genome rearrangements are responsible for immunoglobulin and T-cell receptor diversities 278
11.2.2 Changes in chromatin structure 280
11.2.3 Genome regulation by feedback loops 281
11.3 Regulation of Genome Activity During Development 282
11.3.1 Sporulation in *Bacillus* 282
Sporulation involves coordinated activities in two distinct cell types 282
Special σ subunits control genome activity during sporulation 284
11.3.2 Vulval development in *Caenorhabditis elegans* 285
C. elegans is a model for multicellular eukaryotic development 285
Determination of cell fate during development of the *C. elegans* vulva 286
11.3.3 Development in *Drosophila melanogaster* 287
Maternal genes establish protein gradients in the *Drosophila* embryo 289
A cascade of gene expression converts positional information into a segmentation pattern 289
Segment identity is determined by the homeotic selector genes 291
Homeotic selector genes are universal features of higher eukaryotic development 292
Box 11.1 The genetic basis of flower development 292

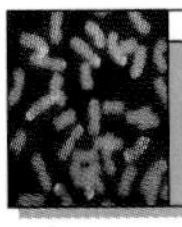
PART 3

How Genomes Replicate and Evolve **296**

Chapter 12 Genome Replication **299**

12.1 The Issues Relevant to Genome Replication 300
12.2 The Topological Problem 301
12.2.1 Experimental proof of the Watson–Crick scheme for DNA replication 301
The Meselson–Stahl experiment 302
Box 12.1 Variations on the semiconservative theme 303
12.2.2 DNA topoisomerases provide a solution to the topological problem 305
12.3 The Replication Process 306
Box 12.2 The diverse functions of DNA topoisomerases 306
12.3.1 Initiation of genome replication 307
Initiation at the *E. coli* origin of replication 307
Origins of replication in yeast have also been clearly defined 308
Replication origins in higher eukaryotes have been less easy to identify 309
12.3.2 The elongation phase of replication 311
The DNA polymerases of bacteria and eukaryotes 311
Discontinuous strand synthesis and the priming problem 314
Events at the bacterial replication fork 314
The eukaryotic replication fork: variations on the bacterial theme 317
Box 12.3 Genome replication in the archaea 318
12.3.3 Termination of replication 319
Replication of the *E. coli* genome terminates within a defined region 319
Little is known about termination of replication in eukaryotes 321
Box 12.4 Telomeres in *Drosophila* 322
12.3.4 Maintaining the ends of a linear DNA molecule 322
Telomeric DNA is synthesized by the telomerase enzyme 322
12.4 Regulation of Eukaryotic Genome Replication 323
12.4.1 Coordination of genome replication and cell division 323
Establishment of the pre-replication complex enables genome replication to commence 325
Regulation of pre-RC assembly 326
12.4.2 Control within S phase 326

Early and late replication origins 326
Checkpoints within S phase 326

Chapter 13 The Molecular Basis of Genome Evolution 329

13.1 Mutations 331
13.1.1 The causes of mutations 331
Box 13.1 Terminology for describing point mutations 332
Errors in replication are a source of point mutations 332
Replication errors can also lead to insertion and deletion mutations 335
Mutations are also caused by chemical and physical mutagens 337
13.1.2 The effects of mutations 340
The effects of mutations on genomes 340
The effects of mutations on multicellular organisms 343
The effects of mutations on microorganisms 344
13.1.3 Hypermutation and the possibility of programmed mutations 345
13.1.4 DNA repair 346
Box 13.2 DNA repair and human disease 348
Direct repair systems fill in nicks and correct some types of nucleotide modification 348
Base excision repairs many types of damaged nucleotide 348
Box 13.3 Cell cycle checkpoints for monitoring DNA damage 349
Nucleotide excision repair is used to correct more extensive types of damage 349
Mismatch repair: correcting errors of replication 351
13.2 Recombination 353
13.2.1 Homologous recombination 353
The Holliday model for homologous recombination 353
Proteins involved in homologous recombination in *E. coli* 355
The double-strand break model for recombination in yeast 356
Box 13.4 Double-strand break repair 357
13.2.2 Site-specific recombination 357
Integration of λ DNA into the *E. coli* genome involves site-specific recombination 357
Box 13.5 The RecE and RecF recombination pathways of *E. coli* 359
13.2.3 Transposition 359
Replicative and conservative transposition of DNA transposons 359
Tranposition of retroelements 363

Chapter 14 Patterns of Genome Evolution 367

14.1 Genomes: The First Ten Billion Years 368
14.1.1 The origins of genomes 369
The first biochemical systems were centered on RNA 369
The first DNA genomes 370
Box 14.1 How unique is life? 371
14.2 Acquisition of New Genes 372
14.2.1 Acquisition of new genes by gene duplication 373
Whole genome duplications can result in sudden expansions in gene number 374
Duplications of genes and groups of genes have occurred frequently in the past 375
Box 14.2 Gene duplication and genetic redundancy 378
The sequences of some duplicated genes do not diverge 380
Genome evolution also involves rearrangement of existing genes 380
14.2.2 Acquisition of new genes from other species 383
Box 14.3 The origins of the eukaryotic cell 383
14.3 Noncoding DNA and Genome Evolution 384
Box 14.4 The origin of a microsatellite 384
14.3.1 Transposable elements and genome evolution 384

14.3.2 The origins of introns 385
'Introns early' and 'introns late': two competing hypotheses 385
The current evidence disproves neither hypothesis 386
Box 14.5 The role of noncoding DNA 386
14.4 The Human Genome: The Last Five Million Years 387

Chapter 15 Molecular Phylogenetics **391**
15.1 The Origins of Molecular Phylogenetics 392
Box 15.1 Phenetics and cladistics 393
15.2 The Reconstruction of DNA-based Phylogenetic Trees 394
15.2.1 The key features of DNA-based phylogenetic trees 394
Gene trees are not the same as species trees 395
15.2.2 Tree reconstruction 397
Box 15.2 Terminology for molecular phylogenetics 397
Sequence alignment is the essential preliminary to tree reconstruction 397
Box 15.3 Multiple substitutions 399
Converting the alignment data into a phylogenetic tree 399
Assessing the accuracy of a reconstructed tree 401
Molecular clocks enable the time of divergence of ancestral sequences to be estimated 401
15.3 The Applications of Molecular Phylogenetics 402
15.3.1 Examples of the use of phylogenetic trees 402
DNA phylogenetics has clarified the evolutionary relationships between humans and other primates 402
The oldest life on Earth 402
The origins of AIDS 404
Problems with prions 405
15.3.2 Molecular phylogenetics as a tool in the study of human prehistory 406
Intraspecific studies require highly variable genetic loci 406
Box 15.4 Haplotypes 406
The origins of modern humans - out of Africa or not? 406
Box 15.5 Genes in populations 408
The patterns of more recent migrations into Europe are also controversial 410
Prehistoric human migrations into the New World 410

Appendix – Keeping up to Date **415**
Glossary **419**
Index **453**

Additional information

RESEARCH BRIEFINGS

2.1	*The human genetic map*	33
3.1	*The human physical map*	56
5.1	*Using genetic footprinting to assign functions to yeast genes*	99
5.2	*Patterns of gene expression during yeast sporulation*	108
6.1	*Supercoiled domains in the* E. coli *nucleoid*	130
7.1	*Genes are made of DNA*	148
8.1	*Similarities between TFIID and the histone core octamer*	183
9.1	*Probing the structure of the transcription elongation complex*	205
9.2	*Transcription through nucleosomes*	211
10.1	*Elucidation of the genetic code*	238
10.2	*Peptidyl transferase is a ribozyme*	249
11.1	*Unraveling a signal transduction pathway*	274
11.2	*A* Drosophila *mutant with an extended lifespan*	288
12.1	*Identification of mammalian origins of replication*	310
13.1	*Adaptive mutations?*	347
14.1	*An ancient duplication of the yeast genome*	376
14.2	*Directed evolution in the design of better genes*	379
15.1	*Neandertal DNA*	409

TECHNICAL NOTES

2.1	*Southern hybridization*	19
2.2	*Polymerase chain reaction*	20
2.3	*DNA chips*	23
3.1	*Restriction endonucleases*	40
3.2	*Gel electrophoresis*	43
3.3	*DNA labeling*	46
3.4	*DNA cloning*	52
4.1	*Polyacrylamide gel electrophoresis*	62
4.2	*Enzymes for DNA manipulations*	64
4.3	*Cloning vectors based on bacteriophage* λ	74
5.1	*Techniques for studying RNA*	91
5.2	*Site-directed mutagenesis*	101
6.1	*Ultracentrifugation techniques*	123
13.1	*Mutation detection*	334
15.1	*Phylogenetic analysis*	399

Abbreviations

AIDS	acquired immunodeficiency syndrome
AP	apurinic/apyrimidinic
ARS	autonomously replicating sequence
ASO	allele-specific oligonucleotide
ATP	adenosine triphosphate
BAC	bacterial artificial chromosome
bp	base pairs
BSE	bovine spongiform encephalopathy
cAMP	cyclic AMP
CASP	CTD-associated SR-like protein
CHEF	contour-clamped homogeneous electric fields
CPSF	cleavage and polyadenylation specificity factor
CRM	chromatin remodeling machine
CstF	cleavage stimulation factor
DAG	diacylglycerol
DASH	dynamic allele-specific hybridization
DBS	double-stranded DNA binding site
DNA	deoxyribonucleic acid
dsRAD	double-stranded RNA adenosine deaminase
dsRBD	double-stranded RNA binding domain
EEO	electroendosmosis value
EMS	ethylmethane sulfonate
ERV	endogenous retrovirus
ES	embryonic stem
EST	expressed sequence tag
FEN	flap endonuclease
FIGE	field inversion gel electrophoresis
FISH	fluorescent *in situ* hybridization
GABA	γ-aminobutyric acid
GAP	GTPase activating protein
GNRP	guanine nucleotide releasing protein
GTF	general transcription factor
GTP	guanosine triphosphate
HAT	hypoxanthine + aminopterin + thymidine
HBS	heteroduplex binding site
HIV	human immunodeficiency virus
hnRNA	heterogenous nuclear RNA
HPRT	hypoxanthine phosphoribosyl transferase
IF	initiation factor
Inr	initiator
IRES	internal ribosome entry site
ITF	integration host factor
ITR	inverted terminal repeat
JAK	Janus kinase
kb	kilobase pairs
LCR	locus control region
LINE	long interspersed nuclear element
LTR	long terminal repeat
MAP	mitogen activated protein
MAR	matrix-associated region
Mb	megabase pairs
MeCP	methyl-CpG-binding protein
MGMT	O^6-methylguanine-DNA methyltransferase
mRNA	messenger RNA
NHEG	nonhomologous end joining
OFAGE	orthogonal field alternation gel electrophoresis
ORC	origin recognition complex
ORF	open reading frame
PAC	P1-derived artificial chromosome
PADP	polyadenylate binding protein
PCNA	proliferating cell nuclear antigen
PCR	polymerase chain reaction
PTRF	polymerase I and transcript release factor
RACE	rapid amplification of cDNA ends
RBS	RNA binding site
RC	replication complex
RF	release factor
RFC	replication factor C
RFLP	restriction fragment length polymorphism
RLF	replication licensing factor
RNA	ribonucleic acid
RNP	ribonucleoprotein
RPA	replication protein A
rRNA	ribosomal RNA
RTVL	retroviral-like element
RT-PCR	reverse transcriptase polymerase chain reaction
SAP	stress activated protein
SAR	scaffold attachment region
scRNA	small cytoplasmic RNA
SINE	short interspersed nuclear element
SIV	simian immunodeficiency virus
snoRNA	small nucleolar RNA
SNP	single nucleotide polymorphism
snRNA	small nuclear RNA
snRNP	small nuclear ribonucleoprotein
SRF	serum response factor
SSB	single-strand binding protein
SSLP	simple sequence length polymorphism
STAT	signal transducer and activator of transcription
STR	simple tandem repeat
STS	sequence tagged site
TAF	TBP-associated factor
TBP	TATA-binding protein
TF	transcription factor
TK	thymidine kinase
T_m	melting temperature
tmRNA	transfer-messenger RNA
TPA	tissue plasminogen activator
tRNA	transfer RNA
VNTR	variable number of tandem repeats
YAC	yeast artificial chromosome

Preface

Genomes attempts to bring a fresh approach to the teaching of undergraduate molecular biology. It starts with the premise that the syllabus for a university course in molecular biology should reflect the major research issues of the new millennium rather than those topics that were in vogue during the 1970s and 1980s. The book is therefore centered on genomes, not genes, in recognition of the fact that today's molecular biology is driven less by research into the activities of individual genes and more by genome sequencing and functional analysis. Many of today's molecular biology undergraduates will be involved in genome research when they begin their graduate careers and all of them will find their work influenced in one way or another by genome projects. If the objective of undergraduate teaching is to prepare students for their future careers then they must be taught about genomes!

It would of course be foolish to suggest that genes are no longer important. The major challenge that I faced when writing *Genomes* was to combine the essential elements of the traditional molecular biology syllabus with the new material relating to genomes. It is not yet possible to describe adequately the events leading from DNA to protein entirely in terms of 'genome to proteome', hence a substantial part of *Genomes* is devoted to the expression pathways of individual genes. This book differs from many others in that it attempts to describe these expression pathways in the context of the activity and function of the genome as a whole. Similarly, DNA replication, mutation and recombination are dealt with largely in terms of their effects on the genome, and not simply as processes responsible for the replication and alteration of genes.

My belief that molecular biology teaching should be centered on genomes grew as I wrote this book and discovered how much more satisfying and informative the approach is compared with the traditional syllabus. A number of topics that in the past have seemed to me to be of peripheral interest have fallen into place and taken on new relevance. I hope that at least some of the excitement that I felt while writing *Genomes* is conveyed to the reader.

T.A. Brown
Manchester

Acknowledgements

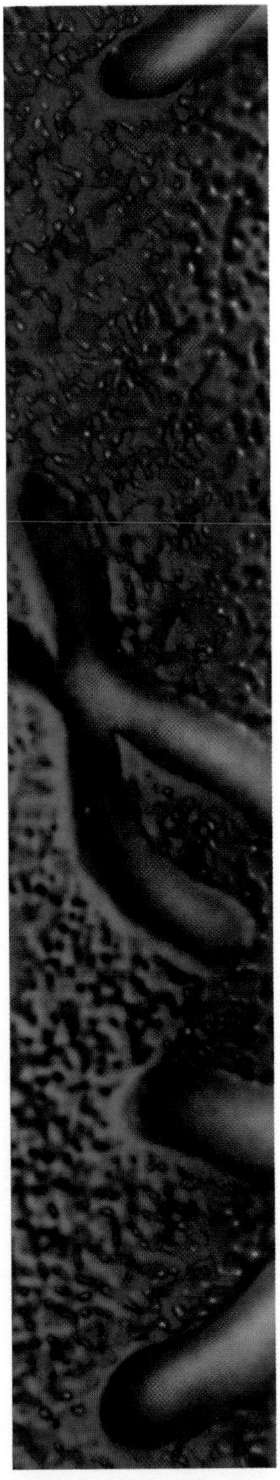

Genomes has not been an easy book to write. I began with the conviction that if the book was going to provide a new approach to molecular biology then it must not be derivative in any way, and so I made very little use of existing textbooks during the period when I was writing *Genomes*. My first acknowledgement is therefore to the editors of *Nature, Science, Trends in Genetics* and *Trends in Biochemical Sciences*, whose publications over the last three years have formed the bulk of the source material from which this book has been written. The initial stimulus that persuaded me to tackle such a massive task came from Jonathan Ray at BIOS, and I thank Jonathan, Lisa Mansell, Fran Kingston and the other production staff for the superb work that they have done at all stages from the initial idea to the finished book. I especially want to thank Phil Welbourn of J&L Composition whose ability to convert my semi-coherent doodlings into professional diagrams has been a major factor in making the science in *Genomes* accessible to the reader.

A number of my colleagues at UMIST read parts of *Genomes* and provided comments that have made the book better than it would otherwise have been. I thank Andrew Doig, Jeremy Derrick and Robin Allaby for clarifying a number of points about NMR, X-ray crystallography and molecular phylogenetics, respectively. I am also grateful to R Slater, John Mattick, Raymond Dalgleish, Gudmundur Eggertsson, Andrew Lambertsson, R Krumlauf, John Quackenbush, Julian Blow, Anne Donaldson, Professor GA Dover and Mike Jackson for their more general comments on drafts of various chapters – these comments have had an important influence on the quality of the book.

At a personal level, I am grateful to Rob Beynon and John McCarthy, my Heads of Department during the period when I was writing *Genomes*, for allowing me the time and freedom to produce the book. I must also thank my current research group – Robin Allaby, Robert Sallares, Jayne Threadgold, Christine Flaherty, Sarah Lindsay, Lisa Nencioni and Susan Gomzi – for maintaining their research output despite having to put up with an absentee supervisor for most of 1998. Last, but by no means least, my wife and research colleague Keri Brown convinced me that I could write this book and kept her faith in me at times when I was ready to give up. She also suffered, without complaint, a virtual cessation of our social life and lengthy periods when 'The Book' dominated everything else. If you find this book useful then you should thank Keri, not me, because she is the one who ensured that it was written.

An Introduction to GENOMES

I have tried to make *Genomes* as user-friendly as possible. The book therefore includes a number of devices intended to help the reader and make it a more effective teaching aid.

ORGANIZATION OF THE BOOK

Genomes is divided into three parts:

- **Part 1 How Genomes are Studied** does what its title says by describing and explaining the techniques used to study genomes. It begins with a brief chapter that introduces the reader to some typical genomes and sets out the issues that will be covered elsewhere in the book. The techniques for studying genomes are then described in the order in which they would be used in a genome project: methods for constructing genetic (Chapter 2) and physical maps (Chapter 3); DNA sequencing methodology and the strategies used to assemble a contiguous genome sequence (Chapter 4); methods for identifying genes in a genome sequence and determining the functions of those genes in the cell (Chapter 5). The Human Genome Project forms a continuous thread throughout Part 1, but this is not to the exclusion of all else and I have tried to give adequate coverage to the strategies that have been used, and are being used, to understand the genomes of other organisms.
- **Part 2 How Genomes Function** covers the material that in the past has been described (inadequately in my opinion) as 'DNA goes to RNA goes to protein'. The first two chapters set the scene: Chapter 6 surveys genome anatomies – the structures and contents of the genomes of different organisms – and Chapter 7 emphasizes the importance of the part played by DNA-binding proteins in expressing the genome, a theme that recurs when the events responsible for initiation of transcription (Chapter 8) and synthesis of RNA (Chapter 9) are subsequently described. Chapter 10 covers the genetic code, protein synthesis and protein processing, and Chapter 11 surveys the regulation of genome activity. Keeping Chapter 11 to a manageable length proved difficult, as many different topics are relevant to genome regulation, and I hope that I have achieved a satisfactory balance between conciseness and breadth of coverage.
- **Part 3 How Genomes Replicate and Evolve** links DNA replication, mutation and recombination with the gradual evolution of genomes over time. In Chapters 12 and 13 the molecular processes responsible for replication, mutation, repair and recombination are described, and in Chapter 14 the ways in which these processes are thought to have shaped the structures and genetic contents of genomes over evolutionary time are considered. Finally, Chapter 15 is devoted to the increasingly informative use of molecular phylogenetics as a means of inferring the evolutionary relationships between DNA sequences.

FIGURES

A good diagram is certainly worth a thousand words but a bad one can confuse the reader and a superfluous one is merely distracting. I have therefore tried to ensure that every figure is necessary and fulfils a purpose beyond simply breaking up the text and making the book look pretty. I have also tried to make figures reproducible because in my opinion this makes them much more useful as a learning aid for the student. I have never understood the penchant for making textbook diagrams into works of art because if the student cannot redraw a diagram then it is merely an illustration and does not help the student learn the information that it is designed to convey. The figures in *Genomes* are as clear, simple and uncluttered as possible.

BOXES, TECHNICAL NOTES AND RESEARCH BRIEFINGS

The main text in each chapter is supported and extended by additional information, separated into three distinct categories:

- **Boxes** contain discrete packages of information that I have taken out of the main text, either for emphasis or to avoid disrupting the flow of the main text. Some boxes summarize the key points regarding a topic that is described at length in the text, or provide a pointer towards a later topic that has a bearing on the issues being discussed. Other boxes are used to give a more extended coverage of interesting topics, and some describe current speculation regarding areas that have not yet been resolved.
- Each **Technical Note** is a self-contained description of a technique or a group of techniques important in the study of genomes. The Technical Notes are designed to be read in conjunction with the main

text, each one being located at the place in the book where an application of that technique is described for the first time. Alternatively, all of the Technical Notes could be read at once, before the book as a whole is tackled, so the reader gains a good grounding in experimental procedures at the very outset of their program of study.

- **Research Briefings** are designed to illustrate some of the strategies that are used to study genomes. Each Briefing is based on one or a few research papers and explains the background and rationale of a research project, describes how the resulting data were analyzed, and summarizes the conclusions that were drawn. The objective is to illustrate the way in which real research is conducted and how research into molecular biology has established the 'facts' about genomes. All of the papers described in the Research Briefings are 'classics' in that each one constitutes a major advance in our understanding of genomes, but they are not intended to be a 'Hall of Fame' and readers should be aware that the pace of genome research is such that many other equally important papers are published every year.

READING LISTS

The reading lists at the end of each chapter are divided into two sections:

- **References** are lists of articles that are cited in the text. *Genomes* is not itself a research publication and the text is not referenced in the way that would be appropriate for a review or scientific paper. Many points and facts are not referenced at all, and those citations that are given are often review articles rather than the relevant primary research papers. In several cases, for example, I have referred to a *Science* Perspective or *Nature* News and Views article, rather than a research paper, because these general articles are usually more helpful in explaining the context and relevance of a piece of work. My intention throughout *Genomes* has been that the reference lists should be as valuable as possible to students writing extended essays or dissertations on particular topics.
- **Further Reading** contains books and review articles that are not referred to directly in the main text but which are useful sources of additional material. In most cases I have appended a one-line summary stating the particular value of each item to help the reader decide which ones he or she wishes to seek out. The lists are not all-inclusive and I encourage readers to spend some time searching the shelves of their own libraries for other books and articles. Browsing is an excellent way to discover interests that you never realized you had!

APPENDIX – KEEPING UP TO DATE

The Appendix gives the reader advice regarding the best way to keep up to date with the latest research discoveries. It is divided into two sections. The first section covers the various journal and other publications that include reviews and news articles on genome research, and the second section contains a list of some of the many Internet sites that contain relevant information.

GLOSSARY

I am very much in favor of glossaries as learning aids and I have provided an extensive one for *Genomes*. Every term that is highlighted in bold in the text is defined in the Glossary, along with a number of additional terms that the reader might come across when referring to books or articles in the reading lists. Each term in the Glossary also appears in the index, so the reader can quickly gain access to the relevant pages where the Glossary term is covered in more detail.

PART

1 How Genomes are Studied

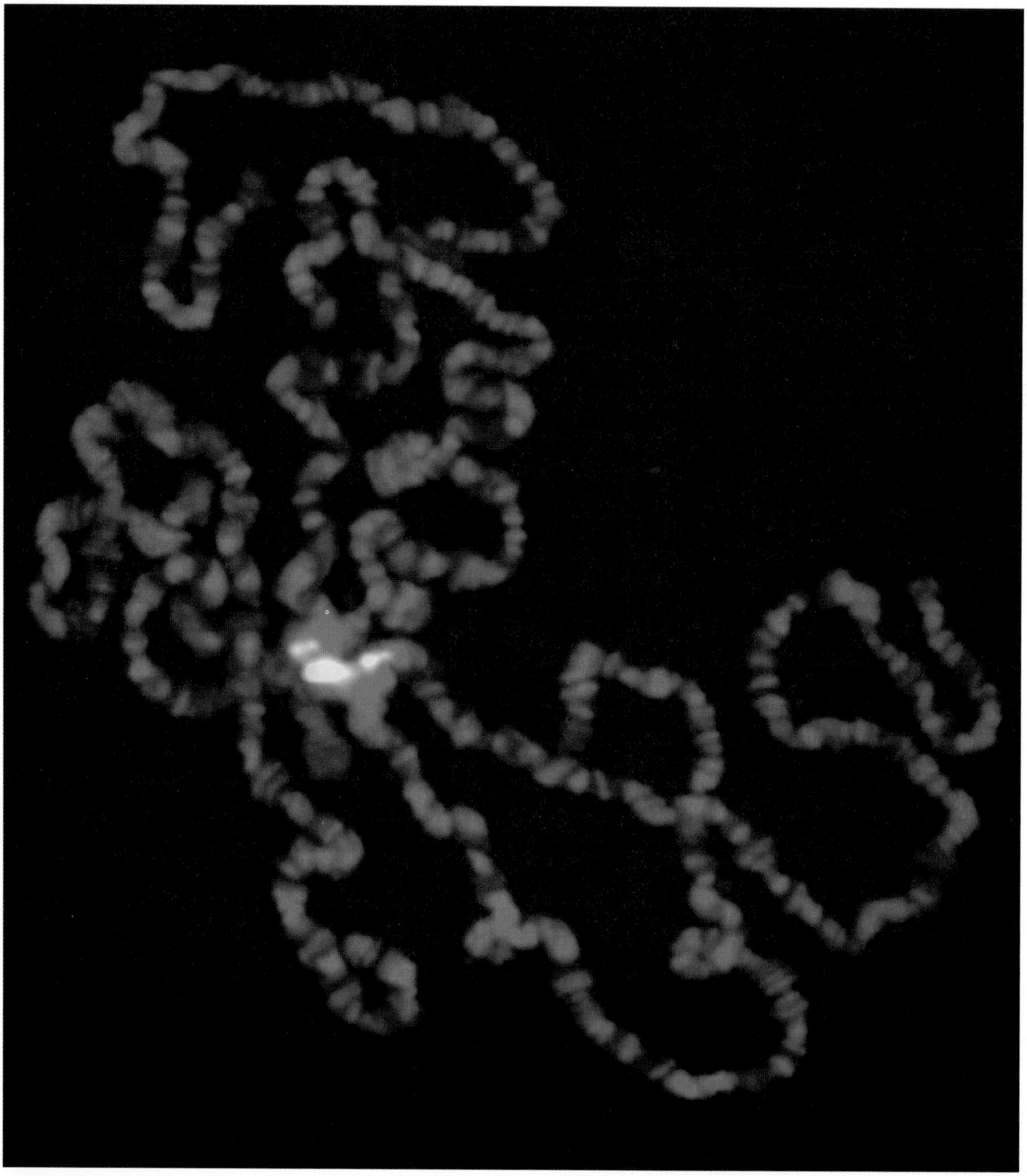

Chapter 1
What is a Genome?

Chapter 2
Mapping Genomes by Genetic Techniques

Chapter 3
Mapping Genomes by Physical Techniques

Chapter 4
Sequencing Genomes

Chapter 5
Understanding a Genome Sequence

Opposite page: A single chromosome of the fruit fly *Drosophila melanogaster*, examined by fluorescent *in situ* hybridization (FISH) with a probe specific for the repetitive DNA sequences of the centromere. The chromatids are seen in red with bright yellow fluorescence located at the centromere. See Section 3.2 for a description of FISH and Sections 6.1.1 and 6.3.1 for information on centromeres.

Acknowledgement: Reproduced with the kind permission of P.A. Coelho and C.E. Sunkel (Instituto de Biologia Molecular e Celular da Universidade do Porto, Portugal).

The image was supplied by Bio-Rad Microscience Ltd.

CHAPTER

1 What is a Genome?

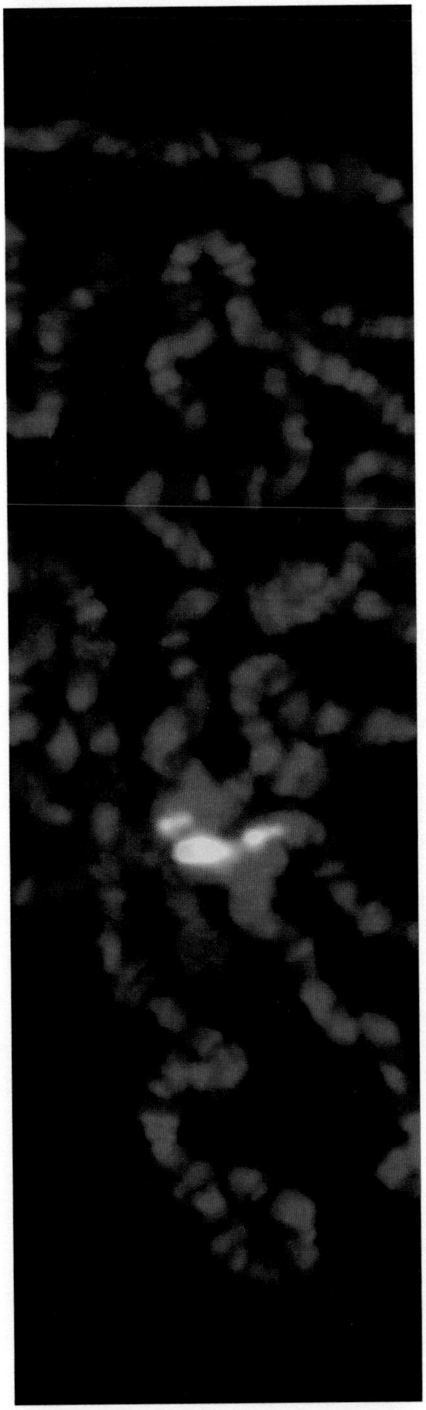

Contents

1.1 *The Human Genome* **3**
1.1.1 The physical structure of the human genome 3
1.1.2 The genetic content of the human genome 6
1.2 *Genomes of Other Organisms* **7**
1.2.1 Genomes of eukaryotes 8
1.2.2 Genomes of prokaryotes 9
1.3 *Why are Genome Projects Important?* **11**

Concepts

- *The genome is the entire DNA content of a cell, including all of the genes and all of the intergenic regions*
- *The human genome contains approximately 80 000 genes but the coding regions of these genes take up only 3% of the genome*
- *The yeast genome contains 6000 genes and has a more compact organization*
- *The genomes of some plants are dominated by repetitive DNA sequences*
- *Prokaryotic genomes are small with very little space between genes*
- *Understanding the information contained in genome sequences will be the major challenge of the early 21st century*

LIFE AS WE KNOW IT is specified by **genomes**. Every organism possesses a genome that contains the **biological information** needed to construct and maintain a living example of that organism. Most genomes, including those for all cellular lifeforms, are made of **DNA** (deoxyribonucleic acid) but a few viruses have **RNA** (ribonucleic acid) genomes. DNA and RNA are polymeric molecules made up of linear, unbranched chains of monomeric subunits called **nucleotides**. Each nucleotide has three parts: a sugar, a phosphate group, and a base (*Figure 1.1*). In DNA, the sugar is 2′-deoxyribose and the bases are adenine (A), cytosine (C), guanine (G) and thymine (T). Nucleotides are linked to one another by **phosphodiester bonds** to form a DNA polymer, or **polynucleotide**, which might be several million nucleotides in length. DNA in living cells is **double-stranded**, two polynucleotides being wound around one another to form the **double helix**. The double helix is held together by **hydrogen bonds** between the base components of the nucleotides in the two strands. The **base-pairing** rules are that A base-pairs with T, and G base-pairs with C. The two DNA molecules in a double helix therefore have **complementary** sequences.

In an RNA nucleotide the sugar is ribose rather than 2′-deoxyribose, and thymine is replaced by the related base uracil (U). RNA polymers are rarely more than a few thousand nucleotides in length, and RNA in the cell is usually **single-stranded**, although base pairs might form between different parts of a single molecule.

The biological information contained in a genome is encoded in the nucleotide sequence of its DNA or RNA molecules and is divided into discrete units called **genes**. The information contained in a gene is read by proteins that attach to the genome at the appropriate positions and initiate a series of biochemical reactions referred to as **gene expression**. For organisms with DNA genomes, this process was originally looked on as comprising two stages, **transcription** and **translation**, the first producing an RNA copy of the gene and the second resulting in synthesis of a protein whose amino acid sequence is determined, via the **genetic code**, by the nucleotide sequence of the RNA transcript (*Figure 1.2A*). This is still an accurate description of gene expression in simple organisms such as bacteria, but it gives an incomplete picture of the events involved in conversion of genomic information into functional proteins in higher organisms (*Figure 1.2B*). A particular weakness of the transcription–translation interpretation is that it can result in attention being drawn away from the key points in the gene expression pathway at which information flow is regulated.

A complete copy of the genome must be made every time a cell divides. **DNA replication** has to be extremely accurate in order to avoid the introduction of **mutations** into the genome copies. Some mutations do, however,

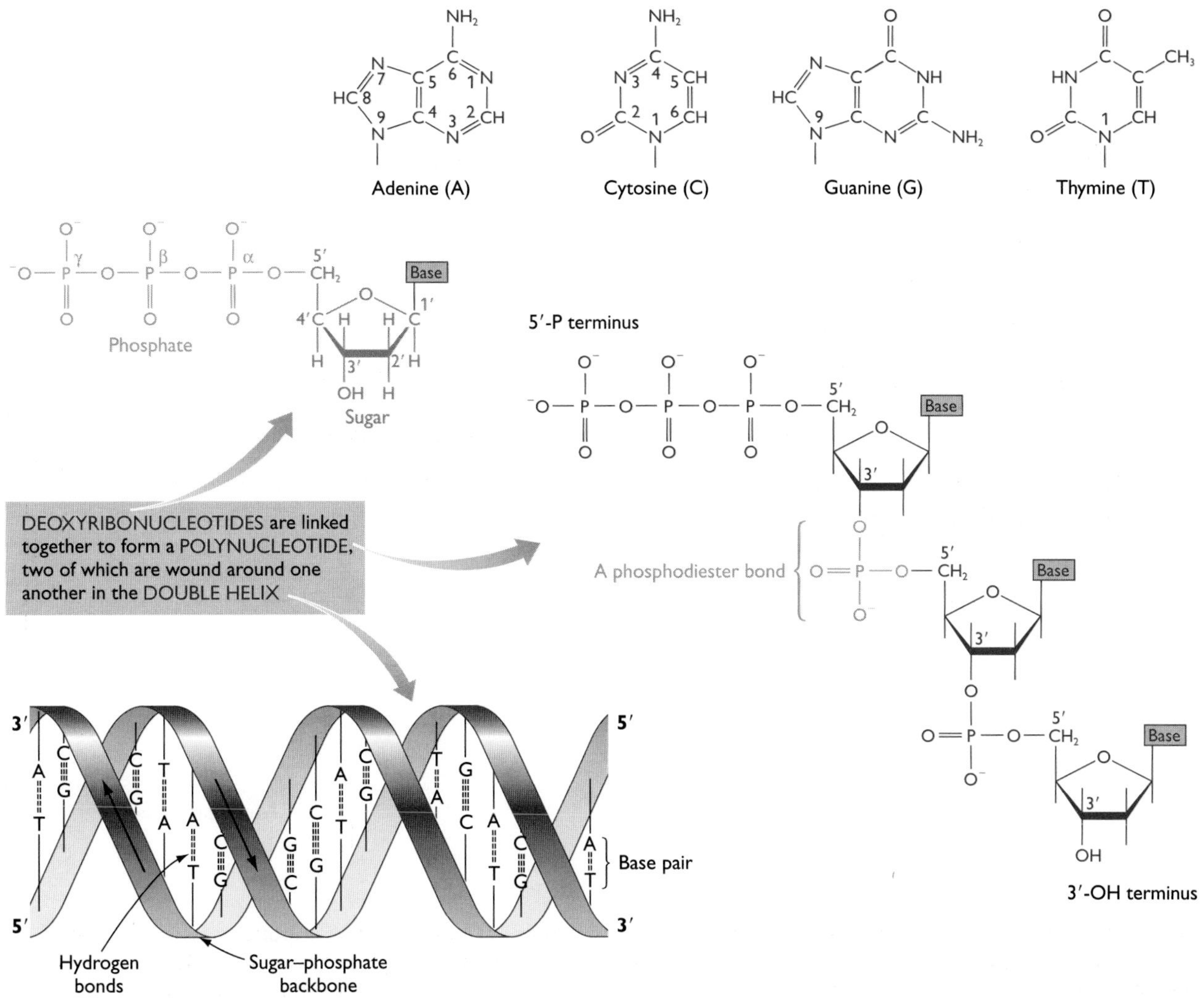

Figure 1.1 The structure of DNA.

Each deoxyribonucleotide consists of the sugar 2′-deoxyribose linked to a phosphate group and one of four bases: adenine, cytosine, guanine or thymine. Nucleotides are linked together by phosphodiester bonds to give a polynucleotide, the two ends of which are chemically different. In the double helix, two polynucleotides, running in different directions, are wound around one another and held together by hydrogen bonds between pairs of bases. See Section 7.1 for a more detailed description of DNA and RNA structure.

occur, either as errors in replication or due to the effects of chemical and physical mutagens that directly alter the chemical structure of DNA. **DNA repair** enzymes correct many of these errors; those that escape the repair processes become permanent features of the lineage descending from the original mutated genome. These events, along with genome rearrangements resulting from **recombination**, underlie **molecular evolution**, the driving force behind the evolution of living organisms.

1.1 THE HUMAN GENOME

Of all the genomes in existence our own is quite naturally the one that interests us the most. We will therefore begin our exploration with an overview of the structure and organization of the human genome. We will then, in the second part of this chapter, consider how similar or different the human genome is to the genomes of other organisms.

1.1.1 The physical structure of the human genome

The human genome is made up of two distinct components (*Figure 1.3*):

- The **nuclear genome**, which comprises approximately 3 000 000 000 bp (**base pairs**) of DNA. This

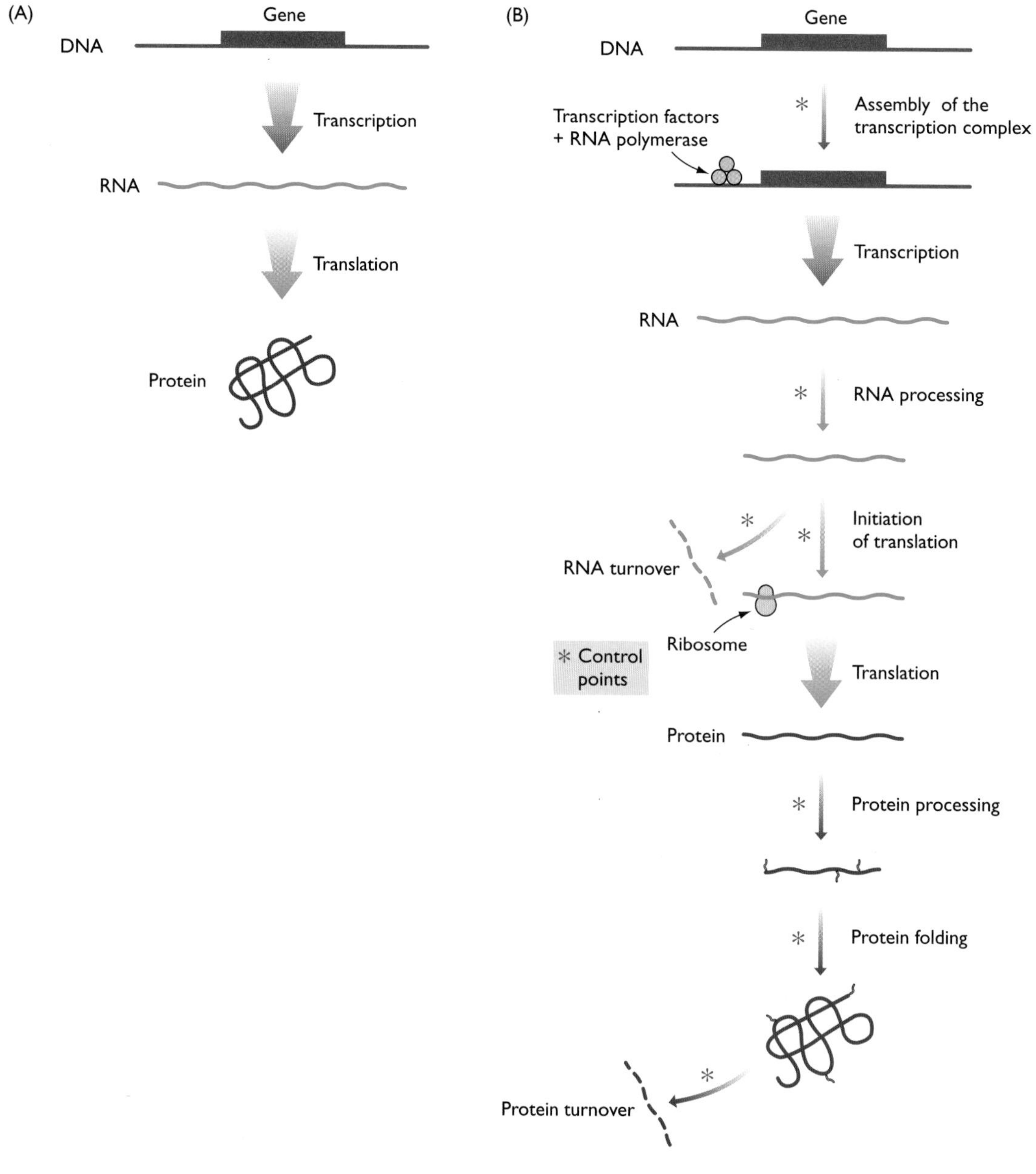

Figure 1.2 Two views of gene expression.

(A) shows the traditional depiction of gene expression, summarized as 'DNA makes RNA makes protein', the two steps being called transcription and translation. (B) gives a more accurate outline of the events involved in gene expression in higher organisms. Key regulatory points are highlighted. Note that these schemes apply only to protein-coding genes. Some genes give rise to noncoding RNAs, such as ribosomal RNA and transfer RNA: these genes are transcribed and processed as shown but the RNAs are not translated (see Section 9.1.1). Full details of all the steps in gene expression are given in Part 2 of this book.

figure is the same as 3 000 000 kb (**kilobase pairs**) or 3000 Mb (**megabase pairs**). The nuclear genome is divided into 24 linear DNA molecules, the shortest 55 Mb in length and the longest 250 Mb, each contained in a different **chromosome**. These 24 chromosomes consist of 22 **autosomes** and the two **sex chromosomes**, X and Y.

- The **mitochondrial genome**, a circular DNA molecule of 16 569 bp, many copies of which are located in the energy-generating organelles called **mitochondria**.

Each of the approximately 10^{13} cells in the adult human body has its own copy or copies of the genome, the only exceptions being those few cell types, such as red blood

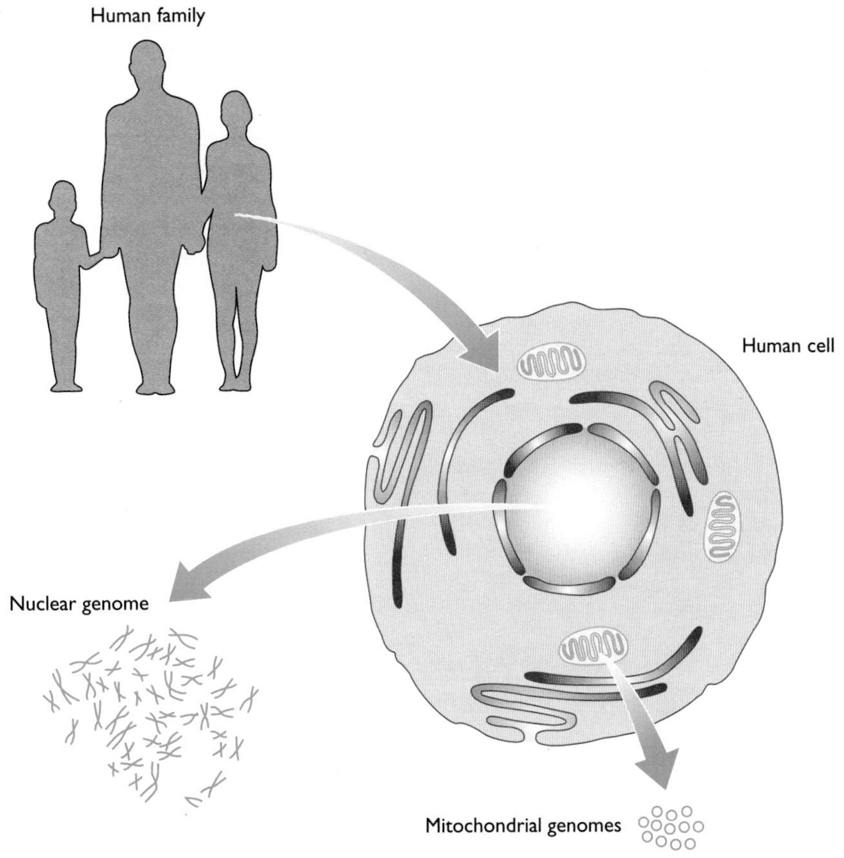

Figure 1.3 The nuclear and mitochondrial components of the human genome.

For more details on the anatomy of the human genome, see Section 6.1.

cells, that lack nuclei in their fully differentiated state. The vast majority of cells are **diploid** and so have two copies of each autosome, plus two sex chromosomes, XX for females or XY for males – 46 chromosomes in all. These are called **somatic cells**, in contrast to **sex cells** or **gametes**, which are **haploid** and have just 23 chromosomes, comprising one of each autosome and one sex chromosome. Both types of cell have about 8000 copies of the mitochondrial genome, 10 or so in each mitochondrion.

Before we go any further we should try to gain some appreciation of the immensity of the human genome. Three billion base pairs is such a large number of nucleotides that it is difficult to grasp the scale that it represents; an analogy is helpful. The typeface used for the text of this book enables approximately 60 nucleotides of DNA sequence to be written in a line 10 cm in length. If printed out in this format, the human genome sequence would stretch for 5000 km, the distance from Montreal to London, Los Angeles to Panama, Tokyo to Calcutta, Cape Town to Addis Ababa, or Auckland to Perth (*Figure 1.4*). The sequence would fill about 3000 books the size of this one. Even the genome of the simplest bacterium would take up a kilometer of this typeface. Such is the enormity of the task facing us if we hope to understand how genomes are constructed and how they work.

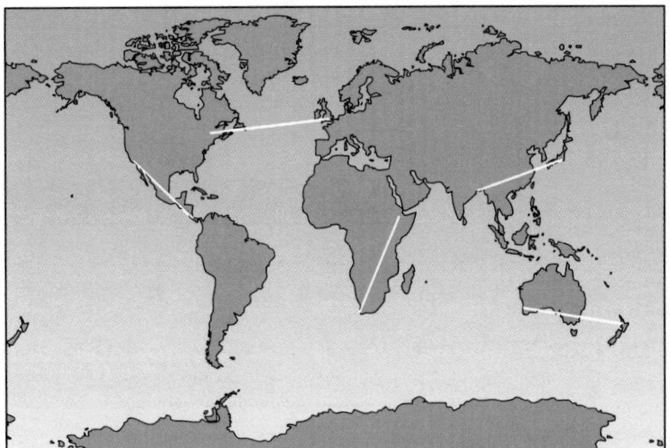

Figure 1.4 The immense length of the human genome.

The map illustrates the distance that would be covered by the human genome sequence if it was printed in the typeface used in this book.

1.1.2 The genetic content of the human genome

Understanding how the human nuclear genome is constructed is one of the goals of the **Human Genome Project**. First conceived in 1984 and begun in earnest in 1990, the Project aims to complete the sequence of the human genome by 2003. What will the sequence reveal?

The current consensus is that the human genome contains approximately 80 000 genes, though numbers as low as 50 000 and as high as 150 000 have been suggested by scientists working on the Human Genome Project (Cohen, 1997). In Chapter 6 the evidence on which these estimates are based will be examined (see Box 6.2, p. 122), and we will also learn that the information contained in these 80 000 or so genes takes up only 3% of the nuclear genome. This is illustrated by *Figure 1.5*, which shows the genetic organization of a 50-kb segment of chromosome 7, this segment forming part of the 'human β T-cell receptor locus', a much larger (685 kb) region of the genome that specifies proteins involved in the immune response (Rowen *et al.*, 1996). Our 50-kb segment contains the following genetic features:

- **One gene**. This gene is called TRY4 and it codes for trypsinogen, the inactive precursor of the digestive enzyme trypsin. TRY4 is one of a family of trypsinogen genes present in two clusters at either end of the β T-cell receptor locus. These genes have nothing to do with the immune response, they simply share this part of chromosome 7 with the β T-cell receptor locus.
- **Two gene segments**. These are V28 and V29-1 and they code for a part of the β T-cell receptor protein after which the locus is named. V28 and V29-1 are quite unusual as they are not complete genes, only segments of a gene, and before being expressed they must be linked to other gene segments from elsewhere in the locus. This occurs in T lymphocytes and is an example of how a permanent change in the activity of the genome can arise during cellular differentiation (see Section 11.2.1). Note that TRY4, V28

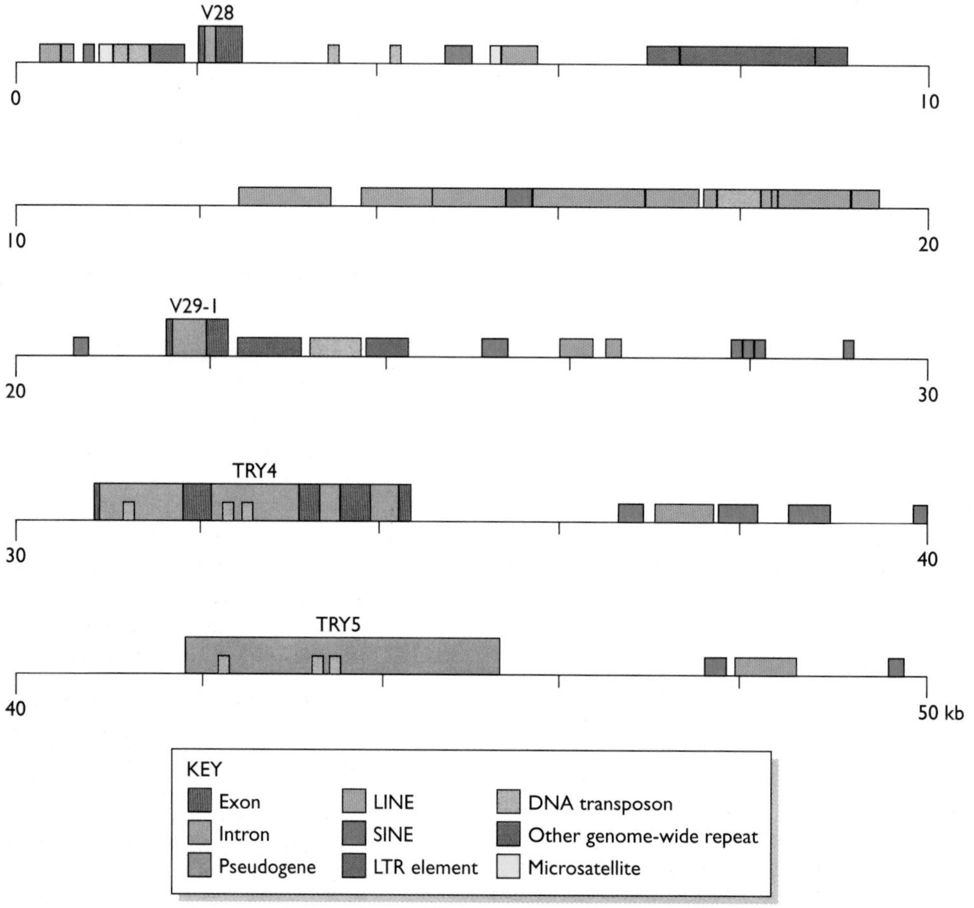

Figure 1.5 A segment of the human genome.

This map shows the location of genes, gene segments, pseudogenes, genome-wide repeats and microsatellites in a 50-kb segment of the human β T-cell receptor locus on chromosome 7. Redrawn from Rowen *et al.* (1996).

Box 1.1: Some key facts about genes

A gene is a segment of the genome that is transcribed into RNA. If the RNA is a transcript of a protein-coding gene then it is called a **messenger RNA (mRNA)** and is translated into protein. If the RNA is noncoding, such as **ribosomal RNA (rRNA)** and **transfer RNA (tRNA)** then it is not translated. Noncoding RNAs have various functions in the cell (Section 9.1.1).

The part of a protein-coding gene that is translated into protein is called the **open reading frame (ORF)**. Each triplet of nucleotides in the ORF is a **codon** that specifies an amino acid in accordance with the rules of the **genetic code** (Section 10.1.2). The ORF is read in the 5′ to 3′ direction along the mRNA. The ORF starts with an **initiation codon** and ends with a **termination codon.** The part of the mRNA before the ORF is called the **leader** segment, and that following the ORF is the **trailer** segment.

Many genes in eukaryotes are discontinuous, being split into **exons** and **introns**. The introns are removed from the primary transcript by **splicing** to produce the functional RNA molecule (Section 9.2.3).

'Upstream' refers to the region of DNA before a gene; **'downstream'** is after the gene.

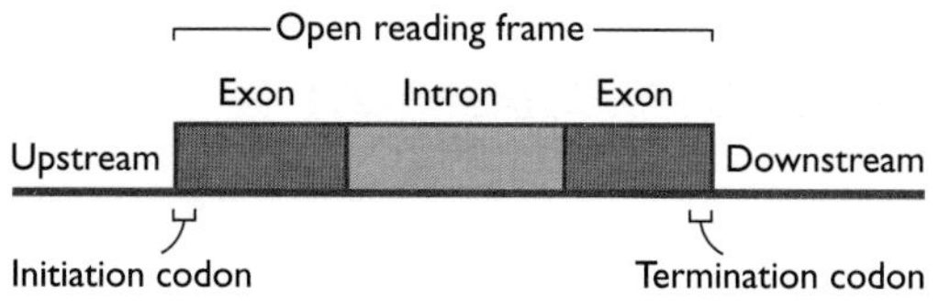

and V29-1, like most human genes, are **discontinuous**, being made up of **exons**, containing protein-coding information, separated by noncoding **introns**. The exons, when added together, make up a total of 1414 bp, or 2.8% of the 50-kb segment. The coding capacity of this segment is therefore fairly typical of the genome as a whole.

- **One pseudogene**. A **pseudogene** is a nonfunctional copy of a gene, usually one that has mutated so that its biological information has become unreadable (Section 6.1.1). This particular pseudogene is called TRY5 and it is closely related to the functional members of the trypsinogen gene family.
- **Fifty-two genome-wide repeat sequences**. These are sequences that recur at many places in the genome. There are four main types of genome-wide repeat, called **LINEs** (long interspersed nuclear elements), **SINEs** (short interspersed nuclear elements), **LTR** (long terminal repeat) **elements** and **DNA transposons**, and examples of each type are seen in this segment, together making up 39.1% of the sequence. We will look at genome-wide repeat sequences in more detail in Section 6.3.2.
- **Two microsatellites**. These are sequences in which a short motif is repeated in tandem (Section 6.3.1). One of the microsatellites seen here has the motif GA repeated sixteen times, giving the sequence

  5′–GAGAGAGAGAGAGAGAGAGAGAGAGAGAGAGAGA–3′
  3′–CTCTCTCTCTCTCTCTCTCTCTCTCTCTCTCTCT–5′

  The second microsatellite comprises six repeats of TATT. Many microsatellites are polymorphic, the number of repeats being variable in different individuals. Microsatellites are useful marker points in the genome and in Chapters 2 and 3 we will see how they have been used in the construction of genome maps.
- Finally, approximately 50% of our 50-kb segment of the human genome is made up of stretches of nongenic, nonrepetitive, single-copy DNA of no known function or significance.

No short segment can be considered truly representative of the human genome as a whole. In some respects the 50-kb sequence shown in *Figure 1.5* is probably fairly atypical, but it illustrates the range of genetic features contained in the human genome. The next question that we will address is how similar the human genome is to the genomes of other organisms.

1.2 GENOMES OF OTHER ORGANISMS

Biologists divide the living world into two types of organism (*Figure 1.6*):

- **Eukaryotes**, whose cells contain membrane-bound compartments, including a nucleus and organelles such as mitochondria and, in the case of plant cells, chloroplasts. Eukaryotes include animals, plants, fungi and protozoa.
- **Prokaryotes**, whose cells lack extensive internal compartments. There are two very different groups of prokaryotes, distinguished from one another by characteristic genetic and biochemical features:
  (i) the **bacteria**, which include most of the commonly encountered prokaryotes such as the gram-negatives (e.g. *Escherichia coli*), the gram-positives (e.g. *Bacillus subtilis*), the cyanobacteria (e.g. *Anabaena*) and many more;
  (ii) the **archaea**, which are less well-studied, and have mostly been found in extreme environments such as hot springs, brine pools and anaerobic lake bottoms.

Eukaryotes and prokaryotes have quite different types of genome and we must therefore consider them separately.

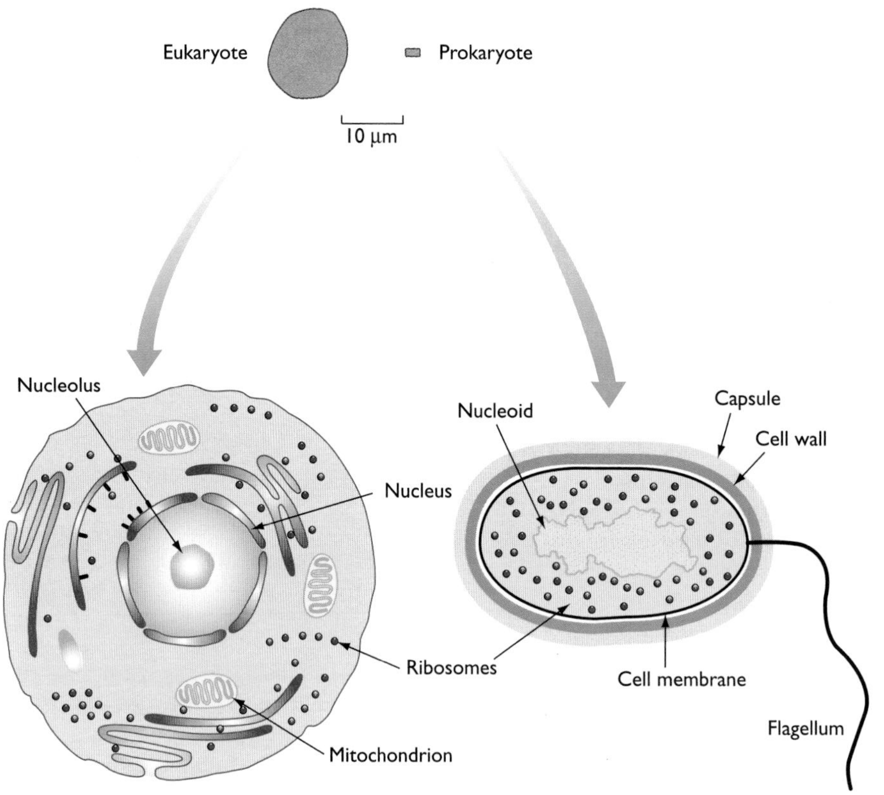

Figure 1.6 Cells of eukaryotes (left) and prokaryotes (right).

The top part of the figure shows a typical human cell and typical bacterium drawn to scale. The human cell is 10 μm in diameter and the bacterium is rod-shaped with dimensions of 1×2 μm. The lower drawings show the internal structures of eukaryotic and prokaryotic cells. Eukaryotic cells are characterized by their membrane-bound compartments, which are absent in prokaryotes. The bacterial DNA is contained in the structure called the nucleoid.

1.2.1 Genomes of eukaryotes

Humans are fairly typical eukaryotes and the human genome is in many respects a good model for eukaryotic genomes in general. All of the eukaryotic nuclear genomes that have been studied are, like the human version, divided into two or more linear DNA molecules, each contained in a different chromosome, and all eukaryotes also possess smaller, usually circular, mitochondrial genomes. The only general eukaryotic feature not illustrated by the human genome is the presence in plants and other photosynthetic organisms of a third genome, located in the chloroplasts. The two organellar genomes – mitochondrial and chloroplast – will be examined in our detailed survey of genome organization in Chapter 6.

Although the basic physical structures of all eukaryotic nuclear genomes are similar, one important feature is very different in different organisms. This is genome size, the smallest eukaryotic genomes being less than 10 Mb in length, and the largest over 100 000 Mb. As can be seen in *Table 1.1*, this size range coincides to a certain extent with the complexity of the organism, the simplest eukaryotes such as fungi having the smallest genomes, and higher eukaryotes such as vertebrates and flowering plants having the largest ones. This makes sense as one would expect the complexity of an organism to be related to the number of genes in its genome – higher eukaryotes need larger genomes to accommodate the extra genes. However, the correlation is not precise: if it was, then the nuclear genome of the yeast *Saccharomyces cerevisiae*, which at 12 Mb is 0.004 times the size of the human nuclear genome, would be expected to contain $0.004 \times 80\,000$ genes, which is just 320. In fact the *S. cerevisiae* genome contains about 6000 genes.

For many years the lack of precise correlation between the complexity of an organism and the size of its genome was looked on as a bit of a puzzle, the so-called C-value paradox. In fact the answer is quite simple: space is saved in the genomes of less complex organisms because the genes are more closely packed together. The *S. cerevisiae* genome, the sequence of which was completed in 1996, illustrates this point, as we can see from the top two parts of *Figure 1.7*, where the 50-kb segment of the human

Table 1.1 Sizes of genomes

Organism	Genome size (Mb)
Prokaryotes	
Mycoplasma genitalium	0.58
Escherichia coli	4.64
Bacillus megaterium	30
Eukaryotes	
Fungi	
Saccharomyces cerevisiae (yeast)	12.1
Aspergillus nidulans	25.4
Protozoa	
Tetrahymena pyriformis	190
Invertebrates	
Caenorhabditis elegans (nematode worm)	100
Drosophila melanogaster (fruit fly)	140
Bombyx mori (silkworm)	490
Strongylocentrotus purpuratus (sea urchin)	845
Locusta migratoria (locust)	5000
Vertebrates	
Fugu rubripes (pufferfish)	400
Homo sapiens (humans)	3000
Mus musculus (mouse)	3300
Plants	
Arabidopsis thaliana (vetch)	100
Oryza sativa (rice)	565
Pisum sativum (pea)	4800
Zea mays (maize)	5000
Triticum aestivum (wheat)	17000
Fritillaria assyriaca (fritillary)	120000

Data taken from Brown (1998).

genome that we looked at earlier in the chapter is compared with a 50-kb segment of the yeast genome. The yeast genome segment, which comes from chromosome III (the first eukaryotic chromosome to be completely sequenced; Oliver *et al*., 1992), has the following distinctive features:

- ***It contains more genes than the human segment.*** This region of yeast chromosome III contains 26 genes thought to code for proteins and two coding for **transfer RNAs (tRNAs)**, short molecules involved in reading the genetic code during translation (Section 10.1). These 28 genes take up 66.4% of the 50-kb sequence.
- ***Relatively few of the yeast genes are discontinuous.*** In this segment of chromosome III none of the genes are discontinuous. In the entire yeast genome there are only 239 introns, which is a remarkably small number compared with the genomes of higher eukaryotes, in which some individual genes can have more than 100 introns.
- ***There are fewer genome-wide repeats.*** This part of chromosome III contains a single LTR element, called *Ty2*, and four truncated LTR elements called delta sequences. These five genome-wide repeats make up 13.5% of the 50-kb segment, but this figure is not entirely typical of the yeast genome as a whole. When all 16 yeast chromosomes are considered the total amount of sequence taken up by genome-wide repeats is only 3.4% of the total.

The picture that emerges is that the genetic organization of the yeast genome is much more economical than that of the human version. The genes themselves are more compact, having fewer introns, and the spaces between the genes are relatively short with much less space taken up by genome-wide repeats and other noncoding sequences. We will see in Chapter 6 that the hypothesis that the more complex eukaryotes have less compact genomes holds when other species are examined. We will also see that the genome-wide repeats appear to play an intriguing role in dictating the compactness or otherwise of a genome, a general rule being that the smaller genomes lack extensive repetition, but the larger ones display a proliferation in the number of repeat sequences that are present. This is strikingly illustrated by the maize genome, which at 5000 Mb is larger than the human genome but still relatively small for a flowering plant. Only a few limited regions of the maize genome have been sequenced, but some remarkable results have been obtained, revealing a genome dominated by repetitive elements. *Figure 1.7C* shows a 50-kb segment of this genome, either side of one member of a family of genes coding for the alcohol dehydrogenase enzymes (SanMiguel *et al*., 1996). This is the only gene in this 50-kb region, though there is a second one, of unknown function, approximately 100 kb beyond the right-hand end of the sequence shown here. Instead of genes, the dominant feature of this genome segment is the genome-wide repeats. The majority of these are of the LTR element type, which comprise virtually all of the noncoding part of the segment, and on their own are estimated to make up approximately 50% of the maize genome.

These examples show that some of the features of eukaryotic genomes are surprisingly unusual. We will fill in the details concerning the contents and organizations of eukaryotic genomes in Chapter 6, and in Chapter 14 we will examine what little is known about the evolutionary events that led to these diverse designs.

1.2.2 Genomes of prokaryotes

Prokaryotic genomes are very different from eukaryotic ones. There is some overlap between the largest prokaryotic and smallest eukaryotic genomes, but on the whole prokaryotic genomes are much smaller (see *Table 1.1*). The *E. coli* genome, for example, is just 4639 kb, two-fifths the size of the yeast genome. The physical organizations are also different. In prokaryotes, most if not all of the genome is contained in a single DNA molecule, and this molecule is circular rather than linear. In addition to

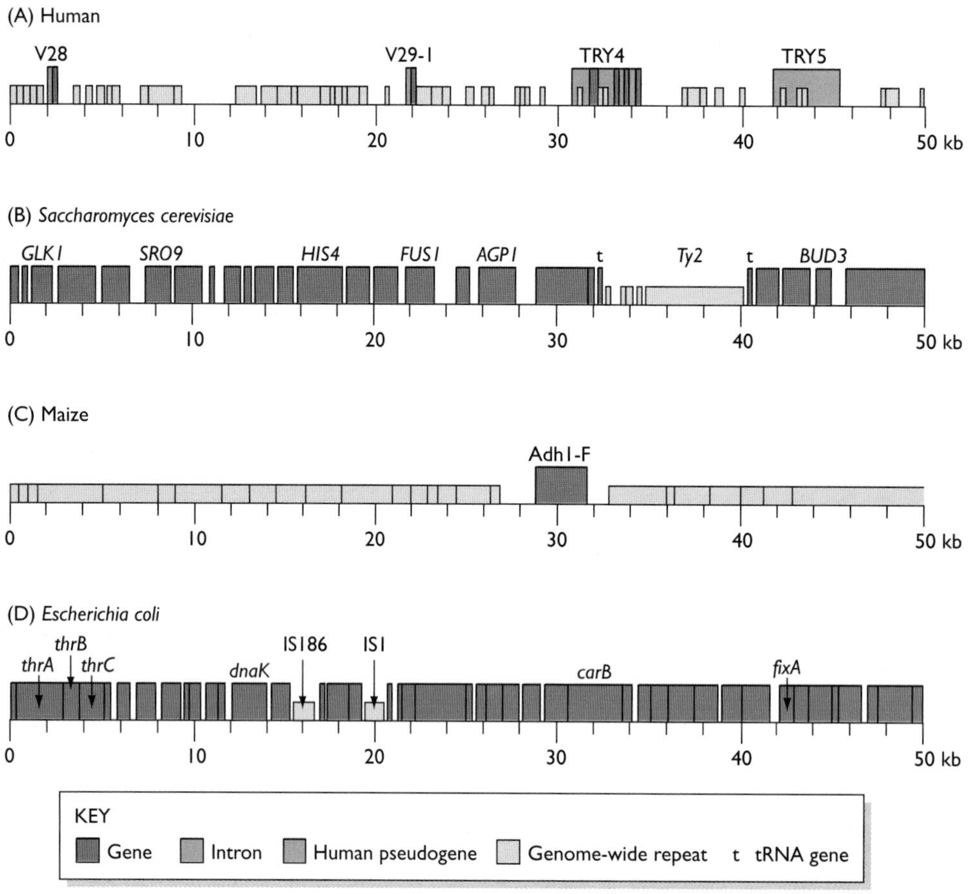

Figure 1.7 Comparison of the genomes of humans, yeast, maize and *E. coli*.

(A) is the 50-kb segment of the human β T-cell receptor locus shown in *Figure 1.5*. This is compared with 50-kb segments from the genomes of (B) *S. cerevisiae* (chromosome III; redrawn from Oliver *et al.*, 1992), (C) maize (redrawn from SanMiguel *et al.*, 1996) and (D) *E. coli* (redrawn from Blattner *et al.*, 1997). See the text for more details.

their single 'chromosome', prokaryotes may also have additional genes on independent, smaller, circular or linear DNA molecules called **plasmids** (*Figure 1.8*). Genes carried by plasmids are useful, coding for properties such as antibiotic resistance or the ability to utilize complex compounds such as toluene as a carbon source; but plasmids appear to be dispensable – a prokaryote can exist quite effectively without them. For this reason, a bacterial or archaeal 'genome' is usually defined as just the main DNA molecule, with plasmids looked on as ancillary components, not parts of the genome itself (for exceptions to this rule, see Section 6.2.1).

As well as having smaller genomes, prokaryotes generally have fewer genes than eukaryotes. *E. coli* has just 4397. After our discussion regarding eukaryotic gene organization, it will probably come as no surprise to learn that prokaryotic genomes are even more compact than those of lower eukaryotes. If we return to *Figure 1.7* we can see this illustrated by part *D*, which shows a 50-kb segment of the *E. coli* genome (Blattner *et al.*, 1997). It is immediately obvious that there are more genes and less space between them, with 43 genes taking up 85.9% of the segment. Some genes have virtually no space between them: *thrA* and *thrB*, for example, are separated by a single nucleotide, and *thrC* begins at the nucleotide immediately following the last nucleotide of *thrB*. These

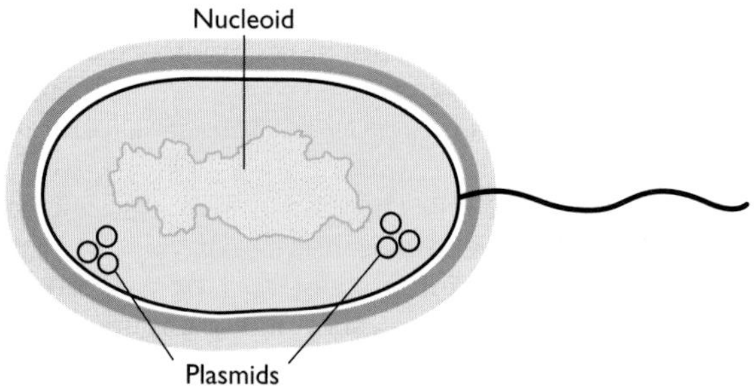

Figure 1.8 Plasmids are small circular DNA molecules that are found inside some prokaryotic cells.

three genes are an example of an **operon**, a group of genes involved in a single biochemical pathway (in this case, synthesis of the amino acid threonine) and expressed in conjunction with one another. Operons have been used as model systems for understanding how gene expression is regulated (Section 8.3.1).

Two other features of prokaryotic genomes can be deduced from *Figure 1.7D*. First, there are no introns in the genes present in this segment of the *E. coli* genome. In fact *E. coli* has no discontinuous genes at all and it is generally believed that this type of gene structure is absent in prokaryotes with a few exceptions, mainly among the archaea. The second feature is the infrequency of repetitive sequences. Prokaryotic genomes do not have anything equivalent to the high-copy-number, genome-wide repeat families found in eukaryotic genomes. They do, however, possess certain sequences that might be repeated elsewhere in the genome, examples being the **insertion sequences** IS1 and IS186 that can be seen in the 50-kb segment shown in *Figure 1.7D*. These are examples of **transposable elements**, sequences that have the ability to move around the genome and to transfer from one organism to another, even sometimes between two different species (Section 6.3.2). The positions of the IS1 and IS186 elements shown in *Figure 1.7D* therefore refer only to the particular *E. coli* strain from which this sequence was obtained: if a different strain is examined then the IS sequences could well be in different positions or might be entirely absent from the genome.

1.3 WHY ARE GENOME PROJECTS IMPORTANT?

As we begin the new millennium, the major goal of molecular biology is to obtain the complete sequences of as many genomes as possible. Considerable effort is being devoted not only to the Human Genome Project, but also to equivalent projects aimed at the genomes of a wide range of other organisms (*Table 1.2*). Why all this activity devoted to genome sequences? There are many reasons.

First, our current perception is that genome sequences are the key to the continued development of not only molecular biology and genetics, but also those areas of biochemistry, cell biology and physiology now described as the **molecular life sciences**. A catalog containing a description of the sequence of every gene in a genome is

Table 1.2 Some of the organisms for which complete genome sequences should be available by 2005

Organism	Genome size (Mb)	Internet address for latest news
Archaea†		
Methanococcus jannaschii	1.66	http://www.tigr.org/tdb/mdb/mjdb/mjdb.html
Methanobacterium thermoautotrophicum	1.75	http://www.genomecorp.com/htdocs/sequences/methanobacter/abstract.html
Archaeoglobus fulgidus	2.18	ftp://ftp.tigr.org/pub/data/a_fulgidus
Bacteria†		
Mycoplasma genitalium	0.58	http://www.tigr.org/tdb/mdb/mgdb/mgdb.html
Mycoplasma pneumoniae	0.81	http://www.zmbh.uni-heidelberg.de/M-pneumoniae/MP_Home.html
Treponema pallidum	1.14	http://www.tigr.org/tdb/mdb/tpdb/tp_bg.html
Borrelia burgdorferi	1.44	ftp://ftp.tigr.org/pub/data/b_burgdorferi
Aquifex aeolicus	1.55	
Helicobacter pylori	1.66	http://www.tigr.org/tdb/mdb/hpdb/hpdb.html
Haemophilus influenzae	1.83	http://www.tigr.org/tdb/mdb/mdb.html
Synechocystis sp.	3.57	http://kazusa.or.jp/cyano/cyano.html
Bacillus subtilis	4.20	http://www.pasteur.fr/Bio/SubtiList.html
Mycobacterium tuberculosis	4.40	http://www.sanger.ac.uk/Projects/M_tuberculosis/
Escherichia coli	4.64	http://www.genetics.wisc.edu:80/index.html
Eukaryotes		
Saccharomyces cerevisiae	12.1	http://www.mips.biochem.mpg.de/
Arabidopsis thaliana	100	http://genome-www.stanford.edu/Arabidopsis/
Caenorhabditis elegans	100	http://moulon.inra.fr/acedb/acedb.html
Drosophila melanogaster	140	http://flybase.bio.indiana.edu/
Oryza sativa	565	http://www.staff.or.jp/
Homo sapiens	3000	http://gdbwww.gdb.org/
Mus musculus	3300	http://www.informatics.jax.org/

† Most of these archaeal and bacterial sequences have already been completed. See also Appendix – Keeping Up to Date.

immensely valuable, even if at first the functions of many of the genes are unknown. Not only will the catalog contain the sequences of the coding parts of every gene, it will also include the regulatory regions for these genes. The genome sequence therefore opens the way to a comprehensive description of the molecular activities of living cells and the ways in which these activities are controlled.

Gene catalogs also aid the isolation and utilization of important genes, such as those human genes responsible for inherited disease, or bacterial genes whose protein products have industrial value. These genes can be isolated from a genome even if the complete genome sequence is not known, but the process is time-consuming and costly, and a different project has to be devised for each gene that is sought. It will be very much easier to obtain a copy of the desired gene if the sequence of its genome is already known, so that the gene can simply be withdrawn from the catalog.

Genome sequences will have additional benefits that at present can only be guessed at. We have seen that the human genome, in common with the genomes of all higher eukaryotes, contains extensive amounts of noncoding DNA. We assume that most of the noncoding DNA has no function, but perhaps this is because we do not know enough about it. Could the noncoding DNA have a role, but one that at present is too subtle for us to grasp? The first step in addressing this possibility is to obtain a complete description of the organization of the noncoding DNA in different genomes, so that common features, which might indicate a role for some or all of these sequences, can be identified.

There is one final reason for genome projects. The work stretches current technology to its limits. Genome sequencing therefore represents the frontier of molecular biology, territory that was inaccessible just a few years ago and which still demands innovative approaches and a lot of sheer hard work. Scientists have always striven to achieve the almost impossible, and the motivation for many molecular biologists involved in genome projects is, quite simply, the challenge of the unknown.

REFERENCES

Blattner FR, Plunkett G, Bloch CA, *et al.* (1997) The complete genome sequence of *Escherichia coli* K-12. *Science*, **277**, 1453–1462.

Brown TA (1998) *Molecular Biology Labfax*, 2nd edition, Volume 1. Academic Press, London.

Cohen J (1997) How many genes are there? *Science*, **275**, 769.

Oliver SG, van der Aart QJM, Agostini-Carbone ML, *et al.* (1992) The complete DNA sequence of yeast chromosome III. *Nature*, **357**, 38–46.

Rowen L, Koop BF and Hood L (1996) The complete 685-kilobase DNA sequence of the human β T cell receptor locus. *Science*, **272**, 1755–1762.

SanMiguel P, Tikhonov A, Jin Y-K, *et al.* (1996) Nested retrotransposons in the intergenic regions of the maize genome. *Science*, **274**, 765–768.

FURTHER READING

Alberts B, Bray D, Lewis J, Raff M, Roberts K and Watson JD (1994) *Molecular Biology of the Cell*, 3rd edition. Garland Publishing, New York. — *The best source of general information on eukaryotic cell biology.*

McKusick VA (1989) The Human Genome Organisation: history, purposes and membership. *Genomics*, **5**, 385–387. — *Describes the goals of the Human Genome Project.*

Prescott LM, Harley JP and Klein DA (1996) *Microbiology*, 3rd edition. Wm C. Brown, Dubuque. — *One of the best books on prokaryotic biology.*

CHAPTER

2 Mapping Genomes by Genetic Techniques

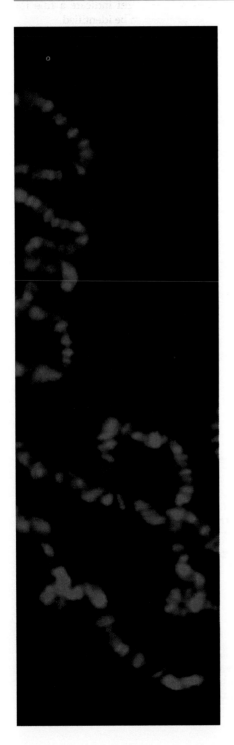

Contents

2.1	*Genetic and Physical Maps*	**15**
2.2	*Markers for Genetic Maps*	**15**
	2.2.1 Genes were the first markers to be used	15
	2.2.2 DNA markers for genetic mapping	18
2.3	*Approaches to Genetic Mapping*	**22**
	2.3.1 Linkage analysis is the basis of genetic mapping	22
	2.3.2 Carrying out linkage analysis with different types of organism	26

Concepts

- *Genetic mapping uses genetic techniques such as crossbreeding and pedigree analysis to construct genome maps*
- *Genes were originally used as markers but these are now supplemented with DNA markers such as RFLPs, SSLPs and SNPs*
- *The basis of genetic mapping is the calculation of recombination frequency by linkage analysis*
- *If planned breeding experiments are possible then two-point or multipoint test crosses are used*
- *Genetic mapping with humans is complicated by the limitations of pedigree analysis*
- *Genetic analysis of bacteria requires gene transfer from one cell to another*

THE NEXT THREE CHAPTERS describe the techniques and strategies used to obtain genome sequences. These techniques involve much more than just methods for sequencing DNA molecules. Those methods are obviously of paramount importance but they have one major limitation: even with the most sophisticated technology it is rarely possible to obtain a sequence of more than about 750 bp in a single experiment. This means that the sequence of a long DNA molecule has to be constructed from a series of shorter sequences. One approach is to break the molecule into fragments, determine the sequence of each one, and use a computer to search for overlaps and build up the master sequence (*Figure 2.1*). This **shotgun method** is the standard approach for sequencing small prokaryotic genomes, but the required data analysis becomes disproportionately more complex as the number of fragments increases (for n fragments the number of possible overlaps is given by $2n^2 - 2n$). A second problem with the shotgun method is that it can lead to errors when repetitive regions of a genome are analyzed. When a repetitive sequence is broken into fragments many of the resulting pieces contain the same, or very similar, sequence motifs. It would be very easy to reassemble these sequences so that a portion of the repetitive regions was left out or even so that two pieces of the same or different chromosomes were mistakenly connected together (*Figure 2.2*).

The shotgun approach on its own is therefore inappropriate for larger, more complex genomes. Instead, a genome **map** must first be generated. A genome map provides a guide for the sequencing experiments by showing the positions of genes and other distinctive features.

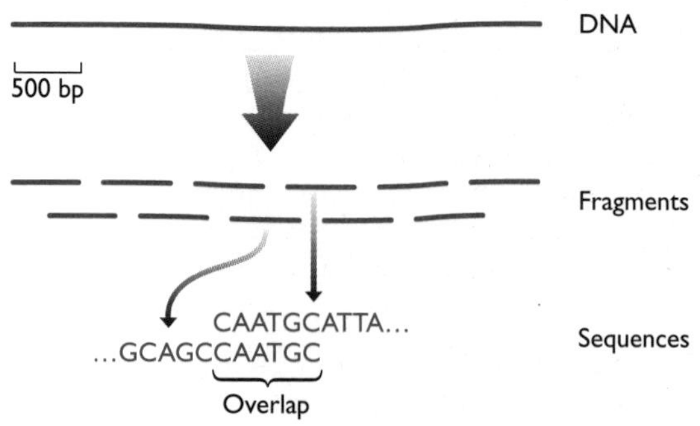

Figure 2.1 The shotgun approach to sequence assembly.

The DNA molecule is broken into small fragments, each of which is sequenced. The master sequence is assembled by searching for overlaps between the sequences of individual fragments. In practice, an overlap of several tens of base pairs would be needed to establish that two sequences should be linked together.

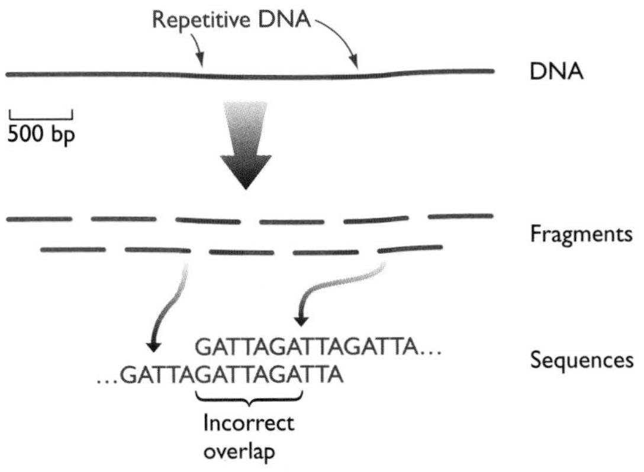

Figure 2.2 Problems with the shotgun approach.

In this example the DNA molecule contains two segments of repetitive DNA, each made up of many copies of the motif GATTA. When the sequences are examined, two fragments from different parts of the DNA appear to overlap. If the error is not recognized then the central segment of the DNA molecule will be left out of the master sequence. If the two repetitive DNA segments were on different chromosomes, then the sequences of these chromosomes might be mistakenly assembled.

Once a genome map is available, the sequencing phase of the project can proceed in either of two ways (*Figure 2.3*):

- By the **clone contig** approach (Section 4.2.2). The genome is broken into manageable segments, each a few hundred kb or a few Mb in length, which are then sequenced individually, usually by the shotgun method. Once the sequence of a segment has been completed, it can be positioned at its correct location on the map.
- By a **directed shotgun** approach (Section 4.2.3). This strategy uses the distinctive features on a genome map as landmarks to aid assembly of the master sequence from the huge numbers of short sequences obtained by the shotgun approach. The map is also used to check that errors have not been made when assembling sequences from repetitive regions. In theory, the entire human genome could be sequenced this way (Venter *et al.*, 1998).

With both approaches, the map provides the framework for carrying out the sequencing phase of the project. Some types of map also indicate the positions of genes, enabling the initial part of a sequencing project to be directed at the interesting regions of a genome so that the sequences of important genes are obtained as quickly as possible. For these reasons, the first 6 years of the Human Genome Project were devoted almost exclusively to mapping the human genome, rather than sequencing it.

2.1 GENETIC AND PHYSICAL MAPS

The convention has been to divide genome mapping methods into two categories:

- **Genetic mapping** is based on the use of genetic techniques to construct maps showing the positions of genes and other sequence features on a genome. Genetic techniques include cross-breeding experiments or, in the case of humans, the examination of family histories (**pedigrees**). Genetic mapping is described in this chapter.
- **Physical mapping** uses molecular biology techniques to examine DNA molecules directly in order to construct maps showing the positions of sequence features including genes. Physical mapping is described in the next chapter.

2.2 MARKERS FOR GENETIC MAPS

As with any type of map, a genetic map must show the positions of distinctive features. In a geographical map these **markers** are recognizable components of the landscape, such as rivers, roads and buildings. What markers can we use in a genetic landscape?

2.2.1 Genes were the first markers to be used

The first genetic maps, constructed in the early decades of the 20th century for organisms such as the fruit fly, used genes as markers. This was many years before it was understood that genes are segments of DNA molecules. Instead, genes were looked on as abstract entities responsible for the transmission of heritable characteristics from parent to offspring. To be useful in genetic analysis, a heritable characteristic had to exist in two alternative forms or **phenotypes**, an example being tall or short stems in the pea plants originally studied by Mendel. Each phenotype is specified by a different **allele** of the corresponding gene. To begin with, the only genes that could be studied were those specifying phenotypes that were distinguishable by visual examination. So, for example, the first fruit fly maps showed the positions of genes for body color, eye color, wing shape and suchlike, all of these phenotypes being visible simply by looking at the flies with a low power microscope or the naked eye. This approach was fine in the early days but geneticists soon realized that there are only a limited number of visual phenotypes whose inheritance can be studied, and in many cases the analysis is made less easy because more than one gene affects a single physical feature. For example, by 1922 over 50 genes had been mapped on the four fruit fly chromosomes but nine of these were for eye color, and any newcomer wishing to make a contribution to this field first had to learn to distinguish between fly eyes that were colored red, light red, vermilion, garnet, carnation, cinnabar, sepia, scarlet, pink, cardinal or claret. To make

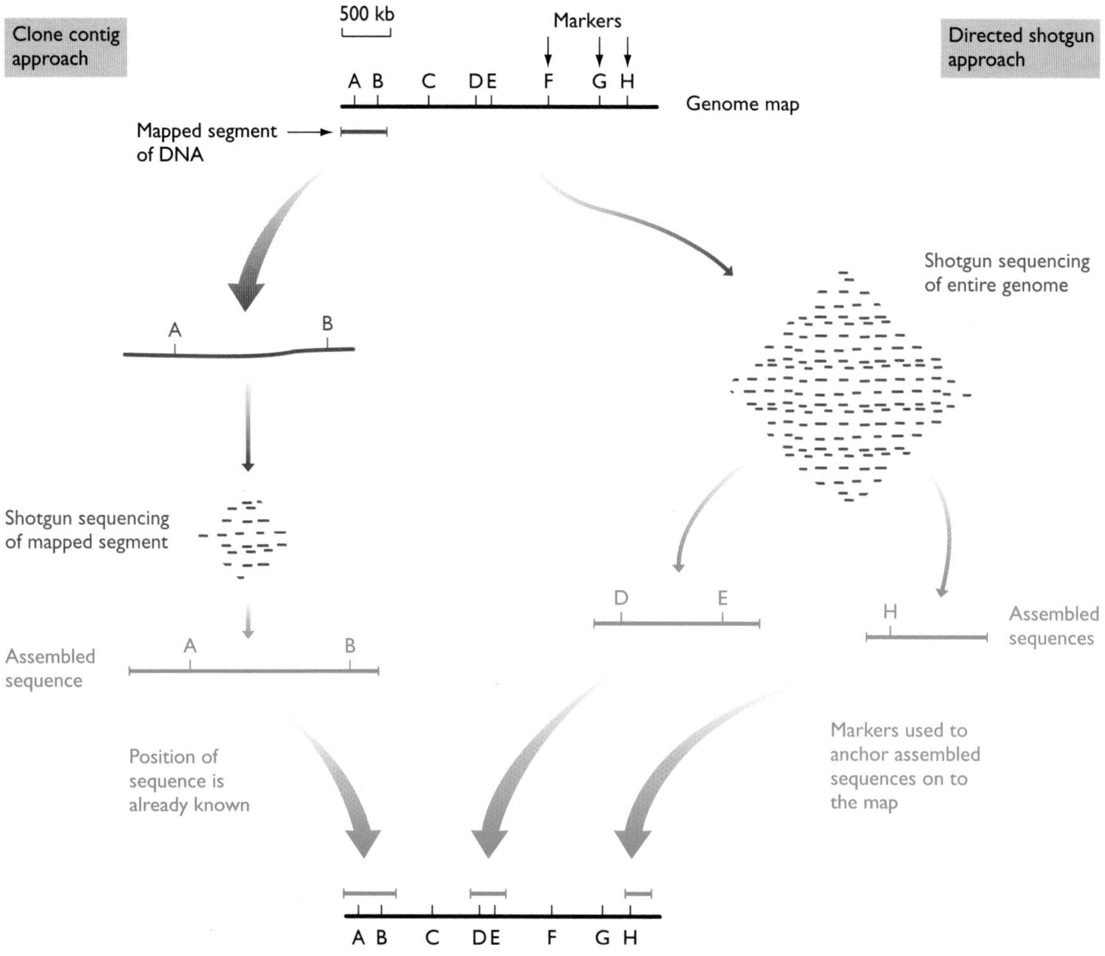

Figure 2.3 Alternative approaches to genome sequencing.

A genome, consisting of a linear DNA molecule of 2.5 Mb, has been mapped and the positions of eight markers (A–H) are known. On the left, the clone contig approach starts with a segment of DNA whose position on the genome map has been identified because it contains markers A and B. The segment is sequenced by the shotgun method and the master sequence placed at its known position on the map. On the right, the directed shotgun approach involves random sequencing of the entire genome. This results in pieces of contiguous sequence, possibly hundreds of kb in length. If a contiguous sequence contains a marker, then it can be positioned on the genome map. Note that, with either method, the more markers there are on the genome map the better. For more details of these sequencing strategies, see Section 4.2.

gene maps more comprehensive it would be necessary to find characteristics that were more numerous, more distinctive and less complex than visual ones.

The answer was to use biochemistry to distinguish phenotypes. This has been particularly important with two types of organisms – microbes and humans. Microbes, such as bacteria and yeast, have very few visual characteristics so gene mapping with these organisms has to rely on biochemical phenotypes such as those listed in *Table 2.1*. Humans have visual characteristics, but biochemical phenotypes scorable by blood typing have been studied since the 1920s, not only the standard blood groups such as the ABO series (Yamamoto *et al.*, 1990), but also allelic variants of blood serum proteins and immunological proteins such as the human leukocyte antigens (the HLA system). A big advantage of these markers over visual phenotypes is that many of the relevant genes have **multiple alleles** (Mori *et al.*, 1997). For example, the gene called *HLA-DRB1* has at least 59 alleles and *HLA-B* has at least 60. This is relevant because of the way in which gene mapping is carried out with humans (Section 2.3.2). Rather than setting up many breeding experiments, which is the procedure with an experimental organism such as fruit flies or mice, data on inheritance of human genes have to be gleaned by examining the phenotypes displayed by members of a single family. If all the family members are homozygous for the gene being studied then no useful information can be obtained. This will frequently be the case if the gene has just two alleles, because the relevant marriages occurred, by chance, between individuals who were homozygous for the same allele. This is much less likely if the gene being studied has 60 rather than two alleles.

Box 2.1: A quick guide to Mendelian genetics

Genetic mapping is based on the principles of inheritance as first described by Gregor Mendel in 1865 (Orel, 1995). From the results of his breeding experiments with peas, Mendel concluded that each pea plant possesses two alleles for each gene, but displays only one phenotype. This is easy to understand if the plant is pure-breeding, or **homozygous**, for a particular characteristic, as then it possesses two identical alleles and displays the appropriate phenotype. However, Mendel showed that if two pure-breeding plants with different phenotypes are crossed then all the progeny (the F_1 generation) display the same phenotype. These F_1 plants must be **heterozygous** for the phenotypes being studied, as each inherits one allele from either parent. The phenotype expressed in the F_1 plants is therefore **dominant** over the second, **recessive** phenotype.

Gene	Alleles	
Height	Tall *T*	Short *t*
Pea shape	Round *R*	Wrinkled *r*

Parents	Tall *TT* × *tt* Short	Round *RR* × *rr* Wrinkled
	F_1 plants inherit one allele from each parent	
F_1 genotypes	All *Tt*	All *Rr*
F_1 phenotypes	All tall	All round
Conclusion	Tall is dominant	Round is dominant

Mendel carried out additional crosses that enabled him to establish two Laws of Genetics. The First Law states that ***alleles segregate randomly***. In other words, if the parent's alleles are *A* and *a*, then a member of the F_1 generation has the same chance of inheriting *A* as it has of inheriting *a*. The Second Law is that ***pairs of alleles segregate independently***, so that inheritance of the alleles of gene A is independent of inheritance of the alleles of gene B. Because of these laws, the outcomes of genetic crosses are predictable, as shown in the example at the top of the next column.

Mendel's great legacy to modern genetics was the fact that, to all intents and purposes, he was absolutely correct in the picture he painted of genes and the way they are inherited. We now know that the two alleles of a gene are

Monohybrid cross

Parents Tall *Tt* × *Tt* Tall

F_1 genotypes

	T	*t*
T	*TT*	*Tt*
t	*Tt*	*tt*

F_1 phenotypes 3 tall : 1 short

Dihybrid cross

Parents Tall round *TtRr* × *TtRr* Tall round

F_1 genotypes

	TR	*Tr*	*tR*	*tr*
TR	*TTRR*	*TTRr*	*TtRR*	*TtRr*
Tr	*TTRr*	*TTrr*	*TtRr*	*Ttrr*
tR	*TtRR*	*TtRr*	*ttRR*	*ttRr*
tr	*TtRr*	*Ttrr*	*ttRr*	*ttrr*

F_1 phenotypes 9 tall round : 3 tall wrinkled : 3 short round : 1 short wrinkled

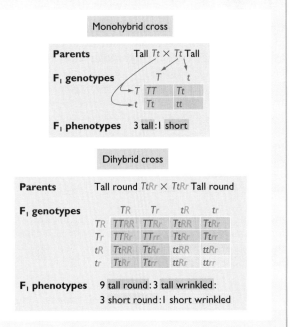

carried by the two homologous chromosomes in a diploid cell (see p. 24) and that the way in which alleles are transmitted from parent to offspring, as deduced by Mendel, corresponds with the events that occur when haploid gametes are produced during meiosis (see *Figure 2.10*). We also appreciate that the simple dominant–recessive rule proposed by Mendel can be complicated by situations that he did not encounter with his pea plants. One of these is **incomplete dominance**, where the heterozygous phenotype is intermediate between the two homozygous forms. An example is when red carnations are crossed with white ones, the F_1 heterozygotes being pink. Another complication is **codominance**, when both alleles are detectable in the heterozygote. Codominance is the typical situation for DNA markers, whose alleles can be detected directly by DNA methods (Section 2.2.2). Mendel made only one major error: his Second Law does not allow for **linkage**, the possibility that the alleles of two genes will be inherited together because they are on the same chromosome. As we will see in Section 2.3.1, the discovery of linkage by Mendel's followers led to the methods used for genetic mapping.

Table 2.1 Typical biochemical markers used for genetic analysis of *Saccharomyces cerevisiae*

Marker	Phenotype	Method by which cells carrying the marker are identified
ADE2	Requires adenine	Grows only when adenine is present in the medium
CAN1	Resistant to canavanine	Grows in the presence of canavanine
CUP1	Resistant to copper	Grows in the presence of copper
CYH1	Resistant to cycloheximide	Grows in the presence of cycloheximide
LEU2	Requires leucine	Grows only when leucine is present in the medium
SUC2	Able to ferment sucrose	Grows if sucrose is the only carbohydrate in the medium
URA3	Requires uracil	Grows only when uracil is present in the medium

2.2.2 DNA markers for genetic mapping

Genes are very useful markers but they are by no means ideal. One problem, especially with larger genomes such as those of vertebrates and flowering plants, is that a map based entirely on genes is not very detailed. This would be true even if every gene could be mapped because, as we saw in Chapter 1, in the genomes of these organisms the genes are widely spaced out with large gaps between them (see *Figure 1.7*, p. 10). The problem is made worse by the fact that only a fraction of the total number of genes exist in allelic forms that can be distinguished conveniently. Gene maps are therefore not very comprehensive. We need other types of marker.

Mapped features that are not genes are called **DNA markers**. As with gene markers, to be useful a DNA marker must exist in at least two allelic forms. There are three types of DNA sequence feature that satisfy this requirement.

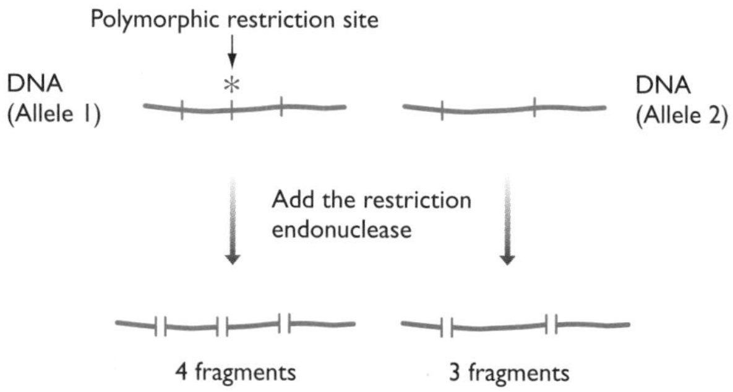

Figure 2.4 A restriction fragment length polymorphism (RFLP).

The DNA molecule on the left has a polymorphic restriction site (marked with the asterisk) that is not present in the molecule on the right. The RFLP is revealed after treatment with the restriction enzyme because one of the molecules is cut into four fragments whereas the other is cut into three fragments.

Restriction fragment length polymorphisms (RFLPs)

RFLPs were the first type of DNA marker to be studied. Restriction fragments are produced when a DNA molecule is treated with a **restriction endonuclease**, which is a type of enzyme that cuts DNA molecules at defined sequences (see Technical Note 3.1, p. 40). For example, the restriction endonuclease *Eco*RI cuts double-stranded DNA at:

5′–G A A T T C–3′
3′–C T T A A G–5′

and at no other sequence. This sequence specificity means that treatment of a DNA molecule with a restriction enzyme should always produce the same set of fragments. This is not always the case with genomic DNA molecules because some restriction sites are polymorphic, existing as two alleles, one allele displaying the correct sequence for the restriction site and therefore being cut when the DNA is treated with the enzyme, the second allele with a sequence alteration so the restriction site is no longer recognized. The result of the latter is that the two adjacent restriction fragments remain linked together after treatment with the enzyme, leading to a length polymorphism (*Figure 2.4*). This is an RFLP and its position on a genome map can be worked out by following the inheritance of its alleles, just as is done when genes are used as markers. There are thought to be about 10^5 RFLPs in the human genome, but of course for each RFLP there can only be two alleles (with and without the site). The value of RFLPs in human gene mapping is therefore limited by the high possibility that any human family will be entirely homozygous for an individual RFLP.

In order to score an RFLP, it is necessary to determine the size of just one or two individual restriction fragments against a background of many irrelevant fragments. This is not a trivial problem: an enzyme such as *Eco*RI, with a 6-bp recognition sequence, should cut DNA approximately once every $4^6 = 4096$ bp (Section 3.1.2) and so would give almost 750 000 fragments when used with human DNA. Fortunately, there are techniques that enable individual fragments to be detected:

- **Southern hybridization** (Technical Note 2.1) was the first technique used to type RFLPs. The DNA fragments resulting from restriction are separated from one another by **agarose gel electrophoresis** and the ones being studied detected by **hybridization probing**.
- The **polymerase chain reaction** or **PCR** (Technical Note 2.2) is now more commonly used because it is faster than Southern analysis. In PCR, the region of DNA containing the RFLP is amplified into multiple copies before the DNA is treated with the restriction enzyme. Now the RFLP can be visualized by staining an agarose gel of the restriction fragments with ethidium bromide, which is a less sensitive detection method than hybridization probing. The background of unamplified DNA does not show up, the only fragments that are seen being those derived from the amplified DNA.

Note that these techniques directly detect both alleles of an RFLP: this means that (as with other DNA markers) the alleles are codominant.

Simple sequence length polymorphisms (SSLPs)

SSLPs are arrays of repeat sequences that display length variations, different alleles containing different numbers of repeat units (*Figure 2.5A*). Unlike RFLPs, SSLPs can be multiallelic as each SSLP can have a number of different length variants. There are two types of SSLP:

- **Minisatellites**, also known as **variable number of tandem repeats** (**VNTRs**), in which the repeat unit is a few tens of nucleotides in length.
- **Microsatellites** or **simple tandem repeats** (**STRs**), whose repeats are much shorter, usually di-, tri- or

TECHNICAL **2.1** NOTES

Southern hybridization

Detection of a specific restriction fragment against a background of many other restriction fragments.

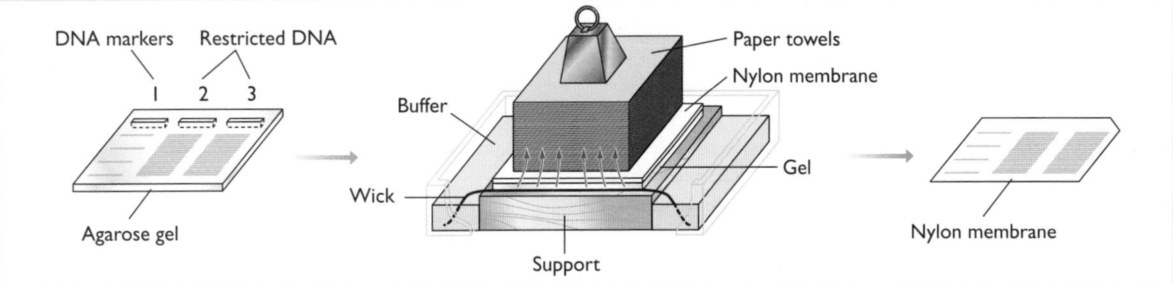

In Southern hybridization, a collection of restriction fragments is transferred from an agarose gel to a nylon membrane and the specific ones being studied are detected by hybridization probing. As well as its use in RFLP mapping (Section 2.2.2) it has many applications in other procedures that require identification of specific restriction fragments, for example to check the identity of a cloned fragment (Technical Note 3.4, p. 52).

Agarose gel electrophoresis (Technical Note 3.2, p. 43) separates linear DNA molecules, such as restriction fragments, according to their sizes. If there are only a few fragments in the digest then each one forms a separate band that can be seen after staining the gel with ethidium bromide. If there are more than about 50 fragments then the bands merge together into a smear and individual fragments cannot be seen. Instead, the fragments are transferred to a nylon membrane, simply by placing the membrane on the gel and allowing buffer to soak through, taking the DNA from the gel to the membrane where it becomes bound (top figure). You can buy an expensive machine to do this, but it is simpler to use a pile of paper towels.

A hybridization probe is a labeled DNA molecule whose sequence is complementary to the target DNA that we wish to detect. Because the probe and target DNAs are complementary they can base pair or **hybridize**. The position of the target restriction fragment on the membrane is therefore identified by detecting the signal given out by the label attached to the probe. The probe could be a synthetic oligonucleotide (see *Figure 2.7*) or it could be a cloned DNA fragment (Technical Note 3.4, p. 53). For detecting an RFLP, the probe is usually a cloned DNA fragment that spans the polymorphic restriction site.

To carry out the hybridization, the membrane is placed in a glass bottle with the labeled probe and some buffer, and the bottle gently rotated for several hours so that the probe has plenty of opportunity to hybridize to its target DNAs. The membrane is then washed to remove any probe that has not become hybridized, and the signal from the label is detected. Details of labeling and detection are given in Technical Note 3.3, p. 46; in the example shown below the probe is radioactively labeled and the signal is detected by **autoradiography**. The bands that are seen on the autoradiograph are those that are complementary to the probe. We see that Lane 2 has just one hybridizing band, but Lane 3 has two bands. The DNA in Lane 3 has therefore been cut at the polymorphic restriction site, whereas the DNA in lane 2 has not.

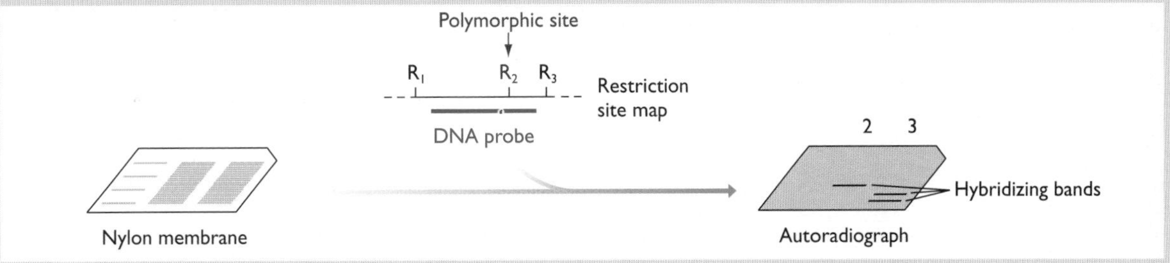

TECHNICAL 2.2 NOTES

Polymerase chain reaction

Exponential amplification of a selected region of a DNA molecule.

The polymerase chain reaction (PCR) results in amplification of a selected region of a DNA molecule. Regions of 5 kb and more can be amplified without difficulty, and longer amplifications – up to 40 kb – are possible using modifications to the standard technique. The technique has many applications in molecular biology, for example in DNA mapping (e.g. Section 2.2.2), DNA sequencing (e.g. Sections 4.1.1 and 4.2.1) and molecular phylogenetics (Chapter 15). A modified version of PCR can be used to amplify DNA copies of RNA molecules (Technical Note 5.1, p. 91). PCR is also used extensively in specialist areas such as forensic science and clinical diagnosis, where its ability to work with very few starting molecules is particularly useful.

The reaction is carried out by mixing together the target DNA molecule, which can be present in extremely small amounts, with nucleotides, two synthetic oligonucleotide **primers**, and a thermostable DNA polymerase (see Technical Note 4.2, p. 64) that is resistant to denaturation by heat treatment. Usually *Taq* polymerase from the bacterium *Thermus aquaticus*, which lives in hot springs, is used. The two primers must anneal to the target DNA on either side of the region to be amplified, which means that the sequences of these borders must be known so that the appropriate oligonucleotides can be made. The oligonucleotides prime the synthesis of new complementary polynucleotides which, as with all enzymatic DNA syntheses, are made in the 5′ to 3′ direction by copying the template DNA in the 3′ to 5′ direction (see *Figure 1.1*, p. 3 and Section 12.3.2).

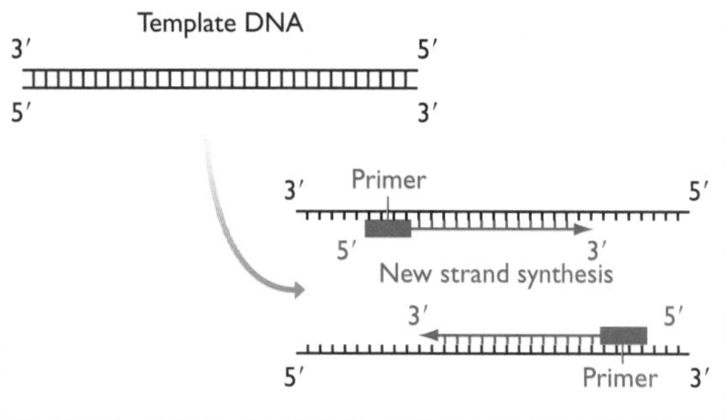

Because the *Taq* polymerase is thermostable, the reaction mixture can be heated to 90°C without destroying the enzyme activity. At this temperature, the new strands detach from the template DNA. When the mixture is cooled down again, more primers anneal to the template DNA and also to the new strands, and the *Taq* polymerase carries out a second cycle of DNA synthesis.

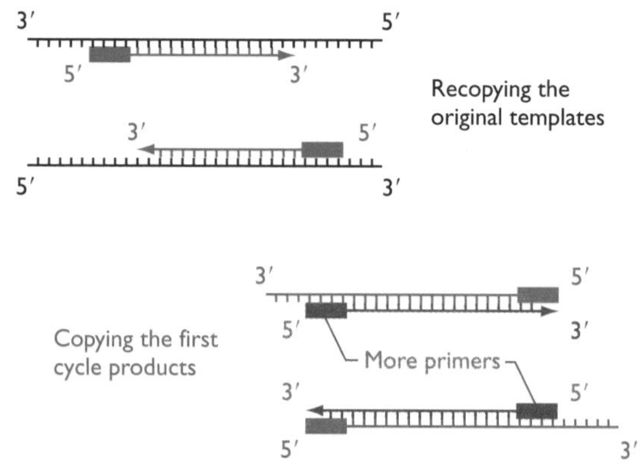

The PCR can be continued for 30–40 cycles before the enzyme eventually becomes inactivated or the primers or nucleotides are used up. A single starting molecule can be amplified into tens of millions of identical fragments, representing a few micrograms of DNA. The presence of a polymorphic restriction site in the amplified region can be tested by digesting the PCR product with the restriction enzyme and then running a sample on an agarose gel.

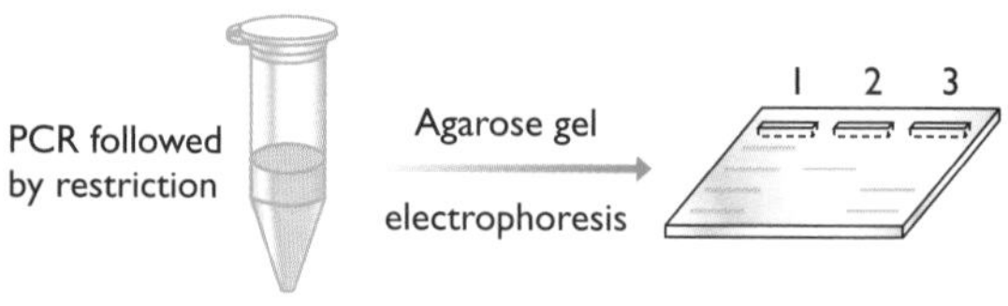

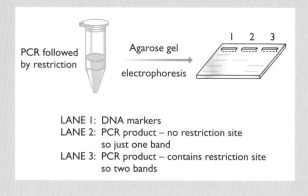

LANE 1: DNA markers
LANE 2: PCR product – no restriction site so just one band
LANE 3: PCR product – contains restriction site so two bands

tetranucleotide units. We met two microsatellites in Chapter 1 when we looked at a typical 50-kb segment of the human genome (see *Figure 1.5*, p. 6).

Microsatellites are more popular than minisatellites as DNA markers for two reasons. First, minisatellites are not spread evenly around the genome but tend to be found more frequently near the ends of chromosomes. In geographical terms, this is equivalent to trying to use a map of lighthouses to find one's way around the middle of an island. Microsatellites are more conveniently spaced throughout the genome. Second, the quickest way to type a length polymorphism is by PCR (*Figure 2.5B*), but PCR typing is much quicker and more accurate with sequences less than 300 bp in length. Most minisatellite alleles are longer than this because the repeat units are relatively long and there tend to be many of them in a single array. Typical microsatellites consist of 10–30 copies of a repeat that is no longer than four bp in length, and so they are more amenable to PCR typing. About 10^4 microsatellites have so far been discovered in the human genome, but there are probably many more (Section 6.3.1).

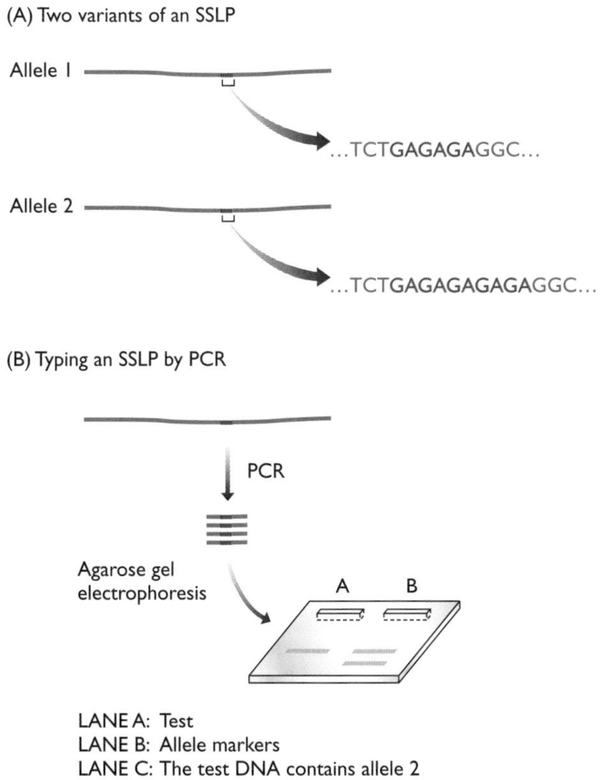

Figure 2.5 SSLPs and how they are typed.

(A) Two alleles of a microsatellite SSLP. In allele 1 the motif 'GA' is repeated three times, and in allele 2 it is repeated five times. (B) How the SSLP could be typed by PCR. The region surrounding the SSLP is amplified and the products examined by agarose gel electrophoresis. The PCR products give a band that corresponds in size to the larger of the two allele sequences, showing that the DNA that was tested contained allele 2. For a description of gel electrophoresis, see Technical Note 3.2, p. 43.

Single nucleotide polymorphisms (SNPs)

These are individual **point mutations** in the genome (*Figure 2.6*). There are vast numbers of these, some of which also give rise to RFLPs, but many of which do not because the sequence in which they lie is not recognized by any restriction enzyme. In the human genome there are thought to be over 200 000 SNPs that lie in genes, and probably ten times this number, possibly more, in non-genic DNA (Collins *et al.*, 1997).

Each SNP has just two alleles, so these markers suffer from the same drawback as RFLPs with regards to human genetic mapping: there is a high possibility that all members of a human family will be homozygous for an individual SNP. The advantages of SNPs are their abundant numbers and the fact that they can be typed by methods that do not involve gel electrophoresis. This is important because gel electrophoresis has proved difficult to automate so any detection method that uses it will be relatively slow and labor intensive. SNP detection is more rapid because it is based on **oligonucleotide hybridization analysis**. An oligonucleotide is a short, single-stranded DNA molecule, usually less than 50 nucleotides in length, that is synthesized in the test tube. If the conditions are just right, then an oligonucleotide will hybridize with another DNA molecule only if the oligonucleotide forms a completely base-paired structure with the second molecule. If there is a single mismatch, a single position within the oligonucleotide that does not form a base pair, then hybridization does not occur (*Figure 2.7*). Oligonucleotide hybridization can therefore discriminate between the two alleles of a SNP. Screening strategies include:

- **DNA chip** technology (Technical Note 2.3). A DNA chip is a wafer of silicon, 2 cm^2 or less in area, carrying many different oligonucleotides in a high density array. The DNA to be tested is labeled with a fluorescent marker and pipetted on to the surface of the chip. Hybridization is detected by examining the chip with a fluorescence microscope, the positions at which the fluorescent signal is emitted indicating which oligonucleotides have hybridized with the test DNA. Many SNPs can therefore be scored in a single experiment (Wang *et al.*, 1998).

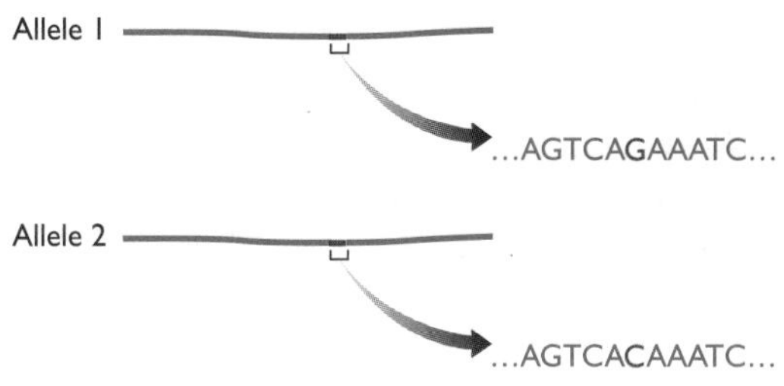

Figure 2.6 A single nucleotide polymorphism.

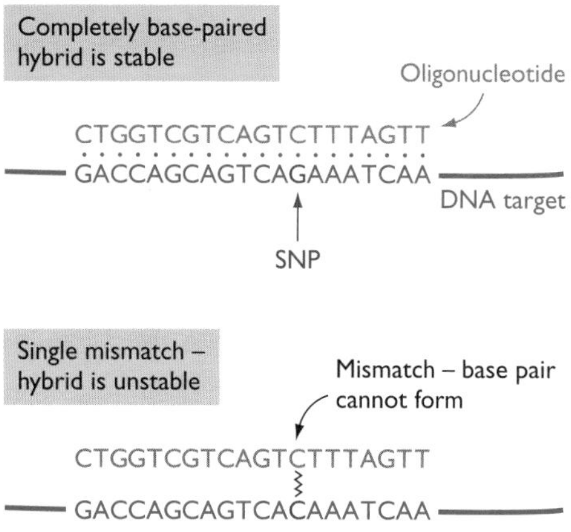

Figure 2.7 Oligonucleotide hybridization is very specific.

Under highly stringent hybridization conditions, a stable hybrid occurs only if the oligonucleotide is able to form a completely base-paired structure with the target DNA. If there is a single mismatch then the hybrid does not form. To achieve this level of stringency the incubation temperature must be just below the **melting temperature** or $\mathbf{T_m}$ of the oligonucleotide. At temperatures above the T_m, even the fully base-paired hybrid is unstable. At more than 5°C below the T_m, mismatched hybrids might be stable. The T_m for the oligonucleotide shown in the figure would be about 58°C. This is calculated from the formula $T_m = (4 \times \text{number of G and C nucleotides}) + (2 \times \text{number of A and T nucleotides})$ °C.

- **Dynamic allele-specific hybridization** (DASH; Pennisi, 1998). In this technique, hybridization occurs in solution, for example in one of the wells of a 96-well microtiter tray, and is detected by a fluorescent marker that binds only to double-stranded DNA and so emits a signal only when hybridization takes place. The hybridization is initially carried out under conditions that allow mismatched hybrids to form. At this stage of the experiment, the oligonucleotide and test DNA hybridize regardless of which SNP allele the latter contains. Discrimination between alleles is achieved by raising the temperature, because mismatched hybrids are less stable than complete hybrids and break down at a lower temperature. Which allele is present in the test DNA can therefore be determined from the temperature at which the hybridization-dependent fluorescent signal disappears.

2.3 APPROACHES TO GENETIC MAPPING

2.3.1 *Linkage analysis is the basis of genetic mapping*

Now that we have assembled a set of markers with which to construct a genetic map we can move on to look at the mapping techniques themselves. These techniques are all based on **genetic linkage**, which was first discovered by Bateson, Saunders and Punnett in 1905 but not fully understood until Thomas Hunt Morgan began his groundbreaking work with fruit flies in 1910–11. The early geneticists of the 20th century recognized that genes are located in chromosomes and that each individual chromosome is inherited as an intact unit. In other words, if a pair of genes is located in the same chromosome then they are physically linked together and therefore should be inherited together. This appeared to be a secure hypothesis, but when the early geneticists carried out genetic crosses similar to those published 40 years earlier by Mendel, they discovered that very few pairs of genes displayed complete linkage. Pairs of genes were either inherited independently, as expected for genes in different chromosomes, or if they showed linkage then it was only **partial linkage**: sometimes they were inherited together and sometimes they were not (*Figure 2.8*). The resolution of this contradiction between theory and observation was the critical step in the development of genetic mapping techniques.

The behavior of chromosomes during meiosis explains why genes display partial and not complete linkage

The critical breakthrough was made by Thomas Hunt Morgan, who made the conceptual leap between partial linkage and the behavior of chromosomes when the nucleus of a cell divides. Cytologists in the late 19th century had distinguished two types of nuclear division: **mitosis** and **meiosis**. Mitosis is more common, being the process by which the diploid nucleus of a somatic cell divides to produce two daughter nuclei, both of which are still diploid (*Figure 2.9*). Approximately 10^{17} mitoses are needed to produce all the cells required during a human lifetime. Before mitosis begins, each chromosome in the nucleus is replicated, but the replicants do not immediately break away from one another. To begin with they remain attached at their **centromeres**, and they do not separate until later in mitosis when the chromosomes are distributed between the two new nuclei. It is obviously important that each daughter nucleus receives a complete set of chromosomes, and most of the intricacies of mitosis appear to be devoted to achieving this end.

Mitosis illustrates the basic events occurring during nuclear division but it is not directly relevant to genetic mapping. Instead, it is the distinctive features of meiosis that interest us. Meiosis occurs only in reproductive cells, and results in a diploid cell giving rise to four haploid gametes, each of which can subsequently fuse with a gamete of the opposite sex during sexual reproduction. The fact that meiosis results in four haploid cells whereas mitosis gives two diploid cells is easy to explain: meiosis involves two nuclear divisions, one after the other, whereas mitosis is just a single nuclear division. The critical difference between mitosis and meiosis is more subtle. Recall that in a diploid cell there are two separate

TECHNICAL 2.3 NOTES

DNA chips

High density arrays of DNA molecules for parallel hybridization analyses.

DNA chips are designed to allow many hybridization experiments to be performed in parallel. Their main applications have been in the screening of polymorphisms such as SNPs (Section 2.2.2) and comparing the RNA populations of different cells (Section 5.4.1). They also have potential in novel DNA sequencing methodologies (Section 4.1.2).

A DNA chip carries a large number of DNA probes, each one with a different sequence and each at a defined position on the surface of the chip. The probes can be synthetic oligonucleotides or other short DNA molecules such as **complementary DNAs (cDNAs**; see *Figure 3.11*, p. 50). In the earliest technology, the oligonucleotides or cDNAs were spotted on to a glass microscope slide or a piece of nylon membrane to form a **microarray**. With this approach, only a relatively low density can be achieved, typically 6400 spots as an 80 × 80 array in an area of 18 mm by 18 mm. To prepare really high density arrays a more sophisticated technology was needed. The answer has been to synthesize oligonucleotides *in situ* on the chip surface, which in turn has required an advance in the methodology used for preparing oligonucleotides, as shown in the figure opposite. The normal method involves adding nucleotides one by one to the growing end of an oligonucleotide, the sequence determined by the order in which the dNTP substrates are added to the reaction mixture. If used for synthesis on a chip, this method would result in each oligonucleotide having the same sequence. Instead, modified dNTP substrates are used, ones that have to be light activated before they will attach to the end of a growing oligonucleotide. The dNTPs are added one after another to the chip surface, photolithography being used to direct pulses of light at individual positions in the array and hence determine which of the growing oligonucleotides will be extended by addition of the particular dNTP added at each step.

A density of up to one million oligonucleotides per cm^2 is possible (Ramsay, 1998), so if used for SNP screening, half a million polymorphisms can be typed in a single experiment, presuming there are oligonucleotides for both alleles of each SNP. A DNA chip is not complicated to use (see figure below). The chip is incubated with labeled target DNA to allow hybridization to take place. Which oligonucleotides have hybridized to the target DNA is determined by scanning the surface of the chip and recording the positions at which the signal emitted by the label is detectable. Radioactive labels can be used with a low density microarray, signals being detected electronically by **phosphorimaging**. This does not provide enough resolution for a high density chip, so with these it is necessary to use a fluorescent label (Technical Note 3.3, p. 46). The fluorescent signal is detected by laser scanning or, more routinely, by fluorescent confocal microscopy.

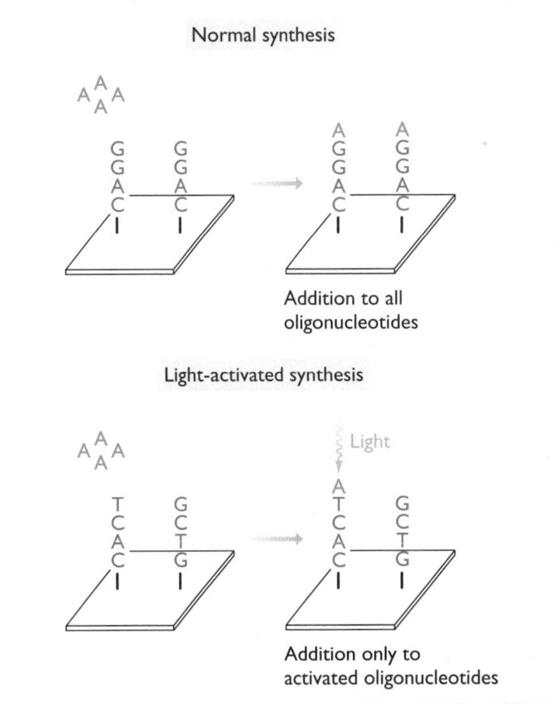

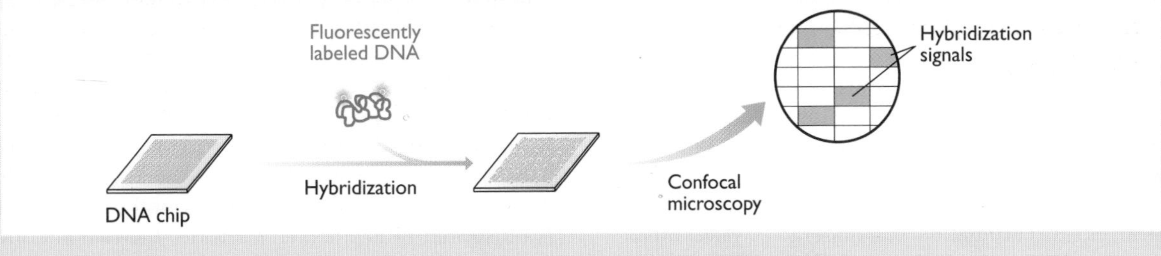

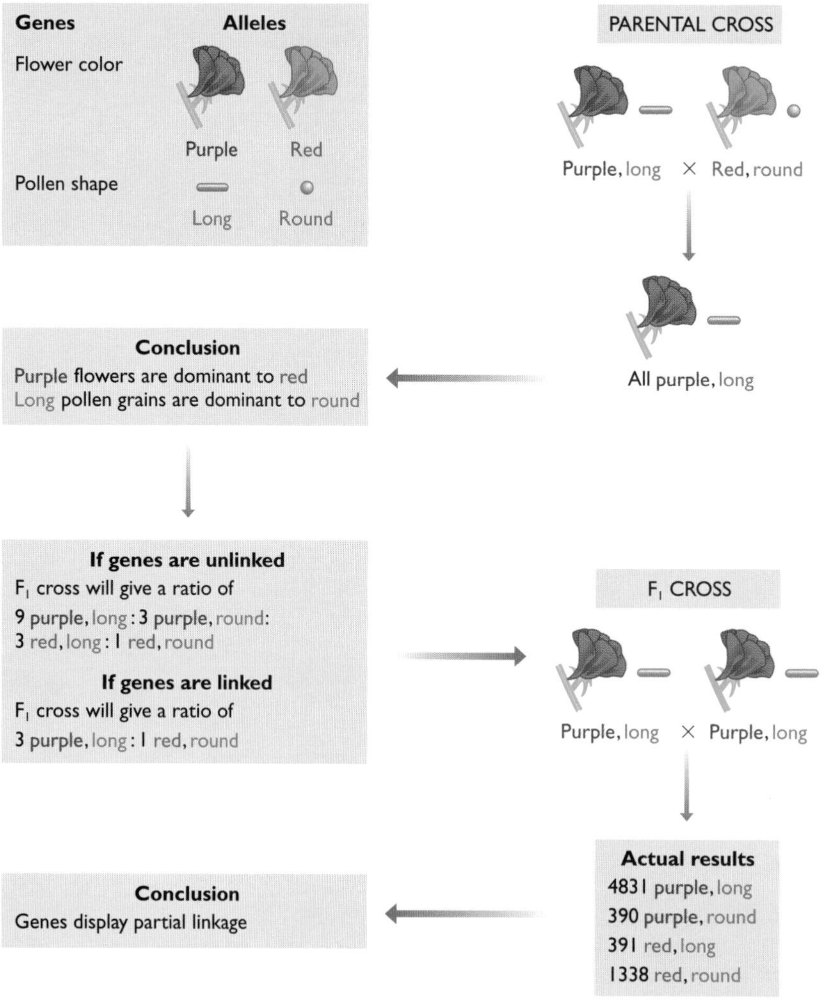

Figure 2.8 Partial linkage.

Partial linkage was discovered in the early 20th century. The cross shown here was carried out by Bateson, Saunders and Punnett in 1905 with sweet peas. The parental cross gives the typical dihybrid result (see Box 2.1, p. 17) with all the F_1 plants displaying the same phenotype, indicating that the dominant alleles are purple flowers and long pollen grains. The F_1 cross gives unexpected results as the progeny show neither a 9:3:3:1 ratio (expected for genes on different chromosomes) nor a 3:1 ratio (expected if the genes are completely linked). An unusual ratio is typical of partial linkage.

copies of each chromosome (Section 1.1.1). We refer to these as pairs of **homologous chromosomes**. During mitosis, homologous chromosomes remain separate from one another, each member of the pair replicating and being passed to a daughter nucleus independently of its homolog. In meiosis, however, the pairs of homologous chromosomes are by no means independent. During meiosis I, each chromosome lines up with its homolog to form a **bivalent** (*Figure 2.10*). This occurs after each chromosome has replicated, but before the replicated structures split, so the bivalent in fact contains four chromosome copies, each of which is destined to find its way into one of the four gametes that will be produced at the end of the meiosis. Within the bivalent, the chromosome arms (the **chromatids**) can undergo physical breakage and exchange of segments of DNA. This is called **crossing-over** or **recombination** and was discovered by the Belgian cytologist Janssens in 1909, just 2 years before Morgan started to think about partial linkage.

How did the discovery of crossing-over help Morgan explain partial linkage? To understand this we need to think about the effect that crossing-over can have on the inheritance of genes. Let us consider two genes, each of which has two alleles. We will call the first gene A and its alleles *A* and *a*, and the second gene B with alleles *B* and *b*. Imagine that the two genes are located on chromosome number 2 of *Drosophila melanogaster*, the species of fruit fly studied by Morgan. We are going to follow the meiosis of a diploid nucleus in which one copy of chromosome 2 has alleles *A* and *B*, and the second has *a* and *b*. This situation is illustrated in *Figure 2.11*. Consider the two alternative scenarios:

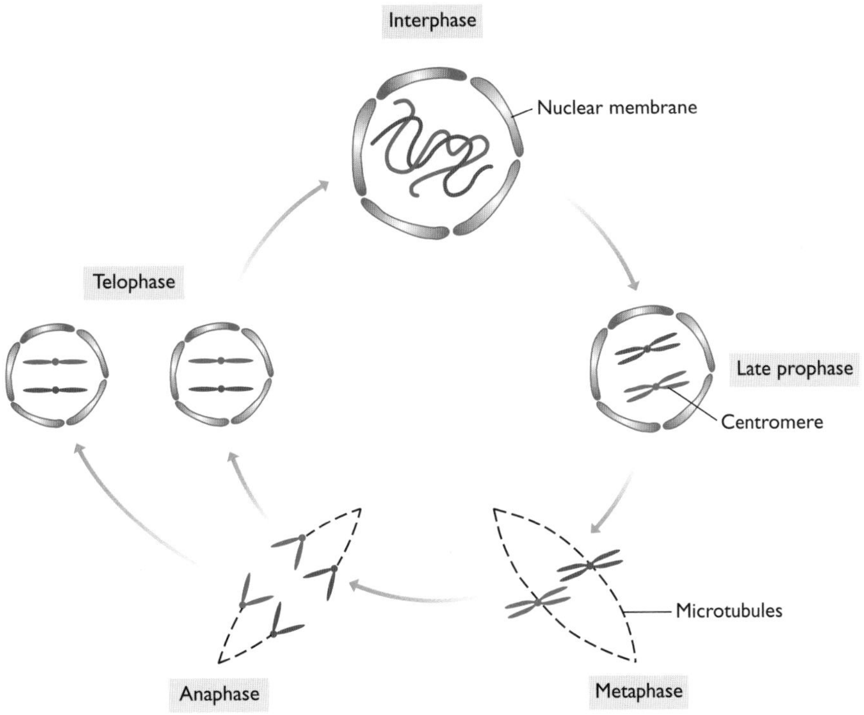

Figure 2.9 Mitosis.

During interphase, the period between nuclear divisions, the chromosomes are in their extended form (Section 6.1.1). At the start of mitosis the chromosomes condense and by late prophase have formed structures that are visible with the light microscope. Each chromosome has already undergone DNA replication but the two daughter chromosomes are held together by the centromere. During metaphase the nuclear membrane breaks down (in most eukaryotes) and the chromosomes line up in the center of the cell. Microtubules now draw the daughter chromosomes towards either end of the cell. In telophase, nuclear membranes reform around each collection of daughter chromosomes. The result is that the parent nucleus has given rise to two identical daughter nuclei. For simplicity, just one pair of homologous chromosomes is shown; one member of the pair is red, the other is blue.

(i) ***A crossover does not occur between genes A and B.*** If this is what happens then two of the resulting gametes will contain chromosome copies with alleles *A* and *B*, the other two will contain *a* and *b*. In other words, two of the gametes have the **genotype** *AB* and two have the genotype *ab*.

(ii) ***A crossover does occur between genes A and B.*** This could lead to segments of DNA containing gene B being exchanged between homologous chromosomes. The eventual result is that each gamete has a different genotype: 1 *AB*, 1 *aB*, 1 *Ab*, 1 *ab*.

Now think about what would happen if we looked at the results of meiosis in 100 identical cells. If crossovers never occur then the resulting gametes will have the following genotypes:

200 *AB*
200 *ab*

This is complete linkage: genes A and B behave as a single unit during meiosis. But if (as is more likely) crossovers occur between A and B in some of the nuclei, then the allele pairs will not be inherited as single units. Let us say that crossovers occur during 40 of the 100 meioses. The following gametes will result:

160 *AB*
160 *ab*
40 *Ab*
40 *aB*

The linkage is not complete, it is only partial. As well as the two **parental** genotypes (*AB*, *ab*) we see gametes with **recombinant** genotypes (*Ab*, *aB*).

From partial linkage to genetic mapping

Once Morgan had understood how partial linkage could be explained by crossing over during meiosis he was able to devise a way of mapping the relative positions of genes on a chromosome. In fact the key breakthrough was made not by Morgan himself, but by an undergraduate working in his lab, Arthur Sturtevant (Sturtevant, 1913). Sturtevant assumed that crossing-over was a random event, there being an equal chance of it occurring at any position along a pair of lined up chromatids. If this assumption is correct then two genes that are close together will be separated by crossovers less frequently than two genes that

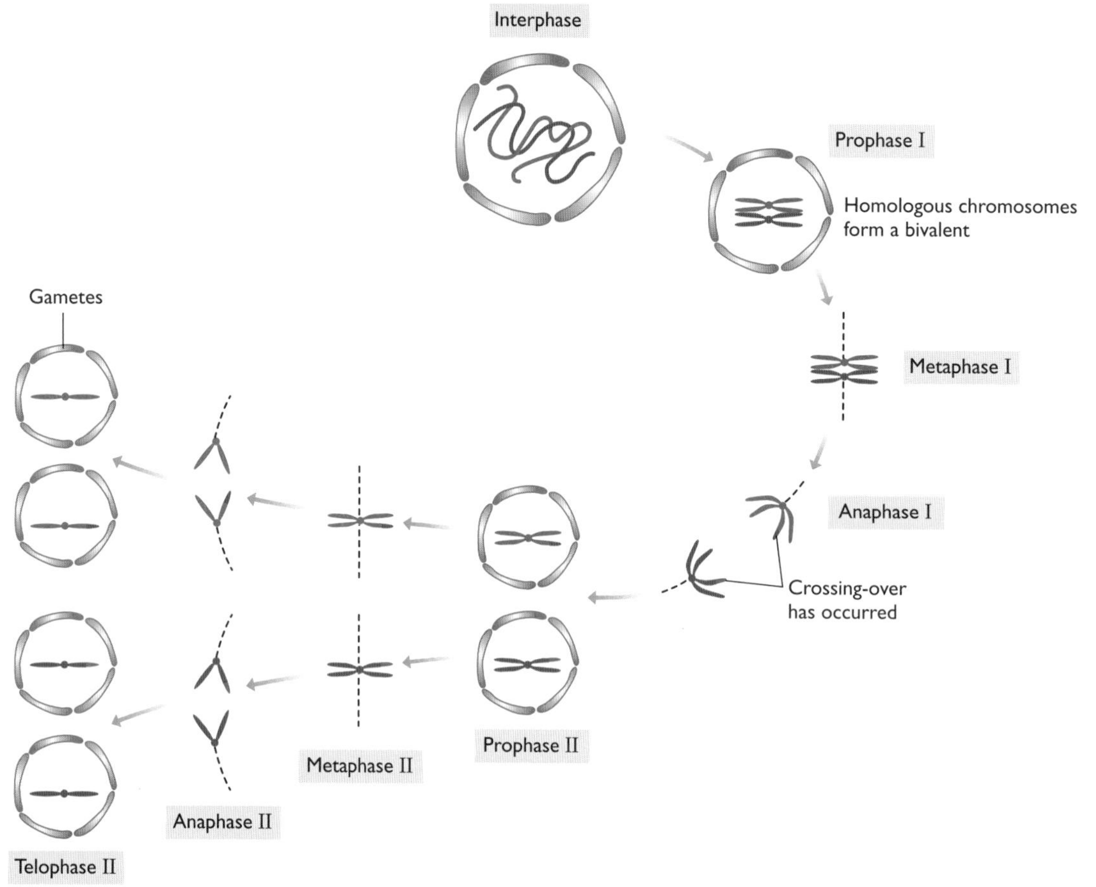

Figure 2.10 Meiosis.

The events involving one pair of homologous chromosomes are shown; one member of the pair is red, the other is blue. At the start of meiosis the chromosomes condense and each homologous pair lines up to form a bivalent. Within the bivalent, crossing-over might occur, involving breakage of chromosome arms and exchange of DNA. Meiosis then proceeds by a pair of mitotic nuclear divisions that result initially in two nuclei, each with two copies of each chromosome still attached at their centromeres, and finally in four nuclei, each with a single copy of each chromosome. These final products of meiosis, the gametes, are therefore haploid. The molecular basis of recombination will be described in Section 13.2.

are more distant from one another. Furthermore, the frequency with which the genes are unlinked by crossovers will be directly proportional to how far apart they are on their chromosome. The **recombination frequency** is therefore a measure of the distance between two genes. If you work out the recombination frequencies for different pairs of genes, you can construct a map of their relative positions on the chromosome (*Figure 2.12*).

It turns out that Sturtevant's assumption about the randomness of crossovers was not entirely justified. Comparisons between genetic maps and the actual positions of genes on DNA molecules, as revealed by physical mapping and DNA sequencing, have shown that some regions of chromosomes, called **recombination hotspots**, are more likely to be involved in crossovers than others. This means that a genetic map distance does not necessarily indicate the physical distance between two markers. Also, we now realize that a single chromatid can participate in more than one crossover at the same time, but that there are limitations on how close together these crossovers can be, leading to more inaccuracies in the mapping procedure. Despite these qualifications, **linkage analysis** usually makes correct deductions about gene order, and distance estimates are sufficiently accurate to generate genetic maps that are of value as frameworks for genome sequencing projects.

2.3.2 Carrying out linkage analysis with different types of organism

To see how linkage analysis is actually carried out, we need to consider three quite different situations:

- Linkage analysis with species such as fruit flies and mice, with which we can carry out planned breeding experiments.
- Linkage analysis with humans, with which we cannot carry out planned experiments but instead make use of family pedigrees.

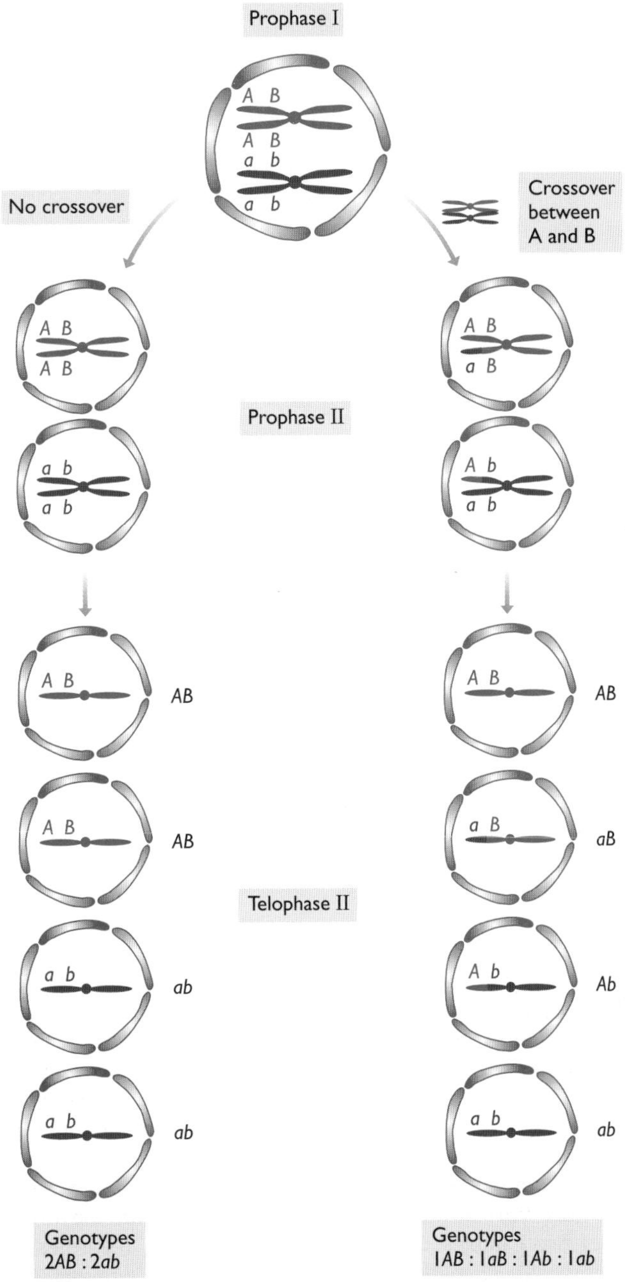

Figure 2.11 The effect of a crossover on linked genes.

The drawing shows a pair of homologous chromosomes, one red and the other blue. A and B are linked genes with alleles *A*, *a*, *B* and *b*. On the left is shown a meiosis with no crossover between A and B. Two of the resulting gametes have the genotype *AB* and the other two are *ab*. On the right, a crossover occurs between A and B: the four gametes display all of the possible genotypes: *AB*, *aB*, *Ab* and *ab*.

- Linkage analysis with bacteria, which do not undergo meiosis.

Linkage analysis when planned breeding experiments are possible

The first type of linkage analysis is the modern counterpart of the method developed by Morgan and his colleagues. The method is based on analysis of the progeny of experimental crosses set up between parents of known genotypes and is, at least in theory, applicable to all eukaryotes. Ethical considerations preclude its use with humans, and practical problems such as the length of the gestation period and the time taken for the newborn to reach maturity (and hence to participate in subsequent

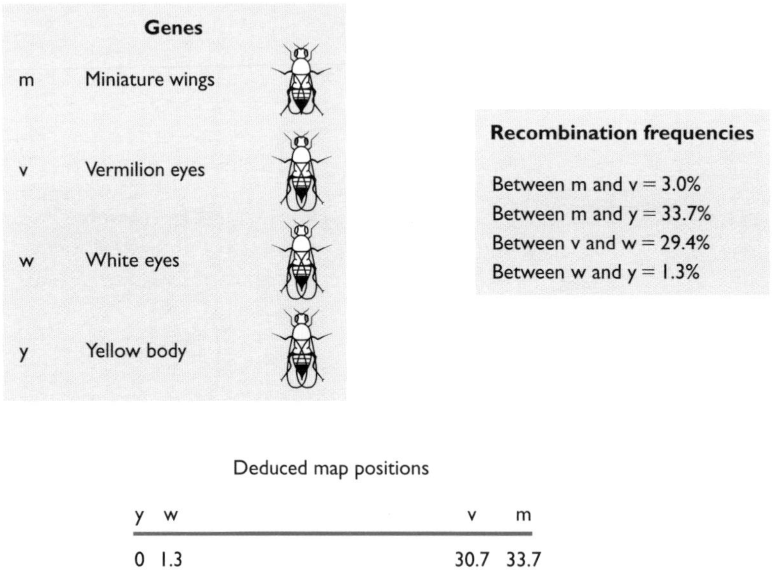

Figure 2.12 Working out a genetic map from recombination frequencies.

The example is taken from the original experiments with fruit flies carried out by Arthur Sturtevant. All four genes are on the fruit fly X chromosome. Recombination frequencies between the genes are shown, along with their deduced map positions.

crosses) limits the effectiveness of the method with some animals and plants.

If we return to *Figure 2.11* we see that the key to gene mapping is the ability to determine the genotypes of the gametes resulting from meiosis. In a few situations this is possible by directly examining the gametes. For example, the gametes produced by some microbial eukaryotes, including the yeast *Saccharomyces cerevisiae*, can be grown into colonies of haploid cells, whose genotypes can be determined by biochemical tests. Direct genotyping of gametes is also possible with higher eukaryotes if DNA markers are used, as PCR can be carried out with the DNA from individual spermatozoa, enabling RFLPs, SSLPs and SNPs to be typed. Unfortunately, sperm typing is laborious. Routine linkage analysis with higher eukaryotes is therefore carried out not by examining the gametes directly but by determining the genotypes of the diploid progeny resulting from fusion of two gametes, one from each of a pair of parents. In other words, a genetic cross is performed.

The complication with a genetic cross is that the resulting diploid progeny are the product not of one meiosis but of two (one in each parent), and in most organisms crossover events are equally likely to occur during production of both the male and female gametes. Somehow we have to be able to disentangle from the genotypes of diploid progeny the crossover events that occurred in each of these two meioses. This means that the cross has to be set up with care. The standard procedure is to use a **test cross**. This is illustrated in *Figure 2.13*, Scenario 1, where we have set up a test cross to map the two genes we met earlier, gene A (alleles *A* and *a*) and gene B (alleles *B* and *b*), both on chromosome 2 of the fruit fly. The critical feature of a test cross is the genotype of the two parents:

- One parent is a **double heterozygote**. This means that all four alleles are present in this parent: its genotype is *AB*/*ab*. Double heterozygotes can be obtained by crossing two pure-breeding strains, for example *AB*/*AB* × *ab*/*ab* (see Box 2.1, p. 17).
- The second parent is a pure-breeding **double homozygote**. In this parent both homologous copies of chromosome 2 are the same: in the example shown in Scenario 1 both have alleles *a* and *b* and the genotype of the parent is *ab*/*ab*.

The double heterozygote has the same genotype as the cell whose meiosis we followed in *Figure 2.11*. Our objective is therefore to infer the genotypes of the gametes produced by this parent and to calculate the fraction that are recombinants. Note that all the gametes produced by the second parent, the double homozygote, will have the genotype *ab*, regardless of whether they are parental or recombinant gametes. As a result of this, meiosis in this parent is, in effect, invisible when the genotypes of the progeny are examined. This means that, as shown in *Figure 2.13*, Scenario 1, the genotypes of the diploid progeny can be unambiguously converted into the genotypes of the gametes from the double heterozygous parent. The test cross therefore enables us to make a direct examination of a single meiosis and hence to calculate a recombination frequency and map distance for the two genes being studied.

Just one additional point needs to be considered. If, as in *Figure 2.13*, Scenario 1, gene markers displaying dominance and recessiveness are used, then the double homozygous parent must have alleles for the two recessive phenotypes; but if codominant DNA markers are used, then the double homozygous parent can have any combination of homozygous alleles (i.e. *AB*/*AB*, *Ab*/*Ab*, *aB*/*aB* and *ab*/*ab*). Senario 2 in *Figure 2.13* shows the reason for this.

Gene mapping by human pedigree analysis

With humans it is of course impossible to preselect the genotypes of parents and set up crosses designed specifically to be useful for mapping purposes. Instead, data for the calculation of recombination frequencies have to be obtained by examining the genotypes of the members of successive generations of existing families. This means that only limited data are available and their interpretation is often difficult, as a human marriage rarely results in a convenient test cross, and often the genotypes of one or more family members are unobtainable because those individuals are dead or unwilling to cooperate.

The problems are illustrated by *Figure 2.14*. In this example we are studying a genetic disease present in a family of two parents and six children. Genetic diseases are frequently used as gene markers in humans, the disease state being one allele and the healthy state being a second allele. The pedigree in *Figure 2.14A* shows us that the mother is affected by the disease as are four of her children. We know from family accounts that the maternal grandmother also suffered from this disease, but both she and her husband, the maternal grandfather, are now dead. We can include them in the pedigree, with slashes indicating that they are dead, but we cannot obtain any further information on their genotypes. Our aim is to map the position of the gene for the genetic disease, and for this purpose we are studying its linkage to a microsatellite marker M, four alleles of which – M_1, M_2, M_3 and M_4 – are present in the living family members. The question is, how many of the children are recombinants?

If we look at the genotypes of the six children we see that numbers 1, 3 and 4 have the disease allele and the microsatellite allele M_1. Numbers 2 and 5 have the healthy allele and M_2. We can therefore construct two alternative hypotheses. The first is that the two copies of the relevant pair of homologous chromosomes in the mother have the genotypes disease-M_1 and healthy-M_2, and therefore

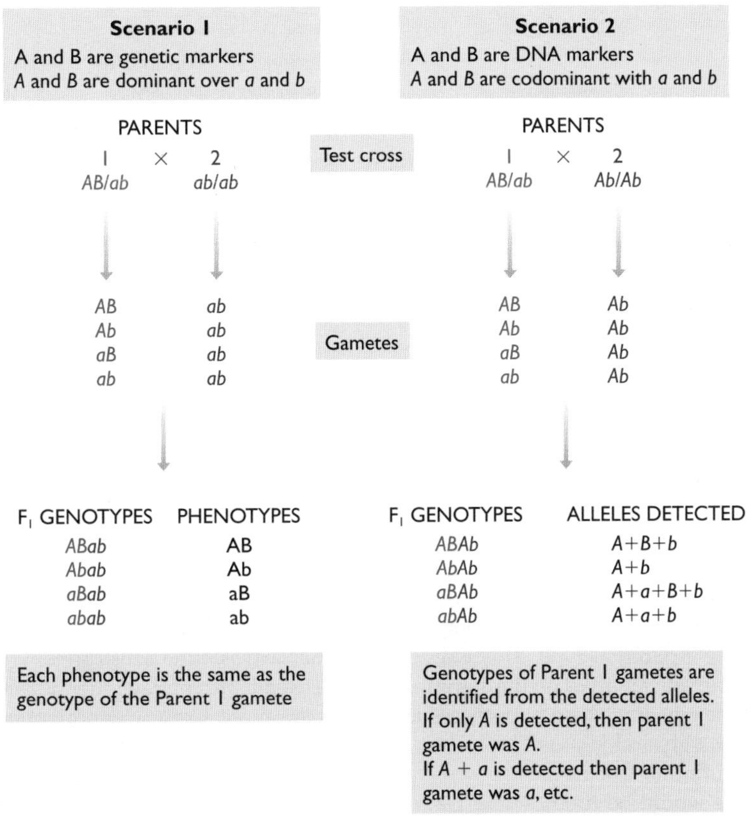

Figure 2.13 Two examples of the test cross.

In Scenario 1, A and B are genetic markers with alleles *A*, *a*, *B* and *b*. The resulting progeny are scored by examining their phenotypes. Because the double homozygous parent (Parent 2) has both recessive alleles, *a* and *b*, it effectively makes no contribution to the phenotypes of the progeny. The phenotype of each individual in the F_1 is therefore the same as the genotype of the gamete from Parent 1 that gave rise to that individual. In Scenario 2, A and B are DNA markers whose allele pairs are codominant. In this particular example, the double homozygous parent has the genotype *Ab/Ab*. The alleles present in each F_1 individual are directly detected, for example by PCR. These allele combinations enable the genotype of the Parent 1 gamete that gave rise to each individual to be deduced.

Box 2.2: Multipoint crosses

The power of linkage analysis is enhanced if more than two markers are followed in a single cross. Not only does this generate recombination frequencies more quickly, it enables the relative order of markers on a chromosome to be determined by simple inspection of the data. This is because two recombination events are required to unlink the central marker from the two outer markers in a series of three. Either of the two outer markers can be unlinked by just a single recombination.

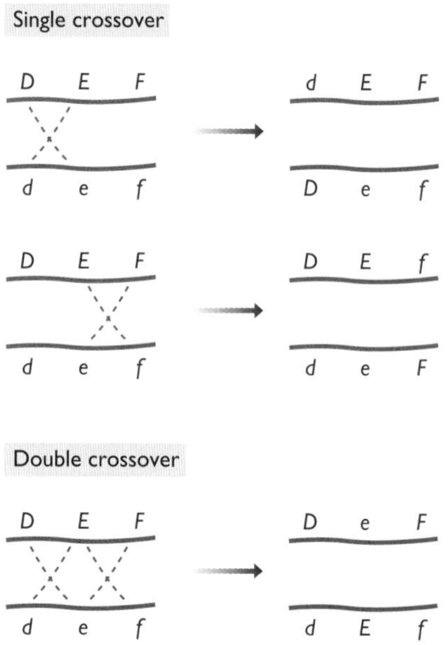

A double recombination is less likely than a single one, so unlinking of the central marker will occur relatively infrequently. An example of the data obtained from a three point cross is shown in the Table. A cross has been set up between a triple heterozygote (*ABC/abc*) and a triple homozygote (*abc/abc*). The most frequent class of progeny are those with one of the two parental genotypes, resulting from an absence of recombination events in the region containing the markers A, B and C. Two other classes of progeny are relatively frequent (51 and 63 progeny in the example shown). Each of these is presumed to arise from a single recombination. Inspection of their genotypes shows that in the first of these two classes marker A has become unlinked from B and C, and in the second class marker B has become unlinked from A and C. The implication is that A and B are the outer markers. This is confirmed by the number of progeny in which marker C has become unlinked from A and B. There are only two of these, showing that a double recombination is needed to produce this genotype. Marker C is therefore between A and B.

Genotypes of progeny	Number of progeny	Inferred recombination events
ABC/abc *abc/abc*	987	None (parental genotypes)
aBC/abc *Abc/abc*	51	One, between A and B, C
AbC/abc *aBc/abc*	63	One, between B and A, C
ABc/abc *abC/abc*	2	Two, one between C and A and one between C and B

children 1, 2, 3, 4 and 5 have parental genotypes and that child six is the one and only recombinant (*Figure 2.14B*). This would suggest that the disease gene and the microsatellite are relatively closely linked and crossovers between them occur infrequently. The alternative hypothesis is that the mother's chromosomes have the genotypes healthy-M_1 and disease-M_2, and children 1–5 are recombinants with number 6 being a parental type. This would mean that the genes are relatively far apart on the chromosome. We cannot determine which of these hypotheses is correct: the data are frustratingly ambiguous.

The most satisfying solution to the problem posed by the pedigree in *Figure 2.14* would be to know the genotype of the grandmother. Let us pretend that this is a soap opera family and the grandmother was not really dead and to everyone's surprise reappears just in time to save the declining audience ratings. Her genotype for microsatellite M turns out to be M_1M_5 (*Figure 2.14C*). This tells us that the disease allele is on the same chromosome as M_1. We can therefore conclude with certainty that Hypothesis 1 is correct and only child 6 is a recombinant.

Resurrection of key individuals is not usually an option open to real life geneticists, though DNA can be obtained from old pathology specimens such as slides and Guthrie cards. Imperfect pedigrees are analyzed statistically, using a measure called the **lod score** (Morton, 1955). This stands for logarithm of the odds that the genes are linked and is used primarily to determine if the two markers being studied lie on the same chromosome, in other words if the genes are linked or not. If the lod analysis establishes linkage then it can also provide a measure of the most likely recombination frequency. Ideally the available data will derive from more than just one pedigree, increasing the confidence in the result. The analysis is less ambiguous for families with larger numbers of children, and, as we saw in *Figure 2.14*, it is important that the members of at least three generations can be genotyped. For this reason, family collections have been established, such as the one maintained by the Centre d'Études du Polymorphisme Humaine (CEPH) in Paris (Dausset *et al.*, 1990). The CEPH collection contains cultured cell lines from families in which all four grandparents as well as at least eight second generation children could be sampled. This collection is available for DNA marker

(A) The pedigree

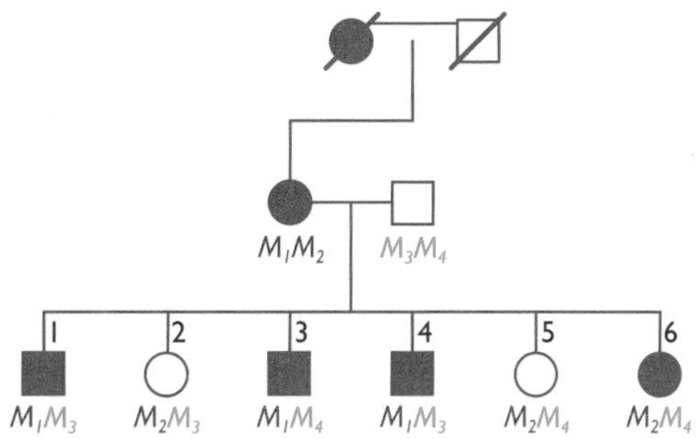

(B) Possible interpretations of the pedigree

MOTHER'S CHROMOSOMES

		Hypothesis 1	Hypothesis 2
		Disease M_1 / *Healthy M_2*	*Healthy M_1* / *Disease M_2*
CHILD 1	*Disease M_1*	Parental	Recombinant
CHILD 2	*Healthy M_2*	Parental	Recombinant
CHILD 3	*Disease M_1*	Parental	Recombinant
CHILD 4	*Disease M_1*	Parental	Recombinant
CHILD 5	*Healthy M_2*	Parental	Recombinant
CHILD 6	*Disease M_2*	Recombinant	Parental
	Recombination frequency	1/6 = 16.7%	5/6 = 83.3%

(C) Resurrection of the maternal grandmother

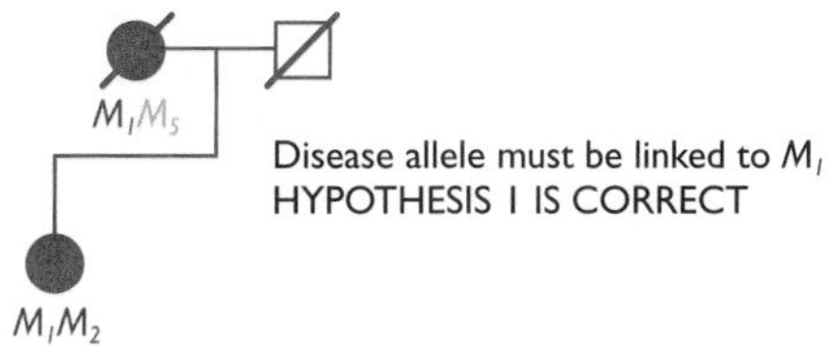

Figure 2.14 An example of human pedigree analysis.

(A) The pedigree shows inheritance of a genetic disease in a family of two living parents and six children, with information about the maternal grandparent available from family records. The disease allele (closed symbols) is dominant over the healthy allele (open symbols). The objective is to determine the degree of linkage between the disease gene and the microsatellite M by typing the alleles for this microsatellite (M_1, M_2, etc.) in living members of the family. (B) The pedigree can be interpreted in two different ways, Hypothesis 1 giving a low recombination frequency and indicating that the disease gene is tightly linked to microsatellite M, and Hypothesis 2 suggesting that the gene and microsatellite are much less closely linked. In (C), the issue is resolved by the reappearance of the maternal grandmother, whose microsatellite genotype is consistent only with Hypothesis 1. See the text for more details.

mapping by any researcher who agrees to submit the resulting data to the central CEPH database (see Research Briefing 2.1).

Genetic mapping in bacteria

The final type of genetic mapping that we must consider is the strategy used with bacteria. The main difficulty that geneticists faced when trying to develop genetic mapping techniques for bacteria is that these organisms, normally being haploid, do not undergo meiosis. Some other way therefore has to be devised to induce crossovers between homologous segments of bacterial DNA. The answer is to make use of one of three methods that exist for transferring pieces of DNA from one bacterium to another (*Figure 2.15*):

(i) **Conjugation**, in which two bacteria come into physical contact and one bacterium (the donor) transfers DNA to the second bacterium (the recipient). The transferred DNA can be a copy of some or possibly all of the donor cell's chromosome, or a segment of chromosomal DNA, up to 1 Mb in length, integrated in a plasmid (Section 1.2.2). The latter is called **episome transfer**.
(ii) **Transduction**, which involves transfer of a small segment of DNA, up to 50 kb or so, from donor to recipient via a **bacteriophage**.
(iii) **Transformation**, whereby the recipient cell takes up from its environment a fragment of DNA, rarely longer than 50 kb, released from a donor cell.

After transfer, a double crossover must occur so that the DNA from the donor bacterium is integrated into the recipient cell's chromosome (*Figure 2.16A*). If this does not occur then the transferred DNA is lost when the recipient cell divides. The only exception is after episome transfer, plasmids being able to propagate independently of the host chromosome.

Biochemical markers are invariably used, the dominant or **wild type** phenotype being possession of a biochemical characteristic (e.g. ability to synthesize tryptophan; sensitivity to an antibiotic) and the recessive phenotype being the complementary characteristic (e.g. inability to synthesize tryptophan; resistance to the antibiotic). The gene transfer is usually set up between a donor strain that possesses the wild-type alleles and a recipient with the recessive alleles, transfer into the recipient strain being monitored by looking for acquisition of the biochemical

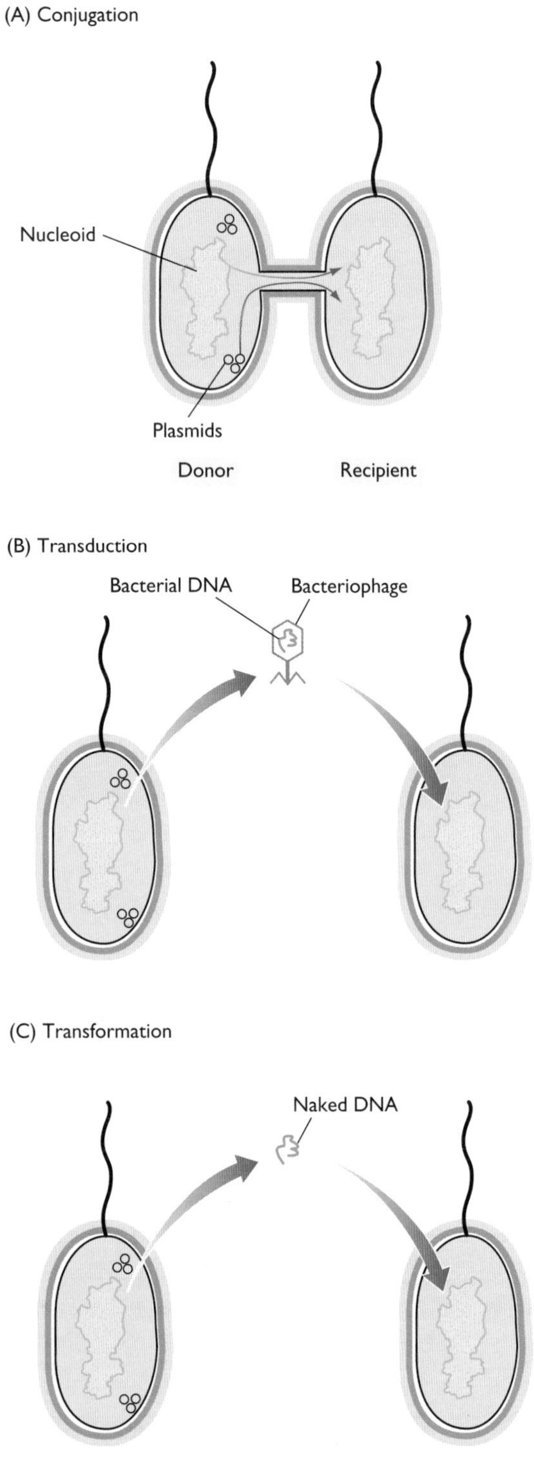

Figure 2.15 Three ways of achieving DNA transfer between bacteria.

(A) Conjugation can result in transfer of chromosomal or plasmid DNA from the donor bacterium to the recipient. Conjugation involves physical contact between the two bacteria, with transfer thought to occur through a narrow tube called the **pilus**. (B) Transduction is the transfer of a small segment of the donor cell's DNA via a bacteriophage. (C) Transformation is similar to transduction but 'naked' DNA is transferred. The events illustrated in (B) and (C) are often accompanied by death of the donor cell. In (B), death occurs when the bacteriophages emerge from the donor cell; in (C), release of DNA from the donor cell is usually a consequence of the cell's death through natural causes.

RESEARCH
2.1
BRIEFING

The human genetic map

One of the initial goals of the Human Genome Project was a genetic map with a density of at least one marker for every 1 Mb of the genome. This objective was reached in 1994. The paper described here, published two years later, presents a genetic map with a marker density averaging one per 0.6 Mb.

The map is based on 5264 microsatellites (Section 2.2.2) of the AC/TG type:

5′-ACACACACACAC-3′
3′-TGTGTGTGTGTG-5′

There were two reasons why microsatellites with this particular sequence were chosen. First, they are frequent in the genome with several found in every Mb of DNA, and so provide the degree of coverage needed for high density mapping work. Second, this type of microsatellite is very variable, with several alleles of each one in the population as a whole. In practical terms, this means that they display high **heterozygosity**, there being a high probability that any person chosen at random from the population will be heterozygous for a particular microsatellite. The average heterozygosity for all the markers included in the study was 0.7 (a 7 in 10 chance that an individual will be heterozygous) and only 7% of the markers had a heterozygosity of less than 0.5.

Eight of the families from the CEPH collection were used for the bulk of the mapping work. These families comprised 134 individuals and allowed a total of 186 meioses to be examined. The resulting data were used to generate the maps of chromosomes 1 to 22. To map the X chromosome, 12 more families (an extra 170 people providing 105 more meioses) were included, the additional data being needed because recombination events between X-linked markers are less common in a pedigree because males only have one X chromosome.

The map

The linkage analysis enabled the 5264 markers to be assigned to 2335 chromosomal positions. The number of positions was less than the number of markers because some microsatellites were too close together to be separated and so were mapped to the same position, or **bin** in mapping jargon. The average density was one marker per 599 kb for the map as a whole, ranging from one per 495 kb for chromosome 17 to one per 767 kb for chromosome 9. Only at three places in the genome were the closest markers apparently separated by more than 4 Mb of DNA. It is suspected that these gaps are not in fact as large as they seem, probably being due to recombination hotspots (see p. 26). Two markers either side of a hotspot will appear to be further apart than they really are because of the high frequency of recombination occurring between them.

Shortly after publication of the genetic map, a physical map with marker density approaching one per 100 kb was completed (see Research Briefing 3.1, p. 56). This and other physical maps include positions for 7000 of the microsatellites present on the genetic map, enabling the two types of map to be integrated into a single, comprehensive human map.

The human microsatellite map.

Each red dash indicates one of the 2335 positions to which the microsatellite markers were mapped. Reprinted with permission from Dib C *et al.*, *Nature*, **380**, 152–154. Copyright 1996 Macmillan Magazines Limited.

Reference

Dib C, Fuaré S, Fizames C, et al. (1996) A comprehensive genetic map of the human genome based on 5,264 microsatellites. *Nature*, **380**, 152–154.

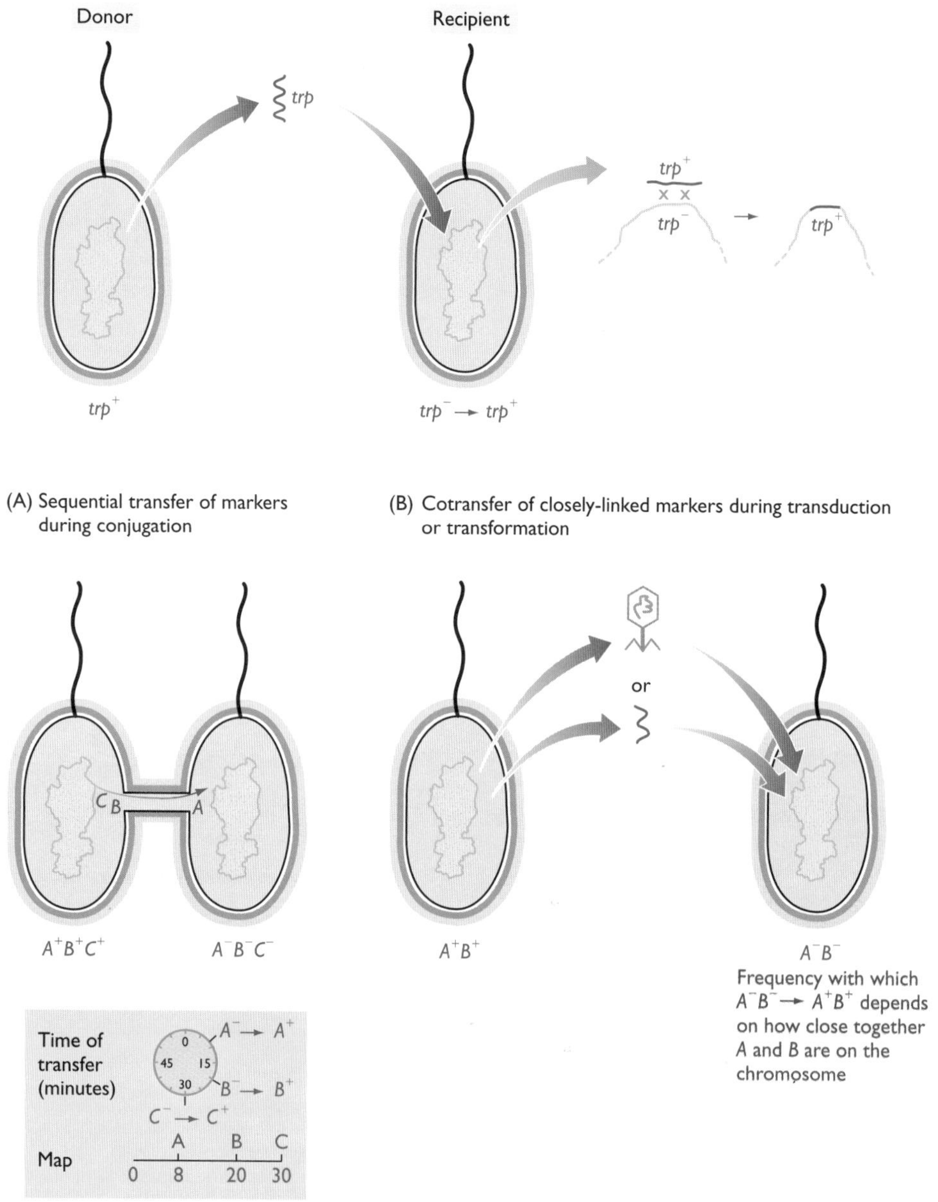

Figure 2.16 The basis of gene mapping in bacteria.

The top drawing shows transfer of a functional gene for tryptophan biosynthesis from a wild-type bacterium (genotype described as trp^+) to a recipient that lacks a functional copy of this gene (trp^-). The recipient is called a tryptophan **auxotroph** (the word used to describe a mutant bacterium that can survive only if provided with a nutrient – in this case, tryptophan – not required by the wild type; see Section 13.1.2). After transfer, two crossovers are needed to integrate the transferred gene into the recipient cell's chromosome, converting the recipient from trp^- to trp^+. (A) During conjugation, DNA is transferred from donor to recipient in the same way that a string is pulled through a tube. The relative positions of markers on the DNA molecule can therefore be mapped by determining the times at which the markers appear in the recipient cell. In the example shown, markers *A*, *B* and *C* are transferred 8, 20 and 30 minutes after the beginning of conjugation, respectively. The entire *E. coli* chromosome takes approximately 100 minutes to transfer. (B) To be cotransferred during transduction and transformation, two or more markers must be closely linked, because these processes usually result in less than 50 kb of DNA being passed from donor to recipient. Transduction and transformation mapping are used to determine the relative positions of markers that are too close together to be mapped precisely by conjugation analysis. For more details on bacterial gene mapping see Freifelder (1987).

functions specified by the genes being studied (*Figure 2.16B*). The precise details of the mapping procedure depend on the type of gene transfer that is being used. In conjugation mapping the donor DNA is transferred as a continuous thread into the recipient, and gene positions are mapped by timing the entry of the wild-type alleles into the recipient. Transduction and transformation mapping enable genes that are relatively close together to be mapped, because the transferred DNA segment is short (<50 kb), so the chances of two genes being transferred together depends on how close together they are on the bacterial chromosome.

REFERENCES

Collins FS, Guyer MS and Chakravarti A (1997) Variations on a theme: cataloging human DNA sequence variation. *Science,* **278**, 1580–1581.

Dausset J, Cann H, Cohen D, *et al.* (1990) Program description: Centre d'Étude du Polymorphisme Humain – collaborative genetic mapping of the human genome. *Genomics,* **6**, 575–577.

Freifelder D (1987) *Microbial Genetics*. Jones & Bartlett, Boston.

Mori M, Beatty PG, Graves M, Boucher KM and Milford EL (1997) HLA gene and haplotype frequencies in the North American population. *Transplantation,* **65**, U1.

Morton NE (1955) Sequential tests for the detection of linkage. *Am. J. Hum. Genet.,* **7**, 277-318.

Orel V (1995) *Gregor Mendel: The First Geneticist*. Oxford University Press, Oxford.

Pennisi E (1998) Sifting through and making sense of genome sequences. *Science,* **280**, 1692–1693.

Ramsay G (1998) DNA chips: state of the art. *Nature Biotechnol.,* **16**, 40–44.

Sturtevant AH (1913) The linear arrangement of six sex-linked factors in *Drosophila* as shown by mode of association. *J. Exp. Zool.,* **14**, 39–45.

Venter JC, Adams MD, Sutton GG, Kerlavage AR, Smith HO and Hunkapiller M (1998) Shotgun sequencing of the human genome. *Science,* **280**, 1540–1542.

Wang DG, Fan J-B, Siao C-J, *et al.* (1998) Large-scale identification, mapping, and genotyping of single-nucleotide polymorphisms in the human genome. *Science,* **280**, 1077–1082.

Yamamoto F, Clausen H, White T, Marken J and Hakamori S (1990) Molecular genetic basis of the histo-blood group ABO system. *Nature,* **345**, 229–233.

FURTHER READING

Alberts B, Bray D, Lewis J, Raff M, Roberts K and Watson JD (1994) *Molecular Biology of the Cell*, 3rd edition. Garland Publishing, New York. — *Contains full details of mitosis and meiosis.*

Fincham JRS, Day PR and Radford A (1979) *Fungal Genetics,* 4th edition. Blackwell, London. — *The bible for gene mapping in microbial eukaryotes.*

Griffiths AJF, Miller JH, Suzuki DT, Lewontin RC and Gelbart WM (1996) *An Introduction to Genetic Analysis*, 6th edition. W. H. Freeman, New York. — *Chapters 5 and 6 are particularly good for gene mapping by experimental crosses. Chapter 10 deals with mapping in bacteria.*

Strachan T and Read AP (1996) *Human Molecular Genetics*. BIOS Scientific Publishers, Oxford. — *Chapter 12 covers human genetic mapping.*

Sturtevant AH (1965) *A History of Genetics*. Harper and Row, New York. — *Describes the early gene mapping work carried out by Morgan and his colleagues.*

CHAPTER

3 Mapping Genomes by Physical Techniques

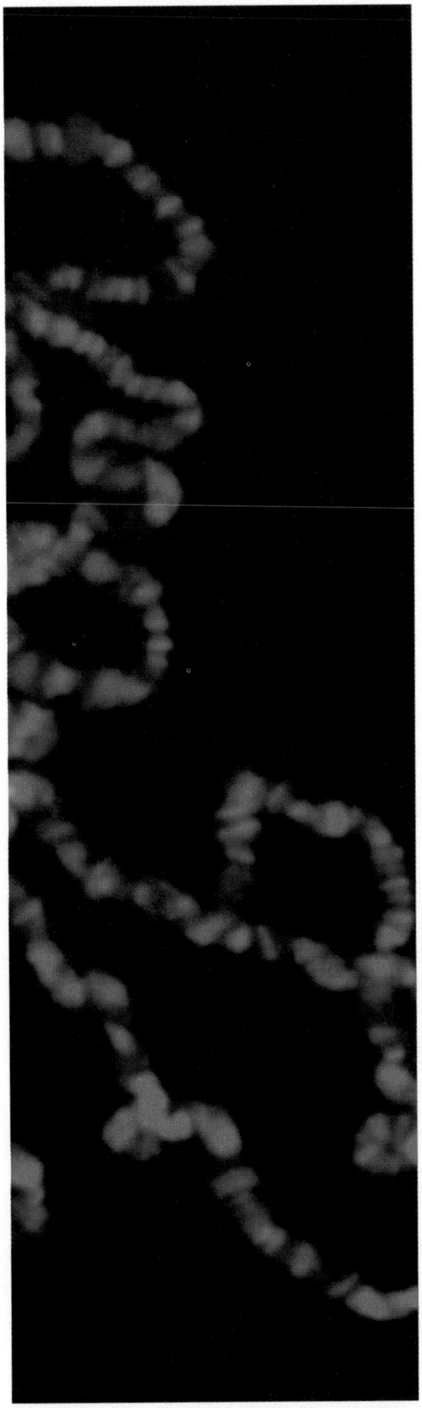

Contents

3.1 *Restriction Mapping* 39
- 3.1.1 The basic methodology for restriction mapping 39
- 3.1.2 The scale of restriction mapping is limited by the sizes of the restriction fragments 41

3.2 *FISH – Fluorescent* In Situ *Hybridization* 45
- 3.2.1 *In situ* hybridization with radioactive or fluorescent probes 45
- 3.2.2 FISH in action 47

3.3 *Sequence Tagged Site (STS) Mapping* 48
- 3.3.1 Any unique DNA sequence can be used as an STS 49
- 3.3.2 Fragments of DNA for STS mapping 49

3.4 *Progress Towards the Human Genome Map* 54

Concepts

- *Physical mapping uses molecular biology techniques such as hybridization analysis and PCR to examine DNA molecules for the positions of distinctive sequence features*
- *Restriction mapping is a useful method for physical mapping of small genomes*
- *FISH is a technique that enables the positions of DNA markers in a chromosome to be visualized by fluorescent labeling*
- *STS mapping with radiation hybrid panels and clone libraries is the most powerful type of physical mapping*
- *Considerable progress in mapping the human genome was made in the mid-1990s, culminating in a combined genetic and physical map with a marker density approaching one per 100 kb*

IN THE PREVIOUS CHAPTER we saw how genetic analysis can be used to construct genome maps showing the chromosomal locations of genes and DNA markers such as RFLPs, SSLPs and SNPs. In this chapter we will look at the alternative genome mapping techniques, those based not on genetic analysis but on direct physical examination of DNA molecules. We will see how physical and genetic maps complement one another to provide the framework from which a complete genome DNA sequence can be obtained.

The first question to ask is why do we need physical mapping techniques? Why are the maps provided by genetic techniques not in themselves sufficient for directing the sequencing phase of a genome project? There are two main reasons:

- ***A genetic map has a limited degree of resolution.*** This is not a major problem for microorganisms, because these can be studied in huge numbers, enabling many crossovers to be scored, and resulting in a highly detailed genetic map in which the markers are just a few kb apart. For example, when the *Escherichia coli* genome sequencing project began in 1990 the latest genetic map for this organism comprised over 1400 markers, an average of one per 3.3 kb. This was sufficient detail to direct the sequencing programme without the need for extensive physical mapping. Similarly, the *Saccharomyces cerevisiae* project was supported by a fine-scale genetic map (approximately 1150 genetic markers, on average one per 10 kb). The problem is that with humans and most other higher eukaryotes it is simply not possible to obtain large numbers of progeny, so relatively few meioses can be studied and the resolving power of linkage analysis is restricted. The human genetic map described in Research Briefing 2.1, p. 33 has a marker density of only one per 599 kb. This is not to say that this map is an under-achievement, in fact the reverse is true. In the mid-1980s, when the idea of the human genome project was first discussed seriously, it was decided that a human genetic map with a marker density of one per 1 Mb was desirable, but that a density of one per 2–5 Mb was probably the best that could be hoped for. The current map therefore exceeds all expectations, but is still short of the total marker density of one per 100 kb set as a prerequisite for the sequencing phase of the human project. This degree of density cannot be achieved by genetic mapping alone: it will require additional mapping by nongenetic means.
- ***Genetic maps have limited accuracy.*** We touched on this point in Section 2.3.1 when we assessed Sturtevant's assumption that crossovers occur at random along chromosomes. This assumption is only partly correct because the presence of recombination hotspots means that crossovers are more likely to occur at some points rather than at others. The effect that this can have on the accuracy of a genetic map

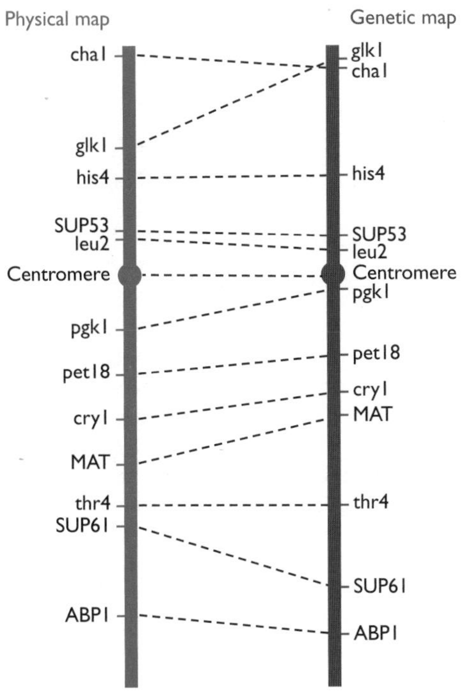

Figure 3.1 Comparison between the genetic and physical maps of *S. cerevisiae* chromosome III.

The comparison shows the discrepancies between the genetic and physical maps, the latter determined by DNA sequencing. Note that the order of the upper two markers, glk1 and cha1, is incorrect on the genetic map, and that there are also differences in the relative positioning of other pairs of markers. Reprinted with permission from Oliver SG, *et al.*, *Nature*, **357**, 38–46. Copyright 1992 Macmillan Magazines Limited.

was illustrated in 1992 when the complete sequence for *S. cerevisiae* chromosome III was published, enabling the first direct comparison to be made between a genetic map and the actual positions of markers as shown by DNA sequencing (*Figure 3.1*). There were quite considerable discrepancies, even to the extent that one pair of genes had been ordered incorrectly by genetic analysis. Bear in mind that *S. cerevisiae* is one of the two eukaryotes (fruit fly is the second) whose genomes have been subjected to really intensive genetic mapping. If the yeast genetic map is inaccurate then how precise are the genetic maps of organisms subjected to less detailed analysis?

These two limitations of genetic mapping mean that with most eukaryotes a genetic map must be checked and supplemented by alternative mapping procedures before large scale DNA sequencing begins. A plethora of physical mapping techniques have been developed but the most important of these fall into three categories:

- **Restriction mapping,** which locates the relative positions on a DNA molecule of the cut sites for restriction endonucleases.
- **Fluorescent *in situ* hybridization (FISH),** in which marker locations are mapped by hybridizing the marker to intact chromosomes.
- **Sequence tagged site (STS) mapping,** in which the positions of short sequences are mapped by PCR and/or hybridization analysis of genome fragments.

3.1 RESTRICTION MAPPING

The description of RFLPs in Section 2.2.2 introduced us to restriction endonucleases, the enzymes that cut DNA molecules at defined sequences. There are many different restriction enzymes with many different target sequences. The majority of these target sites are four-, five- or six-nucleotide sequences, but there are also examples that recognize sequences of seven, eight or more nucleotides (Technical Note 3.1). Very few of the restriction sites in genomic DNA display the polymorphism needed for RFLP mapping, so many sites are not mapped by this technique (*Figure 3.2*). Could we increase the marker density on a genome map by using an alternative method to locate the positions of some of the nonpolymorphic restriction sites? This is what restriction mapping achieves, though in practice the technique has limitations that mean that it is applicable only to relatively small DNA molecules. We will look first at the technique and then consider its relevance to genome mapping.

3.1.1 The basic methodology for restriction mapping

The simplest way to construct a restriction map is to compare the fragment sizes produced when a DNA molecule is digested with two different restriction enzymes, ones

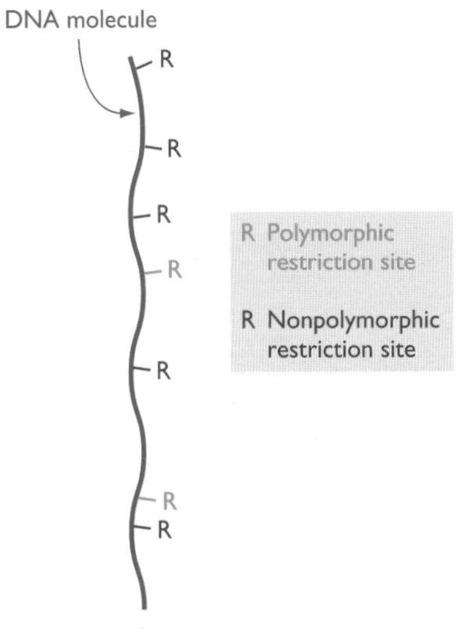

Figure 3.2 Not all restriction sites are polymorphic.

TECHNICAL 3.1 NOTES

Restriction endonucleases

Enzymes for cutting DNA molecules at defined sequences.

A restriction endonuclease is an enzyme that binds to a DNA molecule at a specific sequence and makes a double-stranded cut at or near that sequence. Restriction enzymes have specialized applications in techniques such as RFLP analysis (Section 2.2.2) and restriction mapping (Section 3.1) and are also used extensively to manipulate DNA molecules during procedures such as cloning (Technical Note 3.4, p. 52).

There are three types of restriction endonuclease. With Types I and III there is no strict control over the position of the cut relative to the recognition sequence, but with Type II enzymes the cut is always at the same place, either within the recognition sequence or very close to it. For example, the Type II enzyme called *Eco*RI (isolated from *Escherichia coli*) cuts DNA only at the hexanucleotide 5′-GAATTC-3′. Digestion of DNA with a Type II enzyme therefore gives a reproducible set of fragments, the sizes of these fragments being predictable if the sequence of the target DNA molecule is known.

Over 2500 Type II enzymes have been isolated and more than 300 are available for use in the laboratory (Brown, 1998). Many enzymes have hexanucleotide target sites, but others recognize shorter or longer sequences. There are also examples of enzymes with degenerate recognition sequences, meaning that they cut DNA at any of a family of related sites. *Hin*fI (from *Haemophilus influenzae*), for example, recognizes 5′-GANTC-3′ and so cuts at 5′-GAATC-3′, 5′-GATTC-3′, 5′-GAGTC-3′ and 5′-GACTC-3′. Restriction enzymes cut DNA in two different ways. Many make a simple double-stranded cut giving a **blunt** or flush end, but others cut the two DNA strands at different positions, usually two or four nucleotides apart, so that the resulting DNA fragments have short single-stranded overhangs at each end. These are called **sticky** or **cohesive** ends, as base-pairing between them can stick the DNA molecule back together again. The resulting base-paired structure is likely to fall apart quite quickly, but it can be stabilized if **DNA ligase** is added. This is an enzyme that synthesizes missing phosphodiester bonds (Section 12.3.2) and so can join together two restriction fragments that are transiently base-paired via their sticky ends. It can also join blunt ended molecules but with much less efficiency.

The Table gives details of some commonly used restriction enzymes. Note that most but not all recognition sequences have inverted symmetry: when read in the 5′→3′ direction, the sequence is the same on both strands. Some sticky end cutters give 5′ overhangs (e.g. *Sau*3AI, *Hin*fI), others leave 3′ overhangs (e.g. *Bgl*I). Some pairs of restriction enzymes have different recognition sequences but give the same sticky ends (e.g. *Sau*3AI and *Bam*HI).

Enzyme	Recognition sequence	Type of ends	End sequences
*Alu*I	5′-AGCT-3′ 3′-TCGA-5′	Blunt	5′-AG CT-3′ 3′-TC GA-5′
*Sau*3AI	5′-GATC-3′ 3′-CTAG-5′	Sticky, 3′ overhang	5′- GATC-3′ 3′-CTAG -5′
*Hin*fI	5′-GANTC-3′ 3′-CTNAG-5′	Sticky, 3′ overhang	5′-G ANTC-3′ 3′-CTNA G-5′
*Bam*HI	5′-GGATCC-3′ 3′-CCTAGG-5′	Sticky, 3′ overhang	5′-G GATCC-3′ 3′-CCTAG G-5′
*Bsr*BI	5′-CCGCTC-3′ 3′-GGCGAG-5′	Blunt	5′- NNNCCGCTC-3′ 3′- NNNGGCGAG-5′
*Eco*RI	5′-GAATTC-3′ 3′-CTTAAG-5′	Sticky, 3′ overhang	5′-G AATTC-3′ 3′-CTTAA G-5′
*Not*I	5′-GCGGCCGC-3′ 3′-CGCCGGCG-5′	Sticky, 3′ overhang	5′-GC GGCCGC-3′ 3′-CGCCGG CG-5′
*Bgl*I	5′-GCCNNNNNGGC-3′ 3′-CGGNNNNNCCG-5′	Sticky, 5′ overhang	5′-GCCNNN NNGGC-3′ 3′-CGGNN NNNCCG-5′

N = any nucleotide. For details of all known restriction enzymes see REBASE <http://www.neb.com/rebase/rebase.html>.

that recognize different target sequences. An example using the restriction enzymes *Eco*RI and *Bam*HI is shown in *Figure 3.3*. First, the DNA molecule is digested with one of the two enzymes on its own and the sizes of the resulting fragments measured by agarose gel electrophoresis (Technical Note 3.2). Next, the molecule is digested with the second enzyme and the resulting fragments again sized in an agarose gel. The results so far enable the number of restriction sites for each enzyme to be worked out, but do not allow their relative positions to be determined. Additional information is therefore obtained by cutting the DNA molecule with both enzymes at once. In the example shown in *Figure 3.3*, this **double restriction** enables three of the sites to be mapped, but a problem arises with the larger *Eco*RI fragment, as this contains two *Bam*HI sites and there are two alternative possibilities for the map location of the outer one of these. The problem is solved by going back to the original DNA molecule and treating it again with *Bam*HI, on its own, but this time preventing the digestion from going to completion by, for example, incubating the reaction for only a short time or using a suboptimal incubation temperature. This is called a **partial restriction** and leads to a more complex set of products, the complete restriction products now being supplemented with partially restricted fragments that still contain one or more uncut *Bam*HI sites. In the example shown in *Figure 3.3*, the size of one of the partial restriction fragments is diagnostic and the correct map can be identified.

Usually, a partial restriction gives the information needed to complete a map, but if there are many restriction sites then this type of analysis becomes unwieldy, simply because there are so many different fragments to consider. An alternative strategy is simpler as it enables the bulk of the fragments to be ignored. This is achieved by attaching a radioactive or other type of marker to each end of the starting DNA molecule before carrying out the partial digestion. The result is that many of the partial restriction products become 'invisible', as they do not contain an end-fragment and so do not show up when the agarose gel is screened for labeled products. The sizes of the partial restriction products that are visible enables the unmapped site to be positioned relative to the end of the starting molecule.

3.1.2 The scale of restriction mapping is limited by the sizes of the restriction fragments

Restriction maps are easy to generate if there are relatively few cut sites for the enzymes being used. But as the number of cut sites increases, so also do the numbers of single, double and partial restriction products whose sizes must be determined and compared in order for the map to be constructed. Computer analysis can be brought into play but problems still eventually arise. A stage will be reached when a digest contains so many fragments that individual bands merge on the agarose gel, increasing the chances of one or more fragments being measured incorrectly or missed out entirely. Even if all the fragments can be identified then if several have similar sizes it may not be possible to assemble them into an unambiguous map.

Restriction mapping is therefore more applicable to small rather than large molecules, with the upper limit for the technique dependent on the frequency of the restriction sites in the molecule being mapped. In practice, if a DNA molecule is less than 50 kb in length it is usually possible to construct an unambiguous restriction map for a selection of enzymes with six-nucleotide recognition sequences. Fifty kb is of course way below the minimum size for bacterial or eukaryotic chromosomes, although it does cover a few viral and organellar genomes, and whole-genome restriction maps have indeed been important in directing sequencing projects with these small molecules. Restriction maps are equally useful after bacterial or eukaryotic genomic DNA has been cloned, as the cloned fragments are often less than 50 kb, meaning that a detailed restriction map can be built up as a preliminary to sequencing the cloned region. This is an important application of restriction mapping in sequencing projects with large genomes, but is there any possibility of using restriction analysis for the more general mapping of entire genomes larger than 50 kb? The next section will show that the answer is a qualified yes.

Restriction mapping with rare cutters

The limitations of restriction mapping can be eased slightly by choosing enzymes expected to have infrequent cut sites in the target DNA molecule. These 'rare cutters' fall into two categories:

- ***Enzymes with seven- or eight-nucleotide recognition sequences.*** A few restriction enzymes cut at seven- or eight-nucleotide recognition sequences. Examples are *Sap*I (5′-GCTCTTC-3′) and *Sgf*I (5′-GCGATCGC-3′). The seven-nucleotide enzymes would be expected to cut a DNA molecule with a **GC content** of 50%, on average, once every $4^7 = 16384$ bp. The eight-nucleotide enzymes should cut once every $4^8 = 65536$ bp. These figures compare with $4^6 = 4096$ bp for six-nucleotide enzymes such as *Bam*H1 and *Eco*R1. Seven- and eight-nucleotide cutters are often used in restriction mapping of large molecules but the approach is not as useful as it might be simply because not many of these enzymes are known.
- ***Enzymes whose recognition sequences contain motifs that are rare in the target DNA.*** Genomic DNA molecules do not have random sequences and some are significantly deficient in certain motifs. For example, the sequence 5′-CG-3′ is rare in human DNA, because human cells possess an enzyme that adds a methyl group to carbon 5 of the C nucleotide in this sequence. The resulting 5-methylcytosine is unstable and tends to undergo deamination to give thymine (*Figure 3.4*). The consequence is that during

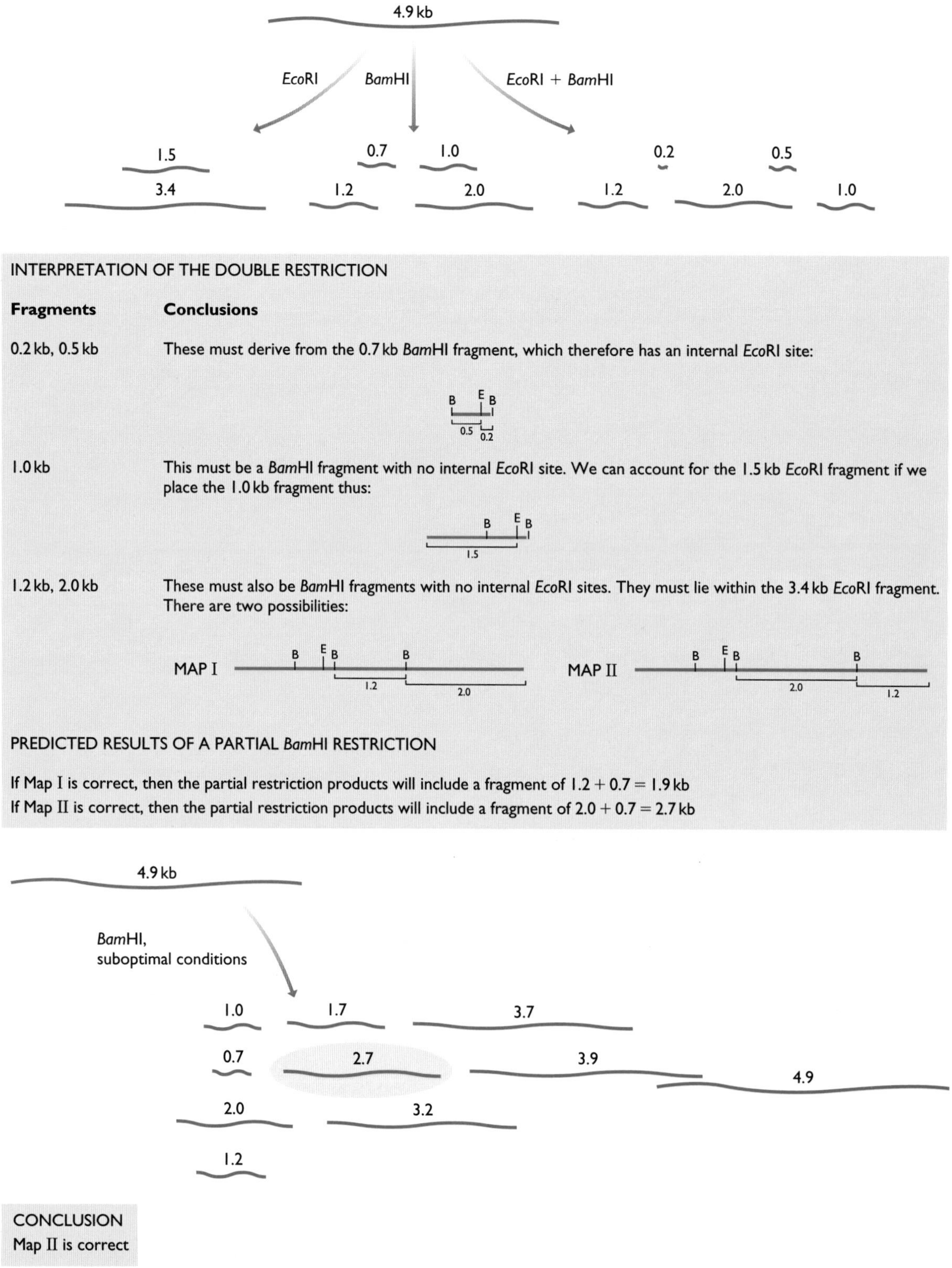
4.9 kb
EcoRI
BamHI
EcoRI + BamHI
1.5
3.4
0.7
1.0
1.2
2.0
0.2
0.5
1.2
2.0
1.0
INTERPRETATION OF THE DOUBLE RESTRICTION
Fragments
Conclusions
0.2 kb, 0.5 kb
These must derive from the 0.7 kb BamHI fragment, which therefore has an internal EcoRI site:
B E B
0.5 0.2
1.0 kb
This must be a BamHI fragment with no internal EcoRI site. We can account for the 1.5 kb EcoRI fragment if we place the 1.0 kb fragment thus:
B E B
1.5
1.2 kb, 2.0 kb
These must also be BamHI fragments with no internal EcoRI sites. They must lie within the 3.4 kb EcoRI fragment. There are two possibilities:
MAP I
B E B B
1.2
2.0
MAP II
B E B B
2.0
1.2
PREDICTED RESULTS OF A PARTIAL BamHI RESTRICTION
If Map I is correct, then the partial restriction products will include a fragment of 1.2 + 0.7 = 1.9 kb
If Map II is correct, then the partial restriction products will include a fragment of 2.0 + 0.7 = 2.7 kb
4.9 kb
BamHI,
suboptimal conditions
1.0
1.7
3.7
0.7
2.7
3.9
4.9
2.0
3.2
1.2
CONCLUSION
Map II is correct

Figure 3.3 Restriction mapping.

The objective is to map the *Eco*RI (E) and *Bam*HI (B) sites in a linear DNA molecule of 4.9 kb. The results of single and double restrictions are shown at the top. The sizes of the fragments given after double restriction enable two alternative maps to be constructed, as explained in the central panel, the unresolved issue being the position of one of the three *Bam*HI sites. The two maps are tested by a partial *Bam*HI restriction (bottom), which shows that Map II is the correct one.

TECHNICAL 3.2 NOTES

Gel electrophoresis

Separation of DNA and RNA molecules of different lengths.

Gel electrophoresis is the standard method for separating DNA molecules of different lengths. It has many applications in size analysis of restriction fragments (e.g. in RFLP analysis [Section 2.2.2] and restriction mapping [Section 3.1]), in purification of individual fragments to be cloned or sequenced, in examination of PCR products (Technical Note 2.2, p. 20) and during DNA sequencing (Section 4.1.1). It can also be used to separate RNA molecules (Technical Note 5.1, p. 91).

Electrophoresis is the movement of charged molecules in an electric field, negatively-charged molecules migrating towards the positive electrode and positively-charged ones towards the negative electrode. The technique was originally carried out in aqueous solution, but this is not particularly useful for DNA separations because the predominant factors influencing migration rate in solution phase are the shape of a molecule and its electric charge. Most DNA molecules are the same shape (linear) and although the charge of a DNA molecule is dependent on its length, the differences in charge are not sufficient to result in effective separation. The situation is different when electrophoresis is carried out in a gel, because now shape and charge are less important and molecular length is the critical determinant of migration rate. This is because the gel is a network of pores through which the DNA molecules have to travel to reach the positive electrode. Shorter molecules are less impeded by the pores than longer molecules and so move through the gel more quickly. Molecules of different lengths therefore form bands in the gel.

Two types of gel are used in molecular biology: **agarose** gels, as described here, and **polyacrylamide** gels, which are covered in Technical Note 4.1, p. 62. Agarose is a polysaccharide that forms gels with pores ranging from 100 to 300 nm in diameter, the size depending on the concentration of agarose in the gel. Gel concentration therefore determines the range of DNA fragments that can be separated, a 0.3% gel, for example, being used for molecules between 5 and 50 kb, and a 5% gel being used for 100–500 bp molecules (Brown, 1998). The separation range is also affected by the electroendosmosis value (EEO) of the agarose, this being a measure of the amount of bound sulfate and pyruvate anions. The greater the EEO, the slower the migration rate for a negatively-charged molecule such as DNA.

An agarose gel is prepared by mixing the appropriate amount of agarose powder in a buffer, heating to dissolve the agarose, and then pouring the molten gel on to a perspex plate with tape around the sides to prevent spillage. A comb is placed in the gel to form wells for the samples. The gel is allowed to set and the electrophoresis then carried out with the gel submerged under buffer. One or two dyes, of known migration rates, are added to the DNA samples before loading so the progress of the electrophoresis can be followed, and the bands of DNA are visualized by soaking the gel in ethidium bromide solution, this compound intercalating between DNA base pairs and fluorescing when activated with ultraviolet radiation.

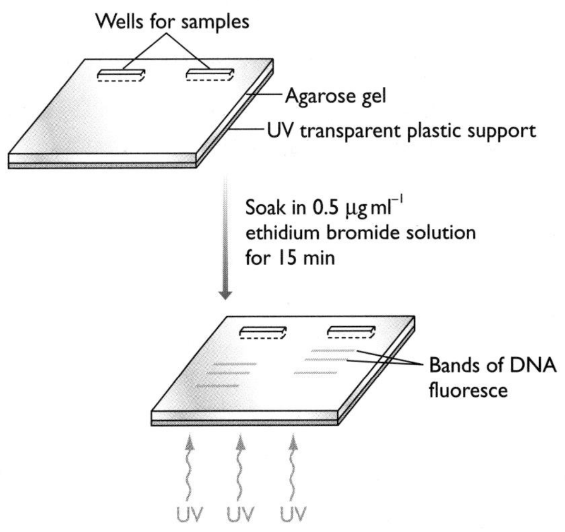

human evolution many of the 5′-CG-3′ sequences that were originally in our genome have become converted to 5′-TG-3′. Restriction enzymes that recognize a site containing 5′-CG-3′ therefore cut human DNA relatively infrequently. Examples are *Sma*I (5′-CCCGGG-3′), which cuts human DNA on average once every 78 kb, and *Bss*HII (5′-GCGCGC-3′) which cuts once every 390 kb. Note that *Not*I, an eight-nucleotide cutter, also targets 5′-CG-3′ sequences (recognition sequence 5′-GCGGCCGC-3′) and cuts human DNA very rarely, approximately once every 10 Mb.

The potential of restriction mapping is therefore increased by using rare cutters. It is still not possible to construct restriction maps of the larger genomes of

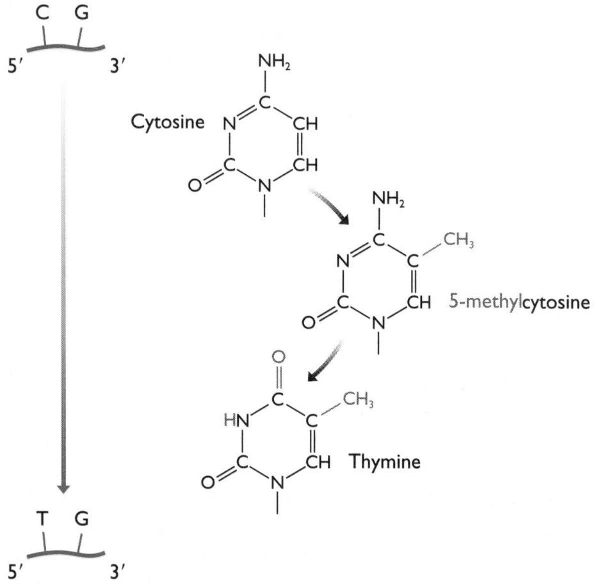

Figure 3.4 The sequence 5′-CG-3′ is rare in human DNA because of methylation of the C, followed by deamination to give T.

animals and plants, but it is feasible to use the technique with the smaller chromosomal DNA molecules of lower eukaryotes such as yeast and fungi.

Monitoring the results of restriction with a rare cutter

If a rare cutter is used then it may be necessary to use a special type of agarose gel electrophoresis to study the resulting restriction fragments. This is because the relationship between the length of a DNA molecule and its migration rate in an electrophoresis gel is not linear, the resolution decreasing as the molecules get longer (*Figure 3.5A*). The result is that it is not possible to separate molecules more than about 50 kb in length because all of these longer molecules run as a single, slowly migrating band in a standard agarose gel. To separate them it is necessary to replace the linear electric field used in conventional gel electrophoresis with a more complex field. An example is provided by **orthogonal field alternation gel electrophoresis** (**OFAGE**), in which the electric field alternates between two pairs of electrodes, each positioned at an angle of 45° to the length of the gel (*Figure 3.5B*). The DNA molecules still move down through the gel, but each change in the field forces the molecules to realign. Shorter molecules realign more quickly than longer ones, so the shorter molecules migrate more rapidly through the gel. The overall result is resolution of molecules much longer than those separated by conventional gel electrophoresis. Related techniques include **CHEF** (**contour clamped homogeneous electric fields**) and **FIGE** (**field inversion gel electrophoresis**).

It is also possible to use methods other than electrophoresis to map restriction sites for rare cutters. With isolated yeast chromosomes, restriction sites have been examined directly by looking at the cut DNA molecules with a microscope, a technique called **optical mapping** (Schwartz *et al.*, 1993). There are two methods for preparing DNA fibers for optical mapping: **gel stretching** and **molecular combing**. We will look at the first method here and return to molecular combing later in the chapter (Section 3.2.2). To prepare gel-stretched DNA fibers (Schwartz *et al.*, 1993), chromosomal DNA is suspended in molten agarose and placed on a microscope slide. As the gel cools and solidifies the DNA molecules become extended (*Figure 3.6*). To utilize gel stretching in optical mapping, the microscope slide onto which the molten agarose is placed is first coated with a restriction enzyme. The enzyme is inactive at this stage because there are no magnesium ions, which the enzyme needs in order to function. Once the gel has solidified it is washed with a solution containing $MgCl_2$, which activates the restriction enzyme, and a fluorescent dye is added, such as DAPI (4,6-diamino-2-phenylindole dihydrochloride) which stains the DNA so the fibers can be seen when the slide is examined with a high power fluorescence microscope. The restriction sites in the extended molecules gradually become gaps as the degree of fiber extension is reduced by the natural springiness of the DNA, enabling the relative positions of the cuts to be recorded.

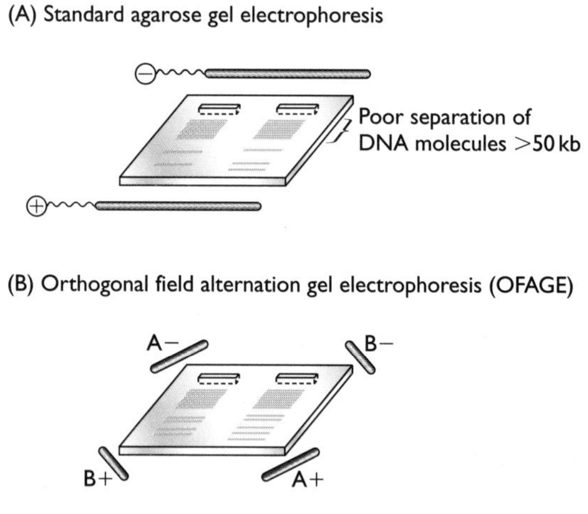

Figure 3.5 Conventional and nonconventional agarose gel electrophoresis.

(A) In standard agarose gel electrophoresis the electrodes are placed at either end of the gel and the DNA molecules migrate directly towards the positive electrode. This does not enable molecules longer than about 50 kb to be separated from one another. (B) In OFAGE, the electrodes are placed at the corners of the gel, with the field pulsing between the A pair and the B pair. OFAGE enables molecules up to 2 Mb to be separated. There are a variety of other techniques using different electrode orientations: for details see Andrews (1991). For more information on agarose gel electrophoresis, see Technical Note 3.2.

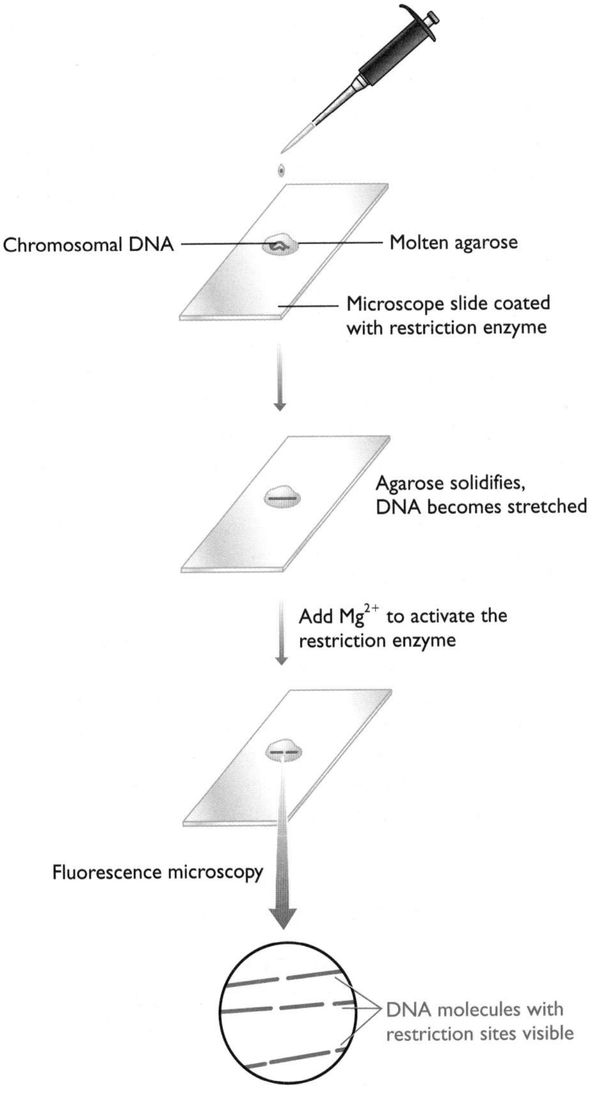

Figure 3.6 Gel stretching.

Molten agarose containing chromosomal DNA molecules is pipetted on to a microscope slide coated with a restriction enzyme. As the gel solidifies, the DNA molecules become stretched. It is not understood why this happens but it is thought that fluid movement on the glass surface during gelation might be responsible. Addition of $MgCl_2$ activates the restriction enzyme, which cuts the DNA molecules. As the molecules gradually coil up, the gaps representing the cut sites become visible.

3.2 FISH – FLUORESCENT *IN SITU* HYBRIDIZATION

The optical mapping method described at the end of the previous section provides a link to the second type of physical mapping procedure that we will consider – FISH (Heiskanen *et al.*, 1996). As in optical mapping, FISH enables the position of a marker on a chromosome or extended DNA molecule to be directly visualized. In optical mapping the marker is a restriction site and it is visualized as a gap in an extended DNA fiber. In FISH, the marker is a DNA sequence and it is visualized by hybridizing a fluorescent probe to it.

3.2.1 In situ *hybridization with radioactive or fluorescent probes*

A number of molecular biology techniques make use of **nucleic acid hybridization**, the ability of two complementary or partly complementary nucleic acid molecules to form a stable base-paired hybrid. We encountered two of these techniques – Southern hybridization and oligonucleotide probing – in Chapter 2. *In situ* hybridization is a version of hybridization analysis in which an intact chromosome is examined by probing it with a labeled DNA molecule. The position on the chromosome at which hybridization occurs provides information about the map location of the DNA sequence used as the probe (*Figure* 3.7). For the method to work, the DNA in the chromosome must be made single-stranded by

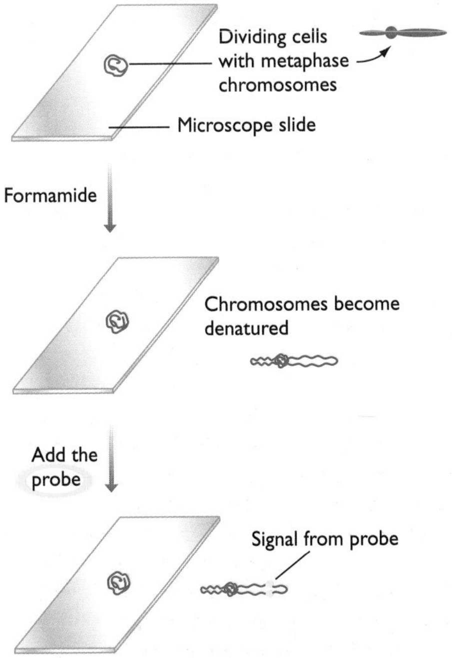

Figure 3.7 Fluorescent *in situ* hybridization.

A sample of dividing cells is dried on to a microscope slide and treated with formamide so that the chromosomes become denatured but do not lose their characteristic metaphase morphologies (see Section 6.1.1). The position at which the probe hybridizes to the chromosomal DNA is visualized by detecting the fluorescent signal emitted by the labeled DNA.

TECHNICAL 3.3 NOTES

DNA labeling

Attachment of radioactive, fluorescent or other types of marker to DNA molecules.

DNA labeling is a central part of many molecular biology procedures including, for example, Southern hybridization (Technical Note 2.1, p. 19), FISH (Section 3.2) and DNA sequencing (Section 4.1.1). It enables the location of a particular DNA molecule – on a nylon membrane, in a chromosome or in a gel – to be determined by detecting the signal emitted by the marker. Labeled RNA molecules are also used in some applications (Technical Note 5.1, p. 91).

Radioactive markers are frequently used for labeling DNA molecules. Nucleotides can be synthesized with one of the phosphate atoms replaced with ^{32}P or ^{33}P, one of the oxygen atoms in the phosphate group replaced with ^{35}S, or one or more of the hydrogen atoms replaced with ^{3}H (see *Figure 1.1*, p. 3). Radioactive nucleotides still act as substrates for DNA polymerases and so are incorporated into a DNA molecule by any strand synthesis reaction catalyzed by a DNA polymerase (Technical Note 4.2, p. 64). Labeled nucleotides or individual phosphate groups can also be attached to one or both ends of a DNA molecule by reactions catalyzed by T4 polynucleotide kinase or terminal deoxynucleotidyl transferase. The radioactive signal can be detected by scintillation counting, but for most molecular biology applications positional information is needed, so detection is by exposure of an X-ray sensitive film (**autoradiography**) or a radiation-sensitive phosphorescent screen (**phosphorimaging**). The choice between the various radioactive labels depends on the requirements of the procedure. High sensitivity is possible with ^{32}P because this isotope has a high emission energy, but sensitivity is accompanied by low resolution because of signal scattering. Low emission isotopes such as ^{35}S or ^{3}H give less sensitivity but greater resolution.

Health and environmental issues have meant that radioactive markers have become less popular in recent years and for many procedures they are now largely superseded by nonradioactive alternatives. The most useful of these are fluorescent markers, which are central components of techniques such as FISH (Section 3.2) and automated DNA sequencing (Section 4.1.1). Fluorophores with various emission wavelengths (i.e. of different colors) are incorporated into nucleotides or attached directly to DNA molecules, and are detected with a suitable film, by fluorescence microscopy, or with a fluorescence detector. Other types of nonradioactive labeling make use of chemiluminescent emissions, but these have the disadvantage that the signal is not generated directly by the label, but instead must be 'developed' by treatment of the labeled molecule with chemicals. For example, one method involves labeling the DNA with the enzyme alkaline phosphatase, which is detected by applying dioxetane, which is dephosphorylated by the enzyme to produce the chemiluminescence.

breaking the base pairs that hold the double helix together (this is called 'denaturing' the DNA). Only then will the chromosomal DNA be able to hybridize with the probe. The standard method for denaturing chromosomal DNA without destroying the morphology of the chromosome is to dry the preparation on to a glass microscope slide and then treat with formamide.

In the early versions of *in situ* hybridization the probe was radioactively labeled (Technical Note 3.3) but this procedure was unsatisfactory because it is difficult with a radioactive label to achieve both sensitivity and resolution, two critical requirements for successful *in situ* hybridization. Sensitivity requires that the radioactive label has a high emission energy (an example of such a radiolabel is ^{32}P), but if the radiolabel has a high emission energy then it scatters its signal and so gives poor resolution. High resolution is possible if a radiolabel with low emission energy, such as ^{3}H, is used, but these have such low sensitivity that lengthy exposures are needed, leading to a high background and difficulties in discerning the genuine signal.

These problems were solved in the late 1980s by the development of nonradioactive, fluorescent DNA labels. These labels combine high sensitivity with high resolution and are ideal for *in situ* hybridization. Fluorolabels with different colored emissions have been designed, so it is possible to hybridize a number of different probes to a single chromosome and distinguish their individual hybridization signals, enabling the relative positions of the probe sequences to be mapped. To maximize sensitivity, the probes must be as heavily labeled as possible, which in the past has meant that they must be quite lengthy DNA molecules, usually cloned DNA fragments of at least 40 kb such as cosmid clones (see Technical Note 4.3, p. 74). This requirement is less important now as techniques for achieving heavy labeling with shorter molecules have been developed. As far as the construction of a physical map is concerned, a cloned DNA fragment can be looked on as simply another type of marker, though in practice the use of clones as markers adds a second dimension as the cloned DNA is the material from which the DNA sequence is determined. Mapping the positions of clones therefore provides a direct link between a genome map and its DNA sequence.

If the probe is a long fragment of DNA then one potential problem, at least with higher eukaryotes, is that it is likely to contain examples of repetitive DNA sequences (see *Figures 1.5*, p. 6, and *1.7*, p. 10, and Section 6.3) and so may hybridize to many chromosomal positions, not just the specific point to which it is perfectly matched. To reduce this nonspecific hybridization the probe, before use, is mixed with unlabeled DNA from the organism being studied. This DNA can simply be total nuclear DNA (i.e. representing the entire genome) but it is better if a fraction enriched for repeat sequences is used. The idea is that the unlabeled DNA hybridizes to the repetitive DNA sequences in the probe, blocking these so that the subsequent *in situ* hybridization is driven wholly by the unique sequences (Lichter *et al.*, 1990). Nonspecific hybridization is therefore reduced or eliminated entirely (*Figure 3.8*).

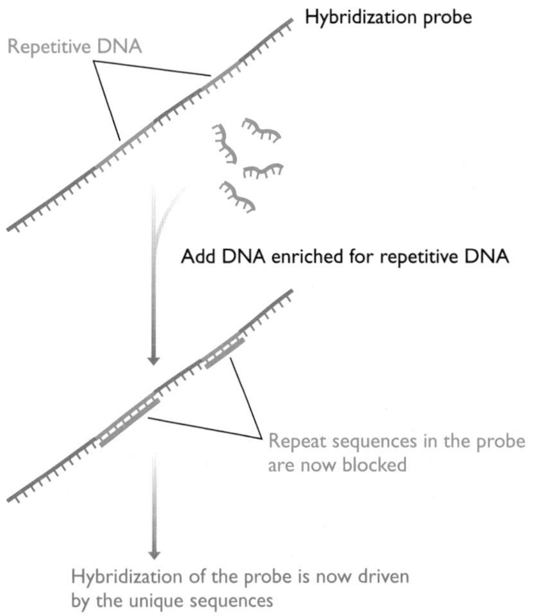

Figure 3.8 A method for blocking repetitive DNA sequences in a hybridization probe.

In this example the probe molecule contains two genome-wide repeat sequences (shown in green). If these sequences are not blocked then the probe will hybridize nonspecifically to any copies of these genome-wide repeats in the target DNA. To block the repeat sequences, the probe is prehybridized with a DNA fraction enriched for repetitive DNA.

3.2.2 FISH in action

Low-resolution FISH with metaphase chromosomes

FISH was originally used with **metaphase chromosomes**. These chromosomes, prepared from nuclei that are undergoing division (Section 6.1.1), are highly condensed with each chromosome in a set taking up a recognizable appearance, characterized by the position of its centromere and the banding pattern that emerges after the chromosome preparation is stained (see *Figure 6.4*, p. 121; Craig and Bickmore, 1993). With metaphase chromosomes, a fluorescent signal obtained by FISH is mapped by measuring its position relative to the end of the short arm of the chromosome (the **FLpter value**). A disadvantage is that the highly condensed nature of metaphase chromosomes means that only low-resolution mapping is possible, two markers needing to be at least 1 Mb apart to be resolved as separate hybridization signals (Trask *et al.*, 1991). This degree of resolution is insufficient for the construction of useful chromosome maps, and the main application of metaphase FISH has been in determining which chromosome a new marker is located on, and providing a rough idea of its map position, as a preliminary to finer scale mapping by other methods. For several years these 'other methods' did not involve any form of FISH, but since 1995 a range of higher resolution FISH techniques have been developed.

High-resolution FISH with extended chromosomes and DNA fibers

The resolution of FISH is determined not by the hybridization technique itself but by the nature of the chromosomal preparation being studied. If metaphase chromosomes are too condensed for fine scale mapping then we must use chromosomes that are more extended. There are two ways of doing this (Heiskanen *et al.*, 1996):

- **Mechanically stretched chromosomes** can be obtained by modifying the preparative method used to isolate chromosomes from metaphase nuclei. Centrifugation generates shear forces which can result in the chromosomes becoming stretched up to 20 times their normal length. Individual chromosomes are still recognizable and FISH signals can be mapped in the same way as with normal metaphase chromosomes. The resolution is significantly improved and markers that are 200–300 kb apart can be distinguished.
- **Nonmetaphase chromosomes** can be used because it is only during metaphase that chromosomes are highly condensed, at other stages of the cell cycle the chromosomes are naturally unpacked. Attempts have been made to use prophase nuclei (see *Figure 2.9*, p. 25) as in these the chromosomes are still sufficiently condensed for individual ones to be identified. In practice these preparations provide no advantage over mechanically stretched chromosomes. Interphase chromosomes are more useful, as this stage of cell cycle (between nuclear divisions) is when the chromosomes are most unpacked. Resolution down to 25 kb is possible, but chromosome morphology is lost so there are no external reference

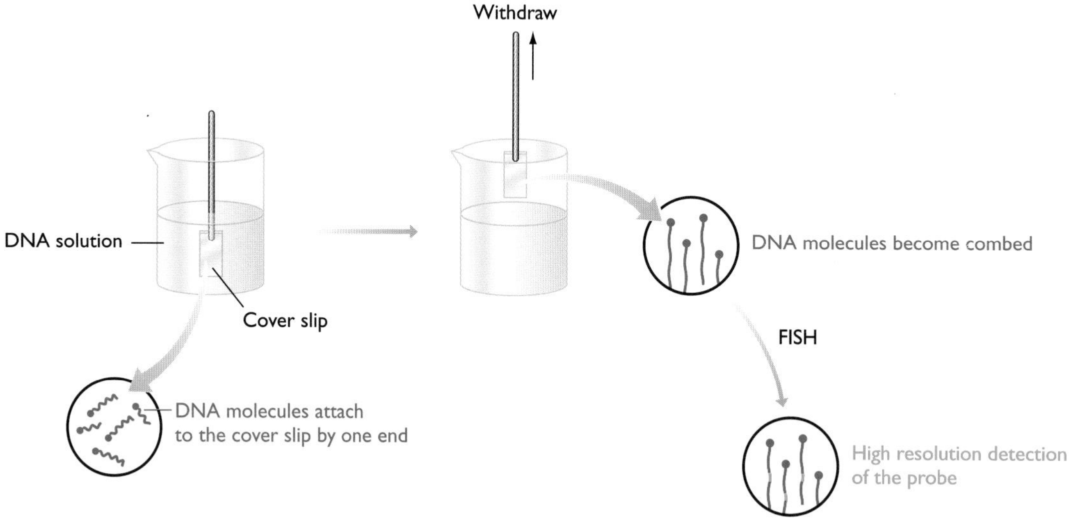

Figure 3.9 Molecular combing enables FISH to be carried out at high resolution.

DNA molecules from the solution attach to the cover slip by their ends. The slip is withdrawn from the solution at a rate of 0.3 mm s^{-1}, which results in a 'comb' of parallel DNA molecules suitable for high resolution FISH.

points against which to map the position of the probe. This technique is therefore used after preliminary map information has been obtained, usually as a means of determining the order of a series of markers in a small region of a chromosome.

The most significant improvement in the resolution of FISH has been achieved by abandoning intact chromosomes and instead using purified DNA. The gel stretching technique described in Section 3.1.2 (see *Figure 3.6*) can be used to prepare DNA for **fiber-FISH**, as can molecular combing (Michalet *et al.*, 1997). The latter is remarkably simple in outline. A molecular comb of DNA fibers is prepared by dipping a silicone-coated microscope cover slip in a solution of DNA, leaving it for 5 minutes (during which time the DNA molecules attach to the slip by their ends), and then removing the slip at a constant speed of 0.3 mm s^{-1} (*Figure 3.9*). The force required to pull the DNA molecules through the meniscus causes them to line up, and once in the air the cover slip surface dries, retaining the DNA molecules as an array of parallel fibers. When used as a target for FISH, markers that are less than 10 kb apart can be distinguished.

3.3 SEQUENCE TAGGED SITE (STS) MAPPING

To generate a detailed physical map of a large genome we need, ideally, a high-resolution mapping procedure that is rapid and not technically demanding. Neither of the two techniques that we have considered so far – restriction mapping and FISH – meets these requirements. Restriction mapping is rapid, easy and provides detailed information, but it cannot be applied to large genomes. FISH can be applied to large genomes and modified versions such as fiber-FISH can give high resolution data, but FISH is difficult to carry out and data accumulation is slow, map positions for no more than three or four markers being obtained in a single experiment. If detailed physical maps are to become a reality then we need a more powerful technique.

At present the most powerful physical mapping technique, and the one that has been responsible for generation of the most detailed maps of large genomes, is STS mapping. A **sequence tagged site** or **STS** is simply a short DNA sequence, generally between 100 and 500 bp in length, that is easily recognizable and occurs only once in the chromosome or genome being studied. To map a set of STSs, a collection of overlapping DNA fragments from a single chromosome or from the entire genome is needed. In the example shown in *Figure 3.10* a fragment collection has been prepared from a single chromosome, with each point along the chromosome represented on average five times in the collection. The data from which the map will be derived are obtained by taking each STS in turn and noting which fragments contain which STSs. This can be done by hybridization analysis but PCR is generally used because PCR is quicker and has proven to be more amenable to automation. The chances of two STSs being present on the same fragment will, of course, depend on how close together they are in the genome. If they are very close then there is a good chance that they will always be on the same fragment, but if they are

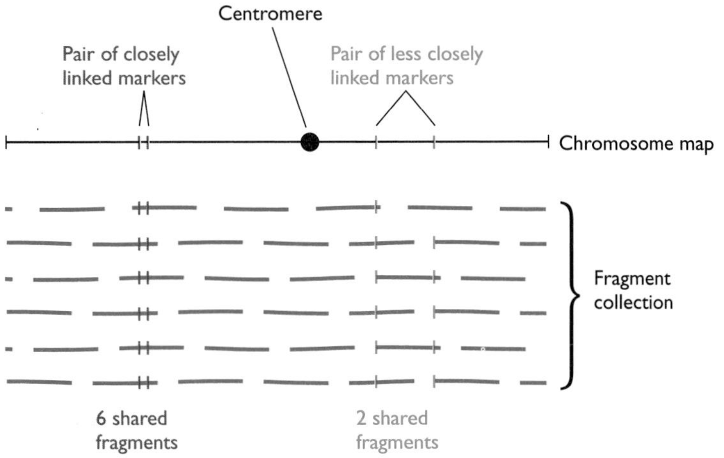

Figure 3.10 A fragment collection suitable for STS mapping.

The fragments span the entire length of a chromosome with each point on the chromosome present in an average of five fragments. The two blue markers are close together on the chromosome map and there is a high probability that they will be found together on the same fragment. The two green markers are more distant from one another and so are less likely to be found on the same fragment.

further apart then sometimes they will be on the same fragment and sometimes they will not (*Figure 3.10*). The data can therefore be used to calculate the distance between two markers, in a manner analogous to the way in which map distances are determined by linkage analysis (Section 2.3). Remember that in linkage analysis a map distance is calculated from the frequency at which crossovers occur between two markers. STS mapping is exactly the same, except that each map distance is based on the frequency at which *breaks* occur between two markers.

The description of STS mapping given above leaves out some critical questions. What exactly is an STS? How is the DNA fragment collection obtained?

3.3.1 Any unique DNA sequence can be used as an STS

To qualify as an STS a DNA sequence must satisfy two criteria. The first is that its sequence must be known, so that a PCR assay can be set up to test for the presence or absence of the STS on different DNA fragments. The second requirement is that the STS must have a unique location in the chromosome being studied, or in the genome as a whole if the DNA fragment set covers the entire genome. If the STS sequence occurs at more than one position then the mapping data will be ambiguous. Care must therefore be taken to ensure that STSs do not include sequences found in repetitive DNA.

These are easy criteria to satisfy and STSs can be obtained in many ways, the most common sources being:

- **Expressed sequence tags (ESTs).** These are short sequences obtained by analysis of **complementary DNA (cDNA)** clones (Marra *et al.*, 1998). Complementary DNA is prepared by converting an mRNA preparation into double-stranded DNA (*Figure 3.11*). Because the mRNA in a cell is derived from protein coding genes, cDNAs represent the sequences of the genes that were being expressed in the cell from which the mRNA was prepared. ESTs are looked on as a rapid means of gaining access to the sequences of important genes, and they are valuable even if their sequences are incomplete. An EST can also be used as an STS, assuming that it comes from a unique gene and not from a member of a gene family in which all the genes have the same or very similar sequences.
- **SSLPs**. In Section 2.2.2 we examined the use of microsatellites and other SSLPs in genetic mapping. SSLPs can also be used as STSs in physical mapping. SSLPs which are polymorphic and have already been mapped by linkage analysis are particularly valuable as they provide a direct connection between the genetic and physical maps.
- **Random genomic sequences** are obtained by sequencing random pieces of cloned genomic DNA, or simply by looking at sequences that have been deposited in the databases.

3.3.2 Fragments of DNA for STS mapping

The second component of an STS mapping procedure is the collection of DNA fragments spanning the chromosome or genome being studied. This collection is sometimes called the **mapping reagent** and at present there are two ways in which it can be assembled.

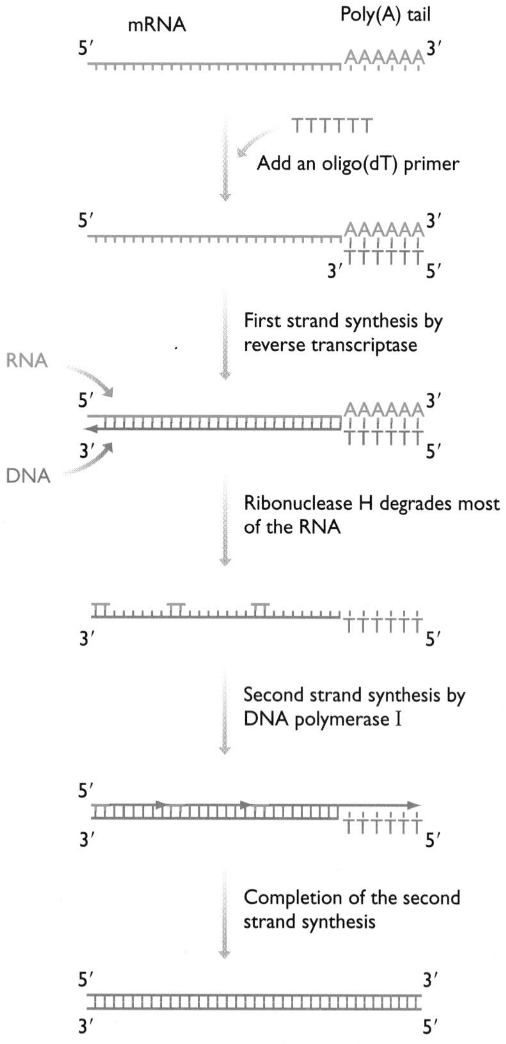

Figure 3.11 One method for preparing cDNA.

Most eukaryotic mRNAs have a poly(A) tail at their 3′-end (Section 9.2.2). This series of A nucleotides is used as the priming site for the first stage of cDNA synthesis, carried out by reverse transcriptase, a DNA polymerase that copies an RNA template. The primer is a short synthetic DNA oligonucleotide, typically 20 nucleotides in length, made up entirely of Ts (an 'oligo(dT)' primer). When the first strand synthesis has been completed, the preparation is treated with ribonuclease H, which specifically degrades the RNA component of an RNA–DNA hybrid. Under the conditions used, the enzyme does not degrade all of the RNA, instead leaving short segments that prime the second DNA strand synthesis reaction, this one catalyzed by DNA polymerase I. See Technical Note 4.2, p. 64 for more details on DNA polymerases and the role of the primer in DNA synthesis.

Radiation hybrids

A radiation hybrid is a rodent cell that contains fragments of chromosomes from a second organism (McCarthy, 1996). The technology was first developed in the 1970s when it was discovered that exposure of human cells to X-ray doses of 3000–8000 rads causes the chromosomes to break up randomly into fragments, larger X-ray doses producing smaller fragments (*Figure 3.12A*). This treatment is of course lethal for the human cells, but the chromosome fragments can be propagated if the irradiated cells are subsequently fused with nonirradiated hamster or other rodent cells. Fusion is stimulated either chemically with polyethylene glycol or by exposure to Sendai virus (*Figure 3.12B*). Not all of the hamster cells take up chromosome fragments so a means of identifying the hybrids is needed. The routine selection process is to use a hamster cell line that is unable to make either thymidine kinase (TK) or hypoxanthine phosphoribosyl transferase (HPRT), deficiencies in either of these two enzymes being lethal when the cells are grown in a medium containing a mixture of hypoxanthine + aminopterin + thymidine (HAT

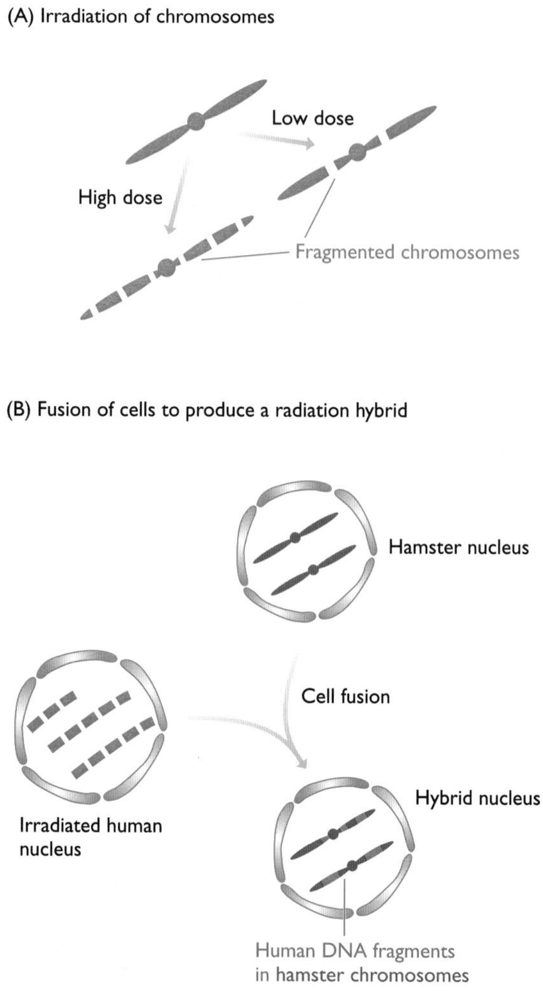

Figure 3.12 Radiation hybrids.

(A) The results of irradiation of human cells: the chromosomes break into fragments, smaller fragments generated by higher X-ray doses. In (B), a radiation hybrid is produced by fusing an irradiated human cell with an untreated hamster cell. For purposes of clarity, only the nuclei are shown.

medium). After fusion, the cells are placed in HAT medium. Those that grow are hybrid hamster cells that have acquired human DNA fragments that include genes for the human TK and/or HPRT enzymes, which are synthesized inside the hybrids enabling these cells to grow in the selective medium. The treatment results in hybrid cells containing a random selection of human DNA fragments, inserted into the hamster chromosomes. Typically the fragments are 5–10 Mb in size with each cell containing fragments equivalent to 15–35% of the human genome. The collection of cells is called a radiation hybrid panel and they can be used as a mapping reagent in STS mapping, providing that the PCR assay used to identify the STS does not amplify the equivalent region of DNA from the hamster genome.

A second type of radiation hybrid panel, containing DNA from just one human chromosome, can be constructed if the cell line that is irradiated is not a human one but a second type of rodent hybrid. Cytogeneticists have developed a number of rodent cell lines in which a single human chromosome is stably propagated in the rodent nucleus. If a cell line of this type is irradiated and fused with hamster cells, then the hybrid hamster cells obtained after selection will contain either human or mouse chromosome fragments, or a mixture of both. The ones containing human DNA can be identified by probing with a human-specific genome-wide repeat sequence, such as the short interspersed nuclear element (SINE) called *Alu* (Section 6.3.2), which has a copy number of almost one million and so occurs on average once every 4 kb in the human genome. Only cells containing human DNA will hybridize to *Alu* sequences, enabling the uninteresting mouse hybrids to be discarded and STS mapping to be directed at the cells containing human chromosome fragments.

Radiation hybrid mapping of the human genome was initially carried out with chromosome-specific rather than whole-genome panels because it was thought that fewer hybrids would be needed to map a single chromosome than would be needed to map the entire genome. It turns out that a high resolution map of a single human chromosome requires a panel of 100–200 hybrids, which is about the most that can be handled conveniently in a PCR screening program. But whole-genome and single-chromosome panels are constructed differently, the former involving irradiation of just human DNA, and the latter requiring irradiation of a mouse cell containing much mouse DNA and relatively little human DNA. This means that in a single-chromosome panel the human DNA content per hybrid is much lower than in a whole-genome panel. It transpires that detailed mapping of the entire human genome is possible with fewer than 100 whole-genome radiation hybrids, so whole-genome mapping is no more difficult than single-chromosome mapping. Once this was realized, whole-genome radiation hybrids became a central component of the mapping phase of the Human Genome Project (see Research Briefing 3.1, p. 56). Standard whole-genome panels are now available, such as the Genebridge4 Panel of 93 human hybrids, and these are used worldwide so that map data from different projects are directly comparable. Whole-genome libraries are also being used for STS mapping of other mammalian genomes and for those of zebrafish and chicken (McCarthy, 1996).

A clone library can also be used as the mapping reagent for STS analysis

A preliminary to the sequencing phase of a genome project is to break the genome or isolated chromosomes into fragments and clone each one in a high-capacity vector, one able to handle large fragments of DNA (see Technical Note 3.4 and Section 4.2.2). This results in a clone library, a collection of DNA fragments which, in this case, have an average size of 1–2 Mb. This type of clone library can be used as a mapping reagent in STS analysis.

As with radiation hybrid panels, a clone library can be prepared from genomic DNA, and so represent the entire genome, or a chromosome-specific library can be made if the starting DNA comes from just one type of chromosome. The latter is possible because individual chromosomes can be separated by **flow cytometry**. To carry out this technique, dividing cells (ones with condensed chromosomes) are carefully broken open so that a mixture of intact chromosomes is obtained. The chromosomes are then stained with a fluorescent dye. The amount of dye that a chromosome binds depends on its size, so larger chromosomes bind more dye and fluoresce more brightly than small ones. The chromosome preparation is diluted and passed through a fine aperture, producing a stream of droplets, each one containing a single chromosome. The droplets pass through a detector which measures the amount of fluorescence, and hence identifies which droplets contain the particular chromosome being sought. An electric charge is applied to these drops, and no others (*Figure 3.13*), enabling the droplets containing the desired chromosome to be deflected and separated from the rest. What if two different chromosomes have similar sizes, as is the case with human chromosomes 21 and 22? These can usually be separated if the dye that is used is not one that binds nonspecifically to DNA, but instead has a preference for AT- or GC-rich regions. Examples of such dyes are Hoechst 33258 and chromomycin A_3, respectively. Two chromosomes that are the same size rarely have identical GC contents, and so can be distinguished by the amounts of AT- or GC-specific dye that they bind.

Compared to radiation hybrid panels, clone libraries have one important advantage for STS mapping. This is the fact that the individual clones can subsequently provide the DNA that is actually sequenced. The data resulting from STS analysis, from which the physical map is generated, can equally well be used to determine which clones contain overlapping DNA fragments, enabling a **clone contig** to be built up (*Figure 3.14*; for other methods for assembling clone contigs see Section 4.2.2). This assembly of overlapping clones can be used as the base material for a lengthy, continuous DNA sequence, and the STS data can later be used to anchor this sequence precisely onto the physical map. If the STSs also include

TECHNICAL 3.4 NOTES

DNA cloning

Part 1. Molecules and manipulations.

Cloning revolutionized biological research in the 1970s by making it possible to study individual genes. Its broad range of applications can be divided into three categories: isolation of genes and other DNA sequences for *in vitro* procedures such as hybridization probing and DNA sequencing; preparation of modified versions of genes to reintroduce in to the original host so that gene expression and its control can be studied; and transfer of genes between organisms to obtain large quantities of the protein coded by the gene or to modify the phenotype of the recipient.

A DNA cloning experiment with *E. coli* is illustrated below. First, the DNA is digested with a restriction enzyme, or randomly sheared by sonication or by some other method (Section 4.2.1). The entire digest can be cloned, or a single fragment can be purified by cutting its band from an agarose gel. The **cloning vector** (see Part 2 of this Technical Note) is also cut with a restriction enzyme, one that has a single recognition site in the vector sequence and so converts the circular molecule into a linear form. The vector and DNA fragments are then ligated together to produce **recombinant DNA molecules.** This is easiest if both the vector and the DNA have the same sticky ends or if both have blunt ends (see Technical Note 3.1); if the ends are different then they can be blunted by enzymatic treatment.

The recombinant DNA molecules are usually introduced into *E. coli* by **transformation**, the natural DNA uptake mechanism possessed by bacteria and enhanced in the laboratory by soaking cells in calcium chloride. An alternative, with certain vectors, is ***in vitro* packaging**, which results in the recombinant molecules being inserted into infectious bacteriophage particles. Within the bacterium, the vector directs multiplication of the recombinant DNA molecule, producing a number of identical copies. The vector replication process is such that only one recombinant DNA molecule can propagate within a single bacterium, so each resulting clone contains multiple copies of just one DNA insert. Cells are plated onto agar medium and the **markers** carried by the vector (see Part 2) used to distinguish colonies containing recombinant molecules from untransformed colonies and from ones containing a vector with no insert DNA.

Cloning vectors for organisms other than *E. coli* are used in a similar way to that described above. The only significant differences concern the ways in which the recombinant DNA molecules are introduced into the host cells. To induce DNA uptake by yeast and plant cells it is normally necessary to remove their cell walls by enzymatic treatment. Animal cells can be transformed by precipitating DNA on to a cell monolayer. More sophisticated methods include bombardment of tissues with gold or tungsten microprojectiles coated in DNA, or direct microinjection of DNA into the host cell's nucleus.

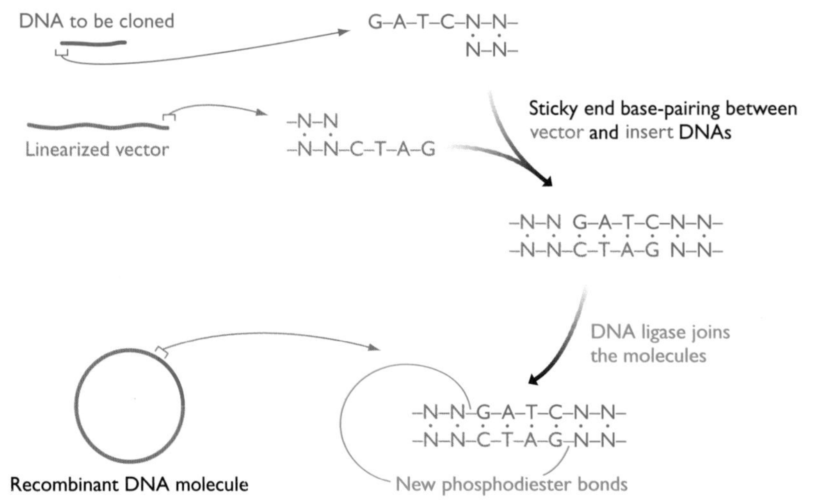

TECHNICAL 3.4 NOTES

DNA cloning

Part 2. Cloning vectors.

A cloning vector has two key requirements. First, it must be a relatively short DNA molecule, ideally less than 10 kb, so that it is easily manipulated in the test tube. During a cloning experiment (see Part 1 of this Technical Note) the vector must be cut at just one position, so it must be short enough to have single recognition sequences for one or more restriction enzymes. The second requirement is that the vector must be efficiently replicated inside the host cell. Plasmids and bacteriophages satisfy these requirements as far as bacterial hosts are concerned, but plasmids are uncommon in eukaryotes, though *Saccharomyces cerevisiae* possesses one that is sometimes used for cloning purposes. Most eukaryotic vectors are therefore based on viruses. Alternatively, with a eukaryotic host the replication requirement can be bypassed by inducing the foreign DNA to be inserted into one of the host chromosomes.

The vector must also carry markers that enable cells containing the vector to be identified and recombinant molecules to be distinguished from nonrecombinant ones. Markers include genes for antibiotic resistance, histochemical markers that result in colonies taking on a distinctive color, and nutritional markers which enable the transformed cell to grow on a medium lacking a nutrient that it normally requires.

Many vectors are based on natural *E. coli* plasmids (Brown, 1998). A typical example is pUC8 (see below), which is 2.7 kb and has unique sites for a range of restriction enzymes. It carries two markers: a gene for ampicillin resistance which is used to determine if a bacterium contains a vector, and the *lacZ′* gene, which codes for part of the β-galactosidase enzyme. The *lacZ′* gene is inactivated if the vector contains a DNA insert. Its activity is monitored by plating cells on agar containing the disaccharide X-gal (5-bromo-4-chloro-3-indolyl-β-D-galactopyranoside), which is similar to lactose, the natural substrate for β-galactosidase (Section 6.2.2), and is split by the enzyme into its two component sugars, one of which is colored blue. This means that colonies containing pUC8 without inserted DNA turn blue, whereas those containing a recombinant vector remain creamy-white.

Other *E. coli* vectors are based on the bacteriophages λ and M13. Phage λ vectors, including cosmids, are described in Technical Note 4.3, p. 74, and the M13 system, which is mainly used in DNA sequencing, is detailed in Section 4.1.1. Special vectors for cloning large pieces of DNA are covered in Section 4.2.2.

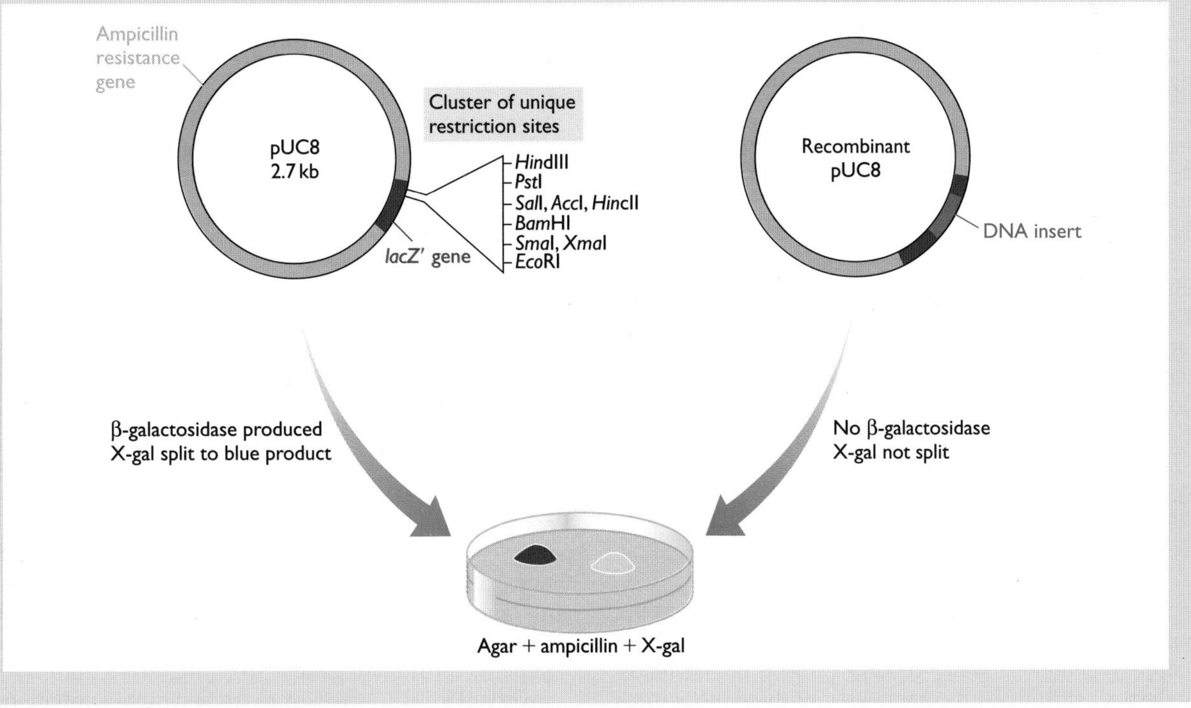

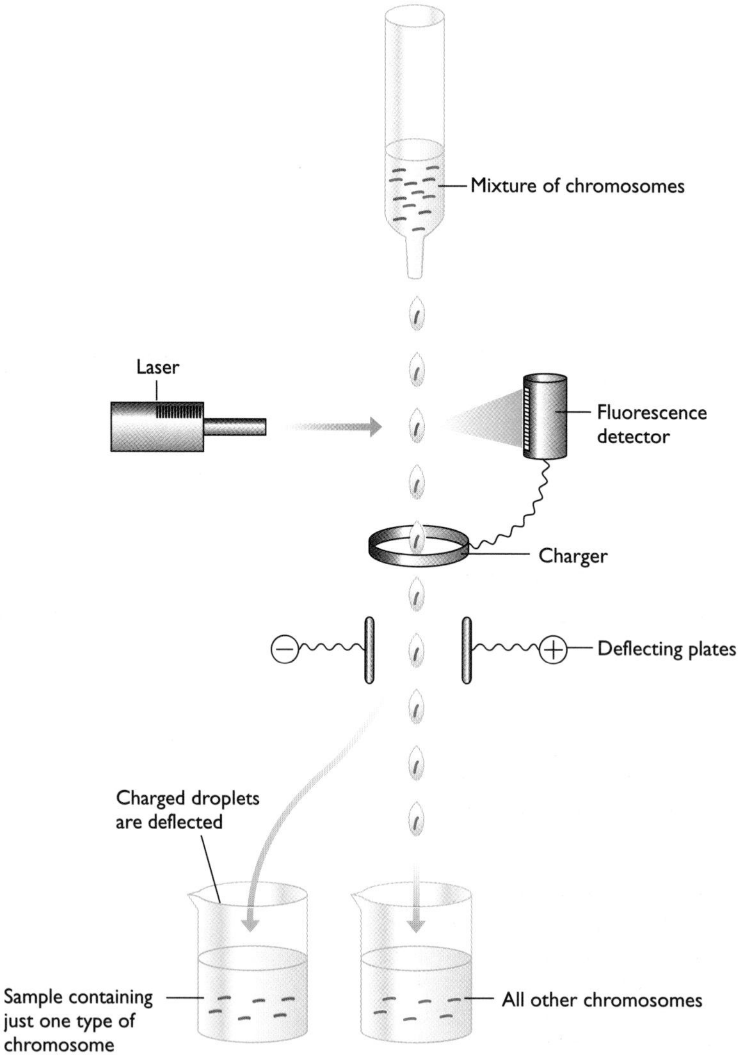

Figure 3.13 Separating chromosomes by flow cytometry.

A mixture of fluorescently-stained chromosomes is passed through a small aperture so that each drop that emerges contains just one chromosome. The fluorescence detector identifies the signal from drops containing the correct chromosome and applies an electric charge to these drops. When the drops reach the electric plates, the charged ones are deflected into their separate beaker. All other drops fall straight through the deflecting plates and are collected in the waste beaker.

SSLPs that have been mapped by genetic linkage analysis then the DNA sequence, physical map and genetic map can all be integrated. This is an ultimate objective of a genome project and we will see in the next section how close we are to achieving it with the human genome.

3.4 PROGRESS TOWARDS THE HUMAN GENOME MAP

In this chapter and the previous one we have covered the important techniques – genetic and physical – used for mapping genomes. Many genome projects are currently underway, each using a combination of these procedures to generate a detailed map to direct and aid interpretation of the DNA sequencing phase of the project. Before moving on to look at DNA sequencing we will briefly review progress in one of these projects, the one for the human genome.

Until the beginning of the 1980s a detailed map of the human genome was looked on as an unattainable objective. Although comprehensive genetic maps had been constructed for fruit flies and a few other organisms, the problems inherent in analysis of human pedigrees (Section 2.3.2) and the relative paucity of polymorphic genetic markers, meant that most geneticists doubted that a human genetic map could ever be achieved. The initial impetus for human genetic mapping came from the discovery of RFLPs, which were the first highly polymorphic

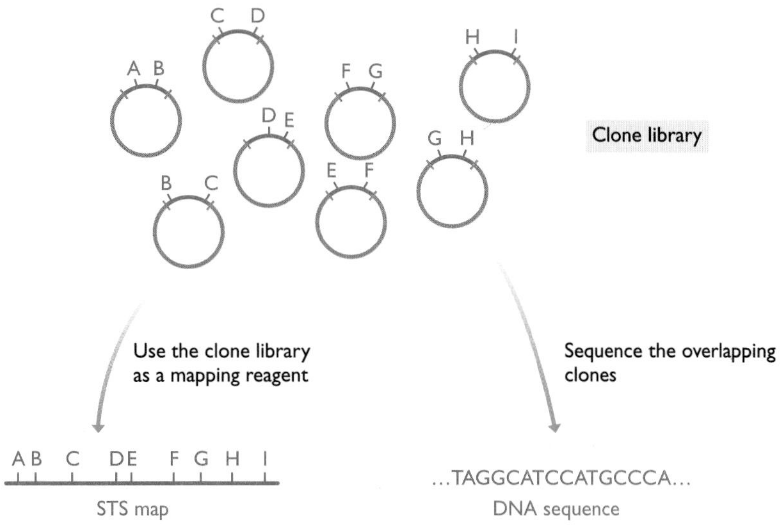

Figure 3.14 The value of clone libraries in genome projects.

The small clone library shown in this example contains sufficient information for an STS map to be constructed, and can also be used as the source of the DNA that will be sequenced.

DNA markers to be recognized in animal genomes. In 1987 the first human RFLP map was published, comprising 393 RFLPs and ten additional polymorphic markers (Donis-Keller *et al.*, 1987). This map, developed from analysis of 21 families, had an average marker density of one per 10 Mb.

In the late 1980s the Human Genome Project became established as a loose but organized collaboration between geneticists in all parts of the world. One of the goals that the Project set itself was a genetic map with a density of one marker per 1 Mb though, as mentioned at the beginning of this chapter, it was thought that a density of one per 2–5 Mb might be the realistic limit. In fact by 1994 an international consortium had met and indeed exceeded the objective, thanks to their use of SSLPs and the large CEPH collection of reference families (Section 2.3.2). The 1994 map contained 5800 markers, of which over 4000 were SSLPs, and had a density of one marker per 0.7 Mb (Murray *et al.*, 1994). The final version of the genetic map, published the following year (Dib *et al.*, 1996; see Research Briefing 2.1, p. 33) took the 1994 map slightly further by inclusion of an additional 1250 SSLPs.

Physical mapping did not lag far behind. In the early 1990s considerable effort was put into the generation of clone contig maps, using STS screening as described in the previous section as well as other clone fingerprinting methods which we will look at in Section 4.2.2. The major achievement of this phase of the physical mapping project was publication of a clone contig map of the entire genome, consisting of 33 000 YACs (yeast artificial chromosomes) containing fragments with an average size of 0.9 Mb (Cohen *et al.*, 1993). However, doubts were raised about the value of YAC contig maps when it was realized that YAC clones can contain two or more pieces of noncontiguous DNA (*Figure 3.15*). The use of these chimeric

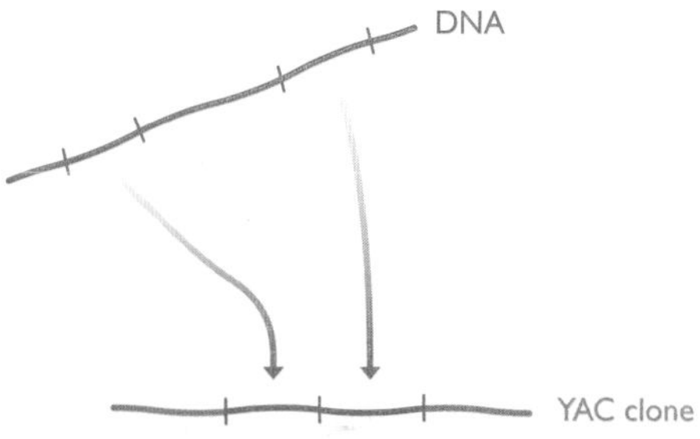

Figure 3.15 Some YAC clones contain segments of DNA from different parts of the human genome.

clones in construction of contig maps could lead to DNA segments that are widely separated in the genome being mistakenly mapped to adjacent positions. These problems led to the adoption of radiation hybrid mapping of STS markers, largely by the Whitehead Institute/MIT Genome Center in Massachusetts, culminating in 1995 with publication of a human STS map containing 15 086 STSs with an average density of one per 199 kb (Hudson *et al.*, 1995). This map was later supplemented with an additional 20 104 STSs, most of these being ESTs and hence positioning protein coding genes on the physical map (Schuler *et al.*, 1996). The resulting map density approached the target of one marker per 100 kb set as the objective for physical mapping at the outset of the Human Genome Project. The construction of the EST map is described in more detail in Research Briefing 3.1.

The combined STS maps include positions for almost 7000 polymorphic SSLPs that have also been mapped onto the genome by genetic means. As a result, the physical and genetic maps can be directly compared, and

RESEARCH

3.1

BRIEFING

The human physical map

A combined genetic and physical map with a density of at least one marker per 100 kb was set by the Human Genome Project as a prerequisite before large scale sequencing could begin. The physical mapping component of the Project culminated in 1996 with this paper.

Previous physical maps had been based on STS markers but the vast majority of these were thought to lie in intergenic regions. One objective of this project was to increase the number of mapped physical markers that corresponded to coding regions of the genome, as this would enable the initial sequencing to be directed at gene-rich regions. The markers used were therefore expressed sequence tags taken from the human EST databases.

The mapping procedure

At the time that the project was begun approximately 450 000 human ESTs were known, but this collection included a great deal of redundancy, many genes represented by two or more ESTs. The initial phase of the project was therefore a computer-based comparison of a subset of sequences from the EST library in order to identify overlaps and redundancy. This resulted in 49 625 EST groups, each thought to represent a different human gene. Attempts were made to map about 30 000 of these, using the G3 and Genebridge4 radiation hybrid panels (83 and 93 hamster cell lines respectively) and one YAC collection of 32 000 clones with an average insert size of approximately 1 Mb. Map positions could be determined for 20 104 ESTs. Most of these, 19 000 in all, were mapped using one or both of the radiation hybrid panels, the remainder were mapped using the YAC collection. The 20 104 ESTs mapped to 16 354 different loci, indicating that some pairs of ESTs were probably derived from a single gene. The average map density was one marker per 183 kb.

The vast amount of work required to generate the EST map was possible only by an international collaboration. Altogether 18 laboratories participated, located at eight centers in five different countries. Consistency throughout the consortium was maintained by ensuring that some markers were independently mapped by more than one group. Of 3114 markers mapped in duplicate, only about 140 were given vastly different map positions (more than 8 Mb apart) by different groups. These 140 errors appeared to fall into two categories. About half of them were relatively minor errors due simply to an inaccuracy by one of the groups. The remainder involved a disagreement as to which chromosome a marker was located on. In some cases this was probably due to a data handling error, but the possibility exists that at least some of these apparent misassignments were due to the presence of two genes with similar sequences at different positions in the genome.

Significance of the map

Because ESTs are expressed sequences the resulting map indicates the density of genes in different parts of the genome. It is clear that genes are not distributed evenly through the genome, each chromosome having gene-dense and gene-light regions (see also Section 6.1.1). Also, some chromosomes (notably numbers 4, 13, 18, 21 and X) have relatively low gene densities, though this may be due in part to the fact that only about one-fifth of all human genes are included on the map. Integration with other physical maps and the genetic one described in Research Briefing 2.1, p. 33 resulted in the comprehensive human map that is acting as the framework for the sequencing phase of the Human Genome Project.

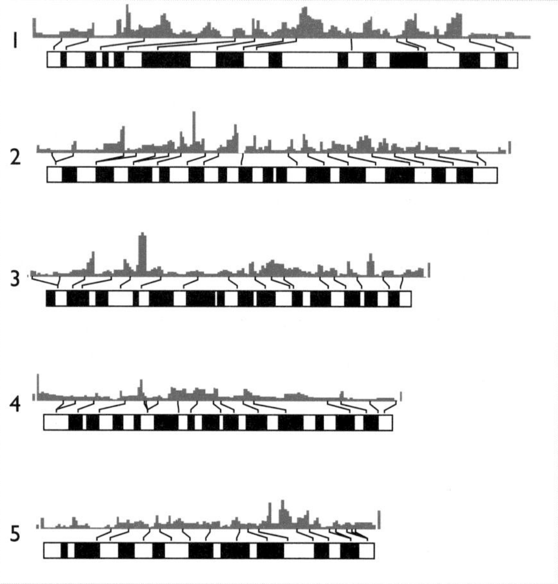

The human EST map.

The maps for chromosomes 1 to 5 are shown. The chromosomes are drawn with the G banding pattern they take up after staining with Giemsa (see *Table 6.2*, p. 120). The red histograms show the distributions of ESTs along each chromosome. As ESTs are derived from genes the histograms show, in effect, the gene densities at different positions. Reprinted with permission from Schuler GD, *et al.* (1996) *Science*, **274**, 540–546. (Updated in 1998.)

Reference

Schuler GD, Boguski MS, Stewart EA, *et al.* (1996) A gene map of the human genome. *Science*, **274**, 540–546.

clone contig maps that include STS data can be anchored onto both maps. The net result is a comprehensive, integrated map (Bentley *et al.*, 1998; Deloukas *et al.*, 1998) that is now being used as the framework for the DNA sequencing phase of the Human Genome Project. It took just 15 years for the unattainable to be achieved.

REFERENCES

Andrews AT (1991) Electrophoresis of nucleic acids. In: TA Brown, ed. *Essential Molecular Biology: A Practical Approach, Volume 1*. IRL Press, Oxford, pp. 89–126.

Bentley DR, Pruitt KD, Deloukas P, Schuler GD and Ostell J (1998) Coordination of human genome sequencing via a consensus framework map. *Trends Genet.*, **14**, 381–384.

Brown TA (1998) *Molecular Biology Labfax*, 2nd edition, Volumes 1 and 2. Academic Press, London.

Cohen D, Chumakov I and Weissenbach J (1993) A first-generation map of the human genome. *Nature*, **366**, 698–701.

Craig JM and Bickmore WA (1993) Chromosome bands – flavors to savour. *Bioessays*, **15**, 349–354.

Deloukas P, Schuler GD, Gyapay G, *et al.* (1998) A physical map of 30,000 genes. *Science*, **282**, 744–746.

Dib C, Fauré S, Fizames C, *et al.* (1996) A comprehensive genetic map of the human genome based on 5,264 microsatellites. *Nature*, **380**, 152–154.

Donis-Keller H, Green P, Helms C, *et al.* (1987) A genetic map of the human genome. *Cell*, **51**, 319–337.

Heiskanen M, Peltonen L and Palotie A (1996) Visual mapping by high resolution FISH. *Trends Genet.*, **12**, 379–382.

Hudson TJ, Stein LD, Gerety SS, *et al.* (1995) An STS-based map of the human genome. *Science*, **270**, 1945–1954.

Lichter P, Tang CJ, Call K, *et al.* (1990) High resolution mapping of human chromosome 11 by *in situ* hybridization with cosmid clones. *Science*, **247**, 64–69.

Marra MA, Hillier L and Waterston RH (1998) Expressed sequence tags – ESTablishing bridges between genomes. *Trends Genet.*, **14**, 4–7.

McCarthy L (1996) Whole genome radiation hybrid mapping. *Trends Genet.*, **12**, 491–493.

Michalet X, Ekong R, Fougerousse F, *et al.* (1997) Dynamic molecular combing: stretching the whole human genome for high-resolution studies. *Science*, **277**, 1518–1523.

Murray JC, Buetow KH, Weber JL, *et al.* (1994) A comprehensive human linkage map with centimorgan density. *Science*, **265**, 2049–2054.

Oliver SG, van der Aart QJM, Agostoni-Carbone ML, *et al.* (1992) The complete DNA sequence of yeast chromosome III. *Nature*, **357**, 38–46.

Schuler GD, Boguski MS, Stewart EA, *et al.* (1996) A gene map of the human genome. *Science*, **274**, 540–546.

Schwartz DC, Li X, Hernandez LI, Ramnarain SP, Huff EJ and Wang Y-K (1993) Ordered restriction maps of *Saccharomyces cerevisiae* chromosomes constructed by optical mapping. *Science*, **262**, 110–114.

Trask BJ, Massa H, Kenwrick S and Gitschier J (1991) Mapping of human chromosome Xq28 by 2-color fluorescence *in situ* hybridization of DNA sequences to interphase cell nuclei. *Am. J. Hum. Genet.*, **48**, 1–15.

FURTHER READING

Brown TA (1995) *Gene Cloning: An Introduction*, 3rd edition. Chapman & Hall, London. — *Describes the basic methodology, such as restriction, gel electrophoresis and PCR, used in physical mapping and other areas of molecular biology research.*

Lichter P (1997) Multicolor FISHing: what's the catch? *Trends Genet.*, **13**, 475–479. — *An interesting review of some of the applications of FISH.*

Primrose SB (1995) *Principles of Genome Analysis*. Blackwell Science, Oxford. — *A little out of date but a good description of mapping strategies.*

Strachan T and Read AP (1996) *Human Molecular Genetics*. BIOS Scientific Publishers, Oxford. — *Chapter 11 covers physical mapping of the human genome.*

Walter MA, Spillett DJ, Thomas P, Weissenbach J and Goodfellow PN (1994) A method for constructing radiation hybrid maps of whole genomes. *Nature Genet.*, **7**, 22–28. — *An excellent review of this approach to mapping.*

Watson JD, Gilman M, Witkowski J and Zoller M (1992) *Recombinant DNA*, 2nd edition. W.H. Freeman, New York. — *Detailed descriptions of basic recombinant DNA methodology.*

CHAPTER

4

Sequencing Genomes

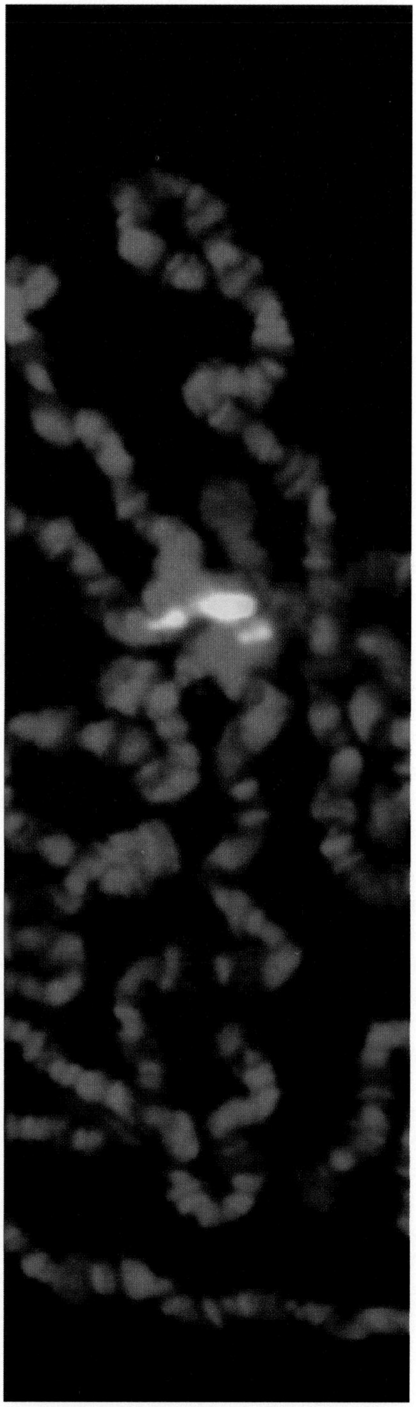

Contents

4.1	*The Methodology for DNA Sequencing*	**60**
	4.1.1 Chain termination DNA sequencing	60
	4.1.2 Radical departures from conventional DNA sequencing	69
4.2	*Assembly of a Contiguous DNA Sequence*	**70**
	4.2.1 Sequence assembly by the shotgun approach	70
	4.2.2 Sequence assembly by the clone contig approach	73
	4.2.3 The directed shotgun approach	79
4.3	*The Sequencing Phase of the Human Genome Project*	**80**
	4.3.1 The Human Genome Project is now largely based on BAC libraries	81
	4.3.2 Rapid sequence acquisition can be directed at interesting regions of the genome	81
	4.3.3 The sequencing phase of the Human Genome Project raises ethical questions	82

Concepts

- *Two methods exist for rapid DNA sequencing but the chain termination procedure is the method of choice for genome sequencing*
- *A single sequencing experiment produces only a few hundred bp of sequence*
- *A contiguous DNA sequence for a small genome can be assembled by random shotgun sequencing*
- *For large genomes, overlapping clone contigs are built up with fragments cloned in high capacity vectors such as BACs*
- *A shotgun sequencing method that uses map data to aid sequence assembly may provide a quicker method for sequencing large genomes*
- *Rapid sequence acquisition from large genomes is possible by EST sequencing and sequence skimming*

THE ULTIMATE OBJECTIVE of a genome project is the complete DNA sequence for the organism being studied, ideally integrated with the genetic and/or physical maps of the genome so that genes and other interesting features can be located within the DNA sequence. This chapter describes the techniques and research strategies that are used during the sequencing phase of a genome project, when this ultimate objective is being directly addressed. Techniques for sequencing DNA are clearly of central importance in this context and we will begin the chapter with a detailed examination of sequencing methodology. This methodology is, however, of little value unless the short sequences that result from individual sequencing experiments can be linked together in the correct order to give the master sequences of the chromosomes that make up the genome. The middle part of this chapter describes the strategies used to ensure that the master sequences are assembled correctly. Finally, we will consider how the sequencing phase of the Human Genome Project is being conducted.

4.1 THE METHODOLOGY FOR DNA SEQUENCING

Rapid and efficient procedures for DNA sequencing were first devised in the mid-1970s. Two different procedures were published at almost the same time:

- **The chain termination method** (Sanger *et al.*, 1977), in which the sequence of a single-stranded DNA molecule is determined by enzymatic synthesis of complementary polynucleotide chains, these chains terminating at specific nucleotide positions.
- **The chemical degradation method** (Maxam and Gilbert, 1977), in which the sequence of a double-stranded DNA molecule is determined by treatment with chemicals that cut the molecule at specific nucleotide positions.

Both methods were equally popular to begin with but the chain termination procedure has gained ascendancy in recent years, especially for genome sequencing. This is partly because the chemicals used in the chemical degradation method are toxic and therefore hazardous to the health of the researchers doing the sequencing experiments, but mainly because it has been easier to automate chain termination sequencing. As we will see later in this chapter, a genome project involves a huge number of individual sequencing experiments and it would take many years to carry all these out by hand. Automated sequencing techniques are therefore essential if the project is to be completed in a reasonable timespan.

4.1.1 Chain termination DNA sequencing

Chain termination DNA sequencing is based on the principle that single-stranded DNA molecules that differ in

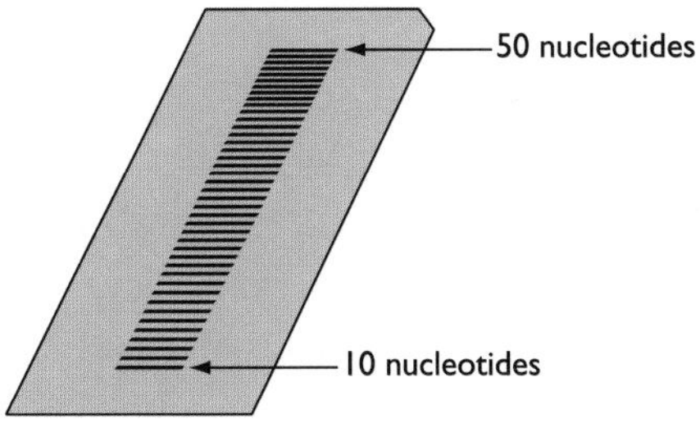

Figure 4.1 Polyacrylamide gel electrophoresis can resolve single-stranded DNA molecules that differ in length by just one nucleotide.

The banding pattern is produced after separation of single-stranded DNA molecules by denaturing polyacrylamide gel electrophoresis. The molecules are labeled with a radioactive marker and the bands visualized by autoradiography. The bands gradually get closer together towards the top of the ladder. In practice, molecules up to about 1500 nucleotides in length can be separated if the electrophoresis is continued for long enough.

length by just a single nucleotide can be separated from one another by **polyacrylamide gel electrophoresis** (Technical Note 4.1). This means that it is possible to resolve a family of molecules, representing all lengths from 10 to 1500 nucleotides, into a series of bands (*Figure 4.1*).

Chain termination sequencing in outline

The starting material for a chain termination sequencing experiment is a preparation of identical single-stranded DNA molecules. The first step is to **anneal** a short oligonucleotide to the same position on each molecule, this oligonucleotide subsequently acting as the **primer** for synthesis of a new DNA strand, complementary to the **template** (*Figure 4.2A*). The strand synthesis reaction, which is catalyzed by a DNA polymerase enzyme (Technical Note 4.2) and requires the four deoxyribonucleotide triphosphates (dATP, dCTP, dGTP and dTTP) as substrates, would normally continue until several thousand nucleotides had been polymerized. This does not occur in

Box 4.1: The chemical degradation sequencing method

The difference between the two sequencing techniques lies in the way in which the A, C, G and T families of molecules are generated. In the chemical degradation procedure these families are produced not by enzymatic synthesis but by treatment with chemicals that cut specifically at a particular nucleotide. The starting material can be double-stranded DNA but before beginning the sequencing procedure the double-stranded molecules must be converted to a single-stranded preparation with each molecule labeled at one end.

To illustrate the procedure, we will follow the 'G' reaction. First, the molecules are treated with dimethyl sulfate, which attaches a methyl group to the purine ring of G nucleotides. Only a limited amount of dimethyl sulfate is added, the objective being to modify, on average, just one G per polynucleotide. At this stage the DNA strands are still intact, cleavage not occurring until a second chemical – piperidine – is added. Piperidine removes the modified purine ring and cuts the DNA molecule at the phosphodiester bond immediately upstream of the baseless site that is created. The result is a set of cleaved DNA molecules, some of which are labeled and some of which are not. The labeled molecules all have one end in common and one end determined by the cut sites, the latter indicating the positions of the G nucleotides in the DNA molecules that were cleaved. In other words, the G family of molecules produced by chemical treatment is equivalent to the G family produced by the chain termination method.

The families of cleaved molecules are electrophoresed in a polyacrylamide gel and the sequence read in a similar way to that described for chain termination sequencing. The only significant difference is that problems have been encountered in developing chemical treatments to cut specifically at A or T, and so the four reactions that are carried out are usually 'G', 'A+G', 'C' and 'C+T'. This does not affect the accuracy of the sequence that is read from the gel.

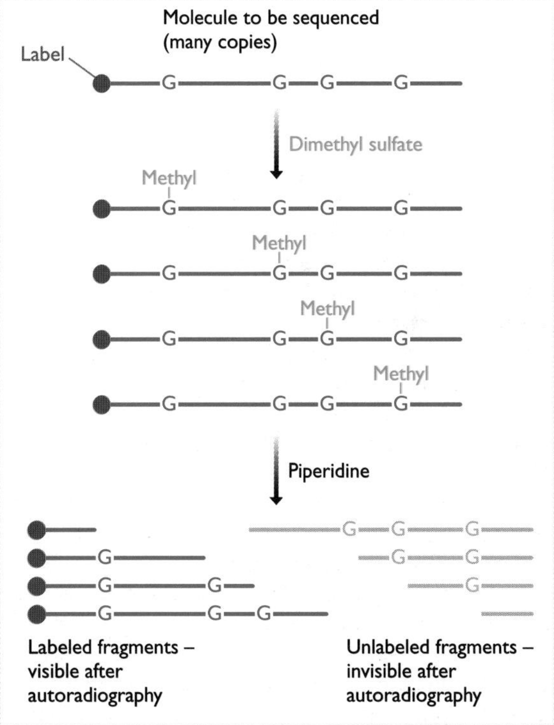

TECHNICAL 4.1 NOTES

Polyacrylamide gel electrophoresis

Separation of DNA molecules differing in length by just one nucleotide.

Polyacrylamide gel electrophoresis is used to examine the families of chain-terminated DNA molecules resulting from a sequencing experiment. Agarose gel electrophoresis (Technical Note 3.2, p. 43) cannot be used for this purpose because it does not have the resolving power needed to separate single-stranded DNA molecules that differ in length by just one nucleotide. Polyacrylamide gels have smaller pore sizes than agarose gels and allow precise separations of molecules from 10–1500 bp. As well as DNA sequencing, polyacrylamide gels are also used for other applications where fine-scale DNA separations are required, for instance in the examination of amplification products from PCRs directed at microsatellite loci, where the products of different alleles might differ in size by just two or three base pairs (see *Figure 2.5*, p. 21).

A polyacrylamide gel consists of chains of acrylamide monomers (CH_2=CH–CO–NH_2) cross-linked with *N, N′*–methylenebisacrylamide units (CH_2=CH–CO–NH–CH_2–NH–CO–CH=CH_2), the latter commonly called 'bis'. The pore size of the gel is determined by both the total concentration of monomers (acrylamide + bis) and the ratio of acrylamide to bis. For DNA sequencing, a 6% gel with an acrylamide : bis ratio of 19:1 is normally used as this allows resolution of single-stranded DNA molecules between 100 and 750 nucleotides in length. About 650 nucleotides of sequence can therefore be read from a single gel. The gel concentration can be increased to 8% in order to read the sequence closer to the primer (resolving molecules 50–400 nucleotides in length) or decreased to 4% to read a more distant sequence (500–1500 nucleotides from the primer). Polymerization of the acrylamide : bis solution is initiated by ammonium persulfate and catalyzed by TEMED (*N, N, N′, N′*-tetramethylethylenediamine). Sequencing gels also contain urea, which is a denaturant that prevents intrastrand base pairs from forming in the chain-terminated molecules. This is important because the change in conformation resulting from base-pairing alters the migration rate of a single-stranded molecule, so the strict equivalence between the length of a molecule and its band position, critical for reading the DNA sequence (see *Figure 4.2D*), is lost.

Polyacrylamide gels are prepared between two glass plates held apart by spacers. This arrangement serves two purposes. First, it enables a very thin (<1 mm) gel to be made, which facilitates sequence reading by improving the sharpness of the bands. Second, it ensures that polymerization, which is inhibited by oxygen, occurs evenly throughout the gel. With such a thin gel the amount of DNA per band is small and the banding pattern is only barely visible after ethidium bromide staining. For this reason, a radioactively-labeled nucleotide is usually included in the sequencing reactions so that the banding pattern can be visualized by autoradiography.

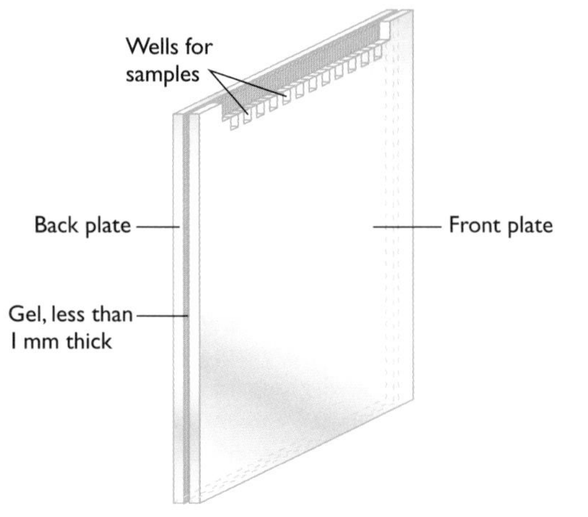

a chain termination sequencing experiment because, as well as the four dNTPs, a small amount of a **dideoxynucleotide** (e.g. ddATP) is added to the reaction. The polymerase enzyme does not discriminate between dNTPs and ddNTPs, so the dideoxynucleotide can be incorporated into the growing chain, but it then blocks further elongation because it lacks the 3′-hydroxyl group needed to form a connection with the next nucleotide (*Figure 4.2B*).

If ddATP is present, chain termination occurs at positions opposite thymidines in the template DNA (*Figure 4.2C*). Because dATP is also present the strand synthesis does not always terminate at the first T in the template, in fact it may continue until several hundred nucleotides have been polymerized before eventually a ddATP is incorporated. The result is therefore a set of new chains, all of different lengths, but each ending in ddATP. Now the polyacrylamide gel comes into play. The family of molecules generated in the presence of ddATP are loaded into one lane of the gel, and the families generated with ddCTP, ddGTP and ddTTP loaded into the three adjacent lanes. After electrophoresis, the DNA sequence can be read directly from the positions of the bands in the gel (*Figure 4.2D*). The band that has moved the furthest

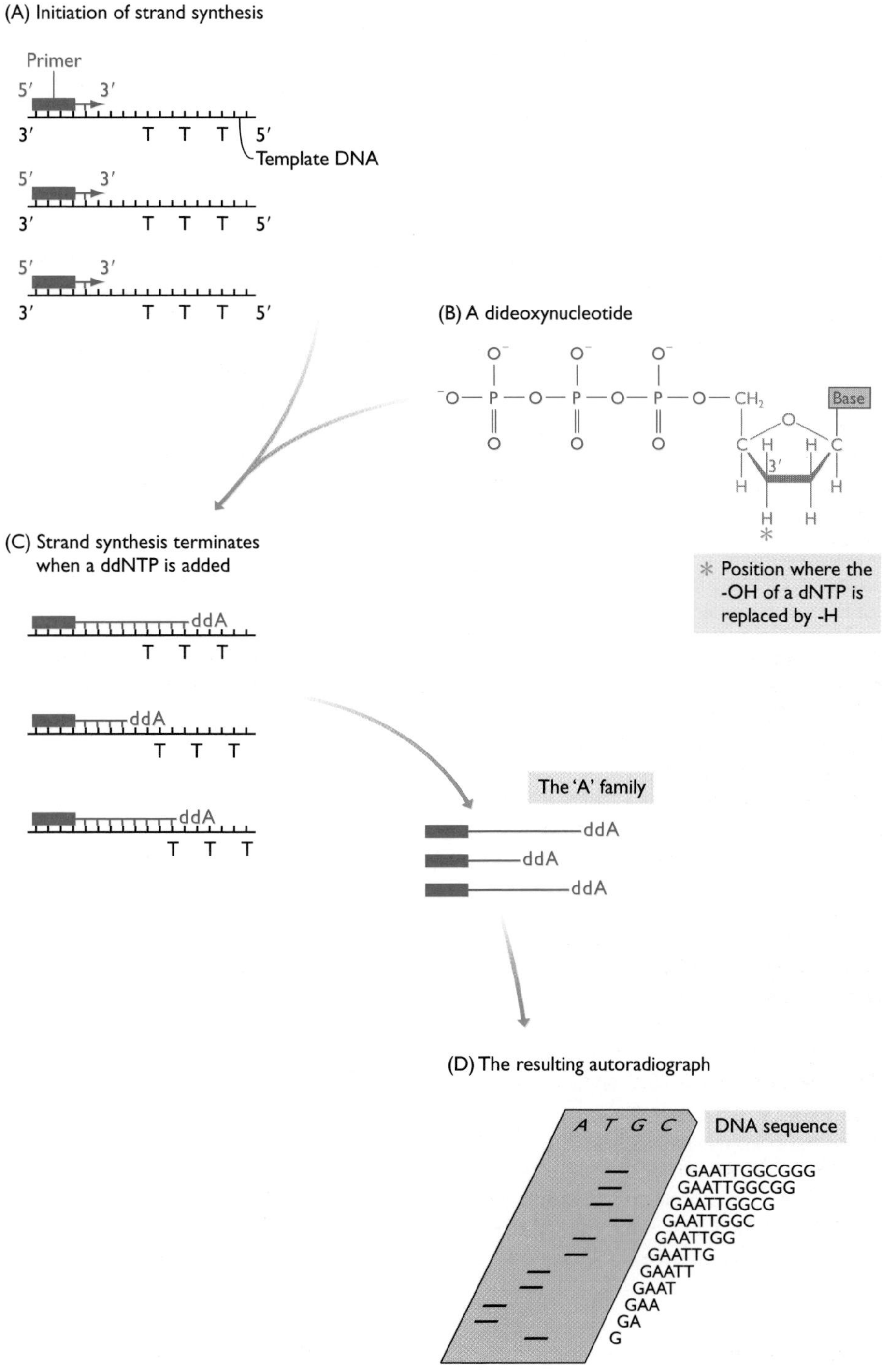

Figure 4.2 Chain termination DNA sequencing.

(A) Chain termination sequencing involves the synthesis of new strands of DNA complementary to a single-stranded template. (B) Strand synthesis does not proceed indefinitely because the reaction mixture contains small amounts of a dideoxynucleotide, which blocks further elongation because it has a hydrogen atom rather than hydroxyl group attached to its 3′-carbon. (C) The result of strand synthesis in the presence of ddATP, chains being terminated opposite T's in the template. This 'A' family of terminated chains is loaded into one lane of a polyacrylamide gel, alongside the families of terminated chains from the T, G and C reactions. (D) In the methodology shown here, the banding pattern is visualized by autoradiography, the terminated chains having become radioactively labeled by inclusion of a labeled dNTP in the strand synthesis reactions. The sequence, shown on the right, is read by noting which lane each band lies in, starting at the bottom of the autoradiograph and moving band by band towards the top.

TECHNICAL 4.2 NOTES

Enzymes for DNA manipulations

A wide variety of purified enzymes are used for test tube manipulation of DNA and RNA molecules.

Molecular biological research makes extensive use of purified enzymes which enable DNA and RNA molecules to be manipulated in the test tube in predetermined ways.

- **DNA polymerases** are enzymes that synthesize new polynucleotides complementary to an existing DNA or RNA template. DNA polymerases cannot begin synthesis on an entirely single-stranded template, so if the template lacks a double-stranded region the new synthesis has to be primed by attachment of a short, synthetic oligonucleotide. In molecular biology techniques, the primer is used to specify the point on the template at which DNA synthesis should begin. For examples of the use of primers see the descriptions of PCR (Technical Note 2.2, p. 20) and DNA sequencing (Section 4.1.1). The DNA-dependent DNA polymerases used in molecular biology research are predominantly versions of DNA polymerase I, one of the two polymerases involved in bacterial DNA replication (Section 12.3.2). These enzymes include *Taq* polymerase (PCR), Sequenase (chain termination DNA sequencing) and *E. coli* DNA polymerase I, the last of these used in various procedures including cDNA synthesis (see *Figure 3.11*, p. 50) and the nick translation method for preparing labeled DNA (Technical Note 3.3, p. 46). Reverse transcriptase, whose main application is also in cDNA synthesis, is an RNA-dependent DNA polymerase.
- **Nucleases** are enzymes that degrade DNA and/or RNA molecules by breaking the phosphodiester bonds that link one nucleotide to the next. Some nucleases have a broad range of activities but most are either **exonucleases**, removing nucleotides from the ends of DNA and/or RNA molecules, or **endonucleases**, making cuts at internal phosphodiester bonds. Some nucleases are specific for DNA and some for RNA, some work only on double-stranded DNA and others only on single-stranded DNA, and some are not fussy what they work on. Restriction endonucleases (Technical Note 3.1, p. 40) are the most specific nucleases but many others have applications in molecular biology. Examples are deoxyribonuclease I, used in nuclease protection experiments (e.g. Section 7.3.1) and S1 nuclease, used in transcript mapping (Section 5.1.2).
- **Ligases** join DNA molecules together by synthesizing phosphodiester bonds between nucleotides at the ends of two different molecules, or at the two ends of a single molecule. The most frequently used DNA ligase is the one coded by T4 bacteriophage, which is obtained from infected *E. coli* cells.
- **End-modification enzymes** are useful in a number of ways. Alkaline phosphatase removes phosphate groups from the 5′ ends of DNA molecules, which prevents these molecules from being ligated to one another. Linear versions of cloning vectors (Technical Note 3.4, p. 52) are often treated with alkaline phosphatase so that the two ends of a vector cannot self-ligate. A phosphatased end can ligate to a nonphosphatased end, so the ability of these vectors to form recombinant DNA molecules is not impaired. The advantage is that the ligation reaction is now biased towards formation of recombinant molecules, because the vector cannot recircularize unless new DNA is inserted. T4 polynucleotide kinase carries out the reverse reaction to alkaline phosphatase, adding phosphates to 5′ ends. This is a useful way of end-labeling a DNA molecule. End-labeling can also be carried out with terminal deoxynucleotidyl transferase, which adds one or more nucleotides to the 3′ ends of DNA molecules.

For a comprehensive review of enzymes use to manipulate DNA and RNA molecules, see Brown (1998).

represents the smallest piece of DNA, this being the strand that terminated by incorporation of a ddNTP at the first position in the template. In the example shown in *Figure 4.2* this band lies in the 'G' lane (i.e. the lane containing the molecules terminated with ddGTP), so the first nucleotide in the sequence is 'G'. The next band, corresponding to the molecule that is one nucleotide longer than the first, is in the 'A' lane, so the second nucleotide is 'A' and the sequence so far is 'GA'. Continuing up through the gel we see that the next band also lies in the 'A' lane (sequence GAA), then we move to the 'T' lane (GAAT), and so on. The sequence can be continued up to the region of the gel where individual bands are not separated.

Chain termination sequencing requires a single-stranded DNA template

The template for a chain termination experiment is a single-stranded version of the DNA molecule to be sequenced. There are several ways in which this can be obtained:

- *The DNA can be cloned in a plasmid vector* (see Technical Note 3.4, p. 52). The resulting DNA will be double-stranded so cannot be used directly in sequencing. Instead, it must be converted into single-stranded DNA by denaturation with alkali or by boiling. This is probably the most common method for obtaining template DNA for DNA sequencing, partly because cloning in a plasmid vector is a routine technique, but also because denaturation results in two complementary, single-stranded DNA molecules. This means that both ends of the cloned DNA molecule can be sequenced. A shortcoming is that it can be difficult to prepare plasmid DNA that is not contaminated with small quantities of bacterial DNA and RNA which can act as spurious templates or primers in the DNA sequencing experiment.
- *The DNA can be cloned in an M13 vector*. Vectors based on M13 bacteriophage are designed specifically for the production of single-stranded templates for DNA sequencing. M13 bacteriophage has a single-stranded DNA genome which, after infection of *Escherichia coli* bacteria, is converted into a double-stranded **replicative form**. The replicative form is copied until over 100 molecules are present in the cell, and when the cell divides the copy number in the new cells is maintained by further replication. At the same time, the infected cells continually secrete new M13 phage particles, approximately 1000 per generation, these phages containing the single-stranded version of the genome (*Figure 4.3*). M13 vectors are double-stranded DNA molecules based on the replicative form of the M13 genome. They can be manipulated in exactly the same way as a plasmid cloning vector. The difference is that cells that have been transfected with a recombinant M13 vector secrete phage particles containing single-stranded DNA, this DNA comprising the vector molecule plus any additional DNA that has been ligated into it. The phages therefore provide the template DNA for chain termination sequencing. Their one disadvantage is that DNA fragments longer than about 3 kb suffer deletions and rearrangements when cloned in an M13 vector, so the system can only be used with short pieces of DNA.
- *The DNA can be cloned in a phagemid*. This is a plasmid cloning vector that contains, as well as its plasmid origin of replication, the origin from M13 or another phage with a single-stranded DNA genome. If an *E. coli* cell contains both a phagemid and a **helper phage**, the latter carrying genes for the phage replication enzymes and coat proteins, then the phage origin of the phagemid becomes activated, resulting in synthesis of phages containing the single-stranded version of the phagemid. The double-stranded plasmid DNA can therefore be converted into single-stranded template DNA for DNA sequencing. This system avoids the instabilities of M13 cloning and can be used with fragments up to 10 kb or more.

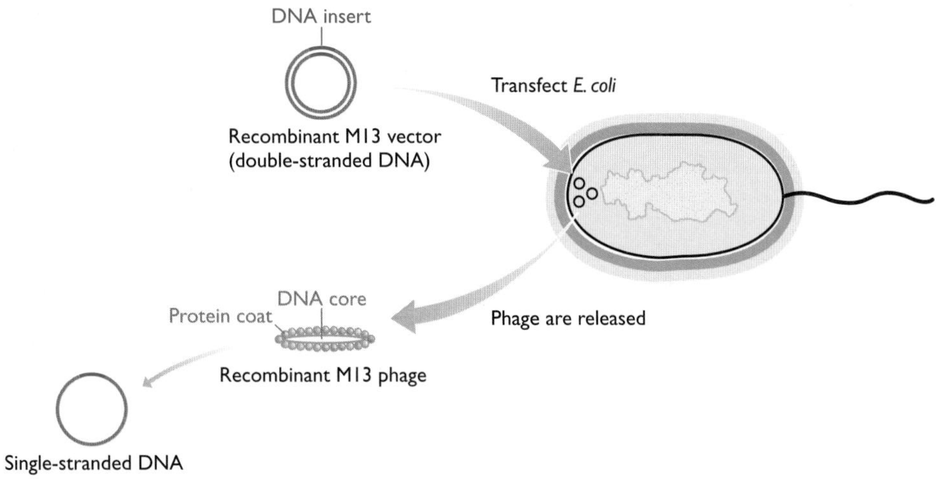

Figure 4.3 Obtaining single-stranded DNA by cloning in a bacteriophage M13 vector.

M13 vectors can be obtained in two forms: the double-stranded replicative molecule and the single-stranded version found in bacteriophage particles. The replicative form can be manipulated in the same way as a plasmid cloning vector (Technical Note 3.4, p. 52) with new DNA inserted by restriction followed by ligation. The recombinant vector is introduced into *E. coli* cells by **transfection**, which is the same as transformation (see Technical Note 3.4) except that phage rather plasmid DNA is involved. Once inside an *E. coli* cell, the double-stranded vector replicates and directs synthesis of single-stranded copies that are packaged into phage particles and secreted from the cell. The phage particles can be collected from the culture medium after centrifuging to pellet the bacteria. The protein coats of the phages are removed by treating with phenol and the single-stranded version of the recombinant vector purified for use in DNA sequencing.

Box 4.2: DNA polymerases for chain termination sequencing

Any template-dependent DNA polymerase is capable of extending a primer that has been annealed to a single-stranded DNA molecule, but not all polymerases do this in a way that is useful for DNA sequencing. Three criteria in particular must be fulfilled by a sequencing enzyme:

- **High processivity**. This refers to the length of polynucleotide that is synthesized before the polymerase terminates through natural causes. A sequencing polymerase must have high processivity so that it does not dissociate from the template before incorporating a chain-terminating nucleotide.
- **Negligible or zero 5′→3′ exonuclease activity**. Most DNA polymerases also have exonuclease activities, meaning that they can degrade DNA polynucleotides as well as synthesize them (see the diagram below). We will deal with the biological reasons for these exonuclease functions when we examine the roles of polymerases in DNA replication in Chapter 12. A 5′→3′ exonuclease activity enables the polymerase to remove a DNA strand that is already attached to the template. This is a disadvantage in DNA sequencing because removal of nucleotides from the 5′ ends of the newly synthesized strands results in the lengths of these strands being altered, making it impossible to read the sequence from the banding pattern in the polyacrylamide gel.
- **Negligible or zero 3′→5′ exonuclease activity** is also desirable so that the polymerase does not remove the chain termination nucleotide once it has been incorporated.

These are stringent requirements and are not entirely met by any naturally occurring DNA polymerase. Instead, artificially modified enzymes are generally used. The first of these to be developed was the Klenow polymerase, which is a version of *E. coli* DNA polymerase I from which the 5′→3′ exonuclease activity of the standard enzyme has been removed, either by cleaving away the relevant part of the protein or by genetic engineering. The Klenow polymerase has relatively low processivity, limiting the length of sequence that can be obtained from a single experiment to about 250 bp and giving nonspecific bands on the sequencing gel, these 'shadow' bands representing strands that have terminated naturally rather than by incorporation of a ddNTP. More recently the Klenow enzyme has been superseded by a modified version of the DNA polymerase encoded by bacteriophage T7, this enzyme going under the tradename 'Sequenase'. Sequenase has high processivity and no exonuclease activity, and also possesses other desirable features such as rapid reaction rate and the ability to use many modified nucleotides as substrates.

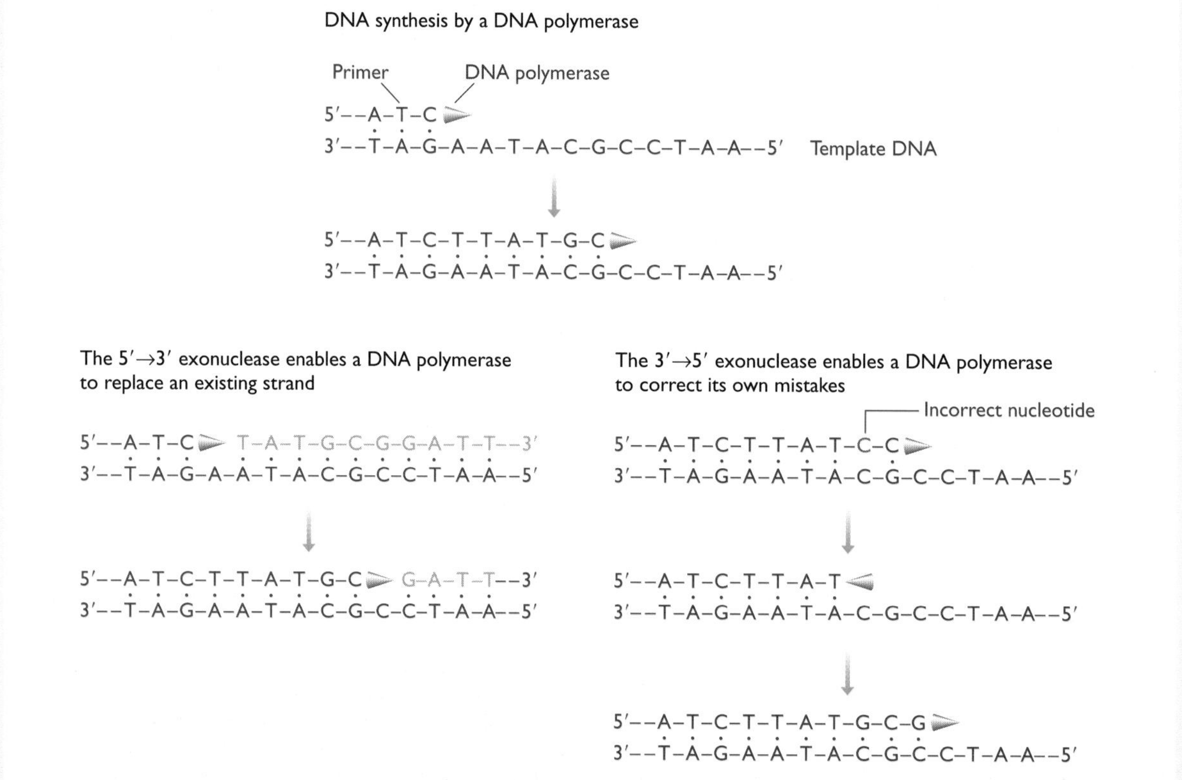

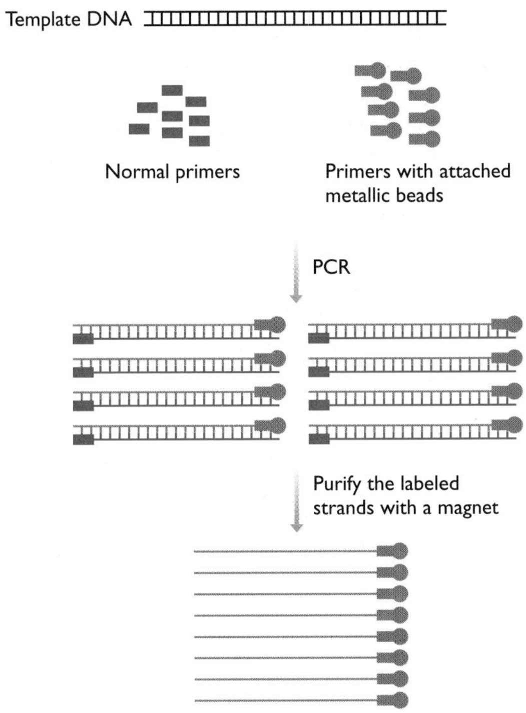

Figure 4.4 One way of using PCR to prepare template DNA for chain termination sequencing.

The PCR is carried out with one normal primer (shown in red), and one primer that is labeled with a metallic bead (shown in brown). After PCR, the labeled strands are purified with a magnetic device. For more details about PCR, see Technical Note 2.2, p. 20.

- ***PCR can be used to generate single-stranded DNA.*** There are various methods, the most effective being to modify one of the two primers so that DNA strands synthesized from this primer are easily purified. One possibility is to attach small metallic beads to the primer and then use a magnetic device to purify the resulting strands (*Figure 4.4*).

The primer determines the region of the template DNA that will be sequenced

To begin a chain termination sequencing experiment an oligonucleotide primer is annealed on to the template DNA. The primer is needed because template-dependent DNA polymerases cannot initiate DNA synthesis on a molecule that is entirely single-stranded: there must be a short double-stranded region to provide a 3′ end onto which the enzyme can add new nucleotides (see Section 12.3.2).

The primer also plays the critical role of determining the region of the template molecule that will be sequenced. For most sequencing experiments a 'universal' primer is used, this being one that is complementary to the part of the vector DNA immediately adjacent to the point into which new DNA is ligated (*Figure 4.5A*). The same universal primer can therefore give the sequence of any piece of DNA that has been ligated into the vector. Of course if this inserted DNA is longer than 750 bp or so then only a part of its sequence will be obtained, but usually this is not a problem because the project as a whole simply requires that a large number of short sequences are generated and subsequently assembled into the contiguous master sequence. It is immaterial whether or not the short sequences are the complete or only partial sequences of the DNA fragments used as templates. If double-stranded plasmid DNA is being used to provide the template then, if desired, more sequence can be obtained from the other end of the insert. Alternatively, it is possible to extend the sequence in one direction by synthesizing a nonuniversal primer, designed to anneal at a position corresponding to the end of the longest sequence so far obtained (*Figure 4.5B*). An experiment with this primer will provide a second short sequence, overlapping the previous one.

Thermal cycle sequencing offers an alternative to the traditional methodology

The discovery of thermostable DNA polymerases, which led to the development of PCR (see Technical Note 2.2, p. 20) has also resulted in new methodologies for chain termination sequencing. In particular, the innovation called **thermal cycle sequencing** (Sears *et al.*, 1992) has two advantages over traditional chain termination sequencing. First, it uses double-stranded rather than single-stranded DNA as the starting material. Second, very little template DNA is needed, so the DNA does not have to be cloned before being sequenced.

Thermal cycle sequencing is carried out in a similar way to PCR but just one primer is used and each reaction

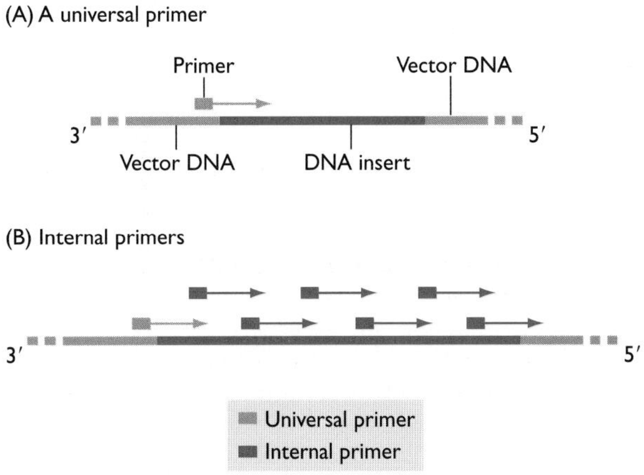

Figure 4.5 Different types of primer for chain termination sequencing.

(A) A universal primer anneals to the vector DNA, adjacent to the position at which new DNA is inserted. A single universal primer can therefore be used to sequence any DNA insert, but only provides the sequence of one end of the insert. (B) One way of obtaining a longer sequence is to carry out a series of chain termination experiments, each with a different internal primer that anneals within the DNA insert.

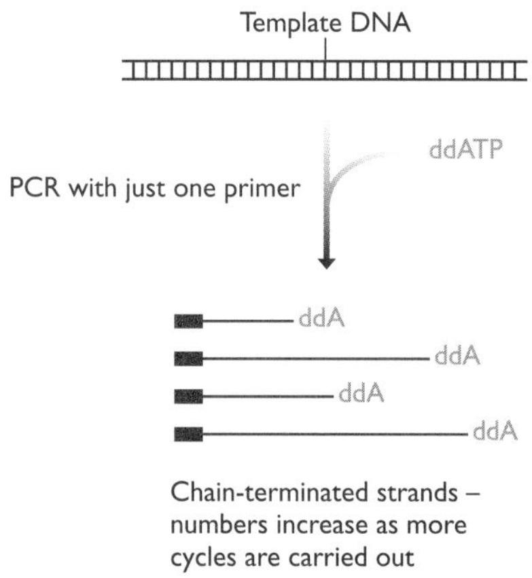

Figure 4.6 Thermal cycle sequencing.

PCR is carried out with just one primer and with a dideoxynucleotide present in the reaction mixture. The result is a family of chain-terminated strands – the 'A' family in the reaction shown. These strands, along with the products of the C, G and T reactions, are electrophoresed as in the standard methodology (see *Figure 4.2D*).

mixture includes one of the ddNTPs (*Figure 4.6*). Because there is only one primer only one of the strands of the starting molecule is copied, and the product accumulates in a linear fashion, not exponentially as is the case in a real PCR. The presence of the ddNTP in the reaction mixture causes chain termination, as in the standard methodology, and the family of resulting strands can be analyzed and the sequence read in the normal manner by polyacrylamide gel electrophoresis.

Fluorescent primers are the basis of automated sequence reading

The standard chain termination sequencing methodology employs radioactive labels with the banding pattern in the polyacrylamide gel visualized by autoradiography. Usually one of the nucleotides in the sequencing reaction is labeled so that the newly synthesized strands contain radiolabels along their lengths, giving a high detection sensitivity. To ensure good band resolution, ^{33}P or ^{35}S is generally used, as the emission energies of these isotopes are relatively low, in contrast to ^{32}P, which has a higher emission energy and gives poorer band resolution because of signal scattering.

In Section 3.2 we saw how the replacement of radioactive labels by fluorescent ones has given a new dimension to *in situ* hybridization techniques. Fluorolabeling has been equally important in the development of sequencing methodology, in particular because the detection system for fluorolabels has opened the way to automated sequence reading (Prober *et al.*, 1987). The label is attached to the ddNTPs, with a different fluorolabel being used for each ddNTP (*Figure 4.7A*). Chains terminated with A are therefore labeled with one fluorophore, chains terminated with C are labeled with a second fluorophore, and so on. Now it is possible to carry out the four sequencing reactions – for A, C, G and T – in a single tube and to load all four families of molecules into just one lane of the polyacrylamide gel, because the fluorescent detector can discriminate between the different labels and hence determine if each band represents an A, C, G or T. The sequence can be read directly as the bands pass in front of the detector and either printed out in a form readable by eye (*Figure 4.7B*) or sent straight to a computer for storage. When combined with robotic devices that prepare the sequencing reactions and load the gel, the fluorescent detection system provides a major increase in

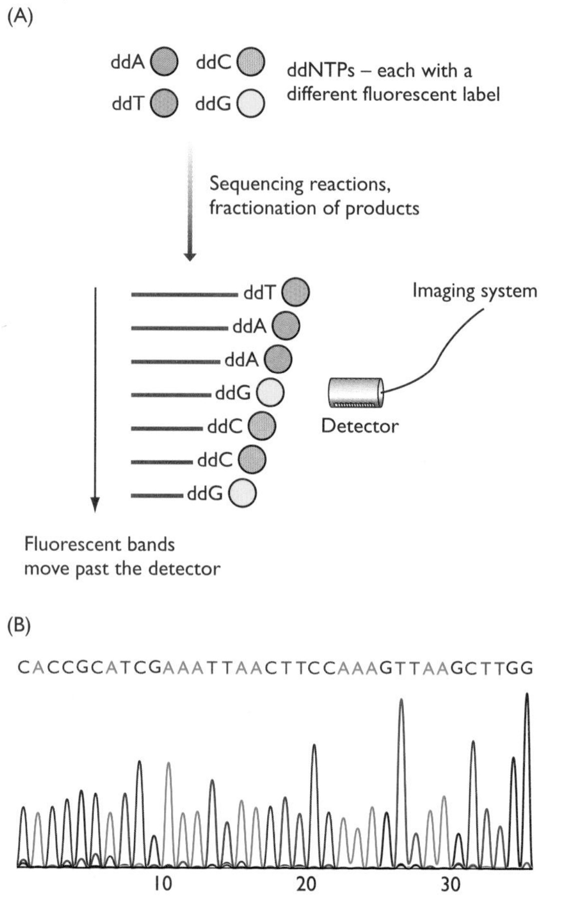

Figure 4.7 Automated DNA sequencing with fluorescently-labeled dideoxynucleotides.

(A) The chain termination reactions are carried out in a single tube, with each dideoxynucleotide labeled with a different fluorophore. In the automated sequencer, the bands in the electrophoresis gel move past a fluorescence detector, which identifies which dideoxynucleotide is present in each band. The information is passed to the imaging system. (B) The printout from an automatic sequencer. The sequence is represented by a series of peaks, one for each nucleotide position. In this example, a green peak is an 'A', blue is 'C', black is 'G', and red is 'T'. Sequence printout provided by Dr C. James, Department of Biomolecular Sciences, UMIST.

throughput and avoids errors that might arise when a sequence is read by eye and then entered manually into a computer. It is only by use of these automated techniques that we can hope to generate sequence data rapidly enough to complete a genome project in a reasonable length of time.

4.1.2 Radical departures from conventional DNA sequencing

In spite of the development of automated techniques, conventional DNA sequencing suffers from the limitation that only a few hundred bp of sequence can be determined in a single experiment. In the context of the Human Genome Project, this means that each experiment provides only one five-millionth of the total genome sequence. Attempts are continually being made to modify the technology so that sequence acquisition is more rapid, a recent example being the introduction of new automated sequencers that use capillary separation rather than a polyacrylamide gel. These have 96 channels so 96 sequences can be determined in parallel, and each run takes less than 2 hours to be completed, enabling up to 1000 sequences to be obtained in a single day.

There have also been attempts to make sequence acquisition more rapid by devising new sequencing methodologies. One possibility is **pyrosequencing**, which does not require electrophoresis or any other fragment separation procedure and so is more rapid than chain termination sequencing (Ronaghi *et al.*, 1998). In pyrosequencing, the template is copied in a straightforward manner without added ddNTPs. As the new strand is being made, the order in which the dNTPs are incorporated is detected, so the sequence can be 'read' as the reaction proceeds. The addition of a nucleotide to the end of the growing strand is detectable because it is accompanied by release of a molecule of pyrophosphate, which can be converted by the enzyme sulfurylase into a flash of chemiluminescence. Of course, if all four dNTPs were added at once then flashes of light would be seen all the time and no useful sequence information would be obtained. Each dNTP is therefore added separately, one after the other, with a nucleotidase enzyme also present in the reaction mixture so that if a dNTP is not incorporated into the polynucleotide then it is rapidly degraded before the next dNTP is added (*Figure 4.8*). This procedure makes it possible to follow the order with which the dNTPs are incorporated into the growing strand. The technique sounds complicated, but it simply requires that a repetitive series of additions be made to the reaction mixture, precisely the type of procedure that is easily automated with the possibility of many experiments being carried out in parallel.

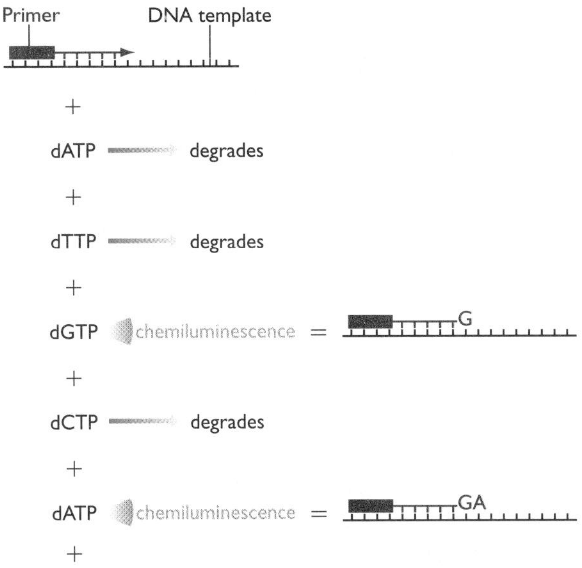

Figure 4.8 Pyrosequencing.

The strand synthesis reaction is carried out in the absence of dideoxynucleotides. Each dNTP is added individually, with a nucleotidase enzyme that degrades the dNTP if it is not incorporated into the strand being synthesized. Incorporation of a nucleotide is detected by a flash of chemiluminescence induced by the pyrophosphate released from the dNTP. The order in which nucleotides are added to the growing strand can therefore be followed.

A very different approach to DNA sequencing might one day be possible through the use of DNA chips (see Technical Note 2.3, p. 23). A chip carrying an array of different oligonucleotides could be used in DNA sequencing by applying the test molecule, the one whose sequence is to be determined, to the array and determining at which positions it hybridizes. Hybridization to an individual oligonucleotide would indicate the presence of that particular oligonucleotide sequence in the test molecule, and comparison of all the oligonucleotides to which hybridization occurs would enable the sequence of the test molecule to be deduced (*Figure 4.9*). The problem with this approach is that the maximum length of the molecule that can be sequenced is given by the square root of the number of oligonucleotides in the array, so if every possible 8-mer oligonucleotide (ones containing eight nucleotides) were attached to the chip – all 65 536 of them – then the maximum length of readable sequence would only be 256 bp (Southern, 1996). Even if the chip carried all the 1 048 576 different 10-mer sequences it could still only be used to sequence a 1 kb molecule. To sequence a 1 Mb molecule (this being the sort of advance in sequence capability that is really needed) the chip would have to carry all of the 1×10^{12} possible 20-mers. This may sound an outlandish proposition but advances in miniaturization, together with the possibility of electronic rather than visual detection of hybridization, could bring such an array within reach in the future.

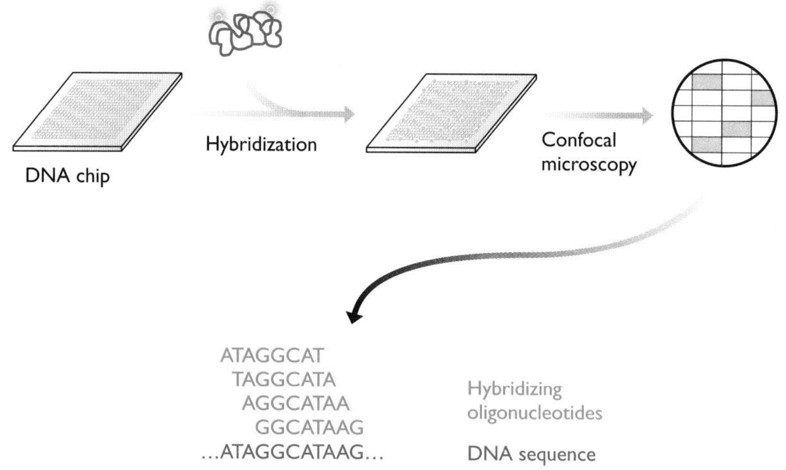

Figure 4.9 A possible way of using chip technology in DNA sequencing.

The chip carries an array of every possible 8-mer oligonucleotide. The DNA to be sequenced is labeled with a fluorescent marker and applied to the chip, and the positions of hybridizing oligonucleotides determined by confocal microscopy. Each hybridizing oligonucleotide represents an 8-nucleotide sequence motif that is present in the probe DNA. The sequence of the probe DNA can therefore be deduced from the overlaps between the sequences of these hybridizing oligonucleotides. See Technical Note 2.3, p. 23 for more information on DNA chips.

4.2 ASSEMBLY OF A CONTIGUOUS DNA SEQUENCE

The next question to address is how the master sequence of a chromosome, possibly several tens of Mb in length, can be assembled from the multitude of short sequences generated by chain termination sequencing. There are three solutions to this problem: the **shotgun** approach, the **clone contig** approach, and the **directed shotgun** approach.

4.2.1 Sequence assembly by the shotgun approach

The straightforward approach to sequence assembly is to build up the master sequence directly from the short sequences obtained from individual sequencing experiments, simply by examining the sequences for overlaps (see *Figure 2.1*, p. 14). This is called the shotgun approach. It does not require any prior knowledge of the genome and so can be carried out in the absence of a genetic or physical map.

The potential of the shotgun approach was proven by the Haemophilus influenzae *sequence*

During the early 1990s there was extensive debate about whether the shotgun approach would work in practice, many molecular biologists being of the opinion that the amount of data handling needed to compare all the mini-sequences and identify overlaps, even with the smallest genomes, would be beyond the capabilities of existing computer systems. These doubts were laid to rest in 1995 when the sequence of the 1830 kb genome of the bacterium *Haemophilus influenzae* was published (Fleischmann *et al.*, 1995).

The *H. influenzae* genome was sequenced entirely by the shotgun approach and without recourse to any genetic or physical map information. The strategy used to obtain the sequence is shown in *Figure 4.10*. The first step was to break the genomic DNA into fragments by sonication, a technique which uses high frequency sound waves to make random cuts in DNA molecules. The fragments were then electrophoresed and those in the range 1.6 to 2.0 kb purified from the agarose gel and ligated into a plasmid vector. From the resulting library, 19 687 clones were taken at random and 28 643 sequencing experiments carried out, the number of sequencing experiments being greater than the number of plasmids because both ends of some inserts were sequenced. Of these sequencing experiments, 16% were considered to be failures because they resulted in less than 400 bp of sequence. The remaining 24 304 sequences gave a total of 11 631 485 bp, corresponding to six times the length of the *H. influenzae* genome, this amount of redundancy being deemed necessary to ensure complete coverage. Sequence assembly required 30 hours of time on a computer with 512 Mb of RAM, and resulted in 140 lengthy, contiguous sequences, each of these **sequence contigs** representing a different, nonoverlapping portion of the genome.

The next step was to join up pairs of contigs by obtaining sequences from the gaps between them (*Figure 4.11*). First, the library was checked to see if there were any clones whose end sequences were located in different contigs. If such a clone could be identified, then additional

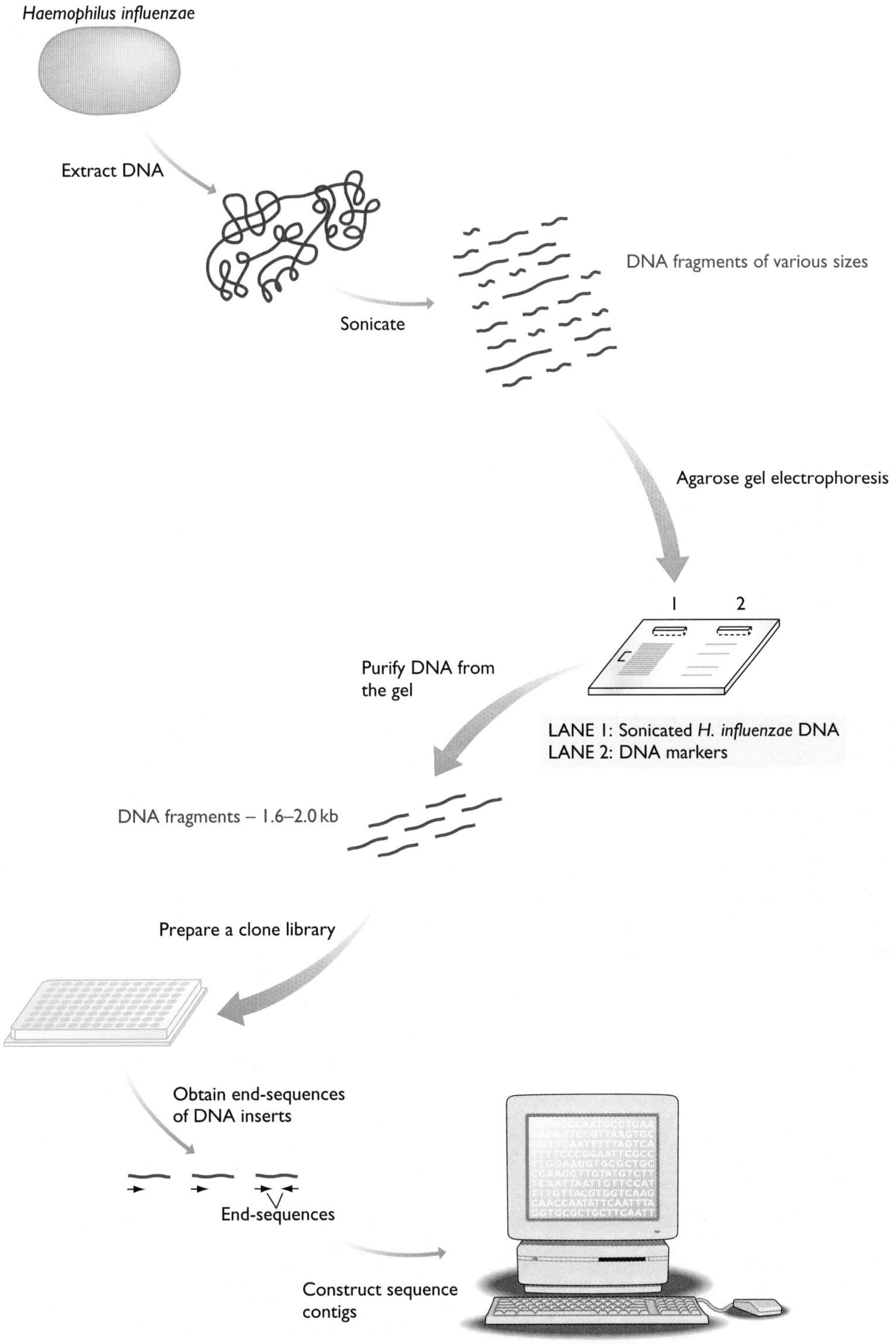

Figure 4.10 The way in which the shotgun approach was used to obtain the DNA sequence of the *Haemophilus influenzae* genome.

H. influenzae DNA was sonicated and fragments with sizes between 1.6 and 2.0 kb purified from an agarose gel and ligated into a plasmid vector to produce a clone library. End-sequences were obtained from clones taken from this library, and a computer used to identify overlaps between sequences. This resulted in 140 sequence contigs, which were assembled into the complete genome sequence as shown in *Figure 4.11*. For further details, see Fleischmann *et al.*, 1995.

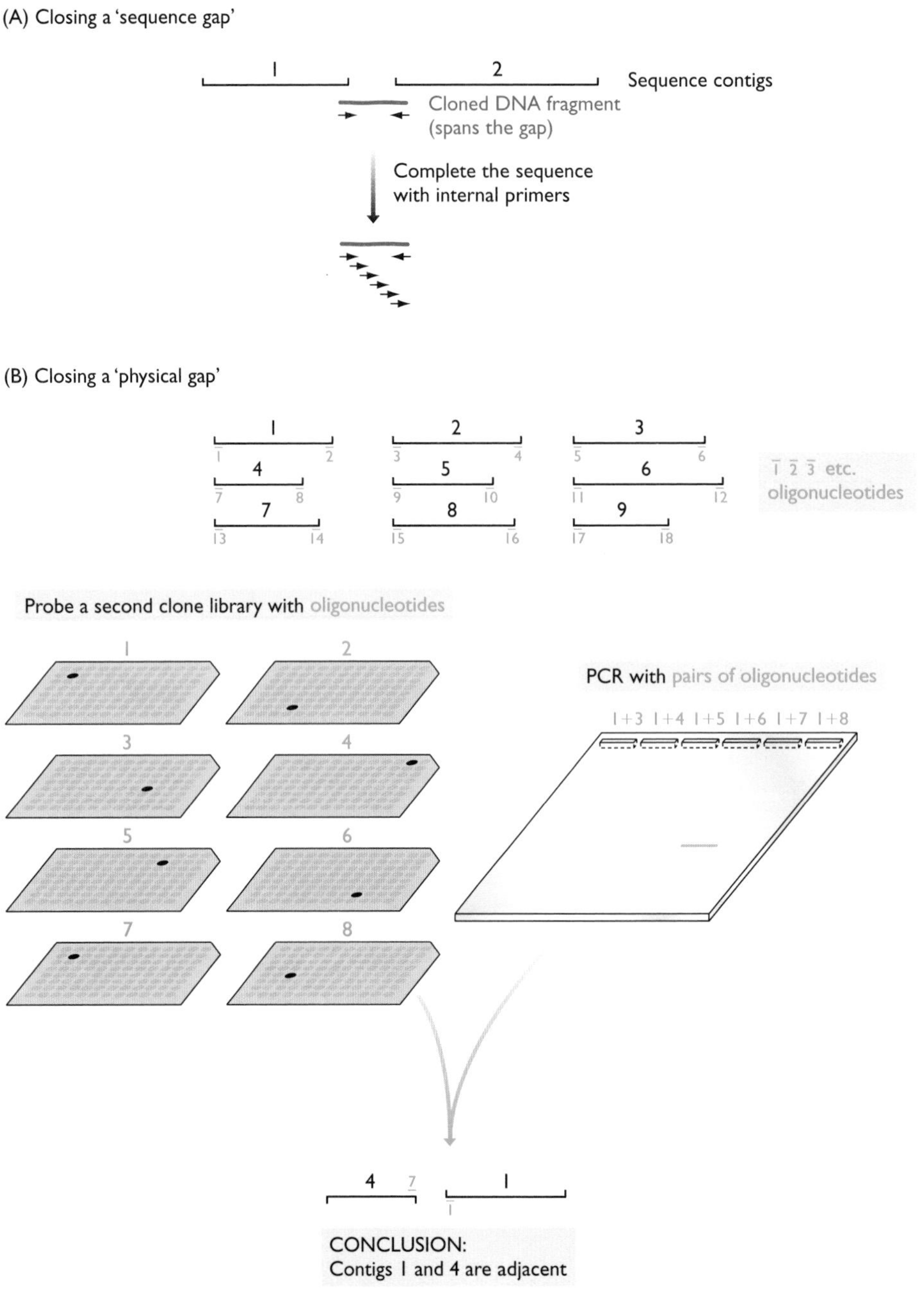

Figure 4.11 Assembly of the complete *Haemophilus influenzae* genome sequence by spanning the gaps between individual sequence contigs.

(A) 'Sequence gaps' are ones which can be closed by further sequencing of clones already identified from the library. In this example, the end-sequences of contigs 1 and 2 lie within the same plasmid clone, so further sequencing of this DNA insert with internal primers (see *Figure 4.5B*) will provide the sequence to close the gap. (B) 'Physical gaps' are stretches of sequence that are not present in the clone library, probably because these regions are unstable in the cloning vector that has been used. Two strategies for closing these gaps are shown. On the left, a second clone library, prepared with a different type of vector, is probed with oligonucleotides corresponding to the ends of the contigs. Oligonucleotides 1 and 7 both hybridize to the same clone, whose insert must therefore contain DNA spanning the gap between contigs 1 and 4. On the right, PCRs are carried out with pairs of oligonucleotides. Only numbers 1 and 7 give a PCR product, confirming that the contig ends represented by these two oligonucleotides are close together in the genome. The PCR product or the insert from the λ clone could be sequenced to close the gap between contigs 1 and 4.

sequencing of its insert would close the 'sequence gap' between the two contigs. In fact, there were 99 clones in this category, so 99 of the gaps could be closed without too much difficulty.

This left 42 gaps which probably consisted of DNA sequences that were unstable in the cloning vector and therefore not present in the library. To close these 'physical gaps' a second clone library was prepared, this one with a different type of vector. Rather than using another plasmid, in which the uncloned sequences would probably still be unstable, the second library was prepared in a bacteriophage λ vector (Technical Note 4.3). This new library was probed with 84 oligonucleotides, one at at time, these 84 oligonucleotides having sequences identical to the sequences at the ends of the unlinked contigs (*Figure 4.11B*). The rationale was that if two oligonucleotides hybridized to the same λ clone then the ends of the contigs from which they were derived must lie within that clone, and sequencing the DNA in the λ clone would therefore close the gap. Twenty three of the 42 physical gaps were dealt with in this way.

A second strategy for gap closure was to use pairs of oligonucleotides, from the set of 84 described above, as primers for PCRs of *H. influenzae* genomic DNA. Some oligonucleotide pairs were selected at random and those spanning a gap identified simply from whether or not they gave a PCR product (see *Figure 4.11B*). Sequencing the resulting PCR products closed the relevant gaps. Other primer pairs were chosen on a more rational basis. For example, oligonucleotides were tested as probes with a Southern blot of *H. influenzae* DNA, and pairs which hybridized to similar sets of restriction fragments identified. The two members of an oligonucleotide pair identified in this way must be contained within the same restriction fragments and so are are likely to lie close together on the genome. This means that the pair of contigs that the oligonucleotides are derived from are adjacent, and the gap between them can be spanned by a PCR using the two oligonucleotides as primers, which will provide the template DNA for gap closure.

The strengths and limitations of the shotgun approach

The demonstration that a small genome can be sequenced relatively rapidly by the shotgun approach led in the late 1990s to a sudden plethora of completed microbial genomes. These projects demonstrated that shotgun sequencing can be set up on a production-line basis, with each team member having his or her individual task in DNA preparation, carrying out the sequencing reactions, or analyzing the data. This strategy resulted in the 580 kb genome of *Mycoplasma genitalium* being sequenced by five people in just 8 weeks (Fraser *et al.*, 1995), and it is now accepted that a year should be ample time to generate the complete sequence of any genome less than about 5 Mb, even if nothing is known about the genome before the project begins. The strengths of the shotgun approach are therefore its speed and its ability to work in the absence of a genetic or physical map.

The main weakness is the complexity and limitations of the data analysis needed to identify overlaps between the starting sequences and construct the sequence contigs. Every sequence that is obtained must be compared with every other sequence in order to identify overlaps. Shortcuts can be taken – for example, once a sequence has been attached to a contig it can be taken out of the analysis, reducing the number of comparisons that have to be made with the remaining mini-sequences, but the amount of data analysis is still huge and pushes existing computer systems to their limits. The data analysis is also dependent on there being zero or at worst very little sequence repetition in the genome being studied, as non-contiguous sequences could mistakenly be joined together by 'jumping' from one repeat unit to another (see *Figure 2.2*, p. 15). There are many interesting genomes that are small and lack significant numbers of repeated sequences, and the shotgun approach will continue to play an important role at this end of the sequencing project spectrum, but we must look at more structured approaches if we wish to sequence larger genomes.

4.2.2 Sequence assembly by the clone contig approach

The clone contig approach is the conventional method for obtaining the sequence of a eukaryotic genome and has also been used with those microbial genomes which have previously been mapped by genetic and/or physical means. The clone contig approach accepts that shotgun sequencing is the most efficient method of generating sequence contigs of up to 5 Mb, and therefore aims to use partial restriction (Section 3.1.1) to break a larger genome down into a library of overlapping, cloned fragments in this size range (see *Figure 2.3*, p. 16). Ideally these overlapping fragments are anchored onto a genetic and/or physical map of the genome, so that the sequence data from individual clones can be checked and interpreted by looking for features (e.g. STSs, SSLPs, genes) known to be present in a particular region.

To understand the sequence contig approach we must first examine the types of cloning vector available for handling large fragments of DNA, and then we must look at the methods used to identify overlapping clones and hence build up the clone contig.

Cloning vectors for large pieces of DNA

The first cloning vectors to be developed, back in the 1970s, were based on bacterial plasmids. These types of plasmid vector are still used as workhorses in general DNA manipulation but they are unstable if their total size is increased by more than a few kb, so they cannot be used to clone large fragments of DNA and are unsuitable for the purpose that we are discussing. The desire to work with larger pieces of DNA led to the development of cloning vectors based on bacteriophage λ, including both regular λ vectors and the λ-plasmid hybrids called

TECHNICAL 4.3 NOTES

Cloning vectors based on bacteriophage λ

Vectors for cloning DNA fragments up to 44 kb in length.

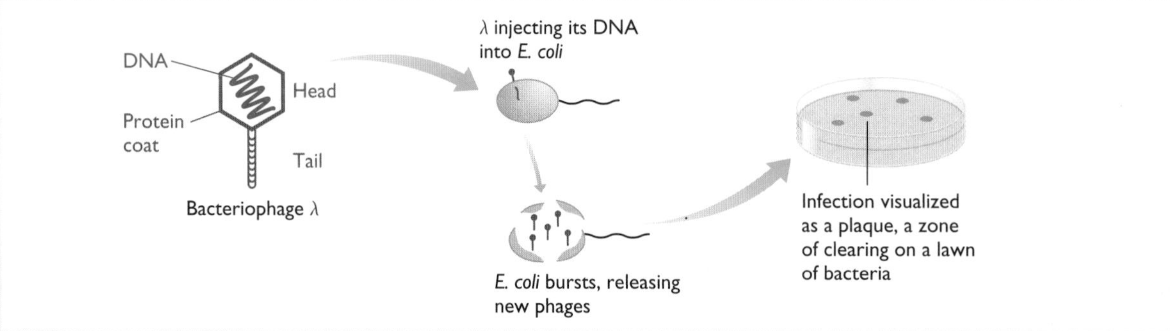

Plasmid vectors (Technical Note 3.4, Part 2, p. 53) are generally unable to clone DNA fragments greater than about 10 kb in size, larger inserts undergoing rearrangements or interfering with the plasmid replication system so that the recombinant DNA molecules are lost from the host cells. The first attempts to develop vectors able to handle larger fragments of DNA centered on bacteriophage λ. The λ genome is 48.5 kb but part of its sequence, some 15 kb or so, is 'optional' in that it contains genes that are only needed for integration of the phage DNA into the *E. coli* chromosome (see Section 13.2.2). These segments can therefore can be deleted without impairing the ability of the phage to infect bacteria and direct synthesis of new λ particles by the lytic cycle, as shown above.

The λ genome is linear, so removal of the central segment results in two arms between which new DNA is inserted. Two types of vector have been developed: insertion vectors, which consist of the two λ arms separated by the restriction site used for insertion of new DNA; and replacement vectors, which contain a **stuffer fragment** that is replaced when the DNA to be cloned is ligated into the vector (see below).

Various methods are used to distinguish recombinant plaques from nonrecombinant ones, including the lactose system described for plasmid vectors in Technical Note 3.4, Part 2.

The λ phage particle can accommodate up to 52 kb of DNA, so if the genome has 15 kb removed then up to 18 kb of new DNA can be cloned. This maximum insert size can be increased further by use of a special type of λ vector called a **cosmid**. A cosmid is basically a plasmid vector that carries a λ *cos* site, this being the critical sequence needed to package a DNA molecule into a λ phage particle. Ligation of the insert DNA into a cosmid is carried out in such a way that recombinant cosmid molecules become linked together to produce linear chains. The recombinant cosmids in these chains are recognized as 'λ genomes' by the components of an *in vitro* packaging mix, a collection of proteins that spontaneously assemble into phage particles. It does not matter that the cosmids do not contain any λ genes as the packaging reaction is dependent solely on the presence of the *cos* sites. The cosmid itself can be 8 kb or less in size, so up to 44 kb of new DNA can be inserted before the packaging limit of the λ phage particle is reached.

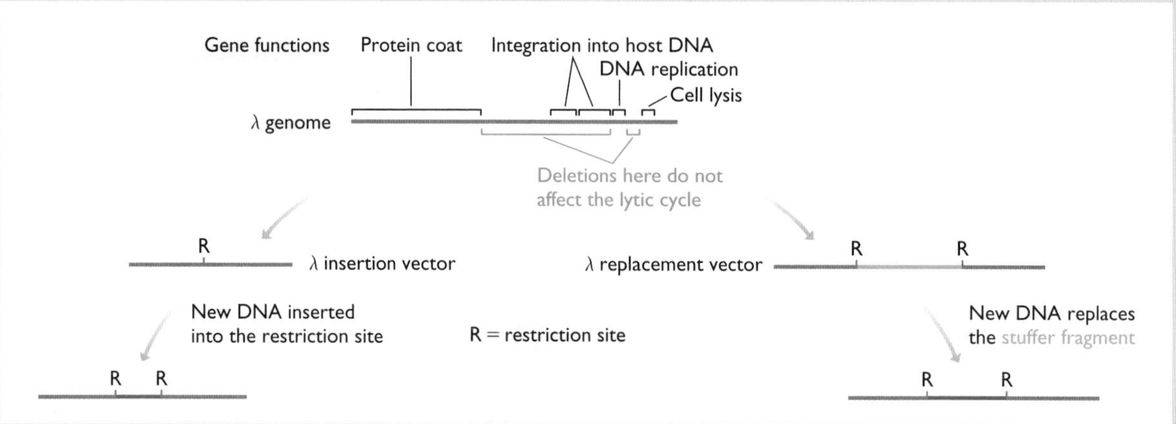

cosmids (Technical Note 4.3). With these vectors, the upper limit for the length of the cloned DNA is set by the space available within the head of the λ phage particle, as the recombinant vector DNA must be packaged within λ phages in order to be transferred to new *E. coli* cells during the cloning process. There is only so much space in the λ phage head so these vectors cannot be improved in any way: their maximum capacities have already been reached. Despite this limitation, some small eukaryotic genomes, notably that of the yeast *Saccharomyces cerevisiae*, have been sequenced mainly or entirely from clone contigs constructed with cosmids (Oliver *et al.*, 1992).

The first major breakthrough in attempts to clone DNA fragments larger than 100 kb came with the invention of **yeast artificial chromosomes** or **YACs** (Burke *et al.*, 1987). These vectors are propagated not in *E. coli* but in *S. cerevisiae* and they are based not on plasmids or viruses but on chromosomes. The first YACs were constructed after studies of natural chromosomes had shown that, in addition to the genes that it carries, each chromosome has three important components:

- The **centromere**, which plays a critical role during nuclear division (see *Figure 2.9*, p. 25).
- The **telomeres**, the special sequences which mark the ends of chromosomal DNA molecules (Section 6.1.1).
- One or more **origins of replication**, which initiate synthesis of new DNA when the chromosome divides (Section 12.3.1).

In a YAC, the DNA sequences that underlie these chromosomal components are linked together with one or more selectable markers and at least one restriction site into which new DNA can be inserted (*Figure 4.12*). All of these components can be contained in a DNA molecule of 10–15 kb in length. As natural yeast chromosomes range from 230 kb to over 1700 kb, the possibility arises that a YAC could be used to clone Mb-sized DNA fragments. This potential has been realized, standard YACs being able to clone 600 kb fragments, with special types able to handle DNA up to 1400 kb in length. Currently this is the highest capacity of any type of cloning vector and several genome projects have made extensive use of YACs. Unfortunately, with some types of YAC there have been problems with insert stability, the cloned DNA sometimes becoming rearranged by recombination (Anderson, 1993; see *Figure 3.15*, p. 55). For this reason there is also great interest in other types of vectors, ones that cannot clone such large pieces of DNA but which do not suffer so much from instability problems. These vectors include:

- **Bacteriophage P1** vectors (Sternberg, 1990), which are very similar to λ vectors, being based on a deleted version of a natural phage genome, the capacity of the cloning vector being determined by the size of the deletion and the space within the phage particle. The P1 genome is larger than the λ genome, and the phage particle is bigger, so a P1 vector can clone larger fragments of DNA than λ, up to 125 kb using current technology.
- **Bacterial artificial chromosomes** or **BACs** (Shizuya *et al.*, 1992) are based on the naturally-occurring F plasmid of *E. coli*. Unlike the plasmids used to construct the early cloning vectors, the F plasmid is relatively large and vectors based on it therefore have a higher capacity for accepting inserted DNA. BACs can be used to clone fragments of 300 kb and greater.
- **P1-derived artificial chromosomes** or **PACs** (Ioannou *et al.*, 1994) combine features of P1 vectors and BACs and have a capacity of up to 300 kb.
- **Fosmids** (Kim *et al.*, 1992) contain the F plasmid origin of replication and a λ *cos* site. They are similar to cosmids but have a lower copy number in *E. coli*, which means that they are less prone to instability problems.

Clone contigs can be built up by chromosome walking, but the method is laborious

The simplest way to build up an overlapping series of cloned DNA fragments is to begin with one clone from a library, identify a second clone whose insert overlaps with the insert in the first clone, then identify a third clone whose insert overlaps with the second clone, and so on. This is the basis of **chromosome walking**, which was the first method devised for assembly of clone contigs.

Chromosome walking was originally used to move relatively short distances along DNA molecules, using clone libraries prepared with λ or cosmid vectors. The most straightforward approach is to use the insert DNA from the starting clone as a hybridization probe to screen all the other clones in the library. Clones whose inserts overlap with the probe give positive hybridization signals, and their inserts can be used as new probes to continue the walk (*Figure 4.13*).

The main problem that arises is that if the probe contains a genome-wide repeat sequence then it will hybridize not only to overlapping clones but also to nonoverlapping clones whose inserts also contain copies of the repeat. The extent of this nonspecific hybridization can be reduced by blocking the repeat sequences by prehybridization with unlabeled genomic DNA (see *Figure 3.8*, p. 47), but this does not completely solve the problem, especially if the walk is being carried out with long inserts from high-capacity vectors such as YACs or BACs. In fact, intact inserts are rarely used for chromosome walks with human and similar DNAs which have a high frequency of genome-wide repeats. Instead, a fragment from the end of an insert is used as the probe, there being less chance of a genome-wide repeat occurring in a short end-fragment compared to the insert as a whole. If complete confidence is required then the end-fragment can be sequenced prior to its use to ensure that no repetitive DNA is present.

If the end-fragment has been sequenced then the walk can be speeded up by using PCR rather that hybridization to identify clones with overlapping inserts. Primers are designed from the sequence of the end-fragment and

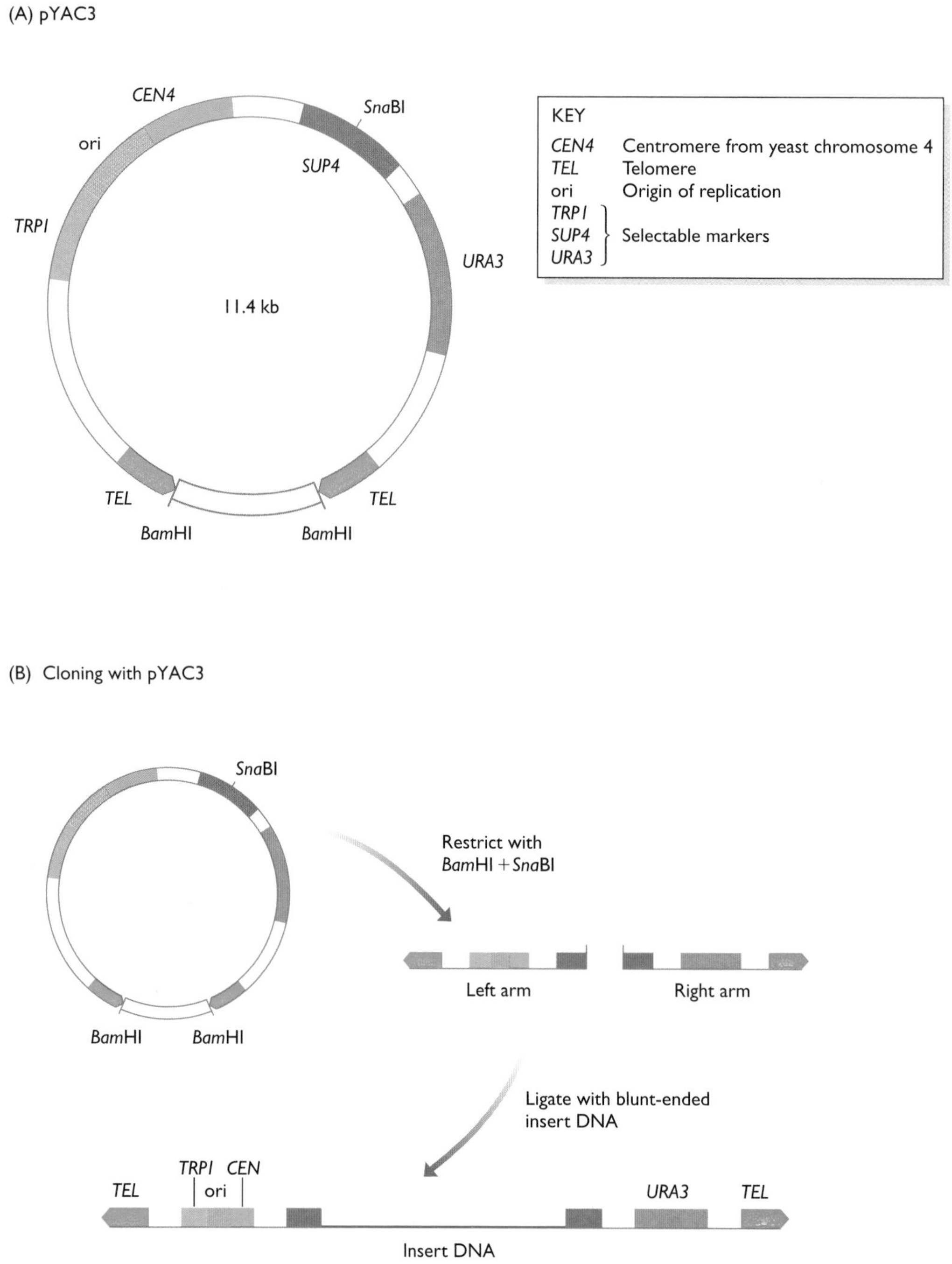

Figure 4.12 Working with a YAC.

(A) The cloning vector pYAC3. (B) To clone with pYAC3, the circular vector is digested with *Bam*HI and *Sna*BI. *Bam*HI restriction removes the 'stuffer fragment' held between the two telomeres in the circular molecule. *Sna*BI cuts within the *SUP4* gene and provides the site into which new DNA will be inserted. Ligation of the two vector arms with new DNA produces the structure shown at the bottom. This structure carries functional copies of the *TRPI* and *URA3* selectable markers. The host strain has inactivated copies of these genes, which means that it requires tryptophan and uracil as nutrients. After transformation, cells are plated onto a minimal medium, lacking tryptophan and uracil. Only cells that contain the vector, and so can synthesize tryptophan and uracil, are able to survive on this medium and produce colonies. Note that if a vector comprises two right arms, or two left arms, then it will not give rise to colonies because the transformed cells will still require one of the nutrients. The presence of insert DNA in the cloned vector molecules is checked by testing for inactivation of *SUP4*, which is carried out by a color test: on the appropriate medium, colonies containing recombinant vectors (i.e. with an insert) are white; nonrecombinants (vector but no insert) are red.

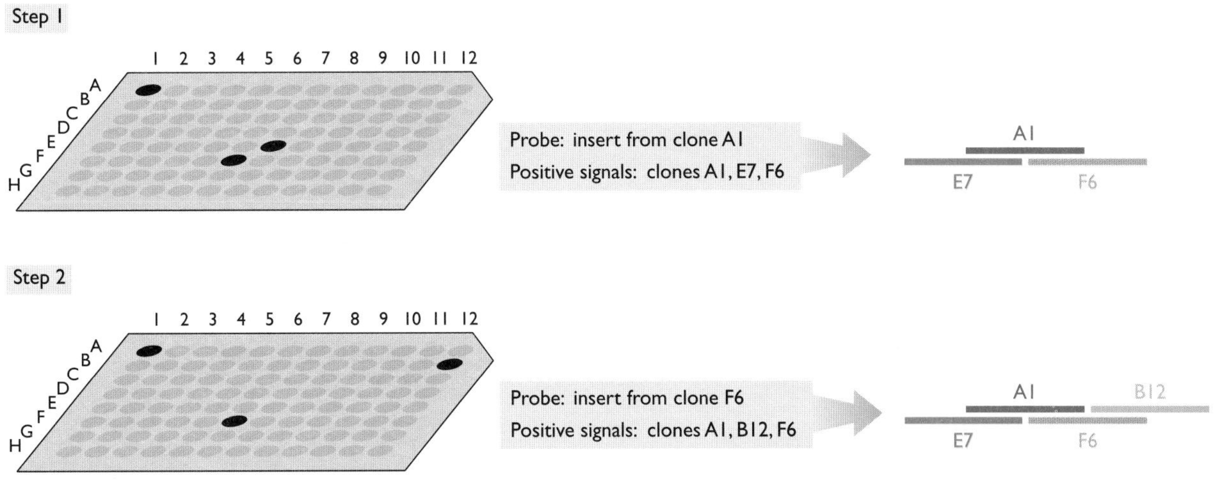

Figure 4.13 Chromosome walking.

The library comprises 96 clones, each containing a different insert. To begin the walk, the insert from one of the clones is used as a hybridization probe against all the other clones in the library. In the example shown, clone A1 is the probe, and it hybridizes to itself and to clones E7 and F6. The inserts from the last two clones must therefore overlap with the insert from clone A1. To continue the walk, the probing is repeated but this time with the insert from clone F6. The hybridizing clones are A1, F6 and B12, showing that the insert from B12 overlaps with the insert from F6.

used in attempted PCRs with all the other clones in the library. A clone that gives a PCR product of the correct size must contain an overlapping insert (*Figure 4.14*). To speed the process up even more, rather than performing a PCR with each individual clone, groups of clones are mixed together in such a way that an unambiguous identification of overlapping ones can still be made. The method is illustrated in *Figure 4.15*, in which a library of 960 clones has been prepared in 10 microtiter trays, each tray comprising 96 wells in an 8 × 12 array, with one clone per well. PCRs are carried out as follows:

(i) Samples of each clone in row A of the first microtiter tray are mixed together and a single PCR carried out. This is repeated for every row of every tray, 80 PCRs in all.
(ii) Samples of each clone in column 1 of the first microtiter tray are mixed together and a single PCR

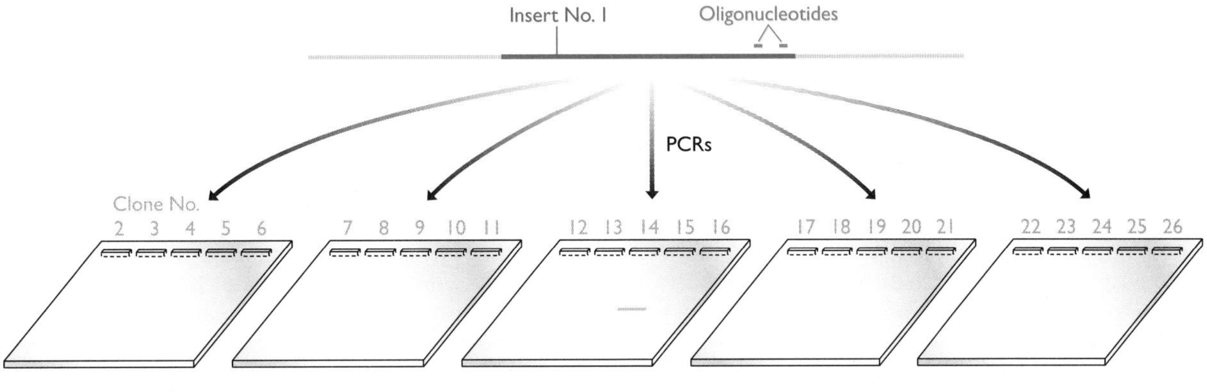

Figure 4.14 Chromosome walking by PCR.

The two oligonucleotides anneal within the end-region of insert number 1. They are used in PCRs with all the other clones in the library. Only clone 15 gives a PCR product, showing that the inserts in clones 1 and 15 overlap. The walk would be continued by sequencing the fragment from the other end of clone 15, designing a second pair of oligonucleotides, and using these in a new set of PCRs with all the other clones.

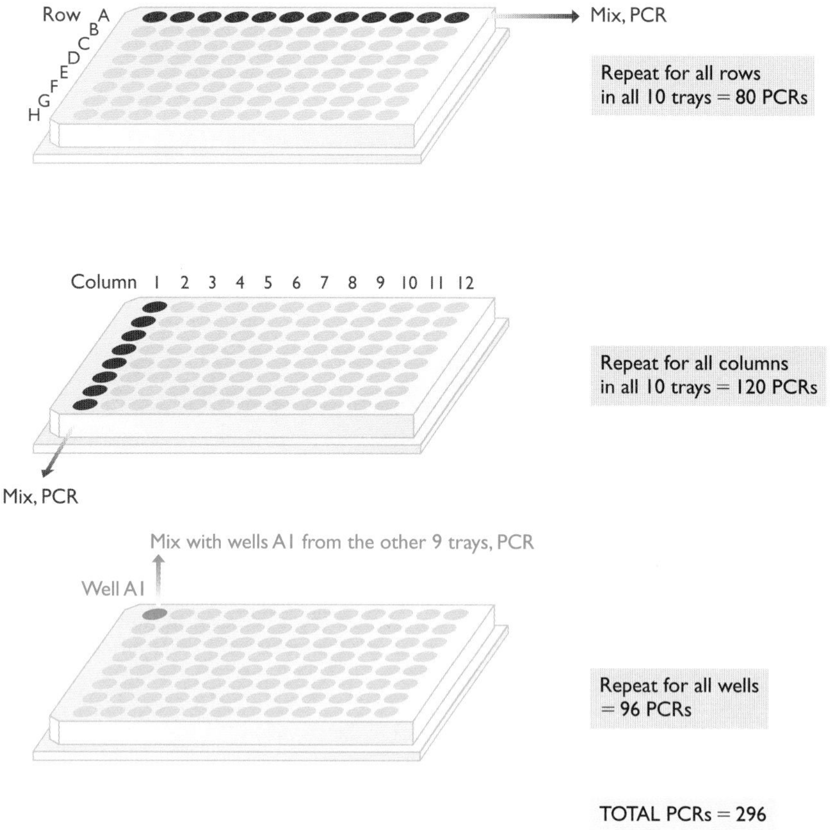

Figure 4.15 Combinatorial screening of clones in microtiter trays.

In this example, a library of 960 clones has to be screened by PCR. Rather than carrying out 960 individual PCRs, the clones are grouped as shown and just 296 PCRs are performed. In most cases, the results enable positive clones to be identified unambiguously. In fact, if there are few positive clones, then sometimes they can be identified by just the 'row' and 'column' PCRs. For example, if positive PCRs are obtained with tray 2 row A, tray 6 row D, tray 2 column 7, and tray 6 column 9, then it can be concluded that there are two positive clones, one in tray 2 well A7 and one in tray 6 well D9. The 'well' PCRs are needed if there are two or more positive clones in the same tray.

carried out. This is repeated for every column of every tray, 120 PCRs in all.

(iii) Clones from well A1 of each of the ten microtiter trays are mixed together and a single PCR carried out. This is repeated for every well, 96 PCRs in all.

As explained in the legend to *Figure 4.15*, these 296 PCRs provide enough information to identify which of the 960 clones give products and which do not. Ambiguities arise only if a substantial number of clones turn out to be positive.

Newer, more rapid methods for clone contig assembly

Even when the screening step is carried out by the combinatorial PCR approach shown in *Figure 4.15*, chromosome walking is a slow process and it is rarely possible to assemble contigs of more than 15–20 clones by this method. The procedure has been extremely valuable in **positional cloning**, where the objective is to walk from a mapped site to an interesting gene that is known to be no more than a few Mb distant. It has been less valuable for assembling clone contigs across entire genomes, especially with the complex genomes of higher eukaryotes. So what alternative methods are there?

The main alternative is to use a **clone fingerprinting** technique. Clone fingerprinting provides limited information on the physical structure of a cloned DNA fragment, this physical information or 'fingerprint' being compared with equivalent data from other clones, enabling those with similarities, possibly indicating overlaps, to be identified. One or a combination of the following techniques are used (*Figure 4.16*):

- **Restriction patterns** can be generated by digesting clones with a variety of restriction enzymes and separating the products in an agarose gel. If two clones contain overlapping inserts then their restriction fingerprints will have bands in common, as both will contain fragments derived from the overlap region.

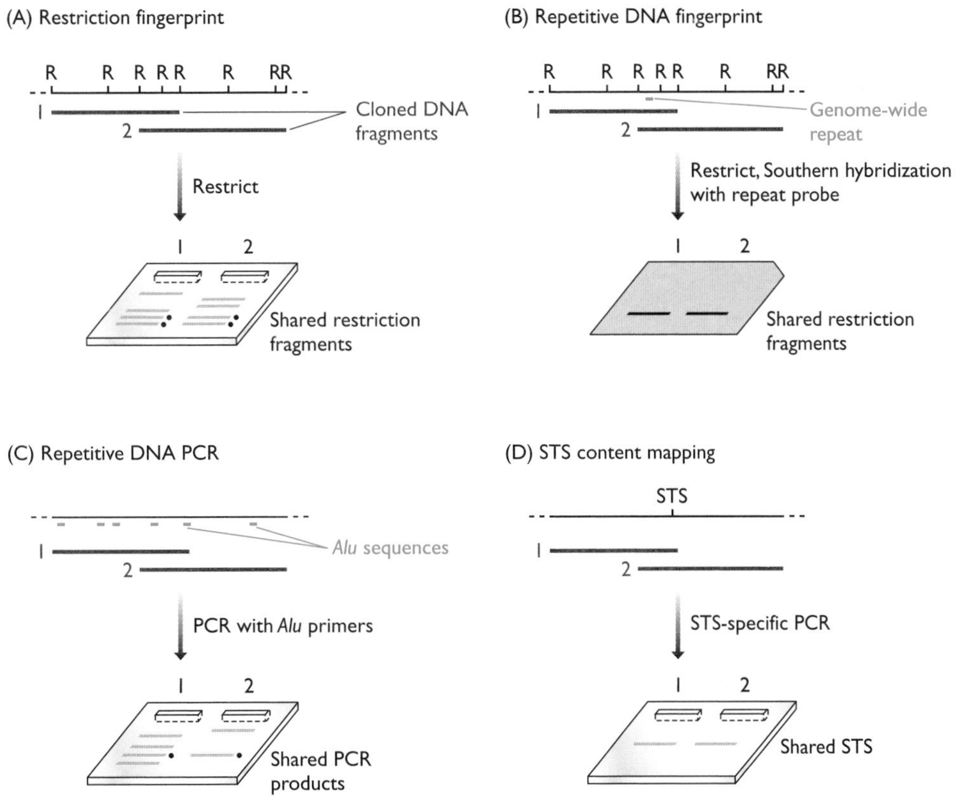

Figure 4.16 Four clone fingerprinting techniques.

- **Repetitive DNA fingerprints** can be prepared by blotting a set of restriction fragments and carrying out Southern hybridization with probes specific for one or more types of genome-wide repeat. As for the restriction fingerprints, overlaps are identified by looking for two clones that have some hybridizing bands in common.
- **Repetitive DNA PCR**, or **interspersed repeat element PCR (IRE-PCR)**, uses primers that anneal within genome-wide repeats and so amplify the single copy DNA between two neighboring repeats. Because genome-wide repeat sequences are not evenly spaced out in a genome, the sizes of the products obtained after repetitive DNA PCR can be used as a fingerprint in comparisons with other clones, in order to identify potential overlaps. With human DNA, the genome-wide repeats called *Alu* elements (Section 6.3.2) are often used as these occur on average once every 4 kb. An ***Alu*-PCR** of a human BAC insert of 150 kb would therefore be expected to give approximately 38 PCR products, producing a detailed fingerprint.
- **STS content mapping** is particularly useful because it can result in a clone contig that is anchored on to a physical map of STS locations. PCRs directed at individuals STSs (Section 3.3) are carried out with each member of a clone library. Presuming the STS is single copy in the genome, then all clones that give PCR products must contain overlapping inserts.

As with chromosome walking, efficient application of these fingerprinting techniques requires combinatorial screening of gridded clones, ideally with computerized methodology for analyzing the resulting data.

4.2.3 The directed shotgun approach

Experience with the conventional shotgun approach to genome sequencing (Section 4.2.1) has shown that if the total length of sequence that is generated is between 6.5 and 8 times the length of the genome being studied, then the resulting sequence contigs will span over 99.8% of the genome sequence (Fraser, 1997), with a few gaps that can be closed by one of the methods developed during the *H. influenzae* project (see *Figure 4.11*). This implies that 70 million individual sequences, each 500 bp or so in length, corresponding to a total of 35 000 Mb, would be sufficient if the random approach were taken with the human genome. Seventy million sequences is not an impossibility: in fact, with 75 automatic sequencers, each performing 1000 sequences per day, the task could be achieved in 3 years.

The big question is whether the 70 million sequences could be assembled correctly. If the conventional shotgun approach is used, which makes no reference to a genome map, then the answer is certainly no. The huge amount of computer time needed to identify overlaps between the sequences, and the errors or at best uncertainties caused by the extensive repetitive DNA content of the human genome, would make the task impossible. But it has been argued that a 'directed shotgun' approach, one which makes full use of the genome map during the sequence assembly phase of the project, would work (Venter *et al.*, 1998).

This project is currently underway with the 70 million sequences being generated as follows:

- Sixty million sequences are being obtained from a plasmid library of human DNA with inserts averaging 2 kb in length. Approximately 500 bp are being sequenced from both ends of the inserts in these clones.
- Ten million sequences are being obtained as end-sequences from a second library, prepared with a different type of plasmid cloning vector, containing 10-kb inserts.

Two libraries are being used because with any cloning vector it is anticipated that some fragments will not be cloned because of incompatibility problems which prevent vectors containing these fragments from being propagated. Different types of vector suffer from different problems so fragments that cannot be cloned in one vector can often be cloned if a second vector is used. Using two different vectors should therefore improve the overall coverage of the genome. The 10-kb library will also help to ensure that errors do not occur when sequencing around the commonest genome-wide repeats. Most of these repeats are 5 kb or so in size, and hence can be entirely contained in a 10-kb fragment but not a 2-kb one. Sequence 'jumps', from one repeat sequence to another, can therefore be avoided by ensuring that the two end-sequences of each 10-kb insert are at their appropriate positions in the master sequence (*Figure 4.17*).

The critical feature of this project is that the master sequence will be assembled with reference to the framework provided by the human genome maps. In particular, the STS map data (Section 3.4), as well as the end sequences of fragments from 300 000 BAC clones whose positions in the genome are known, will be used to anchor the sequence contigs resulting from random sequencing on to the genome map. Whether this will result in a complete and accurate genome sequence, or a 'jumbled collection of paragraphs' (Francis Collins, quoted by Patterson, 1998) is not yet known. Because of these doubts, the directed shotgun project is not looked on as a replacement for the Human Genome Project, the latter continuing to be based firmly on the clone contig approach with the objective of completing a highly accurate sequence by 2003 (Collins *et al.*, 1998). We will therefore close this chapter by looking at how the sequencing phase of the Project is progressing.

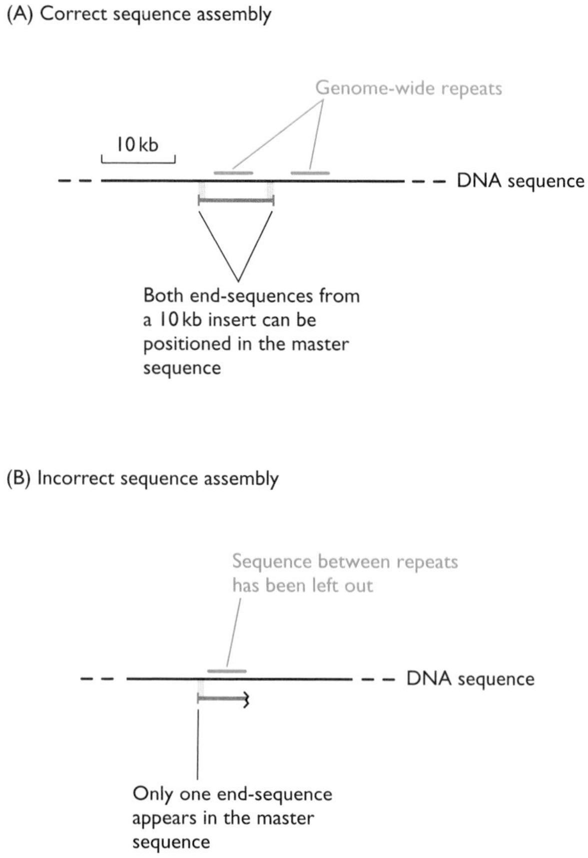

Figure 4.17 Avoiding errors when the directed shotgun approach is used.

In *Figure 2.2*, p. 15, we saw how easy it would be to 'jump' between repeat sequences when assembling the master sequence by the standard shotgun approach. The result of such an error would be to lose all the sequence between the two repeats that had mistakenly been linked together. In the directed shotgun approach, this type of error is avoided by ensuring that the two end-sequences of a 10-kb cloned DNA fragment both appear on the master sequence, at their expected positions in the unique DNA sequences to either side of a genome-wide repeat. If one of the end-sequences is missing, then an error has been made when assembling the master sequence.

4.3 THE SEQUENCING PHASE OF THE HUMAN GENOME PROJECT

In Section 3.4 we surveyed the mapping phase of the Human Genome Project and saw how this culminated in 1995 with an integrated genetic and physical map with a marker density approaching one per 100 kb, the target that had originally been set as a requirement before large-scale sequencing could begin. Since achieving this objective, work on the human genome has been focused more directly on DNA sequencing.

4.3.1 The Human Genome Project is now largely based on BAC libraries

The high capacity of YACs, which means that they can be used with DNA fragments longer than can be handled by any other type of cloning system, led to these vectors being chosen as the source material for construction of the first clone contigs of the entire human genome (Cohen *et al.*, 1993) and of individual chromosomes (e.g Cayanis *et al.*, 1998). The original idea was that the YAC libraries would also be used for the sequencing phase of the Project, but this plan has largely been abandoned because of concerns that some YAC clones (up to half according to some estimates) contain noncontiguous fragments from different parts of the genome or have undergone rearrangements or deletions (see *Figure 3.15*, p. 55). These instabilities would lead to major errors if YAC inserts were used to generate the genome sequence.

To avoid these errors, a library of 300 000 BAC clones has been generated and these clones mapped onto the genome. This forms the 'sequence-ready' map which is acting as the primary foundation for the sequencing phase of the project, the inserts from each BAC being completely sequenced by the shotgun method.

4.3.2 Rapid sequence acquisition can be directed at interesting regions of the genome

It is recognized that the clone contig approach being taken by the Human Genome Project will result in a relatively slow accumulation of sequence data. With the complete genome sequence not due until 2003, a number of alternative strategies, designed to generate sequence data more quickly than is possible by rigorous mapping and contig assembly, have been developed. Bear in mind that although the genome as a whole is interesting, the very interesting bits are the genes, especially those that are relevant to human disease. Rapid sequencing strategies that result specifically in gene sequences are therefore worth pursuing.

EST sequencing is a rapid way of accessing genes

Genes can be directly accessed by analysis of expressed sequence tags (ESTs). These are a type of STS that can be used as markers in physical mapping. ESTs are obtained by sequencing cDNA, which itself is synthesized from the mRNA molecules in a cell. The mRNAs in a cell are copies of the genes that are being expressed, and do not contain sequences from the regions between genes, nor from the noncoding introns that are present within many genes. ESTs are therefore copies of just the very interesting parts of the genome. ESTs also have a number of practical advantages which mean that their sequences can be generated very rapidly:

- mRNA is converted into cDNA by a simple test-tube reaction (see *Figure 3.11*, p. 50), so a cDNA library can be prepared very quickly.
- Only one sequencing experiment is needed per cDNA, because 500 bp of sequence is enough to identify the gene that an EST represents (Section 5.1.1), though in practice both end-sequences are usually obtained.
- The sequence of an EST does not need to be checked by repeating the sequencing experiment, because mistakes arising from the inherent error rate of chain termination sequencing (about one error per 100 nucleotides) do not prevent identification of the gene from which the EST is derived. In contrast, the minimal acceptable error rate for the genome sequence (one per 10 000 nucleotides, as set by the Human Genome Project) requires that each region be sequenced at least three times.

By late 1998, over one million human ESTs had been obtained, representing 50 000 to 60 000 different human genes. In other words, partial sequences are available for well over half of all human genes. The fact that these are only partial sequences, either because the sequencing experiment did not cover the entire cDNA, or the cDNA synthesis did not result in a complete copy of the mRNA, is not a problem. As mentioned above, the sequence information in a single EST is enough to identify the gene from which it is derived, nonidentifications resulting not from a lack of sequence data but from limitations in the procedures used for gene identification (Section 5.1.1). If the complete gene sequence is needed then the EST can be used as a hybridization probe, first to find the appropriate clone in a BAC library, and then to identify the region within the clone that contains the gene sequence (*Figure 4.18*).

Sequence skimming is equally rapid

A second method of rapid sequence acquisition is **sequence skimming** (Claverie, 1994), which in effect is the first stage of a shotgun sequencing project. The rationale is that the first few random sequences obtained from a long fragment will map at various positions within that fragment, so if the fragment contains one or more genes then there is a good chance that at least some of them will be 'hit' by these initial sequences. For example, imagine the fragment being studied is 100 kb in length. One hundred random sequences of 500 bp each will produce about 50 kb of sequence, which even allowing for some overlaps would be anticipated to include over one-third of the 100 kb fragment. These initial sequences will be error prone, but as with ESTs this is not a problem if the initial objective is simply to identify the genes that are present.

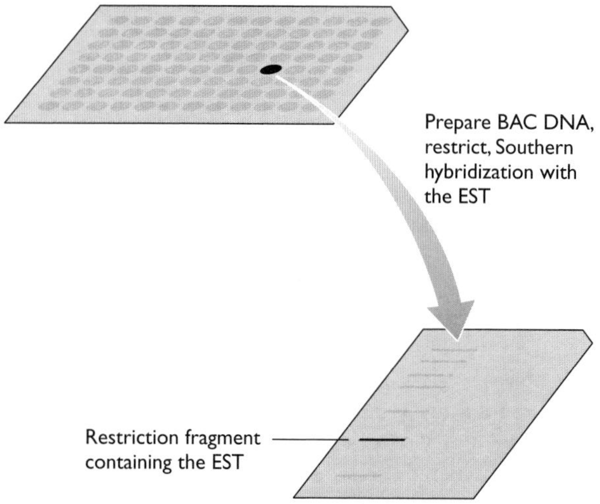

Figure 4.18 From EST to gene sequence.

An interesting EST has been discovered and the corresponding gene sequence is now being sought. A BAC library is therefore probed with the EST and a hybridizing clone identified. DNA from this clone is purified, restricted and the fragments run in an agarose gel. The gel is blotted and Southern hybridization carried out with the EST probe. The hybridizing band is the restriction fragment containing the EST. This restriction fragment will therefore contain part or all of the gene from which the EST is derived.

4.3.3 The sequencing phase of the Human Genome Project raises ethical questions

The Human Genome Project has been controversial since its inception and now that large-scale sequencing is getting underway a number of ethical issues are being raised. One controversy concerns the question of who if anyone will own human DNA sequences. To many, the idea of ownership of a DNA sequence is a peculiar concept, but large sums of money can be made from the information contained in the human genome, for example by using gene sequences to direct development of new drugs and therapies against cancer and other diseases. Pharmaceutical companies involved in genome sequencing naturally want to protect their investments, as they would for any other research enterprise, and currently the only way of doing this is by patenting the DNA sequences that they discover. Unfortunately, in the past, errors have been made in dealing with the financial issues relating to research into human biological material, the individual from whom the material is obtained not always being a party in the profit-sharing. These issues have still to be resolved.

The problems relating to the public usage of human genome sequences are even more contentious. A major concern is the possibility that once the sequence is known, individuals whose sequences are considered 'substandard', for whatever reason, might be discriminated against. The dangers range from increased insurance premiums for individuals whose sequence includes mutations predisposing them to one or other genetic disease, to the possibility that racists might attempt to define good and bad sequence features, with depressingly predictable implications for the individuals unlucky enough to fall into the 'bad' category. Many scientists argue in support of a Human Genome Diversity Project, which will build up a catalog of sequence variability in different populations and ethnic groups, in the expectation that the resulting data will emphasize the unity of the human race by showing that patterns of genetic variability do not reflect the geographical and political groupings that humans have adopted during the last few centuries.

The Human Genome Project, especially in the USA, continues to support research and debate into the ethical, legal and social issues raised by genome sequencing. In particular, great care is being taken to ensure that the genome sequence that results from the project cannot be identified with any single individual. The DNA that is being cloned and sequenced is taken only from individuals who have given full consent for their material to be used in this way and for whom anonymity can be guaranteed. This has required a certain amount of realignment of the research effort, as older clone libraries have been destroyed and the existing physical maps have had to be checked using the new material. It is accepted, however, that this extra work is necessary and that care must be taken to maintain and enhance public confidence in the project.

REFERENCES

Anderson C (1993) Genome shortcut leads to problems. *Science*, **259**, 1684–1687.

Brown TA (1998) *Molecular Biology Labfax*, 2nd edition, Volumes 1 and 2. Academic Press, London.

Burke DT, Carle GF and Olson MV (1987) Cloning of large segments of exogenous DNA into yeast by means of artificial chromosome vectors. *Science*, **236**, 806–812.

Cayanis E, Russo JJ, Kalachikov S, *et al*. (1998) High-resolution YAC-cosmid-STS map of human chromosome 13. *Genomics*, **47**, 26–43.

Claverie JM (1994) A streamlined random sequencing strategy for finding coding exons. *Genomics*, **23**, 575–581.

Cohen D, Chumakov I and Weissenbach J (1993) A first-generation map of the human genome. *Nature*, **366**, 698–701.

Collins FS, Patrinos A, Jordan E, *et al.* (1998) New goals for the US Human Genome Project: 1998–2003. *Science*, **282**, 682–689.

Fleischmann RD, Adams MD, White O, *et al.* (1995) Whole-genome random sequencing and assembly of *Haemophilus influenzae* Rd. *Science*, **269**, 496–512.

Fraser CM (1997) How to sequence a small genome. *Trends Genet.*, **13**, poster insert.

Fraser CM, Gocayne JD, White O, *et al.* (1995) The minimal gene complement of *Mycoplasma genitalium*. *Science*, **270**, 397–403.

Ioannou PA, Amemiya CT, Garnes J, Kroisel PM, Shizuya H, Chen C, Batzer MA and de Jong PJ (1994) P1-derived vector for the propagation of large human DNA fragments. *Nature Genet.*, **6**, 84–89.

Kim U-J, Shizuya H, de Jong PJ, Birren B and Simon MI (1992) Stable propagation of cosmid and human DNA inserts in an F factor based vector. *Nucleic Acids Res.*, **20**, 1083–1085.

Maxam AM and Gilbert W (1977) A new method for sequencing DNA. *Proc. Natl Acad. Sci. USA*, **74**, 560–564.

Oliver SG, van der Aart QJM, Agostini-Carbone ML, *et al.* (1992) The complete DNA sequence of yeast chromosome III. *Nature*, **357**, 38–46.

Patterson M (1998) Politicogenomics takes centre stage. *Trends Genet.*, **14**, 259–260.

Prober JM, Trainor GL, Dam RJ, Hobbs FW, Robertson CW, Zagursky RJ, Cocuzza AJ, Jensen MA and Baumeister K (1987) A system for rapid DNA sequencing with fluorescent chain-terminating dideoxynucleotides. *Science*, **238**, 336–341.

Ronaghi M, Ehleen M and Nyrén P (1998) A sequencing method based on real-time pyrophosphate. *Science*, **281**, 363–365.

Sanger F, Nicklen S and Coulson AR (1977) DNA sequencing with chain terminating inhibitors. *Proc. Natl Acad. Sci. USA*, **74**, 5463–5467

Sears LE, Moran LS, Kisinger C, Creasey T, Peery-O'Keefe H, Roskey M, Sutherland E and Slatko BS (1992) Circum-Vent thermal cycle sequencing and alternative manual and automated DNA sequencing protocols using the highly thermostable Vent (exo^-) DNA polymerase. *Biotechniques*, **13**, 626–633.

Shizuya H, Birren B, Kim UJ, Mancino V, Slepak T, Tachiiri Y and Simon M (1992) Cloning and stable maintenance of 300-kilobase-pair fragments of human DNA in *Escherichia coli* using an F-factor-based vector. *Proc. Natl Acad. Sci. USA*, **89**, 8794–8797.

Southern EM (1996) DNA chips: analysing sequence by hybridization to oligonucleotides on a large scale. *Trends Genet.*, **12**, 110–115.

Sternberg N (1990) Bacteriophage P1 cloning system for the isolation, amplification, and recovery of DNA fragments as large as 100 kilobase pairs. *Proc. Natl Acad. Sci. USA*, **87**, 103–107.

Venter JC, Adams MD, Sutton GG, Kerlavage AR, Smith HO and Hunkapiller M (1998) Shotgun sequencing of the human genome. *Science*, **280**, 1540–1542.

FURTHER READING

Brown TA (1994) *DNA Sequencing: The Basics.* Oxford University Press, Oxford. — *Details of DNA sequencing methodology.*

Monaco AP and Larin Z (1994) YACs, BACs, PACs and MACs – artificial chromosomes as research tools. *Trends Biotechnol.*, **12**, 280–286. — *A good review of high-capacity cloning vectors.*

Wilkie T (1993) *Perilous Knowledge: The Human Genome Project and its Implications.* Faber and Faber, New York. — *A view of the social impact of the Human Genome Project.*

CHAPTER

5 Understanding a Genome Sequence

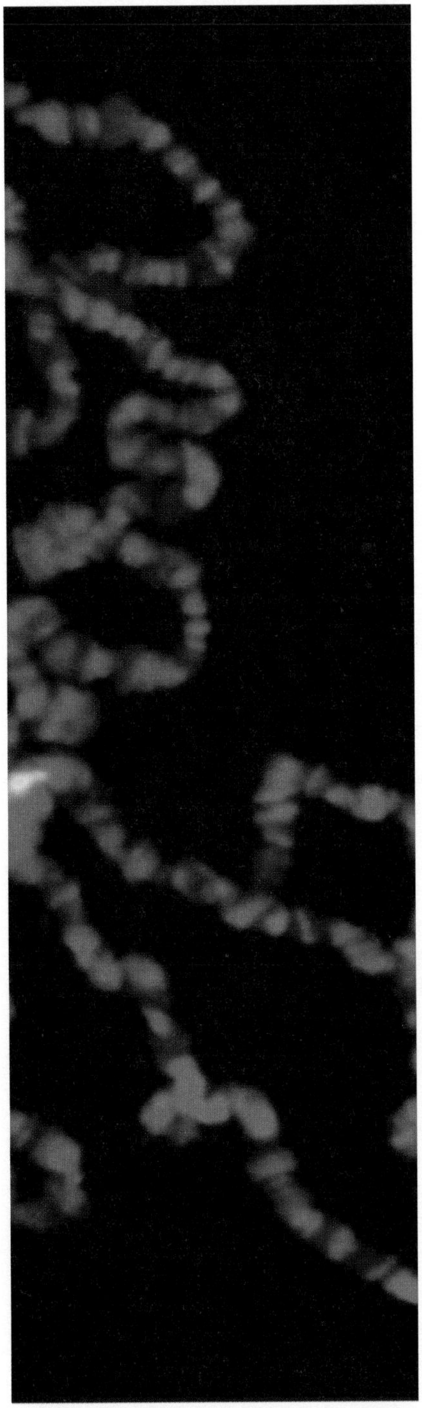

Contents

5.1 *Locating Genes in DNA Sequences* **86**
- 5.1.1 Gene location by sequence inspection 86
- 5.1.2 Experimental techniques for gene location 89

5.2 *Determining the Function of a Gene* **92**
- 5.2.1 Computer analysis of gene function 93
- 5.2.2 Assigning gene function by experimental analysis 96
- 5.2.3 More detailed studies of the activity of a protein coded by an unknown gene 100

5.3 *Comparative Genomics* **103**
- 5.3.1 Comparative genomics as an aid to gene mapping 103
- 5.3.2 Comparative genomics in the study of human disease genes 106

5.4 *From Genome to Cell* **106**
- 5.4.1 Patterns of gene expression 107

Concepts

- *Understanding the information contained in genome sequences will be the major challenge of the early 21st century*
- *Genes can be located in DNA sequences by searching for ORFs, but the procedure is less successful with eukaryotic genomes*
- *Genes can be located experimentally by searching for expressed sequences*
- *Homology analysis is a powerful means of assessing the function of unknown genes*
- *A function can sometimes be assigned to a gene by determining the effect that its inactivation has on the phenotype of the organism*
- *Information on the genetic content of a large genome can be obtained by studying a smaller, related genome*
- *Studies of the transcriptome and proteome hold the key to understanding genome activity*

A GENOME SEQUENCE is not an end in itself. A major challenge still has to be met in understanding what the genome contains and how the genome functions. The former is addressed by a combination of computer analysis and experimentation with the primary aim of locating the genes and their control regions and assigning a role to each gene. The bulk of this chapter will be devoted to these methods. The second question, understanding how the genome functions is, to a certain extent, merely a different way of stating the objectives of molecular biology over the last 30 years. The difference is that in the past attention has been directed at the expression pathways for individual genes, with groups of genes being considered only when the expression of one gene is linked to that of another. Now the issues have become more general and relate to the expression of the genome as a whole. We will familiarize ourselves with the relevant points at the end of this chapter, and return to them constantly as we progress through Part 2 of the book.

5.1 LOCATING GENES IN DNA SEQUENCES

Once a DNA sequence has been obtained, whether it is the sequence of a single cloned fragment or of an entire chromosome, then various methods can be employed to locate the genes that are present. These methods can be divided into those that involve simply inspecting the sequence, by eye or more frequently by computer, to look for the special sequence features associated with genes, and those methods that locate genes by experimental analysis of the DNA sequence.

5.1.1 Gene location by sequence inspection

Sequence inspection can be used to locate genes because genes are not random series of nucleotides but instead have distinctive features. These features determine whether a sequence is a gene or not, and so by definition are not possessed by noncoding DNA. If in the future we can fully understand the exact nature of the specific sequence features that define a gene then sequence inspection will become a foolproof way of locating genes. We are not yet at this stage but sequence inspection is still a powerful tool in genome analysis.

Genes are open reading frames

Genes that code for proteins comprise **open reading frames** (**ORFs**) consisting of a series of **codons** that specify the amino acid sequence of the protein that the gene codes for (see Box 1.1, p. 7). The ORF begins with an **initiation codon**, usually but not always ATG, and ends with a **termination codon**, either TAA, TAG or TGA (Section 10.1.2). Searching a DNA sequence for ORFs that begin with an ATG and end with a termination triplet is therefore one way of looking for genes. The analysis is complicated by the fact that each DNA sequence has six

reading frames, three in one direction and three in the reverse direction on the complementary strand (*Figure 5.1*), but computers are quite capable of scanning all six reading frames for ORFs. How effective is this as a means of gene location?

The key to the success of ORF scanning is the frequency with which termination triplets appear in the DNA sequence. If the DNA has random sequence and a GC content of 50% then each of the three termination triplets – TAA, TAG and TGA – will appear, on average, once every $4^3 = 64$ bp. If the GC content is >50% then the termination triplets, being AT-rich, will occur less frequently but one would still be expected every 100–200 bp. As there are three termination triplets and three reading frames in either direction, random DNA should not show many ORFs longer than 50 triplets in length, especially if the presence of a starting ATG is used as part of the definition of an 'ORF'. Most genes, on the other hand, are longer than 50 codons (the average lengths are 317 codons for *Escherichia coli* and 483 codons for *Saccharomyces cerevisiae*). ORF scanning, in its simplest form, therefore takes a figure of, say, 100 codons as the shortest length of a putative gene and records positive hits for all ORFs longer than this.

How well does this strategy work in practice? With bacterial genomes, simple ORF scanning is an effective way of locating most of the genes in a DNA sequence. This is illustrated by *Figure 5.2*, which shows a segment of the *E. coli* genome with all ORFs longer than 50 triplets highlighted. The real genes in the sequence cannot be mistaken because they are much longer than 50 codons in length. With bacteria the analysis is further simplified by the fact that there is relatively little noncoding DNA in the genome (only 11% for *E. coli*, see Section 6.2.2). If we assume that the real genes do not overlap, and there are no genes-within-genes (see Box 6.3, p. 124), which are valid assumptions for bacterial DNA, then it is only in the noncoding regions that there is a possibility of mistaking a short, spurious ORF for a real gene. So if the noncoding component of a genome is small then there is a reduced chance of making mistakes in interpreting the results of a simple ORF scan.

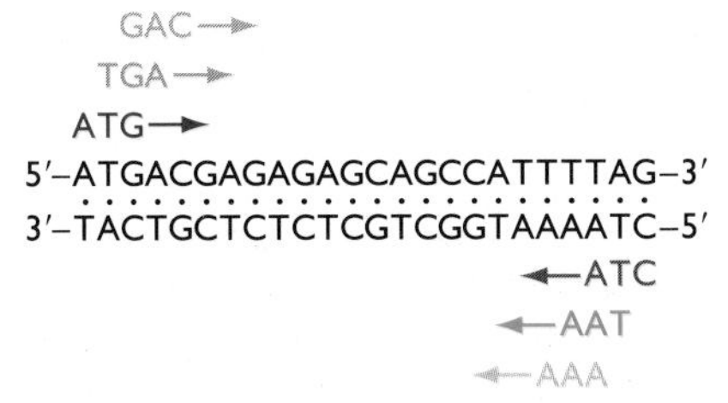

Figure 5.1 A double-stranded DNA molecule has six reading frames.

Both strands are read in the 5′→3′ direction. Each strand has three reading frames, depending on which nucleotide is chosen as the starting position.

Simple ORF scans are less effective with higher eukaryotic DNA

Although ORF scans work well for simple bacterial genomes, they are less effective for locating genes in DNA sequences from higher eukaryotes. This is partly because there is substantially more space between real genes (70% of the human genome is intergenic), increasing the chances of finding spurious ORFs. But the main problem with the human genome and those of higher eukaryotes in general is that their genes are often split by introns (Sections 1.1.2 and 9.2.3), and so do not appear as continuous ORFs in the DNA sequence. Many exons are shorter than 100 codons, some less than 50 codons, and continuing the reading frame into an intron usually leads to a termination sequence that appears to close the ORF (*Figure 5.3*). In other words, the genes of higher eukaryotes do not appear in their DNA sequences as long ORFs and simple ORF scanning cannot locate them.

Solving the problem posed by introns is the main challenge for bioinformaticists writing new software programs for ORF location. Three modifications to the basic procedure for ORF scanning have been adopted (Fickett, 1996):

(i) **Codon bias** is taken into account. 'Codon bias' refers to the fact that not all codons are used equally frequently in the genes of a particular organism. For example, in human genes the amino acid alanine is only infrequently specified by the codon GCG, usually one of the other three codons for alanine (GCA, GCC or GCT) is used. Similarly, threonine is most frequently coded by ACA, ACC or ACT and less often by ACG (*Table 5.1*). The biological reason for codon bias is not understood, but all organisms have a bias, different in different species. Real exons are expected to display the codon bias but chance series of triplets do not. The codon bias of the organism being studied is therefore written into the ORF scanning software.

(ii) **Exon–intron boundaries** can be searched for as these have distinctive sequence features, though unfortunately the distinctiveness of these sequences is not so great as to make their location a trivial task. The sequence of the upstream, exon–intron boundary is usually described as

5′–AG↓GTAAGT–3′

the arrow indicating the precise boundary point. However, only the 'GT' immediately after the arrow is invariable, at the preceding positions nucleotides other than the ones shown are quite often found. In other words, the sequence shown is a **consensus**, the average of a range of variabilities. The downstream, intron–exon boundary is even less well defined:

5′–PyPyPyPyPyPyNCAG↓–3′

where 'Py' means one of the pyrimidine nucleotides, T or C, and 'N' is any nucleotide. Simply searching for the consensus sequences will not locate

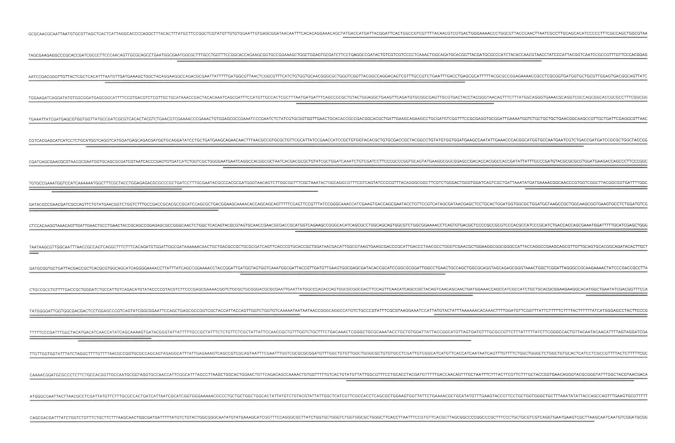

Figure 5.2 ORF screening is an effective way of locating genes in a bacterial genome.

The diagram shows 4522 bp of the lactose operon of *Escherichia coli* with all ORFs longer than 50 codons marked. The sequence contains two real genes – *lacZ* and *lacY* – indicated by the red lines. These real genes cannot be mistaken because they are much longer than the spurious ORFs, shown in blue. See *Figure 6.14A*, p. 133 for the detailed structure of the lactose operon.

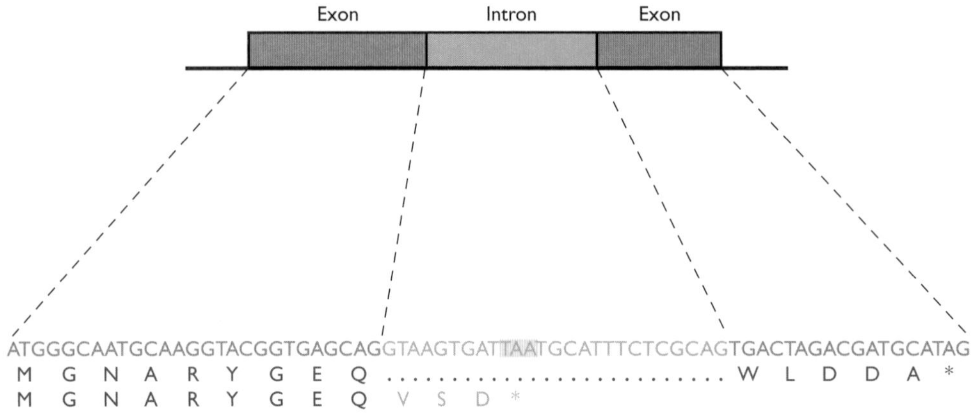

Figure 5.3 ORF scans are complicated by introns.

The nucleotide sequence of a short gene containing a single intron is shown. The correct translation is given immediately below the nucleotide sequence: in this translation the intron has been left out and the amino acid sequence is hence split into two segments. In the lower line, the sequence has been translated without realizing that an intron is present. As a result of this error, the amino acid sequence appears to terminate within the intron. The amino acid sequences have been written using the standard one-letter abbreviations: these are defined in *Table 7.3*, p. 155. The genetic code is described in Section 10.1.2 and introns are covered in detail in Section 9.2.3.

Table 5.1 Examples of codon bias in the human genome

Amino acid	Codons	Codon frequency
Alanine	GCA	22%
	GCC	41%
	GCG	11%
	GCT	26%
Threonine	ACA	27%
	ACC	38%
	ACG	12%
	ACT	23%
Valine	GTA	11%
	GTC	25%
	GTG	48%
	GTT	17%

See Section 10.1.2 for details concerning the genetic code. The data are taken from the internet site maintained by Yasukazu Nakamura, Laboratory of Gene Structure 2, Kazusa DNA Research Institute, Japan: <http://www.dna.affrc.go.jp/~nakamura/codon.html>

more than a few exon–intron boundaries because most have sequences other than the ones shown. Writing software to take account of the known variabilities has proven difficult (Frech *et al.*, 1997), and at present locating exon–intron boundaries by sequence analysis is a hit and miss affair.

(iii) **Upstream control sequences** can be used to locate the regions where genes begin. This is because these control sequences, like exon–intron boundaries, have distinctive sequence features that they possess in order to carry out their role as recognition signals for the DNA-binding proteins involved in gene expression (Chapters 7 and 8). Unfortunately, as with exon–intron boundaries, the control sequences are variable, more so in eukaryotes than prokaryotes, and in eukaryotes not all genes have the same collection of control sequences. Using these to locate genes is therefore problematical.

These three extensions of simple ORF scanning are generally applicable to all higher eukaryotic genomes. Additional strategies are also possible with individual organisms, based on the special features of their genomes. For example, vertebrate genomes contain **CpG islands** upstream of many genes (Bird, 1986), these being sequences of approximately 1 kb in which the GC content is greater than the average for the genome as a whole. For example, some 56% of human genes are associated with an upstream CpG island. These sequences are distinctive and when one is located in vertebrate DNA a strong assumption can be made that a gene begins in the region immediately downstream.

Homology searches give an extra dimension to sequence inspection

The limitations of ORF scanning with higher eukaryotic genomes are offset to a certain extent by the use of a **homology search** to test whether a series of triplets is a real exon or a chance sequence. In this analysis the DNA databases are searched to determine if the test sequence is identical or similar to any genes that have already been sequenced. Obviously, if the test sequence is part of a gene that has already been sequenced by someone else then an identical match will be found, but this is not the point of a homology search. Instead the intention is to determine if an entirely new sequence is *similar* to any known genes, because if it is then there is a chance that the test and match sequences are **homologous**, meaning that they represent genes that are evolutionarily related. The main use of homology searching is to assign functions to newly discovered genes, and we will therefore return to it when we deal with this aspect of genome analysis later in the chapter (Section 5.2.1). At this point, we will note simply that the technique is also central to *gene location*, as it enables tentative exon sequences located by ORF scanning to be tested for functionality. If the tentative exon sequence gives one or more positive matches after a homology search then it probably is a real exon, if it gives no match then its authenticity must remain in doubt until it is assessed by one or other of the experiment-based gene location techniques. The value of this approach is increasing rapidly as the numbers of authentic gene sequences in databases multiply.

5.1.2 Experimental techniques for gene location

It is to the experiment-based gene location techniques that we now turn our attention. These procedures locate genes by examining the RNA molecules that are transcribed from a DNA fragment. All genes are transcribed into RNA, and if the gene is discontinuous then the primary transcript is subsequently processed to remove the introns and link up the exons (Section 9.2.3). Techniques which map the positions of transcribed (or 'expressed') sequences in a DNA fragment can therefore be used to locate exons and entire genes. The only problem to be kept in mind is that the transcript is usually longer than the coding part of the gene because it begins several tens of nucleotides upstream of the initiation codon and continues several tens or hundreds of nucleotides downstream of the termination codon (see Box 1.1, p. 7 and Chapters 8 and 9). Transcript analysis does not therefore give a precise definition of the start and end of the coding region of a gene, but it does tell you that a gene is present in a particular region and it can locate the exon–intron boundaries. Often this is sufficient information to enable the coding region to be delineated.

Hybridization tests can determine if a fragment contains expressed sequences

The simplest procedures for studying expressed sequences are based on hybridization analysis. RNA molecules can be separated by specialized forms of agarose gel electrophoresis and transferred to a nitrocellulose or nylon membrane by the process called **northern blotting** (Technical Note 5.1). This differs from Southern blotting (see Technical Note 2.1, p. 19) only in the precise conditions under which the transfer is carried out and the fact that it was not invented by a Dr Northern and so does not have a capital 'N'. If a northern blot is probed with a labeled DNA fragment, then RNAs expressed from the fragment will be detected (*Figure 5.4*). Northern hybridization is therefore, theoretically, a means of determining the number of genes present in a DNA fragment and the size of each coding region. There are two weaknesses with this approach:

- Some individual genes give rise to two or more transcripts, of different lengths, because some of their exons are optional and may or may not be retained in the mature RNA (Section 9.2.3). If this is the case, then a fragment that contains just one gene could detect two or more hybridizing bands in the northern blot. A similar problem can occur if the gene is a member of a multigene family (Section 6.1.1).
- With many species, it is not practical to make a preparation from an entire organism so the RNA extract is obtained from a single organ or tissue. Consequently any genes not expressed in that organ or tissue will not be represented in the RNA population, and so will not be detected when the RNA is probed with the DNA fragment being studied. Even if the whole organism is used, not all genes will give hybridization signals because many are expressed only at a particular developmental stage, and others are weakly expressed, meaning that their RNA products are present in amounts too low to be detected by hybridization analysis.

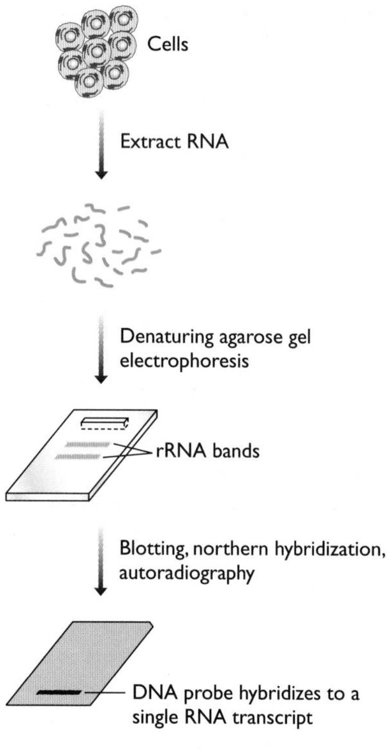

Figure 5.4 Northern hybridization.

An RNA extract is electrophoresed under denaturing conditions in an agarose gel (Technical Note 5.1). After ethidium bromide staining, two bands are seen. These are the two largest ribosomal RNA (rRNA) molecules (Section 10.2.1) which are abundant in most cells. The smaller rRNAs, which are also abundant, are not seen because they are so short that they run out of the bottom of the gel and, in most cells, none of the messenger RNAs (the transcripts of protein-coding genes) are abundant enough to form a band visible after ethidium bromide staining. The gel is blotted on to a nylon membrane and, in this example, probed with a radioactively labeled DNA fragment. A single band is visible on the autoradiograph, showing that the DNA fragment used as the probe contains part or all of one expressed sequence.

A second type of hybridization analysis avoids the problems with poorly expressed and tissue-specific genes by searching not for RNAs but for related sequences in the DNAs of other organisms. This approach, like homology searching, is based on the fact that homologous genes in related organisms have similar sequences, whereas the noncoding DNA is quite different. If a DNA fragment from one species is used to probe a Southern blot of DNAs from related species, and one or more hybridization signals are obtained, then it is likely that the probe contains one or more genes (*Figure 5.5*). This is called **zoo-blotting**.

cDNA sequencing enables genes to be mapped within DNA fragments

Northern hybridization and zoo-blotting enable the presence or absence of genes in a DNA fragment to be determined, but give no positional information relating to the location of those genes in the DNA sequence. The easiest way to obtain this information is to sequence the relevant cDNAs. A cDNA is a copy of an mRNA (see *Figure 3.11*, p. 50) and so corresponds to the coding region of a gene, plus any leader or trailer sequences that are also transcribed. Comparing a cDNA sequence with a genomic DNA sequence therefore delineates the position of the relevant gene and reveals the exon–intron boundaries.

The degree of success of cDNA sequencing as a means of gene location depends on two factors. The first is the frequency of the appropriate cDNAs in the cDNA clone library that has been prepared. As with northern hybridization, the problem relates to the different expression levels of different genes. If the DNA fragment being studied contains one or more poorly expressed genes, then the relevant cDNAs will be rare in the library and it might be necessary to screen many clones before the desired one is identified. To get around this problem various methods of **cDNA capture** or **cDNA selection** have been devised, based around repeated hybridization between the DNA fragment being studied and the pool of cDNAs (Lovett, 1994). Because the cDNA pool contains so many different sequences it is generally not possible to

TECHNICAL 5.1 NOTES

Techniques for studying RNA

Many of the techniques devised for studying DNA molecules can be adapted for use with RNA. Examples include:

- **Agarose gel electrophoresis** of RNA is carried out under denaturing conditions so that the migration rate of a single-stranded DNA molecule is dependent entirely on its length, and is not influenced by the intramolecular base pairs that form in many RNAs (e.g. see *Figure 10.11*, p. 244). The denaturant, usually formaldehyde or glyoxal, is added to the running buffer and possibly also the sample loading buffer.
- **Northern hybridization** refers to the procedure whereby an RNA gel is blotted on to a nylon membrane and hybridized to a labeled probe (e.g. see *Figure 5.4*). This is equivalent to Southern hybridization and is carried out in a similar way.
- **Labeled RNA molecules** are usually prepared by copying a DNA template into RNA in the presence of a labeled ribonucleotide. The RNA polymerase enzymes from SP6, T3 or T7 bacteriophages are used because they can produce up to 30 μg of labeled RNA from 1 μg of DNA in 30 min. RNA can also be end-labeled by treatment with poly(A) polymerase (Section 9.2.2).
- **PCR** of RNA molecules requires a modification to the first step of the normal reaction. *Taq* DNA polymerase cannot copy an RNA molecule so the first step is catalyzed by a reverse transcriptase, which makes a DNA copy of the RNA template. This DNA copy is then amplified by *Taq* polymerase (e.g. see *Figure 5.6*). The technique is called **reverse transcriptase-PCR** or **RT-PCR**. The discovery of thermostable enzymes that make DNA copies of both RNA and DNA templates (e.g. the *Tth* DNA polymerase from the bacterium *Thermus thermophilus*) raises the possibility of carrying out RT-PCR in a single reaction with just one enzyme.
- **RNA sequencing** methods exist but these are difficult and applicable only to small molecules. The methods are similar to chemical degradation sequencing of DNA (Box 4.1, p. 61) but employ sequence-specific endonucleases rather than chemicals to generate the cleaved template molecules. In practice, the sequence of an RNA molecule is usually obtained by converting it into cDNA (see *Figure 3.11*, p. 50) and sequencing this by the chain termination method.
- **Specialist methods** have been developed for mapping the positions of RNA molecules on to DNA sequences, for example to determine the start and end points of transcription and to locate the positions of introns in a DNA sequence. These methods are described in Section 5.1.2.

The only major deficiency in the RNA 'toolkit' is the absence of enzymes with the degree of sequence specificity displayed by the restriction endonucleases that are used routinely in DNA manipulations. Other than this, the only drawback with RNA work is the ease with which RNAs are degraded by ribonucleases that are released when cells are disrupted (as during RNA extraction), which are present on the hands of laboratory workers, and which tend to contaminate glassware and solutions. This means that rigorous laboratory procedures (e.g. cleaning of glassware with ribonuclease inhibitors) have to be adopted to keep RNA molecules intact.

discard all irrelevant clones by these repeated hybridizations, but it is possible to enrich the pool for those clones that specifically hybridize to the DNA fragment. This reduces the size of the library that must subsequently be screened under stringent conditions to identify the desired clones.

A second factor determining success or failure is the completeness of the individual cDNA molecules. Traditionally, cDNAs are made by copying RNA molecules into single-stranded DNA with **reverse transcriptase** and then converting the single-stranded DNA into double-stranded DNA with a DNA polymerase (see *Figure 3.11*, p. 50). There is always a chance that one or other of the strand synthesis reactions will not proceed to completion, resulting in a truncated cDNA. The presence of intramolecular base pairs in the RNA can also lead to incomplete copying. Truncated cDNAs may lack some of the information needed to locate the start and end points of a gene and all its exon–intron boundaries.

Methods are available for precise mapping of the ends of transcripts

The problems with incomplete cDNAs means that more robust methods are needed for locating the precise start and end points of gene transcripts. One possibility is a special type of PCR which uses RNA rather than DNA as the starting material. The first step in this type of PCR is to convert the RNA into cDNA with reverse transcriptase, after which the cDNA is amplified with *Taq* polymerase in the same way as in a normal PCR. These methods go under the collective name of **RT-PCR** (**reverse transcriptase PCR**) but the particular version that interests us at

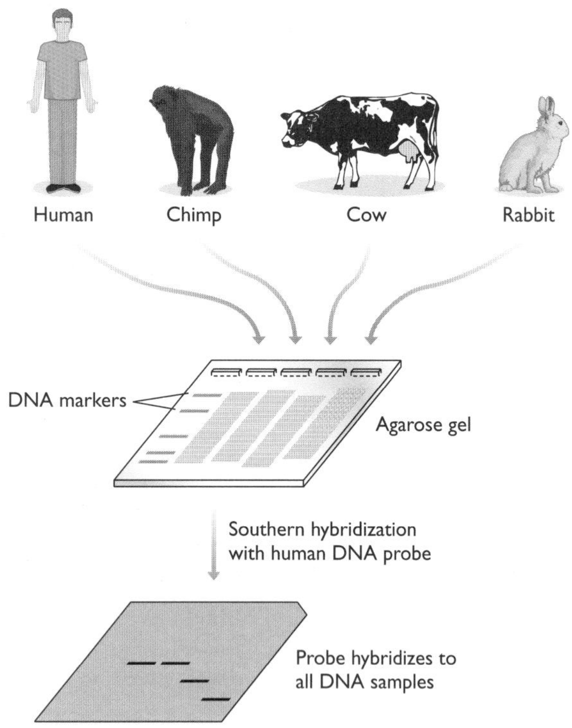

Figure 5.5 Zoo-blotting.

The objective is to determine if a fragment of human DNA hybridizes to DNAs from related species. Samples of human, chimp, cow and rabbit DNAs are therefore prepared, restricted, and electrophoresed in an agarose gel. Southern hybridization is then carried out with a human DNA fragment as the probe. A positive hybridization signal is seen with each of the animal DNAs, suggesting that the human DNA fragment contains an expressed gene. Note that the hybridizing restriction fragments from the cow and rabbit DNAs are smaller than the hybridizing fragments in the human and chimp samples. This indicates that the restriction map around the expressed sequence is different in cows and rabbits, but does not affect the conclusion that a homologous gene is present in all four species.

present is **RACE** (**rapid amplification of cDNA ends**; Frohman *et al.*, 1988). In the simplest form of this method one of the primers is specific for an internal region close to the beginning of the gene being studied. This primer attaches to the mRNA for the gene and directs the first, reverse transcriptase catalyzed, stage of the process, during which a cDNA corresponding to the start of the mRNA is made (*Figure 5.6*). Because only a small segment of the mRNA is being copied the expectation is that the cDNA synthesis will not terminate prematurely, so one end of the cDNA will correspond exactly with the start of the mRNA. Once the cDNA has been made a short linker is attached to its 3′ end. The second primer anneals to this linker and, during the first round of the normal PCR, converts the single-stranded cDNA into a double-stranded molecule, which is subsequently amplified as the PCR proceeds. The sequence of this amplified molecule will reveal the precise position of the start of the transcript.

Other methods for precise transcript mapping involve **heteroduplex analysis**. If the DNA region being studied is cloned as a restriction fragment in an M13 vector (Section 4.1.1) then it can be obtained as single-stranded DNA. When mixed with an appropriate RNA preparation the transcribed sequence in the cloned DNA hybridizes with the equivalent mRNA, forming a double-stranded heteroduplex. In the example shown in *Figure 5.7* the start of this mRNA lies within the cloned restriction fragment, so some of the cloned fragment participates in the heteroduplex, but the rest does not. The single-stranded regions can be digested by treatment with a single-strand-specific nuclease such as S1. The size of the heteroduplex is determined by degrading the RNA component with alkali and electrophoresing the single-stranded DNA in an agarose gel. This size measurement is then used to position the start of the transcript relative to the restriction site at the end of the cloned fragment.

Exon–intron boundaries can also be located with precision

Heteroduplex analysis can also be used to locate exon–intron boundaries. The method is exactly the same as that shown in *Figure 5.7* with the exception that the cloned restriction fragment spans not the start of the transcript but the exon–intron boundary being mapped.

A second method for finding exons in a genome sequence is called **exon trapping** (Church *et al.*, 1994). This requires a special type of vector that contains a **minigene** consisting of two exons flanking an intron sequence, the first exon being preceded by the sequence signals needed to initiate transcription in a eukaryotic cell (*Figure 5.8*). To use the vector the piece of DNA to be studied is inserted into a restriction site located within the vector's intron region. The vector is then introduced into a suitable eukaryotic cell line, where it is transcribed and the RNA produced from it is spliced. The result is that any exon contained in the genomic fragment becomes attached between the upstream and downstream exons from the minigene. RT-PCR with primers annealing within the two minigene exons is now used to amplify a DNA fragment which is sequenced. As the minigene sequence is already known the nucleotide positions at which the inserted exon starts and ends can be determined, precisely delineating this exon.

5.2 DETERMINING THE FUNCTION OF A GENE

Once a new gene has been located, the question of its function has to be addressed. This is turning out to be a difficult area of genomics research, completed sequencing projects revealing that we know rather less than we thought about the content of individual genomes. *E. coli* and *S. cerevisiae*, for example, were intensively studied by conventional genetic analysis prior to the advent of

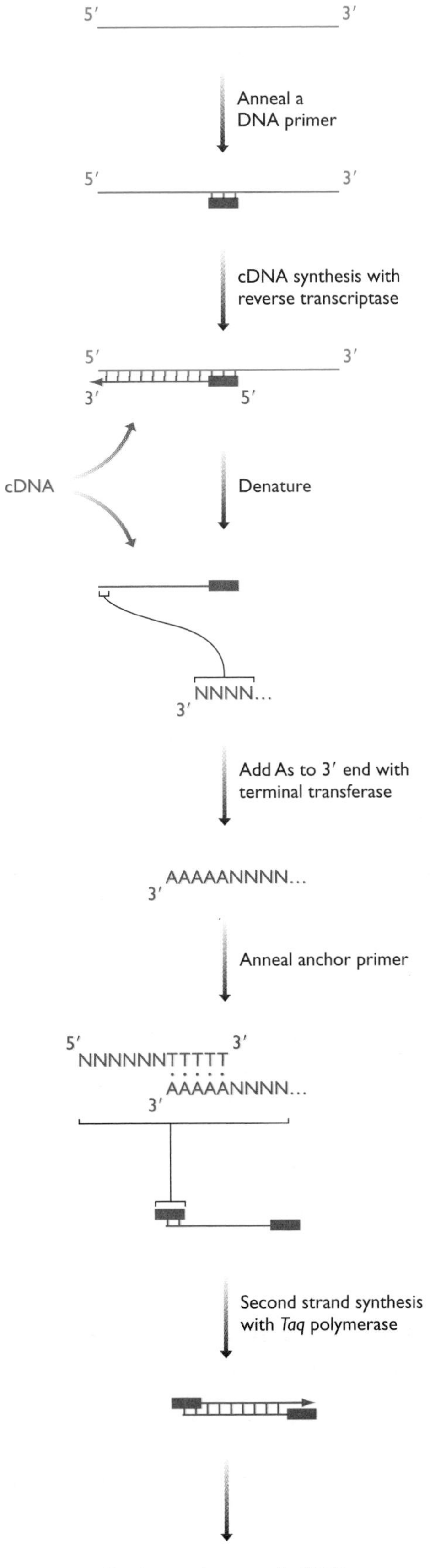

sequencing projects, and geneticists were at one time fairly confident that most of their genes had been identified by mutation studies. The genome sequences revealed that in fact there are large gaps in our knowledge. Of the 4288 protein-coding genes in the *E. coli* genome sequence, only 1853 (43% of the total) had been previously identified (Blattner *et al.*, 1997). For *S. cerevisiae* the figure was only 30% (Dujon, 1996).

As with gene location, attempts to determine the functions of unknown genes are made by computer analysis and by experimental studies.

5.2.1 Computer analysis of gene function

We have already seen that computer analysis plays an important role in locating genes in DNA sequences, and that one of the most powerful tools available for this purpose is homology searching, which locates genes by comparing the DNA sequence under study with all the other DNA sequences in the databases. The basis of homology searching is that related genes have similar sequences and so a new gene can be discovered by virtue of its similarity to an equivalent, already sequenced, gene from a different organism. Now we will look more closely at homology analysis and see how it can be used to assign a function to a new gene.

Homology reflects evolutionary relationships

Homologous genes are ones that share a common evolutionary ancestor, revealed by sequence similarities between the genes. These similarities form the data on which molecular phylogenies are based, as we shall see in Chapter 15. Homologous genes fall into two categories:

- **Orthologous** genes are those homologs that are present in different organisms and whose common ancestor predates the split between the species.
- **Paralogous** genes are present in the same organism, often members of a recognized multigene family (Section 6.1.1), their common ancestor possibly or possibly not predating the species in which the genes are now found.

Figure 5.6 RACE – rapid amplification of cDNA ends.

The RNA being studied is converted into a partial cDNA by extension of a DNA primer that anneals at an internal position not too distant from the 5′-end of the molecule. The 3′-end of the cDNA is further extended by treatment with terminal deoxynucleotidyl transferase (Technical Note 4.2, p. 64) in the presence of dATP, which results in a series of As being added to the cDNA. This series of As acts as the annealing site for the anchor primer. Extension of the anchor primer leads to a double-stranded DNA molecule which can now be amplified by a standard PCR. This is 5′-RACE, so-called because it results in amplification of the 5′-end of the starting RNA. A similar method – 3′-RACE – can be used if the 3′-end sequence is desired.

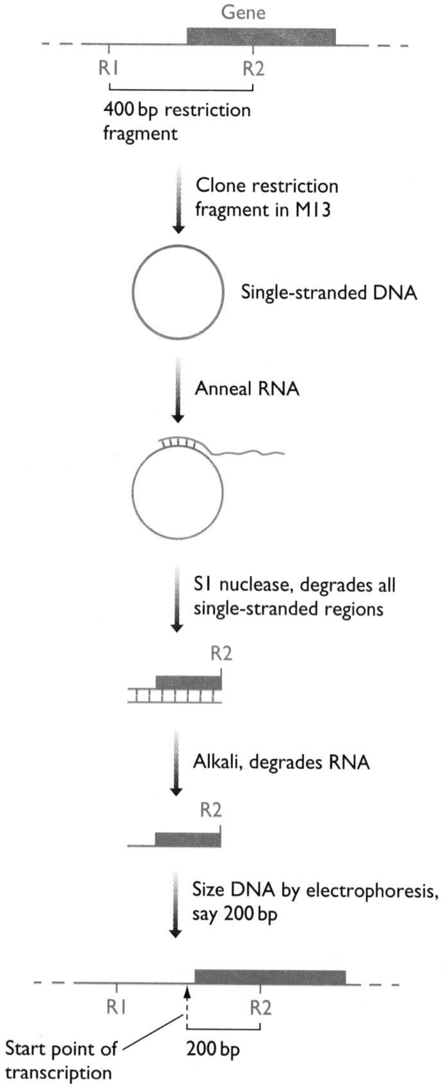

Figure 5.7 S1 nuclease mapping.

This method for transcript mapping makes use of S1 nuclease, an enzyme that degrades single-stranded DNA or RNA polynucleotides, including single-stranded regions in predominantly double-stranded molecules, but has no effect on double-stranded DNA or on DNA–RNA hybrids. In the example shown, a restriction fragment that spans the start of a transcription unit is ligated into an M13 vector and the resulting single-stranded DNA hybridized with an RNA preparation. After S1 treatment, the resulting heteroduplex has one end marked by the start of the transcript and the other by the downstream restriction site (R2). The size of the undigested DNA fragment is therefore measured by gel electrophoresis in order to determine the position of the start of the transcription unit relative to the downstream restriction site.

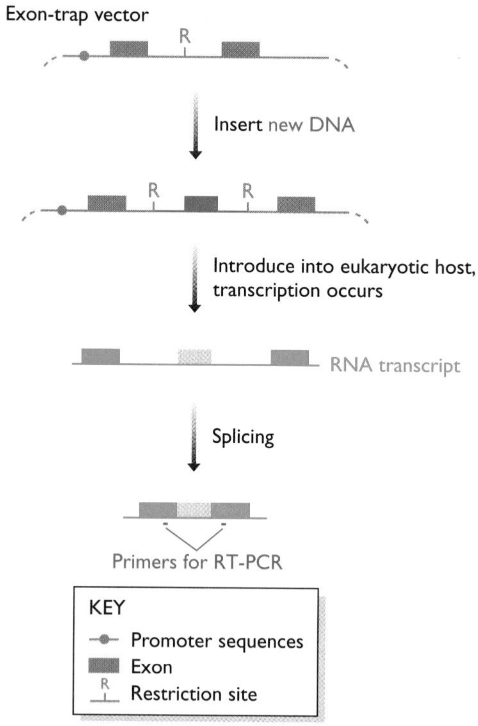

Figure 5.8 Exon trapping.

The exon-trap vector consists of two exon sequences preceded by the signals required for gene expression in a eukaryotic host. New DNA containing an unmapped exon is ligated into the vector and the recombinant molecule introduced into the host cell. The resulting RNA transcript is then examined by RT-PCR to identify the boundaries of the unmapped exon.

A pair of homologous genes do not usually have identical nucleotide sequences, because the two genes undergo different random changes by mutation, but they have similar sequences because these random changes have operated on the same starting sequence, the common ancestral gene. Homology searching makes use of these sequence similarities. The basis of the analysis is that if a newly sequenced gene turns out to be similar to a second, previously sequenced gene, then an evolutionary relationship can be inferred and the function of the new gene is likely to be the same, or at least similar, to the function of the second gene.

It is important not to confuse the words *homology* and *similarity*. It is incorrect to describe a pair of related genes as '80% homologous' if their sequences have 80% nucleotide identity (*Figure 5.9*). A pair of genes are either evolutionarily related or they are not, there are no in between situations and it is therefore meaningless to ascribe a percentage value to homology.

Homology analysis can provide information on the function of an entire gene or of segments within it

A homology search can be conducted with a DNA sequence but usually a tentative gene sequence is converted into an amino acid sequence before the search is

```
Sequence 1  GGTGAGGGTATCATCCCATCTGACTACACCTCATCGGGAGACGGAGCAGT
Sequence 2  GGTCAGGATATGATTCCATCACACTACACCTTATCCCGAGTCGGAGCAGT
Identities  *** *** *** ** *****  ********* ***  *** *********
```

Figure 5.9 Two DNA sequences with 80% sequence identity.

carried out. One reason for this is because there are twenty different amino acids in proteins but only four nucleotides in DNA, so genes that are unrelated usually appear to be more different from one another when their amino acid sequences are compared (*Figure 5.10*). A homology search is therefore less likely to give spurious results if the amino sequence is used. The practicalities of homology searching are not at all daunting. Several software programs exist for carrying out this type of analysis, the most popular being BLAST. The search can be carried out simply by sending a correctly formatted email to the BLAST server at a DNA database. Several hours later the server sends you back an email giving the results of the search. There is currently approximately a 50% chance that a newly sequenced gene will give a positive homology match with a gene already in the databases (Fickett, 1996).

A positive match may give a clear indication of the function of the new gene or the implications of the match might be more subtle. In particular, genes that have no obvious evolutionary relatedness might have short segments that are similar to one another. The explanation of this is often that, although the genes are unrelated, their proteins have similar functions and the shared sequence encodes a domain within each protein that is central to that shared function. Although the genes themselves have no common ancestor, the domains do, but with their common ancestor occurring at a very ancient time, the homologous domains having subsequently evolved not only by single nucleotide changes, but also by more complex rearrangements that have created new genes within which the domains are found (see Section 14.2.1). An interesting example is provided by the tudor domain, an approximately 120-amino-acid motif which was first identified in the sequence of the *Drosophila melanogaster* gene called *tudor* (Ponting, 1997). The protein coded by the *tudor* gene, whose function is unknown, is made up of ten copies of the tudor domain, one after the other (*Figure 5.11*). A homology search using the tudor domain as the test revealed that several known proteins contain this domain. The sequences of these proteins are not highly similar to one another and there is no indication that they are true homologs, but all possess the tudor domain. These proteins include one involved in RNA transport during *Drosophila* oogenesis, a human protein with a role in RNA metabolism, and others whose activity appears to involve RNA in one way or another. The homology analysis therefore suggests that the tudor sequence plays some part in RNA binding, RNA metabolism or some other function involving RNA. The information from the computer analysis is incomplete by itself, but it points the way to the types of experiment that should be done to obtain more clearcut data on the function of the tudor domain.

Homology analysis in the yeast genome project

The *S. cerevisiae* genome project has illustrated both the potential and limitations of homology analysis as a means of assigning functions to new genes. The yeast genome contains approximately 6000 genes, of which 30% had been identified by conventional genetic analysis before the sequencing project got underway. The remaining 70% were studied by homology analysis, giving the following results (*Figure 5.12*; Dujon, 1996):

- Almost another 30% of the genes in the genome could be assigned functions after homology searching of the sequence databases. About half of these were clear homologs of genes whose functions had previously been established, and about half had less striking similarities, including many where the similarities were restricted to discrete domains. For all these genes the homology analysis could be described as successful, but with various degrees of

```
             G  A  P  G  M  W  L  R  L  A  A  G  S  F  E  H  A  G
Sequence 1  GGTGCACCCGGTATGTGACTGCGATTAGCAGCGGGATCATTTCAGCATGCAGGG
            * * ***** **** **** ** *** **** ***** *** ** **** ** *
Sequence 2  GATACACCCCGTATTTGACAGCAATTTGCAGGGGGATGATTGCACCATGGAGCG
             D  T  P  R  I  W  E  E  F  A  G  G  W  L  H  H  G  A
```

Figure 5.10 Lack of homology between two sequences is often more apparent when comparisons are made at the amino acid level.

In this example, the two nucleotide sequences are 76% identical, as indicated by the asterisks. This might be taken as evidence that the sequences are homologous. However, when the sequences are translated into amino acids the identity decreases to 28%, suggesting that the similarity at the nucleotide level was fortuitous. The amino acid sequences have been written using the standard one-letter abbreviations: these are defined in *Table 7.3*, p. 155.

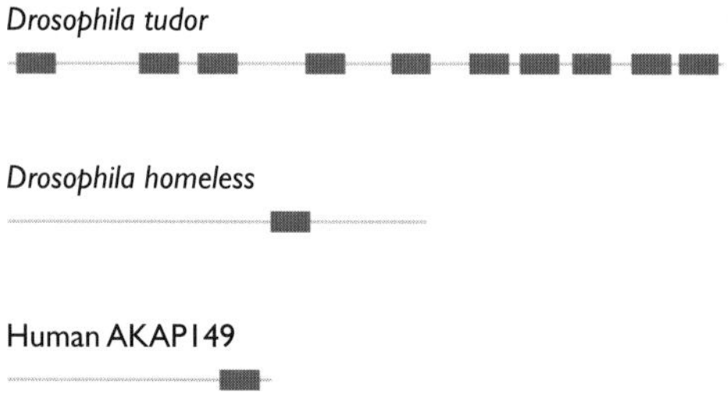

Figure 5.11 The tudor domain.

The top drawing shows the structure of the *Drosophila tudor* protein, which contains ten copies of the tudor domain. The domain is also found in a second *Drosophila* protein, *homeless*, and in the human A-kinase anchor protein (AKAP149). The proteins have dissimilar structures other than the presence of the tudor domains. The activity of each protein involves RNA in one way or another.

usefulness (Oliver, 1996a). For some genes the identification of a homolog enabled the function of the yeast gene to be comprehensively determined: examples included identification of yeast genes for DNA polymerase subunits and aminoacyl-tRNA synthetases. For other genes the functional assignment could only be to a broad category, such as 'gene for a protein kinase', in other words the biochemical properties of the gene product could be inferred but not the exact role of the protein in the cell. Some identifications were initially puzzling, the best example being the discovery of a yeast homolog of a bacterial gene involved in nitrogen fixation. Yeasts do not fix nitrogen so this could not be the function of the yeast gene. In this case, the discovery of the yeast homolog refocused attention on the previously characterized bacterial gene, with the subsequent realization that, although being involved in nitrogen fixation, the primary role of the bacterial gene product was in the

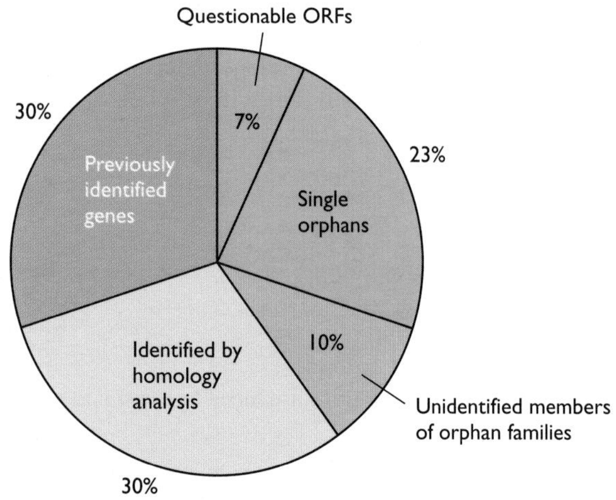

Figure 5.12 Categories of gene in the yeast genome.

synthesis of metal-containing proteins, which have broad roles in all organisms, not just nitrogen-fixing ones.

- About 10% of all the yeast genes had homologs in the databases, but the functions of these homologs were unknown. The homology analysis was therefore unable to help in assigning functions to these yeast genes. These yeast genes and their homologs are called **orphan families**.
- The remaining yeast genes, about 30% of the total, had no homologs in the databases. A small proportion of these (about 7% of the total) were questionable ORFs which might not be real genes, being rather short or having an unusual codon bias. The remainder look like genes but at present are unique. The latter group are called **single orphans**.

5.2.2 Assigning gene function by experimental analysis

It is clear that homology analysis is not a panacea that can identify the functions of all new genes. Experimental methods are therefore needed to complement and extend the results of homology studies. This is proving to be one of the biggest problems in genomics research and most molecular biologists agree that the methodologies and strategies currently in use are not entirely adequate for assigning functions to the vast numbers of unknown genes being discovered by sequencing projects. The problem is that the objective – to plot a course from gene to function – is the reverse of the route normally taken by genetic analysis, in which the starting point is a phenotype and the objective is to identify the underlying gene or genes. The problem we are currently addressing takes us in the opposite direction, starting from a new gene and hopefully leading to identification of the associated phenotype.

Gene inactivation is the key to functional analysis

In conventional genetic analysis, the genetic basis of a phenotype is usually studied by searching for mutant organisms in which the phenotype has become altered. The mutants might be obtained experimentally, for example by treating a population of organisms (e.g. a culture of bacteria) with ultraviolet radiation or a mutagenic chemical (see Section 13.1.1), or the mutants might be present in a natural population. The gene or genes that have been altered in the mutant organism are then studied by genetic crosses (Chapter 2), which can locate the position of a gene in a genome and also determine if the gene is the same as one that has already been characterized. The gene can then be studied further by molecular biology techniques, for example by cloning and sequencing.

The general principle of this conventional analysis is that the genes responsible for a phenotype can be identified by determining which genes are inactivated in organisms that display a mutant version of the phenotype. If the starting point is the gene, rather than the phenotype, then the equivalent strategy would be to mutate the gene and identify the phenotypic change that results. This is the basis of most of the techniques used to assign functions to unknown genes.

Methods for inactivating specific genes depend on homologous recombination

The easiest way to inactivate a specific gene is to disrupt it with an unrelated segment of DNA (*Figure 5.13*). This can be achieved by **homologous recombination** between the chromosomal copy of the gene and a second piece of DNA that shares some sequence identity with the target gene. Homologous, and other types of, recombination are complex events that we will deal with in detail in Section 13.2: for present purposes it is enough to know that if two DNA molecules have similar sequences, then recombination can result in segments of the molecules being exchanged.

How is gene inactivation carried out in practice? We will consider two examples, the first with *S. cerevisiae.* Since completing the genome sequence in 1996, yeast molecular biologists have embarked on a coordinated, international effort to determine the functions of as many orphan genes as possible (Oliver, 1996b). One technique that is being extensively used is shown in *Figure 5.14* (Wach *et al.*, 1994). The central component is the 'deletion cassette', which carries a gene for antibiotic resistance. This gene is not a normal component of the yeast genome but it will work if transferred into a yeast chromosome, giving rise to a transformed yeast cell that is resistant to the antibiotic geneticin. Before using the deletion cassette new segments of DNA are attached as tails to either end. These segments have sequences identical to parts of the yeast gene that is going to be inactivated. After the modified cassette is introduced into a yeast cell, homologous recombination occurs between the DNA tails and the chromosomal copy of the yeast gene, replacing the latter with the antibiotic resistant gene. Cells which have undergone the replacement are therefore selected by plating the culture on to agar medium containing geneticin. The resulting colonies lack the target gene and their phenotypes can be examined to gain some insight into the function of the gene.

The second example of gene inactivation uses an analogous process but with mice rather than yeast. The mouse is frequently used as a **model organism** for humans, because the mouse genome is similar to the human genome, containing many of the same genes. Identifying the functions of unknown human genes is therefore being

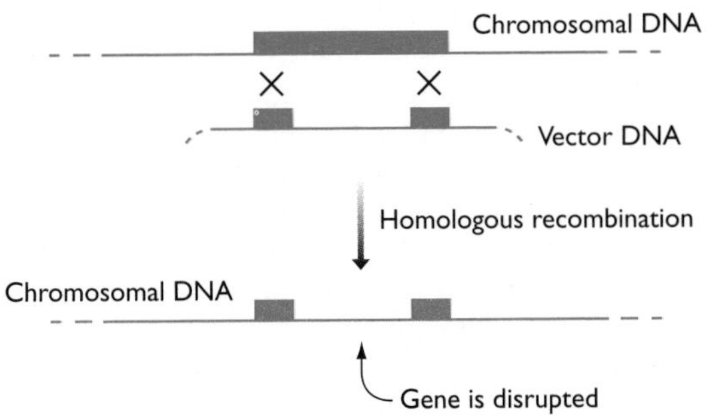

Figure 5.13 Gene inactivation by homologous recombination.

The chromosomal copy of the target gene recombines with a disrupted version of the gene carried by a cloning vector. As a result, the target gene becomes inactivated. For more information on recombination see Section 13.2.

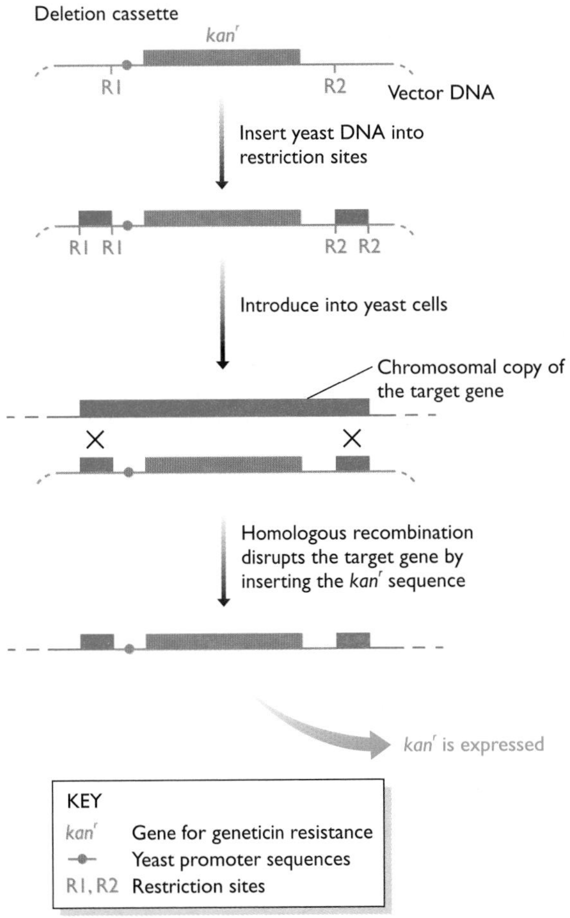

Figure 5.14 The use of a yeast deletion cassette.

The deletion cassette consists of an antibiotic resistance gene preceded by the promoter sequences needed for expression in yeast, and flanked by two restriction sites. The start and end segments of the target gene are inserted into the restriction sites and the vector introduced into yeast cells. Recombination between the gene segments in the vector and the chromosomal copy of the target gene results in disruption of the latter. Cells in which the disruption has occurred are identified because they now express the antibiotic resistance gene and so will grow on an agar medium containing geneticin. The gene designation '*kan*^{r}' is an abbreviation for 'kanamycin resistance', kanamycin being the family name of the group of antibiotics that include geneticin.

carried out largely by inactivating the equivalent genes in the mouse, these experiments being ethically unthinkable with humans. The homologous recombination part of the procedure is identical to that described for yeast and once again results in a cell in which the target gene has been inactivated. The problem is that we do not want just one mutated cell, we want a whole mutant mouse, as only with the complete organism can we make a complete assessment of the effect of the gene inactivation on the phenotype. To achieve this it is necessary to use a special type of mouse cell, an **embryonic stem** or **ES** cell (Evans *et al.*, 1997). Unlike most mouse cells, ES cells are **totipotent**, meaning that they are not committed to a single developmental pathway and can therefore give rise to all types of differentiated cell. The engineered ES cell is therefore injected into a mouse embryo, which continues to develop and eventually give rise to a **chimera**, a mouse whose cells are a mixture of mutant ones, derived from the engineered ES cells, and nonmutant ones, derived from all the other cells in the embryo. This is still not quite what we want, so the chimeric mice are allowed to mate with one another. Some of the offspring result from fusion of two engineered gametes, and will therefore be nonchimeric, as every one of their cells will carry the inactivated gene. These are **knockout mice**, and it is hoped that examination of their phenotypes provides the desired information on the function of the gene being studied. This works well for many gene inactivations but some are lethal and so cannot be studied in a homozygous knockout mouse. Instead, a heterozygous mouse is obtained, the product of fusion between one normal and one engineered gamete, in the hope that the phenotypic effect of the gene inactivation will be apparent even though the mouse still has one correct copy of the gene being studied.

The phenotypic effect of gene inactivation is sometimes difficult to discern

Once a gene-inactivated yeast strain, knockout mouse, or equivalent with any other organism has been obtained the next stage is to examine the phenotype of the mutant in order to assign a function to the unknown gene. This can be much more difficult than it sounds. With any organism the range of phenotypes that must be examined is immense. Even with yeast, the list is quite lengthy (*Table 5.2*) and with higher organisms some phenotypes (e.g. behavioral ones) are difficult if not impossible to assess in a comprehensive fashion. Furthermore, the effects of some gene inactivations are very subtle and may not be recognized when the phenotype is examined. A good example was provided by the longest gene on yeast chromosome III which, at 2167 codons and with typical yeast codon bias, simply had to be a functional gene rather than a spurious ORF. Inactivation of this gene had no apparent effect, the mutant yeast cells appearing to have an identical phenotype to normal yeast. For some time it was thought that perhaps this gene is dispensable, its protein product either involved in some completely nonessential function, or having a function that is duplicated by a second gene. Eventually it was shown that the mutants died when they were grown at low pH in the presence of glucose and acetic acid, which normal yeasts can tolerate, and it was concluded that the gene codes for a protein that pumps acetate out of the cell (Oliver, 1996a). This is definitely an essential function as the gene plays a vital role in protecting yeast from acetic acid-induced damage, but this essentiality was difficult to track down from the phenotype tests.

Table 5.2 List of yeast phenotypes that must be examined in order to assign a function to an unknown gene

Possible biological functions of an unassigned gene
DNA synthesis and the cell cycle
RNA synthesis and processing
Protein synthesis
Stress responses
Cell wall synthesis and morphogenesis
Transport of biochemicals within the cell
Energy and carbohydrate metabolism
Lipid metabolism
Development
DNA repair and recombination
Meiosis
Chromosome structure
Cell architecture
Secretion and protein trafficking

Based on the categories defined by the European Functional Analysis Network (Oliver, 1996b).

These problems, together with the daunting prospect of having to perform a gene inactivation and full phenotype screen for every one of the 2500 yeast orphans, has prompted yeast biologists to pioneer the search for more efficient methods of assigning functions to unknown genes. One possibility is described in Research Briefing 5.1.

Gene overexpression can also be used to assess function

So far we have concentrated on techniques that result in inactivation of the gene being studied ('loss of function'). The complementary approach is to engineer an organism in which the test gene is much more active than normal ('gain of function') and to determine what changes, if any, this has on the phenotype. The results of these experiments must be treated with caution because of the need to distinguish between a phenotype change that is due to the specific function of an overexpressed gene and a less specific phenotype change that reflects the abnormality of the situation where a single gene product is being synthesized in excessive amounts, possibly in tissues in which the gene is normally inactive. Despite this qualification, overexpression has provided some important information on gene function.

To overexpress a gene a special type of cloning vector must be used, one designed to ensure that the cloned gene directs the synthesis of as much protein as possible. The vector is therefore **multicopy**, meaning that inside

RESEARCH

5.1

BRIEFING

Using genetic footprinting to assign functions to yeast genes

Most of the available techniques for functional analysis are designed for use with individual genes. This paper describes a novel approach which starts with a large set of genes, such as an entire chromosome, and enables those with related functions to be identified.

The *Saccharomyces cerevisiae* genome project revealed that yeast has approximately 6000 genes, only 30% of which had been identified by conventional genetic techniques. Almost half of the previously unknown genes could be assigned functions by homology analysis (Section 5.2.1), but this still left some 2500 unidentified genes. Assigning functions by examining each of these orphans one by one would take a great deal of time, so methods were needed for placing genes into functional groups, each group comprising genes with related roles. 'Genetic footprinting' does this by identifying the mutations carried by cells that are unable to grow at the maximal rate in one or more of a variety of different types of culture media.

The technical approach

The first step in the genetic footprinting procedure is to generate a mixed population of yeast cells, each with a single inactivated gene. This can be achieved by *Ty1* mutagenesis. *Ty1* is a retrotransposon (Section 6.3.2), a type of repeat sequence that can move from place to place ('transpose') within a genome via an RNA intermediate. If the retrotransposon moves into a gene then that gene is inactivated because it now has an extra segment of DNA embedded within it. In one particular strain of yeast, transposition of a *Ty1* sequence can be induced by placing the cells in a culture medium containing galactose. The retrotransposon moves to different positions in different cells, so if a large enough population is used (approximately 10^{11} cells) then the resulting mutated culture contains inactivated versions of every yeast gene.

After mutagenesis, the culture is divided into aliquots and each grown under different conditions. The aim is to identify the mutations carried by cells that are unable to grow at the maximal rate. This is done by a series of PCR tests, one PCR for every gene being studied. Each PCR uses a primer that anneals within the *Ty1* sequence, together with a second primer that is specific for the gene being tested. This means that, although the template DNA is prepared from the culture as a whole, each PCR only amplifies DNA from those cells that have an inactivated copy of the target gene. The abundance of these cells in the culture is monitored by comparing the amounts of PCR product obtained at different times: if the cells suffer a growth disadvantage under the conditions being used, then their PCR products become less abundant as time goes on.

Genetic footprinting of yeast chromosome V

Genetic footprinting was first applied to functional analysis of the 268 genes on yeast chromosome V. Most gene inactivations led to a general growth disadvantage that manifested itself under all culture conditions, including growth in a nutrient-rich medium. For 51 genes, inactivation led to a

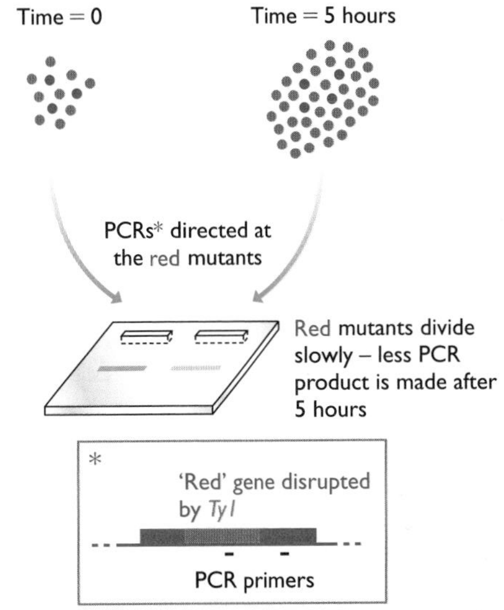

'major' growth defect (>25% reduction in growth) suggesting that these genes code for essential proteins. Most of these genes had previously been identified and included, for example, ones coding for amino acid biosynthesis enzymes and DNA repair proteins. The category also included 14 orphans. Other orphan inactivations resulted in a growth disadvantage only when the cells were grown under special conditions. For example, growth in a lactate-rich medium, which tests for respiratory proficiency, linked three orphans with this function. Similarly, two orphans were found to have roles in tolerance to high-salt conditions and one was associated with growth at high temperature.

Altogether it proved possible to assign broad functional classes to approximately 30% of the orphans on chromosome V. Although the information on each individual gene is imprecise, the results indicate the types of experiment that should be carried out with individual genes in order to pin down their exact functions. If the success rate with chromosome V could be reproduced across the entire yeast genome then some 750 orphans would be assigned a functional class, with the possibility that this number might be increased further if a greater range of culture regimes was tested.

References

Smith V, Botstein D and Brown PO (1995) Genetic footprinting: a genomic strategy for determining a gene's function given its sequence. *Proc. Natl Acad. Sci. USA* **92**, 6479–6483.

Smith V, Chou KN, Lashkari D, Botstein D and Brown PO (1996) Functional analysis of the genes of yeast chromosome V by genetic footprinting. *Science*, **274**, 2069–2074.

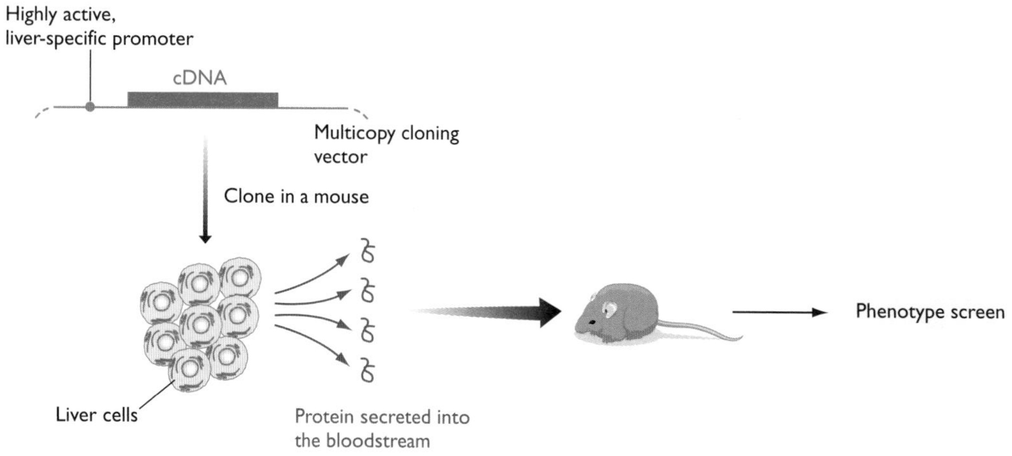

Figure 5.15 Functional analysis by gene overexpression.

The objective is to determine if overexpression of the gene being studied has an effect on the phenotype of a transgenic mouse. A cDNA of the gene is therefore inserted into a cloning vector carrying a highly-active promoter sequence that directs expression of the cloned gene in mouse liver cells.

the host organism it multiplies to 40–200 copies per cell, so there are many copies of the test gene. The vector must also contain a highly active promoter (Section 8.3.1) so that each copy of the test gene is converted into large quantities of mRNA, again ensuring that as much protein as possible is made. An example of the technique used with mice genes is shown in *Figure 5.15* (Simonet *et al.*, 1997). In this project the genes to be studied were identified from EST sequences and so each one was cloned as a cDNA. The genes were selected because their sequences suggested that they code for proteins that are secreted into the bloodstream. The cloning vector that was used contained a highly active promoter that is expressed only in the liver, so each **transgenic mouse** overexpressed the test gene in its liver and then secreted the resulting protein into the blood. The phenotype of each transgenic mouse was examined in the search for clues regarding the functions of the cloned genes. An interesting discovery was made when it was realized that one transgenic mouse had bones that were significantly more dense than those of normal mice. This was important for two reasons: first, it enabled the relevant gene to be identified as one involved in bone synthesis; and second, the discovery of a protein that increases bone density has implications for the development of treatments for human osteoporosis, a fragile-bone disease.

5.2.3 More detailed studies of the activity of a protein coded by an unknown gene

Gene inactivation and overexpression are the primary techniques used by genome researchers to determine the function of a new gene, but these are not the only procedures that can provide information on gene activity. Other methods can extend and elaborate the results of inactivation and overexpression. These can be used to provide additional information that will aid identification of a gene function, or might form the basis of a more comprehensive examination of the activity of a protein whose gene has already been characterized.

Directed mutagenesis can be used to probe gene function in detail

Inactivation and overexpression can determine the overall function of a gene, but they cannot provide detailed information on the activity of a protein coded by a gene. For example, it might be suspected that part of a gene codes for an amino acid sequence that directs its protein product to a particular compartment in the cell, or is responsible for the ability of the protein to respond to a chemical or physical signal. To test these hypotheses it would be necessary to delete or alter the relevant part of the gene sequence, but to leave the bulk unmodified so that the protein is still synthesized and retains the major part of its activity. The various procedures of **site-directed** or ***in vitro* mutagenesis** (Technical Note 5.2) can be used to make these subtle changes. These are important techniques whose applications lie not only with the study of gene activity but also in the area of **protein engineering**, where the objective is to create novel proteins with properties that are better suited for use in industrial or clinical settings.

After mutagenesis the gene sequence must be introduced into the host cell so that homologous recombination can replace the existing copy of the gene with the modified version. This presents a problem because we must have a way of knowing which cells have undergone homologous recombination. Even with yeast this will only be a fraction of the total, and with mice the fraction will be very small. Normally we would solve this problem by placing a marker gene (e.g. one coding for

TECHNICAL 5.2 NOTES

Site-directed mutagenesis

Methods for making a precise alteration in a gene sequence in order to change the structure and possibly the activity of a protein.

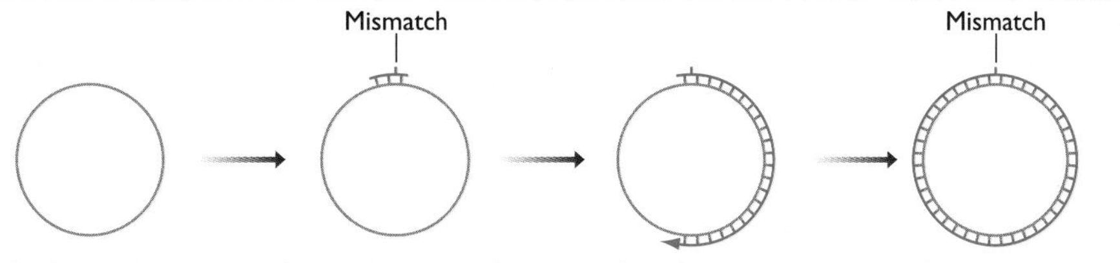

Changes in protein structure can be engineered by site-directed mutagenesis techniques, which result in defined alterations being made to the nucleotide sequence of the gene coding for a protein of interest. These techniques have applications in general research as a means of examining the functions of different parts of a protein (Section 5.2.3), and also have widespread importance in the development of new enzymes for biotechnological purposes.

Conventional mutagenesis is a random process that introduces changes at unspecified positions in a DNA molecule. Screening of large numbers of mutated organisms is necessary to find a mutation of interest. Even with microbes, which can be screened in huge numbers, the best that can be hoped for is a range of mutations in the correct gene, one of which might affect the part of the protein being studied. Site-directed mutagenesis offers a means of making much more specific mutations.

An almost unlimited variety of DNA manipulations can be used to introduce mutations into cloned genes. The simplest are to delete a restriction fragment, to open the DNA at a restriction site and remove a few nucleotides, or to insert new DNA at a restriction site. These are relatively large-scale changes. To alter a single nucleotide at a specified position the technique called **oligonucleotide-directed mutagenesis** is used, one version of which is described here. First, a single-stranded version of the gene is obtained by cloning in an M13 vector (Section 4.1.1). A short oligonucleotide is then synthesized, complementary to the relevant region of the gene, but containing the desired nucleotide alteration. This oligonucleotide is hybridized to the DNA and used as the primer for a strand synthesis reaction that is allowed to continue all the way around the circular template molecule, as shown above.

After introduction into *E. coli*, DNA replication produces numerous copies of this recombinant DNA molecule. Half the resulting double-stranded molecules are copies of the original strand of DNA, and half are copies of the strand that contains the mutated sequence. The double-stranded molecules direct synthesis of phage particles so about half the phage released from the infected bacteria carry a single-stranded version of the mutated molecule. The phages are plated on to solid agar so that plaques are produced, and the mutant ones identified by hybridization probing with the original oligonucleotide (see below). The mutated gene can then be placed back in its original host by homologous recombination, as described in Section 5.2.3, or transferred to an *E. coli* vector designed for synthesis of protein from cloned DNA, so that a sample of the mutated protein can be obtained and the effect of the mutation determined.

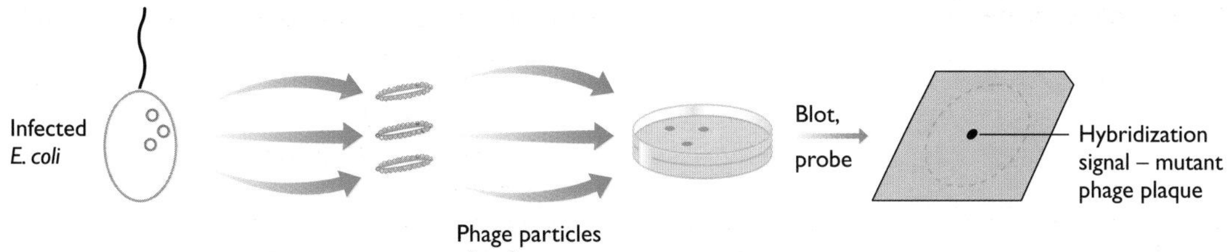

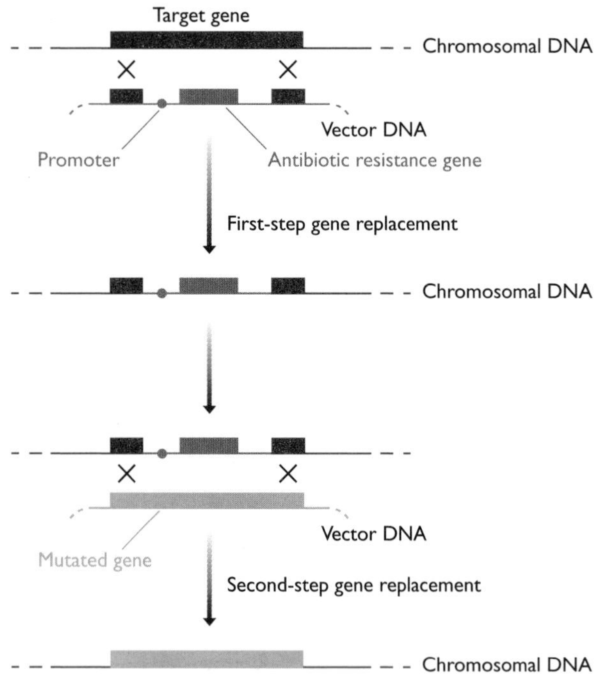

Figure 5.16 Two-step gene replacement.

antibiotic resistance) next to the mutated gene and looking for cells that take on the phenotype conferred by this marker. In most cases, cells that insert the marker gene into their genome also insert the closely attached, mutated gene and so are the ones we want. The problem is that in a site-directed mutagenesis experiment we must be sure that any change in the activity of the gene being studied is due to the specific mutation that was introduced into the gene rather than to the indirect result of changing its environment in the genome by inserting a marker gene next to it. The answer is to use a more complex, two-step gene replacement (*Figure 5.16*). In this procedure the target gene is first replaced with the marker gene, on its own, the cells in which this recombination

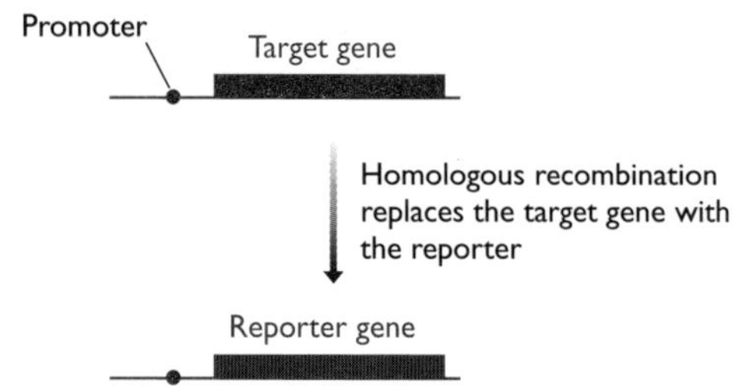

Figure 5.17 A reporter gene.

The open reading frame of the reporter gene replaces the open reading frame of the gene being studied. The result is that the reporter gene is placed under control of the regulatory sequences that usually dictate the expression pattern of the test gene. For more information on these regulatory sequences, see Sections 8.2 and 8.3.

takes place being identified by selecting for the marker gene phenotype. These cells are then used in the second stage of the gene replacement, when the marker gene is replaced by the mutated gene, success now being monitored by looking for cells that have lost the marker gene phenotype. These cells contain the mutated gene and their phenotypes can be examined to determine the effect of the directed mutation on the activity of the protein product.

Reporter genes and immunocytochemistry can be used to locate where and when genes are expressed

Clues to the function of a gene can often be obtained by determining where and when the gene is expressed. If gene expression is restricted to a particular organ or tissue of a multicellular organism, or to a single set of cells within an organ or tissue, then this positional information can be used to infer the general role of the gene product. The same is true of information relating to the developmental stage at which a gene is expressed, this type of analysis having proved particularly useful in understanding the activities of genes involved in the earliest stages of *Drosophila* development (Section 11.3.3) and increasingly being used to unravel the genetics of mammalian development. It is also applicable to those unicellular organisms, such as yeast, which have distinctive developmental stages in their life cycle.

Determining the pattern of gene expression within an organism is possible with a **reporter gene**. This is a gene whose expression can be monitored in a convenient way, ideally by visual examination (*Table 5.3*), cells that express the reporter gene becoming blue, fluorescing or giving off some other visible signal. For the reporter gene to give a reliable indication of where and when a test gene is expressed the reporter must be subject to the same regulatory signals as the test gene. This is achieved by replacing the ORF of the test gene with the ORF of the reporter gene, the regulatory signals being contained in the region of DNA upstream of an ORF (*Figure 5.17*). The reporter gene will now display the same expression pattern as the test gene, this expression pattern being determined by examining the organism for the reporter signal.

As well as knowing in which cells a gene is expressed, it is often useful to locate the position within the cell where the protein coded by the gene is found. For example, key data regarding gene function can be obtained by showing that the protein product is located in mitochondria, in the nucleus, or on the cell surface. Reporter genes

Table 5.3 Examples of reporter genes

Gene	Gene product	Assay
lacZ	β-galactosidase	Histochemical test
uidA	β-glucuronidase	Histochemical test
lux	Luciferase	Bioluminescence
GFP	Green fluorescent protein	Fluorescence

cannot help here as the DNA sequence upstream of the gene, the sequence to which the reporter gene is attached, is not involved in targeting the protein product to its correct intracellular location. Instead it is the amino acid sequence of the protein itself that is important. Therefore the only way to determine where the protein is located is to search directly for it. This is done by **immunocytochemistry**, which makes use of an antibody that is specific for the protein of interest and so binds to this protein and no other. The antibody is labeled so that its position in the cell, and hence the position of the target protein, can be visualized. For low resolution studies, fluorescent labeling and light microscopy are used; alternatively, high resolution immunocytochemistry can be carried out by electron microscopy with an electron dense label such as colloidal gold.

Protein–protein interactions can be examined by phage display and the yeast two-hybrid system

Important data pertaining to gene function can often be obtained by determining if the protein coded by the gene interacts with other, known proteins. If an interaction to a well characterized protein is identified, then it might be possible to infer the function of the unknown protein. For example, an interaction between an unknown protein and one known to be located on the cell surface might indicate that the unknown protein is involved in cell–cell signaling (Section 11.1.2). There are several methods for studying protein–protein interactions, the two most useful being **phage display** and the **yeast two-hybrid system**.

In phage display a special type of cloning vector is used (Clackson and Wells, 1994). This vector is based on λ bacteriophage or one of the filamentous bacteriophages such as M13. It is designed so that a new gene that is cloned into it is expressed in such a way that its protein product is synthesized as a fusion with one of the phage coat proteins (*Figure 5.18A*). The phage protein therefore carries the foreign protein into the phage coat, where it is 'displayed' in a form that enables it to interact with other proteins that the phage encounters. There are several ways in which phage display can be used to study protein interactions. In one method, the test protein is displayed and interactions sought with a series of purified proteins or protein fragments of known function. This approach is limited as it takes time to carry out each test, so is feasible only if some prior information has been obtained about likely interactions. A more powerful strategy is to prepare a phage display library, a collection of clones displaying a range of proteins, and identifying which members of the library interact with the test protein (*Figure 5.18B*).

The yeast two-hybrid system detects protein interactions in a more complex way (Fields and Sternglanz, 1994). In Section 8.3.2 we will see that proteins called **transcription factors** are responsible for controlling the expression of genes in eukaryotes. To carry out this function a transcription factor must bind to a DNA sequence upstream of a gene and activate the RNA polymerase enzyme that copies the gene into RNA. These two activities are specified by different parts of the transcription factor's protein structure, and some factors will work even after the protein has been cleaved into two segments, one segment containing the DNA-binding domain and one the activation domain. In the cell, the two segments interact to form the functional transcription factor.

The two-hybrid system makes use of an *S. cerevisiae* strain that lacks a transcription factor for a reporter gene. This gene is therefore switched off. The two segments of the transcription factor are coded by separate genes, both of which have been cloned. One of these genes, usually the one for the DNA-binding domain, is ligated to the gene for the protein whose interactions we wish to study. This protein can come from any organism, not just yeast: in the example shown in *Figure 5.19A* it is a human protein, which for convenience we will call 'Protein X'. After introduction into yeast, this engineered gene will specify synthesis of a fusion protein, made up of the DNA-binding domain of the transcription factor attached to Protein X. The second gene, the one for the activation domain of the transcription factor, is ligated with a mixture of DNA fragments so that many different constructs are made. The engineered genes are mixed together with yeast cells, some of which become cotransformed with an engineered DNA-binding gene and an engineered activation gene (*Figure 5.19B*). If the fusion proteins coded by these two genes can cooperate then the reporter gene will be switched on. If this happens then the protein attached to the activation domain of the transcription factor must be one that can interact with Protein X.

5.3 COMPARATIVE GENOMICS

We have already seen how similarities between homologous genes from different organisms provide one way of assigning a function to an unknown gene (Section 5.2.1). This is an example of how knowledge about the genome of one organism can help in understanding the genome of a second organism. The possibility that **comparative genomics** might be a valuable means of deciphering the human genome was recognized when the Human Genome Project was planned in the late 1980s, and the Project has actively stimulated the development of genome projects for model organisms such as the mouse and fruit fly. In this section we will explore the extent to which comparisons between different genomes are proving useful.

5.3.1 Comparative genomics as an aid to gene mapping

The basis of comparative genomics is that the genomes of related organisms are similar. The argument is the same one that we considered when looking at homologous genes (Section 5.2.1). Two organisms with a relatively recent common ancestor will have genomes that display species-specific differences built onto the common plan

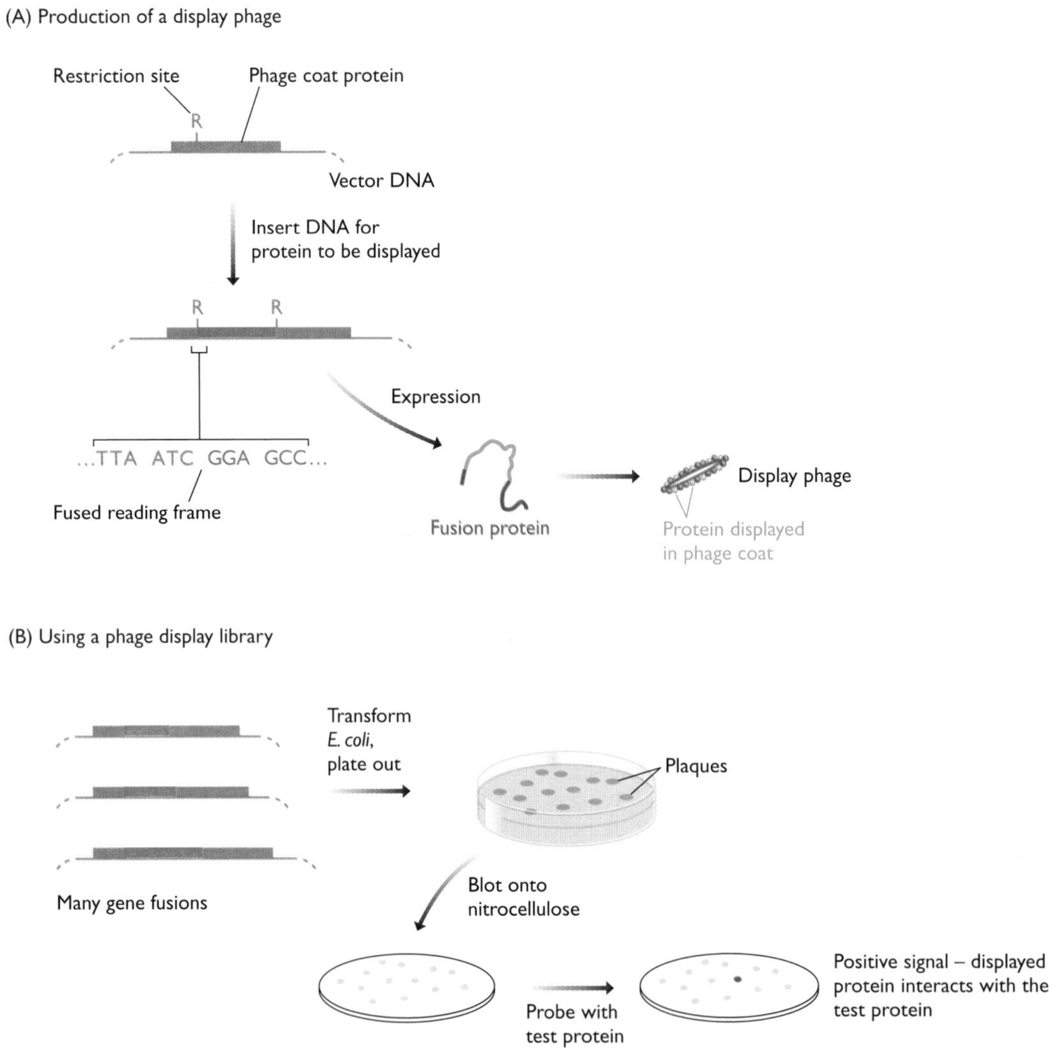

Figure 5.18 Phage display.

(A) The cloning vector used for phage display is a bacteriophage genome with a unique restriction site located within a gene for a coat protein. The technique was originally carried out with the gene III coat protein of the filamentous phage called f1, but has now been extended to other phages including λ. To create a display phage, the DNA sequence coding for the test protein is ligated into the restriction site so that a fused reading frame is produced – one in which the series of codons continues unbroken from the coat protein gene into the test gene. After transformation of *E. coli*, this recombinant molecule directs synthesis of a hybrid protein, made up of the test protein fused to the coat protein. Phage particles produced by these transformed bacteria therefore display the test protein in their coats. (B) A phage display library is produced by ligating many different DNA fragments into the cloning vector. After transformation, the bacteria are plated on to an agar medium so that plaques are produced. Each plaque contains phage displaying a different protein. The lawn of plaques is blotted on to a nitrocellulose membrane and the test protein applied. If the test protein interacts with any of the displayed proteins then it will bind to the membrane. Binding is detected either because the test protein is itself labeled (e.g. with a fluorescent marker) or by treating the membrane with a labeled antibody specific for the test protein. Note that a nitrocellulose membrane is used, not a nylon one as in Southern hybridization (Technical Note 2.1, p. 19), because nylon membranes give a high background with protein probes.

possessed by the ancestral genome. The closer two organisms are on the evolutionary scale, the more related their genomes will be (Nadeau and Sankoff, 1998).

If the two organisms are sufficiently closely related then their genomes might display **synteny**, the partial or complete conservation of gene order. Then it is possible to use map information from one genome to locate genes in the second genome. At one time it was thought that mapping the genomes of mouse and other mammals, which are at least partially syntenic with the human genome, might provide valuable information that could be used in construction of the human genome map. The problem with this approach is that all the close relatives of humans have equally large genomes that are just as difficult to

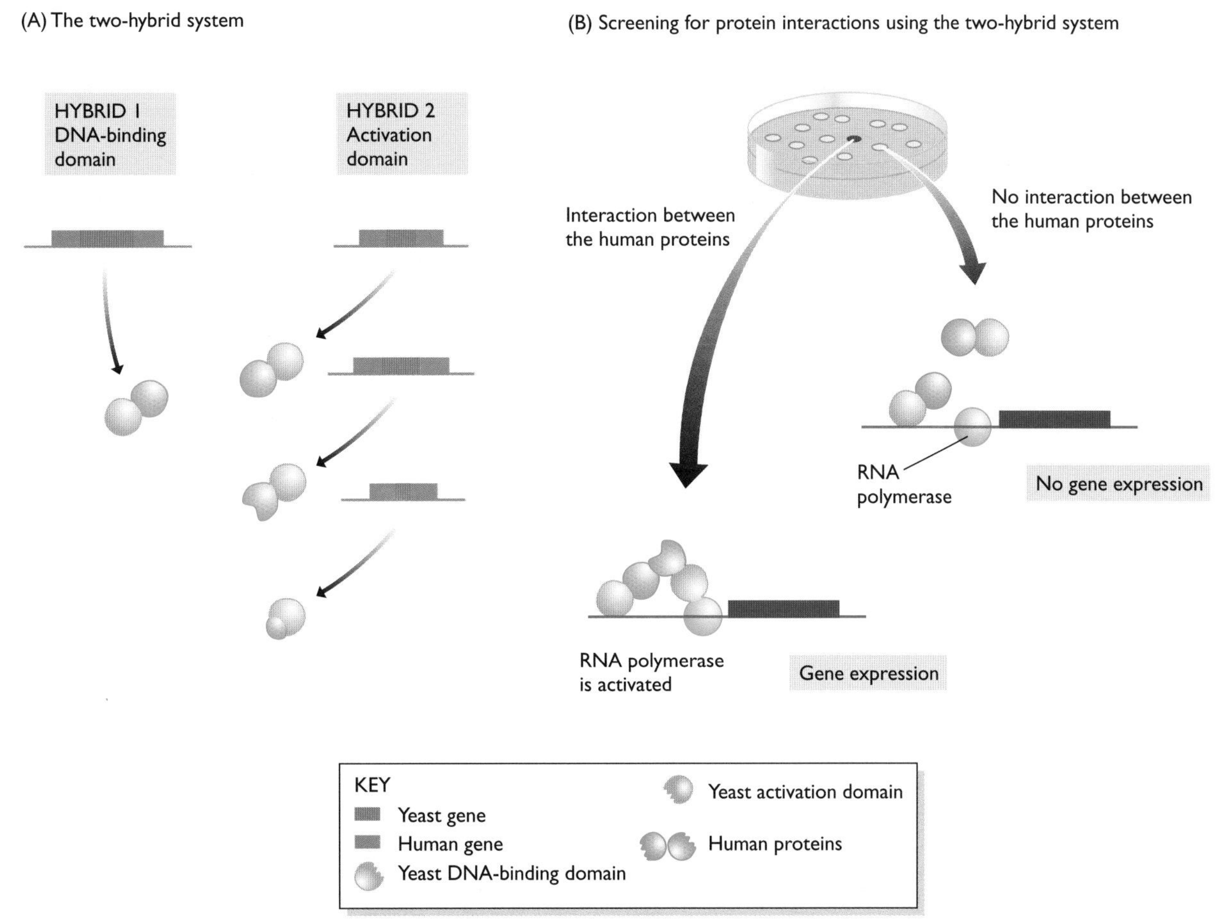

Figure 5.19 The yeast two-hybrid system.

(A) On the left, a gene for a human protein has been ligated to the gene for the DNA-binding domain of a yeast transcription factor. After transformation of yeast, this construct specifies a fusion protein, part human protein and part yeast transcription factor. On the right, various human DNA fragments have been ligated to the gene for the activation domain of the transcription factor: these constructs specify a variety of fusion proteins. (B) The two sets of constructs are mixed and cotransformed into yeast. A colony in which the reporter gene is expressed contains fusion proteins whose human segments interact, thereby bringing the DNA-binding and activation domains of the transcription factor into proximity and enabling the RNA polymerase to be activated. See Section 8.3.2 for more information on transcription factors.

study, the only advantage being that a genetic map is easier to construct with an animal which, unlike humans, can be subjected to experimental breeding programs (Section 2.3.2). Despite the limitations of human pedigree analysis, progress has been more rapid in mapping the human genome than those of any of our close relatives, so in this respect comparative genomics is proving more useful in mapping the animal genomes rather than our own. This is in itself a useful corollary to the Human Genome Project because it is revealing animal homologs of human genes involved in diseases, providing animal models in which these diseases can be studied.

Mapping is significantly easier with a small genome compared with a large one. This means that if one member of a pair of syntenic genomes is substantially smaller than the other, then mapping studies with this small genome is likely to provide a real boost to equivalent work with the larger genome. The pufferfish, *Fugu rubripes*, has been proposed in this capacity with respect to the human genome. The pufferfish genome is just 400 Mb, less than one-seventh the size of the human genome but containing approximately the same number of genes. The mapping work carried out to date with the pufferfish indicates that there is some similarity with the human gene order, at least over short distances. This means that it should be possible, to a certain extent, to use the pufferfish map position to find human homologs of sequenced pufferfish genes, and *vice versa*. This may be useful in locating unmapped human genes, but holds greatest promise in locating essential sequences such as promoters and other regulatory signals upstream of human genes. This is because these signals are likely to be similar in the two genomes, and recognizable because they are surrounded with noncoding DNA that has diverged quite considerably by random mutations (Elgar *et al.*, 1996).

One area where comparative genomics has a definite advantage is in the mapping of plant genomes. Wheat provides a good example. Wheat is the most important food plant in the human diet, being responsible for approximately 20% of the human calorific intake, and is therefore one of the crop plants that we most wish to study and possibly manipulate in the quest for improved crops. Unfortunately, the wheat genome is huge at 17 000 Mb, more than five times larger than even the human genome. A small model genome with a gene order similar to that of wheat would therefore be useful as a means of mapping desirable genes which might then be obtained from their equivalent positions in the wheat genome. Wheat, and other cereals such as rice, are members of the *Gramineae*, a large and diverse family of grasses. The rice genome is only 400 Mb, substantially smaller than that of wheat, and there are probably other grasses with even smaller genomes. Comparative mapping of the rice and wheat genomes has revealed many similarities, and the possibility therefore exists that genes from the wheat genome might be isolated by first mapping the positions of the equivalent genes in a smaller *Gramineae* genome (Gale and Devos, 1998).

5.3.2 Comparative genomics in the study of human disease genes

One of the main reasons for sequencing the human genome is to gain access to the sequences of genes involved in human disease. The hope is that the sequence of a disease gene will provide an insight into the biochemical basis of the disease and hence indicate a way of preventing or treating the disease. Comparative genomics has an important role to play in the study of disease genes because the discovery of a homolog of a human disease gene in a second organism is often the key to understanding the biochemical function of the human gene. If the homolog has already been characterized then the information needed to understand the biochemical role of the human gene may already be in place; if it is not then the necessary research can be directed at the homolog.

To be useful in the study of disease-causing genes, the second genome does not need to be syntenic with the human genome, nor even particularly closely related. *Drosophila* holds great promise in this respect, as the phenotypic effects of many *Drosophila* genes are already well known, so the data already exist for inferring the mode of action of human disease genes that have homologs in the *Drosophila* genome (Guffanti *et al.*, 1997). But the greatest success has been with yeast. Several human disease genes have homologs in the *S. cerevisiae* genome (*Table 5.4*). These disease genes include ones involved in cancer, cystic fibrosis and neurological syndromes, and in several cases the yeast homolog has a known function that provides a clear indication of the biochemical activity of the human gene. In some cases it has even been possible to demonstrate a physiological similarity between the gene activity in humans and yeast. For example, the yeast gene *SGS1* is a homolog of a human gene involved in the premature aging disease called Werner's syndrome. Yeasts with a mutant *SGS1* gene live for shorter periods than normal yeasts and display accelerated onset of aging indicators such as sterility (Sinclair *et al.*, 1997). In yeast the underlying cause of the premature aging, and possibly the direct effect of the mutation, appears to be a change in the distribution of proteins in the nucleolus, a subcompartment of the nucleus. The link between *SGS1* and the Werner's syndrome gene, provided by comparative genomics, means that these findings are directly relevant to study of the human disease.

5.4 FROM GENOME TO CELL

Even if every gene in a genome can be identified and assigned a function, a challenge still remains. This is to understand how the genome as a whole operates within the cell, specifying and coordinating the various biochemical activities that take place. Describing and explaining this aspect of genome biology will be a mammoth task that will occupy researchers for many decades

Table 5.4 Examples of human disease genes that have homologs in *S. cerevisiae*

Human disease gene	Yeast homolog	Function of the yeast gene
Amyotrophic lateral sclerosis	*SOD1*	Protection against superoxide (O_2^-)
Ataxia telangiectasia	*TEL1*	Codes for a protein kinase
Colon cancer	*MSH2, MLH1*	DNA repair
Cystic fibrosis	*YCF1*	Metal resistance
Myotonic dystrophy	*YPK1*	Codes for a protein kinase
Type I neurofibromatosis	*IRA2*	Codes for a regulatory protein
Werner's syndrome	*SGS1*	Nucleolus protein distribution
Wilson's disease	*CCC2*	Copper transport?

Data taken from Bassett *et al.* (1996) and Sinclair *et al.* (1997).

of the 21st century. Attempts are currently being made to chart the pattern of gene expression – which genes are switched on and which are switched off – in different tissues, during different developmental stages and, in the case of humans, in different disease states.

5.4.1 Patterns of gene expression

All organisms regulate expression of their genes, so that those genes whose protein products are not needed at a particular time are switched off. Understanding which genes are active in particular tissues at particular times, and assessing the relative degrees of activity of the genes that are switched on, is the first step in understanding how the genome as a whole functions within the cell.

Gene expression is assayed by measuring levels of RNA transcripts

The first stage of gene expression involves copying the gene into an RNA transcript. Identifying if a transcript of a gene is present in a cell or tissue is therefore the most direct way of determining whether the gene is switched on. This can be done by immobilizing the DNA copy of the gene on a solid support and hybridizing it with a sample of RNA extracted from the cells being studied. In fact the RNA is usually converted into cDNA prior to being used as the hybridization probe, as cDNA synthesis is the most convenient way of obtaining labeled sequences representing RNA molecules. If transcripts of the gene are present in the RNA extract then a hybridization signal will be seen (*Figure 5.20*).

Determining if a gene is being expressed is therefore a routine task. The complications become apparent when one appreciates that, in order to assay the entire mRNA content of a cell (referred to as its **transcriptome**), this basic procedure has to be repeated for every protein-coding gene in the genome. The answer is to use a DNA chip (Technical Note 2.3, p. 23) or other kind of microarray. In an early experiment, all 6000 *S. cerevisiae* genes were immobilized as an 80 × 80 array in a 18 mm by 18 mm square on a glass slide, and changes in gene expression that accompany the transition from aerobic to anaerobic respiration assayed, using fluorescently labeled cDNA with hybridization of the cDNA to the microarray being visualized by scanning confocal microscopy (*Figure 5.21*; DeRisi *et al.*, 1997). Twofold or greater changes in expression level, either up or down, were recorded for 1740 genes, over a quarter of the total, including many orphans whose functions are not known. A more recent example of this approach to studying the yeast transcriptome is described in Research Briefing 5.2.

Not surprisingly, applying these and equivalent techniques to studies of human gene expression patterns has proved more difficult. Most effort has been directed at cancerous tissues because understanding the abnormal features of gene expression in these cells might lead to a means of diagnosing the disease state and monitoring progress during treatment, and might even point the way towards new therapies. One study discovered 289 genes whose expression patterns differed significantly when normal colon epithelial cells were compared with colon cancer cells, about half of these genes also showing abnormal expression levels in pancreatic cancer cells (Zhang *et al.*, 1997). The implication is that some genes are abnormally expressed in a range of cancers and some are specific for a particular type of cancerous tissue.

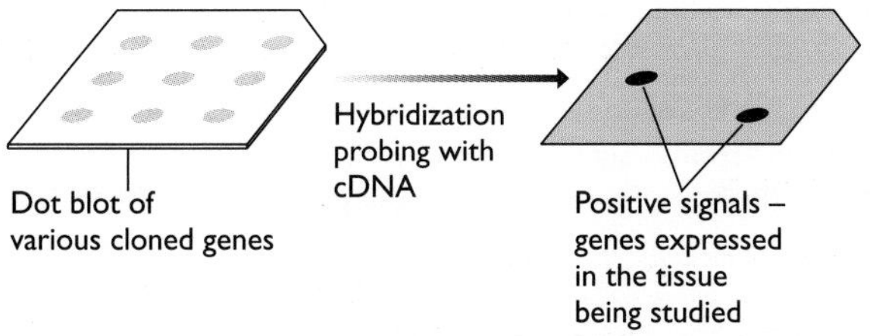

Figure 5.20 Assaying gene expression by dot blot hybridization analysis.

The objective is to determine which genes are expressed in a particular cell or tissue. Clones of the genes being tested are spotted on to a nylon membrane by the procedure called **dot blotting**. The RNA sample is converted into labeled cDNA and hybridized to the membrane. Positive signals indicate which genes are expressed.

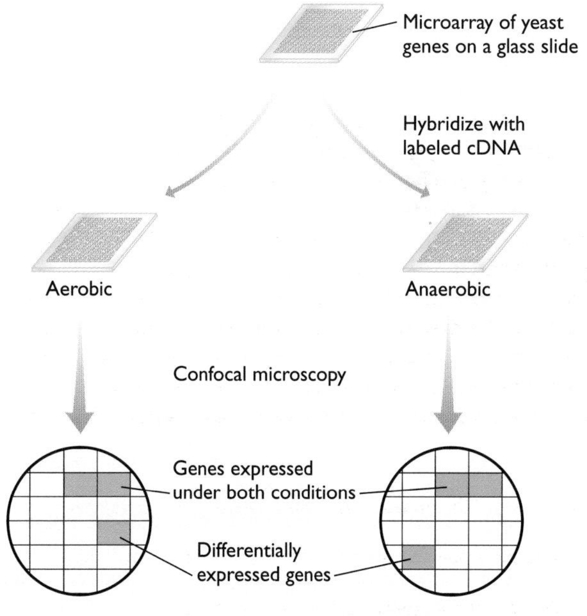

Figure 5.21 Assaying gene expression with a microarray.

As in Figure 5.20, the objective is to determine which genes are expressed in a particular cell or tissue. This time the gene sequences are immobilized as a microarray, which enables many more genes to be tested in one experiment. The cDNA is fluorescently labeled and hybridization detected by confocal microscopy. Compare this analysis with the example of the use of a DNA chip described in Technical Note 2.3, p. 23.

RESEARCH 5.2 BRIEFING

Patterns of gene expression during yeast sporulation

DNA microarrays have been used to follow the expression pattern of nearly every gene in the *Saccharomyces cerevisiae* genome during the developmental pathway that results in formation of spores.

Spore formation, or sporulation, is the key developmental pathway during the life cycle of the yeast *Saccharomyces cerevisiae*. It begins when the environmental conditions induce a diploid cell to undergo meiosis and produce four haploid spores (see *Figure 2.10*, p. 26), two of the a mating type and two α. After germination, each spore is able to fuse with a partner of the opposite mating type to produce another diploid cell. This is the sexual reproduction cycle for yeast.

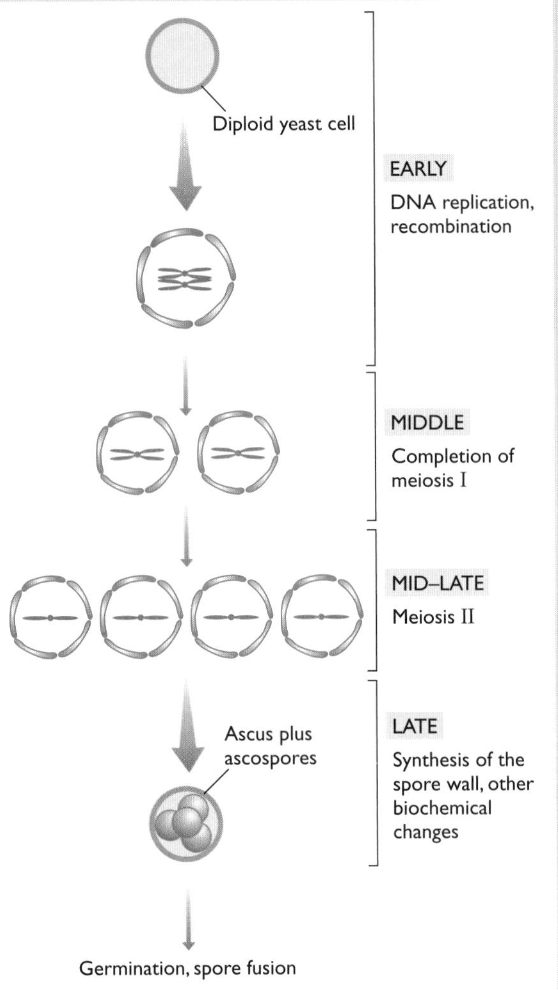

Gene expression during spore formation

Sporulation was extensively studied in yeast in the years before the complete genome sequence was available. The process had been divided into four stages – early, middle, mid–late and late, as shown in the diagram – with each stage characterized by expression of a different set of genes. Various methods had been used to identify some of these sporulation-specific genes, including examination of mutant yeast cells that are unable to form spores and the use of northern hybridization (Technical Note 5.1, p. 91) to determine which genes are transcribed into mRNA during sporulation. Altogether, approximately 150 sporulation-specific genes had been found.

The yeast genome contains 6000 protein-coding genes, about 2400 of which have no known function. Is it possible that some of these unidentified genes are involved in sporulation? And could some of the genes whose protein products are known have a previously unsuspected role at some stage in spore formation? The development of DNA microarray technology (Technical Note 2.3, p. 23), provides a new approach to answering these questions, one that is much more powerful than the traditional methods used during the pregenome era of yeast biology.

DNA microarrays and gene expression

Messenger RNA preparations were made from yeasts at various times after the onset of sporulation (induced by transferring cells to a nitrogen-deficient medium), and each preparation was labeled and then hybridized to a microarray containing the 6000 yeast genes. The hybridization signal associated with each gene was then used to assess the amount of that particular mRNA in each preparation, and hence to follow the expression pattern of each gene during sporulation.

As expected, the microarray approach resulted in a significantly more detailed description of gene expression during yeast sporulation than had previously been possible. The patterns of gene activity indicated that the early stage of sporulation is subdivided into three distinct phases – the early (I), early (II) and early–middle phases – each with its own characteristic set of expressed genes. In total, 256 genes specific to these early periods were identified, several of these possessing the sequence (5′-GGCGGC-3′) approximately 600 bp upstream of their initiation codons. This 'URS1 consensus sequence' is thought to be a recognition signal for a transcription factor (Section 8.3) involved in regulating the changeover from vegetative growth to sporulation. Another 158 genes were shown to be active during the middle phase of sporulation, 70% of these genes having an 'MSE consensus sequence', again suggesting coregulation by a single transcription factor. A further 61 genes were expressed during the mid–late phase and five during the late phase. An additional 600 genes that are repressed during sporulation were identified, these presumably coding for proteins that are needed during vegetative growth but whose synthesis must be switched off when spores are being formed.

The future for microarray and chip technology

As well as its importance in delineating the pattern of gene expression during yeast sporulation, this project is also significant as a demonstration of the potential of microarray technology in the functional analysis of genomes. We can expect DNA microarrays and chips to be used in similar ways in analysis of the recently completed genome of the nematode worm *Caenorhabditis elegans*, and in future years these technologies will also play important roles in the post-sequencing phase of the Human Genome Project.

Reference

Chu S, DeRisi J, Eisen M, Mulholland J, Botstein D, Brown PO and Herskowitz I (1998) The transcriptional program of sporulation in budding yeast. *Science*, **282**, 699–705.

The proteome

In the second stage of gene expression, mRNA transcripts are translated into protein, the entire protein content of the cell being referred to as the **proteome**. Determining the identities and relative amounts of the proteins in a cell or tissue is more difficult than transcript studies because techniques for protein characterization are less well advanced than the equivalent RNA procedures. Delineating the proteome is, however, one of the keys to understanding how the genome as a whole is expressed. Techniques at present are based around two-dimensional polyacrylamide gel electrophoresis to separate proteins, or peptides derived from proteins by proteolytic cleavage, followed by mass spectrometry to determine the amino acid compositions of the protein or peptide in each spot on the two-dimensional gel (Yates, 1998). This information can be related to the sequences of all the protein-coding genes in the genome in order to identify which gene specifies which spot. A change in the proteome due, for example, to cellular differentiation, is revealed by a change in the pattern of spots.

REFERENCES

Bassett DE, Boguski MS and Hieter P (1996) Yeast genes and human disease. *Nature*, **379**, 589–590.

Bird A (1986) CpG-rich islands and the function of DNA methylation. *Nature*, **321**, 209–213.

Blattner FR, Plunkett G, Bloch CA, et *al.* (1997) The complete genome sequence of *Escherichia coli* K-12. *Science*, **277**, 1453–1462.

Church DM, Stotler CJ, Rutter JL, Murrell JR, Trofatter JA and Buckler AJ (1994) Isolation of genes from complex sources of mammalian genomic DNA using exon amplification. *Nature Genet.*, **6**, 98–105.

Clackson T and Wells JA (1994) *In vitro* selection from protein and peptide libraries. *Trends Biotechnol.*, **12**, 173–184.

DeRisi JL, Iyer VR and Brown PO (1997) Exploring the metabolic and genetic control of gene expression on a genomic scale. *Science*, **278**, 680–686.

Dujon B (1996) The yeast genome project: what did we learn? *Trends Genet.*, **12**, 263–270.

Elgar G, Sandford R, Aparicio S, Macrae A, Vekatesh B and Brenner S (1996) Small is beautiful: comparative genomics with the pufferfish (*Fugu rubripes*). *Trends Genet.*, **12**, 145–150.

Evans MJ, Carlton MBL and Russ AP (1997) Gene trapping and functional genomics. *Trends Genet.*, **13**, 370–374.

Fickett JW (1996) Finding genes by computer: the state of the art. *Trends Genet.*, **12**, 316–320.

Fields S and Sternglanz R (1994) The two-hybrid system: an assay for protein–protein interactions. *Trends Genet.*, **10**, 286–292.

Frech K, Quandt K and Werner T (1997) Finding protein-binding sites in DNA sequences: the next generation. *Trends Biochem. Sci.*, **22**, 103–104.

Frohman MA, Dush MK and Martin GR (1988) Rapid production of full-length cDNAs from rare transcripts: amplification using a single gene-specific oligonucleotide primer. *Proc. Natl Acad. Sci. USA*, **85**, 8998–9002.

Gale MD and Devos KM (1998) Plant comparative genetics after 10 years. *Science*, **282**, 656–659.

Guffanti A, Banfi S, Simon G, Ballabio A and Borsani G (1997) DRES search engine: of flies, men and ESTs. *Trends Genet.*, **13**, 79–80.

Lovett M (1994) Fishing for complements: finding genes by direct selection. *Trends Genet.*, **10**, 352–357.

Nadeau JH and Sankoff D (1998) Counting on comparative maps. *Trends Genet.,* **14**, 495–501.

Oliver SG (1996a) From DNA sequence to biological function. *Nature,* **379**, 597–600.

Oliver SG (1996b) A network approach to the systematic analysis of yeast gene function. *Trends Genet.,* **12**, 241–242.

Ponting CP (1997) Tudor domains in proteins that interact with RNA. *Trends Biochem. Sci.,* **22**, 51–52.

Simonet WS, Lacey DL, Dunstan CR, *et al.* (1997) Osteoprotegrin: a novel secreted protein involved in the regulation of bone density. *Cell,* **89**, 309–319.

Sinclair DA, Mills K and Guarente L (1997) Accelerated aging and nucleolar fragmentation in yeast *sgs1* mutants. *Science,* **277**, 1313–1316.

Wach A, Brachat A, Pohlmann R and Philippsen P (1994) New heterologous modules for classical or PCR-based gene disruptions in *Saccharomyces cerevisiae. Yeast,* **10**, 1793–1808.

Yates JR (1998) Mass spectrometry and the age of the proteome. *J. Mass Spectrom.,* **33**, 1–19.

Zhang L, Zhou W, Velculescu VE, Kern SE, Hruban RH, Hamilton SR, Vogelstein B and Kinzler KW (1997) Gene expression in normal and cancer cells. *Science,* **276**, 1268–1272.

FURTHER READING

Lander ES (1996) The new genomics: global views of biology. *Science,* **274**, 536–539. — *A commentary on 'post-genome' research including RNA transcript analysis and proteome studies.*

O'Brien SJ, Weinberg J and Lyons LA (1997) Comparative genomics: lessons from cats. *Trends Genet.,* **13**, 393–399.— *A review that draws out the importance of comparative methods in studies of mammalian genomes.*

Oliver SG, Winson MK, Kell DB and Baganz F (1998) Systematic functional analysis of the yeast genome. *Trends Biotechnol.,* **16**, 373–378. — *Describes the various approaches being taken to assign functions to yeast genes.*

Smith TF (1998) Functional genomics – bioinformatics is ready for the challenge. *Trends Genet.,* **14**, 291–293. — *A review of the applications of computer-based studies in analysis of genome sequences.*

Tang CM, Hood DW and Moxon ER (1997) *Haemophilus* influence: the impact of whole genome sequencing on microbiology. *Trends Genet.,* **13**, 399–404. — *Summarizes research areas that have been stimulated by the completion of microbial genome sequences.*

PART

2 How Genomes Function

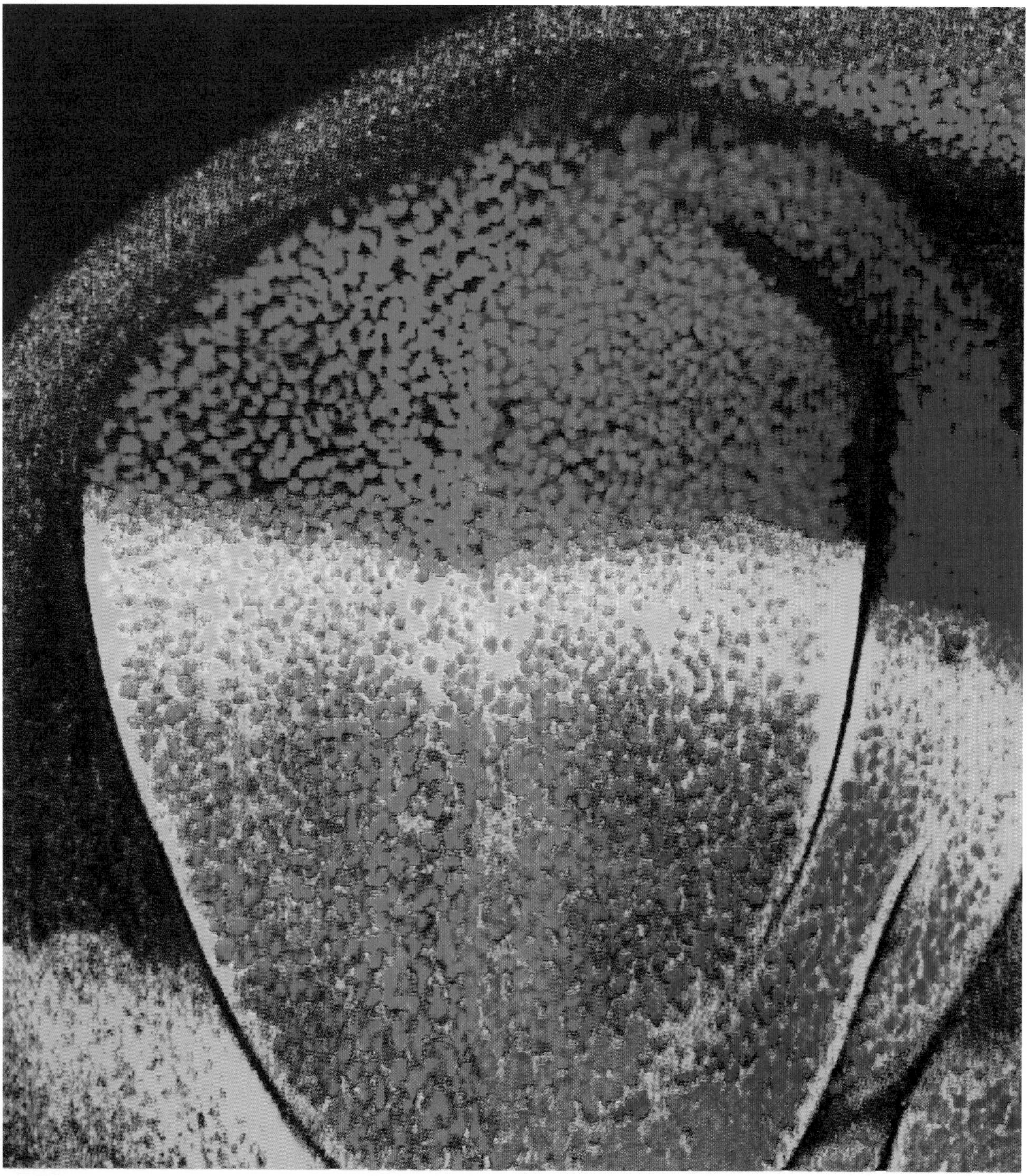

Chapter 6
Genome Anatomies

Chapter 7
The Role of DNA-binding Proteins

Chapter 8
Initiation of Transcription: the First Step in Gene Expression

Chapter 9
Synthesis and Processing of RNA

Chapter 10
Synthesis and Processing of the Proteome

Chapter 11
Regulation of Genome Activity

Opposite page: Gene expression during wing development in the fruit fly *Drosophila melanogaster*. Cells expressing the developmental genes *apterous* (blue), *vestigial* (red) and *CiD* (green) have been visualized by staining with fluorescently-labeled antibodies that bind to the proteins coded by these genes. The purple regions show the overlap between the expression regions for *apterous* and *vestigial*. See Section 11.3.3 for more on *Drosophila* development.

Acknowledgement: Reproduced with kind permission of J. Williams, S. Paddock and S. Carroll (Howard Hughes Medical Institute, University of Wisconsin, USA).

The image was supplied by Bio-Rad Microscience Ltd.

CHAPTER

6 Genome Anatomies

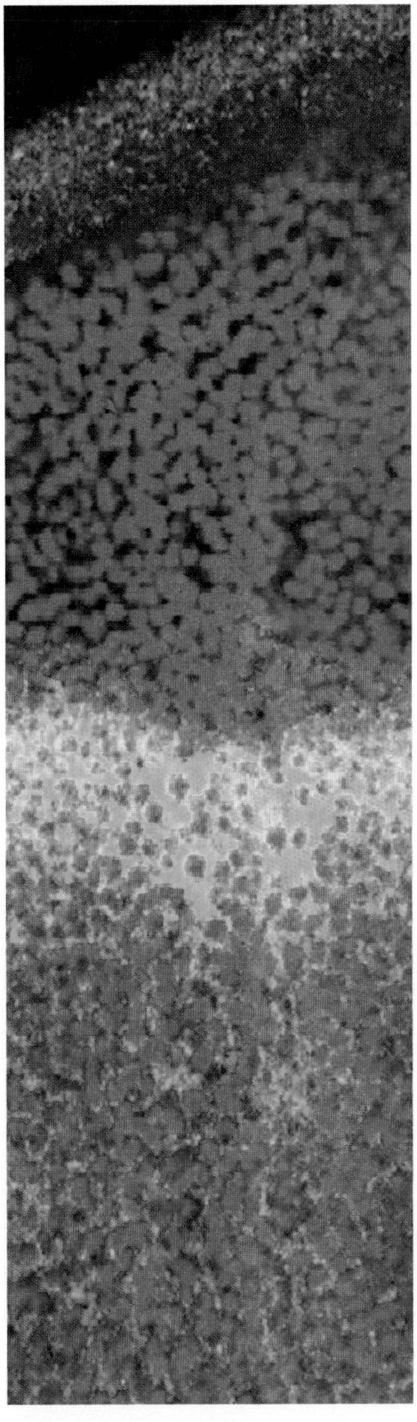

Contents

6.1 The Anatomy of the Eukaryotic Genome **116**
6.1.1 Eukaryotic nuclear genomes 117
6.1.2 Eukaryotic organelle genomes 125
6.2 The Anatomy of the Prokaryotic Genome **128**
6.2.1 The physical structure of the prokaryotic genome 128
6.2.2 The genetic organization of the prokaryotic genome 132
6.3 The Repetitive DNA Content of Genomes **134**
6.3.1 Tandemly repeated DNA 136
6.3.2 Interspersed genome-wide repeats 137
6.3.3 Repetitive DNA and eukaryotic chromosome architecture 140

Concepts

- *Eukaryotic nuclear genomes vary in length from less than 10 Mb to over 100 000 Mb*
- *DNA is packaged in the nucleus by association with histone proteins in chromatin*
- *Many nuclear genes are members of multigene families*
- *The mitochondria and chloroplasts of eukaryotes have their own genomes, but they do not contain many genes*
- *Most, but not all, bacterial genomes are circular DNA molecules*
- *Gene families called operons are common in bacterial genomes*
- *Eukaryotic genomes contain clustered and dispersed repetitive DNA sequences*
- *Many dispersed repeats, in eukaryotes and bacteria, are active or ancient transposable elements*

IN PART 1 we looked at the various methods that are used to study genomes. We discovered that there are many ways of mapping genomes and that these, combined with DNA sequencing, have already resulted in complete nucleotide-by-nucleotide descriptions for several species. But by the end of Chapter 5 we had realized that mapping and sequencing projects can only take us so far. A sequence provides the foundation for understanding a genome, but it is not an end in itself.

To understand how genomes function – what they do and how they do it – the results of mapping and sequencing experiments must be combined with information about the activities of genomes in living cells. In Part 2 we will examine these activities. We will draw on a great deal of information that was obtained during the pre-genome era, the period up to the mid-1990s when molecular biologists, in the absence of complete genome sequences, were forced to study the activities of individual genes or, at most, small groups of genes that work together. These studies expanded the simple description of 'DNA makes RNA makes protein', first stated by Crick in 1958, into the more complex series of events that we now recognize as being involved in the gene expression pathway (see *Figure 1.2*, p. 4). Work continues in the realms of biochemistry and molecular cell biology to describe gene expression in ever increasing detail and many exciting new discoveries will undoubtedly be made in the future, especially in the area of gene regulation. But the major challenge for molecular biology has now moved on from description of individual gene activities to the more general questions regarding expression and regulation of the genome as a whole. These are the issues that will dominate research during the early decades of the new millennium.

In Part 2 we will discover that genome expression depends on interactions between DNA-binding proteins and specific nucleotide sequences within the genome (Chapter 7), these interactions determining which genes are expressed at a particular time (Chapter 8). In Chapters 9 and 10 we will see how genome expression leads to synthesis of RNA and protein, and Chapter 11 will describe how changes in genome expression, and the resulting changes in cellular activities, underlie the processes of differentiation and development. First, however, we must examine what mapping, sequencing and other studies have revealed about the anatomies – the organizations and contents – of the genomes of different organisms.

6.1 THE ANATOMY OF THE EUKARYOTIC GENOME

We have already learnt that the human genome is split into two components: the nuclear genome and the mitochondrial genome (see *Figure 1.3*, p. 5). This is the typical pattern for most eukaryotes, the bulk of the genome being contained in the chromosomes in the cell nucleus with a much smaller part located in the mitochondria and, in the case of photosynthetic organisms, in the chloroplasts. We will look first at the nuclear genome.

6.1.1 Eukaryotic nuclear genomes

Eukaryotic nuclear genomes range in size from less than 10 Mb to over 100 000 Mb (see *Table 1.1*, p. 9). Genome size broadly coincides with organism complexity, the genomes of higher eukaryotes being larger than those of lower eukaryotes, but size is determined not only by the number of genes in the genome but also by the amount of repetitive DNA, the larger genomes tending to be ones in which the copy numbers of the repeat sequences are highest (see *Figure 1.7*, p. 10). We will examine the various types of repetitive DNA later in this chapter (Section 6.3).

The nuclear genome is split into a set of linear DNA molecules, each contained in a chromosome. No exceptions to this pattern are known: all eukaryotes that have been studied have at least two chromosomes and the DNA molecules are always linear. This point is worth noting because later we will learn that organelle and bacterial DNA molecules are more variable in structure. The only variability at this level of eukaryotic genome structure lies with chromosome number, which appears to be unrelated to the biological features of the organism (*Table 6.1*). For example, the yeast *Saccharomyces cerevisiae*, a relatively simple eukaryote, has four times as many chromosomes as the fruit fly *Drosophila melanogaster*. Nor is chromosome number linked to genome size: some salamanders have genomes 30 times bigger than the human version but split into half the number of chromosomes. These comparisons are interesting but at present do not tell us anything useful about the genomes themselves: they are more a reflection of the nonuniformity of the evolutionary events that have shaped genome architecture in different organisms.

Packaging of DNA into chromosomes

Chromosomes are much shorter than the DNA molecules that they contain: the average human chromosome has just under 5 cm of DNA. A highly organized packaging system is therefore needed to fit a DNA molecule into its chromosome. We must understand this packaging system before we start to think about how genomes function because the nature of the packaging has an influence on the processes involved in expression of individual genes (Section 8.1).

The important breakthroughs in understanding DNA packaging were made in the early 1970s by a combination of biochemical analysis and electron microscopy. It was already known that nuclear DNA is associated with DNA-binding proteins called **histones** but the exact nature of the association had not been delineated. In 1973–74 several groups carried out **nuclease protection experiments** on **chromatin** (DNA–histone complexes) that had been gently extracted from nuclei by methods designed to retain as much of the chromatin structure as possible. In a nuclease protection experiment the complex is treated with an enzyme that cuts the DNA at positions that are not 'protected' by attachment to a protein. The sizes of the resulting DNA fragments indicate the positioning of the protein complexes on the original DNA molecule (*Figure 6.1*). After limited nuclease treatment of purified chromatin the bulk of the DNA fragments had lengths of approximately 200 bp and multiples thereof, suggesting a regular spacing of histone proteins along the DNA.

In 1974 these biochemical results were supplemented by electron micrographs of purified chromatin, which enabled the regular spacing inferred by the protection experiments to be visualized as beads of protein on the string of DNA (*Figure 6.2A*). Further biochemical analysis indicated that each bead, or **nucleosome**, contains eight histone protein molecules, these being two each of histones H2A, H2B, H3 and H4. Structural studies have shown that these eight proteins form a barrel-shaped octamer with the DNA wound twice around the outside (*Figure 6.2B*). Between 140 and 150 bp of DNA are associated with the nucleosome particle depending on species, and each nucleosome is separated by 50–70 bp of linker DNA, giving the repeat length of 190–220 bp previously shown by the nuclease protection experiments.

As well as the proteins of the core octamer, there is a group of additional histones, all closely related to one another and collectively called **linker histones**. In vertebrates these include histones H1a to H1e, H1°, H1t and H5. A single linker histone is attached to each nucleosome, but its precise positioning is not known. Recent structural studies support the traditional model in which the linker histone acts as a clamp, preventing the coiled DNA from detaching from the nucleosome (*Figure 6.2C*; Zhou *et al.*, 1998), but other results suggest that, at least in some organisms, the linker histone is not located on the extreme surface of the nucleosome–DNA assembly, as expected if it really is a clamp, but instead is inserted between the core octamer and the DNA (Pruss *et al.*, 1995; Pennisi, 1996).

The beads-on-a-string structure shown in *Figure 6.2A* is thought to represent an unpacked form of chromatin that occurs only infrequently in living nuclei. Very gentle cell

Table 6.1 Chromosome numbers for various organisms

Organism	Genome size (Mb)	Number of chromosomes
Saccharomyces cerevisiae	12.1	16
Drosophila melanogaster	140	4
Human	3 000	23
Maize	5 000	10
Salamander	90 000	12

The number of chromosomes refers to a haploid cell, which contains a single copy of the genome.

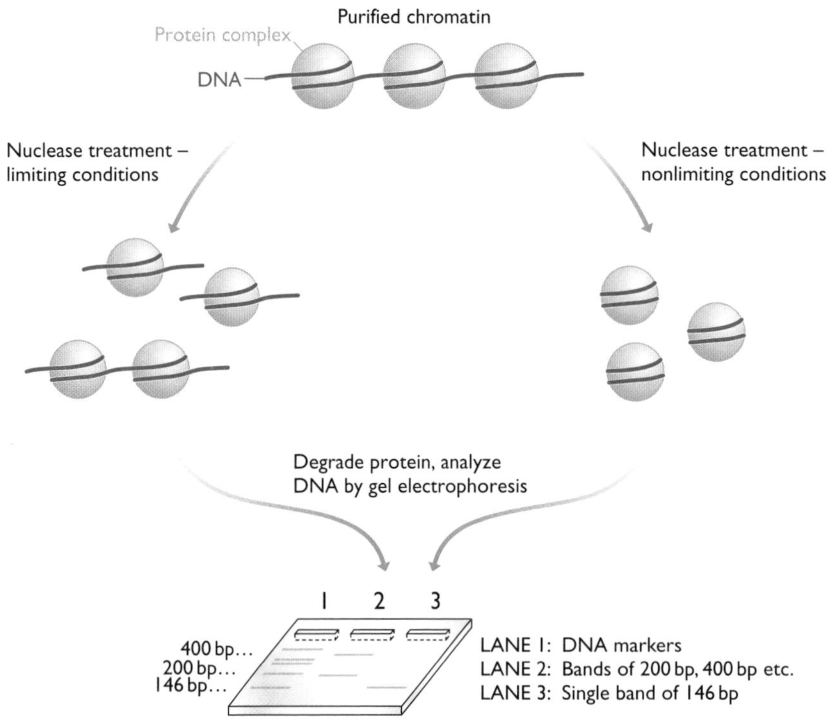

Figure 6.1 Nuclease protection analysis of chromatin from human nuclei.

Chromatin is purified from nuclei under low salt conditions (<1 mM NaCl) and treated with a nuclease enzyme. On the left, the nuclease treatment is carried out under limiting conditions so that the DNA is cut, on average, just once in each of the linker regions between the bound proteins. After removal of the protein the DNA fragments are analyzed by agarose gel electrophoresis and found to be 200 bp in length, or multiples thereof. On the right, the nuclease treatment proceeds to completion, so all the DNA in the linker regions is digested. The remaining DNA fragments are all 146 bp in length. The results show that in this form of chromatin, protein complexes are spaced along the DNA at regular intervals, one for each 200 bp, with 146 bp of DNA closely attached to each protein complex.

breakage techniques developed in the mid 1970s resulted in a more condensed version of the complex called the **30 nm fiber** (it is approximately 30 nm in width). The exact way in which nucleosomes associate to form the 30 nm fiber is not known, but several models have been proposed, the most popular of which is the solenoid structure shown in *Figure 6.3*. The individual nucleosomes within the 30 nm fiber may be held together by interactions between the linker histones, or the attachments might involve the core histones, whose protein 'tails' extend outside the nucleosome (see *Figure 8.5*, p. 176). The latter hypothesis is attractive because chemical modification of these tails results in the 30 nm fiber opening up, enabling genes contained within it to be activated (Section 8.1).

The special features of metaphase chromosomes

The 30 nm fiber is probably the major type of chromatin in the nucleus during interphase, the period between divisions (Section 8.1). When the nucleus divides, the DNA adopts a more compact form of packaging, resulting in the highly condensed **metaphase chromosomes** that can be seen with the light microscope and which have the appearance generally associated with the word 'chromosome' (see *Figures 2.9*, p. 25 and *2.10*, p. 26). For many years it has been known that the individual metaphase chromosomes in a nucleus do not all look the same. As well as obvious structural differences, such as size and the position of the centromere relative to the two ends, more subtle differences are revealed when chromosomes are stained. There are a number of different staining techniques (*Table 6.2*), each resulting in a banding pattern that is characteristic for the individual chromosome. This means that the set of chromosomes possessed by an organism can be represented as a **karyogram**, in which the banded appearance of each one is depicted. The human karyogram is shown in *Figure 6.4*. The karyogram is, in essence, a physical map, though a very low resolution one.

Two components of the metaphase chromosome are functionally important. The first is the centromere, which not only holds the daughter chromosomes together after DNA replication (see *Figure 2.9*, p. 25), but also acts as the attachment point for microtubules which draw the daughters into their respective nuclei when the cell divides. The DNA from centromeres has been sequenced and shown to be made up principally of repetitive units, which in humans are 171 bp in length and called **alphoid**

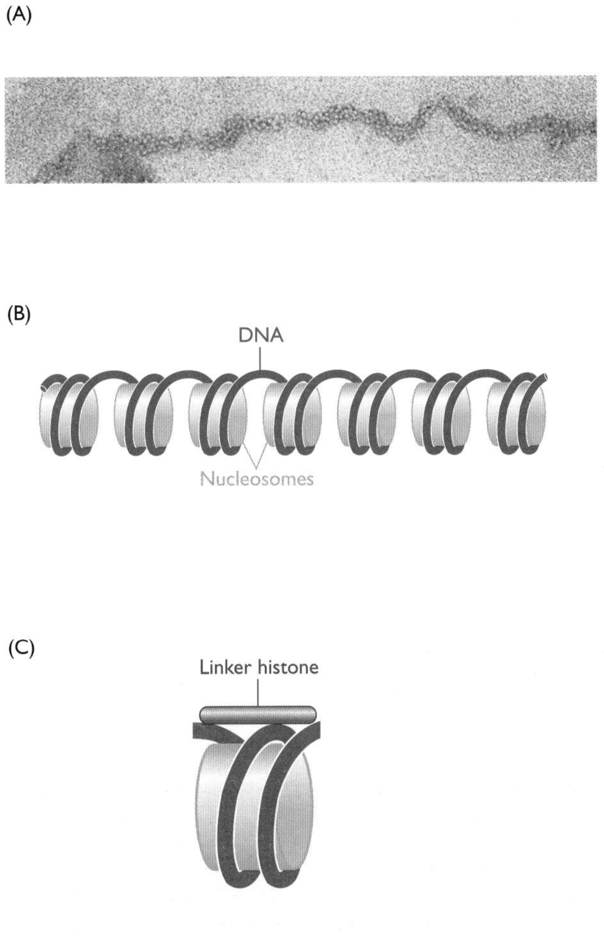

Figure 6.2 Nucleosomes.

(A) Electron micrograph of a purified chromatin strand showing the beads-on-a-string structure. Courtesy of Dr Barbara Hamkalo, University of California, Irvine. (B) The model for the beads-on-a-string structure in which each bead is a barrel-shaped nucleosome with the DNA wound twice around the outside. Each nucleosome is made up of eight proteins: a central tetramer of two histone H3 and two histone H4 subunits, plus a pair of H2A–H2B dimers, one above and one below the central tetramer (see *Figure 8.5*, p. 176). (C) The precise position of the linker histone relative to the nucleosome is not known, but as shown here the linker histone may act as a clamp, preventing the DNA from detaching from the outside of the nucleosome.

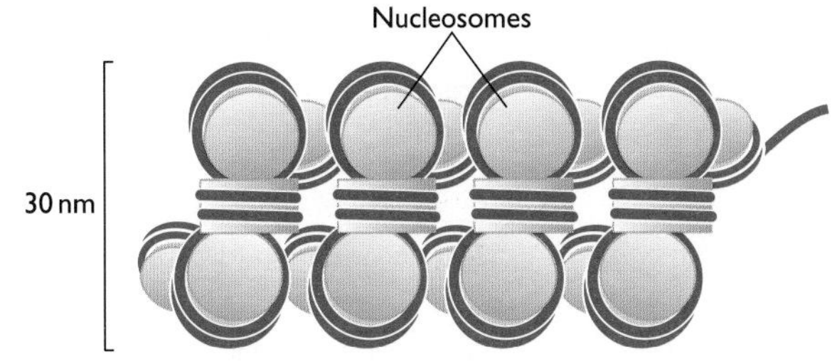

Figure 6.3 The solenoid model for the 30 nm chromatin fiber.

In this model, the beads-on-a-string structure of chromatin is condensed by winding the nucleosomes into a helix with six nucleosomes per turn. Higher levels of chromatin packaging will be described in Section 8.1.

DNA. Studies of abnormal chromosomes in which segments of centromeric DNA have been deleted suggest that about 500 kb of DNA, mainly comprising these alphoid repeats, is needed for complete centromere function. The individual nature of this DNA is reflected by the presence in the centromere of at least seven proteins not found elsewhere in the chromosome. One of these proteins, CENP-A, is very similar to histone H3 and is thought to replace this histone in the centromeric nucleosomes. It is assumed that the small distinctions between CENP-A and H3 confer special properties on centromeric nucleosomes, but exactly what these properties might be, and how they relate to the function of the centromere, is not yet known. Part of the function of the centromere is revealed by the electron microscope, which shows that in a dividing cell a pair of plate-like **kinetochores** are present on the surface of the chromosome in the centromeric region. These structures act as the attachment points for the microtubules that radiate from the spindle pole bodies located at the nuclear surface and which draw the divided chromosomes into the daughter nuclei. Part of the kinetochore is made up of alphoid DNA plus CENP-A and other proteins, but its structure has not been described in detail (Vafa and Sullivan, 1997).

The second important chromosomal structure is the **telomere**, the terminal region of each chromatid. Telomeres are important because they mark the ends of chromosomes and therefore enable the cell to distinguish a real end from an unnatural end caused by chromosome breakage, an essential requirement as the cell must repair the latter but not the former. Telomeric DNA is made up of hundreds of copies of a repeated motif, 5′-TTAGGG-3′ in humans, with a short extension of the 3′ terminus of the double-stranded molecule (*Figure 6.5*). Special telomeric proteins are believed to bind to the repeat sequences and, as in the centromere region, the telomeric nucleosomes may have special structures. The result is thought to be a proteinaceous 'cap' that protects the chromosome ends.

Where are the genes in a genome?

When one looks at a set of human chromosomes under the microscope the inevitable question that springs to mind is 'Where exactly are the genes?'. The EST map produced by Schuler *et al.* (1996; see Research Briefing 3.1, p. 56) largely answers this question for the human genome and shows that the genes are not distributed evenly along the chromosomes but that each chromosome is made up of regions in which genes are abundant interspersed with regions of lower gene content.

This uneven gene distribution was suspected for several years before the EST map was published, based on

Table 6.2 Staining techniques used to produce chromosome banding patterns

Technique	Procedure	Banding pattern
G-banding	Mild proteolysis followed by Giemsa staining	Dark bands are AT-rich Pale bands are GC-rich
R-banding	Heat denaturation followed by Giemsa staining	Dark bands are GC-rich Pale bands are AT-rich
Q-banding	Stain with quinacrine	Dark bands are AT-rich Pale bands are GC-rich
C-banding	Denature with barium hydroxide then stain with Giemsa	Dark bands are constitutive heterochromatin (Section 8.1)

two related lines of evidence. The first of these is the fact that banding patterns are produced when chromosomes are stained. The dyes used in these procedures (see *Table 6.2*) bind to DNA molecules, but in most cases with preferences for certain base pairs. Giemsa, for example, has a greater affinity for DNA regions that are rich in A and T nucleotides. The dark G-bands in the human karyogram (see *Figure 6.4*) are therefore thought to be AT-rich regions of the genome. The base composition of the genome as a whole is 59.7% A+T so the dark G-bands must have AT contents substantially greater than 60%. This means that it is unlikely that there are many genes in dark G-bands, because genes generally have AT contents of 45–50%.

The second line of evidence pointing to uneven gene distribution derives from the **isochore** model of genome organization (Gardiner, 1996). According to this model, the genomes of vertebrates and plants (and possibly of lower eukaryotes also) are mosaics of segments of DNA, each at least 300 kb in length, each segment having a uniform base composition that differs from that of the adjacent segments. Support for the isochore model comes from experiments in which genomic DNA is broken into fragments of approximately 100 kb, treated with dyes that bind specifically to AT- or GC-rich regions, and the pieces separated by density gradient centrifugation (Technical Note 6.1). When this experiment is carried out with human DNA, five fractions are seen, each representing a different isochore type with its distinctive base composition: two AT-rich isochores called L1 and L2, and three GC-rich classes, H1, H2 and H3. The last of these, H3, is the least abundant in the human genome, making up only 3% of the total, but contains over 25% of the genes. This is a clear indication that genes are not distributed evenly through the human genome.

Do the light G-bands and H3 isochore regions indeed correspond to the gene-rich areas revealed by the EST map of Schuler *et al.* (1996)? This question is quite difficult to answer because the EST map has a much higher resolution than either the G-banding karyogram or the H3 isochore distribution map (Saccone *et al.*, 1992). Several of the chromosome regions with high EST contents appear to coincide with light G-bands and H3 areas but the correspondence is not precise. This does not invalidate the use of chromosome banding and isochore distribution as a means of identifying gene rich regions of genomes for which EST maps are not available, but it shows that the results are likely to be imprecise.

Box 6.1: Unusual chromosome types

The karyograms of some organisms display unusual features not illustrated by the human version. These include:

- **Minichromosomes**, which are relatively short in length but rich in genes. The chicken genome, for example, is split into 39 chromosomes: six **macrochromosomes** containing 66% of the DNA but only 25% of the genes, and 33 minichromosomes containing the remaining one-third of the genome and 75% of the genes. The gene density in the minichromosomes is therefore some six times greater than that in the macrochromosomes (McQueen *et al.*, 1998).
- **B chromosomes**, which are additional chromosomes possessed by some individuals in a population, but not all. They are common in plants (Jones, 1995) and also known in fungi, insects and animals. B chromosomes appear to be fragmentary versions of normal chromosomes that result from unusual events during nuclear division. Some contain genes, often for ribosomal RNAs, but it is not clear if these genes are active. The presence of B chromosomes can affect the phenotype of the organism, particularly in plants where they are associated with reduced viability. It is presumed that B chromosomes are gradually lost from cell lineages due to irregularities in their inheritance pattern.

Families of genes

As more and more information is obtained from sequencing projects we are beginning to realize that **multigene families**, groups of genes of identical or similar sequence,

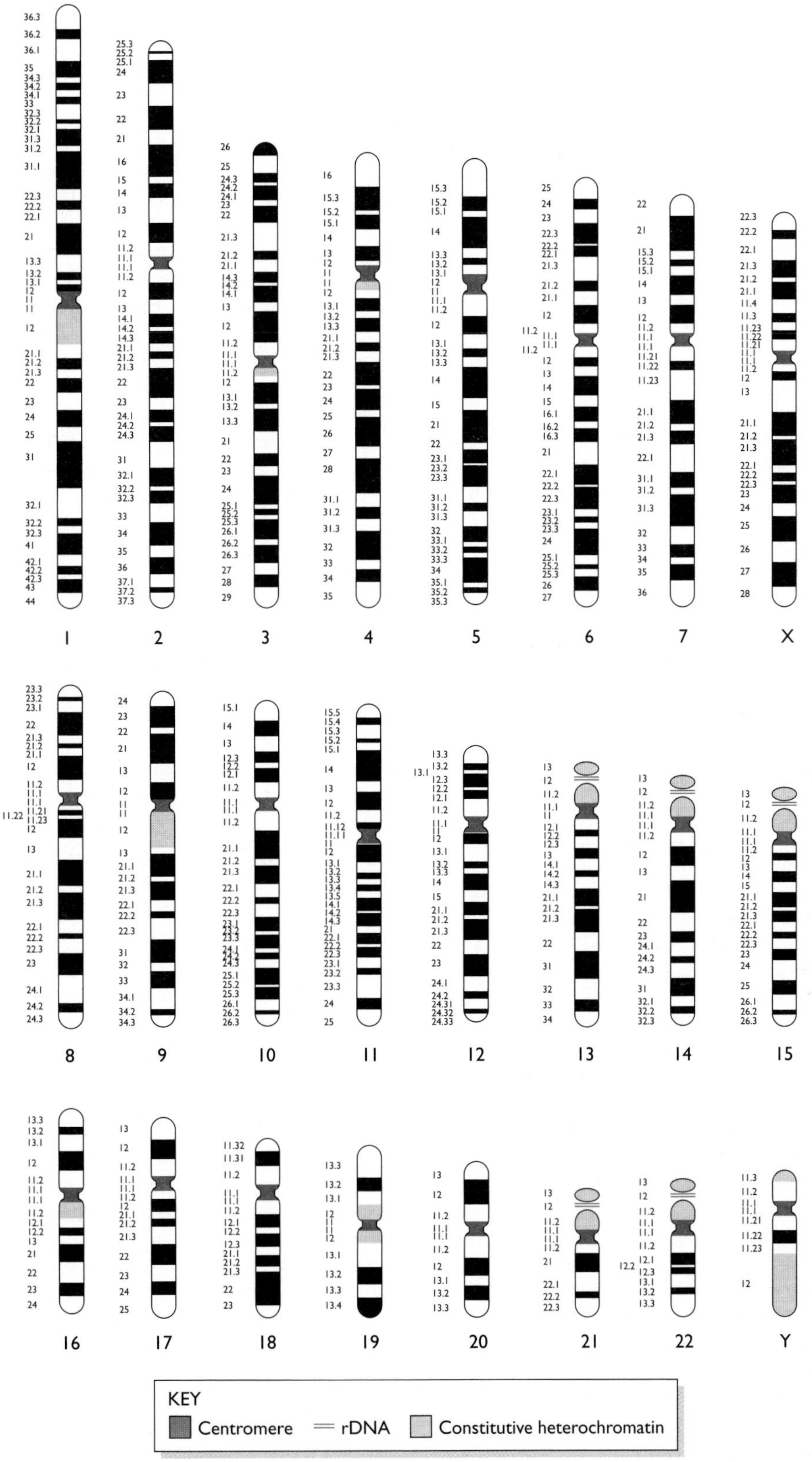

Figure 6.4 The human karyogram.

The chromosomes are shown with the G-banding pattern obtained after Giemsa staining. Chromosome numbers are given below each structure and the band numbers to the left. 'rDNA' is a region containing a cluster of repeat units for the 28S, 5.8S and 18S rRNA genes (see p. 122). 'Constitutive heterochromatin' is very compact chromatin that lacks genes (Section 8.1.1). Redrawn from Strachan and Read (1996).

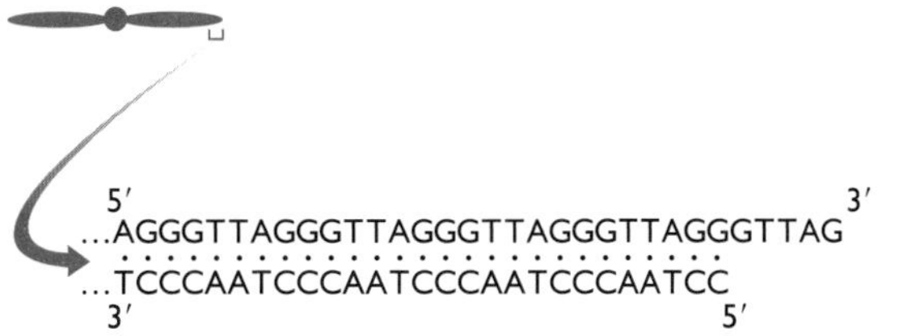

Figure 6.5 Telomeric DNA.

The length of the 3′-extension is different in each telomere. For more details about telomeric DNA, see Section 12.3.4.

Box 6.2: How many genes in the human genome?

The question of how many genes there are in the human genome has intrigued molecular biologists for over a decade. During the early 1990s the 'best guess' was approximately 50 000, but this figure gradually increased as DNA sequencing showed that more genes than previously suspected were members of multigene families. Extrapolations from the gene contents of large sequenced regions of the human genome give an estimate of 60 000–70 000 genes, a figure that agrees with the estimated gene number for the pufferfish genome, which is thought to be a good guide to the human genome (Section 5.3.1; Cohen, 1997). An alternative approach has been to estimate the number of CpG islands in human DNA (Section 5.1.1), on the basis that each CpG island is associated with a single gene and that 56% of all genes are associated with CpG islands. As there are about 45 000 CpG islands, this line of reasoning suggests there are 80 000 or so genes.

Several attempts have been made to estimate the gene number from the contents of EST libraries (Section 3.3.1) but the resulting figures vary widely, from as low as 60 000 to as high as 150 000 (Cohen, 1997). This is because it is difficult to be sure what fraction of the total gene number is represented in a single library, the calculation being further complicated by the possibility that some genes are represented by two or more nonoverlapping ESTs.

are common features of many genomes. For example, every eukaryote that has been studied (as well as all but the simplest bacteria) has multiple copies of the genes for the ribosomal RNAs (Section 9.1.1). This is illustrated by the human genome, which contains approximately 2000 genes for the 5S rRNA (so-called because it has a sedimentation coefficient of 5S: see Technical Note 6.1), all located in a single cluster on chromosome 1, and about 280 copies of a repeat unit containing the 28S, 5.8S and 18S rRNA genes, these units grouped into five clusters, each of 50–70 repeats, one on each of chromosomes 13, 14, 15, 21 and 22 (see *Figure 6.4*). Ribosomal RNAs are components of the protein-synthesizing particles called **ribosomes**, and it is presumed that their genes are present in multiple copies because of the heavy demand for rRNA synthesis during cell division, when several tens of thousands of new ribosomes must be assembled.

The rRNA genes are examples of 'simple' or 'classical' multigene families, in which all the members have identical or nearly identical sequences. These families are believed to have arisen by gene duplication with the sequences of the individual members kept identical by an evolutionary process that, as yet, has not been fully described (but which we will return to in Section 14.2.1). Other multigene families, more common in higher eukaryotes than lower eukaryotes, are called 'complex' because the individual members, although similar in sequence, are sufficiently different for the gene products to have distinctive properties. One of the best examples of this type of multigene family is provided by the mammalian globin genes. The globins are the blood proteins that combine to make hemoglobin, each molecule of the latter being made up of two α-type and two β-type globins. In humans the α-type globins are coded by a small multigene family on chromosome 16 and the β-type globins by a second family on chromosome 11 (*Figure 6.6*). These genes were among the first to be sequenced, back in the late 1970s (Fritsch *et al.*, 1980). The sequence data showed that the genes in each family are similar to one another, but by no means identical. In fact the nucleotide sequences of the two most different genes in the β-type cluster, coding for the β- and ε-globins, are only 79.1% identical. This is still similar enough for the two proteins both to be β-type globins, but sufficiently different for them to have distinctive biochemical properties. Similar variations are seen in the α-cluster.

Why are the members of the globin gene families so different from one another? The answer was revealed when the expression patterns of the individual genes were studied. It was discovered that the genes are expressed at different stages in human development: for example, in the β-type cluster ε is expressed in the early embryo, G_γ and A_γ (which differ by just one amino acid) in the fetus, and δ and β in the adult (*Figure 6.6*). The different biochemical properties of the resulting globin proteins are thought to reflect slight changes in the physiological role that hemoglobin plays during the course of human development.

With some multigene families, the individual members are clustered, as with the globin genes, but in others the genes are dispersed around the genome. An example of the latter is the five aldolase genes of the human genome, which are located on chromosomes 3, 9, 10, 16 and 17. The important point is that, even though dispersed, the members of the multigene family have sequence similarities that point to a common evolutionary origin. When these sequence comparisons are made it is sometimes possible to see relationships not only within a single gene family but also between different families. All of the genes in

TECHNICAL 6.1 NOTES

Ultracentrifugation techniques

Methodology for separation of cell components and large molecules.

The development of high-speed centrifuges in the 1920s led to techniques for separating organelles and other fractions from disrupted cells. The first technique to be used was **differential centrifugation**, in which pellets of successively lighter cell components are collected by centrifuging cell extracts at different speeds. Intact nuclei, for example, are relatively large and can be collected from a cell extract by centrifugation at 1000 g for 10 minutes; mitochondria, being lighter, require centrifugation at 20 000 g for 20 minutes. By careful manipulation of the centrifugation parameters, different cell components can be obtained in fairly pure form.

Density gradient centrifugation was used for the first time in 1951. In this procedure, the cell fraction is not centrifuged in a normal aqueous solution. Instead, a sucrose solution is layered into the tube in such a way that a density gradient is formed, the solution being more concentrated and hence denser towards the bottom of the tube. The cell fraction is now placed on the top of the gradient and the tube centrifuged at a very high speed: at least 500 000 g for several hours. Under these conditions, the rate of migration of a cell component through the gradient depends on its **sedimentation coefficient**, which in turn depends on its molecular mass and shape. For example, eukaryotic ribosomes have a sedimentation coefficient of 80S (S stands for Svedberg units, Svedberg being the Swedish scientist who pioneered the biological applications of ultracentrifugation), whereas bacterial ribosomes, being smaller, have a sedimentation coefficient of 70S.

In a second type of density gradient centrifugation a solution of 8 M cesium chloride is used, substantially denser than the sucrose solution used to measure S values. The starting solution is uniform, the gradient being established during the centrifugation. Cellular components migrate down through the centrifuge tube but molecules such as DNA and proteins do not reach the bottom: instead each one comes to rest at a position where the density of the matrix equals its own **buoyant density** (see *Figure 6.16*, p. 136). This technique has many applications in molecular biology, being able to separate DNA fragments of different base compositions (Sections 6.1.1 and 6.3.1) and DNA molecules with different conformations (e.g. supercoiled, circular and linear DNA). It is also able to distinguish between normal DNA and DNA labeled with a heavy isotope of nitrogen (Section 12.2.1).

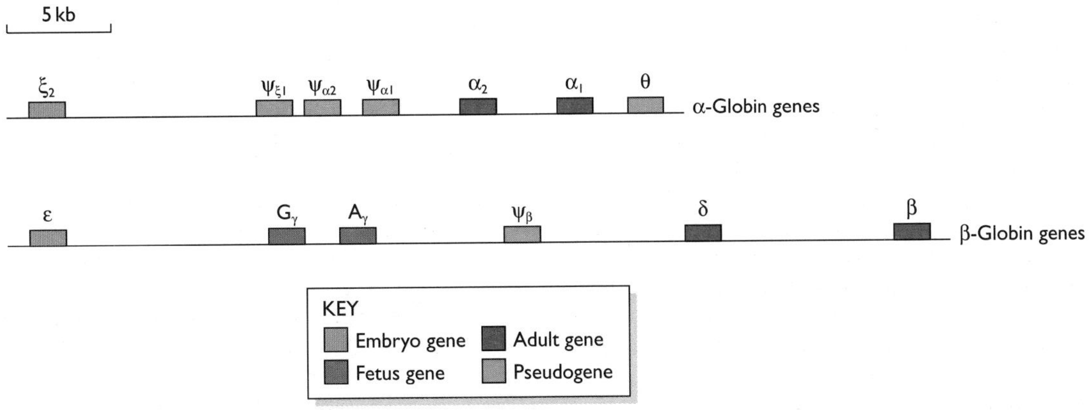

Figure 6.6 The human α- and β-globin gene clusters.

The α-cluster is located on chromosome 16 and the β-globin cluster on chromosome 11. Both clusters contain genes that are expressed at different developmental stages and each includes at least one pseudogene. Note that expression of the α-type gene ξ_2 begins in the embryo and continues during the fetal stage: there is no fetal-specific α-type globin. The θ pseudogene is expressed but its protein product is inactive. None of the other pseudogenes are expressed because their promoter regions are dysfunctional. For more information on the developmental regulation of the β-globin genes, see Section 8.1.1.

Box 6.3: Two examples of unusual gene organization

- **Overlapping genes** are occasionally found in small compact genomes such as those of viruses. Usually the amino acid sequences of the proteins coded by a pair of overlapping genes are not similar because the mRNAs are translated from different reading frames, as illustrated by the *D* and *E* genes of the *E. coli* bacteriophage φX174.

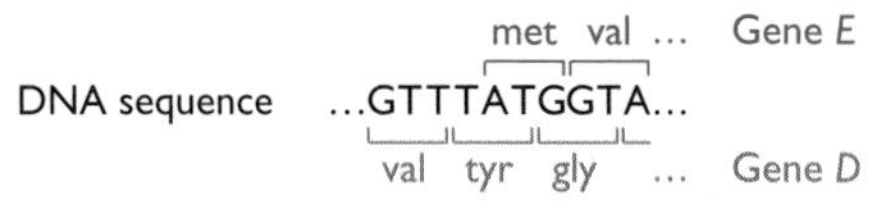

Overlapping genes are very rare in the nuclear genomes of higher organisms though there are examples in the compact mitochondrial genomes of some animals, including humans (see *Figure 6.9A*).

- **Genes-within-genes** are relatively common features of nuclear genomes, one gene being contained within an intron of a second gene. An example in the human genome is the neurofibromatosis type I gene, which has three short genes (called *OGMP*, *EVI2A* and *EVI2B*) within one of its introns. Each of these internal genes is also split into exons and introns.

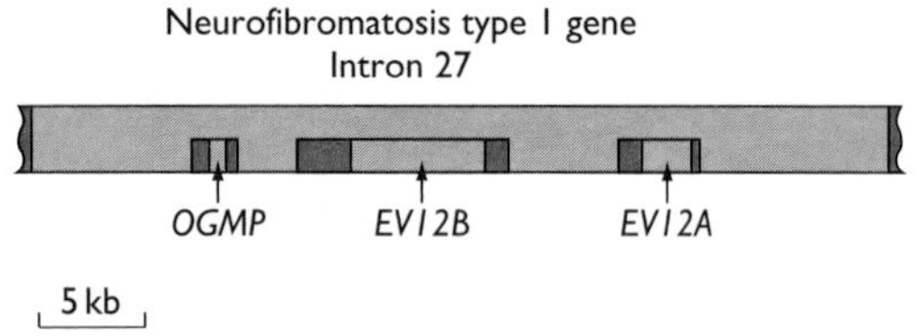

Recently, it has been discovered that many snoRNAs, which are involved in chemical modification of rRNAs (Section 9.4.1), are specified by genes within introns.

α- and β-globin families, for example, have some sequence similarity and are thought to have evolved from a single ancestral globin gene. We therefore refer to these two multigene families as comprising a single globin **gene superfamily**, and from the similarities between the individual genes we can chart the duplication events that have given rise to the series of genes that we see today (Section 14.2.1).

Pseudogenes and other gene relics

Our ideas regarding the evolution of complex multigene families are based on the assumption that after a gene duplication, mutations occur so that the nucleotide sequences of the two new genes gradually become more and more different. What if a mutation results in a sequence change that inactivates one of the genes? This appears to be a common occurrence because when we examine the genomes of higher eukaryotes we see many examples of **pseudogenes**, genes which for one reason or another have become nonfunctional.

Inactivation by mutation results in a **conventional pseudogene**, usually by creating a termination codon (Section 10.1.2) within the coding region of the gene, meaning that its protein product will be truncated and hence unable to carry out its biochemical role. Five conventional pseudogenes are present in the human globin gene families (see *Figure 6.6*), four in the α-cluster ($\psi_{\xi 1}$, $\psi_{\alpha 1}$, $\psi_{\alpha 2}$ and θ) and one in the β-group (ψ_{β}).

A second type of pseudogene arises not by evolutionary decay but by an abnormal adjunct to gene expression. As we have not yet looked at gene expression in detail we cannot yet fully appreciate how these **processed pseudogenes** arise. At this point all we need note is that a processed pseudogene is derived from the RNA transcript of a gene by synthesis of a DNA copy which subsequently reinserts into the genome (*Figure 6.7*). Insertion occurs at random positions so processed pseudogenes are only infrequently found with their functional cousins in clustered multigene families, they are more usually at dispersed positions. We will fill in the details of the process that results in a processed pseudogene, and look at their characteristic features, in Chapter 9 (see Box 9.2, p. 214).

As well as pseudogenes, genomes also contain other gene relics in the form of **truncated genes**, which lack a

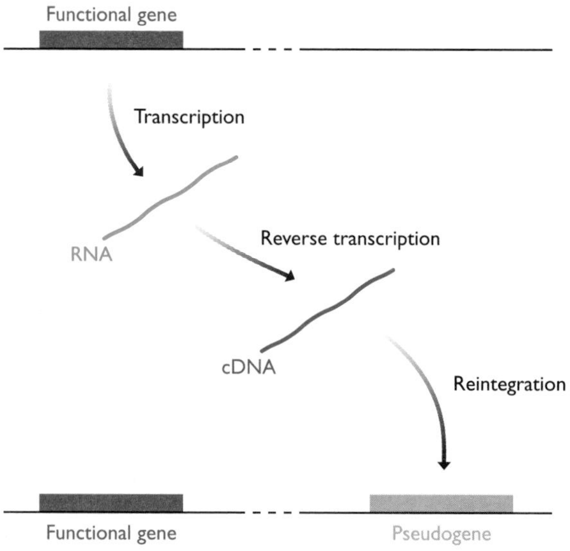

Figure 6.7 The origin of a processed pseudogene.

A processed pseudogene is thought to arise by reverse transcription of the RNA transcript of a functional gene. The resulting complementary DNA (cDNA) can reintegrate into the genome, possibly into the same chromosome as its functional parent, or possibly into a different chromosome. More details about processed pseudogenes are given in Box 9.2, p. 214.

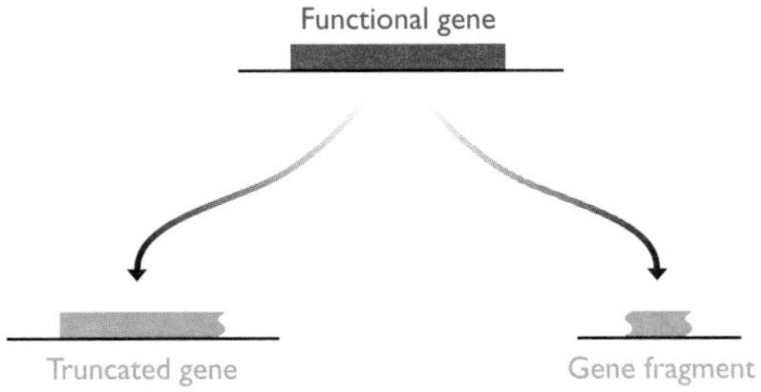

Figure 6.8 A truncated gene and a gene fragment.

greater or lesser stretch from one end of the complete gene, and **gene fragments**, which are short isolated regions from within a gene (*Figure 6.8*). Both types of relic are generally found within multigene families, rather than at dispersed positions, and are thought to arise by deletion or rearrangement of the DNA.

6.1.2 Eukaryotic organelle genomes

Now we move out of the nucleus to examine the organelle genomes present in the mitochondria and chloroplasts of eukaryotic cells. The possibility that some genes might be located outside of the nucleus – **extra-chromosomal genes** as they were initially called – was first raised in the 1950s as a means of explaining the unusual inheritance patterns of certain phenotypes observed in the fungus *Neurospora crassa*, the yeast *Saccharomyces cerevisiae* and the photosynthetic alga *Chlamydomonas reinhardtii*. Electron microscopy and biochemical studies at about the same time provided hints that DNA molecules might be present in mitochondria and chloroplasts. Eventually, in the early 1960s, these various lines of evidence were brought together and the existence of mitochondrial and chloroplast genomes, independent of and distinct from the nuclear genome, was accepted.

Physical features of organelle genomes

Almost all eukaryotes have mitochondrial genomes and all photosynthetic eukaryotes have chloroplast genomes. Initially, it was thought that virtually all organelle genomes were circular DNA molecules. This was because genetic maps for mitochondrial and chloroplast genomes were circular, as were physical maps obtained by restriction analysis (Section 3.1). Electron microscopy studies had observed both circular and linear DNA in some organelles, but it was assumed that the linear molecules were simply fragments of circular genomes that had become broken during preparation for electron microscopy. We still believe that most mitochondrial and chloroplast genomes are circular, but we now recognize that in many eukaryotes the circular genomes coexist in their organelles with linear versions and, in the case of chloroplasts, with smaller circles that contain subcomponents of the genome as a whole. We also now realize that the mitochondrial genomes of some microbial eukaryotes (e.g. *Paramecium*, *Chlamydomonas* and several yeasts) are always linear (Nosek *et al.*, 1998).

Copy numbers for organelle genomes are not particularly well understood. Each human mitochondrion contains about 10 identical molecules, which means that there are about 8000 per cell, but in *S. cerevisiae* the total number is probably smaller (less than 6500) even though there may be over 100 genomes per mitochondrion. Photosynthetic microorganisms such as *Chlamydomonas* have approximately 1000 chloroplast genomes per cell, about one fifth the number present in a higher plant cell. One mystery that dates back to the 1950s, and has never been satisfactorily solved, is that when organelle genes are studied in genetic crosses the results suggest that there is just one copy of a mitochondrial or chloroplast genome per cell. This is clearly not the case but indicates that our understanding of the transmission of organelle genomes from parent to offspring is less than perfect.

Mitochondrial genome sizes are variable (*Table 6.3A*) and unrelated to the complexity of the organism. Most multicellular animals have small mitochondrial genomes with a compact genetic organization, the genes being close together with little space between them. The human mitochondrial genome (*Figure 6.9A*), at 16 569 bp, is typical of this type. Lower eukaryotes such as *Saccharomyces cerevisiae* (*Figure 6.9B*) and flowering plants have larger and less compact mitochondrial genomes, with a number of the genes containing introns. Chloroplast genome sizes are less variable (*Table 6.3B*) and all appear to be organized along similar lines (*Figure 6.10*).

The genetic content of organelle genomes

Organelle genomes are much smaller than their nuclear counterparts and we therefore anticipate that their gene contents are much more limited. This is indeed the case. Again, mitochondrial genomes display the greater variability, gene contents ranging from 12 for the green alga *Chlamydomonas reinhardtii* to 92 for the protozoan *Reclinomonas americana* (*Table 6.4*; Lang *et al.*, 1997; Palmer, 1997a). All mitochondrial genomes contain genes for the mitochondrial rRNAs and at least some of the protein components of the respiratory chain, the latter being the main biochemical feature of the mitochondrion. The more gene-rich genomes also code for transfer RNAs, ribosomal proteins and proteins involved in transcription, translation and transport of other proteins into the mitochondrion from the surrounding cytoplasm (*Table 6.4*). Most chloroplast genomes appear to possess the same set of 200 or so genes, again coding for ribosomal and transfer RNAs, as well as ribosomal proteins and proteins involved in photosynthesis (see *Figure 6.10*).

A general feature of organelle genomes emerges from *Table 6.4*. These genomes specify some of the proteins found in the organelle, but not all of them. These other proteins are coded by nuclear genes, synthesized in the cytoplasm, and transported into the organelle. If the cell has mechanisms for transporting proteins into mitochondria and chloroplasts, then why not have all the organelle proteins specified by the nuclear genome? We do not yet

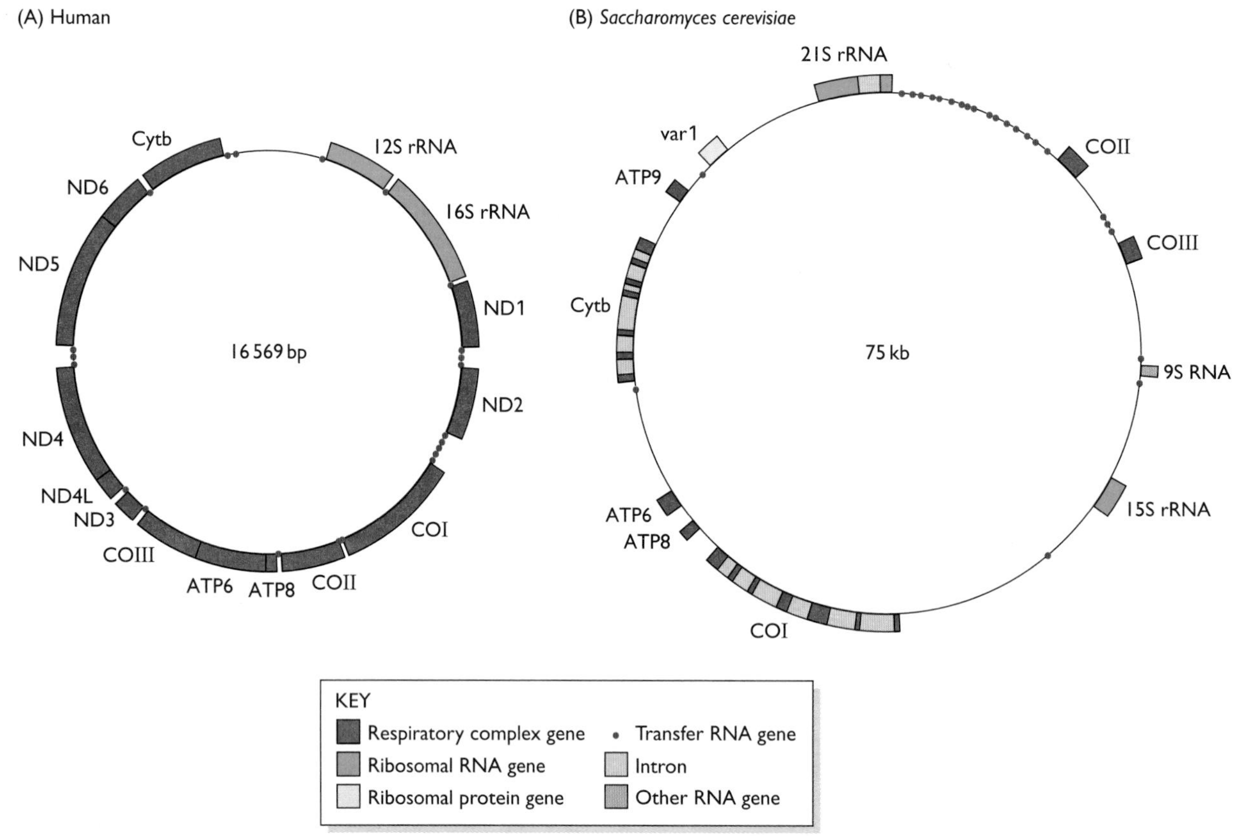

Figure 6.9 The (A) human and (B) *Saccharomyces cerevisiae* mitochondrial genomes.

Because of their relatively small sizes, many mitochondrial genomes have been completely sequenced. The human and yeast genomes illustrate two types of organization, the human genome being small and compact with little wasted space, typical of multicellular animals, and the yeast mitochondrial genome having longer spaces between genes and possessing some introns, the typical organization for lower eukaryotes and plants. In the human mitochondrial genome the ATP6 and ATP8 genes overlap. The yeast genome contains five additional open reading frames, not shown on this map, that have not yet been shown to code for functional gene products, and there are also several genes located within the introns of the split genes. Most of the latter code for maturase proteins that are involved in splicing the introns from the transcripts of these genes (Section 9.3.3). The two genomes are drawn at different scales. Abbreviations: ATP6, ATP8, ATP9, genes for ATPase subunits 6, 8 and 9; COI, COII, COIII, genes for cytochrome *c* oxidase subunits I, II and III; Cytb, gene for apocytochrome *b*; ND1–ND6, genes for NADH hydrogenase subunits 1–6; var1, yeast mitochondrial gene for a ribosome-associated protein. The 9S RNA gene in the yeast genome specifies the RNA component of the enzyme ribonuclease P (Section 9.3.2).

have a convincing answer to this question although it has been suggested that at least some of the proteins coded by organelle genomes are extremely hydrophobic and cannot be transported through the membranes that surround mitochondria and chloroplasts, and so simply cannot be moved into the organelle from the cytoplasm (Palmer, 1997b). The only way in which the cell can get them into the organelle is to make them there in the first place.

The origins of organelle genomes

The discovery of organelle genomes led to many speculations about their origins. Today most biologists accept that the **endosymbiont theory** is correct, at least in outline, even though it was considered quite unorthodox when first proposed in the 1960s. The endosymbiont theory is based on the observation that the gene expression processes occurring in organelles are similar in many respects to equivalent processes in bacteria. In addition, when nucleotide sequences are compared organelle genes are found to be more similar to equivalent genes from bacteria than they are to eukaryotic nuclear genes. The endosymbiont theory therefore holds that mitochondria and chloroplasts are the relics of free-living bacteria that formed a symbiotic association with the precursor of the eukaryotic cell, way back at the very earliest stages of evolution.

Support for the endosymbiont theory has come from the discovery of organisms which appear to exhibit stages of endosymbiosis that are less advanced than seen with mitochondria and chloroplasts. For example, an early stage in endosymbiosis is displayed by the protozoan *Cyanophora paradoxa*, whose photosynthetic structures, called **cyanelles**, are different from chloroplasts and

Table 6.3 Sizes of mitochondrial and chloroplast genomes

Species	Type of organism	Genome size (kb)
A. Mitochondrial genomes		
Chlamydomonas reinhardtii	Green alga	16
Mus musculus	Vertebrate (mouse)	16
Homo sapiens	Vertebrate (human)	17
Metridium senile	Invertebrate (sea anemone)	17
Drosophila melanogaster	Invertebrate (fruit fly)	19
Chondrus crispus	Red alga	26
Aspergillus nidulans	Ascomycete fungus	33
Reclinomonas americana	Protozoa	69
Saccharomyces cerevisiae	Yeast	75
Suillus grisellus	Basidiomycete fungus	121
Brassica oleracea	Flowering plant (cabbage)	160
Arabidopsis thaliana	Flowering plant (vetch)	367
Zea mays	Flowering plant (maize)	570
Cucumis melo	Flowering plant (melon)	2500
B. Chloroplast genomes		
Pisum sativum	Flowering plant (pea)	120
Marchantia polymorpha	Liverwort	121
Oryza sativa	Flowering plant (rice)	136
Nicotiana tabacum	Flowering plant (tobacco)	156
Chlamydomonas reinhardtii	Green alga	195

Data taken from Brown (1998).

instead resemble ingested cyanobacteria. Similarly, the *Rickettsia*, which live inside eukaryotic cells, might be modern versions of the bacteria that gave rise to mitochondria (Andersson *et al.*, 1998). It has also been suggested that the hydrogenosomes of trichomonads (unicellular microbes, many of which are parasites), some of which have a genome, but most of which do not, represent an advanced type of mitochondrial endosymbiosis (Palmer, 1997b; Akhmanova *et al.*, 1998).

If mitochondria and chloroplasts were once free-living bacteria, then since the endosymbiosis was set up there must have been a transfer of genes from the organelle into the nucleus. We do not understand how this occurred, or indeed whether there was a mass transfer of many genes at once, or a gradual trickle from one site to the other. But we do know that DNA transfer from organelle to nucleus, and indeed between organelles, still occurs. This was first discovered in the early 1980s, when the first partial sequences of chloroplast genomes were obtained. It was found that in some plants the chloroplast genome contains segments of DNA, often including entire genes, that are copies of parts of the mitochondrial genome. The implication is that this **promiscuous DNA**, as it is called, has been transferred from one organelle to the other. We

Table 6.4 Features of mitochondrial genomes

Feature	*Chlamydomonas reinhardtii*	*Homo sapiens*	*Saccharomyces cerevisiae*	*Arabidopsis thaliana*	*Reclinomonas americana*
Total number of genes	12	37	35	52	92
Types of genes					
Protein-coding genes	7	13	8	27	62
Respiratory complex	7	13	7	17	24
Ribosomal proteins	0	0	1	7	27
Transport proteins	0	0	0	3	6
RNA polymerase	0	0	0	0	4
Translation factor	0	0	0	0	1
Ribosomal RNA genes	2	2	2	3	3
Transfer RNA genes	3	22	24	22	26
Other RNA genes	0	0	1	0	1
Number of introns	1	0	8	23	1
Genome size (kb)	16	17	75	367	69

Based on Palmer (1997a).

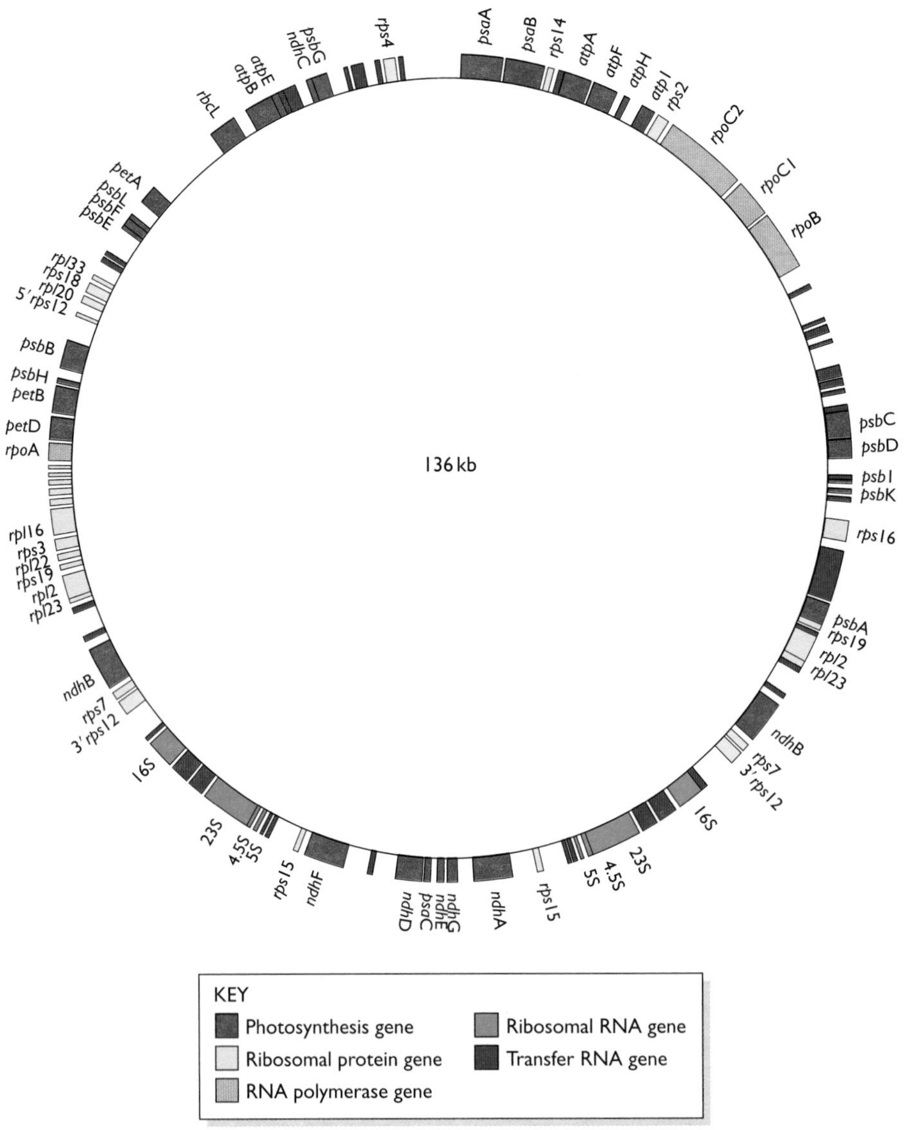

Figure 6.10 The rice chloroplast genome.

Only those genes with known functions are shown. A number of the genes contain introns which are not indicated on this map. These split genes include several of those for tRNAs, which is why the tRNA genes are of different lengths even though the tRNAs that they specify are all of similar size.

now know that this is not the only type of transfer that can occur, both mitochondrial and chloroplast DNA segments being present in plant nuclear genomes, and mammalian chromosomes containing sequences derived from mitochondrial DNA.

6.2 THE ANATOMY OF THE PROKARYOTIC GENOME

The relatively small sizes of prokaryotic genomes, which means that they are amenable to rapid sequence analysis by the shotgun approach (Section 4.2.1), has resulted in the frequent publication over the last few years of complete genome sequences for members of the bacteria and archaea. As a consequence we are beginning to understand a great deal about the anatomies of prokaryotic genomes, and in many respects we know more about these organisms than we do about eukaryotes.

6.2.1 The physical structure of the prokaryotic genome

Most prokaryotic genomes are less than 5 Mb in size, although there are a few that are substantially larger than this (see *Table 1.1*, p. 9). The traditional view has been that in a typical prokaryote the genome is contained in a single circular DNA molecule, localized within the **nucleoid**, the lightly staining region of the otherwise featureless prokaryotic cell (see *Figure 1.6*, p. 8). This is certainly true for *Escherichia coli* and many of the other commonly studied bacteria but, as we will see, our increasing knowledge of prokaryotic genomes is leading us to question several of the preconceptions that became established during the pre-genome era of microbiology. These preconceptions relate both to the physical structure of the prokaryotic genome and its genetic organization.

The traditional view of the bacterial 'chromosome'

As with eukaryotic chromosomes, a prokaryotic genome has to squeeze into a relatively tiny space (the circular *E. coli* chromosome has a circumference of 1.6 mm whereas an *E. coli* cell is just $1.0 \times 2.0\,\mu m$) and, as with eukaryotes, this is achieved with the help of DNA-binding proteins that package the genome in an organized fashion. The resulting structure has no substantial similarities with a eukaryotic chromosome, but we still use 'bacterial chromosome' as a convenient term to describe it.

Most of what we know about the organization of DNA in the nucleoid comes from studies of *E. coli*. The first feature to be recognized was that the circular *E. coli* genome is **supercoiled**. Supercoiling occurs when additional turns are introduced into the DNA double helix (positive supercoiling) or if turns are removed (negative supercoiling). With a linear molecule, the torsional stress introduced by over- or underwinding is immediately released by rotation of the ends of the DNA molecule, but a circular molecule, having no ends, cannot reduce the strain in this way. Instead the circular molecule responds by winding around itself to form a more compact structure (*Figure 6.11*). Supercoiling is therefore an ideal way of packaging a circular molecule into a small space. Evidence that supercoiling is involved in packaging the circular *E. coli* genome was first obtained in the 1970s from examination of isolated nucleoids, and subsequently confirmed as a feature of DNA in living cells in 1981 (Research Briefing 6.1). In *E. coli*, the supercoiling is thought to be generated and controlled by two enzymes, DNA gyrase and DNA topoisomerase I, which we will look at in more detail in Section 12.2.2 when we examine the roles of these enzymes in DNA replication.

Studies of isolated nucleoids and of living cells have shown that the *E. coli* DNA molecule does not have unlimited freedom to rotate once a break is introduced (see Research Briefing 6.1). The most likely explanation is that the bacterial DNA is attached to proteins that restrict its ability to relax, so that rotation at a break site results only in loss of supercoiling from a small segment of the molecule (*Figure 6.12*). The current model has the *E. coli* DNA attached to a protein core from which 40–50 supercoiled loops radiate out into the cell. Each loop contains approximately 100 kb of supercoiled DNA, the amount of DNA that becomes unwound from a single break.

The protein component of the nucleoid includes DNA gyrase and DNA topoisomerase I, the two enzymes that are primarily responsible for maintaining the supercoiled state, as well as a set of at least four proteins believed to have a more specific role in packaging the bacterial DNA. The most abundant of these packaging proteins is HU, which is structurally very different to eukaryotic histones but acts in a similar way, forming a tetramer around which approximately 60 bp of DNA becomes wound. There are some 60 000 HU proteins per *E. coli* cell, enough to cover about one-fifth of the DNA molecule, but it is not known if the tetramers are evenly spread along the DNA or restricted to the core region of the nucleoid.

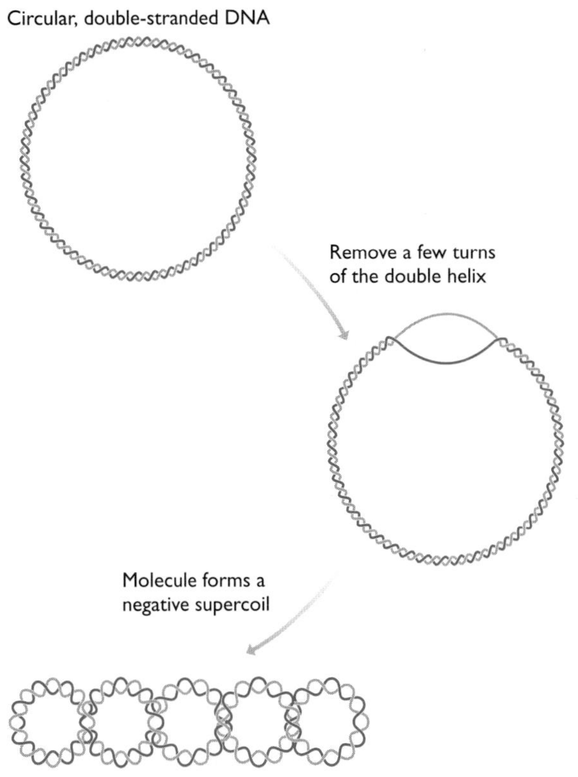

Figure 6.11 Supercoiling.

The diagram shows how underwinding a circular, double-stranded DNA molecule results in negative supercoiling.

Complications on the *E. coli* theme

In recent years it has become clear that the straightforward view of prokaryotic genome anatomy developed from studies of *E. coli* is an over-simplification. Although the vast majority of bacterial and archaeal chromosomes are indeed circular, an increasing number of linear ones are being found. The first of these, for *Borrelia burgdorferi*, was described in 1989 (Ferdows and Barbour, 1989) and during the following years similar discoveries were made for *Streptomyces* and other bacteria (Chen, 1996).

A second complication concerns the precise status of **plasmids** with regard to the prokaryotic genome. A plasmid is a small piece of DNA, often but not always circular, that coexists with the main chromosome in a bacterial

RESEARCH 6.1 BRIEFING

Supercoiled domains in the *E. coli* nucleoid

Studies with extracted nucleoids suggested that the *E. coli* genome is organized into a series of supercoiled domains. Was this model correct for living bacteria?

Early attempts to understand the organization of bacterial DNA made use of isolated nucleoids from *E. coli* cell extracts. These experiments showed that *E. coli* DNA is supercoiled, and also suggested that relaxation of supercoils after strand breakage is limited to the single region in which the break occurs, more extensive relaxation being prevented by the presence of proteins or other structures that bind to the DNA and separate the molecule into independent domains. This model was considered unproven because domains were observed only when the cell extracts contained contaminating RNA molecules. If these RNA molecules were degraded by treating with ribonuclease then the DNA molecule lost all its supercoiling. The supercoiled domains might therefore be an artefact of the extraction procedure. To resolve this problem it was necessary to devise a way of directly examining the structure of the nucleoid in living cells.

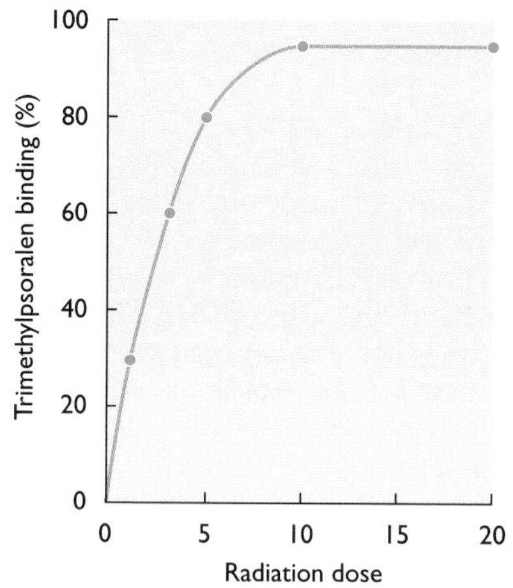

The experiment

One approach was to exploit the ability of trimethylpsoralen to distinguish between supercoiled and relaxed DNA. This compound, when photoactivated by a pulse of light at 360 nm, binds to double-stranded DNA at a rate that is directly proportional to the degree of torsional stress possessed by the molecule. The degree of supercoiling can therefore be assayed by measuring the amount of trimethylpsoralen that binds to the molecule in unit time. The technique was devised by Richard Sinden and David Pettijohn of the University of Colorado Health Sciences Center, and used by them in 1981 to probe the structure of the nucleoids in living *E. coli* bacteria. Cells were exposed to varying doses of radiation to introduce single-strand breaks into the DNA molecule, and then suspended in a buffer containing radioactively-labeled trimethylpsoralen. After photoactivation, DNA was extracted from each batch of cells and the amount of bound trimethylpsoralen measured by radioactive counting. Typical results are shown in the graph.

The conclusion

The amount of trimethylpsoralen bound by the nucleoids of nonirradiated cells confirms that DNA is supercoiled in living *E. coli*. The results also support the domain model. If the *E. coli* nucleoid is not organized into domains then one break in the DNA molecule would lead to complete loss of supercoiling: irradiation would therefore have an all-or-nothing effect on trimethylpsoralen binding. This was not observed. Instead, the results show that supercoiling is reduced in a gradual, incremental fashion as the irradiation dose increases. This is the response predicted by the domain model, in which the overall supercoiling of the molecule is gradually relaxed as greater doses of radiation cause breaks within an increasing number of domains. The trimethylpsoralen method therefore confirms that the domain model is correct.

Can these results also be used to estimate the number of domains in the *E. coli* nucleoid? The number of single-strand breaks caused by a given dose of radiation can be determined by purifying the DNA, denaturing with alkali to separate the strands, and measuring the average length of the polynucleotides that remain. The number of breaks needed to produce a complete loss of supercoiling can therefore be determined, but this figure is not the same as the number of domains. Radiation 'hits' occur at random, so some domains may accumulate two or more breaks before all of the others have received their first hits. However, 'randomness' is quantifiable statistically, so the total number of domains can be calculated from the numbers of strand breaks needed to produce different degrees of reduction in the overall amounts of supercoiling. Using this approach, Sinden and Pettijohn calculated that the *E. coli* nucleoid contains 43 ± 10 domains.

Reference

Sinden RR and Pettijohn DE (1981) Chromosomes in living *Escherichia coli* cells are segregated into domains of supercoiling. *Proc. Natl Acad. Sci. USA*, **78**, 224–228.

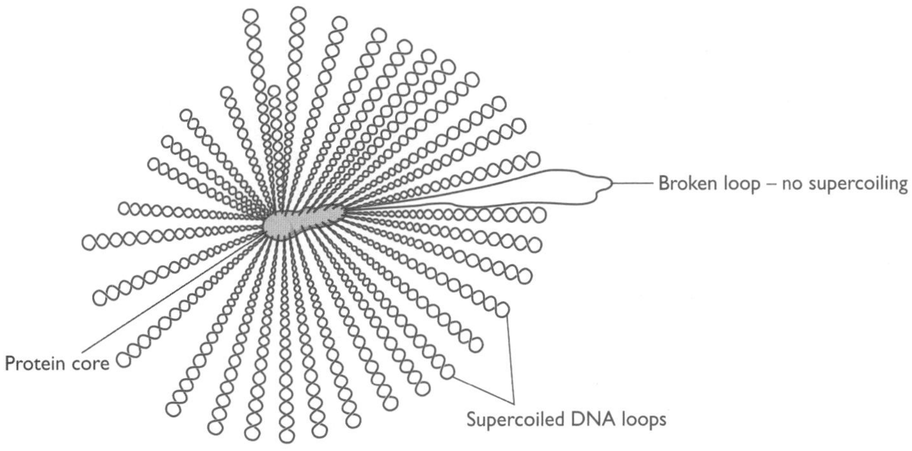

Figure 6.12 A model for the structure of the *E. coli* nucleoid.

Between 40 and 50 supercoiled loops of DNA radiate from the central protein core. One of the loops is shown in circular form, indicating that a break has occurred in this segment of DNA, resulting in a loss of the supercoiling.

cell (see *Figure 1.8*, p. 10). Some types of plasmid are able to integrate into the main genome, but others are thought to be permanently independent. Plasmids carry genes that usually are not present in the main chromosome, but in many cases these genes are nonessential to the bacterium, coding for phenotypes such as antibiotic resistance which the bacterium can live without if the environmental conditions are amenable (*Table 6.5*). As well as this apparent dispensability, many plasmids are able to transfer from one cell to another, and the same plasmids are sometimes found in bacteria that belong to different species. These various features of plasmids suggest that they are independent entities and that in most cases the plasmid content of a prokaryotic cell should not be included in the definition of its genome.

With a bacterium such as *E. coli*, which has a 4.6 Mb chromosome and can harbour various combinations of plasmids, none of which is more than a few kb in size and all of which are dispensable, it is acceptable to define the main chromosome as the 'genome'. With other prokaryotes it is not so easy. The linear chromosome of *Borrelia burgdorferi*, 910 kb in length and carrying 853 genes, is accompanied by at least 17 linear and circular plasmids which together contribute another 533 kb and at least 430 genes (Fraser *et al.*, 1997). Although most of these genes are orphans, those that have been identified include several which would not normally be considered dispensable, such as genes for membrane proteins and purine biosynthesis. The implication is that at least some of the 17 *Borrelia* plasmids are essential components of the genome, leading to the possibility that some prokaryotes have multipartite genomes, comprising a number of separate DNA molecules, more akin to what we see in the eukaryotic nucleus rather than the 'typical' prokaryotic arrangement. However, this interpretation of the *Borrelia* genome is still controversial, and is complicated by the fact that the related bacterium *Treponema pallidum*, whose genome is a single, circular DNA molecule of

Table 6.5 Features of typical plasmids

Type of plasmid	Gene functions	Examples
Resistance	Antibiotic resistance	Rbk of *E. coli* and other bacteria
Fertility	Conjugation and DNA transfer between bacteria	F of *E. coli*
Killer	Synthesis of toxins that kill other bacteria	Col of *E. coli*, for colicin production
Degradative	Enzymes for metabolism of unusual molecules	TOL of *Pseudomonas putida*, for toluene metabolism
Virulence	Pathogenicity	Ti of *Agrobacterium tumefaciens*, conferring the ability to cause crown gall disease on dicotyledonous plants

1138 kb containing 1041 genes (Fraser *et al.*, 1998), lacks homologs for any of the genes present on the *Borrelia* plasmids.

The final complication regarding the physical structures of prokaryotic genomes concerns differences between the packaging systems for bacterial and archaeal DNA molecules. One of the reasons why the archaea are looked on as a distinct group of organisms, different from the bacteria, is because archaea do not possess packaging proteins such as HU but instead have proteins that are much more similar to histones. Currently we have no information on the structure of the archaeal nucleoid, but the assumption is that these histone-like proteins play a central role in DNA packaging. The intriguing questions raised by this and other similarities between archaeons and eukaryotes is one of the areas of molecular evolution that we will explore in Chapter 14 (see Box 14.3, p. 383).

6.2.2 The genetic organization of the prokaryotic genome

We have already learnt that bacterial genomes have compact genetic organizations with very little space between genes (see *Figure 1.7*, p. 10). To re-emphasize this point, the complete, circular gene map of the *E. coli* genome is shown in *Figure 6.13*. There *is* noncoding DNA in the *E. coli* genome, but it accounts for only 11% of the total and it is distributed around the genome in small segments that do not show up when the map is drawn at this scale. In this regard, *E. coli* is typical of all prokaryotes whose genomes have so far been sequenced: prokaryotic genomes have very little wasted space. There are theories that this compact organization is beneficial to prokaryotes, for example by enabling the genome to be replicated

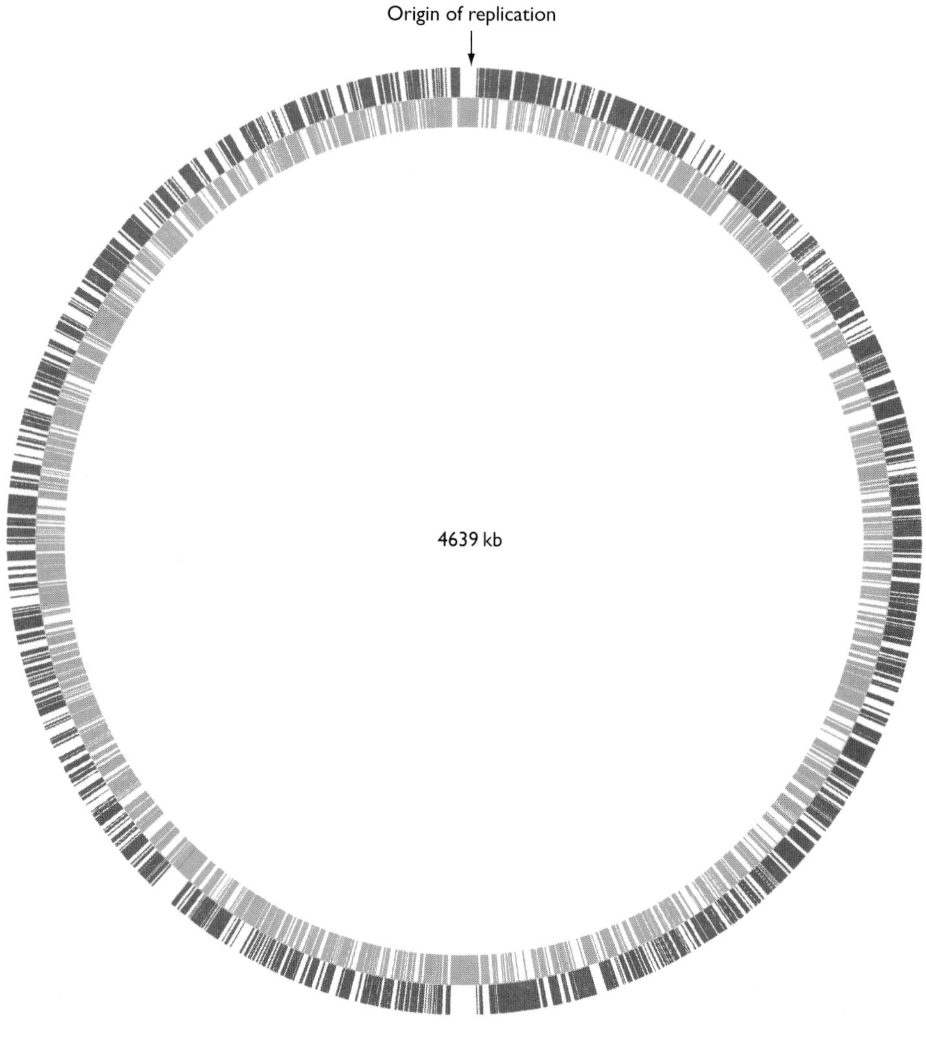

Figure 6.13 The *E. coli* genome.

The map is shown with the origin of replication (Section 12.3.1) positioned at the top. Genes on the outside of the circle are transcribed in the clockwise direction and those on the inside are transcribed in the anticlockwise direction. Image supplied courtesy of Dr FR Blattner, Laboratory of Genetics, University of Wisconsin-Madison. Reproduced with permission.

relatively quickly, but these ideas have never been supported by hard experimental evidence.

Operons are characteristic features of prokaryotic genomes

A second characteristic feature of prokaryotic genomes illustrated by *E. coli* is the presence of **operons**. In the years before genome sequences we thought we understood operons very well; now we are not so sure.

An operon is a group of genes that are located adjacent to one another in the genome, with perhaps just one or two nucleotides between the end of one gene and the start of the next. All the genes in an operon are expressed as a single unit. This type of arrangement is common in prokaryotic genomes, a typical *E. coli* example being the **lactose operon**, the first operon to be discovered (Jacob and Monod, 1961), which contains three genes involved in conversion of the disaccharide sugar lactose into its monosaccharide units glucose and galactose (*Figure 6.14A*). The monosaccharides are substrates for the energy-generating glycolytic pathway, so the function of the genes in the lactose operon is to convert lactose into a form that is utilizable by *E. coli* as an energy source. Lactose is not a common component of *E. coli*'s natural environment, so most of the time the operon is not expressed and the enzymes for lactose utilization are not made by the bacterium. When lactose becomes available it

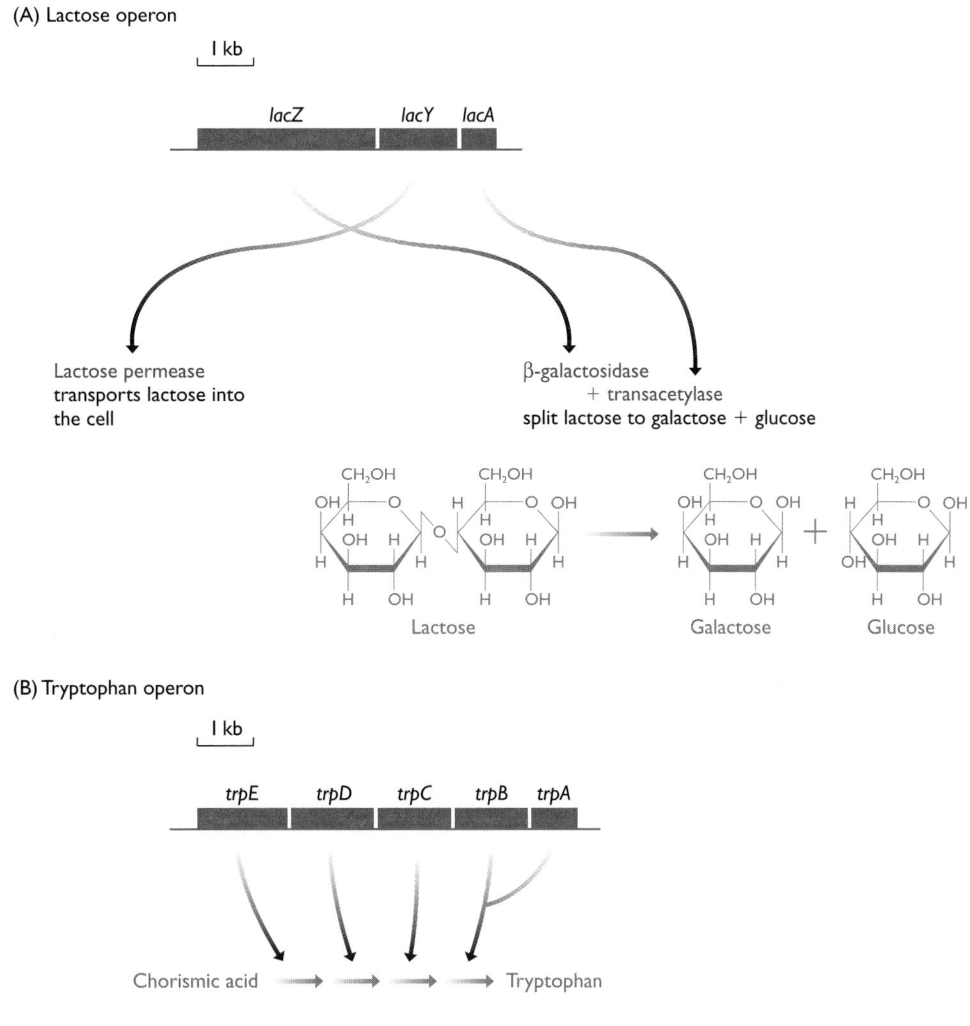

Figure 6.14 Two operons of *E. coli*.

(A) The lactose operon. The three genes are called *lacZ*, *lacY* and *lacA*, the first two separated by 52 bp and the second two by 64 bp. All three genes are expressed together, *lacY* coding for the lactose permease which transports lactose into the cell, and *lacZ* and *lacA* coding for enzymes that split lactose into its component sugars, galactose and glucose. (B) The tryptophan operon, which contains five genes coding for enzymes involved in the multistep biochemical pathway that converts chorismic acid into the amino acid tryptophan. The genes in the tryptophan operon are closer together than those in the lactose operon: *trpE* and *trpD* overlap by one bp, as do *trpB* and *trpA*; *trpD* and *trpC* are separated by 4 bp and *trpC* and *trpB* by 12 bp. For more details on the regulation of these operons, see Sections 8.3.1 and 11.1.1.

switches on the operon, all three genes being expressed together, resulting in coordinated synthesis of the lactose-utilizing enzymes. This is the classical example of gene regulation in bacteria, and we will examine it in detail in Sections 8.3.1 and 11.1.1.

Altogether there are almost 600 operons in the *E. coli* genome, each containing two or more genes, and a similar number are present in *Bacillus subtilis*. In most cases the genes in an operon are functionally related, coding for a set of proteins that are involved in a single biochemical activity such as utilization of a sugar as an energy source or synthesis of an amino acid. An example of the latter is the tryptophan operon of *E. coli* (*Figure 6.14B*). Microbial geneticists have become very attracted to the simplicity of the system whereby a bacterium is able to control its various biochemical activities by regulating the expression of groups of related genes linked together in operons. This may be a correct interpretation of the function of operons in *E. coli*, *B. subtilis* and many other prokaryotes, but in at least some species the picture is less straightforward. Both the archaeon *Methanococcus jannaschii* and the bacterium *Aquifex aeolicus* have operons, but the genes in an individual operon rarely have any biochemical relationship. For example, one of the operons in the *A. aeolicus* genome contains six linked genes, these genes coding for two proteins involved in DNA recombination (Section 13.2), an enzyme used in protein synthesis, a protein required for motility, an enzyme involved in nucleotide synthesis, and an enzyme for lipid synthesis (*Figure 6.15*; Deckert *et al.*, 1998). This is typical of the operon structure in the *A. aeolicus* and *M. jannaschii* genomes. In other words, the notion that expression of an operon leads to coordinated synthesis of enzymes required for a single biochemical pathway does not hold for these species.

Genome projects have therefore confused our understanding of operons. It is certainly too early to abandon the belief that operons play a central role in biochemical regulation in many bacteria, but we need to explain the unexpected features of the operons in *A. aeolicus* and *M. jannaschii*. It has been pointed out that both *A. aeolicus* and *M. jannaschii* are autotrophs, which means that, unlike many prokaryotes, they are able to synthesize organic compounds from carbon dioxide (Deckert *et al.*, 1998), but how this similarity between the species might be used to explain their operon structures is not clear.

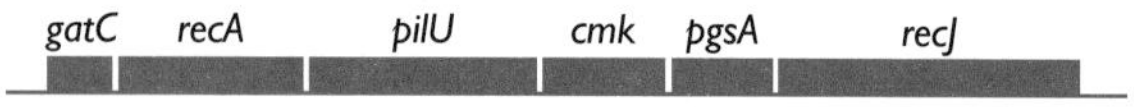

Figure 6.15 A typical operon in the genome of *Aquifex aeolicus*.

The genes code for the following proteins: *gatC*, glutamyl-tRNA aminotransferase subunit C, which plays a role in protein synthesis (Section 10.1.1); *recA*, recombination protein RecA; *pilU*, twitching mobility protein; *cmk*, cytidylate kinase, required for synthesis of cytidine nucleotides; *pgsA*, phosphotidylglycerophosphate synthase, an enzyme involved in lipid biosynthesis; *recJ*, single-strand-specific endonuclease RecJ, which is another recombination protein.

Speculations regarding the content of the minimal genome and the identity of distinctiveness genes

Even though a number of prokaryotic genome sequences have now been published, it is not yet possible to describe a complete catalog of the gene content of any one species, for the simple reason that the functions of many of the genes are unknown. The *E. coli* genome, for example, contains 4288 protein-coding genes, but over 1500 of these are orphans. Despite the incompleteness of the information, it is still interesting to examine the roles of the genes whose functions are known, and to appreciate the number of different genes involved in each of the various biochemical activities that a bacterium such as *E. coli* is able to carry out (*Table 6.6*).

Gene catalogs are even more interesting when comparisons are made between different species. We see, for example, that whereas 243 of the identified genes in the *E. coli* genome are involved in energy metabolism, *Haemophilus influenzae* has only 112 genes in this category and *Mycoplasma genitalium* just 31. These comparisons have led to speculations about the smallest number of genes needed to specify a free-living cell. Most attempts to answer this question have been based on examination of the two smallest genomes sequenced so far, those of *M. genitalium* and its cousin *M. pneumoniae*, which contain 470 and 679 genes, respectively. Theoretical considerations have led to the suggestion that 256 genes are the minimum required (Mushegian and Koonin, 1996), but experiments in which increasing numbers of *Mycoplasma* genes have been mutated suggest that at least 300 are needed. There has been similar interest in searching for 'distinctiveness' genes, ones that distinguish one species from another. Of the 470 genes in the *M. genitalium* genome, 350 are also present in the distantly-related bacterium *Bacillus subtilis* (Doolittle, 1997), which suggests that the biochemical and structural features that distinguish a *Mycoplasma* from a *Bacillus* are encoded in the 120 or so genes that are unique to the former. Unfortunately, the identities of these supposed distinctiveness genes do not provide any obvious clues about what makes a bacterium a *Mycoplasma* rather than anything else.

6.3 THE REPETITIVE DNA CONTENT OF GENOMES

Repetitive DNA is the one aspect of genome structure that we have not touched on so far. *Figure 1.7*, p. 10 showed us that repetitive DNA is found in all organisms and that in some, including humans, it makes up a substantial fraction of the entire genome. In this section we will look at repetitive DNA in more detail.

There are various types of repetitive DNA and several classification systems have been devised. The scheme that

Table 6.6 Partial gene catalogs for *Escherichia coli*, *Haemophilus influenzae* and *Mycoplasma genitalium*

Category	Number of genes in		
	E. coli	***H. influenzae***	***M. genitalium***
Total protein-coding genes	4288	1727	470
Biosynthesis of amino acids	131	68	1
Biosynthesis of cofactors	103	54	5
Biosynthesis of nucleotides	58	53	19
Cell envelope proteins	237	84	17
Energy metabolism	243	112	31
Intermediary metabolism	188	30	6
Lipid metabolism	48	25	6
DNA replication, recombination, repair	115	87	32
Protein folding	9	6	7
Regulatory proteins	178	64	7
Transcription	55	27	12
Translation	182	141	101
Uptake of molecules from the environment	427	123	34

Taken from Fraser *et al.* (1995) and Blattner *et al.* (1997). The numbers refer only to genes whose functions are known and are therefore approximate, because each genome also contains many genes whose functions have not yet been identified.

Box 6.4: The organization of the human genome

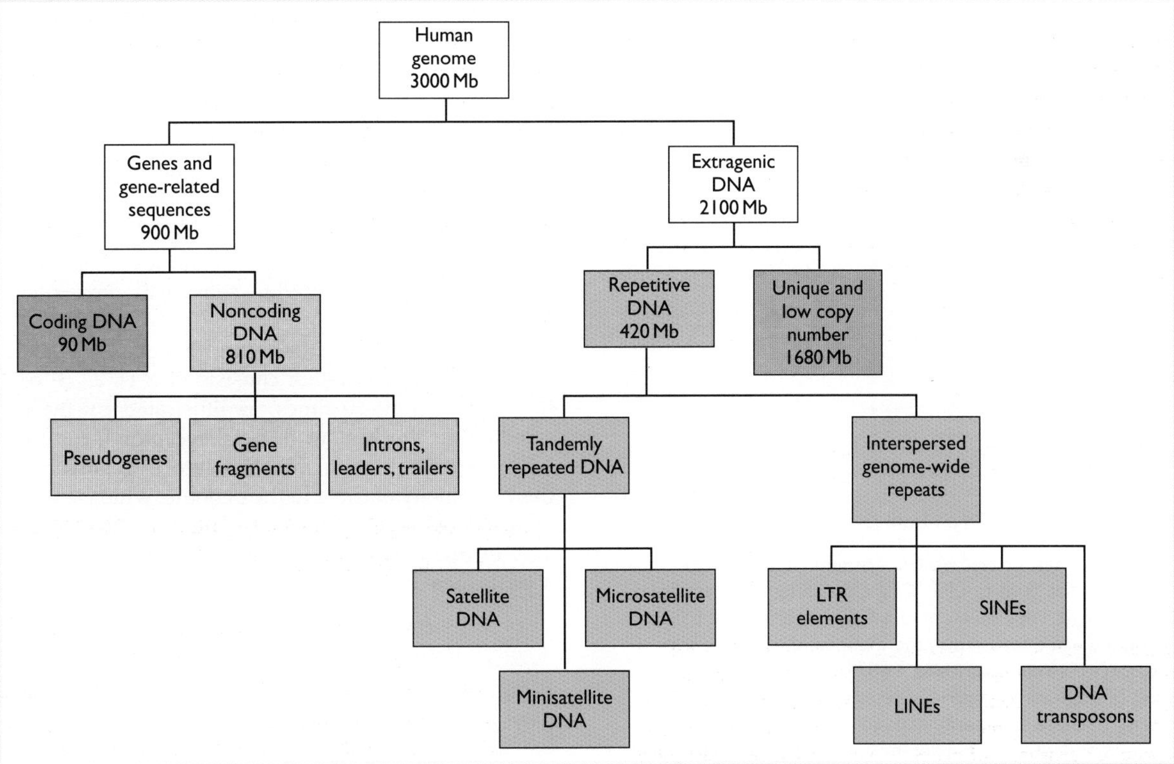

Based on Strachan and Read (1996).

we will use (see Box 6.4) begins by dividing the repeats into those that are clustered into tandem arrays and those which are dispersed around the genome.

6.3.1 Tandemly repeated DNA

Tandemly repeated DNA is a common feature of eukaryotic genomes but is virtually unknown in prokaryotes. This type of repeat is also called **satellite DNA**, because DNA fragments containing tandemly repeated sequences form 'satellite' bands when genomic DNA is fractionated by density gradient centrifugation (see Technical Note 6.1). Human DNA, for example, when broken into fragments 50–100 kb in length, forms a main band (buoyant density 1.701 g cm^{-3}) and three satellite bands (1.687, 1.693 and 1.697 g cm^{-3}). The main band contains DNA fragments made up mostly of single copy sequences with GC compositions close to 40.3%, the average value for the human genome. The satellite bands contain fragments of repetitive DNA, and hence have GC contents and buoyant densities that are atypical of the genome as a whole (*Figure 6.16*).

Satellite DNA is found at centromeres and elsewhere in eukaryotic chromosomes

The satellite bands in density gradients of eukaryotic DNA are made up of fragments that are composed of long series of tandem repeats, possibly hundreds of kb in length. A single genome can contain several different types of satellite DNA, each with a different repeat unit, these units being anything from <5 to >200 bp. The three satellite bands in human DNA include at least four different repeat types.

We have already encountered one type of human satellite DNA, the alphoid DNA repeats found in the centromere regions of chromosomes. Although some satellite DNA is scattered around the genome, most is located in the centromeres, supporting the hypothesis that repetitive DNA plays a structural role in the centromere (see p. 118). An alternative suggestion is that the repetitive DNA content of the centromere reflects the fact that this is the last region of the chromosome to be replicated. In order to delay its replication until the very end of the cell cycle, the centromere DNA must lack sequences that can act as origins of replication. The repetitive nature of centromeric DNA may be a means of ensuring that origins are absent (Csink and Henikoff, 1998).

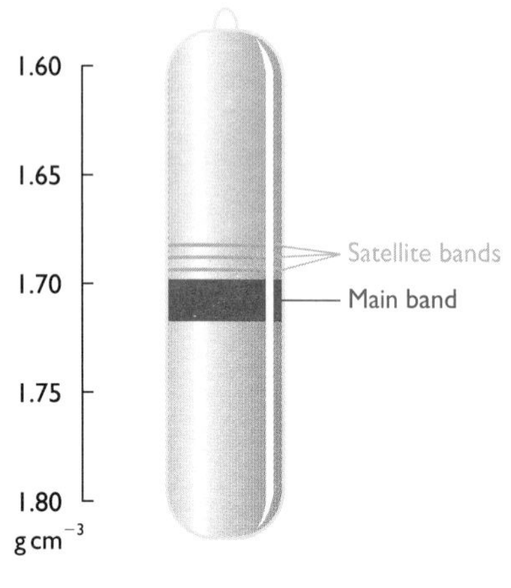

Figure 6.16 Satellite DNA from the human genome.

Human DNA has an average GC content of 40.3% and average buoyant density of 1.701 g cm^{-3}. Fragments made up mainly of single copy DNA have a GC content close to this average and are contained in the main band in the density gradient. The satellite bands at 1.687, 1.693 and 1.697 g cm^{-3} consist of fragments containing repetitive DNA. The GC contents of these fragments depend on their repeat motif sequences and so can be different from the genome average, meaning that these fragments have different buoyant densities to single copy DNA and migrate to different positions in the density gradient.

Mini- and microsatellites

Although not appearing in satellite bands on density gradients, two other types of tandemly-repeated DNA are also classed as 'satellite' DNA. These are **minisatellites** and **microsatellites**. Minisatellites form clusters up to 20 kb in length, with repeat units up to 25 bp; microsatellite clusters are shorter, usually <150 bp, and the repeat unit is usually 4 bp or less.

Minisatellite DNA is a second type of repetitive DNA that we are already familiar with because of its association with structural features of chromosomes. Telomeric DNA, which in humans comprises hundreds of copies of the motif 5′-TTAGGG-3′ (see *Figure 6.5*, p. 122), is an example of a minisatellite. We know a certain amount about how telomeric DNA is formed, and we know that it has an important function in DNA replication. These are topics that we will return to in Section 12.3.4. In addition to telomeric minisatellites, some eukaryotic genomes contain various other clusters of minisatellite DNA, many though not all near the ends of chromosomes. Functions for these other minisatellite sequences have not been identified.

The function of microsatellites is equally mysterious. The typical microsatellite consists of a 1-, 2-, 3- or 4-bp unit repeated 10–20 times, as illustrated by the microsatellites in the human β T-cell receptor locus (Section 1.1.2). Although each microsatellite is relatively short, there are many of them in the genome, which is why they are used as markers on genome maps (Sections 2.2.2 and 3.3). In humans, for example, microsatellites with a CA repeat, such as

5′-CACACACACACACACA-3′
3′-GTGTGTGTGTGTGTGT-5′

make up 0.5% of the genome, 15 Mb in all. Single base-pair repeats of the type

5′-AAAAAAAAAAAAAAAA-3′
3′-TTTTTTTTTTTTTTTT-5′

make up another 0.3%.

Although their function is unknown, microsatellites have proved very useful to geneticists, and not simply because of their value as genetic and physical markers in genome mapping work. Many microsatellites are variable, meaning that the number of repeat units in the array is different in different members of a species. This is because 'slippage' sometimes occurs when a microsatellite is copied during DNA replication, leading to insertion or, less frequently, deletion of one or more repeat units (see *Figure 13.5*, p. 336). The variations are such that no two humans alive today have exactly the same combination of microsatellite alleles: if enough microsatellites are examined then a unique **genetic profile** can be established for every person. Genetic profiling is well known as a tool in forensic science (*Figure 6.17*), but identification of criminals is a fairly trivial application of microsatellite variability. More sophisticated methodology makes use of the fact that a person's genetic profile is inherited partly from the mother and partly from the father. This means that microsatellites can be used to establish population affinities, not only for humans but for other animals and for plants also (see Section 15.3.2).

6.3.2 Interspersed genome-wide repeats

Tandemly repeated DNA sequences are thought to have arisen by expansion of a progenitor sequence, either by replication slippage as described for microsatellites, or by DNA recombination processes (Section 13.2). Both of these events are likely to result in a series of linked repeats, rather than individual repeat units scattered around the genome. Interspersed repeats must therefore have arisen by a different mechanism, one that can result in a copy of a repeat unit appearing in the genome at a position distant from the location of the original sequence. The most frequent way in which this occurs is by **transposition**, and most interspersed repeats have inherent transpositional activity.

Transposition via an RNA intermediate

The precise mechanics of transposition need not worry us until we deal with recombination and related rearrangements to the genome in Section 13.2. All that we need to know at this point is that there are two alternative modes of transposition, one that involves an RNA intermediate and one that does not. The version that involves an RNA intermediate is called **retrotransposition**. The basic mechanism involves three steps (*Figure 6.18*):

1. An RNA copy of the **transposon** is synthesized by the normal process of transcription.
2. The RNA transcript is copied into DNA, which initially exists as an independent molecule outside of the genome. This conversion of RNA to DNA, the reverse of the normal transcription process, requires a special enzyme called **reverse transcriptase**. Often the reverse transcriptase is coded by a gene within the transposon and is translated from the RNA copy synthesized in step 1.
3. The DNA copy of the transposon integrates into the genome, possibly back into the same chromosome

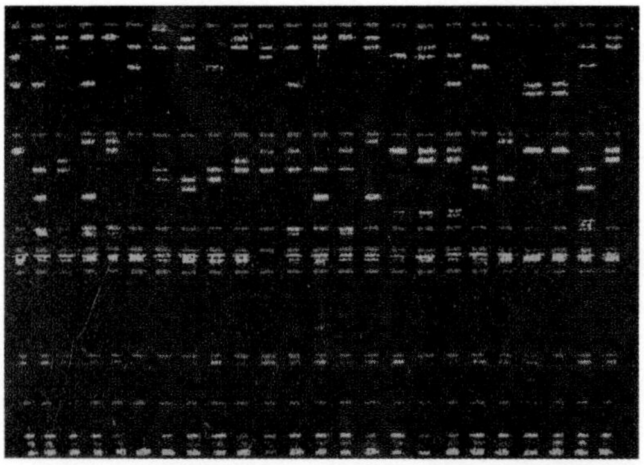

Figure 6.17 The use of microsatellite analysis in genetic profiling.

In this example, microsatellites located on the short arm of chromosome 6 have been amplified by PCR. The PCR products are labeled with a blue or green fluorescent marker and run in a polyacrylamide gel, each lane showing the genetic profile of a different individual. No two individuals have the same genetic profile because each person has a different set of microsatellite alleles, the alleles giving rise to bands of different sizes after PCR. The red bands are DNA size markers. Image supplied courtesy of PE Biosystems, Warrington, UK, and reproduced with permission.

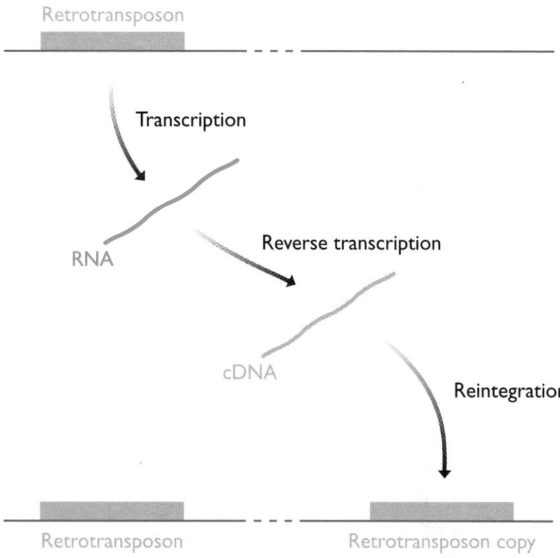

Figure 6.18 Retrotransposition.

Compare with *Figure 6.7*, p. 124, and note that the events are essentially the same as those that result in a processed pseudogene.

occupied by the original unit, or possibly into a different chromosome.

The end result is that there are now two copies of the transposon, at different points in the genome.

RNA transposons or **retroelements** are features of eukaryotic genomes but have not so far been discovered in prokaryotes. They have attracted a great deal of attention because there are clear similarities between some types of retroelement and free-living viruses called **retroviruses**, which include many benign forms but also virulent types such as the human immunodeficiency viruses which cause AIDS. These structural relationships are illustrated in *Figure 6.19* and can be summarized as follows:

- **Retroviruses** are viruses whose genomes are made of RNA and have the genetic organization shown in *Figure 6.19A*. They infect many types of vertebrate. Once inside a cell, the RNA genome is copied into DNA by the reverse transcriptase specified by the viral *pol* gene and the DNA copy integrates into the host genome. New viruses can be produced by copying the integrated DNA into RNA and packaging this into virus coat proteins, the latter coded by the *env* genes on the virus genome.
- **Endogenous retroviruses (ERVs)** are retroviral genomes integrated into vertebrate chromosomes. Some are still active and might, at some stage in a cell's lifetime, direct synthesis of exogenous viruses, but most are decayed relics that no longer have the capacity to form viruses (Patience *et al.*, 1997). These inactive sequences are genome-wide repeats but they are not capable of additional proliferation. There are some 1000 in the human genome. Truncated versions, sometimes called **retroviral-like elements** or **RTVLs**, are more common, with approximately 20 000 in human DNA.
- **Retrotransposons** have sequences similar to ERVs but are features of nonvertebrate eukaryotic genomes (i.e. plants, fungi, invertebrates and microbial eukaryotes) rather than vertebrates. Retrotransposons have very high copy numbers in some genomes, with many different types present. Most of the genome-wide repeats in maize are retrotransposons and in this plant these elements probably make up almost half the genome (see *Figure 1.7*, p. 10). There are two types of retrotransposon: the *Ty3/gypsy* family (*Ty3* and *gypsy* are examples of this class in yeast and fruit fly respectively), which possess the same set of genes as an ERV; and the *Ty1-copia* family, members of which lack the *env* gene (*Figure 6.19B*). Both types are able to transpose, via the mechanism depicted in *Figure 6.18*, but the absence of the *env* gene means that the *Ty1-copia* group cannot form infectious virus particles. In fact, despite the presence of *env* in the *Ty3/gypsy* genome, it has only recently been recognized that some of these elements can form viruses and hence should be looked on as nonvertebrate retroviruses. Although

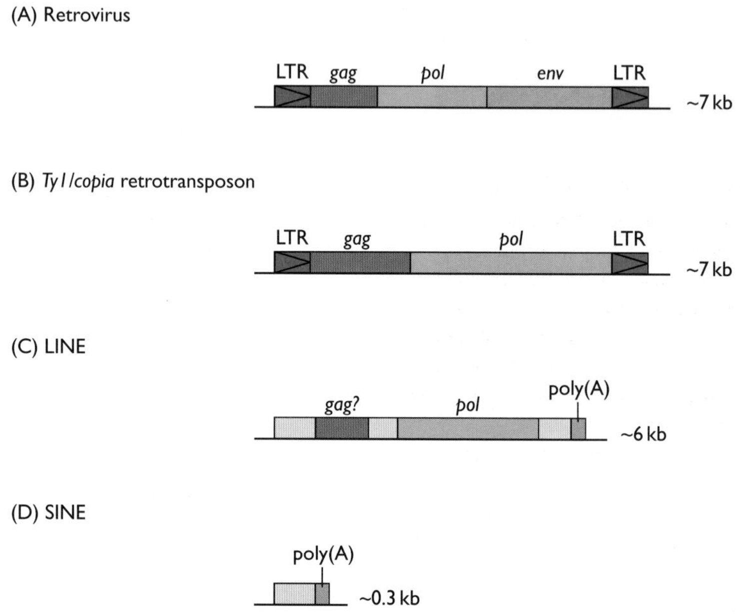

Figure 6.19 Retroelements.

A comparison of the structures of four types of retroelement. Retroviruses and retrotransposons are LTR elements that possess long terminal repeats at each end. The *gag* gene codes for a series of proteins located in the virus core, *pol* codes for the reverse transcriptase and other enzymes involved in replication of the element, and *env* codes for coat proteins. LINEs and SINEs are non-LTR retroelements or retroposons. Both have a poly(A) region (a long series of A nucleotides) at one end.

technically interspersed, retrotransposons are sometimes found in clusters in a genome sequence due to the presence of preferred integration sites for transposing elements.

The three types of retroelement described so far are **LTR elements**, as they have long terminal repeats at either end which play a role in the transposition process (see *Figure 6.19*). Other retroelements do not have LTRs. These are called **retroposons** and in mammals include:

- **LINEs (long interspersed nuclear elements)**, which contain a reverse transcriptase-like gene probably involved in the retrotransposition process (*Figure 6.19C*). An example is the human element LINE-1, which is 6.1 kb and has a copy number of 3500 for the full length element, with several hundred thousand truncated copies also present.
- **SINEs (short interspersed nuclear elements)**, which do not have a reverse transcriptase gene but can still transpose, probably by 'borrowing' reverse transcriptase enzymes that have been synthesized by other retroelements (*Figure 6.19D*). The commonest SINE in the human genome is ***Alu*** which has a copy number approaching one million. *Alu* seems to be derived from the gene for the 7SL RNA, which is involved in movement of proteins around the cell. The first *Alu* element may have arisen by the accidental reverse transcription of a 7SL RNA molecule and integration of the DNA copy into the human genome. Because *Alu* elements are actively transcribed (though the RNA does not seem to have a function) there are relatively large amounts of *Alu* RNA in the cell, providing the opportunity for proliferation of the element.

DNA transposons

Not all transposons require an RNA intermediate. Many are able to transpose in a more direct, DNA to DNA manner. With these elements we are aware of two distinct transposition mechanisms (*Figure 6.20*), one involving direct interaction between the donor transposon and the target site, resulting in copying of the donor element (**replicative transposition**), and the second involving excision of the element and reintegration at a new site (**conservative transposition**). Both mechanisms require enzymes which are usually coded by genes within the transposon.

In eukaryotes, DNA transposons are less common than retrotransposons (there are probably less than 1000 DNA transposon sequences in the human genome), but they have a special place in genetics because a family of plant DNA transposons, the *Ac/Ds* elements of maize, were the first transposable elements to be discovered, by Barbara McClintock in the 1950s. Her conclusions – that some genes are mobile and can move from one position to another in a chromosome – were based on exquisite genetical experiments, but were widely disbelieved by other biologists until the late 1970s when the molecular basis of transposition was first appreciated. Today, eukaryotic DNA transposons are continuing to shake the paradigms of genetics by providing the most convincing evidence for **horizontal gene transfer**, the direct movement of DNA sequences from the genome of one species to that of another. This appears to have occurred on several occasions with the DNA transposon that has rather romantically been called *mariner*. This element, about 1250 bp in length, was first discovered in a *Drosophila* species (Jacobson *et al.*, 1986) but is now known to be present in many animals including humans (Hartl, 1997). A number of examples of horizontal transmission involving *mariner* can be deduced by comparing the nucleotide sequences of *mariner* elements in different species, including two cases of horizontal transmission into humans (Robertson *et al.*, 1996).

Although relatively uncommon in eukaryotes, DNA transposons are an important component of prokaryotic genome anatomy. The **insertion sequences**, IS1 and IS186, present in the 50-kb segment of *E. coli* DNA that we examined in Chapter 1 (see *Figure 1.7*, p. 10), are examples of DNA transposons, and a single *E. coli* genome may contain as many as 20 of these, of various types. Most of the sequence of an IS is taken up by one or two genes that specify the **transposase** enzyme that catalyzes its transposition (*Figure 6.21A*). IS elements can transpose either replicatively or conservatively. Other kinds of DNA transposon known in *E. coli*, and fairly typical of prokaryotes in general, are:

- **Composite transposons**, which are basically a pair of IS elements flanking a segment of DNA usually containing one or more genes, often ones coding for antibiotic resistance (*Figure 6.21B*). The transposition of a composite transposon is catalyzed by the transposase coded by one or both of the IS elements. Composite transposons use the conservative mechanism of transposition.
- **Tn3-type transposons** have their own transposase gene and so do not require flanking IS elements in order to transpose (*Figure 6.21C*). Tn3 elements transpose replicatively.

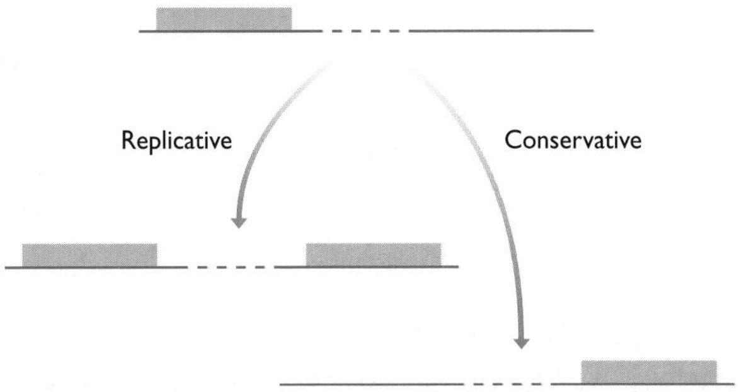

Figure 6.20 Two mechanisms of transposition used by DNA transposons. For more details see Section 13.2.3

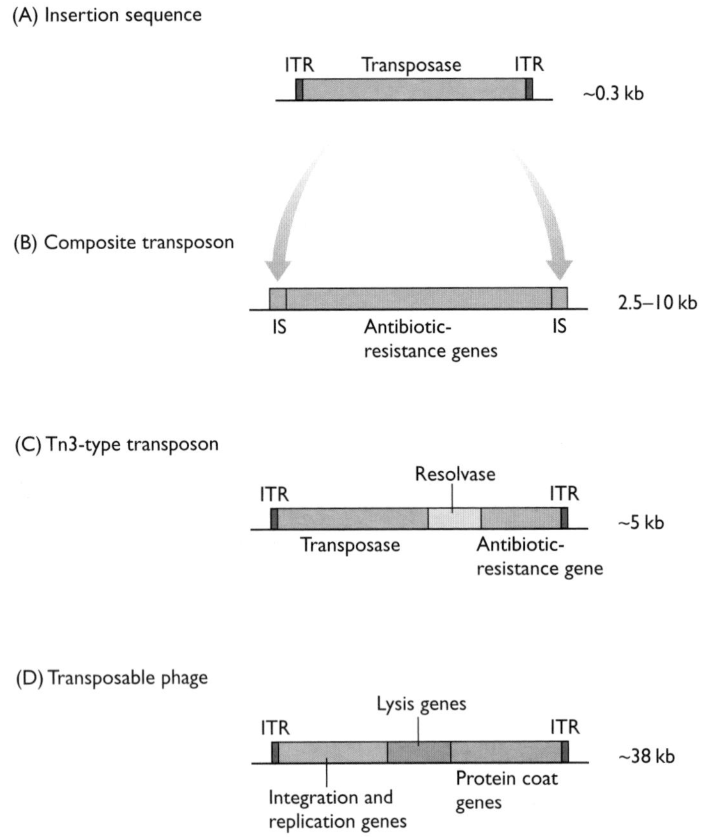

Figure 6.21 DNA transposons of prokaryotes.

Four types are shown. Insertion sequences, Tn3-type transposons and transposable phages are flanked by short (< 50 bp) inverted terminal repeat (ITR) sequences. The resolvase gene of the Tn3-type transposon codes for a protein involved in the transposition process.

- **Transposable phages** are bacterial viruses which transpose replicatively as part of their normal infection cycle (*Figure 6.21D*).

6.3.3 Repetitive DNA and eukaryotic chromosome architecture

The complete prokaryotic sequences that are already available show the precise positions of repeated sequences such as IS elements and transposons in these genomes. For eukaryotes, our knowledge of exactly how the repetitive DNA is organized in a chromosome is much less well advanced. We know the relative amounts of single copy and repetitive DNA in the human genome (see Box 6.4, p. 135), we know that certain classes of satellite DNA occur in the centromere and telomere regions, and we have snapshots of the organization of genome-wide repeats in short segments of plant and human DNA (see *Figure 1.7*, p. 10; Rowen *et al.*, 1996; Voytas, 1996; Andersson *et al.*, 1998). But we lack a comprehensive understanding of how genome-wide repeats are distributed within chromosomal DNA molecules.

An interesting attempt to bring together various lines of evidence and infer the overall architecture of a typical plant chromosome has been made by Schmidt and Heslop-Harrison (1998). The picture they paint emphasizes how insignificant the genes are within the chromosome as a whole (*Figure 6.22*). The bulk of the plant chromosome is made up of 30 or so different repeat sequences, predominantly the tandemly-repeated satellite DNA sequences at the centromere, telomeres and elsewhere, together with long swathes of retrotransposon sequences, the latter technically interspersed in the genome but at such a high copy number that they form clusters that stretch for megabases along the chromosomal DNA (as we saw for maize DNA in *Figure 1.7*, p. 10). Microsatellites and LINEs make a lesser, but still significant, contribution. To what extent this prediction of plant chromosome architecture holds true for animal genomes is not yet known, but it is clear that the human genome sequence will be valuable not only for what it tells us about the genes it contains, but also because of the insights it will provide into the arrangement of repetitive DNA within each chromosomal DNA molecule.

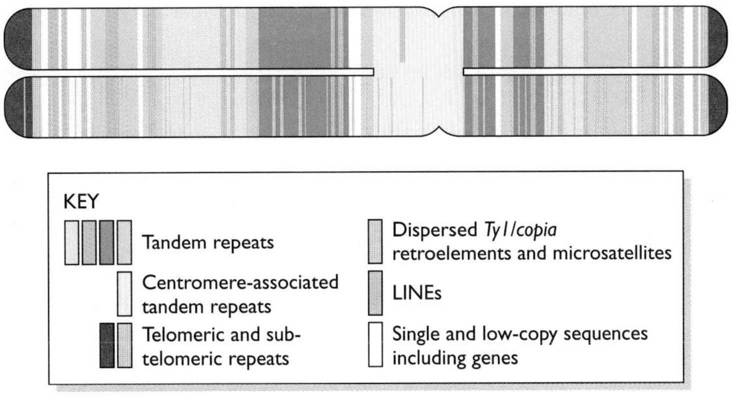

Figure 6.22 Repetitive DNA in a plant chromosome.

The drawing shows a model for the positioning of different types of repetitive DNA in a typical plant chromosome. Note that the positions of the dispersed *Ty1/copia* retroelements and microsatellites are shown in the lower pair of chromatids only. Reprinted from *Trends in Plant Science*, **3**, Schmidt T and Heslop-Harrison JS, pp. 195–199, copyright 1998, with permission from Elsevier Science.

REFERENCES

Akhmanova A, Voncken F, van Allen T, van Hoek A, Boxma B, Vogels G, Veenhuis M and Hackstein JHP (1998) A hydrogenosome with a genome. *Nature*, **396**, 527–528.

Andersson G, Svensson A-C, Setterblad N and Rask L (1998) Retroelements in the human MHC class II region. *Trends Genet.*, **14**, 109–114.

Andersson SGE, Zomorodipour A, Andersson JO, *et al.* (1998) The genome sequence of *Rickettsia prowazekii* and the origin of mitochondria. *Nature*, **396**, 133–140.

Blattner FR, Plunkett G, Bloch CA, *et al.* (1997) The complete genome sequence of *Escherichia coli* K-12. *Science*, **277**, 1453–1462.

Brown TA (1998) *Molecular Biology Labfax*, 2nd edn, Vol. 1. Academic Press, London.

Chen CW (1996) Complications and implications of linear bacterial chromosomes. *Trends Genet.*, **12**, 192–196.

Cohen J (1997) How many genes are there? *Science*, **275**, 769.

Csink AK and Henikoff S (1998) Something from nothing: the evolution and utility of satellite repeats. *Trends Genet.*, **14**, 200–204.

Deckert G, Warren PV, Gaasterland T, *et al.* (1998) The complete genome of the hyperthermophile bacterium *Aquifex aeolicus*. *Nature*, **392**, 353–358.

Doolittle RF (1997) Microbial genomes opened up. *Nature*, **392**, 339–342.

Ferdows MS and Barbour AG (1989) Megabase-sized linear DNA in the bacterium *Borrelia burgdorferi*, the Lyme disease agent. *Proc. Natl Acad. Sci. USA*, **86**, 5969–5973.

Fraser CM, Gocayne JD, White O, *et al.* (1995) The minimal gene complement of *Mycoplasma genitalium*. *Science*, **270**, 397–403.

Fraser CM, Casjens S, Huang WM, *et al.* (1997) Genomic sequence of a Lyme disease spirochaete, *Borrelia burgdorferi*. *Nature*, **390**, 580–586.

Fraser CM, Norris SJ, Weinstock GM, *et al.* (1998) Complete genome sequence of *Treponema pallidum*, the syphilis spirochaete. *Science*, **287**, 375–388.

Fritsch EF, Lawn RM and Maniatis T (1980) Molecular cloning and characterization of the human α-like globin gene cluster. *Cell*, **19**, 959–972.

Gardiner K (1996) Base composition and gene distribution: critical patterns in mammalian genome organization. *Trends Genet.*, **12**, 519–524.

Hartl DL (1997) *Mariner* sails into *Leishmania*. *Science*, **276**, 1659–1660.

Jacob F and Monod J (1961) Genetic regulatory mechanisms in the synthesis of proteins. *J. Mol. Biol.*, **3**, 318–356.

Jacobson JW, Medhora MM and Hartl DL (1986) Molecular structure of a somatically unstable transposable element in *Drosophila*. *Proc. Natl Acad. Sci.* USA, **83**, 8684–8688.

Jones RN (1995) B chromosomes in plants. *New Phytol.*, **131**, 411–434.

Lang BF, Burger G, O'Kelly CJ, Cedergren R, Golding GB, Lemieux C, Sankoff D, Turmel M and Gray MW (1997) An ancestral mitochondrial DNA resembling a eubacterial genome in miniature. *Nature*, **387**, 493–497.

McQueen HA, Siriaco G and Bird AP (1998) Chicken minichromosomes are hyperacetylated, early replicating, and gene rich. *Genome Res.*, **8**, 621–630.

Mushegian AR and Koonin EV (1996) A minimal gene set for cellular life derived by comparison of complete bacterial genomes. *Proc. Natl Acad. Sci. USA*, **93**, 10268–10273.

Nosek J, Tomáska L, Fukuhara H, Suyama Y and Kovác L (1998) Linear mitochondrial genomes: 30 years down the line. *Trends Genet.*, **14**, 184–188.

Palmer JD (1997a) The mitochondrion that time forgot. *Nature*, **387**, 454–455.

Palmer JD (1997b) Organelle genomes: going, going, gone! *Science*, **275**, 790–791.

Patience C, Wilkinson DA and Weiss RA (1997) Our retroviral heritage. *Trends Genet.*, **13**, 116–120.

Pennisi E (1996) Linker histones, DNA's protein custodians, gain new respect. *Science*, **274**, 503–504.

Pruss D, Hayes JJ and Wolffe AP (1995) Nucleosomal anatomy – where are the histones? *Bioessays*, **17**, 161–170.

Robertson HM, Zumpano KL, Lohe AR and Hartl DL (1996) Reconstructing the ancient mariners of humans. *Nature Genet.*, **12**, 360–361.

Rowen L, Koop BF and Hood L (1996) The complete 685-kilobase DNA sequence of the human β T-cell receptor locus. *Science*, **272**, 1755–1762.

Saccone S, Desario A, Dellaville G and Bernardi G (1992) The highest gene concentrations in the human genome are in telomeric bands of metaphase chromosomes. *Proc. Natl Acad. Sci. USA*, **89**, 4913–4917.

Schmidt T and Heslop-Harrison JS (1998) Genomes, genes and junk: the large-scale organization of plant chromosomes. *Trends Plant Sci.*, **3**, 195–199.

Schuler GD, Boguski MS, Stewart EA, *et al.* (1996) A gene map of the human genome. *Science*, **274**, 540–546.

Strachan T and Read AP (1996) *Human Molecular Genetics*. BIOS Scientific Publishers, Oxford.

Vafa O and Sullivan KF (1997) Chromatin containing CENP-A and alpha-satellite DNA is a major component of the inner kinetochore plate. *Curr. Biol.*, **7**, 897–900.

Voytas DF (1996) Retroelements in genome organization. *Science*, **274**, 737–738.

Zhou Y-B, Gerchman SE, Ramakrishnan V, Travers A and Muyldermans S (1998) Position and orientation of the globular domain of linker histone H5 on the nucleosome. *Nature*, **395**, 402–405.

FURTHER READING

Drlica K and Riley M (1990) *The Bacterial Chromosome*. American Society for Microbiology, Washington, DC. — *A source of information on all aspects of bacterial DNA.*

Federoff NV (1983) Controlling elements in maize. In: J Shapiro, ed. *Mobile Genetic Elements*. Academic Press, New York, pp. 1–63. — *Includes a summary of the experiments carried out by McClintock that led to the discovery of transposons.*

Gill P, Ivanov PL, Kimpton C, Piercy R, Benson N, Tully G, Evett I, Hagelberg E and Sullivan K (1994) Identification of the remains of the Romanov family by DNA analysis. *Nature Genet.* **6**, 130–135. — *An example of the use of genetic fingerprinting based on microsatellite variability.*

Margulis L (1970) *Origin of Eukaryotic Cells*. Yale University Press, New Haven. — *The first description of the endosymbiotic theory for the origin of mitochondria and chloroplasts.*

Ramakrishnan V (1997) Histone H1 and chromatin higher-order structure. *Crit. Rev. Eukaryotic Gene Expression*, **7**, 215–230. — *Detailed descriptions of models for the 30 nm chromatin fiber.*

Strachan T and Read AP (1996) *Human Molecular Genetics*. BIOS Scientific Publishers, Oxford. — *The organization of the human genome is described in Chapter 7.*

CHAPTER

7 The Role of DNA-binding Proteins

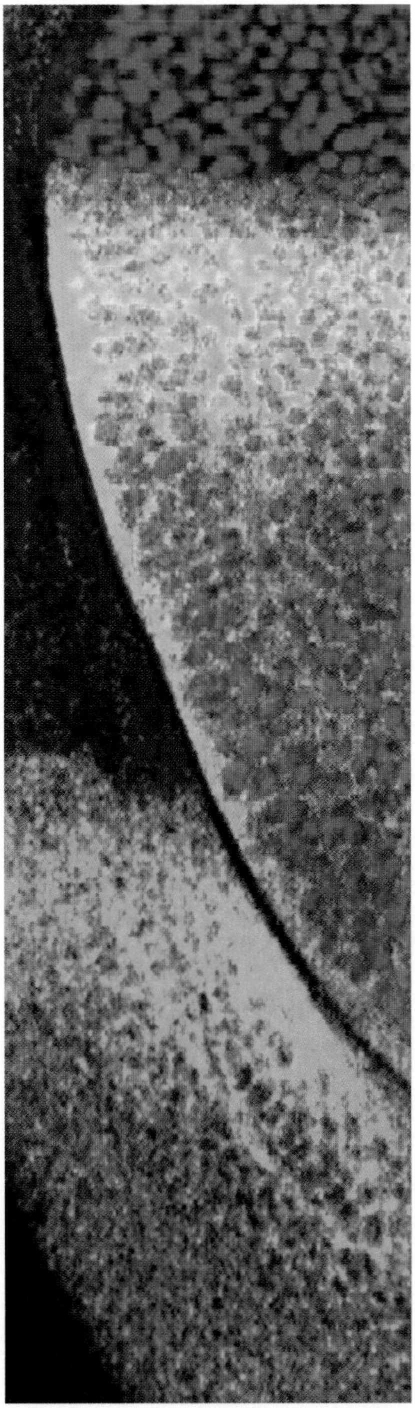

Contents

7.1 *The Structure of DNA* 144
- 7.1.1 Nucleotides and polynucleotides 145
- 7.1.2 The double helix 145

7.2 *Proteins* 150
- 7.2.1 The four levels of protein structure 150

7.3 *Methods for Studying DNA-binding Proteins* 156
- 7.3.1 Locating a protein-binding site on a DNA molecule 156
- 7.3.2 Purification of a DNA-binding protein 157
- 7.3.3 Techniques for studying protein and DNA–protein structures 160

7.4 *Interactions between DNA and DNA-binding Proteins* 162
- 7.4.1 The DNA partner in the interaction 163
- 7.4.2 The protein partner 164
- 7.4.3 DNA and protein in partnership 167

Concepts

- *DNA and proteins are linear, unbranched polymers*
- *The DNA double helix can adopt a variety of conformations*
- *The functional, three-dimensional structure of a protein is determined by its amino acid sequence*
- *Various techniques can be used to locate protein binding sites on DNA molecules*
- *X-ray crystallography and NMR are powerful methods for determining protein structures*
- *Binding proteins are thought to locate their attachment sites mainly by direct sequence readout of the double helix*
- *DNA-binding proteins have special structural motifs that enable them to form a close attachment with the double helix*

ON ITS OWN the genome is inactive. It is a store of information but it does not itself possess the means of releasing that information to the cell. Which genes are expressed at a particular time is determined by the activities of proteins that interact with the genome, either by binding to specific DNA sequences or by attaching in a less specific manner. These DNA-binding proteins include enzymes that copy genes into RNA molecules, and proteins that specify which genes are switched on and which are switched off. Understanding the activities of DNA-binding proteins is therefore one of the keys to understanding how the genome works.

Our current knowledge in this area is based on structural studies that have described the detailed interactions between DNA molecules and binding proteins. This work is itself based on the studies of DNA and proteins that were carried out in the first half of the 20th century and which culminated in the discovery of the double helix structure of DNA (Watson and Crick, 1953) and elucidation of the main structural units in proteins (Pauling *et al.*, 1951; Pauling and Corey, 1953). We begin this chapter by reviewing these classic structural studies and examining how they have been extended by research since the 1950s. We will then survey the techniques used to study interactions between DNA and proteins and finally explore what we currently know about how binding proteins make attachments with DNA molecules. This will set the scene for Chapters 8 and 9, in which we will look at the various roles of DNA-binding proteins in genome expression. In Chapters 9 and 10 we will see that related proteins, ones that bind to RNA rather than DNA, also play an important part in genome expression, and in Chapters 12 and 13 we will discover that DNA-binding proteins are also responsible for replication of the genome and for correcting sequence alterations caused by chemical and physical attack.

7.1 THE STRUCTURE OF DNA

To understand the structure of DNA we must address two questions. The first concerns the structure of the **polynucleotide**, the DNA polymer composed of linked nucleotide subunits that encodes genomic information. To all intents and purposes, this aspect of DNA structure was fully worked out in the years before 1940. The second question concerns the way in which two polynucleotides combine to form the double-stranded DNA molecules found in living cells. This was the question asked by Watson and Crick in the early 1950s. Their discovery of the **double helix** was a landmark in genetics but it was not a complete description of the structure of DNA in living cells, more recent studies having shown that double-stranded DNA can take up a range of conformations.

7.1.1 Nucleotides and polynucleotides

DNA is a linear, unbranched polymer in which the monomeric subunits are four chemically-distinct nucleotides that can be linked together in any order in chains hundreds, thousands or even millions of units in length.

The chemical structures of the four nucleotides

Each nucleotide in a DNA polymer is made up of three components (see *Figure 1.1*, p. 3):

1. **2′-deoxyribose**, which is a **pentose**, a type of sugar composed of five carbon atoms. These five carbons are numbered 1′ (spoken as 'one-prime'), 2′, etc. The name '2′-deoxyribose' indicates that this particular sugar is a derivative of ribose, one in which the hydroxyl (–OH) group attached to the 2′-carbon of ribose has been replaced by a hydrogen (–H) group.
2. **A nitrogenous base**, either **cytosine** or **thymine** (single-ring **pyrimidines**) or **adenine** or **guanine** (double-ring **purines**). The base is attached to the 1′-carbon of the sugar by a **β-*N*-glycosidic bond** attached to nitrogen number 1 of the pyrimidine or number 9 of the purine.
3. **A phosphate group**, comprising one, two or three linked phosphate units attached to the 5′-carbon of the sugar. The phosphates are designated α, β and γ, with the α-phosphate being the one directly attached to the sugar.

A molecule made up of just the sugar and base is called a **nucleoside**; addition of the phosphates converts this to a nucleotide. Although cells contain nucleotides with one, two or three phosphate groups, only the nucleoside triphosphates act as substrates for DNA synthesis. The full chemical names of the four nucleotides that polymerize to make DNA are:

2′-deoxyadenosine 5′-triphosphate
2′-deoxycytidine 5′-triphosphate
2′-deoxyguanosine 5′-triphosphate
2′-deoxythymidine 5′-triphosphate

The abbreviations of these four nucleotides are dATP, dCTP, dGTP and dTTP, respectively, or, when referring to a DNA sequence, A, C, G and T.

Polynucleotides

In a polynucleotide, individual nucleotides are linked together by **phosphodiester bonds** between their 5′- and 3′-carbons (see *Figure 1.1*). From the structure of this linkage we can see that the polymerization reaction (*Figure 7.1*) involves removal of the two outer phosphates (the β- and γ-phosphates) from one nucleotide and replacement of the hydroxyl group attached to the 3′-carbon of the second nucleotide. Note that the two ends of the polynucleotide are chemically distinct, one having an unreacted triphosphate group attached to the 5′-carbon (the **5′** or **5′-P terminus**) and the other having an unreacted hydroxyl attached to the 3′-carbon (the **3′** or **3′-OH terminus**). This means that the polynucleotide has a chemical direction, expressed as either 5′→3′ (down in *Figure 7.1*) or 3′→5′.

An important consequence of the polarity of the phosphodiester bond is that the chemical reaction needed to extend a DNA polymer in the 5′→3′ direction is different to that needed to make a 3′→5′ extension. All natural DNA polymerase enzymes are only able to carry out 5′→3′ synthesis, which adds significant complications to the process by which double-stranded DNA is replicated (Section 12.3.2). The same limitation applies to RNA polymerases, the enzymes which make RNA copies of DNA molecules (Section 8.2.1).

RNA

Although our attention is firmly on DNA, the structure of the RNA polynucleotide is so similar to that of DNA that it makes sense to deal with it here. RNA is also a polynucleotide but with two differences compared with DNA (*Figure 7.2*). First, the sugar in an RNA nucleotide is ribose; and second, RNA contains uracil instead of thymine. The four nucleotide substrates for synthesis of RNA are therefore:

adenosine 5′-triphosphate
cytidine 5′-triphosphate
guanosine 5′-triphosphate
uridine 5′-triphosphate

which are abbreviated to ATP, CTP, GTP and UTP, or A, C, G and U.

As with DNA, RNA polynucleotides contain 3′–5′ phosphodiester bonds, but these phosphodiester bonds are less stable than those in a DNA polynucleotide, because of the indirect effect of the hydroxyl group at the 2′-position of the sugar. This may be one reason why the biological functions of RNA do not require the polynucleotide to be more than a few thousand nucleotides in length, at most. There are no RNA counterparts of the Mb sized DNA molecules found in prokaryotic and eukaryotic chromosomes.

7.1.2 The double helix

The polynucleotide structure of DNA was elucidated before it was discovered that DNA is the genetic material. This discovery came in two stages, first through the demonstration, published in 1945, that the transfer of DNA from one bacterium to another can result in the recipient acquiring new genetic information, and second by the realization in 1952 that the genes of the bacteriophage called T2 are made of DNA (Research Briefing 7.1). Various lines of evidence had shown that cellular DNA molecules are comprised of two or more polynucleotides assembled together in some way, and the possibility that unravelling this structure might provide insights into how genes work prompted Watson and Crick, among others, to try to solve the structure. According to Watson in his book *The Double Helix* (see Further Reading), their work was a desperate race against Linus Pauling who initially proposed an incorrect triple helix model, giving

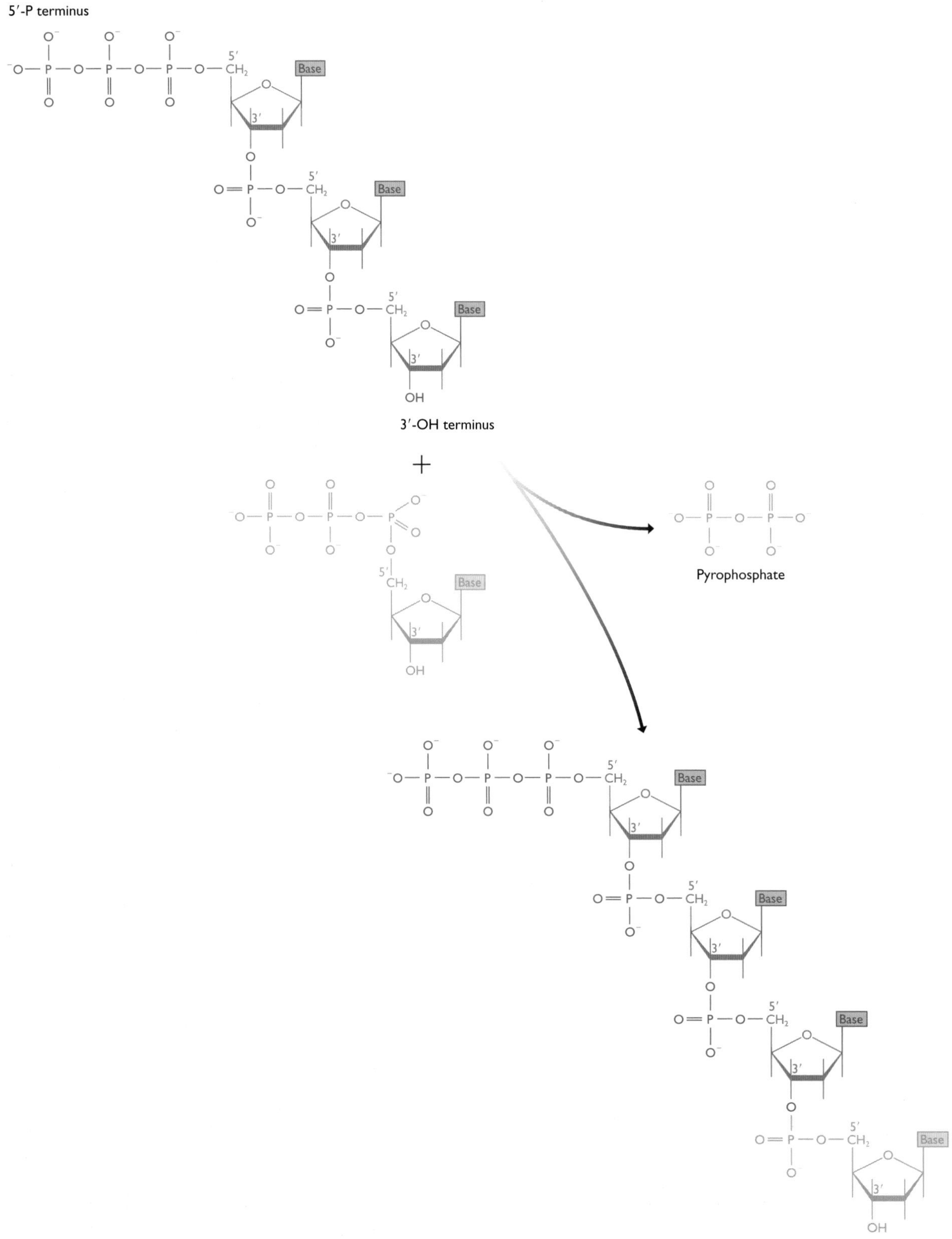

Figure 7.1 The polymerization reaction that results in synthesis of a DNA polynucleotide.

Synthesis occurs on the 5′→3′ direction with the new nucleotide being added to the 3′carbon at the end of the existing polynucleotide. The β- and γ-phosphates of the nucleotide are removed as a pyrophosphate molecule.

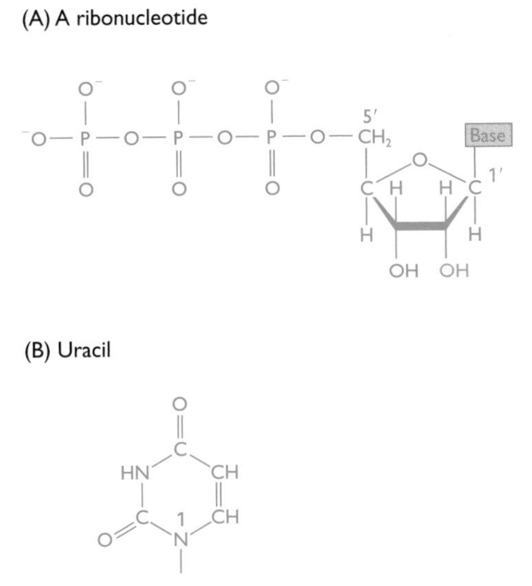

Figure 7.2 The chemical differences between DNA and RNA.

(A) RNA contains ribonucleotides, in which the sugar is ribose rather than 2′-deoxyribose. The difference is that a hydroxyl group rather than hydrogen atom is attached to the 2′ carbon. (B) RNA contains the pyrimidine called uracil instead of thymine. See *Figure 1.1*, p. 3, for the structure of the deoxyribonucleotides found in DNA.

Watson and Crick the time they needed to complete the double helix structure. It is now difficult to separate fact from fiction, especially regarding the part played by Rosalind Franklin, whose X-ray diffraction studies provided the bulk of the experimental data in support of the double helix and who was herself very close to solving the structure. The one thing that is clear is that the double helix, discovered by Watson and Crick on Saturday, 7 March 1953, was the single most important breakthrough in biology during the 20th century.

The key features of the double helix

The double helix comprises two polynucleotides wound around one another with the two strands running in opposite directions (*Figure 7.3A*). The helix is stabilized by two types of chemical interaction:

- **Base-pairing** between the two strands, which involves the formation of hydrogen bonds (see Box 7.2, p. 153) between an adenine on one strand and a thymine on the other strand, or between a cytosine and a guanine (*Figure 7.3B*). These two base-pair combinations – A base-paired with T and G base-paired with C – are the only ones that are permissible. This is partly because of the geometries of the nucleotide bases and the relative positions of the groups that are able to participate in chemical bonds, and partly because the pair must be between a purine and a pyrimidine because a purine–purine pair would be too big to fit within the helix, and a pyrimidine–pyrimidine pair would be too small.
- **Base-stacking**, sometimes called **π–π interactions**, which involves hydrophobic interactions between adjacent base pairs and adds stability to the double helix once the strands have been brought together by base-pairing.

Both base-pairing and base-stacking are important in holding the two polynucleotides together, but base-pairing has added significance because of its biological implications. The limitation that A can only base-pair with T, and G can only base-pair with C, means that DNA replication can result in perfect copies of a parent molecule through the simple expedient of using the sequences of the pre-existing strands to dictate the sequences of the new strands. This is **template-dependent DNA synthesis** and it is the system used by all cellular DNA polymerases (Chapter 12). Its counterpart, **template-dependent RNA synthesis**, is used by RNA polymerases to make RNA copies of genes, these copies preserving the biological information contained in the sequence of the genomic DNA molecule (Chapter 9). The only difference between DNA and RNA syntheses is that when RNA is made the adenines in the DNA template do not specify thymines in the RNA copy, because RNA does not contain thymine: instead adenine pairs with uracil in DNA–RNA hybrids and in double-stranded RNA structures.

The double helix has structural flexibility

The double helix described by Watson and Crick, and shown in *Figure 7.3*, is called the B-form of DNA. Its characteristic features lie in its dimensions: a helical diameter of 2.37 nm, a rise of 0.34 nm per base pair, and a pitch (i.e. distance taken up by a complete turn of the helix) of 3.4 nm, this corresponding to ten base pairs per turn. The DNA inside living cells is thought to be predominantly in this B-form, but it is now clear that genomic DNA molecules are not entirely uniform in structure. This is mainly because each nucleotide in the helix has the flexibility to take up slightly different molecular shapes. To adopt these different conformations, the relative positions of the

Box 7.1: Base-pairing in RNA

RNA can also form base pairs, A pairing with U and G pairing with C. RNA can base-pair with a DNA molecule (as occurs during transcription: Sections 9.2.1 and 9.2.2) or with a second RNA molecule (for example, during intron splicing: Section 9.2.3) but the predominant form of RNA base-pairing is intramolecular. This is seen with both ribosomal and transfer RNAs, resulting in folded structures that enable these molecules to carry out their functions in protein synthesis (Sections 10.1.1 and 10.2.1). Within these structures, base-paired regions adopt helical conformations, but usually these extend for only a few tens of base pairs at most.

RESEARCH 7.1 BRIEFING

Genes are made of DNA

The discovery that genes are made of DNA was the key breakthrough that led to the emergence of modern molecular biology in the 1950s.

Cytochemical studies of cell nuclei carried out in the 1920s had shown that chromosomes contain roughly equal amounts of DNA and protein. Unfortunately, at that time DNA was thought to be a small molecule lacking the variability expected of the genetic material and it was therefore concluded that genes must be made of protein. Two experiments in the middle decades of the 20th century challenged this assumption.

Experimental evidence that genes are made of DNA

The first of these experiments – in reality a major research project – was completed by Avery, MacLeod and McCarty of Columbia University, New York, in 1944. Frederick Griffiths, a British medical officer, had previously shown that an unknown component (the 'transforming principle') of dead *Streptococcus pneumoniae* bacteria can convert living but harmless *S. pneumoniae* into a virulent strain capable of causing pneumonia. Avery and his colleagues carried out a meticulous series of experiments that indicated that, contrary to expectations, the transforming principle is DNA. This did not, however, immediately lead to acceptance of DNA as the genetic material, partly because it was not clear in the minds of all microbiologists that transformation really is a genetic phenomenon, and partly because it was possible to criticize the veracity of some aspects of Avery's work. In particular, doubts were raised about the purity of the deoxyribonuclease that Avery used to inactivate the transforming principle. This latter result, a central part of the evidence for the transforming principle being DNA, would be invalid if, as seemed possible, the enzyme contained even trace amounts of a contaminating protease.

The second experiment, carried out by Alfred Hershey and Martha Chase at Cold Spring Harbor, New York, in 1952, was more conclusive. Primarily designed as simply another step in elucidation of the infection cycle of bacteriophages (or 'phages', viruses that infect bacteria), the Hershey–Chase experiment succeeded in demonstrating that only the DNA component of a bacteriophage possesses genetic activity. A year earlier, James Watson and Ole Maaløe had tried to use radiolabeling to trace the fate of phage DNA during the infection cycle, but their experiments (and similar ones carried out by other groups) had been inconclusive because the remains of the original phages, which attach themselves to the bacteria, could not be distinguished from the new phages produced at the end of the infection cycle. Hershey and Chase adopted the radiolabeling strategy but with an important modification. They infected *Escherichia coli* bacteria with radiolabeled T2 phages, left the culture for a few minutes to allow the phage to inject their genes into the bacteria, and then agitated the culture in a blender. The blending detached the empty phages from the surface of the bacteria, enabling the bacteria plus phage genes to be collected by centrifugation. They discovered that 70% of the phage DNA but only 20% of the protein was present in the bacteria. After completion of the infection cycle, the new phages contained almost half the original DNA, but less than 1% of the protein. These results indicated that injection of phage genes into the bacterium is associated with injection of DNA and that DNA is the component of the infecting phages that is inherited by new phages. Both observations lead to the conclusion that phage genes are made of DNA.

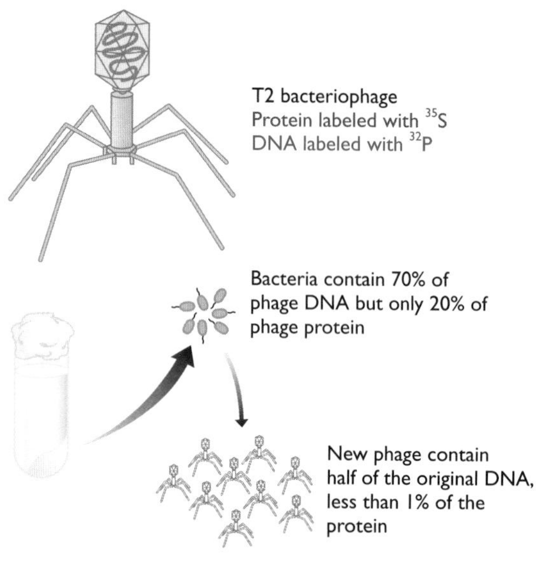

T2 bacteriophage
Protein labeled with ^{35}S
DNA labeled with ^{32}P
Bacteria contain 70% of phage DNA but only 20% of phage protein
New phage contain half of the original DNA, less than 1% of the protein

The impact of the Hershey–Chase experiment

The Hershey–Chase experiment has been assigned a central place in the history of molecular biology. In fact, it is not particularly compelling evidence for the general conclusion that genes are made of DNA, for the simple reason that a bacteriophage is very different from a cell and it might therefore be erroneous to extrapolate from one to the other (indeed, we know that some phage genomes are made of RNA). The Hershey–Chase experiment is important not because of what it tells us but because it alerted biologists, Watson in particular, to the fact that DNA *might* be the genetic material and was therefore worth studying. The notion that DNA might hold the secret to life led directly to the double helix structure of DNA which, with its inbuilt mechanism for replication via complementary base-pairing (see Chapter 12), is the real experimental proof that genes are made of DNA.

References

Avery OT, MacLeod CM and McCarty M (1944) Studies on the chemical nature of the substance inducing transformation of pneumococcal types. *J. Exp. Med.*, **79**, 137–158.

Hershey AD and Chase M (1952) Independent functions of viral protein and nucleic acid in growth of bacteriophage. *J. Gen. Physiol.*, **36**, 39–56.

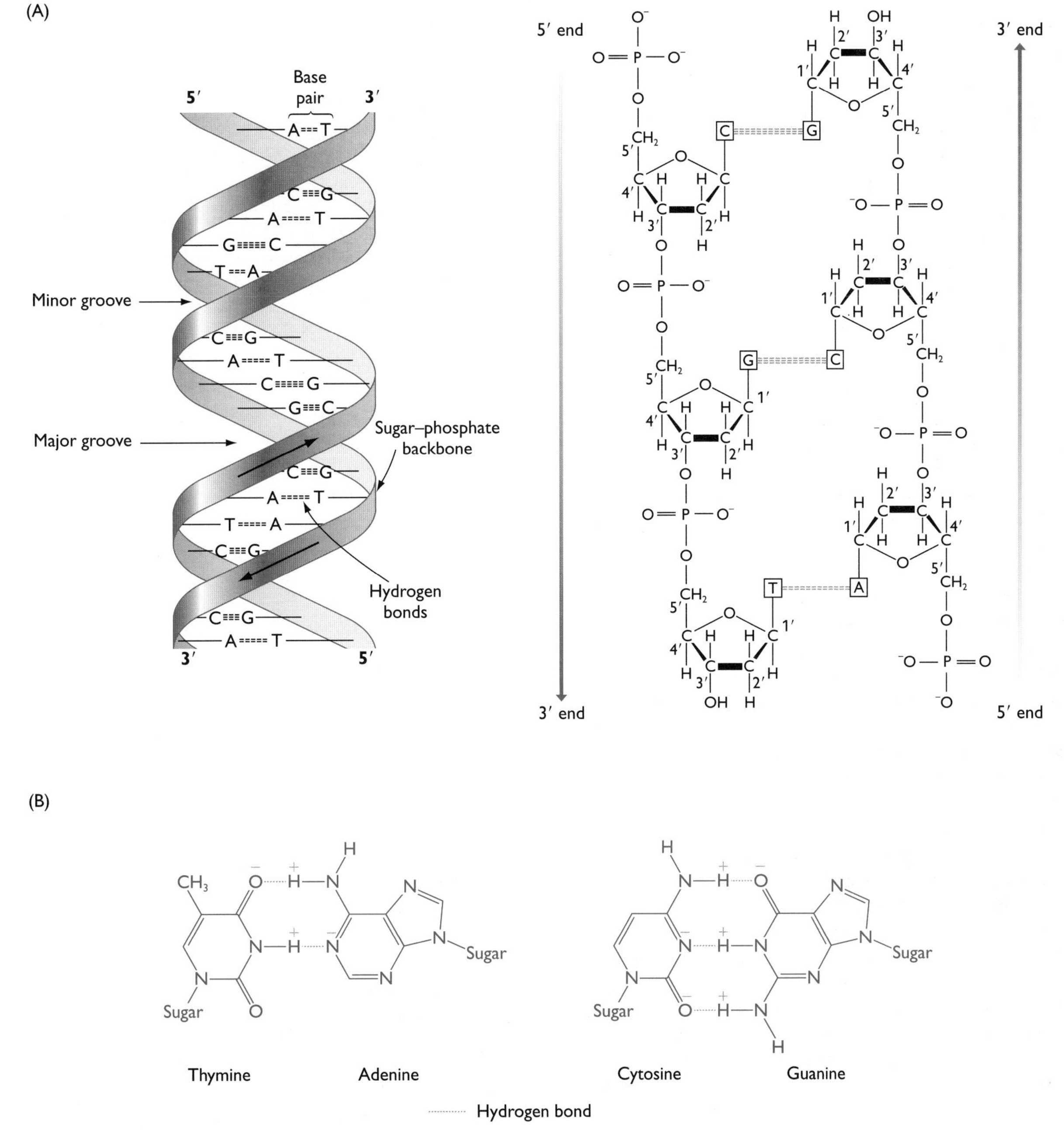

Figure 7.3 The double helix structure of DNA.

(A) Two representations of the double helix. On the left the structure is shown with the sugar–phosphate 'backbones' of each polynucleotide drawn as a red ribbon with the base-pairs in black. On the right the chemical structure for three base pairs is given. (B) A base-pairs with T and G base-pairs with C. The bases are drawn in outline with the hydrogen bonding indicated by dotted lines. Note that a G–C base pair has three hydrogen bonds whereas an A–T base pair has just two. The structures in part (A) are redrawn from Turner *et al.* (1997) (left drawing) and Strachan and Read (1996) (right structure).

atoms in the nucleotide must change slightly. There are a number of possibilities but the most important conformational changes involve rotation around the β-*N*-glycosidic bond, changing the orientation of the base relative to the sugar, and rotation around the bond between the 3′- and 4′-carbons. Both rotations have a significant effect on the double helix: changing the base orientation influences the relative positioning of the two polynucleotides, and rotation around the 3′–4′ bond affects the conformation of the sugar–phosphate backbone.

Rotations within individual nucleotides therefore lead to major changes in the overall structure of the helix. It has been recognized since the 1950s that changes in the dimensions of the double helix occur when fibers containing DNA molecules are exposed to different relative humidities. The modified version of the double helix called the A-form (*Figure 7.4*), for example, has a diameter of 2.55 nm, a rise of 0.29 nm per base pair and a pitch of 3.2 nm, corresponding to 11 base pairs per turn (*Table 7.1*). Other variations include B′-, C-, C′-, C″-, D-, E- and T-DNAs. All these are right-handed helices like the B-form, but a more drastic reorganization is also possible leading to the left-handed Z-DNA (*Figure 7.4*), a slimmer version of the double helix with a diameter of only 1.84 nm.

The bare dimensions of the various forms of the double helix do not reveal what are probably the most significant differences between them. These relate not to diameter and pitch, but the extent to which the internal regions of the helix are accessible from the surface of the structure. As shown in *Figures 7.3* and *7.4*, the B-form of DNA does not have an entirely smooth surface: instead two grooves spiral along the length of the helix. One of these grooves is relatively wide and deep and is called the **major groove**, the other is narrow and less deep and is called the **minor groove**. A-DNA also has two grooves (*Figure 7.4*), but with this conformation the major groove is even deeper, and the minor groove shallower and broader. Z-DNA is different again, with one groove virtually nonexistent but the other very narrow and deep. In each form of DNA, part of the internal surface of at least one of the grooves is formed by chemical groups attached to the nucleotide bases. It is therefore possible for a DNA-binding protein to 'read' the nucleotide sequence, with a greater or lesser degree of ambiguity, simply by reaching down into a groove, and without having to open up the helix by breaking the base pairs. A corollary of this is that a DNA-binding protein whose structure enables it to recognize a specific nucleotide sequence within, say, B-DNA might not be able to recognize that sequence if the DNA has taken up a different conformation. As we will see later in this chapter, conformational variations along the length of a DNA molecule, together with other structural polymorphisms caused by the nucleotide sequence, could be important in determining the specificity of the interactions between the genome and its DNA-binding proteins.

7.2 PROTEINS

A protein, like a DNA molecule, is a linear, unbranched polymer. In proteins the monomeric subunits are called **amino acids** (*Figure 7.5*) and the resulting polymers, or **polypeptides**, are rarely more than 2000 units in length. As with DNA, the key features of protein structure were determined in the first half of the 20th century, this phase of protein biochemistry culminating in the 1940s and early 1950s with elucidation by Pauling and Corey of the major conformations or **secondary structures** taken up by polypeptides (Pauling *et al.*, 1951; Pauling and Corey, 1953). In recent years, interest has focused on how these secondary structures combine to produce the complex three-dimensional shapes of proteins.

Proteins have a tremendous amount of chemical and structural diversity and this enables them to carry out a vast range of different functions (*Table 7.2*). The basis of this chemical diversity will be the theme running through our investigation of protein structure.

7.2.1 The four levels of protein structure

Proteins are traditionally looked on as having four distinct levels of structure. These levels are hierarchical, the protein being built up stage by stage with each level of structure dependent on the one below it:

1. **The primary structure** of the protein is formed by joining amino acids into a polypeptide by **peptide bonds**. These are synthesized by a condensation reaction between the carboxyl group of one amino acid and the amino group of a second amino acid (*Figure 7.6*). In passing, note that, as with a polynucleotide, the two ends of the polypeptide are chemically distinct: one has a free amino group and is called the **amino, NH_2-,** or **N terminus**; the other has a free carboxyl group and is called the **carboxyl, COOH–,** or **C terminus**. The direction of the polypeptide can therefore be expressed as either N→C (left to right in *Figure 7.6*) or C→N.
2. **The secondary structure** refers to the different conformations that can be taken up by the polypeptide.

Table 7.1 Features of different conformations of the DNA double helix

Feature	Conformation B-DNA	A-DNA	Z-DNA
Type of helix	Right-handed	Right-handed	Left-handed
Helical diameter (nm)	2.37	2.55	1.84
Rise per base pair (nm)	0.34	0.29	0.37
Distance per complete turn (= pitch)(nm)	3.4	3.2	4.5
Number of base pairs per complete turn	10	11	12
Topology of major groove	Wide, deep	Narrow, deep	Flat
Topology of minor groove	Narrow, shallow	Broad, shallow	Narrow, deep

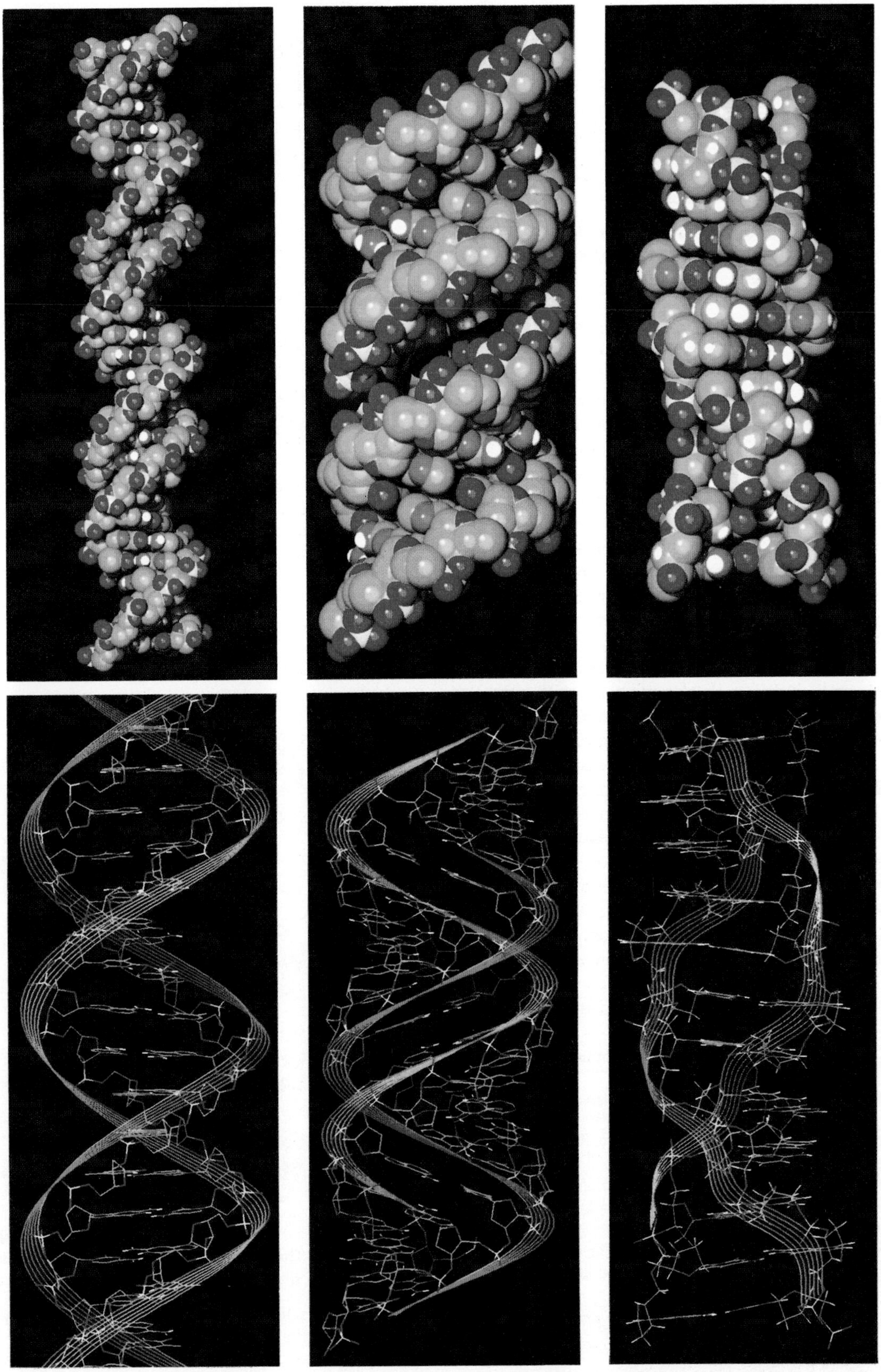

Figure 7.4 Computer-generated images of B-DNA (left), A-DNA (center) and Z-DNA (right).

Reprinted with permission from Kendrew A (ed.), *The Encyclopaedia of Molecular Biology,* Plate 1. Copyright 1994 Blackwell Science.

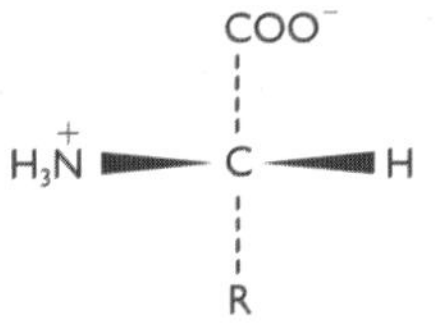

Figure 7.5 The general structure of an amino acid.

All amino acids have the same general structure, comprising a central α-carbon attached to a hydrogen atom, carboxyl group, amino group and R group. The R group is different for each amino acid (see *Figure 7.9*).

The two main types of secondary structure are the **α-helix** and **β-sheet** (*Figure 7.7*), both of which are stabilized by hydrogen bonds that form between different amino acids in the polypeptide. Most polypeptides are long enough to be folded into a series of secondary structures, one after another along the molecule.

3. **The tertiary structure** results from folding the secondary structural components of the polypeptide into a three-dimensional configuration (*Figure 7.8*). The tertiary structure is stabilized by various chemical forces (Box 7.2), notably hydrogen bonding between individual amino acids and hydrophobic forces which dictate that amino acids with nonpolar (i.e. 'water-hating') side groups must be shielded from water by embedding within the internal regions of the protein. There may also be covalent linkages called **disulfide bridges** between cysteine amino acids at various places in the polypeptide.
4. **The quaternary structure** involves the association of two or more polypeptides, each folded into its tertiary structure, into a multisubunit protein. Not all proteins form quaternary structures, but it is a feature of many proteins with complex functions, including several involved in genome expression.

Table 7.2 The functional diversity of proteins

Function	Examples in humans
Biochemical catalysis (enzymes)	DNA polymerase, RNA polymerase
Structure	Collagen in bone and tendons
Movement	Actin and myosin in muscles
Transport	Hemoglobin
Regulation	Hormones such as insulin
Protection	Antibodies, blood clotting proteins
Storage	Ferritin – stores iron in the liver

DNA-binding proteins include examples of three of these functional classes: enzymes (e.g. RNA polymerase: Section 8.2.1), structural proteins (e.g. histones: Section 6.1.1) and regulatory proteins (e.g. transcription factors: Section 8.3.2).

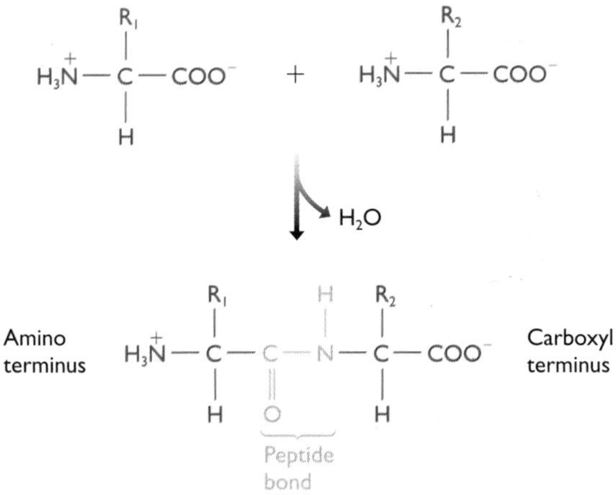

Figure 7.6 In polypeptides, amino acids are linked by peptide bonds.

The drawing shows the chemical reaction that results in two amino acids becoming linked together by a peptide bond. The reaction is called a condensation because it results in elimination of water.

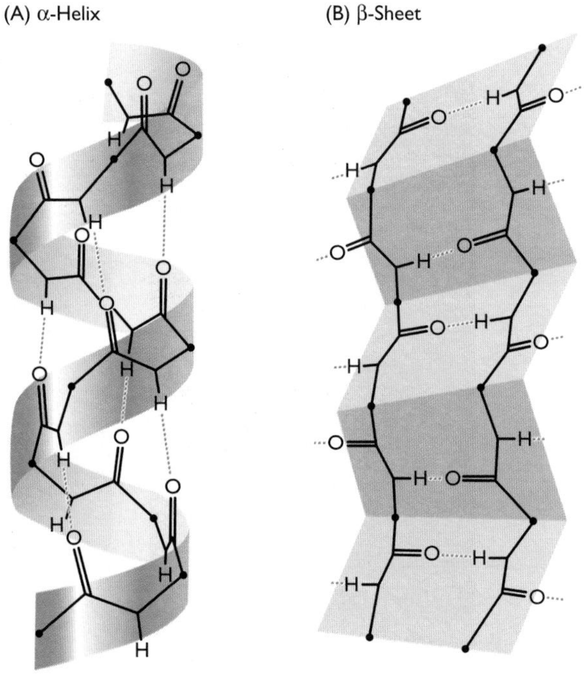

Figure 7.7 The two main secondary structural units found in proteins: (A) the α-helix, and (B) the β-sheet.

The polypeptide chains are shown in outline with the positions of the α-carbons indicated by small dots. The R groups are omitted for clarity. Each structure is stabilized by hydrogen bonds between the C=O and N–H groups of different peptide bonds. The β-sheet conformation that is shown is antiparallel, the two chains running in opposite directions. Parallel β-sheets also occur.

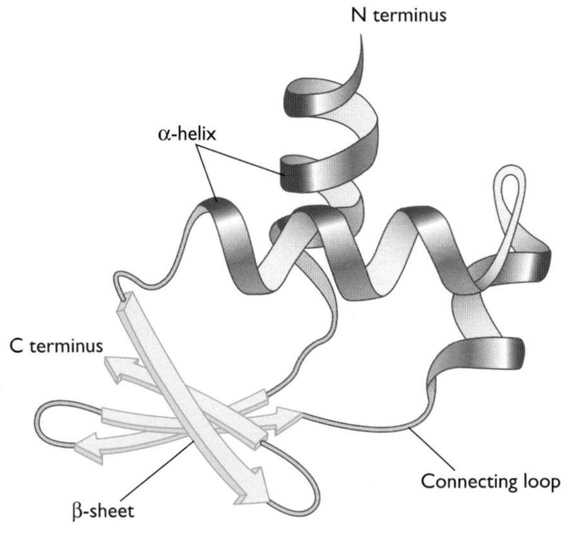

Figure 7.8 The tertiary structure of a protein.

This imaginary protein structure comprises three α-helices, shown as coils, and a four-stranded β-sheet, indicated by the arrows. Redrawn from Turner *et al.* (1997).

Some quaternary structures are held together by disulfide bridges between different polypeptides, but many proteins comprise looser associations of subunits stabilized by hydrogen bonding and hydrophobic effects, and can revert to their component polypeptides, or change their subunit composition, according to the functional requirements.

Amino acid diversity underlies protein diversity

Proteins are functionally diverse because the amino acids from which proteins are made are themselves chemically diverse. Different sequences of amino acids therefore result in different combinations of chemical reactivities, these combinations dictating not only the overall structure of the resulting protein but also the positioning on the surface of the structure of reactive groups that determine the chemical properties of the protein.

Amino acid diversity derives from the R group because this part is different in each amino acid and varies greatly in structure. Proteins are made up from a set of 20 amino acids (*Figure 7.9*; *Table 7.3*). Some of these have R groups that are small, relatively simple structures such as a single hydrogen atom (in the amino acid called glycine) or a methyl group (alanine). Others are large, complex aromatic side chains (phenylalanine, tryptophan and tyrosine). Most are uncharged, but two are negatively charged (aspartic acid and glutamic acid) and three are positively charged (arginine, histidine and lysine). Some are polar (e.g. glycine, serine and threonine), others are nonpolar (e.g. alanine, leucine and valine).

The set of 20 amino acids shown in *Figure 7.9* are the ones that are conventionally looked on as being specified by the genetic code (Section 10.1.2). They are therefore the amino acids encoded in the DNA sequences of genes and linked together, following the instructions present in a messenger RNA copy of the gene, when polypeptides are assembled during protein synthesis. But these 20 amino acids do not on their own represent the limit of the chemical diversity of proteins. The diversity is even greater because of two factors:

- At least one additional amino acid, selenocysteine (*Figure 7.10*), can be inserted into a polypeptide chain during protein synthesis, its insertion directed by a modified reading of the genetic code (Section 10.1.2).
- After synthesis, some proteins are modified by the

Box 7.2: Noncovalent bonds in proteins

The secondary and higher levels of protein structure are stabilized by various types of noncovalent bonds and interactions:

- **Hydrogen bonds** are weak electrostatic attractions between an electronegative atom (such as oxygen or nitrogen) and a hydrogen atom attached to a second electronegative atom. Hydrogen bonds are longer than covalent bonds and much weaker, typical bond energies being 1–10 kcal mol^{-1} at 25°C, compared to up to 90 kcal mol^{-1} for a covalent bond. As well as their role in the DNA double helix, hydrogen bonds stabilize protein secondary structures.
- **van der Waals forces** can be attractive or repulsive. Attactive forces are due to fluctuations between the electron charge densities of adjacent atoms and have energies of 0.1–0.2 kcal mol^{-1}. They promote formation of secondary structures. Repulsive van der Waals forces occur as a result of repulsion between electrons when two atoms come too close together. These repulsive forces place limitations on the degree of packing that can be achieved within a protein.
- **Electrostatic interactions** or ionic bonds form between charged groups. In aqueous environments they are relatively weak – approximately 3 kcal mol^{-1} – because of the shielding effect of water. Electrostatic interactions occur between the R groups of charged amino acids within proteins and are also important on the protein surface, in particular in attachment of a protein to DNA (Section 7.4.3).
- **Hydrophobic effects** arise because the hydrogen bonded structure of water forces hydrophobic groups into the internal parts of a protein. Hydrophobic effects are not true bonds but they are the main determinants of protein tertiary structure.

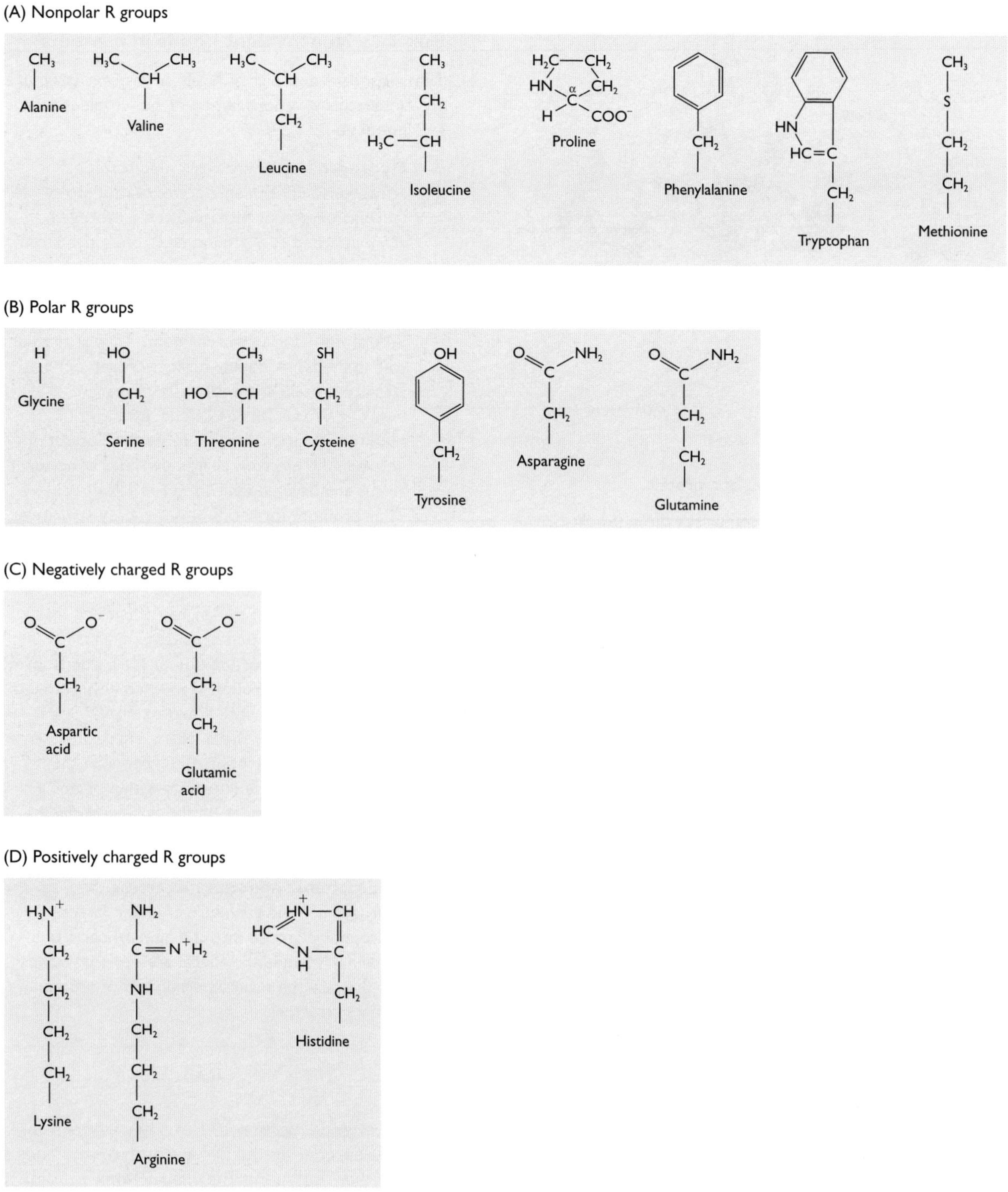

Figure 7.9 Amino acid R groups.

These 20 amino acids are the ones that are conventionally looked on as being specified by the genetic code (Section 10.1.2). Proteins also contain other amino acids, as described in the text. The classification into nonpolar, polar, etc. is as described in Lehninger (1970).

addition of new chemical groups, for example by acetylation or phosphorylation of certain amino acids, or by attachment of large side chains made up of sugar units (Section 10.3.3).

Primary structure dictates protein function

One of the fundamental premises of genetics is that genes specify biological functions. They do this by encoding information, as a sequence of nucleotides, which can be

Table 7.3 Amino acid abbreviations

	Abbreviation	
Amino acid	**Three-letter**	**One-letter**
Alanine	ala	A
Arginine	arg	R
Asparagine	asn	N
Aspartic acid	asp	D
Cysteine	cys	C
Glutamic acid	glu	E
Glutamine	gln	Q
Glycine	gly	G
Histidine	his	H
Isoleucine	ile	I
Leucine	leu	L
Lysine	lys	K
Methionine	met	M
Phenylalanine	phe	F
Proline	pro	P
Serine	ser	S
Threonine	thr	T
Tryptophan	trp	W
Tyrosine	tyr	Y
Valine	val	V

converted by gene expression into a sequence of amino acids in a polypeptide (see *Figure 1.2*, p. 4). But synthesis of an amino acid sequence results in only the primary structure of a protein. For the link between gene and biological function to be completed, the amino acid sequence must itself dictate the function of the protein. This link is implicit in the hierarchical nature of the four levels of protein structure, and before leaving proteins we must examine it in more detail.

The link is established by the ability of the amino acid sequence to determine the series of secondary structures that will be adopted by the polypeptide. Because of the nature of their R groups, certain amino acids are more commonly found in α-helices while others have a predisposition for β-sheets. A secondary structure therefore forms around a group of amino acids that favor that secondary structure and therefore initiate its formation, and then extends to include adjacent amino acids that either favor the structure or have no strong disinclination towards it, and finally terminates when one or more blocking amino acids, which cannot participate in that particular type of structure, are reached (*Figure 7.11*). This process of nucleation, extension and delimitation is repeated along the polypeptide until each part of the chain has adopted its preferred secondary structure. By identifying which amino acids are most frequently located in which secondary structures, and by studying the folding of small polypeptides of known sequence, biochemists have been able to deduce rules for this level of **protein folding**, and to a certain extent can predict which secondary structures will be adopted by a polypeptide simply by examining its primary sequence (Barton, 1995).

The processes occurring during the next two stages of protein folding, which result in the secondary structural units becoming arranged into the tertiary structure, and tertiary units associating to form quaternary multisubunit structures, are not well understood. Most proteins are made up of two or more structural **domains**, possibly with little interaction between them, and these domains are thought to fold independently of one another. Understanding how this occurs is complicated by the fact that many domains include secondary structural units from quite different regions of a polypeptide. For several proteins the spontaneity of the process has been demonstrated by unfolding the protein in the test tube, for example by heat treatment, and observing that it refolds into its correct structure when the treatment is reversed. However, this does not occur with all proteins and is particularly difficult to achieve with larger ones, the problem appearing to be that at various stages of the folding process the protein can adopt alternative partially-folded

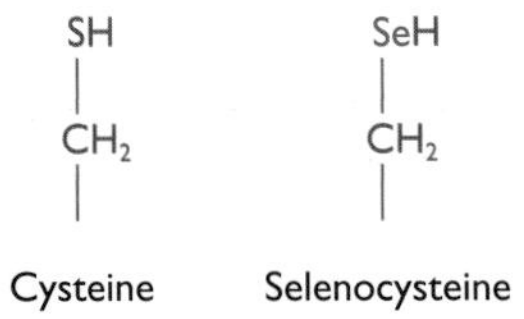

Figure 7.10 The R group of selenocysteine.

Selenocysteine is the same as cysteine but with the sulfur replaced with a selenium atom.

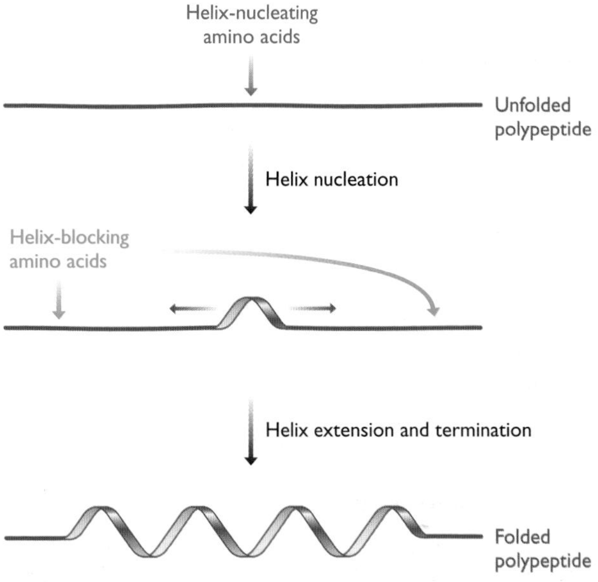

Figure 7.11 Formation of a secondary structure in a polypeptide.

An α-helix is shown nucleating at a position containing amino acids that favor helix formation, and extending in either direction until groups of amino acids that block helix formation are reached.

structures, only one of which leads to the correctly folded tertiary configuration. If the protein makes the wrong 'choice' it can end up at a dead end, where it is partially folded in an incorrect manner from which it cannot escape. In cells, proteins called **molecular chaperones** aid the folding of other proteins, probably by reducing the likelihood that the protein being folded adopts the wrong intermediate structure. We will return to this topic in Section 10.3.1.

The existence of alternative folding pathways, and proteins that aid the folding of other proteins, does not contradict the premise that a protein's functional, folded structure is dictated by its amino acid sequence and hence by the gene that encodes it. Protein folding is simply one stage in the lengthy pathway, summarized in *Figure 1.2B*, p. 4, that converts the biological information of a gene into a protein function.

7.3 METHODS FOR STUDYING DNA-BINDING PROTEINS

As in all areas of molecular biology and genetics, the amount we know about a topic depends on the range and effectiveness of the methods available for its study. With regards to DNA-binding proteins we are fortunate in having a number of powerful techniques that can provide information on the interaction between a protein and the DNA sequence or sequences that it binds to. These techniques can be divided into three categories:

- Methods for identifying the region(s) of a DNA molecule to which a protein binds.
- Methods for purifying a DNA-binding protein.
- Methods for studying the tertiary structure of a DNA-binding protein, including the complex formed when the protein is bound to DNA.

7.3.1 Locating a protein-binding site on a DNA molecule

Often the first thing that is discovered about a DNA-binding protein is not the identity of the protein itself but the features of the DNA sequence that the protein recognizes. This is because genetic and molecular biological experiments, which we will deal with in the next chapter, have shown that many of the proteins that are involved in genome expression bind to short DNA sequences immediately upstream of the genes on which they act (*Figure 7.12*). This means that the sequence of a newly-discovered gene, presuming it includes not only the coding DNA but also the regions upstream of it, provides immediate access to the binding sites of at least some of the proteins responsible for expression of that gene. Because of this, a number of methods have been developed for locating protein binding sites within DNA fragments up to several kb in length, these methods working perfectly well even if the relevant DNA-binding proteins have not been identified.

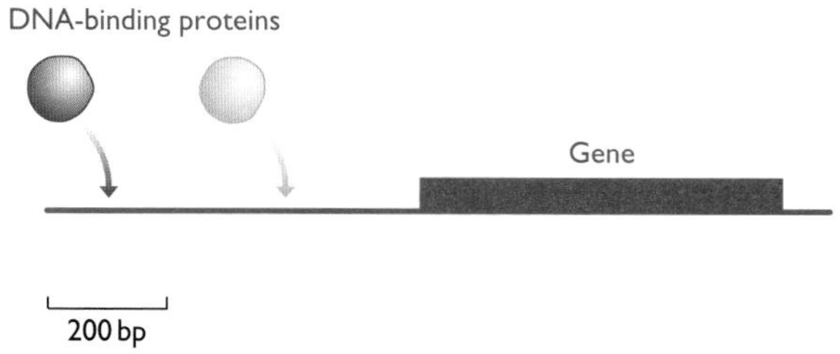

Figure 7.12 Attachment sites for DNA-binding proteins are located immediately upstream of a gene.

See Chapter 8 for more information on the location and function of these protein attachment sites.

Gel retardation identifies DNA fragments that bind to proteins

The first of these methods makes use of the substantial difference between the electrophoretic properties of a 'naked' DNA fragment and one that carries a bound protein. Recall that DNA fragments are separated by agarose gel electrophoresis because smaller fragments migrate through the pore-like structure of the gel more quickly than larger fragments (see Technical Note 3.2, p. 43). If a DNA fragment has a protein bound to it then its mobility through the gel will be impeded: the DNA–protein complex therefore forms a band at a position nearer to the starting point (*Figure 7.13*). This is called **gel retardation** (Garner and Revzin, 1981). In practice the technique is carried out with a collection of restriction fragments that span the region thought to contain a protein-binding site.

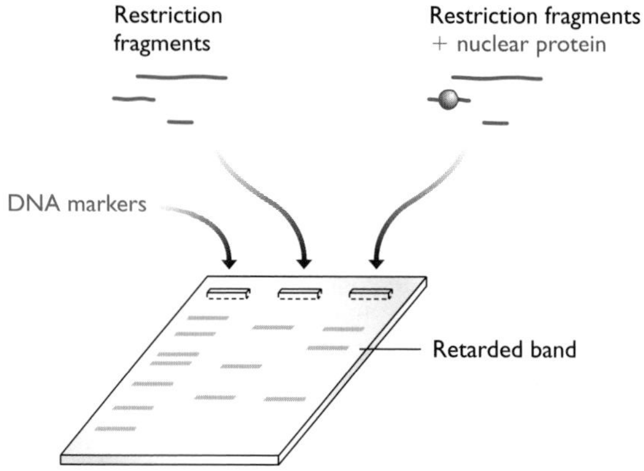

Figure 7.13 Gel retardation analysis.

A set of restriction fragments are mixed with a nuclear extract resulting in attachment of a DNA-binding protein to one of the fragments. The DNA–protein complex has a larger molecular mass than the 'naked' DNA and so runs more slowly during gel electrophoresis. As a result, the band for this fragment is retarded and can be recognized by comparing with the banding pattern produced by restriction fragments that have not been mixed with the nuclear extract.

The digest is mixed with an extract of nuclear proteins (presuming a eukaryote is being studied) and retarded fragments identified by comparing the banding pattern obtained after electrophoresis with the pattern for restricted fragments that have not been mixed with proteins. A nuclear extract is used because at this stage of the project the DNA-binding protein has usually not been purified. If, however, the protein is available then the experiment can be carried out just as easily with the pure protein as with a mixed extract.

Protection assays pinpoint binding sites with greater accuracy

Gel retardation gives a general indication of the location of a protein-binding site in a DNA sequence, but does not pinpoint the site with great accuracy. Often the retarded fragment is several hundred bp in length, compared to the expected length of the binding site of a few tens of bp at most, and so might contain separate binding sites for several proteins. Alternatively, if the retarded restriction fragment is quite small there is the possibility that the binding site also includes nucleotides on adjacent fragments, ones that on their own do not form a stable complex with the protein and so do not lead to a retarded fragment. Retardation studies are therefore a starting point but other techniques are needed to provide more accurate information.

Modification assays can take over where gel retardation leaves off. The basis of these techniques is that if a DNA molecule carries a bound protein then part of its nucleotide sequence will be protected from modification. There are two ways of carrying out the modification:

- By treatment with a nuclease, which cleaves all phosphodiester bonds except those protected by the bound protein.
- By exposure to a methylating agent, such as dimethyl sulfate which adds methyl groups to G nucleotides. Any Gs protected by the bound protein will not be methylated.

The practical details of these two techniques are shown in *Figure 7.14*. Both utilize an experimental approach called **footprinting**. In nuclease footprinting (Galas and Schmitz, 1978), the DNA fragment being examined is labeled at one end, complexed with binding protein (as a nuclear extract or as pure protein) and treated with deoxyribonuclease I (DNase I). Normally, DNase I cleaves every phosphodiester bond leaving a sample of only the DNA segment protected by the binding protein. This is not very useful because sequencing such a small fragment can be difficult. It is quicker to use the more subtle approach shown in *Figure 7.14*. The nuclease treatment is carried out under limiting conditions, such as a low temperature and/or very little enzyme, so that on average each copy of the DNA fragment suffers a single 'hit' – meaning that just one phosphodiester bond is cleaved along its length. Although each fragment is cut just once, in the entire population of fragments all bonds are cleaved except those protected by the bound protein. The protein is now removed, the mixture electrophoresed and the labeled fragments visualized. Each of these fragments has the label at one end and cleavage site at the other. The result is a ladder of bands corresponding to fragments different in length by one nucleotide, the ladder broken by a blank area in which no labeled bands occur. This blank area, or 'footprint', corresponds to the positions of the protected phosphodiester bonds, and hence of the bound protein, in the starting DNA.

Modification interference identifies nucleotides central to protein binding

Modification protection should not be confused with **modification interference**, a different technique with greater sensitivity in the study of protein binding (Hendrickson and Schleif, 1985). Modification interference works on the basis that if a nucleotide critical for protein binding is altered, for example by addition of a methyl group, then binding may be prevented. One of this family of techniques is illustrated in *Figure 7.15*. The DNA fragment, labeled at one end, is treated with the modification reagent, in this case dimethyl sulfate, under limiting conditions so just one guanine per fragment is methylated. Now the binding protein or nuclear extract is added and the fragments electrophoresed. Two bands are seen, one corresponding to the DNA–protein complex and one containing DNA without bound protein. The latter contains molecules which have been prevented from attaching to the protein because the methylation treatment has modified one or more Gs that are crucial for the binding. To identify which Gs are modified, the fragment is purified from the gel and treated with piperidine, a compound that cleaves DNA at methylguanine nucleotides. The result of this treatment is that each fragment is cut into two segments, one carrying the label. The length(s) of the labeled fragment(s), determined by a second round of electrophoresis, tells us which nucleotide(s) in the original fragment were methylated and hence identifies the position in the DNA sequence of Gs that participate in the binding reaction. Equivalent techniques can be used to identify the A, C and T nucleotides involved in binding.

7.3.2 Purification of a DNA-binding protein

Once a binding site has been identified in a DNA molecule it is possible to use this sequence to purify the DNA-binding protein, as a prelude to more detailed structural studies. The purification techniques utilize the ability of the protein to bind to its target site. One possibility is to use a form of **affinity chromatography** (*Figure 7.16A*), which involves immobilizing a DNA fragment or synthetic oligonucleotide in a chromatography column, usually by attaching one end of the DNA to a silica particle (Kadonaga, 1991). The protein extract is then passed through the column in a low salt buffer which promotes binding of proteins to their DNA target sites. The binding

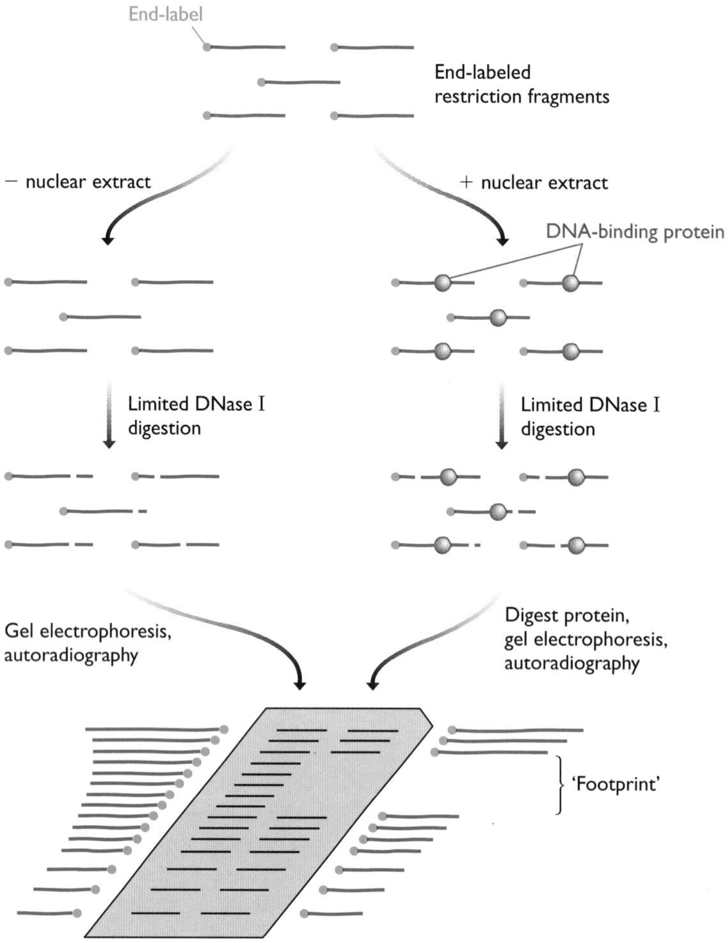

Figure 7.14 DNase I footprinting.

See the description of this method in the text. The restriction fragments used at the start of the procedure must be labeled at just one end. This is usually achieved by treating a set of longer restriction fragments with an enzyme that attaches labels at *both* ends, then cutting these labeled molecules with a second restriction enzyme, and purifying one of the sets of end fragments.

The related dimethyl sulfate (DMS) modification assay is similar to DNase I footprinting. Instead of DNase digestion the fragments are treated with limited amounts of DMS so that a single guanine base is methylated in each fragment. Guanines that are protected by the bound protein cannot be modified. After removal of the protein the DNA is treated with piperidine, which cuts at the modified nucleotide positions. The samples are now examined by electrophoresis. The banding pattern for the control DNA fragments – those not incubated with the nuclear extract – indicates the positions of Gs in the fragment, and the footprint seen in the banding pattern for the test sample shows which Gs were protected.

protein specific for the immobilized sequence is retained in the column while all other proteins pass through. Once these unwanted proteins have been completely washed out the column is eluted with a high salt buffer, which destabilizes the DNA–protein complex. The pure binding protein can then be collected.

An alternative is to use a form of hybridization probing (Singh *et al.*, 1988). A library of cDNA clones, each synthesizing a different cloned protein from the organism being studied, is needed. These clones are blotted onto a nylon membrane in such a way that the protein content of each clone is retained (*Figure 7.16B*). This is slightly different to the types of blotting that we have encountered previously (e.g. *Figure 5.18*, p. 104), in which the objective was to study the DNA content of the clones, but the principle is the same. Now the DNA fragment or oligonucleotide containing the protein binding site is labeled and washed over the membrane. The DNA attaches to a blotted clone only if that clone has been synthesizing the appropriate DNA-binding protein. These clones are identified by detecting where the labeled DNA is located on the membrane. Samples of the clones can then be recovered from the master library and used to produce larger quantities of the binding protein.

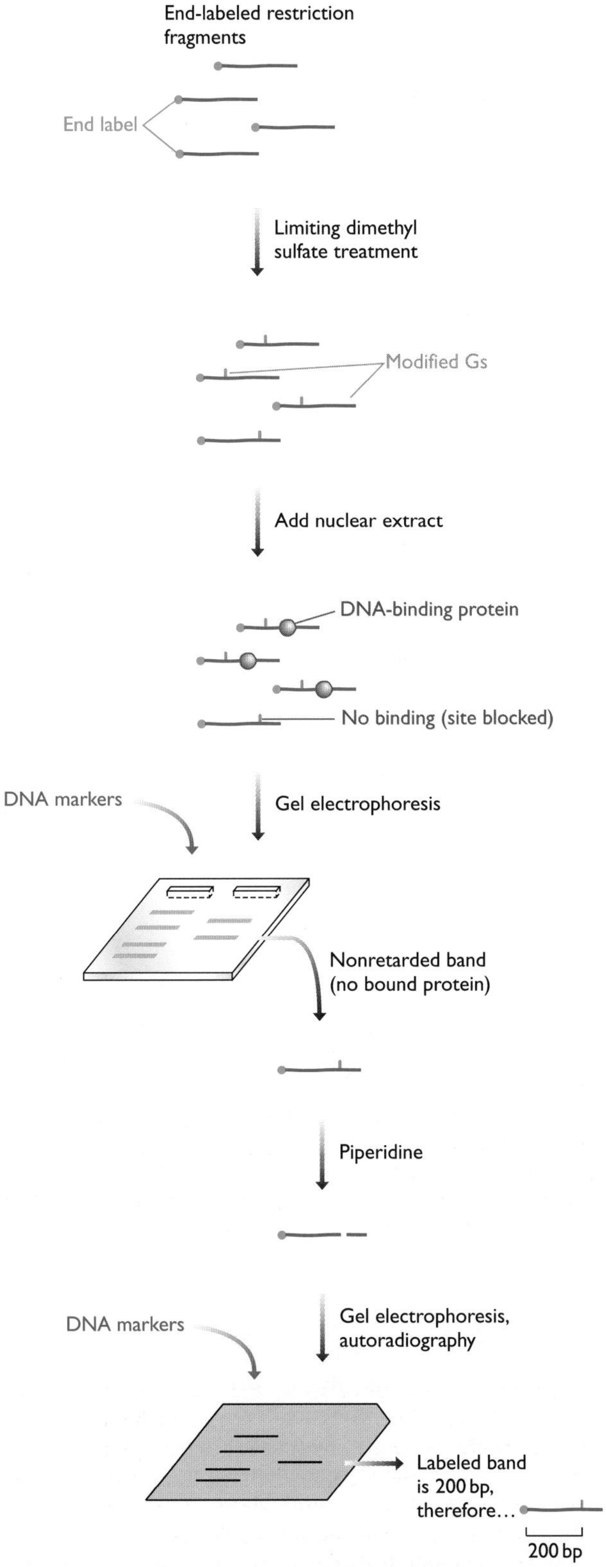

Figure 7.15 Dimethyl sulfate (DMS) modification inteference assay.

The method is described in the text. See the legend to *Figure 7.14* for a description of the procedure used to obtain DNA fragments labeled at just one end.

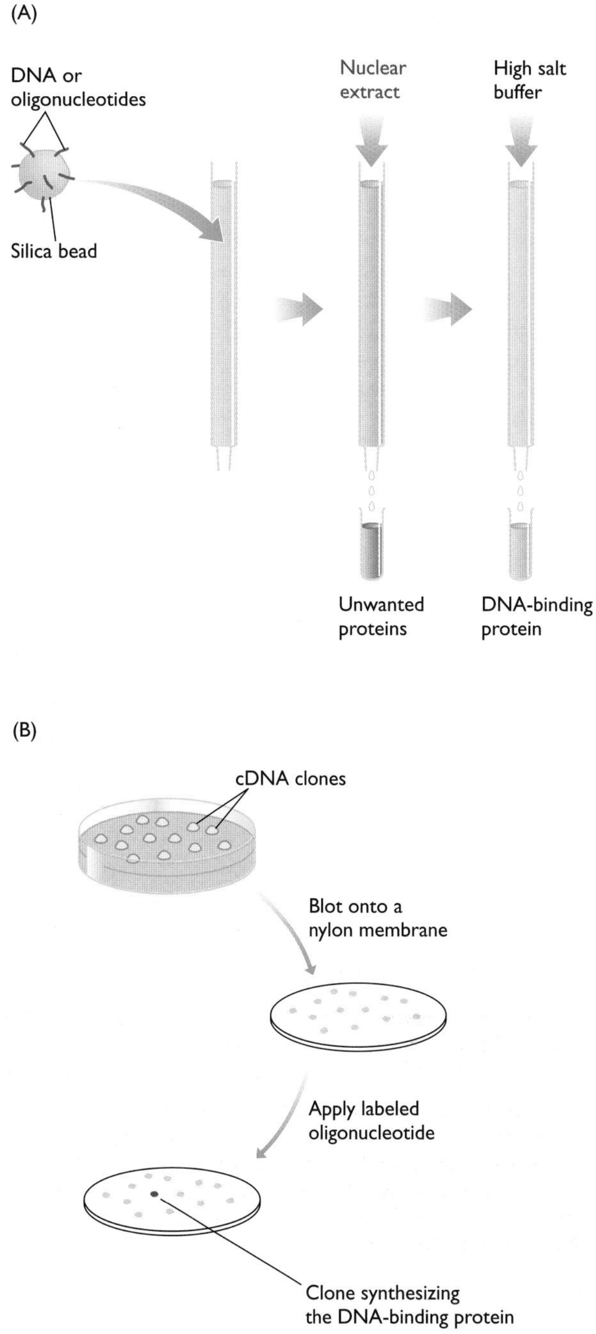

Figure 7.16 Two ways of purifying a DNA-binding protein.

(A) Affinity chromatography. DNA fragments or synthetic oligonucleotides containing the attachment site for the binding protein are attached to silica beads and these packed into a chromatography column. The binding protein is retained in the column when a nuclear extract is passed through, and can subsequently be recovered by eluting with a high salt buffer. (B) Colony hybridization probing. cDNA clones are blotted onto a nylon filter and one that is expressing the binding protein identified through the ability of this colony to bind labeled DNA containing the protein attachment site.

7.3.3 Techniques for studying protein and DNA–protein structures

The availability of a pure sample of a DNA-binding protein makes possible an analysis of its structure, in isolation or attached to its DNA-binding site. This provides the most detailed information on the DNA–protein interaction, enabling the precise structure of the DNA-binding part of the protein to be determined and allowing the identity and nature of the contacts with the DNA helix to be elucidated. Two techniques – **X-ray crystallography** and **nuclear magnetic resonance (NMR) spectroscopy** – are central to this area of research. These techniques and the studies they make possible are one of the major growth areas of the molecular life sciences, not only because of developments in the techniques themselves, but because advances in computation have resulted in new and more powerful ways of analyzing the resulting data.

X-ray crystallography has broad applications in structure determination

X-ray crystallography is a long-established technique whose pedigree stretches back into the late 19th century. Indeed, Nobel Prizes were awarded as early as 1915 to William and Lawrence Bragg, father and son, for working out the basic methodology and using it to determine the crystal structures of salts such as sodium chloride and zinc sulfide. The technique is based on **X-ray diffraction**. X-rays have very short wavelengths, between 0.01 and 10 nm, which is 4000 times shorter than visible light and comparable to the spacings between atoms in chemical structures. When a beam of X-rays is directed on to a crystal some of the X-rays pass straight through, but others are diffracted and emerge from the crystal at a different angle from which they entered (*Figure 7.17A*). If the crystal is comprised of many copies of the same molecule, all positioned in a regular array, then different X-rays are diffracted in similar ways, resulting in overlapping circles of diffracted waves which interfere with one another. An X-ray sensitive photographic film, placed across the beam, reveals a series of spots (*Figure 7.17B*), an **X-ray diffraction pattern**, from which the structure of the molecule in the crystal can be deduced.

The challenge with X-ray crystallography lies with the complexity of the methodology used to deduce the structure of a molecule from its diffraction pattern. The basic principles are that the relative positioning of the spots indicates the arrangement of the molecules in the crystal, and their relative intensities provide information on the structure of the molecule. The problem is that the more complex the molecule, the greater the number of spots and the larger the number of comparisons that must be made between them. Even with computational help the analysis is difficult and time consuming. If successful, the result is an electron density map (*Figure 7.17C* and *D*) which, with a protein, provides a chart of the folded polypeptide from which the positioning of structural features such as α-helices and β-sheets can be determined. If sufficiently detailed, the R-groups of the individual amino acids in the polypeptide can be identified and their orientations relative to one another established, allowing deductions to be made about the hydrogen bonding and other chemical interactions occurring within the protein structure. With luck, these deductions lead to a detailed three-dimensional model of the protein (Rhodes, 1993).

The first protein structures to be determined by X-ray crystallography were for myoglobin and haemoglobin, resulting in more Nobel Prizes, for Perutz and Kendrew in 1962. It still takes several months or longer to complete an X-ray crystallography analysis with a new protein, and there are many pitfalls that can prevent a successful conclusion being reached. In particular, it can often be difficult to obtain a suitable crystal of the protein. Despite these problems, the numbers of completed structures have gradually increased and now include over 30 DNA-binding proteins. An important innovation has been to crystallize DNA-binding proteins in the presence of their target sequences, the resulting protein–DNA structures revealing the precise positioning of the proteins relative to the double helix. It is from this type of information that most of our knowledge about the mode of action of DNA-binding proteins has been obtained.

NMR gives detailed structural information for small proteins

Like X-ray crystallography, NMR traces its origins to the early part of the 20th century, first being described in 1936 with the relevant Nobel Prizes awarded in 1952. The principle of the technique is that rotation of a charged chemical nucleus generates a magnetic moment. When placed in an applied electromagnetic field, the spinning nucleus can orientate in either of two ways, called α and β (*Figure 7.18*), the α-orientation (which is aligned with the magnetic field) having a slightly lower energy. In NMR spectroscopy the magnitude of this energy separation is determined by measuring the frequency of the electromagnetic radiation needed to induce the transition from α to β, the figure being described as the resonance frequency of the nucleus being studied. The critical point is that, although each type of nucleus (e.g. ^{1}H, ^{13}C, ^{15}N) has its own specific resonance frequency, the measured frequency is often slightly different from the standard value (typically by less than 10 parts per million) because electrons in the vicinity of the rotating nucleus shield it to a certain extent from the applied magnetic field. This **chemical shift**, the difference between the observed resonance energy and the standard value for the nucleus being studied, enables the chemical environment of the nucleus to be inferred, and hence provides structural information. Particular types of analysis (called COSY and TOCSY) enable atoms linked by chemical bonds to the spinning nucleus to be identified; other analyses (e.g. NOESY) identify atoms that are close to the spinning nucleus in space but not directly connected to it.

Not all chemical nuclei are suitable for NMR. Most protein NMR projects are ^{1}H studies, the aim being to

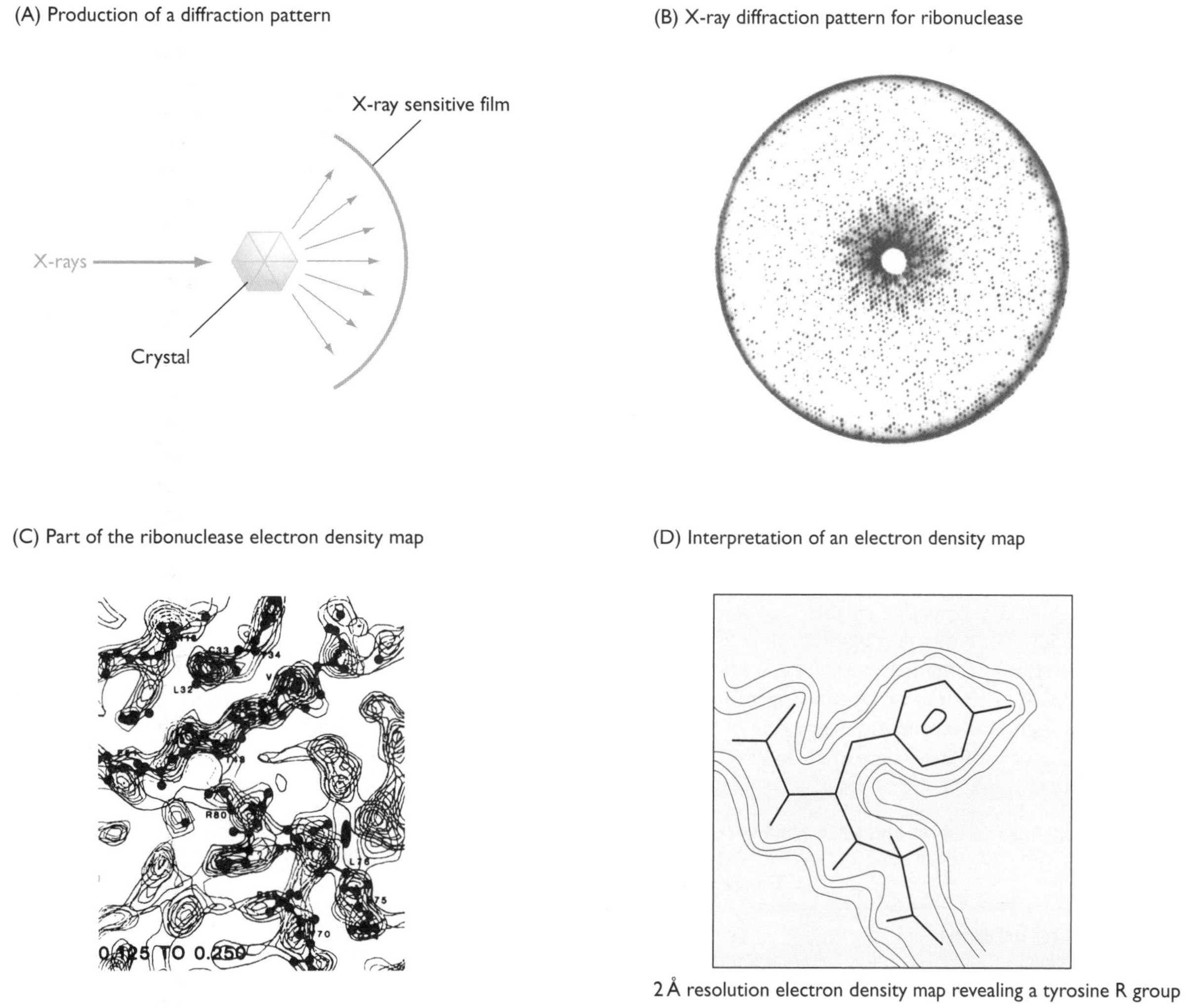

Figure 7.17 X-ray crystallography.

(A) An X-ray diffraction pattern is obtained by passing a beam of X-rays through a crystal of the molecule being studied. (B) shows the diffraction pattern obtained with crystals of ribonuclease and (C) illustrates part of the electron density map derived from this diffraction pattern. (D) If the electron density map has sufficient resolution then it is possible to identify the R groups of individual amino acids, as shown here for tyrosine. (B) and (C) from *Biology*, 3rd edn, by Neil Campbell, copyright 1993 by Benjamin Cummings Publishing Company. Reprinted by permission. (D) reprinted with permission from Zubay G, *Biochemistry*, 4th edn. Copyright 1997 by The McGraw–Hill Companies.

identify the chemical environments and covalent linkages of every hydrogen atom, and from this information to infer the overall structure of the protein. These studies are frequently supplemented by analyses of substituted proteins, in which at least some of the carbon and/or nitrogen atoms have been replaced with the rare isotopes ^{13}C and ^{15}N, these also giving good results with NMR.

When successful, NMR results in the same level of resolution as X-ray crystallography and so provides very detailed information on protein structure (Evans, 1995). The main advantage of NMR is that it works with molecules in solution and so avoids the problems that sometimes occur when attempting to obtain crystals of a protein for X-ray analysis. Solution studies also offer greater flexibility if the aim is to examine changes in protein structure, for example during protein folding or in response to addition of a substrate. The disadvantage of NMR is that it is only suitable for relatively small proteins. There are several reasons for this, one being the need to identify the resonance frequencies for every, or as many as possible, of the ^{1}H or other nuclei being studied. This depends on the various nuclei having different chemical shifts so their frequencies do not overlap. The larger the protein, the greater the number of nuclei and the greater the chances that frequencies overlap and structural information is lost. Although this limits the applicability of NMR, the technique is still very valuable. There are many interesting proteins that are small enough to be studied by NMR, and important information can also be obtained by structural analysis of peptides which, although not complete proteins, can act as models for aspects of protein activity such as nucleic acid binding.

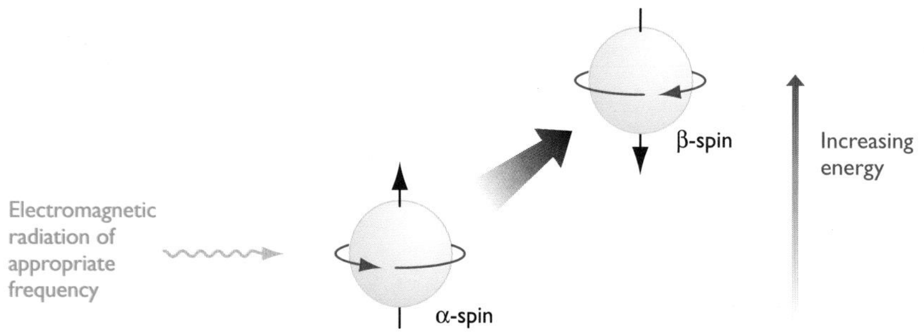

Figure 7.18 The basis of nuclear magnetic resonance spectroscopy.

A rotating nucleus can take up either of two orientations in an applied electromagnetic field. The energy separation between the α and β spin states is determined by measuring the frequency of electromagnetic radiation needed to induce an α→β transition.

7.4 INTERACTIONS BETWEEN DNA AND DNA-BINDING PROTEINS

Although the primary objective of this chapter is to lay the foundations for our detailed examination of genome expression, we should remain aware of the wide range of activities played by different DNA-binding proteins and appreciate that genome expression is just one of the functions of this class of proteins. There are also DNA-binding proteins that have structural roles in eukaryotic and prokaryotic chromosomes, and others that are involved in replication or repair of DNA molecules, as well as the large group of related proteins that bind not to DNA but to RNA (*Table 7.4*). Most of the proteins involved in genome expression recognize specific DNA sequences and bind predominantly to these target sites, whereas others bind nonspecifically at various positions along a DNA molecule. Still others, especially those involved in DNA repair, recognize unusual nucleotide bases and bind

Table 7.4 Functions of DNA- and RNA-binding proteins

Function	Examples	Cross-reference
DNA-binding proteins		
DNA packaging	Eukaryotic histones	Section 6.1.1
	Bacterial nucleoid proteins	Section 6.2.1
DNA recombination	RecA	Section 13.2
DNA repair	DNA glycosylases, nucleases	Section 13.1.4
DNA replication	Origin recognition proteins	Section 12.3.1
	DNA polymerases and ligases	Section 12.3.2
	Single-strand binding proteins	Section 12.3.2
	DNA topoisomerases	Section 12.2.2
Genome expression		
Transcription initiation	Eukaryotic TATA-binding protein	Section 8.2.3
	σ subunit of bacterial RNA polymerase	Section 8.2.3
RNA synthesis	RNA polymerases	Section 8.2.1
Regulation of transcription	Eukaryotic transcription factors	Section 8.3.2
	Bacterial repressors	Section 8.3.1
Others	Prokaryotic restriction enzymes	Section 2.2.2
RNA-binding proteins		
Genome expression		
Intron splicing	snRNP proteins	Section 9.2.3
mRNA polyadenylation	CPSF, CstF	Section 9.2.2
mRNA editing	Adenosine deaminases	Section 9.4.2
rRNA and tRNA processing	Ribonucleases	Section 9.3.2
Translation	Aminoacyl-tRNA synthetases	Section 10.1.1
	Translation factors	Section 10.2.2
RNA turnover	Ribonucleases	Section 9.5
Ribosome structure	Ribosomal proteins	Section 10.2.1

at these positions. In spite of this diversity, all DNA-binding proteins share at least one thing in common – they bind to DNA – and, as we will see, there appears to be only a limited number of ways in which this can be achieved.

7.4.1 The DNA partner in the interaction

In recent years our understanding of the part played by the DNA molecule in the interaction with a binding protein has begun to change. It has always been accepted that the DNA sequence is important, most obviously with those proteins that directly recognize a specific sequence as their binding site. The change in our perception concerns the effect that the nucleotide sequence has on the conformation of the double helix, and the possibility that these conformational changes represent a second, less direct way in which the DNA sequence can influence protein binding.

Direct readout of the nucleotide sequence

It was clear from the double helix structure described by Watson and Crick (Section 7.1.2) that although the nucleotide bases are on the inside of the DNA molecule, they are not entirely buried and some of the chemical groups attached to the purine and pyrimidine bases are accessible from outside of the helix. **Direct readout** of the nucleotide sequence should, therefore, be possible without breaking the base pairs and opening up the molecule.

In order to form chemical bonds with groups attached to the nucleotide bases, a binding protein must form contacts within one or both of the grooves that spiral around the helix (see *Figure 7.3*, p. 149). With the B-form of DNA, the identity and orientation of the exposed parts of the bases within the major groove is such that most sequences can be read unambiguously, whereas within the minor groove it is possible to identify if each base pair is A–T or G–C but difficult to know which nucleotide of the pair is in which strand of the helix (*Figure 7.19*; Kielkopf *et al.*, 1998). Direct readout of the B-form therefore predominantly involves contacts in the major groove. With other DNA types there is much less information on the contacts formed with binding proteins, but the picture is likely to be quite different. In the A-form, for example, the major groove is deep and narrow and less easily penetrated by any part of a protein molecule (see *Table 7.1*, p. 150). The more shallow minor groove is therefore likely to play the major part in direct readout. With Z-DNA the major groove is virtually nonexistent and direct readout is possible to a certain extent without moving beyond the surface of the helix.

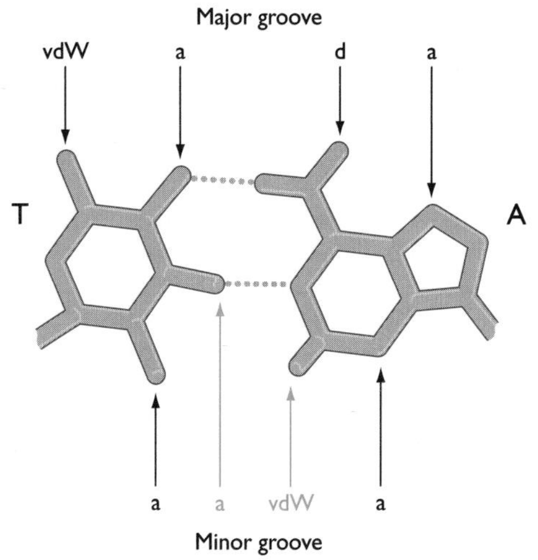

Figure 7.19 Recognition of an A–T base pair in the B-form double helix.

An A–T base pair is shown in outline (see *Figure 7.3B*, p. 149) with arrows indicating the chemical features that can be recognized by accessing the base pair via the major groove (above) and minor groove (below). In the major groove the chemical features are asymmetric and the orientation of the A–T pair can be identified by a binding protein. For some time it was believed that this is not possible in the minor groove, because only the two features shown in black were thought to be present, and these are symmetric. Using these two features, the binding protein could recognize the A–T base pair but not know which nucleotide is in which strand of the helix. The asymmetric features shown in green have recently been discovered, suggesting that the orientation of the pair might in fact be discernable via the minor groove. Abbreviations: a, hydrogen bond acceptor; d, hydrogen bond donor; vdW, van der Waals interaction. Reprinted with permission from Kielkopf CL, *et al.*, *Science*, **282**, 111–115. Copyright 1998 American Association for the Advancement of Science.

The nucleotide sequence has a number of indirect effects on helix structure

The recent change in our view of DNA structure concerns the influence of the nucleotide sequence on the conformation of the helix at different positions along its length. Originally it was thought that cellular DNA molecules have fairly uniform structures, made up mainly of the B-form of the double helix. Some short segments might be in the A-form, and there might be some Z-DNA tracts, especially near the ends of a molecule, but the vast majority of the length of a double helix would be unvarying B-DNA. We now recognize that DNA is much more polymorphic, and that it is possible for the A-, B- and Z-DNA configurations and intermediates between them to coexist within a single DNA molecule, different parts of the molecule having different structures. These conformational variations are sequence dependent, being due largely to the base-stacking interactions that occur between adjacent base pairs. As well as being responsible, along with base-pairing, for the stability of helix, the base-stacking also influences the amount of rotation that occurs around the covalent bonds within individual nucleotides (see p. 149) and hence determines the conformation of the helix at a particular position. The nucleotide sequence therefore indirectly affects the overall conformation of the helix, possibly providing structural information that a binding protein can use to help it locate its appropriate attachment site on a DNA molecule. At present this is just a theoretical possibility as no protein that specifically recognizes a non-B-form of the helix has been identified, but many researchers believe that helix conformation is likely to play some role in the interaction between DNA and protein.

A second type of conformational change is **DNA bending** (Travers, 1995). This does not refer to the natural flexibility of DNA which enables it to form circles and supercoils, but instead to localized positions where the nucleotide sequence causes the DNA to bend. Like other conformational variations, DNA bending is sequence dependent. In particular, a DNA molecule in which one polynucleotide contains two or more groups of repeated adenines, each group comprising 3–5 As, with individual groups separated by 10 or 11 nucleotides, will bend at the 3′ end of the adenine-rich region (Young and Beveridge, 1998). As with helix conformation, it is not yet known to what extent DNA bending influences protein binding, though protein-induced bending at flexible sites has a clearly demonstrated function in regulation of some genes (e.g. Falvo *et al.*, 1995; Section 8.3.2).

7.4.2 The protein partner

Now that we have examined the structural features of DNA relevant to the formation of complexes with binding proteins, we can turn our attention to the proteins themselves. Our main interest lies with those proteins that are able to target a specific nucleotide sequence and hence bind to a limited number of positions on a DNA molecule, as this is the type of interaction that is most important in expression of the genome. To bind in this specific fashion a protein must make contact with the double helix in such a way that the nucleotide sequence can be recognized, which generally requires that part of the protein penetrates into the major and/or minor grooves in order to achieve direct readout of the sequence. This is usually accompanied by more general interactions with the surface of the molecule which may simply stabilize the DNA–protein complex or which may be aimed at accessing the indirect information on nucleotide sequence provided by the conformation of the helix.

When the structures of sequence-specific DNA-binding proteins are compared it is immediately evident that the family as a whole can be divided into a limited number of different groups, based on the structure of the segment of the protein that interacts with the DNA molecule (*Table 7.5*; Luisi, 1995). Each of these **DNA-binding motifs** is present in a range of proteins, often from very different organisms, and at least some of them probably evolved more than once. We will look at two in detail – the **helix-turn-helix** motif and the **zinc finger** – and then briefly survey the others.

The helix-turn-helix motif is present in prokaryotic and eukaryotic proteins

The helix-turn-helix (HTH) motif was the first DNA-binding structure to be identified (Harrison and Aggarwal, 1990). As the name suggests, the motif is made up of

Table 7.5 DNA-binding motifs

Motif	Examples of proteins with this motif
Sequence-specific DNA-binding motifs	
Helix-turn-helix family	
Standard helix-turn-helix	*E. coli* lactose repressor, tryptophan repressor
Homeodomain	*Drosophila* Antennapedia protein
Paired homeodomain	Vertebrate Pax transcription factors
POU domain	Vertebrate regulatory proteins *pit-1*, *oct-1* and *oct-2*
Winged helix-turn-helix	GABP regulatory protein of higher eukaryotes
High mobility group (HMG) domain	Mammalian sex determination protein *SRY*
Zinc finger family	
Cys_2His_2 finger	Transcription factor TFIIIA of eukaryotes
Multicysteine zinc finger	Steroid receptor family of higher eukaryotes
Zinc binuclear cluster	Yeast GAL4 transcription factor
Basic domain	Yeast GCN4 transcription factor
Ribbon-helix-helix	Bacterial MetJ, Arc and Mnt repressors
TBP domain	Eukaryotic TATA-binding protein
β-Barrel dimer	Papillomavirus E2 protein
Rel homology domain (RHB)	Mammalian transcription factor NF-κB
Nonspecific DNA-binding motifs	
Histone fold	Eukaryotic histones
HU/IHF motif*	Bacterial HU and ITF proteins
Polymerase cleft	DNA and RNA polymerases

*The HU/IHF motif is a nonspecific DNA-binding motif in bacterial HU proteins (the nucleoid packaging proteins: Section 6.2.1) but directs sequence-specific binding of the IHF (integration host factor) protein.

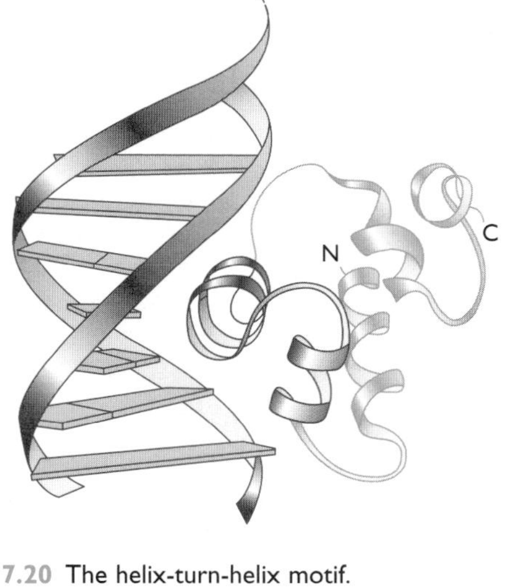

Figure 7.20 The helix-turn-helix motif.

The drawing shows the orientation of the helix-turn-helix motif (in blue) of the *E. coli* bacteriophage 434 repressor in the major groove of the DNA double helix. 'N' and 'C' indicate the N and C termini of the motif, respectively. Reprinted from *DNA–Protein Interactions* by Andrew Travers, published by Chapman & Hall, 1993. Reprinted with kind permission of A Travers.

two α-helices separated by a turn (*Figure 7.20*). The latter is not a random conformation but a specific structure, referred to as a **β-turn**, made up of four amino acids, the second of which is usually glycine. This turn, in conjunction with the first α-helix, positions the second α-helix on the surface of the protein in an orientation that enables it to fit inside the major groove of a DNA molecule. This second α-helix is therefore the **recognition helix** that

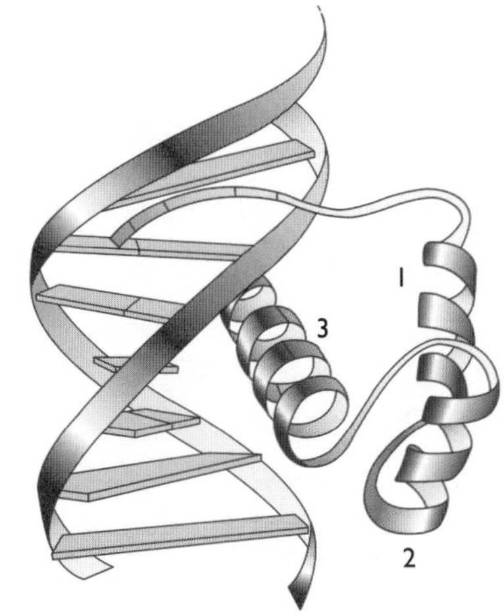

Figure 7.21 The homeodomain motif.

The first three helices of a typical homeodomain are shown with helix 3 orientated in the major groove and helix 1 making contacts in the minor groove. Helices 1–3 run in the N→C direction along the motif. Reprinted from *DNA–Protein Interactions* by Andrew Travers, published by Chapman & Hall, 1993. Reprinted with kind permission of A Travers.

Box 7.3: RNA-binding motifs

The major binding motifs in RNA-binding proteins are as follows (Twyman, 1998):

- The **ribonucleoprotein (RNP) domain** comprises four β-strands and two α-helices in the order β-α-β-β-α-β. The two central β-strands make the critical attachments with the RNA molecule. The RNP domain is the commonest RNA-binding motif and has been found in over 250 proteins.
- The **double-stranded RNA binding domain (dsRBD)** is similar to the RNP domain but with the structure α-β-β-β-α. The RNA-binding function lies between the β and α at the end of the structure. As the name implies, the motif is found in proteins that bind double-stranded RNA.
- The **κ-homology domain** has the structure β-α-α-β-β-α with the binding function between the pair of α-helices. It is relatively uncommon but present in at least one nuclear RNA-binding protein.
- The DNA-binding **homeodomain** may also have RNA-binding activity in some proteins. One ribosomal protein uses a structure similar to a homeodomain to attach to rRNA, and some homeodomain proteins such as Bicoid of *Drosophila melanogaster* (Section 11.3.3) can bind both DNA and RNA.

makes the vital contacts that enable direct readout of the DNA sequence. The HTH structure is usually 20 or so amino acids in length and so is just a small part of the protein as a whole. Some of the other parts of the protein form attachments with the surface of the DNA molecule, primarily to aid the correct positioning of the recognition helix within the major groove.

Many prokaryotic and eukaryotic DNA-binding proteins utilize an HTH motif. In bacteria, HTH motifs are present in some of the best studied regulatory proteins, which switch on and off the expression of individual genes. An example is the **lactose repressor**, which regulates expression of the lactose operon (Sections 8.3.1 and 11.1.1). The various eukaryotic HTH proteins include many whose DNA-binding properties are important in the developmental regulation of genome expression, such as the **homeodomain** proteins, whose roles we will examine in Section 11.3.3. The homeodomain is an extended HTH motif possessed by each of these proteins, made up of 60 amino acids which form four α-helices, numbers 2 and 3 separated by a β-turn, with number 3 acting as the recognition helix and number 1 making contacts within the minor groove (*Figure 7.21*). Other versions of the HTH motif found in eukaryotes include:

- The **POU domain**, which is usually found in proteins that also have a homeodomain, the two motifs

probably working together by binding different regions of a double helix. The name 'POU' comes from the initial letters of the names of the first proteins found to contain this motif (Herr *et al.*, 1988).
- The **winged helix-turn-helix** motif is another extended version of the basic HTH structure, this one with a third α-helix on one side of the HTH motif and a β-sheet on the other side.

Although each of these proteins, prokaryotic and eukaryotic, possesses an HTH motif, the details of the interaction of the recognition helix with the major groove are not exactly the same in all cases. The length of the recognition helix varies, generally being longer in eukaryotic proteins, the orientation of helix in the major groove is not always the same, and the position within the recognition helix of those amino acids that make contacts with nucleotides is different.

Zinc fingers are common in eukaryotic proteins

The second type of DNA-binding motif that we will look at in detail is the zinc finger, which is rare in prokaryotic proteins but very common in eukaryotes (Mackay and Crossley, 1998). There appear to be about 535 different zinc finger proteins in the worm *Caenorhabditis elegans* out of a total 19 000 proteins (Clarke and Berg, 1998) and it is estimated that 1% of all mammalian genes code for zinc finger proteins.

There are at least six different versions of the zinc finger. The first to be studied in detail was the **Cys_2His_2 finger**, which comprises a series of 12 or so amino acids, including two cysteines and two histidines, which form a segment of β-sheet followed by an α-helix. These two structures, which form the 'finger' projecting from the surface of the protein, hold between them a bound zinc atom, coordinated to the two cysteines and two histidines (*Figure 7.22*). The α-helix is the part of the motif that makes the critical contacts within the major groove, its positioning within the groove being determined by the β-sheet, which interacts with the sugar-phosphate backbone of the DNA, and the zinc atom, which holds the sheet and helix in the appropriate positions relative to one another. Other versions of the zinc finger differ in the structure of the finger, some lacking the sheet component and consisting simply of one or more α-helices, and the precise way in which the zinc atom is held in place also varies. For example, the **multicysteine zinc fingers** lack histidines, the zinc atom being coordinated between four cysteines.

An interesting feature of the zinc finger is that multiple copies of the finger are sometimes found on a single protein. Several have two, three or four fingers, but there are examples with many more than this, 37 for one toad protein. In most cases, the individual zinc fingers are thought to make independent contacts with the DNA molecule, but in some cases the relationship between different fingers is more complex. In one particular group of proteins, the nuclear or steroid receptor family, two α-helices containing six cysteines combine to coordinate two zinc atoms in a single DNA-binding domain, larger than a standard zinc finger (*Figure 7.23*). Within this motif it appears that one of the α-helices enters the major groove whereas the second makes contacts with other proteins.

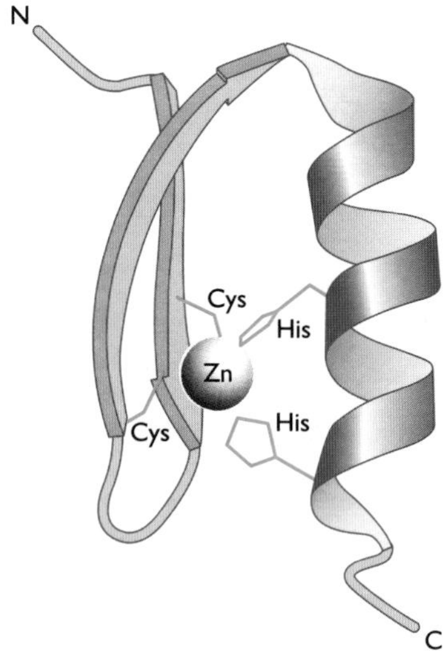

Figure 7.22 The Cys_2His_2 zinc finger.

This particular zinc finger is from the yeast SWI5 protein. The zinc atom is held between two cysteines within the β-sheet part of the motif and two histidines in the α-helix. The solid green lines indicate the R groups of these amino acids. 'N' and 'C' indicate the N- and C-termini of the motif, respectively. Reprinted from *DNA–Protein Interactions* by Andrew Travers, published by Chapman & Hall, 1993. Reprinted with kind permission of A Travers.

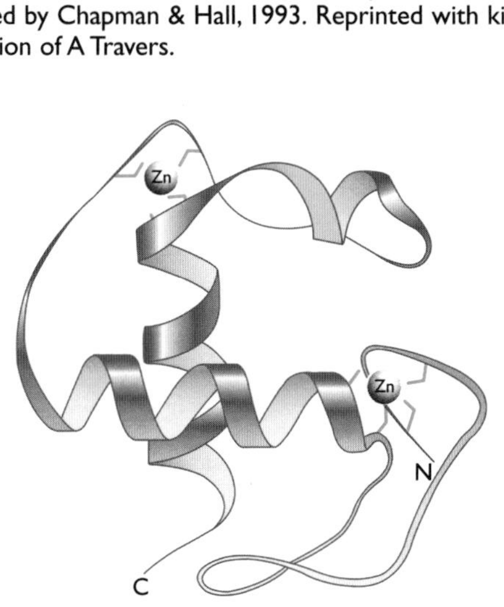

Figure 7.23 The steroid receptor zinc finger.

The R groups of the amino acids involved in the interactions with the zinc atoms are shown as solid green lines. 'N' and 'C' indicate the N and C termini of the motif, respectively. Reprinted from *DNA–Protein Interactions* by Andrew Travers, published by Chapman & Hall, 1993. Reprinted with kind permission of A Travers.

Other DNA-binding motifs

The various other DNA-binding motifs discovered in different proteins include:

- The **basic domain**, in which the DNA recognition structure is an α-helix that contains a high number of basic amino acids (e.g. arginine, serine and threonine). A peculiarity of this motif is that the α-helix only forms when the protein interacts with DNA, in the unbound state the helix has a disorganized structure. Basic domains are found in a number of eukaryotic proteins involved in transcription of DNA into RNA.
- The **ribbon-helix-helix** motif, which is one of the few motifs that achieves sequence-specific DNA binding without making use of an α-helix as the recognition structure. Instead the ribbon (i.e. two strands of a β-sheet) makes contact with the major groove (*Figure 7.24*). Ribbon-helix-helix motifs are found in some gene regulatory proteins in bacteria.
- The **TBP domain** has so far only been discovered in the **TATA-binding protein** (Section 8.2.3) after which it is named (Kim *et al.*, 1993). As with the ribbon-helix-helix motif, the recognition structure is a β-sheet, but in this case the main contacts are with the minor, not major, groove of the DNA molecule.

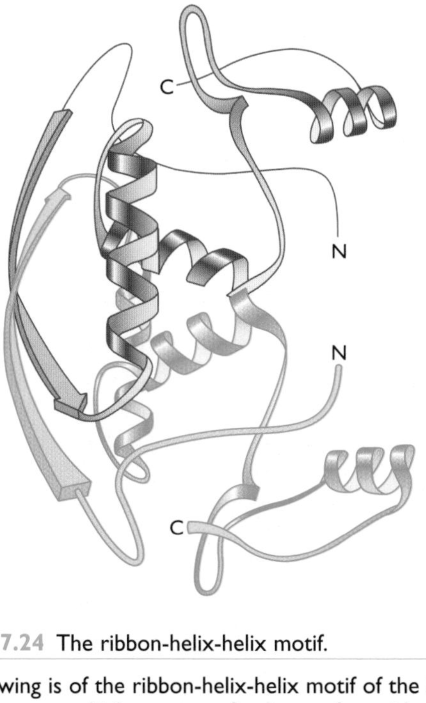

Figure 7.24 The ribbon-helix-helix motif.

The drawing is of the ribbon-helix-helix motif of the *E. coli* MetJ repressor, which consists of a dimer of two identical proteins, one shown in blue and the other in green. The β-strands at the left of the structure make contact with the major groove of the double helix. 'N' and 'C' indicate the N and C termini of the motif, respectively. Reprinted from *DNA–Protein Interactions* by Andrew Travers, published by Chapman & Hall, 1993. Reprinted with kind permission of A Travers.

7.4.3 DNA and protein in partnership

Finally, we will consider the important features of the partnership between DNA and binding proteins.

Contacts between DNA and proteins

The contacts formed between DNA and binding proteins are noncovalent. Within the major groove, hydrogen bonds form between the nucleotide bases and the R-groups of amino acids in the recognition structure of the protein, whereas in the minor groove hydrophobic interactions are more important. On the surface of the helix, the major interactions are electrostatic, between the negative charges on the phosphate component of each nucleotide and the positive charges on the R-groups of amino acids such as lysine and arginine, though some hydrogen bonding also occurs. In some cases, hydrogen bonding, on the surface of the helix or in the major groove, is direct between DNA and protein, in others it is mediated by water molecules. Few generalizations can be made: at this level of DNA–protein interactions each example has its own unique features and the details of the bonding have to be worked out by structural studies rather than comparisons with other proteins.

Most proteins that recognize specific sequences are also able to bind nonspecifically to other parts of a DNA molecule. In fact, it has been suggested that the amount of DNA in a cell is so large, and the numbers of each binding protein so small, that the proteins spend most if not all of their time attached nonspecifically to DNA (Stormo and Fields, 1998). The distinction between the nonspecific and specific forms of binding is that the latter is more favourable in thermodynamic terms. As a result, a protein is able to bind to its specific site even though there are literally millions of other sites to which it could attach nonspecifically. To achieve this thermodynamic favourability the specific binding process must involve the greatest possible number of DNA–protein contacts, which explains in part why the recognition structures of many DNA-binding motifs have evolved to fit snugly into the major groove of the helix, where the opportunity for DNA–protein contacts is greatest. It also explains why some DNA–protein interactions result in conformational changes to one or other partner, increasing still further the complementarity of the interacting surfaces, and allowing additional bonding to occur.

The need to maximize contacts in order to ensure specificity is also one of the reasons why many DNA-binding proteins are dimers, consisting of two proteins attached to one another. This is the case for most HTH proteins and many of the zinc finger type. Dimerization occurs in such a way that the DNA-binding motifs of the two proteins are both able to access the helix, possibly with some degree of cooperativity between them, so that the resulting number of contacts is greater than twice the number achievable by a monomer. As well as their DNA-binding motifs, many proteins contain additional characteristic domains that participate in the protein–protein contacts that result in dimer formation. One of these is the **leucine**

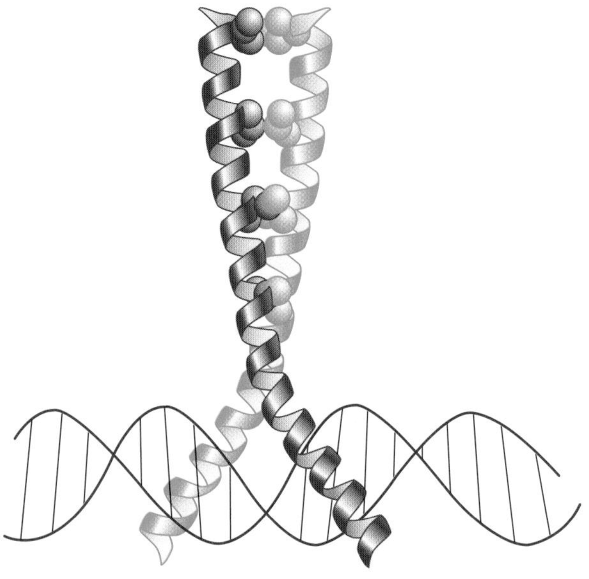

Figure 7.25 A leucine zipper.

This is a bZIP type of leucine zipper. The blue and green structures are parts of different proteins. Each set of spheres represents the R-group of a leucine amino acid. Leucines in the two helices associate with one another via hydrophobic interactions (see Box 7.2, p. 153) to hold the two proteins together in a dimer. In this example, the dimerization helices are extended to form a pair of basic domain DNA-binding motifs, shown making contacts in the major groove. Reprinted from *DNA–Protein Interactions* by Andrew Travers, published by Chapman & Hall, 1993. Reprinted with kind permission of A Travers.

zipper, which is an α-helix that coils more tightly than normal and presents a series of leucines on one of its faces. These can form contacts with the leucines of the zipper on a second protein, forming the dimer (*Figure 7.25*). A second dimerization domain is, rather unfortunately, called the **helix-loop-helix** motif, which is distinct from, and should not be confused with, the helix-turn-helix DNA-binding motif.

Can sequence specificity be predicted from the structure of a recognition helix?

An intriguing question is whether the specificity of DNA binding can be understood in sufficient detail for a prediction to be made, from an examination of the structure of the recognition helix of a DNA-binding motif, of the sequence of a protein's target site. To date, this objective has largely eluded us, but it has been possible to deduce some rules for the interaction involving certain types of zinc finger (Choo and Klug, 1997). In these proteins, four amino acids, three in the recognition helix and one immediately adjacent to it, form critical attachments with the nucleotide bases of the target site. Some of these attachments involve a single amino acid with a single base, others involve two amino acids with one base. By comparing the sequences of amino acids in the recognition helices of different zinc fingers with the sequences of nucleotides at the binding sites, it has been possible to identify a set of rules governing the interaction, these enabling the nucleotide sequence specificity of a new zinc finger protein to be predicted, admittedly with the possibility of some ambiguity, once the amino acid composition of its recognition helix is known.

REFERENCES

Barton GJ (1995) Protein secondary structure prediction. *Curr. Opinion Struct. Biol.*, **5**, 372–376.

Choo Y and Klug A (1997) Physical basis of a protein–DNA recognition code. *Curr. Opinion Struct. Biol.*, **7**, 117–125.

Clarke ND and Berg JM (1998) Zinc fingers in *Caenorhabditis elegans*: finding families and probing pathways. *Science*, **282**, 2018–2022.

Evans JNS (1995) *Biomolecular NMR Spectroscopy*. Oxford University Press, Oxford.

Falvo JV, Thanos D and Maniatis T (1995) Reversal of intrinsic DNA bends in the IFNβ gene enhancer by transcription factors and the architectural protein HMG I(Y). *Cell*, **83**, 1101–1111.

Galas D and Schmitz A (1978) DNase footprinting: a simple method for the detection of protein-DNA binding specificity. *Nucleic Acids Res.*, **5**, 3157–3170.

Garner MM and Revzin A (1981) A gel electrophoretic method for quantifying the binding of proteins to specific DNA regions: application to components of the *Escherichia coli* lactose operon regulatory system. *Nucleic Acids Res.*, **9**, 3047–3060.

Harrison SC and Aggarwal AK (1990) DNA recognition by proteins with the helix-turn-helix motif. *Annu. Rev. Biochem.*, **59**, 933–969.

Hendrickson W and Schleif R (1985) A dimer of AraC protein contacts three adjacent major groove regions at the Ara I DNA site. *Proc. Natl Acad. Sci. USA*, **82**, 3129–3133.

Herr W, Sturm RA, Clerc RG, *et al.* (1988) The POU domain: a large conserved region in the mammalian *pit*-1, *oct*-1, *oct*-2 and *Caenorhabditis elegans unc*-86 gene products. *Genes Develop.*, **2**, 1513–1516.

Kadonaga JT (1991) Purification of sequence-specific DNA binding proteins by DNA affinity chromatography. *Methods Enzymol.*, **208**, 10–23.

Kielkopf CL, White S, Szewczyk JW, Turner JM, Baird EE, Dervan PB and Rees DC (1998) A structural basis for recognition of A·T and T·A base pairs in the minor groove of B-DNA. *Science*, **282**, 111–115.

Kim YC, Geiger JH, Hahn S and Sigler PB (1993) Crystal structure of a yeast TBP/TATA-box complex. *Nature*, **365**, 512–520.

Lehninger A (1970) *Biochemistry*. Worth, New York.

Luisi B (1995) DNA–protein interaction at high resolution. In: DMJ Lilley, ed. *DNA–Protein: Structural Interactions*, pp. 1–48. IRL Press, Oxford.

Mackay JP and Crossley M (1998) Zinc fingers are sticking together. *Trends Biochem. Sci.*, **23**, 1–4.

Pauling L and Corey RB (1953) Configurations of polypeptide chains with favored orientations around single bonds: two new pleated sheets. *Proc. Natl Acad. Sci. USA*, **37**, 729–740.

Pauling L, Corey RB and Branson HR (1951) The structure of proteins: two hydrogen-bonded helical configurations of the polypeptide chain. *Proc. Natl Acad. Sci. USA*, **27**, 205–211.

Rhodes G (1993) *Crystallography Made Crystal Clear*. Academic Press, London.

Singh H, LeBowitz JH, Baldwin AS and Sharp PA (1988) Molecular cloning of an enhancer binding protein: isolation by screening of an expression library with a recognition site DNA. *Cell*, **52**, 415–423.

Stormo GD and Fields DS (1998) Specificity, free energy and information content in protein–DNA interactions. *Trends Biochem. Sci.*, **23**, 109–113.

Strachan T and Read AR (1996) *Human Molecular Genetics*. BIOS Scientific Publishers, Oxford.

Travers AA (1995) DNA bending by sequence and proteins. In: DMJ Lilley ed. *DNA–Protein: Structural Interactions*, pp. 49–75. IRL Press, Oxford.

Turner PC, McLennan AG, Bates AD and White MRH (1997) *Instant Notes in Molecular Biology*. BIOS Scientific Publishers, Oxford.

Twyman RM (1998) *Advanced Molecular Biology: A Concise Reference*. BIOS Scientific Publishers, Oxford.

Watson JD and Crick FHC (1953) Molecular structure of nucleic acids: a structure for deoxyribose nucleic acid. *Nature*, **171**, 737–738.

Young MA and Beveridge DL (1998) Molecular dynamics simulations of an oligonucleotide duplex with adenine tracts phased by a full helix turn. *J. Mol. Biol.*, **281**, 675–687.

FURTHER READING

Bates AD and Maxwell A (1993) *DNA Topology: In Focus*. IRL Press, Oxford. — *A concise description of DNA structure.*

Branden C and Tooze J (1991) *Introduction to Protein Structure*. Garland, New York. — *The best comprehensive text on this subject.*

Lilley DMJ (ed) (1995) *DNA–Protein: Structural Interactions*. IRL Press, Oxford. — *Research-level description of the subject.*

Nagai K and Mattaj IW (eds) (1994) *RNA–Protein Interactions*. IRL Press, Oxford. — *Complements the material in this chapter by providing details of RNA-binding proteins.*

Neidle S (1994) *DNA Structure and Recognition: In Focus*. IRL Press, Oxford. — *Easy to digest information of DNA-binding proteins.*

Travers A (1993) *DNA–Protein Interactions*. Chapman & Hall, London. — *The most accessible of the various books on this topic.*

Watson JD (1968) *The Double Helix*. Atheneum, London. — *The most important discovery of 20th century biology written as a soap opera.*

CHAPTER

8 Initiation of Transcription: the First Step in Gene Expression

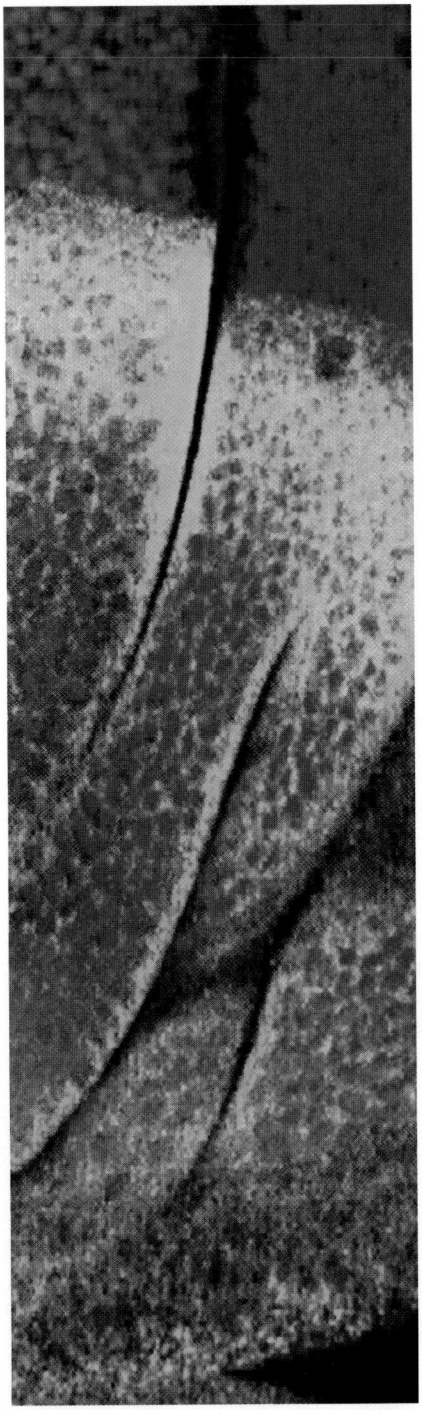

Contents

8.1 Accessing the Genome 173

8.1.1 Effects of chromatin packaging on eukaryotic gene expression 173

8.2 Assembly of the Transcription Initiation Complexes of Prokaryotes and Eukaryotes 178

8.2.1 RNA polymerases 178

8.2.2 Recognition sequences for transcription initiation 179

8.2.3 Assembly of the transcription initiation complex 181

8.3 Regulation of Transcription Initiation 185

8.3.1 Strategies for controlling transcription initiation in bacteria 185

8.3.2 Control of transcription initiation in eukaryotes 188

Concepts

- *Initiation of transcription is the primary control point for gene regulation*
- *Chromatin structure and nucleosome positioning influence patterns of gene expression*
- *DNA-dependent RNA polymerases synthesize RNA on a DNA template*
- *The transcription initiation complex is assembled at the promoter sequence near the start of a gene*
- *Eukaryotic initiation requires general transcription factors and other accessory proteins*
- *In bacteria, initiation rates are controlled by promoter structure and by repressor and activator proteins*
- *In eukaryotes, initiation is controlled by transcription factors that activate or, less frequently, repress the initiation complex*

In order for the cell to utilize the biological information contained within its genome, individual genes, each of which represents a single unit of information, have to be expressed in a coordinated manner. This coordinate gene expression determines the make-up of the cellular RNA, which in turn specifies the nature of the proteome and defines the activities that the cell is able to carry out.

DNA-binding proteins, whose general features we studied in Chapter 7, are responsible for determining which genes are expressed at a particular time and for making RNA transcripts of those genes that are active. For each individual gene, the transcription process can conveniently be divided into two stages (*Figure 8.1*). The first of these, **initiation of transcription**, involves the assembly upstream of the gene of the complex of proteins, including the RNA polymerase enzyme and its various accessory proteins, that will subsequently copy the gene into its RNA transcript. Inherent in this step are the events that determine whether or not the gene is actually expressed. In the second stage of transcription the RNA polymerase moves along the gene and synthesizes the **primary transcript**, which is a direct copy of the gene. This second stage is often accompanied by RNA processing events which include chemical modification of some nucleotides and removal of terminal or internal segments of the transcript.

In this chapter we will study initiation of transcription and in the next we will examine the synthesis and processing of transcripts. We will then look at how the proteome is synthesized before returning to transcription in Chapter 11 as part of our investigation of the ways in which genome activity is modulated by regulation of the expression pathways for individual genes.

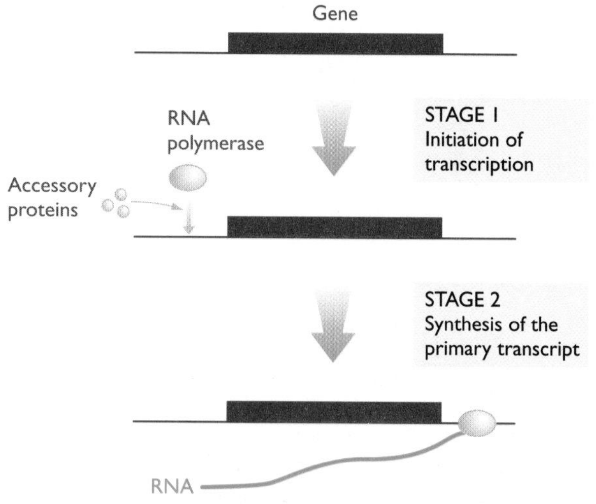

Figure 8.1 The two stages of transcription.

8.1 ACCESSING THE GENOME

When one looks at a genome sequence, written out as a series of As, Cs, Gs and Ts or drawn as a map with the genes indicated by boxes on a string of DNA (as in *Figure 1.5*, p. 6, for example), there is a tendency to imagine that the genes are readily accessible to the DNA-binding proteins responsible for their expression. In reality, the situation is very different. The DNA in the nucleus of a eukaryotic cell or the nucleoid of a prokaryote is attached to a variety of proteins that are not directly involved in gene expression and which must be displaced in order for the RNA polymerase and other expression proteins to gain access to the DNA. In prokaryotes we know very little about these events, a reflection of our generally poor knowledge about the physical organization of the prokaryotic genome (Section 6.2.1), but in eukaryotes we are beginning to understand how the packaging of the genome into chromatin (Section 6.1.1) influences gene expression. This is an exciting area of molecular biology, with recent research indicating that histones and other packaging proteins are not simply inert structures around which the DNA is wound but instead are active participants in the processes that determine which genes are expressed in an individual cell.

8.1.1 Effects of chromatin packaging on eukaryotic gene expression

In Section 6.1.1 we learnt that chromatin is the complex of genomic DNA and chromosomal proteins present in the eukaryotic nucleus. The two lowest levels of DNA packaging in chromatin are the nucleosome and the 30 nm chromatin fiber (see *Figures 6.2* and *6.3*, p. 119). Although there are still gaps in our knowledge, it is becoming clear that chromatin structure influences gene expression in at least two ways (*Figure 8.2*). First, the degree of chromatin packaging displayed by a segment of a chromosome determines whether or not genes within that segment are expressed. Second, if a gene is accessible, then its transcription is influenced by the precise positioning of nucleosomes within and around it. Before studying these effects we must first examine the structure of chromatin in more detail, in particular by considering the higher levels of organization between the 30 nm chromatin fiber and the metaphase chromosome.

Heterochromatin, euchromatin and chromatin loops

The metaphase chromosome represents the most compact form of DNA packaging in eukaryotes and occurs only during nuclear division. After division, the chromosomes become less compact and cannot be distinguished as individual structures. When nondividing nuclei are examined by light microscopy, all that can be seen is a mixture of lightly and darkly staining areas within the nucleus. The dark areas, which tend to be concentrated around the periphery of the nucleus, are called **heterochromatin** and contain DNA that is still in a relatively compact organization, though still less compact than in the metaphase structure. Two types of heterochromatin are recognized:

- **Constitutive heterochromatin** is a permanent feature of all cells and represents DNA that contains no genes and so can always be retained in a compact organization. This fraction includes centromeric and telomeric DNA as well as certain regions of some other chromosomes. For example, most of the human Y chromosome is made of constitutive heterochromatin (see *Figure 6.4*, p. 121).
- **Facultative heterochromatin** is not a permanent feature but is seen in some cells some of the time. Facultative heterochromatin is thought to contain genes that are inactive in some cells or at some periods of the cell cycle. When these genes are inactive, their DNA regions are compacted into heterochromatin.

It is presumed that the organization of heterochromatin is so compact that proteins involved in gene expression simply cannot access the DNA. In contrast, the remaining regions of chromosomal DNA, the parts that contain active genes, are less compact and permit entry of the expression proteins. These regions are called **euchromatin** and they are dispersed throughout the nucleus. The exact organization of the DNA within euchromatin is not known, but with the electron microscope it is possible

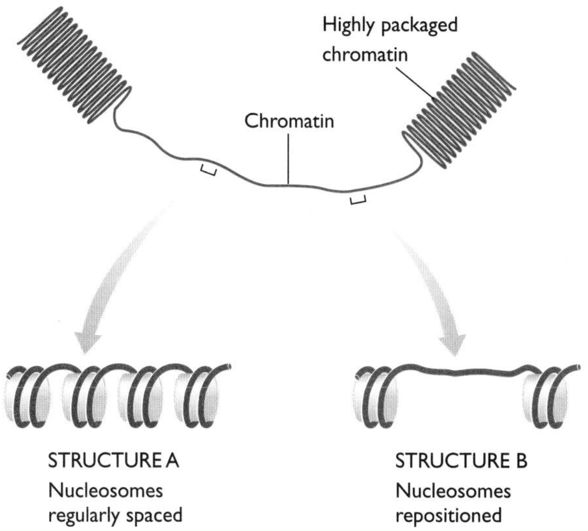

Figure 8.2 Two ways in which chromatin structure can influence gene expression.

A region of unpackaged chromatin in which the genes are accessible is flanked by two more compact segments. Within the unpackaged region, the positioning of the nucleosomes influences gene expression. On the left, the nucleosomes have their regular spacing as displayed by the typical beads-on-a-string structure. On the right, the nucleosome positioning has changed and a short stretch of DNA, approximately 300 bp, is exposed. See *Figures 6.2* and *6.3*, p. 119 for more details on nucleosomes.

to see loops of DNA, each one predominantly in the form of the 30 nm chromatin fiber and between 40 and 100 kb in length. The loops are attached via AT-rich DNA segments called **matrix-associated regions** (**MARs**) or **scaffold attachment regions** (**SARs**) to the **nuclear matrix**, a scaffold-like network that permeates the entire nucleus (*Figure 8.3*).

Structural and functional domains

The loops of DNA attached to the nuclear matrix are called **structural domains**. An intriguing question is the precise relationship between these and the **functional domains** that can be defined when the region of DNA around an expressed gene or set of genes is examined. A functional domain is delineated by treating a region of purified chromatin with deoxyribonuclease I (DNase I) which, being a DNA-binding protein, cannot gain access to the more compacted regions of DNA. Regions sensitive to DNase I span and extend to either side of a gene or set of genes that are being expressed, indicating that in this region the chromatin has a more open organization, though it is not clear whether this organization is the 30 nm fiber or the 'beads-on-a-string' structure (see *Figure 6.2A*, p.119).

The formation and maintenance of an open functional domain is the job, at least for some domains, of a DNA sequence called the **locus control region** or **LCR** (Wolffe, 1995). LCRs were first discovered during a study of the human β-globin genes (Section 6.1.1) and are now thought to be involved in expression of many genes that are active only during specific developmental stages. The globin LCR is contained in a stretch of DNA some 12 kb in length that is positioned upstream of the genes in the 60-kb β-globin functional domain (see *Figure 8.4*). It was

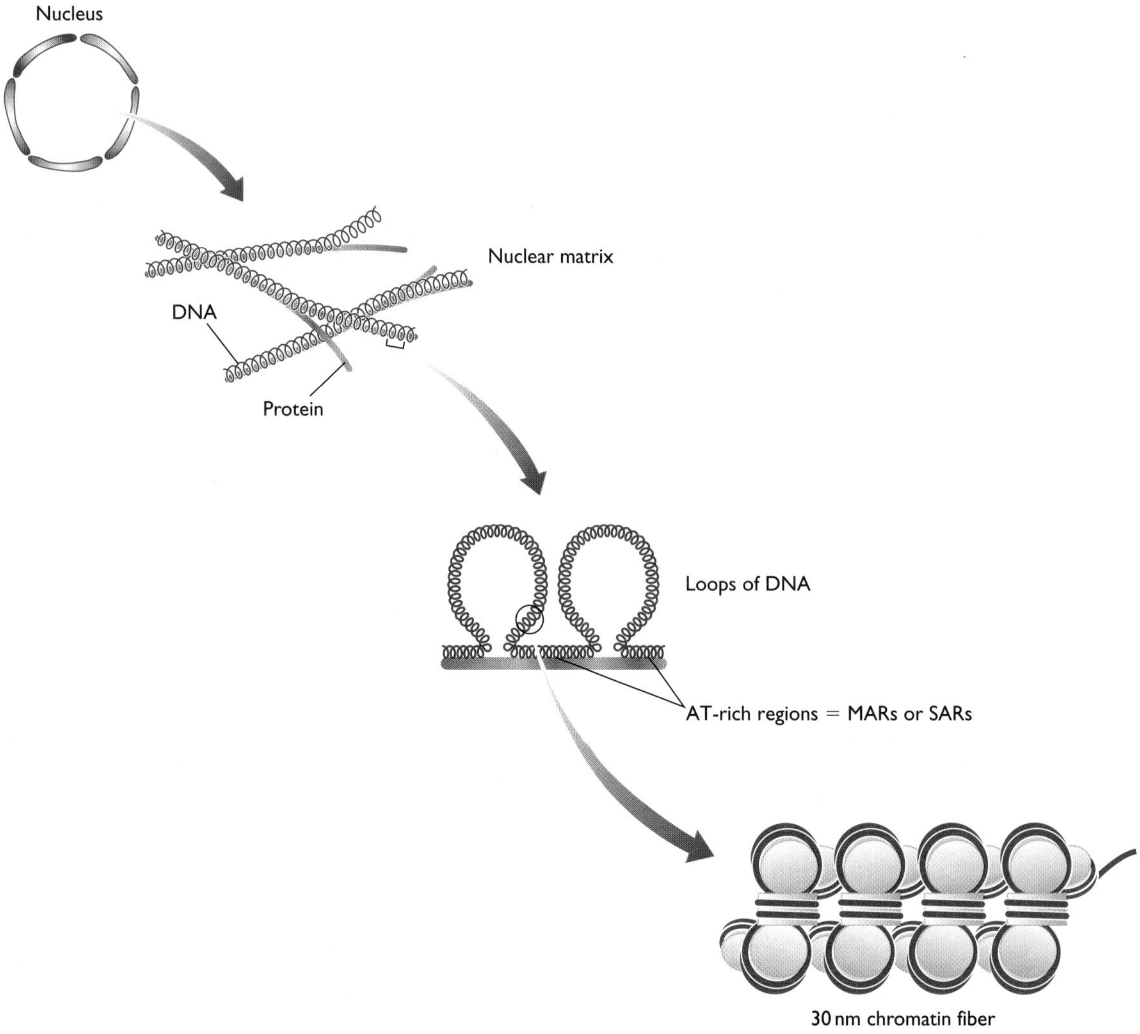

Figure 8.3 A scheme for organization of DNA in the nucleus.

The nuclear matrix is a fibrous protein-based structure whose precise composition and arrangement in the nucleus has not been described. Euchromatin, predominantly in the form of the 30 nm chromatin fiber (see *Figure 6.3*, p. 119) is thought to be attached to the matrix by AT-rich sequences called matrix or scaffold attachment regions (MARs or SARs).

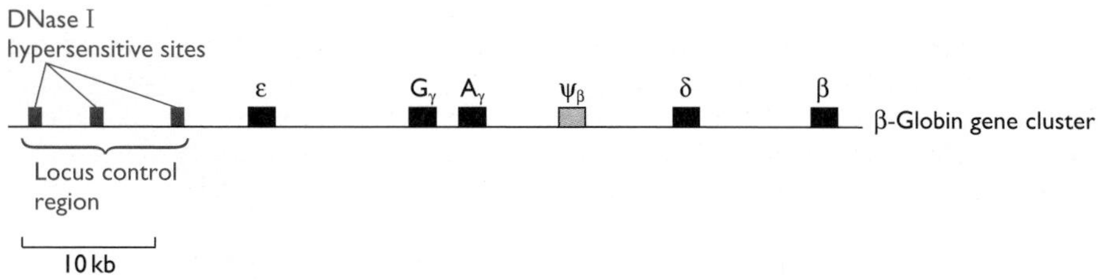

Figure 8.4 DNase I hypersensitive sites indicate the position of the locus control region for the human β-globin gene cluster.

Three hypersensitive sites are located in the 20 kb of DNA upstream of the start of the β-globin gene cluster. These three sites mark the position of the locus control region. Additional hypersensitive sites are seen immediately upstream of each gene, at the position where RNA polymerase attaches to the DNA. These hypersensitive sites are developmental stage specific, being seen only during the phase of development when the adjacent gene is active. The 60 kb region shown here represents the entire β-globin functional domain. See *Figure 6.6*, p. 123 for more information on the developmental regulation of expression of the β-globin gene cluster.

initially discovered during studies of individuals with thalassemia, a blood disease that results from defects in the α- or β-globin proteins. Many thalassemias are due to mutations in the coding regions of the globin genes, but a few map to a 12-kb region upstream of the β-globin gene cluster, the region now called the LCR. The precise function of the LCR was pinned down by experiments in which the human β-globin domain, with and without a complete LCR, was transferred into mouse cells. When DNA is transferred to a new organism it is inserted into one of the recipient's chromosomes, but in a random manner. It was discovered that if the transferred β-globin genes lacked the LCR, then the level of their expression in the mouse cell was variable, presumably because in some cases the genes inserted into a region of open chromatin, and sometimes into a closed region. However, if the intact LCR was present, then the genes were always expressed, because the LCR directed formation of a functional domain within the mouse chromosome. More detailed study of the β-globin LCR has shown that it contains at least three separate sequences, each 200–300 bp in length, that act as recognition sites for DNA-binding proteins. It is these proteins, not the DNA sequence itself, that controls the chromatin structure within the functional domain. Exactly how, and in response to what biochemical signals, is not known.

Is there correspondence between structural and functional domains? Intuition suggests that there should be, and some MARs, which mark the limits of a structural domain, are also located at the boundary of a functional domain. But the correspondence does not seem to be complete as some structural domains contain genes that are not functionally related, and the boundaries of some structural domains lie within genes (Wolffe, 1995).

Nucleosome positioning is important in modulating gene expression

The DNase I sensitivity that characterizes a functional domain is not uniform along the length of that domain. Some short regions are more easily cleaved than others, these **DNase I hypersensitive sites** coinciding with positions where nucleosomes are absent and which are therefore accessible to binding proteins that attach to the DNA. For example, the protein binding positions within the β-globin LCR are hypersensitive sites as are the positions where RNA polymerase attaches to the DNA immediately upstream of the β-globin genes (*Figure 8.4*). These RNA polymerase attachment points illustrate an interesting feature of hypersensitive sites: they are not invariant components of a functional domain. Recall that the different β-type globin genes are expressed at different stages of the human developmental cycle, ε being active in the early embryo, G_γ and A_γ in the fetus, and δ and β in the adult (see *Figure 6.6*, p. 123). Only when the gene is active is the RNA polymerase attachment position marked by a hypersensitive site. Initially it was thought that this was an *effect* of the differential expression of these genes, in other words that in the absence of gene activity it was possible for nucleosomes to cover the RNA polymerase attachment site, presumably to be pushed to one side when it became time to express the gene. Now it is thought that the presence or absence of nucleosomes is a *cause* of gene expression, the gene being switched off if nucleosomes cover the attachment site or on if access to the site is open due to repositioning of the nucleosomes. Nucleosome positioning may even be a key player in the fine tuning of gene expression, regulating the rate of transcription of a gene by continually modulating access of the polymerase (Wade *et al.*, 1997).

This reappraisal of nucleosome function has been prompted by the discovery of two distinct systems for changing the relative positions of nucleosomes and influencing the fine structure of chromatin. The first of these involves **histone acetylation**, the attachment of acetyl groups to lysine amino acids in the N-terminal regions of each of the core histones. These N termini form tails that protrude from the nucleosome core octamer (*Figure 8.5*) and their acetylation reduces the affinity of the histones for DNA and possibly also reduces the interaction between individual nucleosomes that leads to formation

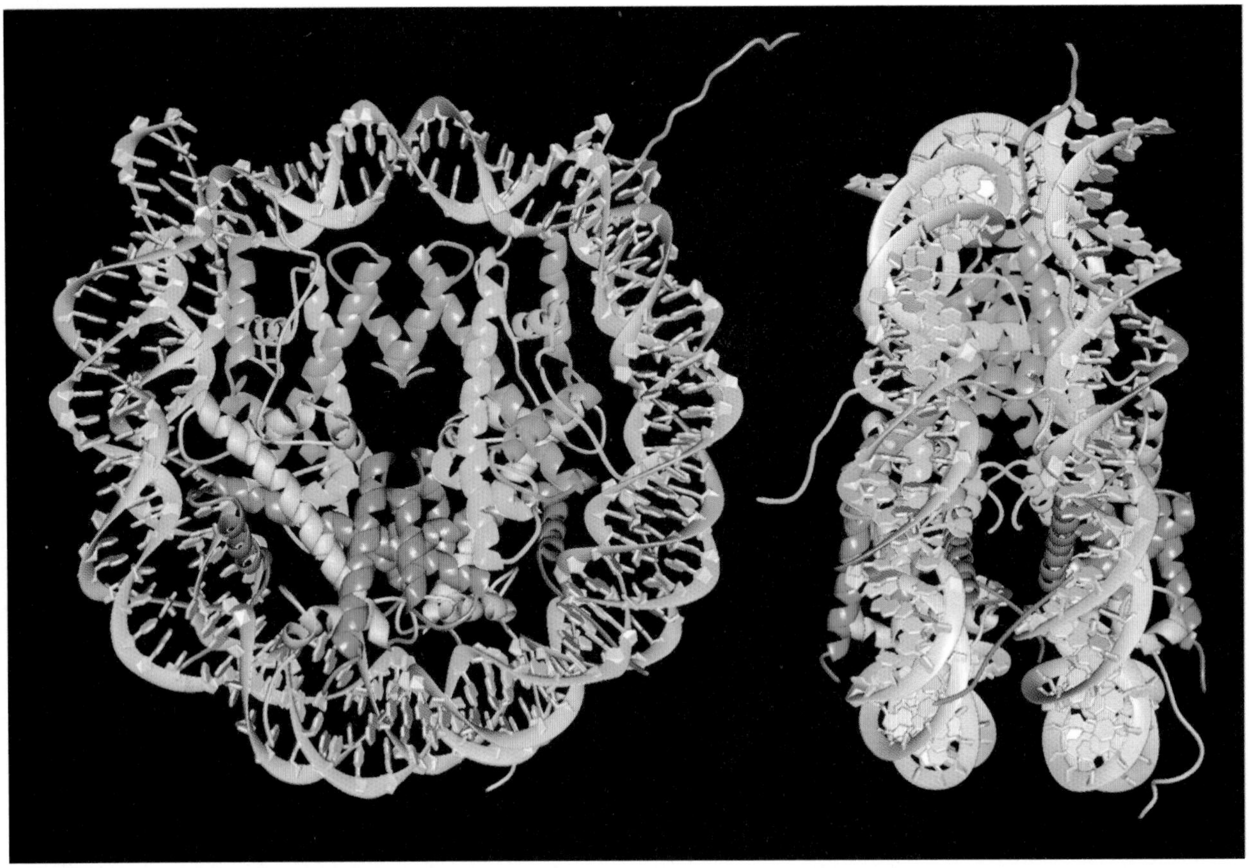

Figure 8.5 Two views of the nucleosome core octamer.

On the left, the view is downward from the top of the barrel-shaped octamer; on the right, the view is from the side. The two strands of the DNA double helix wrapped around the octamer are shown in brown and green. The octamer comprises a central tetramer of two histone H3 (blue) and two histone H4 (bright green) subunits plus a pair of H2A (yellow) – H2B (red) dimers, one above and one below the central tetramer. Note the N-terminal tails of the histone proteins protruding from the core octamer. Reprinted with permission from Luger *et al.*, *Nature*, **389**, 251–260. Copyright 1997 Macmillan Magazines Limited.

of the 30 nm chromatin fiber. The histones in heterochromatin are generally unacetylated whereas those in functional domains are acetylated. The relevance of histone acetylation to gene expression was underlined in 1996 when, after several years of trying, the first examples of histone acetyltransferases and deacetylases, the enzymes that add and remove acetyl groups to histones, were identified (Pennisi, 1997; Wade *et al.*, 1997). It was realized that some proteins that had already been shown to have important influences on gene expression were in fact histone transacetylases or deacetylases. For example, a mammalian protein called p300/CBP, which had been ascribed a clearly defined role in activation of a variety of genes, was found to be a histone transacetylase (Bannister and Kouzarides, 1996). Similarly, it was shown that the retinoblastoma protein, which controls cell proliferation by inhibiting expression of various genes until their activities are required (and which, when mutated, leads to cancer) works in conjunction with a histone deacetylase (Brehm *et al.*, 1998; Magnaghi-Jaulin *et al.*, 1998). These observations, plus the demonstration that different types of cell display different patterns of histone acetylation, are clear indicators that histone acetylation plays a prominent role in regulating gene expression.

The second system for influencing the fine structure of chromatin is less well understood but enough is known about it to show that it too has an important impact on gene expression. This process is called **chromatin remodeling** and involves a set of protein complexes referred to as **chromatin remodeling machines** or **CRMs** (Cairns, 1998). Like the acetylating enzymes, CRMs interact with the histone tails and cause subtle changes in

Box 8.1: DNA methylation and gene expression

The discussion of histone acetylation provides an interesting link with **DNA methylation** (Twyman, 1998), a process whose involvement in the regulation of genome expression has long been recognized but whose mode of action has, until recently, been something of a mystery.

In eukaryotes, cytosine bases in chromosomal DNA molecules are sometimes changed to 5-methylcytosine by the addition of methyl groups by enzymes called DNA methyltransferases. Cytosine methylation is relatively rare in lower eukaryotes but in vertebrates up to 10% of the total number of cytosines in a genome are methylated, and in plants the figure can be high as 30%. The methylation pattern is not random, instead being limited to the cytosine in some copies of the sequences 5′-CG-3′ and, in plants, 5′-CNG-3′. Two types of methylation enzyme can be distinguished. The first type are called **maintenance methylation** enzymes and are responsible for adding methyl groups to a newly-synthesized strand of DNA at positions opposite methylated sites on the parent strand (Section 13.1.4). These enzymes therefore ensure that the two daughter DNA molecules retain the methylation pattern of the parent molecule. The second type are ***de novo* methylation** enzymes, which are presumed to exist but have not yet been isolated from eukaryotes. These enzymes add methyl groups at totally new positions and so can change the pattern of methylation in a localized region of the genome.

Methylation is associated with repression of gene activity. This has been shown by experiments in which methylated or unmethylated genes have been introduced into cells by cloning and their expression levels measured: expression does not occur if the DNA sequence is methylated. The link with gene expression is also apparent when the methylation patterns in chromosomal DNAs are examined, as these show that inactive genes are located in methylated regions. For example, in humans, approximately 56% of all genes are located close to CpG islands (Section 5.1.1). Housekeeping genes, those that are expressed in all tissues, have unmethylated CpG islands, whereas tissue-specific genes are unmethylated only in those tissues in which the adjacent gene is expressed. Note that because the methylation pattern is maintained after cell division, information specifying which genes should be expressed is inherited by the daughter cells, ensuring that in a differentiated tissue the appropriate pattern of gene expression is retained even though the cells in the tissue are being replaced and/or added to by new cells.

How methylation influences gene expression was a puzzle for many years. The isolation of a pair of **methyl-CpG-binding proteins** (**MeCPs**) led to the suggestion that gene silencing might be due to bound MeCPs blocking access to attachment sites for transcription factors (Section 8.3.2). Now it has been shown that methylation is associated with reduced levels of histone acetylation and therefore induces chromatin formation in methylated regions of a chromosomal DNA molecule (Eden *et al.*, 1998). The MeCPs are probably not themselves histone deacetylases; instead it is thought that when bound to methylated sites the MeCPs act as recognition signals for other enzymes with deacetylase activity.

As well as its relevance to the maintenance and inheritance of tissue-specific gene expression, DNA methylation also has at least three other roles in regulation of genome expression:

- In many genomes, transposable elements and other repetitive DNA sequences are hypermethylated (Bender, 1998), probably in order to silence expression of genes such as transposases and hence to prevent transposition of mobile elements (Section 6.3.2), and also to suppress recombination between repeated sequences (Section 13.2). In both cases, the objective is to reduce damage to the genome caused by unwanted DNA rearrangements.
- Methylation underlies **genomic imprinting**, a relatively uncommon but important feature of mammalian genomes in which only one of a pair of genes in a diploid cell is expressed, the second being silenced by methylation. The term 'parental imprinting' is also used, but in most cases it is not clear if the origin of the imprinted gene – from mother or father – is relevant. The number of imprinted genes in a genome has not yet been worked out, but it is not likely to be more than a small fraction. The function of imprinting is not known (Jaenisch, 1997).
- Methylation plays a central role in **X inactivation**, a special form of imprinting that leads to total inactivation of one of the X chromosomes in a female cell. The function of X inactivation is to avoid the situation whereby a female cell possesses two copies of each X-linked gene whereas males only have one. X inactivation is induced by the *Xist* gene, located on the X chromosome. On one of the female X chromosomes the *Xist* gene is expressed, leading to inactivation of that chromosome (Section 11.2.2). On the second chromosome, the activity of *Xist* is repressed by methylation, ensuring that this chromosome remains active.

nucleosome structure that enable the positioning of individual nucleosomes to be changed. Exactly how CRMs achieve this result is not yet known, but they have been shown to facilitate binding to DNA of proteins involved in gene expression. Possible links between CRMs and histone acetyltransferases and deacetylases are being sought (Hampsey, 1997; Tong *et al.*, 1998).

8.2 ASSEMBLY OF THE TRANSCRIPTION INITIATION COMPLEXES OF PROKARYOTES AND EUKARYOTES

Now that we have established that gene expression is influenced by the packaging of DNA in the eukaryotic nucleus, and by implication possibly also in the prokaryotic nucleoid, we can move on to begin our examination of the events involved in the initiation of transcription. We will do this in two stages. First, we will study how the complex of proteins responsible for transcription is assembled on a DNA molecule prior to transcription of a gene. Then, in Section 8.3, we will investigate how the assembly of this complex, and its ability to initiate transcription, can be controlled by various additional proteins that respond to stimuli from inside or outside of the cell and ensure that the correct genes are transcribed at the appropriate times.

8.2.1 RNA polymerases

The enzymes responsible for transcription of DNA into RNA are called **DNA-dependent RNA polymerases**. The name indicates that the enzymatic reaction that they catalyze results in polymerization of RNA from ribonucleotides, and occurs in a DNA-dependent manner, meaning that the sequence of nucleotides in a DNA **template** dictates the sequence of nucleotides in the RNA that is made (*Figure 8.6*). It is permissible to shorten the enzyme name to **RNA polymerase**, as the context in which the name is used means that there is rarely confusion with the **RNA-dependent RNA polymerases** that are involved in replication and expression of some virus genomes. Note that there are also **template-independent RNA polymerases**, including at least one – **poly(A) polymerase** – with an important role in gene expression (Section 9.2.2).

Transcription of eukaryotic nuclear genes requires three different RNA polymerases, called **RNA polymerase I**, **RNA polymerase II** and **RNA polymerase III**. Each is a multisubunit protein (8–12 subunits) with a molecular mass in excess of 500 kDa. Structurally, these polymerases are quite similar to one other, the three largest subunits being closely related and some of the smaller ones being shared by more than one enzyme, but functionally they are quite distinct. Each works on a different set of genes with no interchangeability (*Table 8.1*). Most research attention has been directed at RNA polymerase II, as this is the one that transcribes genes that code for proteins. It also works on a set of genes specifying **small nuclear RNAs** that are involved in RNA processing. RNA polymerase III transcribes other genes for small RNAs, including those for **transfer RNAs (tRNAs)**, and RNA polymerase I transcribes the multicopy repeat units containing the 28S, 5.8S and 18S rRNA genes. The functions of all these RNAs are summarized in Section 9.1.1 and described in detail in Chapters 9 and 10.

Box 8.2: Mitochondrial and chloroplast RNA polymerases

The RNA polymerases that transcribe organellar genes are unlike their counterparts in the nucleus, reflecting the bacterial origins of mitochondria and chloroplasts (Section 6.1.2).

- The mitochondrial RNA polymerase consists of a single subunit with a molecular mass of 140 kDa. Interestingly, this enzyme is more closely related to the RNA polymerases of certain bacteriophages than it is to the standard bacterial version. The polymerase works in conjunction with at least two additional proteins, one of which is a transcription factor, called mtTF1 in mammals and ABF2 in *Saccharomyces cerevisiae*, which binds to mitochondrial promoters and enhances initiation of transcription. These promoters are variable in sequence but most contain an AT-rich region.
- Chloroplast RNA polymerases appear to be much more similar to the bacterial enzymes and their promoters follow the pattern described for *E. coli*, consisting of two boxes at the −35 and −10 positions relative to the start point for transcription.

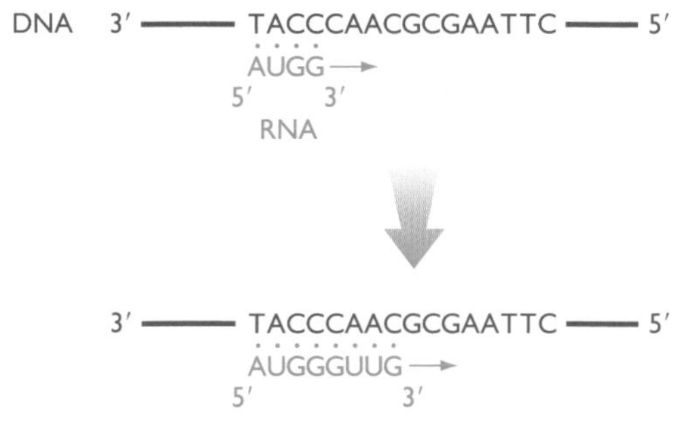

Figure 8.6 Template-dependent RNA synthesis.

The RNA transcript is synthesized in the 5′→3′ direction, reading the DNA in the 3′→5′ direction, with the sequence of the transcript determined by base-pairing to the DNA template.

Table 8.1 Functions of the three eukaryotic RNA polymerases

Polymerase	Genes transcribed
RNA polymerase I	28S, 5.8S and 18S ribosomal RNA genes
RNA polymerase II	Protein-coding genes; most small nuclear RNA (snRNA) genes
RNA polymerase III	Genes for transfer RNAs, 5S ribosomal RNA, U6-snRNA, small nucleolar (sno) RNAs, small cytoplasmic (sc) RNA

See Section 9.1.1 for descriptions of these RNA types.

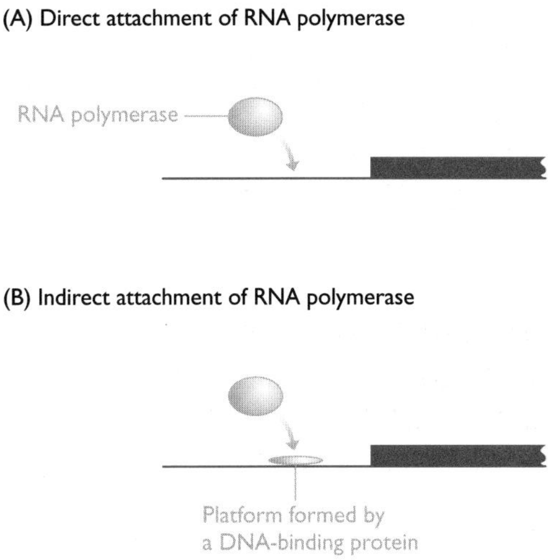

Figure 8.7 Two ways in which RNA polymerases bind to their attachment sites.

(A) shows the direct recognition of the DNA attachment site by the RNA polymerase, as occurs in bacteria. (B) shows recognition of the DNA attachment site by a DNA-binding protein which forms a platform on to which the RNA polymerase binds. This indirect mechanism occurs with eukaryotic and archaeal RNA polymerases.

Archaea, such as *Methanococcus jannaschii*, possess a single RNA polymerase that is very similar to the eukaryotic versions (Bult *et al.*, 1996). But this is not typical of the prokaryotes in general because the bacterial RNA polymerase is very different, consisting of just five subunits described as $\alpha_2\beta\beta'\sigma$ (two α subunits, one each of β and the related β', and one of σ). The α, β and β' subunits are equivalent to the three largest subunits of the eukaryotic RNA polymerases, but the σ subunit has its own special properties, both in terms of its structure and, as we will see in the next section, its function.

8.2.2 Recognition sequences for transcription initiation

It is essential that transcription initiation complexes are constructed at the correct positions on DNA molecules. These positions are marked by target sequences that are recognized either by the RNA polymerase itself or by a DNA-binding protein which, once attached to the DNA, forms a platform to which the RNA polymerase binds (*Figure 8.7*).

Bacterial RNA polymerases bind to promoter sequences

In bacteria, the target sequence for RNA polymerase attachment is called the **promoter**. This term was first used by geneticists in 1964 to describe the function of a locus immediately upstream of the three genes in the lactose operon (*Figure 8.8*). When this locus was inactivated by mutation, the genes in the operon were not expressed: the locus therefore appeared to *promote* expression of the genes. We now know that this is because the locus is the binding site for the RNA polymerase that transcribes the operon.

The sequences that make up the *E. coli* promoter were first identified by comparing the regions upstream of over 100 genes. It was assumed that promoter sequences would be very similar for all genes and so should be recognizable when the upstream regions were compared. These analyses showed that the *E. coli* promoter consists of two segments, both of six nucleotides, described as follows (see *Figure 8.8*):

−35 box	5′-TTGACA-3′
−10 box	5′-TATAAT-3′

These are consensus sequences and so describe the 'average' of all promoter sequences in *E. coli*; the actual sequences upstream of any particular gene might be slightly different (*Table 8.2*). The names of the boxes indicate their positions relative to the point at which transcription begins. The nucleotide at this point is labeled '+1' and is anything between 20 and 600 nucleotides upstream of the start of the coding region of the gene. The spacing between the two boxes is important because it places the two motifs on the same face of the double helix, facilitating their interaction with the DNA-binding component of the RNA polymerase (Section 8.2.3).

Eukaryotic promoters are more complex

In eukaryotes, the term 'promoter' is used to describe all the sequences that are important in initiation of transcription of a gene. For some genes these sequences can be numerous and diverse in their functions, including not only the **core promoter**, sometimes called the **basal promoter**, which is the site at which the initiation complex is assembled, but also one or more **upstream promoter elements** which, as their name implies, lie upstream of the core promoter. Assembly of the initiation complex on the core promoter can usually occur in the absence of the upstream elements, but only in an inefficient way. This indicates that the proteins that bind to the upstream

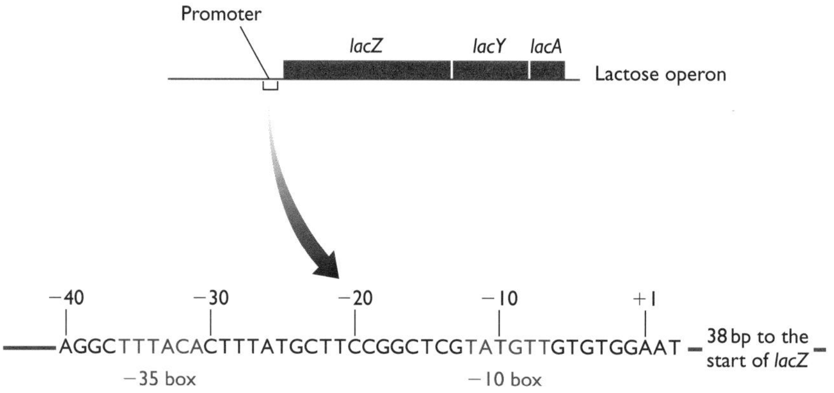

Figure 8.8 The promoter for the lactose operon of *Escherichia coli*.

The promoter is located immediately upstream of *lacZ*, the first gene in the operon. The DNA sequence shows the positions of the −35 and −10 boxes, the two distinct sequence components of the promoter. Compare these sequences with the consensus sequences described in the text. See also *Table 8.2*. For more information on the lactose operon see *Figure 6.14A*, p. 133.

Table 8.2 Sequences of *E. coli* promoters

Promoter	Sequence	
	−35 box	**−10 box**
Consensus	5′-TTGACA-3′	5′-TATAAT-3′
Lactose operon	5′-TTTACA-3′	5′-TATGTT-3′
Tryptophan operon	5′-TTGACA-3′	5′-TTAACT-3′

elements include at least some that are activators of transcription, and which therefore 'promote' gene expression. Inclusion of these sequences in the 'promoter' is therefore justified.

Each of the three types of eukaryotic RNA polymerase recognizes a different type of promoter sequence; indeed, it is the difference between the promoters that defines which genes are transcribed by which polymerases. The details, for vertebrates, are as follows (*Figure 8.9*):

- RNA polymerase I promoters consist of a core promoter spanning the transcription start point, between nucleotides −45 and +20, and an **upstream control element** about 100 bp further upstream.
- RNA polymerase II promoters are variable and can stretch for several kilobases upstream of the transcription start site. The core promoter consists of two segments, the −25 or **TATA box** (consensus 5′-TATAWAW-3′, where W is A or T) and the **initiator (Inr) sequence** (consensus 5′-YYCARR-3′, where Y is C or T, and R is A or G) located around nucleotide +1. Some genes transcribed by RNA polymerase II have only one of these two components of the core promoter, and some, surprisingly, have neither. The latter are called 'null' genes and they are still transcribed, possibly through interactions between the RNA polymerase and a sequence called MED-1 which lies within the gene (Novina and Roy, 1996), though the start position for transcription is more variable than for a gene with a TATA and/or Inr sequence. As well as the core promoter, genes recognized by RNA polymerase II have various upstream promoter elements, the functions of which are described in Section 8.3.2.
- RNA polymerase III promoters are unusual in that they are located within the genes whose transcription they promote. These promoters are variable, falling into at least three categories. Usually the core promoter spans approximately 50–100 bp and comprises two sequence boxes. One category of class III promoter is very similar to the RNA polymerase II type, with a range of upstream promoter elements. Interestingly, this arrangement is seen with the U6

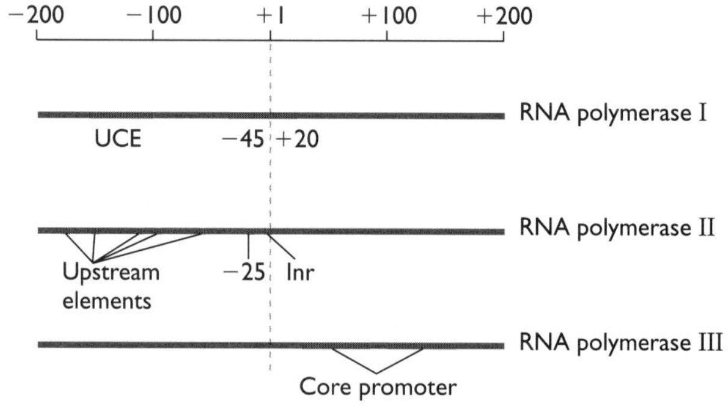

Figure 8.9 Structures of eukaryotic promoters.

Promoter regions are indicated in blue. The RNA polymerase III promoter structure refers to the 5S rRNA genes. Other genes transcribed by RNA polymerase III (see *Table 8.1*) have different promoter structures including some with a TATA box and upstream elements similar to an RNA polymerase II promoter. See the text for more details. Abbreviations: Inr, initiator sequence of the RNA polymerase II promoter; UCE, upstream control element of the RNA polymerase I promoter.

gene, which is one of a family of genes for small nuclear RNAs, all the other members of which are transcribed by RNA polymerase II.

8.2.3 Assembly of the transcription initiation complex

In a general sense, initiation of transcription operates along the same lines with each of the four types of RNA polymerase that we have been considering (*Figure 8.10*). The bacterial polymerase and the three eukaryotic enzymes all begin by attaching, directly or via accessory proteins, to their promoter or core promoter sequences. Next this **closed promoter complex** is converted into an **open promoter complex** by breakage of a limited number of base pairs around the transcription initiation site. Finally, the RNA polymerase moves away from the promoter. This last step is more complicated than it might appear because some attempts by the polymerase to achieve **promoter clearance** are unsuccessful and lead to truncated transcripts that are degraded soon after they are synthesized. The true completion of the initiation stage of transcription is therefore the establishment of a stable transcription complex that is actively transcribing the gene to which it is attached.

Although the scheme shown in *Figure 8.10* is correct in outline for all four polymerases, the details are different for each one. We will begin with the more straightforward events occurring in *E. coli* and other bacteria, and then move on to the ramifications of initiation in eukaryotes.

Transcription initiation in E. coli

In *E. coli*, a direct contact is formed between the promoter and RNA polymerase. The sequence-specificity of the polymerase resides in its σ subunit: the 'core enzyme', which lacks this component, can only make loose and nonspecific attachments to DNA.

Mutational studies of *E. coli* promoters have shown that changes to the sequence of the −35 box affect the ability of RNA polymerase to bind, whereas changes to the −10 box affect the conversion of the closed promoter complex into the open form. These results lead to the model for *E. coli* initiation shown in *Figure 8.11*, where recognition of the promoter occurs by an interaction between the σ subunit and the −35 box, forming a closed promoter complex in which the RNA polymerase spans some 60 bp from upstream of the −35 box to downstream of the −10 box. This is followed by breaking of the base pairs within the −10 box to produce the open complex. The model is consistent with the fact that the −10 boxes of different promoters are comprised mainly or entirely of A–T base pairs, which are weaker than G–C pairs, being linked by just two hydrogen bonds as opposed to three

Box 8.3: Initiation of transcription in the archaea

Initiation of transcription is one of the key areas in which the archaea differ from the bacteria, emphasizing the major distinction between these two types of prokaryote. The archaeal RNA polymerase is made of eleven subunits and closely resembles the RNA polymerases of eukaryotes. Promoters are recognized by a TATA-binding protein and RNA polymerase attachment is mediated by a transcription factor similar to the eukaryotic TFIIB. Archaeal homologs of the other eukaryotic general transcription factors have not been identified and searches of the completed archaeal genomes suggest that they may not be present.

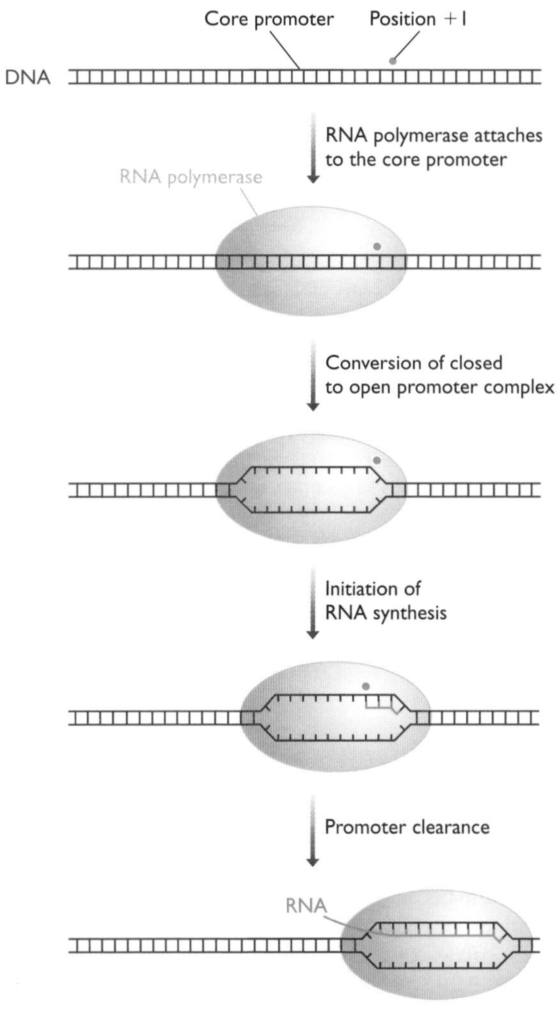

Figure 8.10 Generalized scheme for the events occurring during initiation of transcription.

The core promoter is shown in blue and the transcription initiation site indicated by a green dot. After RNA polymerase attachment the closed complex is converted into the open complex by breakage of base pairs within a short region of the DNA double helix. RNA synthesis begins but successful initiation is not achieved until the polymerase moves away from the promoter region.

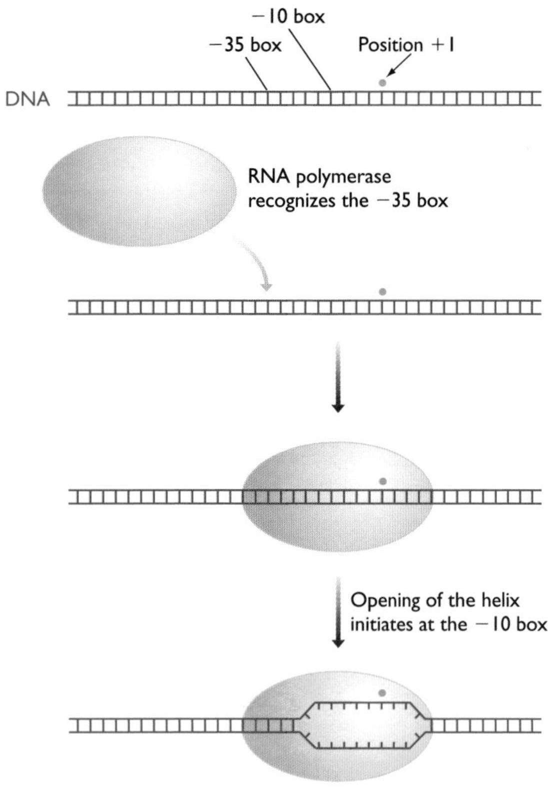

Figure 8.11 Initiation of transcription in *Escherichia coli.*

The *E. coli* RNA polymerase recognizes the −35 box as its binding sequence. After attachment to the DNA, the transition from closed to open complex is initiated by breakage of base pairs in the AT-rich −10 box. Note that although the polymerase is shown as a single structure it is the σ subunit that possesses sequence-specific DNA binding activity and which therefore recognizes the −35 sequence. For subsequent events leading to promoter clearance, refer to *Figure 8.10*.

(see *Figure 7.3B*, p. 149). Opening up of the helix involves contacts between the polymerase and the nontemplate strand (i.e. the one that is not copied into RNA), again with the σ subunit playing a central role (Marr and Roberts, 1997). However, the σ subunit is not all-important because it dissociates soon after initiation is complete, converting the holoenzyme to the core enzyme which carries out the elongation phase of transcription (Section 9.2.1).

Transcription initiation with RNA polymerase II

How do the easily understandable series of events occurring in *E. coli* compare with the equivalent processes in eukaryotes? RNA polymerase II will show us that eukaryotic initiation involves more proteins and has added complexities.

The first difference between initiation of transcription in *E. coli* and eukaryotes is that eukaryotic polymerases do not directly recognize their core promoter sequences. For genes transcribed by RNA polymerase II the initial contact is made by the **general transcription factor (GTF)** TFIID, which is a complex made up of the **TATA-binding protein (TBP)** and at least 12 **TBP-associated factors** or **TAFs**. Most of these are DNA-binding proteins. TBP is a sequence-specific protein which binds to DNA via its unusual TBP domain which makes contact with the minor groove (Section 7.4.2). Several of the TAFs have similarities with core histones and may form a nucleosome-like structure in the promoter region (Research Briefing 8.1; Surridge, 1996). Both TBP and TAFs appear to have important functions in recognition of the core promoter, but both are dispensable, at least for some genes (Chao and Young, 1996; Apone and Green, 1998; Grant and Workman, 1998).

TFIID recognizes the TATA component of the RNA polymerase II core promoter (*Figure 8.12*). It may also recognize the Inr sequence but this is unlikely. It is thought that there is probably an as yet undiscovered Inr binding protein, analogous to but different from TBP, that forms the direct connection with the Inr and then attaches, via protein–protein contacts, to TFIID.

After TFIID has recognized and attached to the core promoter, directly to the TATA sequence and/or indirectly to the Inr sequence, the **preinitiation complex** is formed by binding of additional GTFs in the order TFIIA, TFIIB, TFIIF/RNA polymerase II, TFIIE and TFIIH (see *Figure 8.12*). The specific functions of each of these is described in *Table 8.3*. Some of these GTFs have DNA-binding properties and others interact purely by protein–protein contacts. There are two key steps in this series of events. First, binding of TFIIF, as this step recruits the RNA polymerase into the complex. Second, attachment of TFIIH. This last GTF is a multifunctional protein with roles not only in transcription but also in DNA repair and regulation of the cell cycle (Svejstrup *et al.*, 1996); in transcription it has two functions, acting as a **helicase** which unwinds the DNA, aiding formation of the open promoter complex, and as a kinase that adds phosphate groups to the C-terminal domain of the largest subunit of RNA polymerase II. Once phosphorylated, the polymerase is able to leave the preinitiation complex and begin synthesizing RNA. Most of the GTFs now detach from the core promoter, leaving TFIID (or possibly just TBP) bound to the DNA and ready to initiate a second round of transcription (Roeder, 1996).

Transcription initiation with RNA polymerases I and III

Initiation of transcription at RNA polymerase I and III promoters involves similar events to those seen with RNA polymerase II, but the details are different. One of the most striking similarities is that TBP, first identified as the key sequence-specific DNA-binding component of the RNA polymerase II preinitiation complex, is also involved in initiation of transcription by the two other eukaryotic RNA polymerases.

The RNA polymerase I initiation complex involves four protein complexes in addition to the polymerase

RESEARCH

8.1

BRIEFING

Similarities between TFIID and the histone core octamer

An intriguing insight into the interaction between TFIID and the RNA polymerase II promoter was provided by the discovery that several TAFs have structural similarities with histones.

A key step in initiation of transcription by RNA polymerase II is attachment of the general transcription factor TFIID to the core promoter. The resulting structure forms the platform on which the preinitiation complex is constructed. Intuitively, one anticipates that this platform must be a stable structure in order for the bulky preinitiation complex to be positioned correctly so that productive initiation of transcription can occur. To understand the nature of the attachment, attention has been focused on the TBP-associated factors (TAFs) that, together with the TATA-binding protein, make up the TFIID protein complex.

Homology searching reveals similarities between TAFs and histones

The first link between TAFs and histones was established by homology searching (Section 5.2.1). In 1993–94 a number of TAF genes were cloned, enabling nucleotide sequences to be obtained. The nucleotide sequences were translated into the amino acid sequences of the TAF proteins, and the amino acid sequences used to search the databases for interesting homologies with other proteins. Weak similarities were discovered with the histones of the core octamer of the nucleosome (Section 6.1.1). One of the sequence alignments, between TAF$_{II}$62 and histone H4, is shown above. In this central region of 55 amino acids there are 15 positions at which the same amino acid is present in both sequences, and a further 15 positions where the amino acids are 'similar', these being ones which, although different, have equivalent chemical properties. The same kind of pattern is seen when TAF$_{II}$42 is compared with histone H3. The degree of similarity between these two sets of proteins is not so great that a structural and/or functional relationship can immediately be inferred, but it is sufficiently striking to suggest that more detailed comparisons between TAFs and histones should be made.

Structural comparisons

These more detailed comparisons were carried out by X-ray crystallographic analyses (Section 7.3.3) of TAF$_{II}$42 and TAF$_{II}$62. The results were remarkable, both TAFs being shown to possess a typical histone fold (*Table 7.5*, p. 164), the nonsequence-specific DNA-binding motif, comprising a long α-helix flanked on either side by two shorter α-helices, that is a characteristic feature of histones. Even more striking was the discovery that in the complex formed between TAF$_{II}$42 and TAF$_{II}$62, the two histone folds are orientated in almost exactly the same way as in the histone H3/H4 dimer found in the central tetramer of the nucleosome core particle.

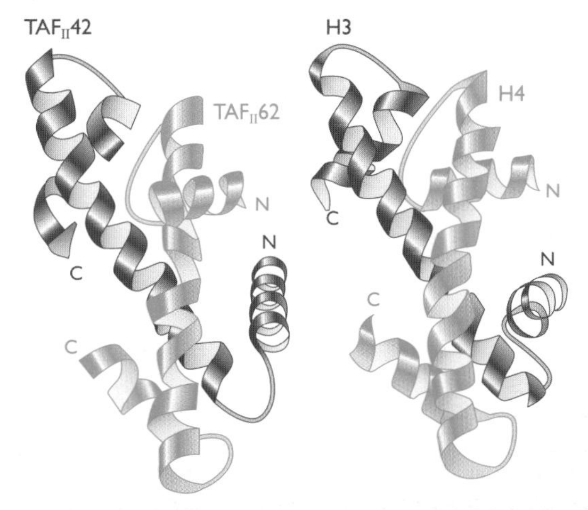

Reprinted with permission from Xie, *et al.*, *Nature*, **380**, 316–322. Copyright 1996 Macmillan Magazines Limited.

Can these results be taken as evidence that TFIID forms a nucleosome-like structure at the core promoter, this pseudonucleosome acting as the platform for assembly of the preinitiation complex? The conclusion is attractive but premature. More research is needed to determine if the similarities between TAFs and histones extend to the details of the attachment between each protein complex and its DNA target. A less exciting but equally possible alternative is that the similarities merely reflect an equivalence in the protein–protein interactions within the two complexes, and are not directly relevant to the DNA attachment.

References

Kokubo T, Gong D-W, Wootton JC, Horikoshi M, Roeder RG and Nakatani Y (1994) Molecular cloning of *Drosophila* TFIID subunits. *Nature*, **367**, 484–487.

Xie X, Kokubo T, Cohen SL, *et al.* (1996) Structural similarity between TAFs and the heterotetrameric core of the histone octamer. *Nature*, **380**, 316–322.

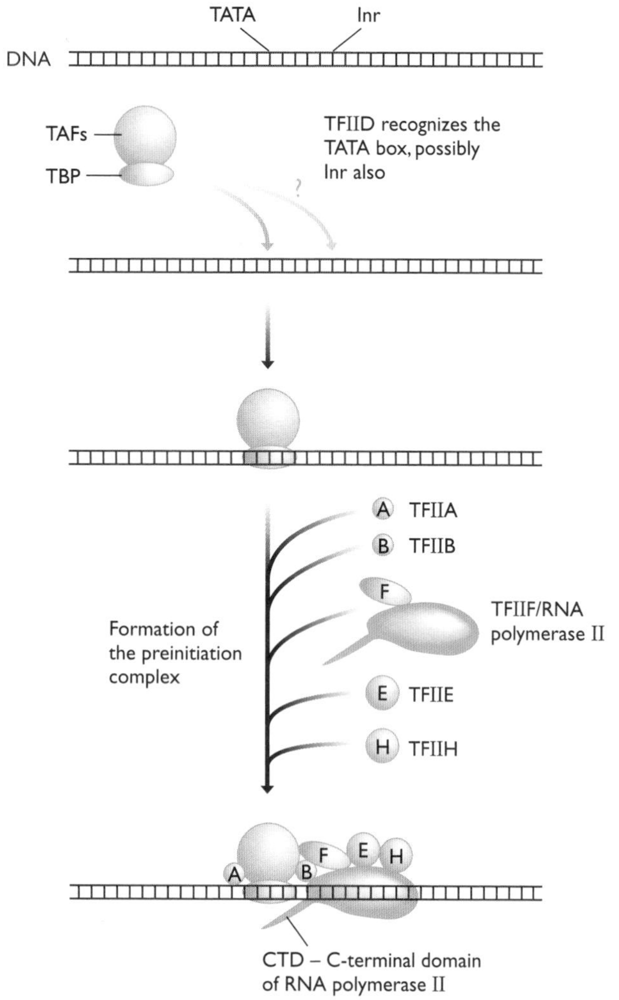

Figure 8.12 Assembly of the RNA polymerase II preinitiation complex.

The first step in assembly of the preinitiation complex is recognition of the TATA box and possibly the Inr sequence by the TATA-binding protein (TBP), probably in conjunction with the TBP-associated factors (TAFs). Other components of the preinitiation complex assemble in the order shown. The drawing is purely schematic with regards to the sizes and shapes of the components and their relative positioning within the preinitiation complex, though account is taken of the protein–protein contacts that are known to occur. The C-terminal domain (CTD) of the largest subunit of RNA polymerase II, which must be phosphorylated before the polymerase can leave the promoter, is depicted as a spike projecting from the bottom of the complex. Based on Roeder (1996), which is recommended reading for more details of initiation by RNA polymerase II.

itself. One of these, UBF, is a dimer of identical proteins that interacts with both the core promoter and the upstream control element (see *Figure 8.9*). UBF is another protein which, like some of the RNA polymerase II TAFs, resembles a histone and may form a nucleosome-like structure in the promoter region (Wolffe, 1994). A second protein complex, called SL1 in humans and TIF-IB in mice, contains TBP and, together with UBF, directs RNA polymerase I and the last two complexes, TIF-IA and TIF-IC, to the promoter. Originally it was thought that the initiation complex was built up in a stepwise fashion but recent results suggest that RNA polymerase I binds the four protein complexes before promoter recognition, the entire assembly attaching to the DNA in a single step (Seither *et al.*, 1998).

RNA polymerase III promoters are variable in structure (see *Figure 8.9*) and this is reflected by a nonuniformity of the processes through which they are recognized. Initiation at the different categories of RNA polymerase III promoter requires different sets of GTFs, but each type of initiation process involves TFIIIB, one of whose subunits is TBP. With promoters of the type seen with the U6 gene, which contain a TATA sequence, TBP probably binds directly to the DNA. In other RNA polymerase III promoters, which have no TATA sequence, binding is probably via a second protein, the latter making the direct DNA contact.

Table 8.3 Functions of human general transcription factors

General transcription factor	Function
TFIID (TBP component)	Recognition of the TATA box and possibly Inr sequence; forms a platform for TFIIB binding
TFIID (TAFs)	Recognition of the core promoter; regulation of TBP binding
TFIIA	Stabilizes TBP and TAF binding
TFIIB	Intermediate in recruitment of RNA polymerase II; influences selection of the start point for transcription
TFIIF	Recruitment of RNA polymerase II
TFIIE	Intermediate in recruitment of TFIIH; modulates the various activities of TFIIH
TFIIH	Helicase activity responsible for the transition from the closed to open promoter complex; possibly influences promoter clearance by the polymerase

Based on Roeder (1996), which is recommended for further details concerning the functions of the human GTFs.

8.3 REGULATION OF TRANSCRIPTION INITIATION

As we progress through the next few chapters we will encounter a number of strategies that organisms use to regulate expression of individual genes. We will discover that virtually every step in the pathway from gene to functioning protein is subject to some degree of control and that regulation of genome expression involves an almost bewildering network of control systems operating on different stages of the expression pathways for different genes. Of all these regulatory systems, it appears that transcription initiation is the step at which the critical controls over gene expression, those that have greatest impact on the biochemical properties of the cell, are exerted. This is perfectly understandable. It makes sense that transcription initiation, being the first step in gene expression, should be the stage at which 'primary' regulation occurs, this being the level of regulation that determines which genes are expressed. Later steps in the pathway might be expected to respond to 'secondary' regulation, the function of which is not to switch genes on or off but to modulate expression of a gene by making small changes to the rate at which the protein product is synthesized, or possibly by changing the nature of the product in some way (*Figure 8.13*).

Earlier in this chapter (Section 8.1) we looked at how chromatin structure can influence gene expression by controlling the accessibility of promoter sequences to RNA polymerase and its associated proteins. This is just one way in which initiation of transcription can be controlled. To obtain a broader picture we will establish some general principles with bacteria, and then examine the events in eukaryotes.

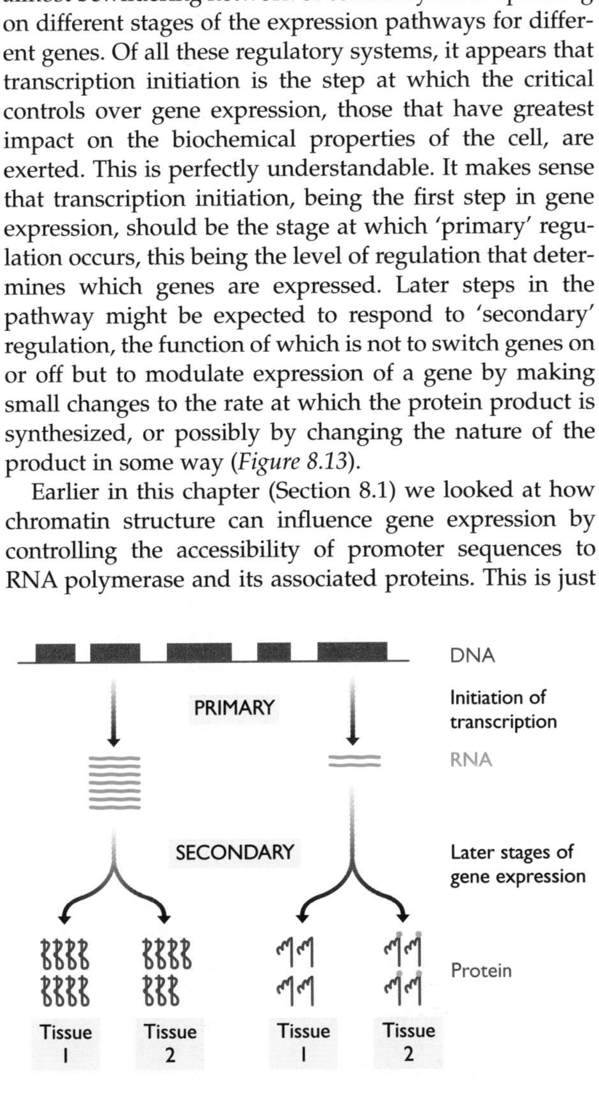

Figure 8.13 Primary and secondary levels of gene regulation.

According to this scheme, 'primary' regulation of gene expression occurs at the level of transcription initiation, this step determining which genes are expressed in a particular cell at a particular time and setting the relative rates of expression of those genes that are switched on. 'Secondary' regulation involves all steps in the gene expression pathway after transcription initiation, and serves to modulate the amount of protein that is synthesized or to change the nature of the protein in some way, for example by chemical modification.

8.3.1 Strategies for controlling transcription initiation in bacteria

In bacteria such as *E. coli*, we recognize two distinct ways in which transcription initiation is controlled:

- **Constitutive control**, which depends on the structure of the promoter.
- **Regulatory control**, which depends on the influence of regulatory proteins.

Promoter structure determines the basal level of transcription initiation

The consensus sequence for the *E. coli* promoter (Section 8.2.2) is quite variable with a range of different motifs being permissible at both the −35 and −10 boxes (see *Table 8.2*, p. 180). These variations, together with less well-defined sequence features around the transcription start site and in the first 50 or so nucleotides of the transcription units, affect the efficiency of the promoter, defined as the number of productive initiations that it promotes per second, a productive initiation being one that results in the RNA polymerase clearing the promoter and beginning synthesis of a full-length transcript. The exact way in which the sequence of the promoter affects initiation is not known, but from our discussion of the events involved in transcription initiation (Section 8.2.3) we might, intuitively, expect that the precise sequence of the −35 box would influence recognition by the σ subunit and hence the rate of attachment of RNA polymerase, that the transition from the closed to open promoter complex might be dependent on the sequence of the −10 box, and that the frequency of abortive initiations (ones that terminate before they progress very far into the transcription unit) might be influenced by the sequence at and immediately downstream of nucleotide +1. All this is speculation but it is a sound 'working hypothesis'. What is clear is that different promoters vary 1000-fold in their efficiencies, the most efficient promoters (called **strong promoters**) directing 1000 times as many productive initiations as the weakest promoters. We refer to these as differences in the **basal rate** of transcription initiation.

Note that the basal rate of transcription initiation for a gene is preprogrammed by the sequence of its promoter and so, under normal circumstances, cannot be changed. It could be changed by a mutation that altered a critical nucleotide in the promoter, and undoubtedly this happens from time to time, but it is not something that the bacterium has control over. The bacterium can, however, determine which promoter sequences are favored by changing the σ subunit of its RNA polymerase. The σ subunit is the part of the polymerase that has the sequence-specific DNA-binding capability (Section 8.2.3), so replacing one version of this subunit with a different version, with a slightly different DNA-binding motif and hence an altered sequence specificity, would result in a different set of promoters being recognized. In *E. coli*, the standard σ subunit, which recognizes the consensus promoter sequence shown on p. 179 and hence directs transcription of most genes, is called σ^{70} (its molecular mass is approximately 70 kDa). *E. coli* also has a second σ subunit, σ^{32}, which is made when the bacterium is exposed to a heat shock. During a heat shock, *E. coli* , in common with other organisms, switches on a set of genes coding for special proteins that help the bacterium withstand the stress (*Figure 8.14*). These genes have special promoter sequences, ones specifically recognized by the σ^{32} subunit. The bacterium is therefore able to switch on a whole range of different genes by making one simple alteration to the structure of its RNA polymerase. This system is common in bacteria: for example, *E. coli* also uses it to control expression of genes involved in nitrogen fixation, this time with the σ^{54} subunit, and *Bacillus* species use a whole range of different σ subunits to switch on and off groups of genes during the changeover from normal growth to formation of spores (Section 11.3.1).

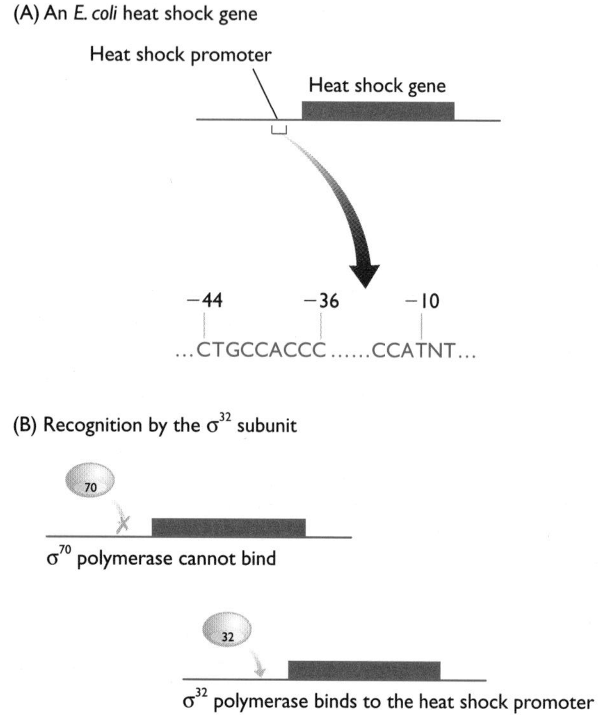

Figure 8.14 Recognition of an *E. coli* heat shock gene by the σ^{32} subunit.

(A) The sequence of the heat shock promoter is different from that of the normal *E. coli* promoter (compare with *Table 8.2*, p. 180). (B) The heat shock promoter is not recognized by the normal *E. coli* RNA polymerase containing the σ^{70} subunit, but is recognized by the σ^{32} RNA polymerase that is active during heat shock. Abbreviation: N, any nucleotide. For more details of the use of novel σ factors by bacteria see Section 11.3.1.

Regulatory control over bacterial transcription initiation

Promoter structure determines the basal level of transcription initiation for a bacterial gene but, with the exception of alternative σ subunits, does not provide any general means by which the expression of the gene can respond to changes in the environment or the biochemical requirements of the cell. Other types of regulatory control are needed.

The foundation of our understanding of regulatory control over transcription initiation in bacteria was laid in the early 1960s by François Jacob, Jacques Monod and other geneticists who studied the lactose operon and other model systems (Brock, 1990). We have already seen how this work led to discovery of the promoter for the lactose operon (Section 8.2.2). It also resulted in identification of the **operator**, a region adjacent to the promoter and which regulates initiation of transcription of the operon (*Figure 8.15A*). The original model envisaged a DNA-binding protein, the **lactose repressor**, attaching to the operator and preventing the RNA polymerase from binding to the promoter, simply by denying it access to the relevant segment of DNA (*Figure 8.15B*). Whether or not the repressor binds depends on the presence in the cell of allolactose, an isomer of lactose, the substrate for the biochemical pathway carried out by the enzymes coded by the three genes in the operon. Allolactose is an **inducer** of the lactose operon. When allolactose is present it binds to the lactose repressor, causing a slight structural change which prevents the helix-turn-helix motifs of the repressor from recognizing the operator as a DNA binding site. The allolactose–repressor complex therefore cannot bind to the operator, enabling the RNA polymerase to gain access to the promoter. When the supply of lactose is used up, the repressor reattaches to the operator and prevents transcription. The operon is therefore expressed only when the enzymes coded by the operon are needed.

The original scheme for regulation of the lactose operon has been confirmed in most details by DNA sequencing of the control region and structural studies of the repressor bound to its operator. The one complication has been the discovery that the repressor has three potential binding sites, at nucleotide positions −82, +11 and +412, and attachment at only one of these, at +11, would be expected to prevent access of the polymerase to the promoter. The repressor is a tetramer or four identical proteins, which work in pairs to attach to a single operator, so it is possible that the repressor has the capacity to bind to two of the three operator sites at once. It is also possible that the repressor can bind to an operator sequence in such a way that it does not block attachment of the polymerase to the promoter, but does prevent a later step in initiation, such as formation of the open promoter complex.

The lactose operon illustrates the basic principle of regulatory control of transcription initiation: attachment of a DNA-binding protein to its specific recognition site can influence the events involved in assembly of the tran-

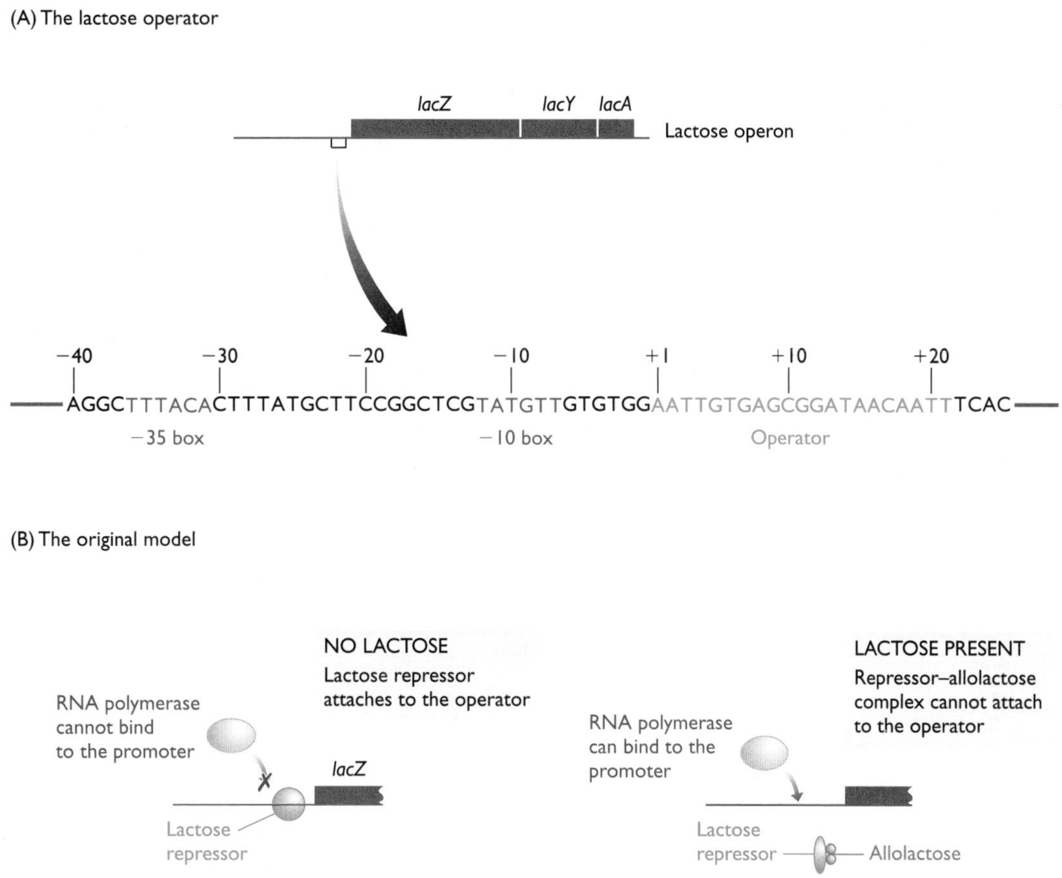

Figure 8.15 Regulation of the lactose operon of *E. coli.*

(A) The operator sequence lies immediately downstream of the promoter for the lactose operon. (B) In the original model for lactose regulation, the lactose repressor is looked on as a simple blocking device that binds to the operator and prevents the RNA polymerase gaining access to the promoter. The three genes in the operon are therefore switched off. This is the situation in the absence of lactose, though transcription is not absolutely blocked because the repressor occasionally detaches allowing a few transcripts to be made. Because of this **basal level** of transcription the bacterium always possesses a few copies of each of the three enzymes coded by the operon (see *Figure 6.14A*, p. 133), probably less than five of each. This means that when the bacterium encounters a source of lactose it is able to transport a few molecules in to the cell and split these into glucose and galactose. An intermediate in this reaction is allolactose, an isomer of lactose, which induces expression of the lactose operon by binding to the repressor, causing a change in the conformation of the latter so it is no longer able to attach to the operator. This allows the RNA polymerase to bind to the promoter and transcribe the three genes. When fully induced approximately 5000 copies of each protein product are present in the cell. When the lactose supply is used up and allolactose is no longer present, the repressor reattaches to the operator and the operon is switched off. Note that the shapes of the repressor and polymerase structures shown here are purely schematic.

scription initiation complex and/or initiation of productive RNA synthesis by an RNA polymerase. Several variations on this theme are seen with other bacterial genes:

- Some repressors respond not to an inducer but to a **co-repressor**. An example is provided by the tryptophan operon of *E. coli*, which codes for a set of genes involved in synthesis of tryptophan (see *Figure 6.14B*, p. 133). In contrast to the lactose operon, the regulatory molecule for the tryptophan operon is not a substrate for the relevant biochemical pathway, but the product, tryptophan itself (*Figure 8.16*). Only when tryptophan is attached to the tryptophan repressor can the latter bind to the operator. The tryptophan operon is therefore switched off in the presence of tryptophan, and switched on when tryptophan is needed. The situation is the reverse of that seen with the lactose operon, but involves the same basic regulatory mechanism.
- Some DNA-binding proteins are **activators** rather than repressors of transcription initiation. In *E. coli* the best example is the **catabolite activator protein,** which binds to sites upstream of several operons, including the lactose operon, and increases the efficiency of transcript initiation, probably by forming a direct contact with the RNA polymerase. The

Box 8.4: *Cis* and *trans*

'*Cis*' and '*trans*' are two important terms relevant to the genetic study of gene regulation in bacteria and other organisms:

- A locus is *cis*-acting on a second locus if it must be on the same DNA molecule in order to have an effect. The operator is a *cis*-acting element because it works only when physically attached to the gene whose expression it regulates.
- A locus is *trans*-acting if it can affect a second locus even when on a different DNA molecule. The gene for the lactose repressor (*lacI*) is *trans*-acting because it can regulate expression of the lactose operon even when removed from the *E. coli* chromosome and placed on a plasmid.

To a molecular biologist, a *cis*-acting regulatory element is usually a target site for a DNA-binding protein, upstream of the gene whose expression is being regulated. A *trans*-acting element is the regulatory protein itself, which can diffuse through the cell from its site of synthesis to its DNA-binding site.

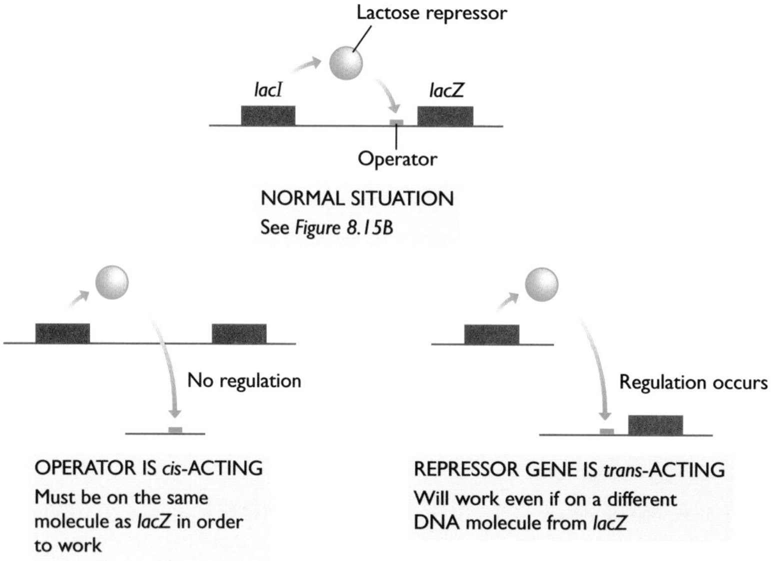

biological role of the catabolite activator protein will be described in Section 11.1.1.

- Other DNA-binding proteins work singly or together to increase or repress transcription of genes to which they are not closely linked. These **enhancers** and **silencers** are not common in bacteria but a few examples are known, including an enhancer that acts on the *E. coli* heat shock genes whose promoters are recognized by the σ^{32} version of the RNA polymerase. Because they are so far from the genes they control they can only form a contact with the RNA polymerase if the DNA forms a loop. A characteristic feature is that a single enhancer or silencer can control expression of more than one gene.

8.3.2 Control of transcription initiation in eukaryotes

With bacteria, it is possible to make a clear distinction between constitutive and regulatory forms of control over transcription initiation. The former depends on promoter structure and determines the basal rate of transcription initiation; the latter depends on the activity of regulatory proteins and changes the rate of transcription initiation if the basal rate is inappropriate for the prevailing conditions. With eukaryotes, categorization of different types of control system is less easy. This is because of a fundamental difference between transcription initiation in bacteria and eukaryotes. In bacteria, the RNA polymerase has a strong affinity for its promoter and the basal rate of transcription initiation is relatively high for all but the weakest promoters. With most eukaryotic genes the reverse is true. The RNA polymerase II and III preinitiation complexes do not assemble efficiently and the basal rate of transcription initiation is therefore very low, regardless of how 'strong' the promoter is. In order to achieve effective initiation, formation of the complex must be activated by additional proteins. Some of these could be defined as 'constitutive' activators, in that they work on many different genes and seem not to respond to any external signals; others could be termed 'regulatory'

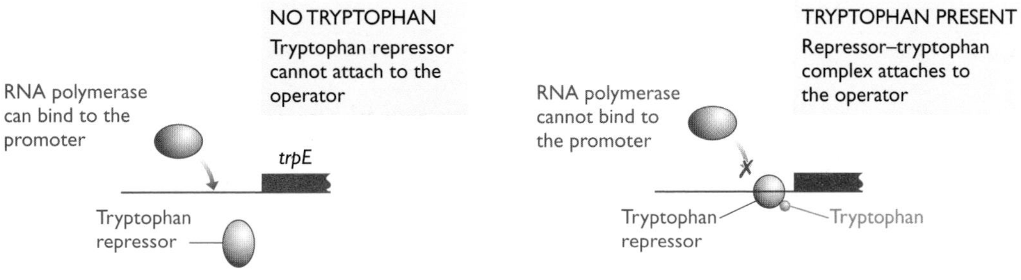

Figure 8.16 Regulation of the tryptophan operon of *E. coli*.

Regulation occurs via a repressor–operator system in a similar way to that described for the lactose operon (*Figure 8.15*) with the difference that the operon is repressed by the regulatory molecule, tryptophan, which is the product of the biochemical pathway specified by the genes in the operon (see *Figure 6.14B*, p. 133). When tryptophan is present, and so does not need to be synthesized, the operon is switched off because the repressor–tryptophan complex binds to the operator. In the absence of tryptophan the repressor cannot attach to the operator, and the operon is expressed.

activators because they target a limited number of genes and do respond to external signals. But there are grey areas between the two types and it is unwise to use this categorization as anything other than a guide to the types of event that occur.

Activators of eukaryotic transcription initiation

Proteins which activate transcription initiation by RNA polymerase II or III are called **transcription factors** (*Table 8.4*). These activators should not be confused with the general transcription factors such as TFIID which play a more central role in assembly of the preinitiation complex.

The traditional view has been that transcription factors are sequence-specific DNA-binding proteins. Some recognize upstream promoter elements (see *Figure 8.9*) and influence transcription initiation only at the promoter to which these elements are attached. Other transcription factors target sites within enhancers and influence transcription of several genes at once. As with bacteria, eukaryotic enhancers can be some distance from the genes that they control but each one is still specific for a certain set of genes: one possibility is that each enhancer controls genes only within the structural domain within which it is located (see Section 8.1.1). Whether bound to an upstream promoter element or a more distant enhancer, the transcription factor, according to the traditional view, activates formation of the preinitiation complex by making contact with it, directly or via another protein, sometimes achieving this by inducing a bend in the DNA between the core promoter and the binding site for the transcription factor (*Figure 8.17*).

This traditional view still holds for the majority of transcription factors that have been identified but cannot be looked on as all-encompassing. We have already seen (Section 8.1.1) that some transcription factors, such as p300/CBP, are able to modify histone proteins and so may function, at least in part, by affecting nucleosome positioning rather than by influencing assembly of the preinitiation complex. Some transcription factors work by bending the DNA into a specific shape that brings into contact other transcription factors that are bound to recognition sites within the bent region, enabling the bound factors to work together in a structure that has been called an **enhanceosome**. An example of a transcription factor that works in this way is SRY, which is the primary protein responsible for determining sex in mammals (Wolffe, 1995). Still other transcription factors have no DNA-binding properties and activate transcription simply by forming protein–protein contacts with the

Box 8.5: Regulation of RNA polymerase I initiation

RNA polymerase I is unusual in that it transcribes just a single set of genes, the multiple copies of the transcription unit containing the 28S, 5.8S and 18S rRNA sequences (Section 6.1.1). These genes are expressed continuously in most cells, but with the rate of transcription varying during the cell cycle and subject to a certain amount of tissue-specific regulation. The regulatory mechanism has not been described in detail but UBF has been implicated as the target for control, due to a clearly defined interaction between this component of the RNA polymerase I initiation complex and the retinoblastoma protein, which limits cell proliferation by repressing transcription initiation by all three polymerases (Voit *et al.*, 1997).

An interesting aspect of RNA polymerase I regulation that is emerging from current research is the role of the **termination factor** for this enzyme. This factor, called TTF-1 in mice and Reb1p in *Saccharomyces cerevisiae*, was first identified as an activator of RNA polymerase II transcription. It appears that the termination factor may also activate RNA polymerase I transcription, because a binding site for it has been located immediately upstream of the promoter for the rRNA transcription unit (Reeder and Lang, 1997). It is not yet clear what biological role the termination factor might play in activation of rRNA synthesis.

Table 8.4 Examples of transcription factors

Factor	DNA-binding site: Name	DNA-binding site: Consensus sequence	Comments
'Constitutive' transcription factors			
CTF/NF1	CAAT box	5′-GCCAATCT-3′	Ubiquitous
CP family	CAAT box	5′-GCCAATCT-3′	Ubiquitous
C/EBP	CAAT box	5′-GCCAATCT-3′	Ubiquitous
	Enhancer core	5′-TGTGGWWWG-3′	
Sp1	GC box	5′-GGGCGG-3′	Ubiquitous
Oct-1	Octamer	5′-ATGCAAAT-3′	Ubiquitous
Oct-2	Octamer	5′-ATGCAAAT-3′	Found in B lymphocytes
Response factors			
Heat shock factor	HSE	5′-CNNGAANNTCCNNG-3′	Responds to heat shock
Serum response factor	SRE	5′-CCATATTAGG-3′	Responds to growth factors in serum
STAT 1	IGRE	5′-TTNCNNNAA-3′	See Section 11.1.2
Cell-specific factors			
GATA-1	–	5′-GATA-3′	Erythroid cell-specific
Pit-1	–	5′-ATATTCAT-3′	Pituitary-specific
MyoD1	E-box	5′-CANNG-3′	Myoblast-specific
NF-κB	κB site	5′GGGACTTTCC-3′	Lymphoid cell-specific
Developmental regulators			
Bicoid	–	5′-TCCTAATCCC-3′	See Section 11.3.3
Antennapedia	–	5′-TAATAATAATAATAA-3′	See Section 11.3.3
Fushi tarazu	–	5′-TCAATTAAATGA-3′	See Section 11.3.3

Based on Twyman RM (1998).
Transcription factors: C/EBP, CAAT/enhancer binding protein; CP, CAAT-binding protein; CTF/NF1, CAAT transcription factor/nuclear factor 1; STAT, signal transducer and activator of transcription. DNA-binding sites: HSE, heat shock element; SRE, serum response element; IGRE, interferon-γ response element.
Nucleotide sequences: N = any nucleotide; W = A or T.

preinitiation complex. As more and more transcription factors are discovered our appreciation of their diversity will undoubtedly grow.

Contacts between transcription factors and the preinitiation complex

A critical feature of transcription factor activity is the contact it forms with the preinitiation complex. The relevant part of the transcription factor is called the **activation domain**. Structural studies have shown that although activation domains are variable most of them fall into one of three categories:

- **Acidic domains** are ones that are relatively rich in acidic amino acids (aspartic acid and glutamic acid). This is the commonest category of activation domain.
- **Glutamine-rich domains** are often found in transcription factors whose DNA-binding domains are of the homeodomain or POU type (Section 7.4.2).
- **Proline-rich domains** are less common.

The type of activation domain present in a transcription factor appears to dictate, at least to a certain extent, the details of the interaction with the preinitiation complex, but this is currently a difficult area with apparently conflicting evidence coming from work with different organisms. In yeast, for example, a protein complex called the **mediator** has been identified, which forms a physical contact between various transcription factors and the C-terminal domain of RNA polymerase II (see *Figure 8.17B*), implying that the polymerase is directly activated by a direct link with the transcription factor (Björklund and Kim, 1996). This mediator is not, however, an essential requirement for activation of all yeast genes (Lee and Lis, 1998). In the fruit fly, searches for the mediator have been unsuccessful; instead several of the TAFs present in TFIID (Section 8.2.3) are implicated as binding partners for different transcription factors. These TAFs are looked on as coactivators that work with the transcription factors to increase the affinity of TFIID for the core promoter. In humans, activation also involves TAFs, but additional coactivators have been discovered which could turn out to be equivalent to the yeast mediator (Kornberg, 1996). There are other lines of evidence suggesting that not only RNA polymerase II, but also TFIIB and TFIIH respond directly to activation from transcription factors.

A few repressors of eukaryotic transcription initiation have been identified

In view of the negligible amount of basal transcription initiation occurring at RNA polymerase II and III promoters, it might seem unlikely that repression of initiation, which is so important in bacteria (Section 8.3.1) plays any major part in control of eukaryotic transcription. In fact a growing number of DNA-binding proteins

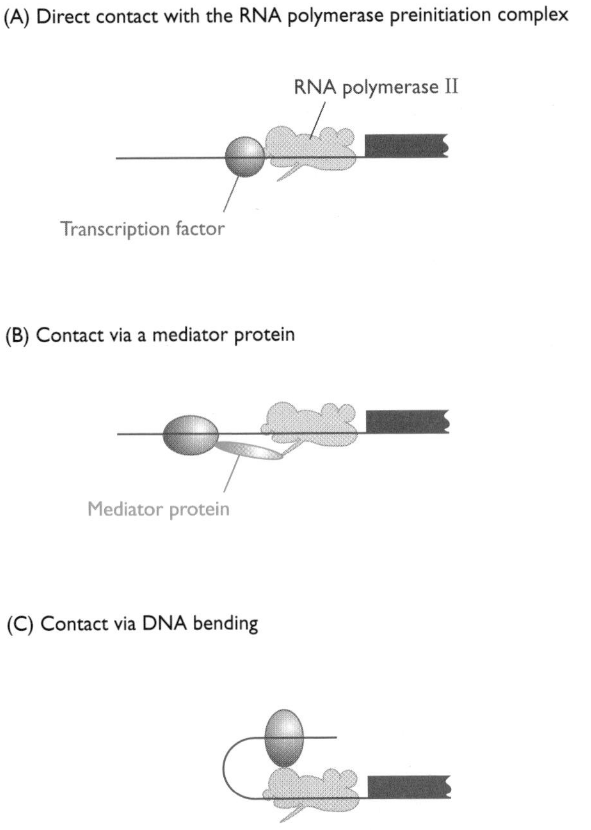

Figure 8.17 Three ways in which a transcription factor could activate initiation of transcription by RNA polymerase II.

(A) Some transcription factors are thought to bind to a target sequence adjacent to the core promoter and then make direct contact with the preinitiation complex. (B) Some factors work in conjunction with a coactivator or a mediator which forms a protein bridge between the transcription factor and preinitiation complex. (C) Some contacts require DNA bending. As described in the text, these three forms of contact are not all-encompassing. It is not yet clear which part of the preinitiation complex makes contact with the transcription factor. In (B), contact with the C-terminal domain of RNA polymerase II is shown, which has been implicated as the mode of action of the yeast mediator protein (Björklund and Kim, 1996).

that repress transcription initiation are being discovered, these proteins binding to upstream promoter elements or more distant sites in silencers.

Repressors have not been comprehensively studied in eukaryotes. A variety of inhibition domains (the converse of an activation domain) have been identified, several of which are rich in prolines, but no general patterns have emerged (Hanna-Rose and Hansen, 1996). Neither have the targets within the preinitiation complex been defined. Several repressors interact with general transcription factors (e.g. TFIIB and TFIID) and are presumed to affect assembly of the preinitiation complex: they may act on

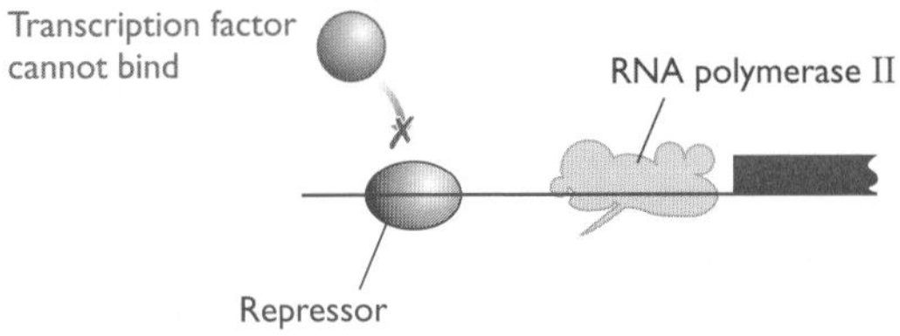

Figure 8.18 Some eukaryotic repressors may compete with transcription activators for the same binding site.

corepressors that have histone deacetylase activity and so induce chromatin formation over the promoter to be silenced. Other repressors target the RNA polymerase, and others bind to the same DNA sites as activator proteins and therefore prevent the activators from attaching and stimulating transcription initiation (*Figure 8.18*).

Control over the activity of transcription factors

The activities of individual transcription factors must be controlled to ensure that the appropriate set of genes is expressed by a cell. We will return to this topic in Chapter 11, when it will form the central theme of our study of the ways in which genome activity is regulated in response to extracellular signals and during differentiation and development.

The activity of a transcription factor can be regulated by controlling either its synthesis or its ability to activate or repress transcription. If the only level of control is over the synthesis of the transcription factor, then rapid changes in expression of the target genes will not be possible. This is because it takes time to accumulate a transcription factor in the cell, or to destroy it when it is not needed. This type of control is therefore associated with transcription factors responsible for maintaining stable patterns of gene expression, for example those underlying cellular differentiation and some aspects of development.

If the activity of the transcription factor is directly controllable then rapid changes in gene expression patterns can occur, enabling the cell to respond to external signals and to transient changes in its biochemical requirements. This type of rapid control over transcription factor activity often depends on the ability of an extracellular signaling compound to influence events occurring inside the cell. This can be achieved in a variety of ways, but all are variations on two themes (*Figure 8.19*):

- **Direct activation** occurs when the extracellular signaling compound is able to enter the cell.
- **Indirect activation** occurs when the extracellular signaling compound is unable to cross the cell membrane and instead binds to a cell surface receptor which transduces the signal to the cell interior.

We will examine the details of these various signaling mechanisms in Section 11.1.

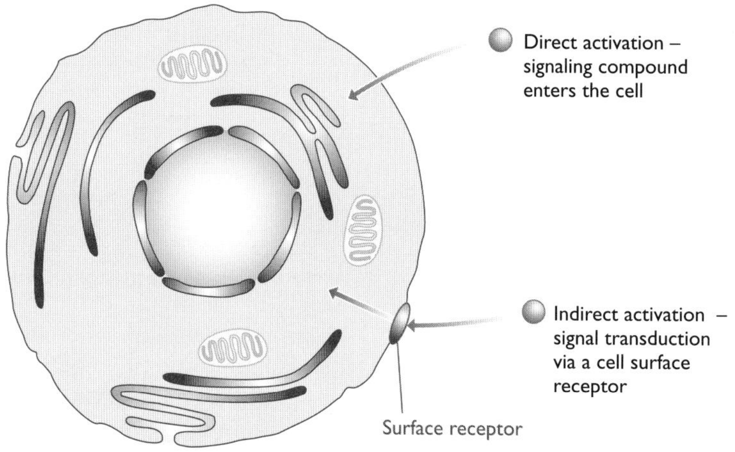

Figure 8.19 Two ways in which an extracellular signaling compound can influence events occurring within a cell. See Section 11.1 for more details on extracellular signaling.

REFERENCES

Apone LM and Green MR (1998) Gene expression: transcription *sans* TBP. *Nature*, **393,** 114–115.

Bannister AJ and Kouzarides T (1996) The CBP co-activator is a histone acetyltransferase. *Nature*, **384**, 641–643.

Bender J (1998) Cytosine methylation of repeated sequences in eukaryotes: the role of DNA pairing. *Trends Biochem. Sci.*, **23**, 252–256.

Björklund S and Kim Y-J (1996) Mediator of transcriptional regulation. *Trends Biochem. Sci.*, **21**, 335–337.

Brehm A, Miiska EA, McCance DJ, Reid JL, Bannister AJ and Kouzarides T (1998) Retinoblastoma protein recruits histone deacetylase to repress transcription. *Nature*, **391**, 597–601.

Brock TD (1990) *The Emergence of Bacterial Genetics*. Cold Spring Harbor Laboratory Press, Cold Spring Harbor, NY.

Bult CJ, White O, Olsen GJ, *et al.* (1996) Complete genome sequence of the methanogenic archaeon, *Methanococcus jannaschii*. *Science*, **273**, 1058–1073.

Cairns BR (1998) Chromatin remodelling machines: similar motors, ulterior motives. *Trends Biochem. Sci.*, **23**, 20–25.

Chao DM and Young RA (1996) Activation without a vital ingredient. *Nature*, **383**, 119–120.

Eden S, Hashimshony T, Keshet I, Cedar H and Thorne AW (1998) DNA methylation models histone acetylation. *Nature*, **394**, 842.

Grant PA and Workman JL (1998) A lesson in sharing? *Nature*, **396**, 410–411.

Hampsey M (1997) A SAGA of histone acetylation and gene expression. *Trends Genet.*, **13**, 427–429.

Hanna-Rose W and Hansen U (1996) Active repression mechanisms of eukaryotic transcription repressors. *Trends Genet.*, **12**, 229–234.

Jaenisch R (1997) DNA methylation and imprinting: why bother? *Trends Genet.*, **13**, 323–329.

Kornberg RD (1996) RNA polymerase II transcription control. *Trends Biochem. Sci.*, **21**, 325–326.

Lee D and Lis JT (1998) Transcriptional activation independent of TFIIH kinase and the RNA polymerase II mediator *in vivo*. *Nature*, **393**, 389–392.

Magnaghi-Jaulin L, Groisman R, Naguibneva I, Robin P, Lorain S, Le Villain JP, Troalen F, Trouche D and Harel-Bellan A (1998) Retinoblastoma protein represses transcription by recruiting a histone deacetylase. *Nature*, **391**, 601–605.

Marr MT and Roberts JW (1997) Promoter recognition as measured by binding of polymerase to nontemplate strand oligonucleotide. *Science*, **276**, 1258–1260.

Novina CD and Roy AL (1996) Core promoters and transcriptional control. *Trends Genet.*, **12**, 351–355.

Pennisi E (1997) Opening the way to gene activity. *Science*, **275**, 155–157.

Reeder RH and Lang WH (1997) Terminating transcription in eukaryotes: lessons learned from RNA polymerase I. *Trends Biochem. Sci.*, **22**, 473–477.

Roeder RG (1996) The role of general initiation factors in transcription by RNA polymerase II. *Trends Biochem. Sci.*, **21**, 327–335.

Seither P, Iben S and Grummt I (1998) Mammalian RNA polymerase I exists as a holoenzyme with associated basal transcription factors. *J. Mol. Biol*, **275**, 43–53.

Surridge C (1996) Transcription: the core curriculum. *Nature*, **380**, 287–288.

Svejstrup JQ, Vichi P and Egly J-M (1996) The multiple roles of transcription factor/repair factor TFIIH. *Trends Biochem. Sci.*, **21**, 346–350.

Tong JK, Hassig CA, Schnitzler GR, Kingston RE and Schreiber SL (1998) Chromatin deacetylation by an ATP-dependent nucleosome remodelling complex. *Nature*, **395**, 917–921.

Twyman RM (1998) *Advanced Molecular Biology: A Concise Reference.* BIOS Scientific Publishers, Oxford.

Voit R, Schafer H and Grummt I (1997) Mechanism of repression of RNA polymerase I transcription by the retinoblastoma protein. *Mol. Cell Biol.*, **17**, 4230–4237.

Wade PA, Pruss D and Wolffe AP (1997) Histone acetylation: chromatin in action. *Trends Biochem. Sci.*, **22**, 128–132.

Wolffe AP (1994) Architectural transcription factors. *Science*, **264**, 1100–1101.

Wolffe AP (1995) Genetic effects of DNA packaging. *Sci. Am.*, **Nov/Dec**, 68–77.

FURTHER READING

Adhya S (1996) The *lac* and *gal* operons today. In: ECC Lin and AS Lynch, eds. *Regulation of Gene Expression in Escherichia coli*, pp. 181–200. Chapman & Hall, New York. — *Comprehensive description of the lactose operon.*

Beebee TJC and Burke J (1992) *Gene Structure and Transcription: In Focus.* IRL Press, Oxford. — *Details on all aspects of transcription.*

Latchman DS (1995) *Gene Regulation: A Eukaryotic Perspective.* Chapman & Hall, London. — *The best general text on this subject.*

Pugh BF (1996) Mechanisms of transcription complex assembly. *Curr. Opinion Cell Biol.*, **8**, 303–311. — *A concise description of the subject.*

CHAPTER

9 Synthesis and Processing of RNA

Contents

9.1 The RNA Content of a Cell **196**
9.1.1 Coding and noncoding RNA 196
9.1.2 Precursor RNA 197
9.1.3 RNA in time and space 198
9.2 Synthesis and Processing of mRNAs **201**
9.2.1 Synthesis of bacterial mRNAs 201
9.2.2 Synthesis of eukaryotic mRNAs by RNA polymerase II 207
9.2.3 Intron splicing 212
9.3 Synthesis and Processing of Noncoding RNAs **219**
9.3.1 Transcript elongation and termination by RNA polymerases I and III 219
9.3.2 Cutting events involved in processing of bacterial and eukaryotic pre-rRNA and pre-tRNA 219
9.3.3 Introns in eukaryotic pre-rRNA and pre-tRNA 221
9.4 Processing of Pre-RNA by Chemical Modification **222**
9.4.1 Chemical modification of pre-rRNAs 225
9.4.2 RNA editing 226
9.5 Turnover of mRNAs **227**

Concepts

- *As well as mRNAs that code for protein, cells also contain various noncoding RNAs*
- *Most RNAs are made as precursors that must be processed to produce the mature molecules*
- *RNA polymerases transcribe genes until a termination signal is reached*
- *Eukaryotic mRNAs are processed by capping, polyadenylation and intron splicing*
- *All pre-rRNAs and pre-tRNAs are processed by cutting events and by chemical modification*
- *Some eukaryotic mRNAs are chemically modified in such a way that their coding properties are changed*
- *mRNA degradation is a specific process that removes coding RNAs whose protein products are no longer required*

INITIATION OF TRANSCRIPTION, culminating with the RNA polymerase leaving the promoter and beginning synthesis of an RNA molecule, is simply the first step in the gene expression pathway. In this chapter and the next we will follow the process onwards and examine how transcription and translation eventually result in synthesis of functioning proteins. Throughout these two chapters our emphasis will be on the expression of individual genes, as it is only by focusing on individual genes that we can gain a detailed understanding of the mechanics of the processes leading to RNA and protein. But we should constantly keep in mind that what is important in the cellular context is not expression of single genes but expression of the genome as a whole. To establish this context, we begin this chapter with an overview of the total RNA content of a living cell.

9.1 THE RNA CONTENT OF A CELL

A typical bacterium contains 0.05–0.10 pg of RNA, making up about 6% of its total weight. A mammalian cell, being much larger, contains more RNA, 20–30 pg in all, but this represents only 1% of the cell as a whole (Alberts *et al.*, 1994). These fractions contain various types of RNA molecule, each with its own function. The best way to understand the RNA content of a cell is therefore to divide it into categories and subcategories depending on function. There are several ways of doing this, the most informative scheme being the one shown in *Figure 9.1.*

9.1.1 Coding and noncoding RNA

The first division is into **coding RNA** and **noncoding RNA**. The coding RNA is made up of just one class of molecule:

- **Messenger** or **mRNAs**, which are transcripts of protein-coding genes and hence are translated into protein in the second stage of genome expression.

Messenger RNAs rarely make up more than 4% of the total RNA, but they are the distinctive component, the **transcriptome**, the part that is continually being restructured by the changing pattern of transcription initiation events occurring throughout the genome. To allow this restructuring, mRNAs are short-lived, being degraded soon after synthesis. Bacterial mRNAs have half-lives of no more than a few minutes and in eukaryotes most mRNAs are degraded a few hours after synthesis. How many different mRNAs are present in a cell at any one time is not known, but studies with *Saccharomyces cerevisiae* suggest that at least three-quarters of the 6200 genes in the yeast genome are transcribed during normal vegetative growth (Velculescu *et al.*, 1997).

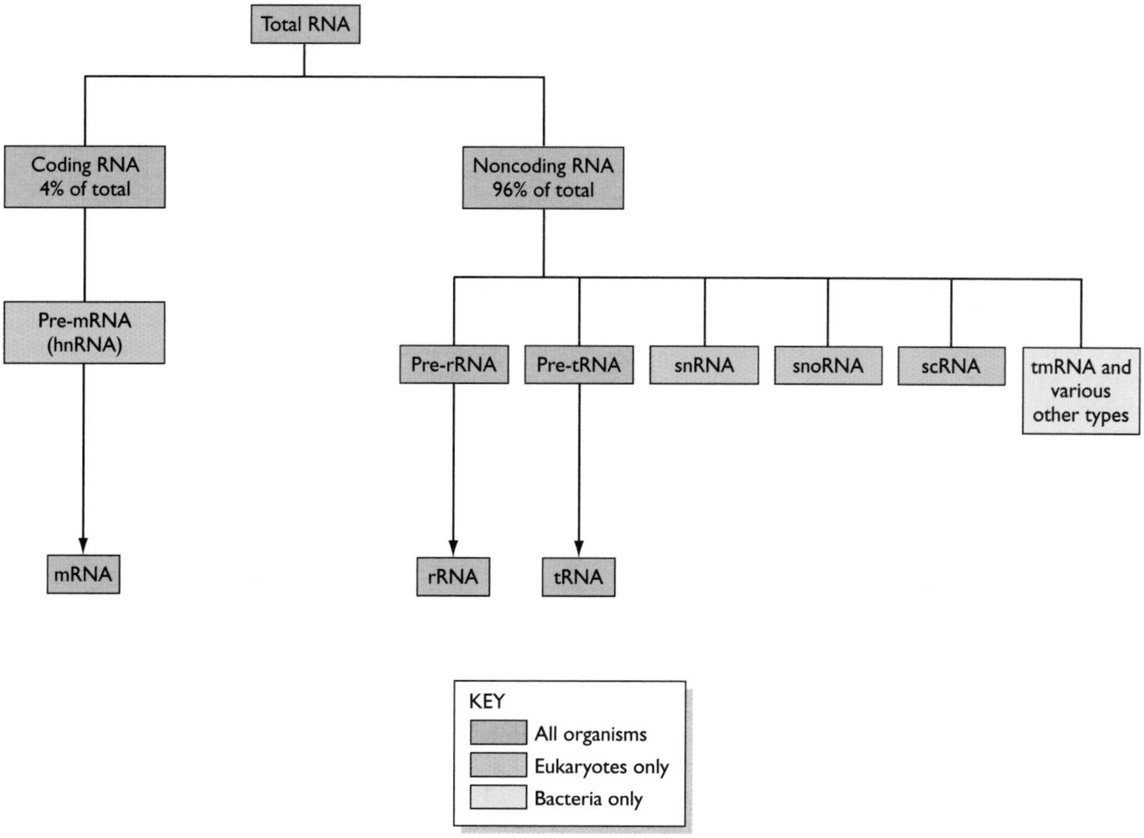

Figure 9.1 The RNA content of a cell.

This scheme shows the types of RNA present in all organisms (eukaryotes, bacteria and archaea) and those categories found only in eukaryotic or bacterial cells. The noncoding RNAs of archaea have not yet been fully characterized and it is not clear which types are present in addition to rRNA and tRNA. For abbreviations, see the text.

The second type of RNA is noncoding. This is more diverse than the coding RNA and comprises transcripts with a number of different functions, all of which are performed by the RNA molecules themselves. In both prokaryotes and eukaryotes the two main types of noncoding RNA are:

- **Ribosomal** or **rRNA**, which is the most abundant RNA in the cell, making up over 80% of the total in actively dividing bacteria. These molecules are components of ribosomes, the structures on which protein synthesis takes place (Section 10.2).
- **Transfer** or **tRNAs**, small molecules that are also involved in protein synthesis, carrying amino acids to the ribosome and ensuring these are linked together in the order specified by the nucleotide sequence of the mRNA that is being translated (Section 10.1).

Ribosomal and transfer RNAs are present in the cells of all organisms. The other noncoding RNA types are specific either to eukaryotes or bacteria (*Figure 9.1*). Eukaryotes contain a variety of short RNAs that are usually divided into three categories, **small nuclear RNA** (**snRNA**, also called **U-RNA** because these molecules are rich in uridine nucleotides), **small nucleolar RNA** (**snoRNA**) and **small cytoplasmic RNA** (**scRNA**), the names indicating their primary locations in the cell. The first two have functions in processing of other RNA molecules and we will meet them shortly (Sections 9.2.3 and 9.4.1). The scRNAs, not all of which are present in all eukaryotes, are themselves a diverse group including molecules with a range of functions, some understood and others still mysterious.

Bacteria also contain noncoding RNAs other than rRNA and tRNA but these molecules do not make up a substantial fraction of the total RNA. They include one interesting RNA type, apparently present in most if not all bacteria, called **transfer-messenger** or **tmRNA**, which looks like a tRNA attached to an mRNA, and which adds short peptide tags onto proteins that have been synthesized incorrectly, labeling them for immediate degradation (Muto *et al.*, 1998).

9.1.2 Precursor RNA

As well as the mature RNAs described above, cells also contain precursor molecules (see *Figure 9.1*). Many RNAs, especially in eukaryotes, are initially synthesized as precursor or **pre-RNA**, which has to be processed before it can carry out its function. The various processing events, all of which are described in detail later in this chapter, include the following (*Figure 9.2*):

- **End-modifications** occur during the synthesis of eukaryotic and archaeal mRNAs, most of which have a single, unusual nucleotide called a **cap** attached at the 5′-end and a **poly(A) tail** attached to the 3′-end.
- **Splicing** is the removal of introns from a precursor RNA. Many eukaryotic protein-coding genes contain introns which are copied when the gene is transcribed. The introns are removed from the **pre-mRNA** by cutting and joining reactions. Unspliced pre-mRNA forms the nuclear RNA fraction called **heterogenous nuclear RNA (hnRNA)**. Some eukaryotic pre-rRNAs and pre-tRNAs also contain introns, as do some archaeal transcripts, but they are extremely rare in bacteria.
- **Cutting events** are particularly important in the processing of rRNA and tRNA, many of which are initially synthesized from transcription units that specify more than one molecule. The **pre-rRNAs** and **pre-tRNAs** must therefore be cut into pieces to produce the mature RNAs. This type of processing occurs in both prokaryotes and eukaryotes.
- **Chemical modifications** are made to rRNAs, tRNAs and mRNAs. The rRNAs and tRNAs of all organisms are modified by addition of new chemical groups, these groups being added to specific nucleotides within each RNA. Chemical modification of mRNA, called **RNA editing**, is uncommon but seen in a diverse group of eukaryotes.

We are not entirely sure of the reason for every type of RNA processing but it is clear that these events have a variety of functions (*Table 9.1*) and that regulation over processing reactions is one way in which expression of individual genes can be modulated (see *Table 11.1*, p. 266). One aspect of mRNA processing that is becoming increasingly apparent is that, at least in some cases, it results in a change in the coding properties of the mRNA. RNA editing, for example, can result in a single pre-mRNA being converted into two different mRNAs coding for quite distinct proteins (*Figure 9.3A*). This does not appear to be particularly common, but **alternative splicing**, in which one pre-mRNA gives rise to two or more mRNAs by assembly of different combinations of exons (*Figure 9.3B*), is fairly widespread. The mRNAs resulting from both editing and alternative splicing often display tissue specificity: the processing events therefore increase the coding capabilities of the genome without the requirement for additional copies of the relevant genes.

9.1.3 RNA in time and space

Before studying the synthesis and processing of individual RNAs, there are two further points to make about the total RNA content of the cell. These relate to the way it changes over time and its distribution within the cell.

The population of RNAs in a cell changes over time

The population of RNA molecules present in a cell at a particular time represents a balance between the synthesis of

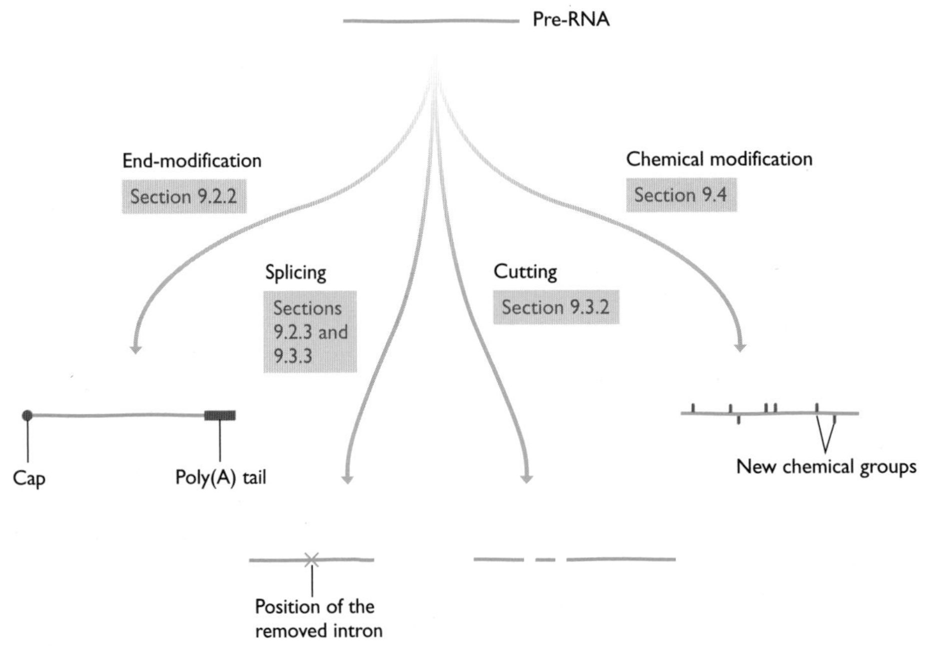

Figure 9.2 Schematic representation of the four types of RNA processing event.

Not all events occur in all organisms: see the text for details.

Table 9.1 Functions of RNA processing events

Processing event	Possible functions	Cross reference
mRNA end modifications		
Capping	Transport of mRNA from nucleus to cytoplasm	Section 9.1.3
	Initiation of translation	Section 10.2.2
Polyadenylation	Termination of transcription by RNA polymerase II	Section 9.2.2
	Initiation of translation	Section 10.2.2
	mRNA turnover	Section 9.5
Splicing		
Pre-mRNA splicing	No role other than removal of introns	Section 9.2.3
Pre-rRNA splicing	No role other than removal of introns	Section 9.3.3
Pre-tRNA splicing	No role other than removal of introns	Section 9.3.3
Cutting events		
Pre-rRNA processing	Releases mature rRNAs from the precursor molecules	Section 9.3.2
Pre-tRNA processing	Releases mature tRNAs from the precursor molecules	Section 9.3.2
Chemical modifications		
rRNA modification	Increases the chemical range of rRNA catalyzed reactions?	Section 9.4.1
tRNA modification	Recognition of tRNAs by aminoacyl-tRNA synthetases	Section 10.1.1
	Recognition of two or more codons by one tRNA ('wobble')	Section 10.1.2
RNA editing	Enables one mRNA to code for two proteins	Section 9.4.2
	Converts abbreviated transcripts into functional RNAs	Box 9.5
	Places termination codons on some mitochondrial mRNAs	Box 9.5

Not all types of processing occur in all organisms: see the text for details.

new RNAs by transcription and the removal of existing RNAs by degradation processes. This is particularly true of the protein-coding RNAs, which constitute the transcriptome that specifies the proteome and through that the biochemical signature of the cell. The transcriptome changes during development and differentiation and in response to external signals that demand an alteration in the biochemical properties of the cell. Most of these changes in the transcriptome result from control over transcription initiation events (see Section 8.3), but some are due to regulatory processes that operate after initiation. The latter will be described at the appropriate places in this and later chapters.

One important point to note is that the transcriptome is never synthesized *de novo*. Every cell receives part of its parent's transcriptome when it is first brought into existence by cell division and maintains a transcriptome throughout its lifetime. Even quiescent cells in bacterial spores or in the seeds of plants have a transcriptome, though expression of that transcriptome into protein may be completely switched off. Transcription does not therefore result in *synthesis* of the transcriptome but instead *maintains* the transcriptome by replacing mRNAs that have been degraded, and brings about *changes* to the composition of the transcriptome via the switching on and off of different sets of genes.

In eukaryotes, RNAs must be transported from nucleus to cytoplasm, and possibly back again

In a typical mammalian cell, about 14% of the total RNA is present in the nucleus (Alberts *et al.*, 1994). About 80% of this nuclear fraction is RNA that is being processed before leaving for the cytoplasm. The other 20% is snRNAs and snoRNAs, playing an active role in the processing events, at least some of these molecules having already been to the cytoplasm where they were coated with protein molecules before being transported back into the nucleus. In other words, eukaryotic RNAs are continually being moved from nucleus to cytoplasm and possibly back to the nucleus again.

The only way for RNAs to leave or enter the nucleus is via one of the many **nuclear pore complexes** that cover the nuclear membrane (*Figure 9.4*). RNAs and most proteins are too large to diffuse through a pore unaided and so have to be transported across by an energy-dependent process. As in many biochemical systems, the energy is obtained by hydrolysis of one of the high-energy phosphate– phosphate bonds in a ribonucleotide triphosphate, in this case by converting GTP to GDP (other processes use ATP→ADP). Energy generation is carried out by a protein called Ran, and transport requires other protein factors called mediators or **exportins** and **importins**. An RNA-specific mediator called exportin-t

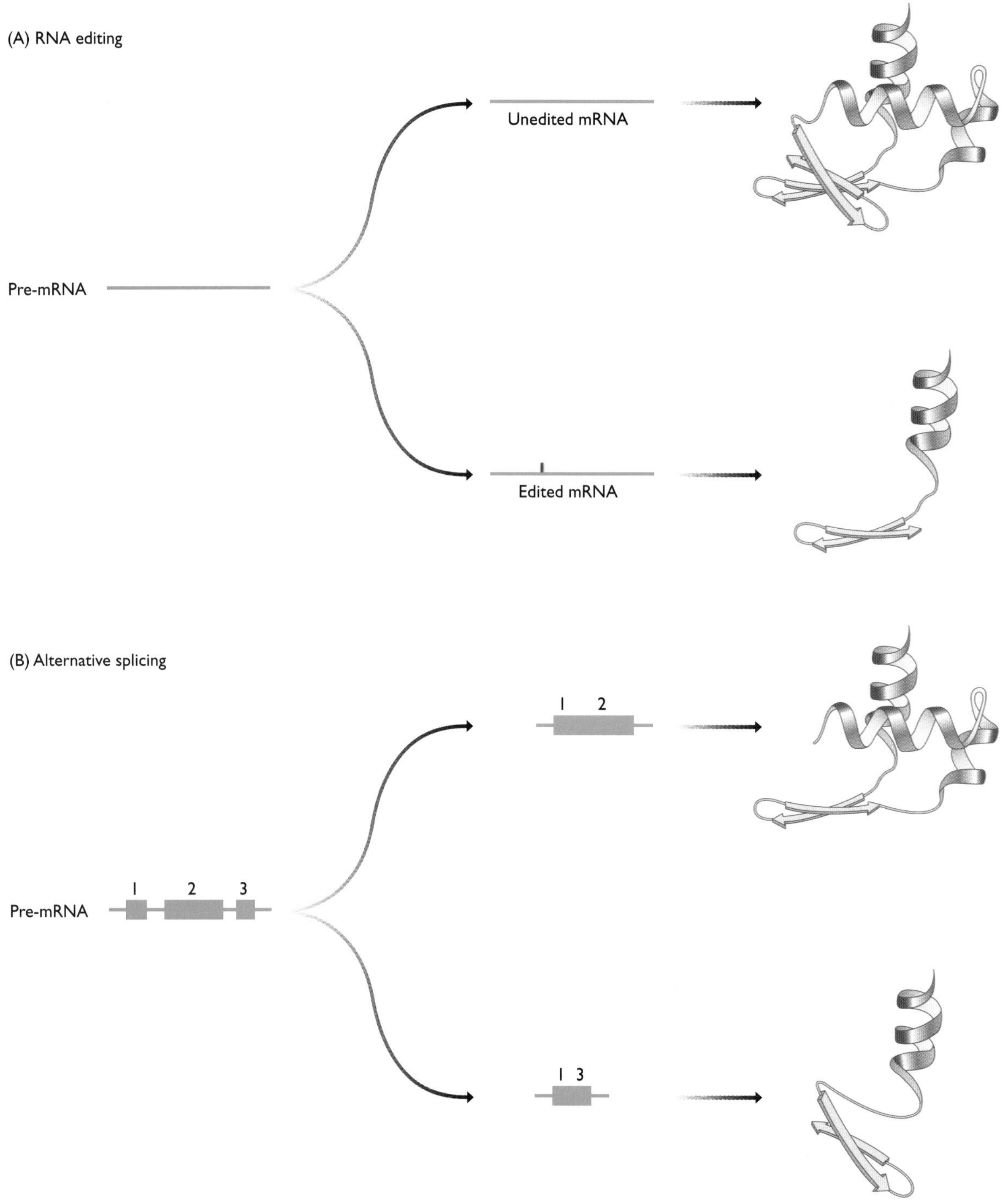

Figure 9.3 Some RNA processing events change the coding properties of an mRNA.

(A) RNA editing can change the sequence of an mRNA, resulting in synthesis of a different protein. An example occurs with the human mRNA for apolipoprotein-B, as shown in *Figure 9.25*, p. 227. (B) Alternative splicing results in different combinations of exons becoming linked together, again resulting in different proteins being synthesized from the same pre-mRNA. *Figure 9.18*, p. 218, shows how alternative splicing underlies sex determination in *Drosophila*.

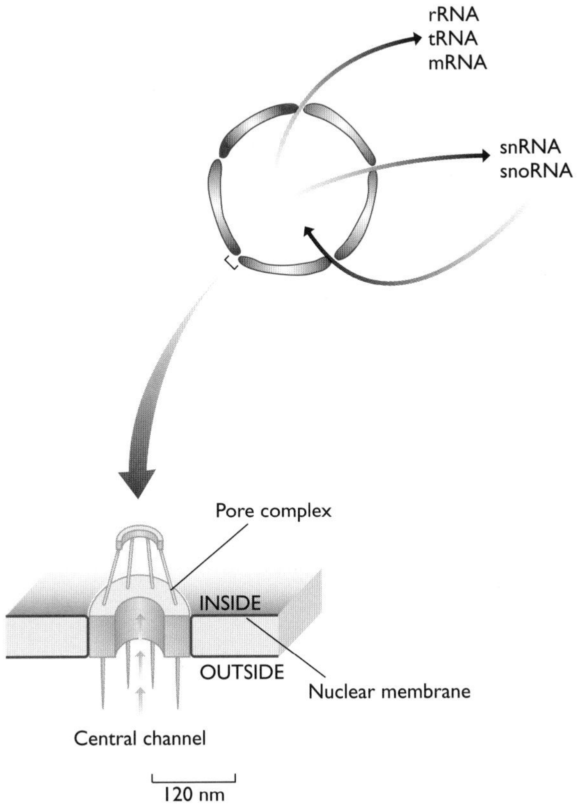

Figure 9.4 Eukaryotic RNAs must be transported through the nuclear pore complexes.

In eukaryotes, rRNAs, tRNAs and mRNAs are transported from nucleus to cytoplasm where these molecules carry out their cellular functions. At least some of the snRNAs and snoRNAs are also transported to the cytoplasm where they are coated with proteins before returning to the nucleus to carry out their roles in RNA processing. The nuclear pore is not simply a hole in the nuclear membrane. It contains a protein assembly comprising a ring embedded in the pore with structures radiating into both the nucleus and cytoplasm. Not shown in this diagram is the central channel complex, a 12 kDa protein that is thought to reside in the channel connecting cytoplasm to nucleus.

has been identified in human and yeast nuclei but this factor seems to be involved only in export of tRNAs. Other types of RNA are probably exported by protein-specific mediators which recognize not the RNA itself but proteins that are bound to it. This also appears to be the case for import of snRNA from cytoplasm to nucleus, which makes use of importin β, a component of one of the protein transport pathways (Nigg, 1997; Weis, 1998).

9.2 SYNTHESIS AND PROCESSING OF mRNAs

We begin our detailed study of transcription by looking at the synthesis and processing of mRNAs. Events in bacteria are different in many respects from those in eukaryotes and so we will deal with the two types of organism in different sections. One aspect of eukaryotic mRNA processing – intron splicing – is so important that it requires a section of its own.

9.2.1 Synthesis of bacterial mRNAs

Bacterial mRNAs undergo no significant forms of processing: the primary transcript that is synthesized by the RNA polymerase is itself the mature mRNA and its translation usually begins before transcription is complete (*Figure 9.5*). This coupling of transcription and translation is important in that it allows special types of control to be applied to the regulation of bacterial mRNA synthesis, as will be described on p. 206.

The elongation phase of bacterial transcription

Because there is just one bacterial RNA polymerase (Section 8.2.1), the general mechanism of transcription is the same for all bacterial genes. The following descriptions of elongation and termination, given in the context of mRNA synthesis, therefore apply equally well to noncoding RNA synthesis.

The chemical basis of the template-dependent synthesis of RNA is illustrated in *Figure 9.6*. Ribonucleotides are added one after another to the growing 3′-end of the RNA transcript, the identity of each nucleotide specified by the base-pairing rules: A base-pairs with T or U; G base-pairs with C. During each nucleotide addition, the β- and γ-phosphates are removed from the incoming nucleotide, and the hydroxyl group is removed from the 3′-carbon of the nucleotide present at the end of the chain, precisely the same as for DNA polymerization (see *Figure 7.1*, p. 146).

During the elongation phase of transcription, the bacterial RNA polymerase moves along the template, maintaining a nonbase-paired **transcription bubble** of some

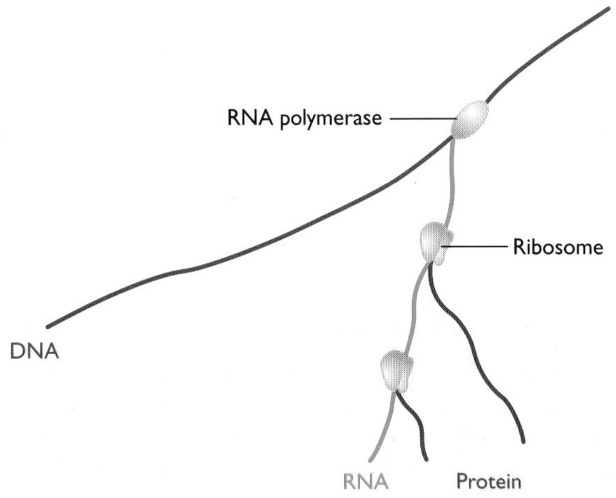

Figure 9.5 In bacteria, transcription and translation are often coupled.

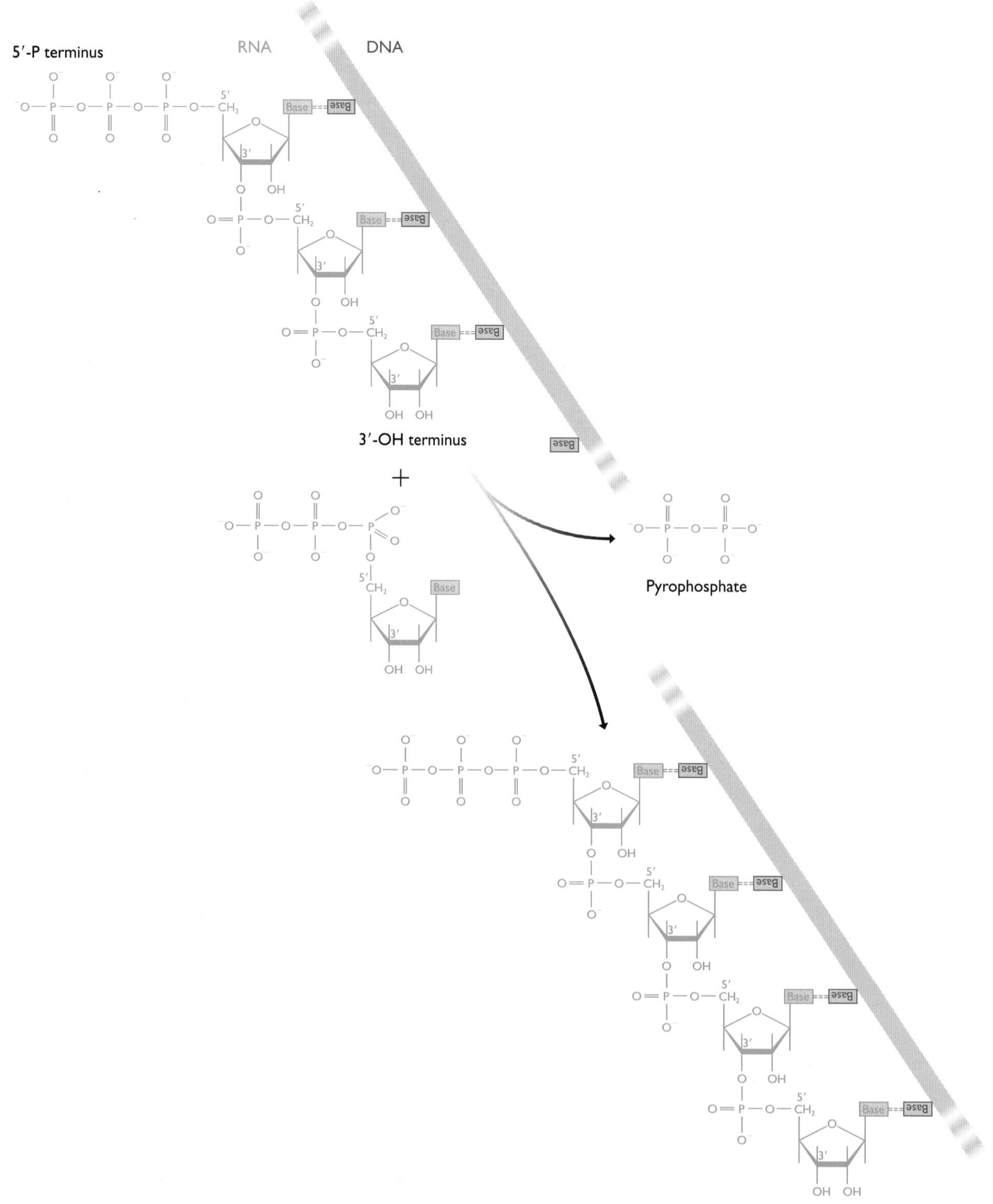

Figure 9.6 Template-dependent synthesis of RNA.

Compare this reaction with polymerization of DNA, illustrated in *Figure 7.1*, p. 146.

15–20 bp, within which the growing transcript is held to the template strand of the DNA by approximately eight RNA–DNA base pairs (*Figure 9.7*). The RNA polymerase covers some 30 bp of DNA, with the stability of the complex depending largely on the β and β′ subunits of the enzyme, which make contacts with short regions of the double-stranded DNA just ahead of the transcription bubble, with the DNA–RNA hybrid within the complex,

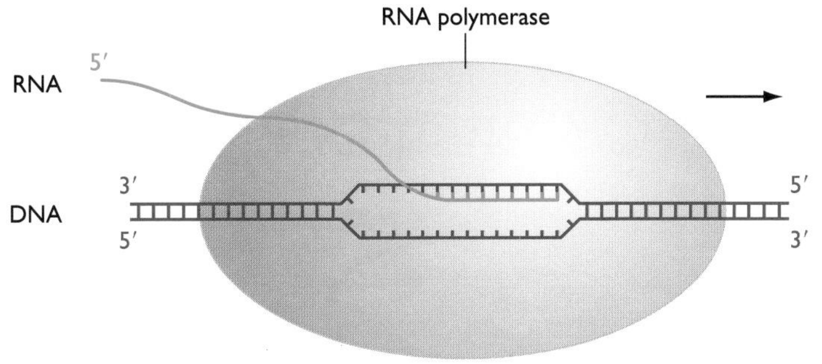

Figure 9.7 Schematic representation of the *E. coli* transcription elongation complex.

The RNA polymerase covers approximately 30 bp of DNA including a transcription bubble of 15–20 bp, with the RNA attached to the template strand of the DNA by eight or so RNA–DNA base pairs. See Research Briefing 9.1 for a more accurate description of the structure of the elongation complex.

and the RNA transcript emerging from it (Research Briefing 9.1; Landick and Roberts, 1996; Nudler *et al.*, 1996, 1998).

Termination of bacterial transcription

Exactly how termination occurs is not known. Current thinking views transcription as a stepwise, nucleotide-by-nucleotide process with the polymerase pausing at each position and making a 'choice' between continuing elongation by adding another ribonucleotide to the transcript, or terminating by dissociating from the template. Which choice is selected depends on which alternative is more favorable in thermodynamic terms (von Hippel, 1998). This model emphasizes that, in order for termination to occur, the polymerase has to reach a position on the template where dissociation is more favorable than continued RNA synthesis.

Bacteria appear to use two distinct strategies for transcription termination. About half the positions at which transcription terminates in *E. coli* correspond to DNA sequences where the template strand contains an inverted palindrome followed by a run of adenosine nucleotides (*Figure 9.8A*). These **intrinsic terminators** are thought to promote dissociation of the polymerase by destabilizing the attachment of the growing transcript to the template, in two ways. First, when the inverted palindrome is transcribed the RNA sequence folds into a stable hairpin, this RNA–RNA base-pairing being favored over the DNA–RNA pairing that normally occurs within the transcription bubble. This reduces the number of contacts made between the template and transcript, weakening the overall interaction and favoring dissociation. The interaction is further weakened when the run of As in the template is transcribed, as the resulting A–U base pairs have only two hydrogen bonds each, compared to three for each G–C pair. The net result is that termination is favored over continued elongation (von Hippel, 1998).

The second type of bacterial termination signal is **Rho dependent**. These signals usually retain the hairpin feature of intrinsic terminators, though the hairpin is less stable and there is no run of As in the template. Termination requires the activity of a protein called Rho, which attaches to the transcript and moves along the RNA towards the polymerase. If the polymerase continues to synthesize RNA then it keeps ahead of the pursuing Rho, but at the termination signal the polymerase stalls (*Figure 9.8B*). Exactly why has not been explained – presumably the hairpin loop that forms in the RNA is responsible in some way – but the result is clear: Rho is able to catch up. Rho is a **helicase**, which means that it actively breaks base pairs, in this case between the template and transcript, resulting in termination of transcription.

Control over the choice between elongation and termination

In bacteria, two mechanisms have evolved for influencing the repeated choice that the polymerase has to make between elongation and termination when copying a template. Both mechanisms are important in regulating the expression of genes contained within operons.

The first process is called **antitermination**. This occurs when the RNA polymerase ignores a termination signal and continues elongating its transcript until a second signal is reached (*Figure 9.9*). It provides a mechanism whereby one or more of the genes at the end of an operon can be switched off or switched on by the polymerase recognizing or not recognizing a termination signal located upstream of those genes. Antitermination is controlled by an **antiterminator protein** which attaches to the DNA near the beginning of the operon and then transfers to the RNA polymerase as it moves past *en route* to the first termination signal. The presence of the antiterminator protein causes the enzyme to ignore the termination signal, presumably by countering the destabilizing properties of an intrinsic terminator or by preventing stalling at a Rho-dependent terminator. Although the mechanics of the process are unclear, the impact that antitermination can

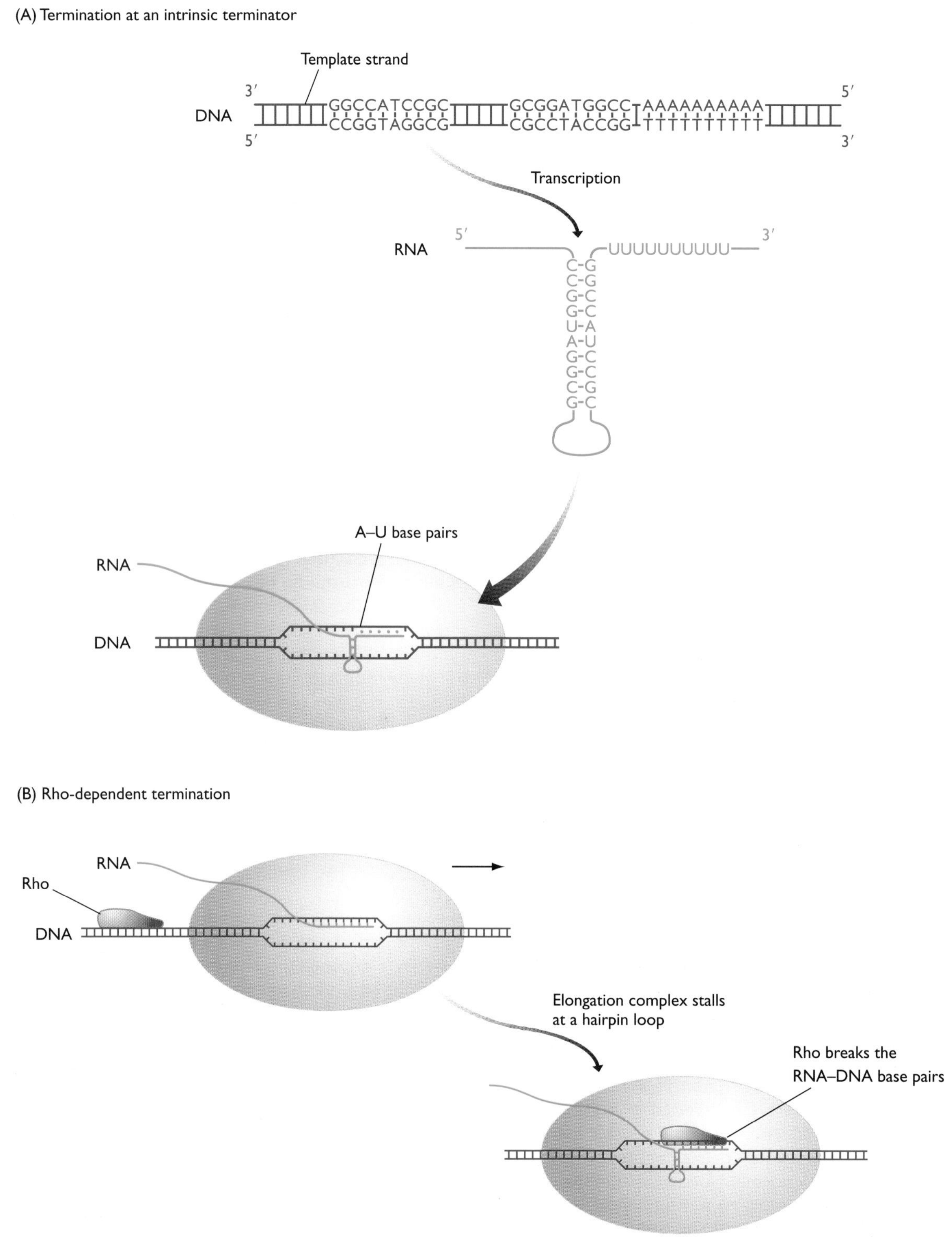

Figure 9.8 Two strategies for termination of transcription in bacteria.

(A) Termination at an intrinsic terminator. The presence of an inverted palindrome in the DNA sequence results in formation of a hairpin loop in the transcript. This hairpin could disrupt the RNA–DNA hybrid within the elongation complex, with the stability of the hybrid further weakened by the run of relatively weak A–U base pairs immediately after the hairpin. (B) Rho-dependent termination. Rho is a helicase that follows the RNA polymerase along the the transcript. When the polymerase stalls at a hairpin, Rho catches up and breaks the RNA–DNA base pairs, releasing the transcript. These two diagrams are schematic and do not reflect the relative sizes of Rho and the RNA polymerase. For alternative ideas regarding termination of transcription, see Research Briefing 9.1.

RESEARCH

9.1

BRIEFING

Probing the structure of the transcription elongation complex

The bacterial RNA polymerase is looked on as a 'sliding clamp' which is able to maintain a tight grip on both the DNA and RNA while retaining its ability to slide along the DNA in a smooth and continuous manner.

During the elongation phase of transcription, the bacterial RNA polymerase comprises four proteins, two relatively small (approximately 35 kDa) α subunits, primarily involved in assembly of the transcription complex during the initiation phase, and one each of the related subunits β and β' (both approximately 150 kDa). The RNA polymerase covers about 30 bp of the template DNA including the transcription bubble of 15–20 bp in which RNA synthesis occurs. Various lines of evidence support a model in which three distinct sites within the RNA polymerase make contact with the DNA/RNA:

- The double-stranded DNA binding site (DBS), which forms a contact with 9 bp of DNA immediately ahead of the transcription bubble.
- The heteroduplex binding site (HBS), which holds on to the short hybrid formed between the RNA transcript and the template strand of the DNA (see *Figure 9.7*).
- The RNA binding site (RBS), which maintains contact with a short region of the RNA transcript, probably no more than 8 bp, as it emerges from the transcription bubble.

The RNA polymerase has to keep a tight grip on both the DNA template and the RNA that it is making in order to prevent the transcription complex from falling apart before the end of the gene is reached, but this grip must not be so tight as to prevent the polymerase from moving along the DNA. These apparently contradictory requirements have led to a model that looks on the polymerase as a 'sliding clamp', the three contacts described above somehow combining the required tightness of grip with the flexibility needed to allow movement of the polymerase.

Identifying DNA- and RNA-binding sites in the polymerase

To test the sliding clamp model it is necessary to understand exactly how the polymerase holds on to the DNA and RNA via the DBS, HBS and RBS. The first requirement is to identify the parts of the β and β' subunits involved in these interactions. This has been achieved by crosslinking experiments in which covalent bonds are formed between the RNA or DNA and the polymerase, these bonds enabling the amino acids closest to the RNA/DNA, presumably those involved in the binding sites, to be identified.

To illustrate the technique we will look at how the RBS has been probed. In these experiments, RNA polymerase enzymes were attached to agarose beads so that transcription could be stalled by rapidly removing the elongation complex from the reaction mixture by centrifugation of the beads. To carry out the crosslinking experiments a modified ribonucleotide, 4-thiouridine, was incorporated at a single position in the growing RNA transcript. Transcription was stopped immediately after incorporation of 4-thiouridine, so that the modified nucleotide would be positioned within the region of the transcript thought to make contact with the RBS. The modified nucleotide is photoreactive and can be induced to form crosslinks with nearby amino acids by UV irradiation. After crosslinking, the β and β' subunits were purified, digested into shorter polypeptides, and the amino acids attached to the RNA identified. The main component of the RBS was shown to be a segment of the β' subunit, near its N-terminus, which contains a zinc finger (Section 7.4.2). Surprisingly, this same region of the β' subunit was also identified as the major component of the DBS, suggesting that the newly synthesized transcript emerges from the elongation complex at a position close to the DNA entry point.

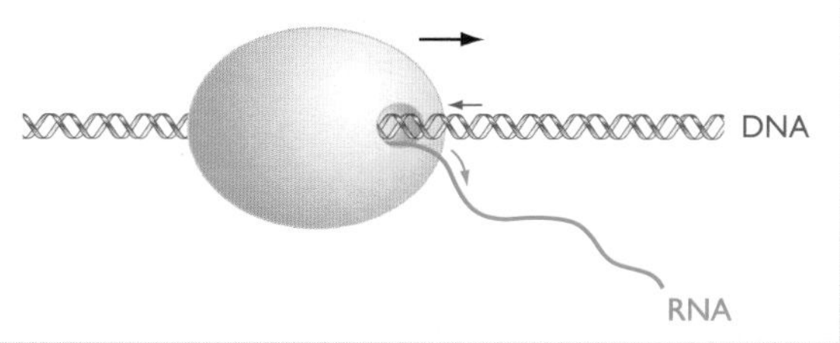

Towards a model of the sliding clamp

Before this work was carried out the RNA polymerase was generally looked on as a bead, the DNA entering one end of the bead and emerging, with the RNA, at the other end. The discovery that the DBS and RBS are in close proximity leads to a different view of the elongation complex, one which is compatible with an interaction of some description between the DNA entering the complex and the RNA emerging from it. This type of interaction could be envisaged as resulting in a clamping effect, and would perhaps be destabilized by a hairpin loop in the RNA, providing a new interpretation of the role of hairpin loops during termination of transcription, slightly different from the conventional view described in the text. The region of the β' subunit that participates in the RBS/DBS is present also in the equivalent subunits of eukaryotic RNA polymerases, suggesting that this model might be universal rather than applying only to bacterial transcription.

Reference

Nudler E, Gusarov I, Avetissova E, Kozlov M and Goldfarb A (1998) Spatial organization of transcription elongation complex in *Escherichia coli*. *Science*, **281**, 424–428.

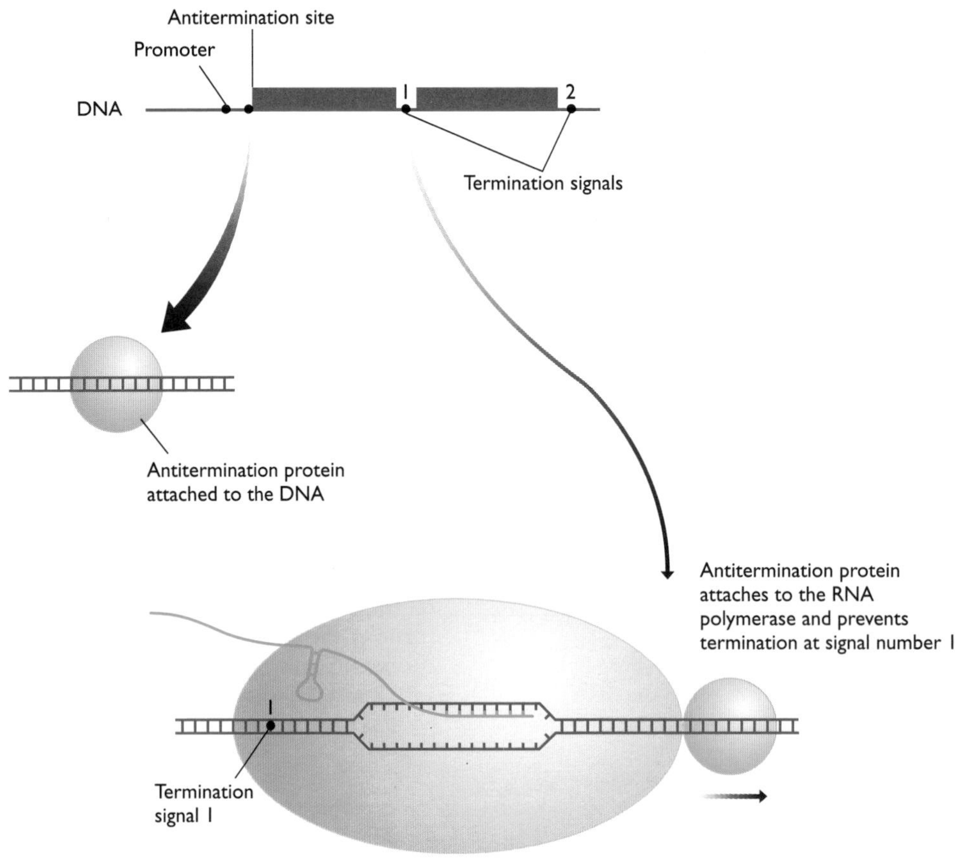

Figure 9.9 Antitermination.

The antiterminator protein attaches to the DNA and transfers to the RNA polymerase as it moves past, subsequently enabling the polymerase to continue transcription through termination signal number 1, so the second of the pair of genes in this operon is transcribed.

have on gene expression has been described in detail, especially during the infection cycle of bacteriophage λ (Box 9.1).

The second type of termination control is called **attenuation**. This system operates primarily with operons that code for enzymes involved in amino acid biosynthesis, but a few other examples are also known. The tryptophan operon (Section 8.3.1) illustrates how it works. In this operon, a termination signal can form in the region immediately after the promoter. Whether or not this hairpin forms depends on the relative positioning between the RNA polymerase and a ribosome which attaches to the 5′-end of the transcript as soon as it is synthesized in order to translate the genes into protein (*Figure 9.10*). If the ribosome is following closely behind the polymerase then the termination signal forms and transcription stops. However, if the ribosome stalls so that it does not keep up with the polymerase, then the termination signal does not form and transcription continues. Ribosome stalling can occur because upstream of the termination signal is a short open reading frame coding for a 14 amino acid peptide that includes two tryptophans. If the amount of free tryptophan is limiting, then the ribosome stalls and the polymerase continues to make its transcript. As this transcript contains copies of the genes coding for the biosynthesis of tryptophan, its continued elongation addresses the requirement that the cell has for this amino acid. When the amount of tryptophan in the cell reaches a satisfactory level, the attenuation system prevents further transcription of the tryptophan operon, because now the ribosome does not stall while it makes the short peptide, and instead keeps pace with the polymerase, allowing the termination signal to form.

The *E. coli* tryptophan operon is controlled not only by attenuation but also by a repressor (Section 8.3.1). Exactly how attenuation and repression work together to regulate expression of the operon is not known, but it is thought that repression provides the basic on–off switch and attenuation modulates the exact level of gene expression which occurs. Other operons, such as those for biosynthesis of histidine, leucine and threonine, make do just with attenuation control.

Box 9.1: Antitermination during the infection cycle of bacteriophage λ

Bacteriophage λ provides the best studied example of the use of antitermination as a means of regulating gene expression (Friedman *et al.*, 1987). Immediately after entering an *Escherichia coli* cell, transcription of the λ genome is initiated by the bacterial RNA polymerase attaching to two promoters, p_L and p_R and synthesizing two 'immediate-early' mRNAs, these terminating at positions t_{L1} and t_{R1}:

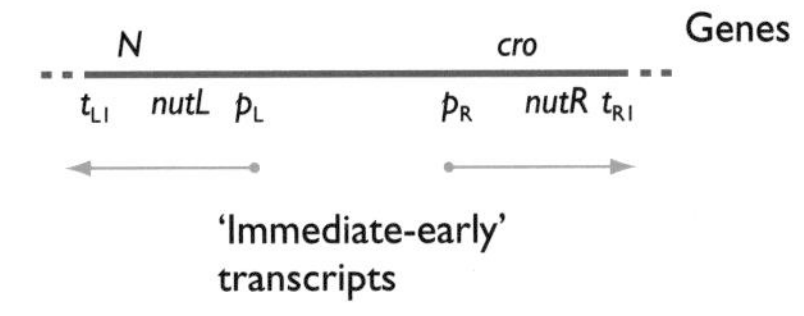

The mRNA transcribed from p_R to t_{R1} codes for a protein called Cro, one of the major regulatory proteins involved in the λ infection cycle. The second mRNA specifies the N protein, which is an antiterminator. The N protein attaches to the λ genome at sites *nutL* and *nutR* and transfers to the RNA polymerase as it passes. Now the RNA polymerase ignores the t_{L1} and t_{R1} terminators and continues transcription downstream of these points. The resulting mRNAs encode the 'delayed-early' proteins:

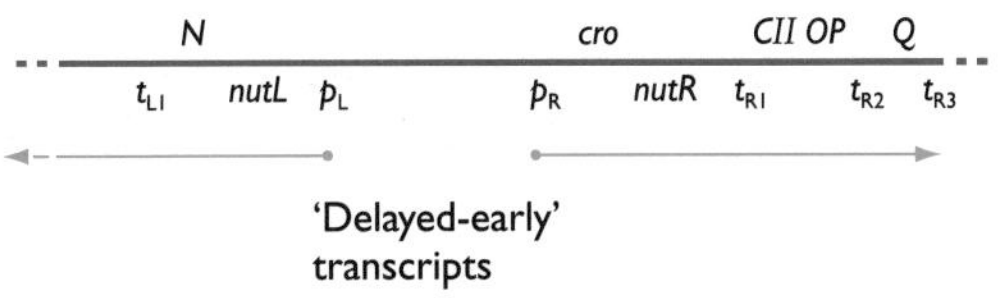

Antitermination controlled by the N protein therefore ensures that the immediate-early and delayed-early proteins are synthesized at the appropriate times during the λ infection cycle. Note that one of the delayed-early proteins, Q, is a second antiterminator that controls the switch to the late stage of the infection cycle.

9.2.2 Synthesis of eukaryotic mRNAs by RNA polymerase II

At the most fundamental level, transcription is very similar in bacteria and eukaryotes. The chemistry of RNA polymerization is identical in all types of organism, and the three eukaryotic RNA polymerases are each structurally related to the *E. coli* RNA polymerase, their three largest subunits being equivalent to the α, β and β′ subunits of the bacterial enzyme. This means that the contacts between a eukaryotic polymerase and its template DNA are probably very similar to the interaction described for bacterial transcription (see Research Briefing 9.1), and the basic principle that transcription is a step-by-step competition between elongation and termination (see p. 203) also holds.

Despite this equivalence, the overall processes for mRNA synthesis in bacteria and eukaryotes are quite dissimilar. The most striking difference is the extent to which eukaryotic mRNAs are processed during transcription. In bacteria, the transcripts of protein-coding genes are not processed at all: the primary transcripts are mature mRNAs and their translation often begins before transcription is completed (see *Figure 9.5*, p. 201). In contrast, all eukaryotic mRNAs have a cap added to the 5′-end, most are also polyadenylated by addition of a series of adenosines to the 3′-end, many contain introns and so undergo splicing, and a few are subject to RNA editing. A function has been assigned to capping, but the reason for polyadenylation largely remains a mystery. With splicing and editing we can appreciate the reasons why the events occur (the former removes introns that block translation of the mRNA, the latter changes the coding properties of the mRNA) but we do not understand why these mechanisms have evolved (why do genes have introns in the first place? why edit an mRNA rather than encoding the desired sequences in the DNA?).

Eukaryotic mRNAs are processed while they are being synthesized. The cap is added as soon as transcription has been initiated, splicing and editing begin while the transcript is still being made, and polyadenylation is an inherent part of the termination mechanism for RNA polymerase II. To deal with all of these events together would, however, lead to confusion with too many different things being described at once. Instead, we will postpone editing until the end of the chapter, which means it can be dealt with in tandem with similar forms of chemical modification occurring during rRNA and tRNA processing, and we will delay splicing until after we have studied capping, elongation and polyadenylation.

Capping of RNA polymerase II transcripts occurs immediately after initiation

Capping is a multistep process that occurs as soon as the 5′-end of the pre-mRNA has emerged from the RNA polymerase II transcription complex, soon after productive initiation has been achieved and the polymerase has moved away from the promoter. The first step is addition of an extra guanosine to the extreme 5′-end of the RNA. Rather than occurring by normal RNA polymerization, capping involves a reaction between the 5′-triphosphate of the terminal nucleotide and the triphosphate of a GTP nucleotide. The γ-phosphate of the terminal nucleotide (the outermost phosphate) is removed, as are the β- and γ-phosphates of the GTP, resulting in a 5′-5′ bond (*Figure 9.11*). The reaction is carried out by the enzyme **guanylyl transferase**. The second step of the capping reaction converts the new terminal guanosine into 7-methylguanosine by attachment of a methyl group to nitrogen number 7 of the purine ring, this modification catalyzed by **guanine methyltransferase**. It is possible that both enzymes

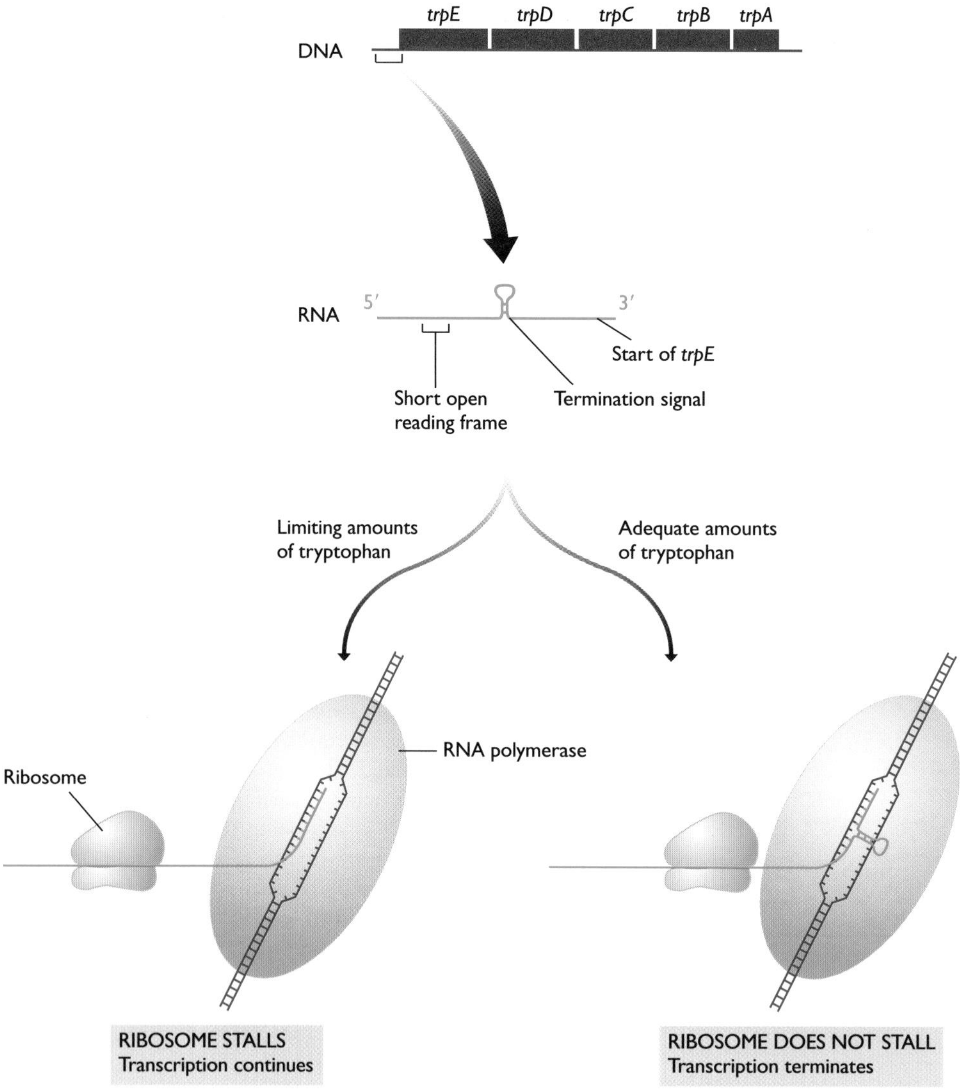

Figure 9.10 Attenuation at the tryptophan operon.

See the text for a description of the basic mechanism. The details of attenuation are complex. The region between the start of the transcript and the beginning of *trpE* can in fact form *two* hairpin loops, but only the smaller of these, the one that acts as a termination signal, is shown in the diagram. The larger hairpin loop is closer to the start of the transcript and is more stable. It also overlaps with the termination hairpin, so only one of the two hairpins can form at any one time. If the ribosome stalls then the larger hairpin forms and transcription continues. However, if the ribosome keeps pace with the RNA polymerase then it disrupts the larger hairpin by attaching to the RNA that forms part of the stem of this hairpin. When this happens the termination hairpin is able to form, and transcription stops. Note that in reality the ribosome is much larger than the RNA polymerase. See Section 8.3.1 for a description of other control systems operating on the tryptophan operon.

are intrinsic components of the RNA polymerase II transcription complex.

The 7-methylguanosine structure is called a **type 0 cap** and is the commonest form in yeast. In higher eukaryotes, additional modifications occur (see *Figure 9.11*):

- A second methylation replaces the hydrogen of the 2′-OH group of what is now the second nucleotide in the transcript. This results in a **type 1 cap**.
- If this second nucleotide is an adenosine, then a methyl group might be added to nitrogen number 6 of the purine.
- Another 2′-OH methylation might occur at the third nucleotide position, resulting in a **type 2 cap**.

All RNAs synthesized by RNA polymerase II are capped in one way or another. This means that as well as mRNAs, the snRNAs that are transcribed by this enzyme are also capped (see *Table 8.1*, p. 179). With snRNAs there is evidence that the cap is one of the features of the molecule that is recognized by the importins that mediate transport of the RNA back into the nucleus after it has travelled to the cytoplasm to collect its binding proteins (Weis, 1998; Section 9.1.3). The cap may also be important for export of snRNAs and mRNAs from the nucleus, but its best defined role is in translation of mRNAs, which we will cover in Section 10.2.2.

Elongation of eukaryotic mRNAs

As mentioned above, the fundamental aspects of transcript elongation are the same in bacteria and eukaryotes. The one major distinction concerns the length of

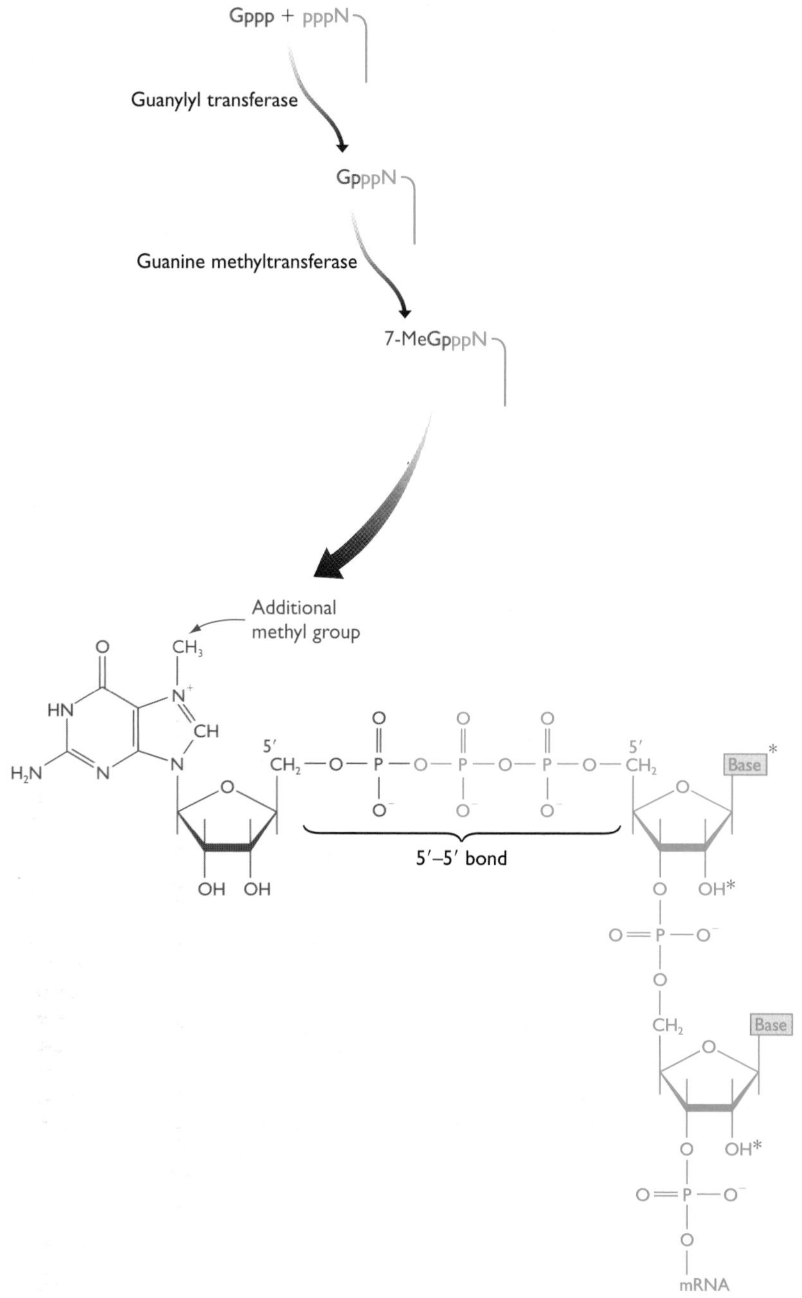

Figure 9.11 Capping of eukaryotic mRNA.

The top part of the diagram shows the capping reaction in outline. A GTP molecule (drawn as Gppp) reacts with the 5′-end of the mRNA to give a triphosphate linkage. In the second step of the process, the terminal G is methylated at nitrogen number 7. The bottom part of the diagram shows the chemical structure of the type 0 cap, with asterisks indicating the positions where additional methylations might occur to produce type I and type 2 cap structures.

transcript that must be synthesized. The longest bacterial genes are only a few kb in length and can be transcribed in a matter of minutes by the bacterial RNA polymerase, which has a polymerization rate of several hundred nucleotides per minute. In contrast, RNA polymerase II can take hours to transcribe a single gene, even though it can work at up to 2000 nucleotides per minute. This is because the presence of multiple introns in many eukaryotic genes (Section 9.2.3) means that considerable lengths of DNA must be copied. For example, the pre-mRNA for the human dystrophin gene is 2400 kb in length and takes about 20 hours to synthesize.

The extreme lengths of eukaryotic genes place demands on the stability of the transcription complex. RNA polymerase II on its own is not able to meet these demands: when the purified enzyme is used in the test tube its polymerization rate is less than 300 nucleotides per minute because the enzyme pauses frequently on the template and sometimes stops altogether. In the nucleus, pausing and stopping is reduced because of the action of a series of **elongation factors**, proteins that associate with the polymerase after it has cleared the promoter and left behind the transcription factors involved in initiation (Reines *et al.*, 1996). Five elongation factors have been studied in detail (*Table 9.2*); the existence of others is suspected but not yet confirmed.

A second difference between bacterial and eukaryotic elongation is that RNA polymerase II, as well as the other eukaryotic nuclear polymerases, has to negotiate the nucleosomes that are attached to the template DNA being transcribed. At first glance, it is difficult to imagine how the polymerase can elongate its transcript through a region of DNA wound round a nucleosome (see *Figure 6.2*, p. 119). One suggestion, based on work with RNA polymerase III but probably equally applicable to other polymerases, is that a short region of DNA detaches from the nucleosome, enabling the polymerase to loop around the structure (Studitsky *et al.*, 1997). The evidence for this model is described in Research Briefing 9.2.

Table 9.2 Elongation factors for RNA polymerase II

Elongation factor	Function
P-TEFb TFIIS (also called SII)	These factors suppress 'pausing' of RNA polymerase II, which can occur when the enzyme transcribes through a region where intrastrand base bairs (e.g. a hairpin loop) can form
ELL Elongin TFIIF	These prevent 'arrest' of RNA polymerase II, which occurs when the RNA polymerase prematurely releases the 3′-end of the transcript, possibly as a result of pausing

Information taken from Reines *et al.* (1996).

Termination of mRNA synthesis is combined with polyadenylation

Virtually all eukaryotic mRNAs have a series of up to 250 adenosines at their 3′-ends. These As are not specified by the DNA and are added to the transcript by a template-independent RNA polymerase called **poly(A) polymerase**. The polymerase does not act at the extreme 3′-end of the transcript, but at an internal site which is cleaved to create a new 3′-end to which the poly(A) tail is added.

The basic features of polyadenylation have been understood for some time. In mammals, polyadenylation is directed by a signal sequence in the mRNA, almost invariably 5′-AAUAAA-3′. This sequence is located between 10 and 30 nucleotides upstream of the polyadenylation site, which is often immediately after the dinucleotide 5′-CA-3′ and followed 10–20 nucleotides later by a GU-rich region. Both the poly(A) signal sequence and the GU-rich region are binding sites for multisubunit protein complexes, which are, respectively, the **cleavage and polyadenylation specificity factor (CPSF)** and the **cleavage stimulation factor (CstF)**. Poly(A) polymerase and at least two other protein factors must associate with bound CPSF and CstF in order for polyadenylation to occur (*Figure 9.12*). These additional factors include **polyadenylate-binding protein**, which helps the polymerase add the adenosines, possibly influences the length of the poly(A) tail that is synthesized, and appears to play a role in maintenance of the tail after synthesis. In yeast, the signal sequences in the transcript are slightly different, but the protein complexes are similar to those in mammals and polyadenylation is thought to occur by more or less the same mechanism (Guo and Sherman, 1996; Manley and Takagaki, 1996).

One interesting feature of polyadenylation has recently emerged. This is that polyadenylation is an inherent part of the transcription process itself, and not a 'post-transcriptional' modification. Evidence for this comes from studies which have shown that the C-terminal domain of the largest subunit of RNA polymerase II is needed for polyadenylation, both in living cells (McCracken *et al.*, 1997) and the test tube (Hirose and Manley, 1998). Interestingly, this C-terminal domain is also implicated in interactions with activation factors during transcription initiation (Section 8.3.2) and splicing factors during intron removal (Section 9.2.3). The implication is that RNA elongation and polyadenylation are closely linked and that polyadenylation is, effectively, the termination process for RNA polymerase II transcription. However, this does not hold true for all genes transcribed by RNA polymerase II as a few are not polyadenylated. The best known examples are the histone mRNAs, whose 3′-ends are produced by a different cleavage process directed not by polyadenylation factors but by U7 snRNA.

Even though polyadenylation can be identified as an inherent part of the termination process, this does not explain why it is necessary to add a poly(A) tail to the transcript. A role for the poly(A) tail has been sought for several years, but no convincing evidence has been found

RESEARCH

9.2

BRIEFING

Transcription through nucleosomes

Eukaryotic nuclear DNA, even in its most open conformation, has nucleosomes attached at frequent intervals. How does transcription progress through nucleosomes without the RNA polymerase being impeded?

In order to transcribe a gene, eukaryotic RNA polymerases have to work around the nucleosomes that are attached to nuclear DNA molecules. An RNA polymerase is larger than a nucleosome, but each nucleosome has 140–150 bp of DNA wrapped around it (see *Figure 6.2*, p. 119) and it is unlikely that the motive force of the RNA polymerase is sufficient to knock a nucleosome off the DNA as it goes past. This scenario is also incompatible with the emerging evidence that nucleosome positioning within transcribed regions is not random, and that nucleosomes are involved in fine tuning the rate of transcription of individual genes (Section 8.1.1).

A model system for transcription through nucleosomes

Progress in understanding how RNA polymerases transcribe nucleosomal DNA has been made with a model system that enables the process to be studied in the test tube. This system comprises a short fragment of DNA, 227 bp in length, that has a single attached nucleosome covering a recognition sequence for the restriction enzyme *Hae*II. The segment of DNA not attached to the nucleosome contains a core promoter for RNA polymerase III, which is able to attach to the DNA and begin synthesis of an RNA transcript before encountering the nucleosome.

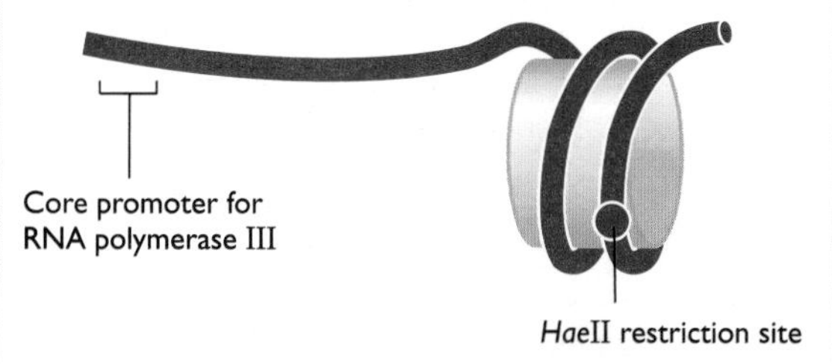

Using this system, it has been possible to show that the presence of the nucleosome does not prevent transcription, confirming that RNA polymerase III can transcribe through a region containing attached nucleosomes, and that the nucleosome is not entirely displaced from the DNA at any stage during the passage of the polymerase. The latter point was demonstrated by mixing radioactively-labeled nucleosomal fragments with a large excess of DNA fragments that were identical except that they were unlabeled and lacked bound nucleosomes. If transcription of the nucleosomal DNA results in complete detachment of the nucleosomes, then it would be anticipated that the freed nucleosomes would rebind to any of the DNA fragments in the reaction mixture, not specifically the ones to which they were originally attached. In fact, after transcription, nucleosomes were still attached to the labeled fragments, indicating that detachment had not occurred.

Other experiments with the model system have started to illuminate the events occurring when an RNA polymerase meets a nucleosome. By using agarose gel electrophoresis (Technical Note 5.1, p. 91) to measure the lengths of the RNA transcripts synthesized after different time periods, it has been shown that the polymerase pauses when it reaches the nucleosome and then moves forward in short bursts, covering 10–11 bp of the DNA before pausing again. The nucleosome therefore impedes the polymerase but only temporarily. The end result is that the polymerase moves past the nucleosome and completes transcription of the DNA, with the nucleosome shifting to a position closer to the promoter end of the fragment. The repositioning of the nucleosome is indicated by the ability of *Hae*II to cut the DNA after transcription. This means that the nucleosome must have moved, because in its original position it blocked access of the restriction enzyme to its cut site.

Looping around a nucleosome

There is still insufficient data to allow anything more than a speculative description of how the RNA polymerase transcribes through a nucleosome, but the mechanism illustrated below is a good working hypothesis. In this scheme, the polymerase approaches the nucleosome and pauses until thermal effects cause the nucleosome to move slightly further along the DNA. The polymerase now moves forward a few more bp, and begins to 'invade' the DNA wound around the nucleosome. By detaching some of the DNA the polymerase exposes a region of the nucleosome surface to which DNA that has already been transcribed can attach. This process continues, enabling the polymerase to loop around the nucleosome and eventually emerge downstream of it.

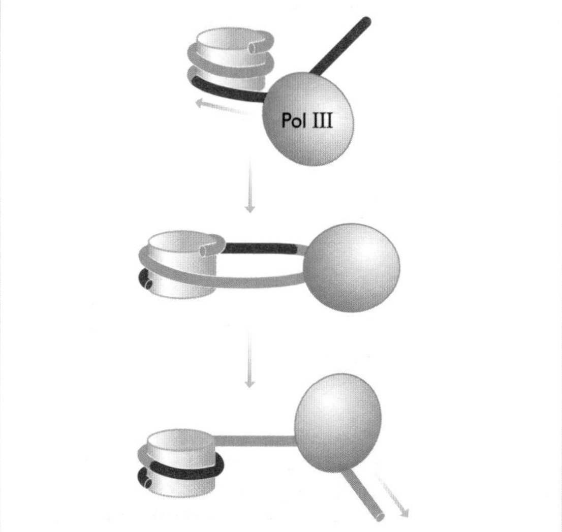

Reference

Studitsky VM, Kassavetis GA, Geiduschek EP and Felsenfeld G (1997) Mechanism of transcription through the nucleosome by eukaryotic RNA polymerase. *Science*, **278**, 1960–1963.

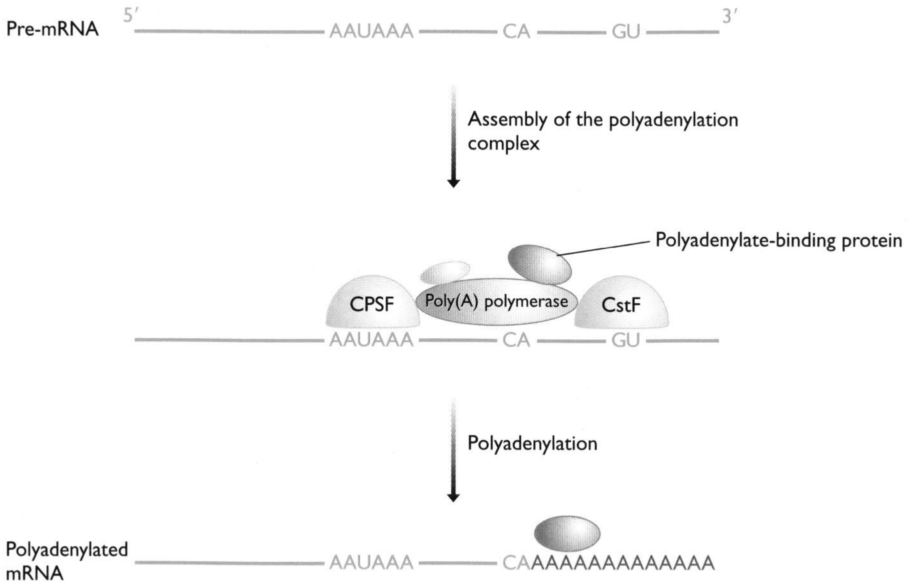

Figure 9.12 Polyadenylation of eukaryotic mRNA.

See the text for details. The diagram is schematic and is not intended to indicate the relative sizes and shapes of the various protein complexes, nor their precise positioning, although CPSF and CstF are thought to bind to the 5′-AAUAAA-3′ and GU-rich sequences, respectively, as shown. Note that 'GU' indicates a GU-rich sequence rather than the dinucleotide 5′-GU-3′.

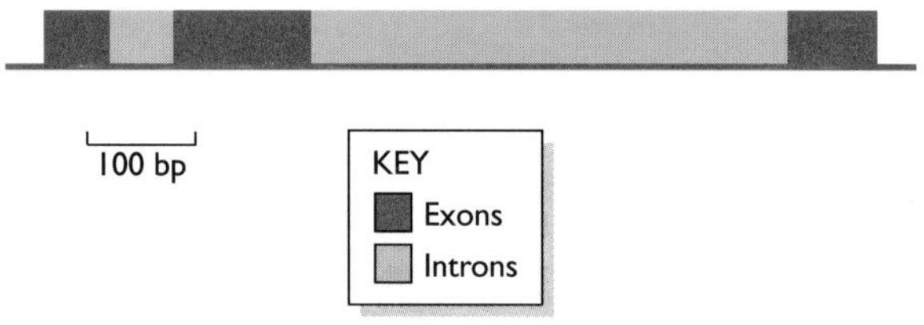

Figure 9.13 Introns.

The structure of the human β-globin gene is shown. This gene is 1423 bp in length and contains two introns, one of 131 bp and one of 851 bp, which together make up 69% of the length of the gene.

for any of the various suggestions that have been made. These include an influence on mRNA stability, which seems unlikely as some stable transcripts have very short poly(A) tails, and a role in initiation of translation. The latter proposal is supported by research showing that poly(A) polymerase is repressed during those periods of the cell cycle when relatively little protein synthesis occurs (Colgan *et al.*, 1996).

9.2.3 Intron splicing

The existence of introns was not suspected until 1977 when DNA sequencing was first applied to eukaryotic genes and it was realized that many of these contain 'intervening sequences' that separate different segments of the coding DNA from one another (*Figure 9.13*). We now recognize seven distinct types of intron in eukaryotes and additional forms in the archaea (*Table 9.3*). Two of these types – the GU–AG and AU–AC introns – are found in eukaryotic protein-coding genes and will be dealt with in this section; the other types will be covered later in the chapter.

Table 9.3 Types of intron

Intron type	Where found	Cross-reference
GU–AG introns	Eukaryotic nuclear pre-mRNA	Section 9.2.3
AU–AC introns	Eukaryotic nuclear pre-mRNA	Section 9.2.3
Group I	Eukaryotic nuclear pre-rRNA, organelle RNAs, few bacterial RNAs	Section 9.3.3
Group II	Organelle RNAs, some prokaryotic RNAs	Box 9.3
Group III	Organelle RNAs	Box 9.3
Twintrons	Organelle RNAs	Box 9.3
Pre-tRNA introns	Eukaryotic nuclear pre-tRNA	Section 9.3.3
Archael introns	Various RNAs	Box 9.3

Table 9.4 Introns in human genes

Gene	Length (kb)	Number of introns	Amount of the gene taken up by the introns (%)
Insulin	1.4	2	67
β-Globin	1.4	2	69
Serum albumin	18	13	89
Type VII collagen	31	117	71
Factor VIII	186	25	95
Dystrophin	2400	78	>99

Adapted from Strachan and Read (1996).

Few rules can be established for the distribution of introns in protein-coding genes, beyond the fact that they are less common in lower eukaryotes: the 6000 genes in the yeast genome contain only 239 introns in total, whereas many individual mammalian genes contain fifty or more introns. When the same gene is compared in related species, it is usually the case that some of the introns are in identical positions but that each species has one or more unique introns. This implies that some introns remain in place for millions of years, retaining their positions while species diversify, while others appear or disappear during this same period. This leads to two competing hypotheses regarding the evolution of introns:

- **'Introns late'** is the hypothesis that introns evolved relatively recently and are gradually accumulating in eukaryotic genomes.
- **'Introns early'** is the alternative hypothesis, that introns are very ancient and are gradually being lost from eukaryotic genomes.

These are issues that we will return to in Section 14.3.2 when we study molecular evolution. For the time being, what is important is that a eukaryotic pre-mRNA may contain many introns, perhaps over 100, taking up a considerable length of the transcript (*Table 9.4*), and that these introns must be excised and the exons joined together in the correct order, before the transcript can function as a mature mRNA.

Conserved sequence motifs indicate the key sites in GU–AG introns

With the vast bulk of pre-mRNA introns the first two nucleotides of the intron sequence are 5′-GU-3′ and the last two 5′-AG-3′. They are therefore called 'GU–AG' introns and all members of this class are spliced in the same way. These conserved motifs were recognized soon after introns were discovered and it was immediately assumed that they must be important in the splicing process. As intron sequences started to accumulate in the databases it was realized that the GU–AG motifs are merely parts of longer consensus sequences that span the 5′ and 3′ splice sites. These consensus sequences vary in different types of eukaryote; in vertebrates they can be described as:

5′ splice site	5′-AG↓GUAAGU-3′
3′ splice site	5′-PyPyPyPyPyPyNCAG↓-3′

In these designations, 'Py' is one of the two pyrimidine nucleotides (U or C), 'N' is any nucleotide, and the arrow indicates the exon–intron boundary. The 5′ splice site is also known as the **donor site** and the 3′ splice site as the **acceptor site**.

Other conserved sequences are present in some but not all eukaryotes. Introns in higher eukaryotes usually have a **polypyrimidine tract**, a pyrimidine-rich region located just upstream of the 3′-end of the intron sequence (*Figure 9.14*). This tract is less frequently seen in yeast introns, but these have an invariant 5′-UACUAAC-3′ sequence,

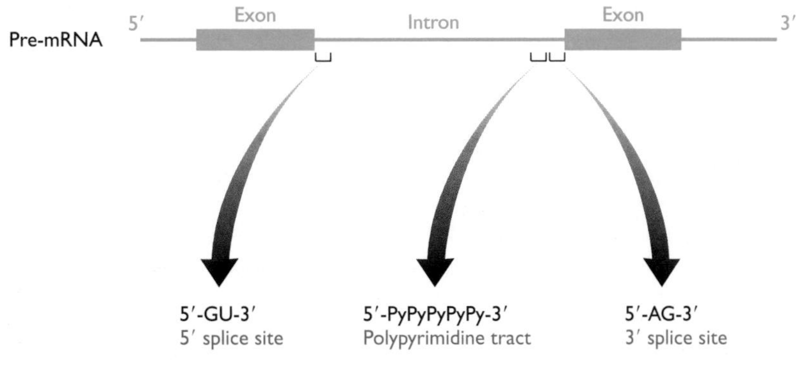

Figure 9.14 Conserved sequences in vertebrate introns.

The longer consensus sequences around the splice sites are given in the text. Abbreviation: Py, pyrimidine nucleotide (U or C).

Box 9.2: Processed pseudogenes

Two types of pseudogene were introduced in Section 6.1.1: the **conventional pseudogene**, which is a functional gene that has become inactivated by the accumulation of deleterious mutations, and the **processed pseudogene**, which arises by reverse transcription of the mRNA of a functional gene, followed by integration of the resulting cDNA back into the genome, as shown in the figure below.

The fact that processed pseudogenes are derived from an mRNA intermediate means that these structures have two important features not displayed by conventional pseudogenes:

- ***Processed pseudogenes are copies of just the transcribed region of a gene***: they lack the upstream and downstream sequences either side of the transcription unit. This means that most processed pseudogenes lack a promoter, because the promoter sequence is not present in the mRNA. Processed pseudogenes therefore cannot be expressed. Exceptions are those processed pseudogenes that are derived from genes transcribed by RNA polymerase III, these having internal promoters that are included in the mRNA sequence (Section 8.2.2). *Alu* elements (Section 6.3.2) fall into this category.
- ***Processed pseudogenes do not contain introns***, even if introns are present in the functional parent copy of the gene. This is because the introns are removed during synthesis of the mRNA and hence are not included in the cDNA sequence that is integrated back into the genome.

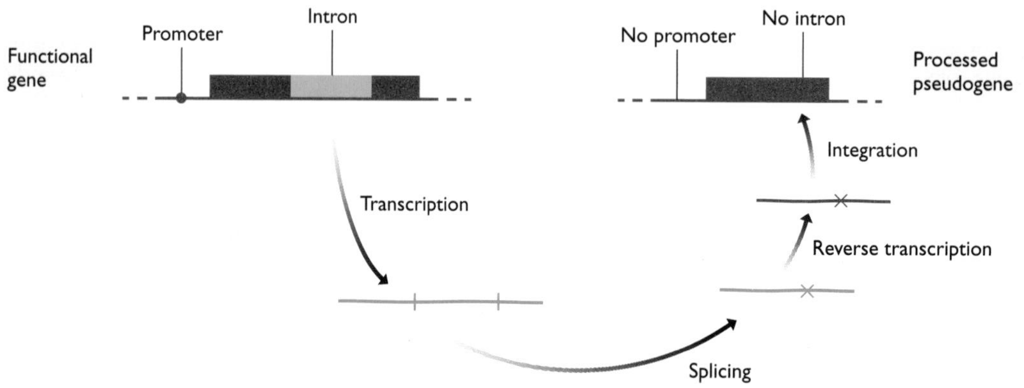

located between 18 and 140 bp upstream of the 3′ splice site, that is not present in higher eukaryotes. However, the yeast sequence is not functionally equivalent to the polypyrimidine tract, as will be described in the next two sections.

Outline of the splicing pathway for GU–AG introns

The conserved sequence motifs indicate important regions of GU–AG introns, regions that we would anticipate either acting as recognition sequences for RNA-binding proteins involved in splicing, or playing some other central role in the process. After several years during which progress was stalled by technical problems (in particular difficulties in developing a cell-free splicing system with which the process could be probed in detail), there has been a recent explosion of information. These results have shown that GU–AG splicing is highly complex. To attempt to simplify it we will first examine the splicing pathway in outline, and then look at the various splicing factors that mediate and control the process.

In outline, the splicing pathway can be divided into two steps (*Figure 9.15*):

1. **Cleavage of the 5′ splice site** occurs by a transesterification reaction promoted by the hydroxyl group attached to the 2′-carbon of an adenosine nucleotide located within the intron sequence. In yeast, this adenosine is the last one in the conserved 5′-UACUAAC-3′ sequence. The result of the hydroxyl attack is cleavage of the phosphodiester bond at the 5′ splice site, accompanied by formation of a new 5′–2′ phosphodiester bond linking the first nucleotide of the intron (the G of the 5′-GU-3′ motif) with the internal adenosine. This means that the intron has now been looped back on itself to create a **lariat** structure.
2. **Cleavage of the 3′ splice site and joining of the exons** result from a second transesterification reaction, this one promoted by the 3′-OH group attached to the end of the upstream exon. This group attacks the phosphodiester bond at the 3′ splice site, cleaving

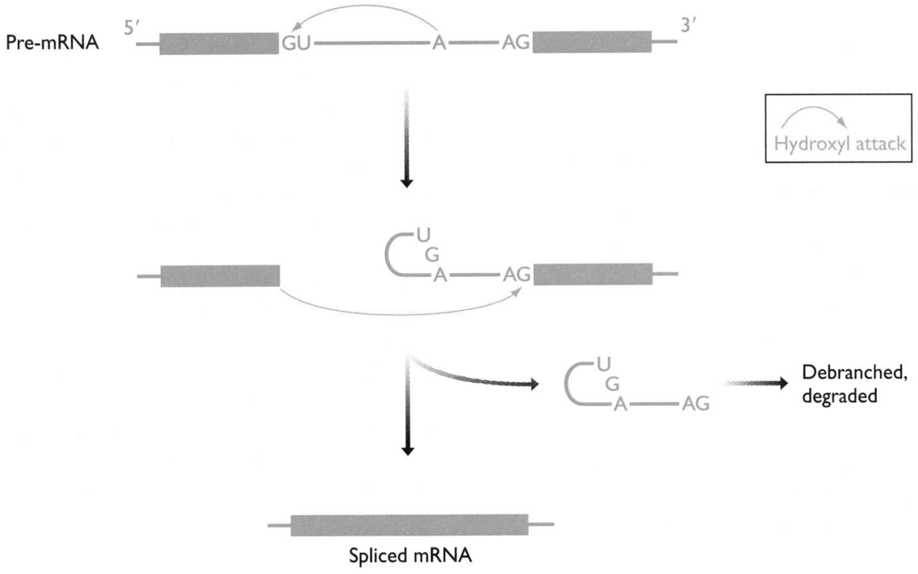

Figure 9.15 Splicing in outline.

Cleavage of the 5′ splice site is promoted by the hydroxyl attached to the 2′-carbon of an adenosine nucleotide within the intron sequence. This results in the lariat structure and is followed by the 3′-OH group of the upstream exon inducing cleavage of the 3′ splice site. This enables the two exons to be ligated with the released intron being debranched and degraded.

it and so releasing the intron as the lariat structure, which is subsequently converted back to a linear RNA and degraded. At the same time, the 3′-end of the upstream exon joins to the newly-formed 5′-end of the downstream exon, completing the splicing process.

In a chemical sense, intron splicing is not a great challenge for the cell. It is simply a double transesterification reaction, no more complicated than many other biochemical reactions that are dealt with by individual enzymes. Why then has such a complex machinery evolved to deal with it? The difficulty lies with the topological problems. The first of these is the substantial distance that might lie between splice sites, possibly a few tens of kb, representing 100 nm or more if the mRNA is in the form of a linear chain. A means is therefore needed of bringing the splice sites into proximity. The second topological problem concerns selection of the correct splice site. All splice sites are similar, so if a pre-mRNA contains two or more introns then there is the possibility that the wrong splice sites could be joined, resulting in **exon skipping**, the loss of an exon from the mature mRNA (*Figure 9.16A*). Equally unfortunate would be selection of a **cryptic splice site**, a site within an intron or exon that has sequence similarity with the consensus motifs of real splice sites (*Figure 9.16B*). Cryptic sites are present in most pre-mRNAs and must be ignored by the splicing apparatus.

snRNAs and their associated proteins are the central components of the splicing apparatus

The central components of the splicing apparatus for GU–AG introns are the snRNAs called U1, U2, U4, U5 and U6. These are short molecules, between 106 nucleotides (U6) and 185 nucleotides (U2) in vertebrates, that associate with protein factors to form **small nuclear ribonucleoproteins (snRNPs)**. These snRNPs, together with other protein factors, form a series of complexes that carry out the two steps of the splicing reaction. There are still gaps in our knowledge, but the process is believed to operate as follows (*Figure 9.17*):

1. **The commitment complex** initiates a splicing activity. This complex comprises U1-snRNP, which binds to the 5′ splice site, partly by RNA–RNA base-pairing, and the protein factors $U2AF^{35}$ and $U2AF^{65}$, which make a protein–RNA contact with the polypyrimidine tract.
2. **Complex A** comprises the commitment complex plus U2-snRNP, the latter attached to the branch site.
3. **Complex B** is formed when U4/U6-snRNP (a single snRNP containing two snRNAs) and U5-snRNP attach to Complex A. U4/U6-snRNP binds to U2-snRNP, and U5–snRNP attaches initially in the downstream exon and then migrates to the 3′ region of the intron. At this stage, an association between U1-snRNP and U2-snRNP brings the 5′ splice site into close proximity with the branch point. U1-snRNP then dissociates from the complex, its role in binding the 5′ splice site now taken over by U6-snRNP.
4. **Complex C** is formed by the departure of U4-snRNA, plus possibly some of the proteins from the U4/U6-snRNP, from Complex B and the repositioning of U5-snRNP onto the 3′ splice site. All three key positions in the intron – the two splice sites and the

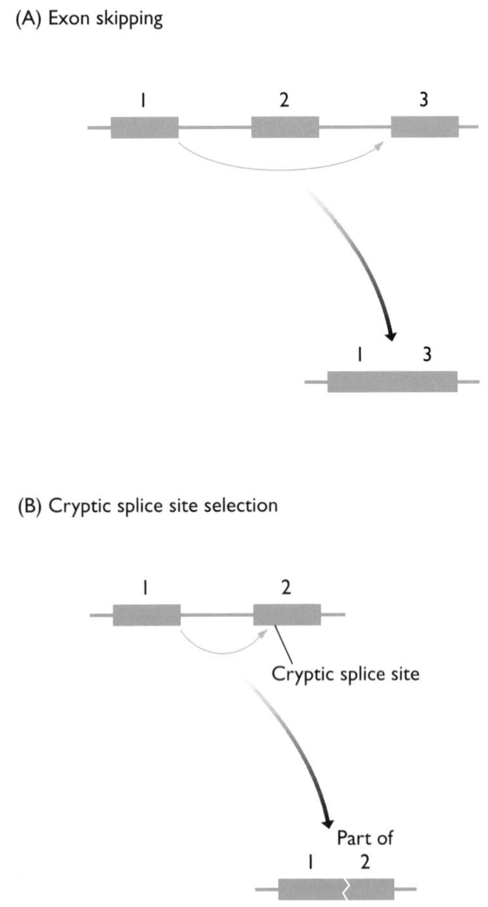

Figure 9.16 Two aberrant forms of splicing.

(A) In exon skipping the aberrant splicing results in an exon being lost from the mRNA. (B) When a cryptic splice site is selected part of an exon might be lost from the mRNA, as shown here, or if the cryptic site lies within an intron then a segment of that intron will be retained in the mRNA.

branch point – are now in proximity and the two transesterifications occur as a linked reaction, possibly catalyzed by U6–snRNP, completing the splicing process.

Although each complex is transient, and none are preformed in the nucleus, the splicing apparatus is often referred to as the **spliceosome**.

SR proteins are mediators in splice site selection

The series of events shown in *Figure 9.17* provides no clues about how the correct splice sites are selected so that exons are not lost during splicing and cryptic sites are ignored. This aspect of splicing is still poorly understood but it has become clear that a set of splicing factors called **SR proteins** are important in splice site selection.

The SR proteins – so-called because their C-terminal domains contain a region rich in serine (abbreviation S) and arginine (R) – were first implicated in splicing when it was discovered that they are components of the spliceosome. They appear to have several roles, including the establishment of a connection between bound U1-snRNP and bound U2AF in the commitment complex (Valcárcel and Green, 1996). This is perhaps the clue to their role in splice site selection as formation of the commitment complex is the critical stage of the splicing process, as this is the event that identifies which sites will be linked. At present the experimental data relevant to the function of SR proteins is patchy, but progress is being made by studying those pre-mRNAs that naturally follow alternative splicing pathways, in particular a series of genes involved in sex determination in the fruit fly (Chabot, 1996). The first gene in this series is *sxl*, whose transcript contains an optional exon which, when spliced to the one preceding it, results in an inactive version of protein SXL. In females the splicing pathway is such that this exon is skipped so that functional SXL is made (*Figure 9.18*). SXL promotes selection of a cryptic splice site in a second transcript, *tra*, by directing $U2AF^{65}$ away from its normal 3′ splice site to a second site further downstream. The resulting female-specific TRA protein is again involved in alternative splicing, this time by interacting with SR proteins to form a multifactor complex that attaches to a site within an exon of a third pre-mRNA, *dsx*, promoting selection of a secondary, female-specific splice site in this pre-mRNA. The resulting male- and female-specific DSX proteins have physiological roles in *Drosophila* sex determination. The binding site within *dsx* that is occupied by TRA and the SR proteins has been called a **splicing enhancer**, and there is evidence that these protein-binding sites might be general features of alternatively spliced pre-mRNAs.

There is one final aspect of SR proteins that we should address. This is the possibility that a subset of them, called **CASPs** (**CTD-associated SR-like proteins**), form a physical connection between the spliceosome and the RNA polymerase II transcription complex, and hence provide a link between transcript elongation and processing. The CTD referred to in the name of these proteins is the C-terminal domain of the largest subunit of RNA polymerase II, the part of the protein also implicated in an interaction with the polyadenylation system (Section 9.2.2) and activation of transcription initiation (Section 8.3.2). Electron microscopic studies have suggested that transcription and splicing occur together and the discovery of splicing factors that have an affinity for the RNA polymerase provides a biochemical basis for this observation (Corden and Patturajan, 1997).

AU–AC introns are similar to GU–AG introns but require a different splicing apparatus

One of the more surprising events of recent years has been the discovery of a few introns in eukaryotic pre-mRNAs that do not fall into the GU–AG category, having different consensus sequences at their splice sites. These are the **AU–AC introns** which, to date, have been found in approximately 20 genes in organisms as diverse as humans, plants and *Drosophila* (Nilsen, 1996; Tarn and Steitz, 1997).

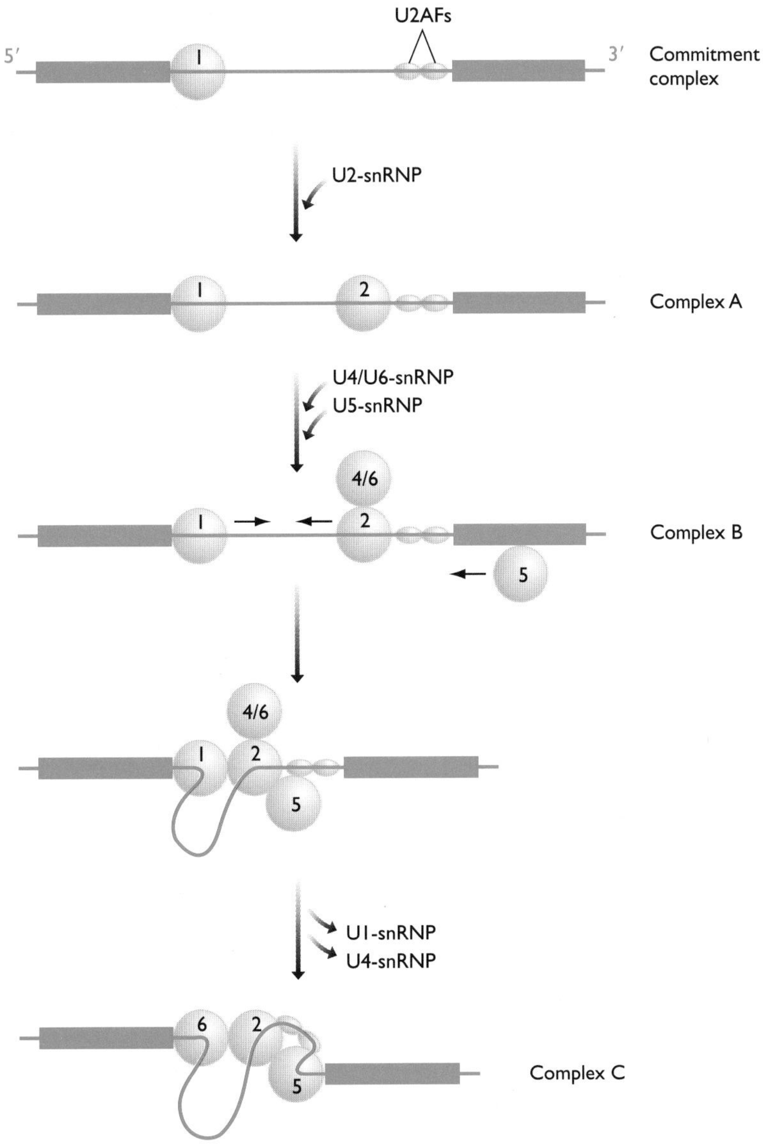

Figure 9.17 The roles of snRNPs and associated proteins in splicing.

See the text for details. There are several unanswered questions about the series of events occurring during splicing and it is unlikely that the scheme shown here is entirely accurate. The key point is that associations between the snRNPs are thought to bring the three critical parts of the intron – the two splice sites and the branch point – into close proximity.

As well as the sequences at their splice sites, AU–AC introns have a conserved (though not invariant) branch site sequence with the consensus 5′-UCCUUAAC-3′, the last adenosine in this motif being the one that participates in the first transesterification reaction. This points us towards the remarkable feature of AU–AC introns: their splicing pathway is virtually identical to that for GU–AG introns, but involves a different set of splicing factors. Only the U5-snRNP is involved in the splicing mechanisms of both types of intron. The roles of U1-snRNP and U2-snRNP are taken by two previously discovered complexes that had never been assigned a function, U11-snRNP and U12-snRNP, and an entirely new U4atac/U6atac-snRNP has subsequently been isolated to complete the picture. The similarities between the two splicing processes extend even to the detailed interaction between the snRNPs in the spliceosome.

AU–AC introns, rather than simply being a curiosity, are proving useful in testing models for interactions occurring during GU–AG intron splicing. The argument is that a predicted interaction between two components of the GU–AG spliceosome can be checked by seeing if the same interaction is possible with the equivalent AU–AC components. This has already been informative

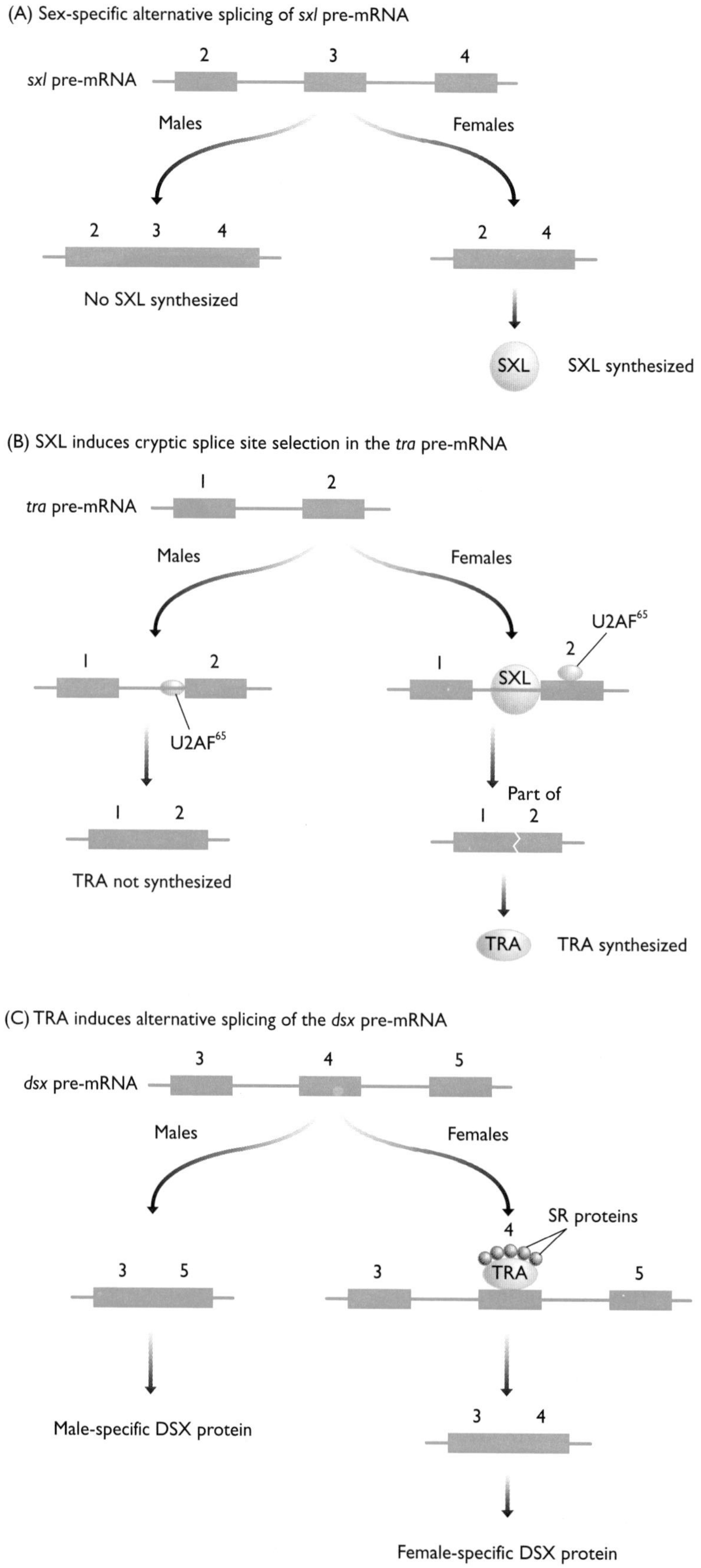
(A) Sex-specific alternative splicing of *sxl* pre-mRNA
2
3
4
sxl pre-mRNA
Males
Females
2 3 4
2 4
No SXL synthesized
SXL
SXL synthesized
(B) SXL induces cryptic splice site selection in the *tra* pre-mRNA
1
2
tra pre-mRNA
Males
Females
U2AF65
SXL
Part of
TRA not synthesized
TRA
TRA synthesized
(C) TRA induces alternative splicing of the *dsx* pre-mRNA
3
4
5
dsx pre-mRNA
Males
Females
SR proteins
TRA
3 5
3 4
Male-specific DSX protein
Female-specific DSX protein

Figure 9.18 Regulation of splicing during expression of genes involved in sex determination in *Drosophila*.

(A) The cascade begins with sex-specific alternative splicing of the *sxl* pre-mRNA. In males all exons are present in the mRNA, but this means that a truncated protein is produced because exon 3 contains a termination codon. In females, exon 3 is skipped, leading to a full-length, functional SXL protein. (B) In females, SXL blocks the 3′ splice site in the first intron of the *tra* pre-mRNA. $U2AF^{65}$ is unable to locate this site and instead directs splicing to a cryptic site in exon 2. This results in an mRNA that codes for a functional TRA protein. In males, there is no SXL so the 3′ splice site is not blocked and a dysfunctional mRNA is produced. (C) In males, exon 4 of the *dsx* pre-mRNA is skipped. The resulting mRNA codes for a male-specific DSX protein. In females, TRA stabilizes the attachment of SR proteins to a splicing enhancer located within exon 4, so this exon is not skipped, resulting in the mRNA that codes for the female-specific DSX protein. The two versions of DSX are the primary determinants of the male and female physiologies. The female *dsx* mRNA ends with exon 4 because the intron between exons 4 and 5 has no 5′ splice site, meaning that exon 5 cannot be ligated to the end of exon 4. Instead a polyadenylation site at the end of exon 4 is recognized in females. See Chabot (1996) for more details. Note that the diagram is schematic and the introns are not drawn to scale.

in helping to define a base-paired structure formed between the U2 and U6 snRNAs in the GU–AG spliceosome (Tarn and Steitz, 1996).

9.3 SYNTHESIS AND PROCESSING OF NONCODING RNAs

In bacteria, the same RNA polymerase synthesizes all types of RNA. The issues that we have already discussed regarding elongation and termination of bacterial mRNA (Section 9.2.1) therefore hold also for rRNA and tRNA synthesis, and the only outstanding areas that we have to cover are the processing of the pre-rRNAs and pre-tRNAs into the mature molecules. This processing involves cutting events and chemical modifications, both types of reaction being similar to equivalent processing events for eukaryotic rRNAs and tRNAs: we will therefore deal with bacterial and eukaryotic processing together, cutting events in Section 9.3.2 and chemical modifications in Section 9.4. The distinctive feature of eukaryotic rRNA and tRNA processing is the presence in some eukaryotic pre-RNAs of introns, different from the pre-mRNA introns described above; these will be covered in Section 9.3.3. First, however, there are issues regarding transcript elongation and termination by RNA polymerases I and III that we must address.

9.3.1 Transcript elongation and termination by RNA polymerases I and III

In general, we know less about transcript elongation and termination by RNA polymerases I and III compared with equivalent processes for RNA polymerase II. The interaction of the polymerase with the template and transcript during elongation appears to be similar with all three enzymes, a reflection of the structural relatedness of the three largest subunits in each polymerase. One difference is the rate of transcription – RNA polymerase I, for example, being much slower than RNA polymerase II, managing a polymerization rate of only 20 nucleotides per minute, compared with 2000 per minute for mRNA synthesis. A second difference is that neither RNA polymerase I nor RNA polymerase III transcripts are capped.

The major differences between the three polymerases are seen when the termination processes are compared. The polyadenylation system for RNA polymerase II termination (Section 9.2.2) is unique to that enzyme and no equivalent has been described for the other two polymerases. Termination by RNA polymerase I involves a DNA-binding protein, called Reb1p in *Saccharomyces cerevisiae* and TTF-I in mice, which attaches to the DNA at a recognition sequence located 12–20 bp downstream of the point at which transcription terminates (*Figure 9.19*). Exactly how the bound protein causes termination is not known, but a model in which the polymerase becomes stalled because of the blocking effect of Reb1p/TTF-I has been proposed (Reeder and Lang, 1997). A second protein, PTRF (polymerase I and transcript release factor) is thought to induce dissociation of the polymerase and the transcript from the DNA template (Jansa *et al.*, 1998). Even less is known about RNA polymerase III termination: a run of adenosines in the template is implicated but the process does not involve a hairpin loop and so is not analogous to termination in bacteria.

9.3.2 Cutting events involved in processing of bacterial and eukaryotic pre-rRNA and pre-tRNA

Bacteria synthesize three different rRNAs, called 5S rRNA, 16S rRNA and 23S rRNA, the names indicating the sizes of the molecules as measured by **sedimentation analysis** (see Technical Note 6.1, p. 123). The three genes for these rRNAs are linked into a single transcription unit (which is usually present in multiple copies, seven for *E. coli*) and so the pre-rRNA contains copies of all three rRNAs. Cutting events are therefore needed to release the mature rRNAs. These cuts are made by various ribonucleases at positions specified by double-stranded regions formed by base-pairing between different parts of the pre-rRNA (*Figure 9.20*). The cut ends are subsequently trimmed by exonucleases.

In eukaryotes there are four rRNAs. One of these, for the 5S rRNA, is transcribed by RNA polymerase III and does not undergo processing. The remaining three (the 5.8S, 18S and 28S rRNAs) are transcribed by RNA polymerase I from a single unit producing a pre-rRNA which, as with the bacterial rRNAs, is processed by cutting and

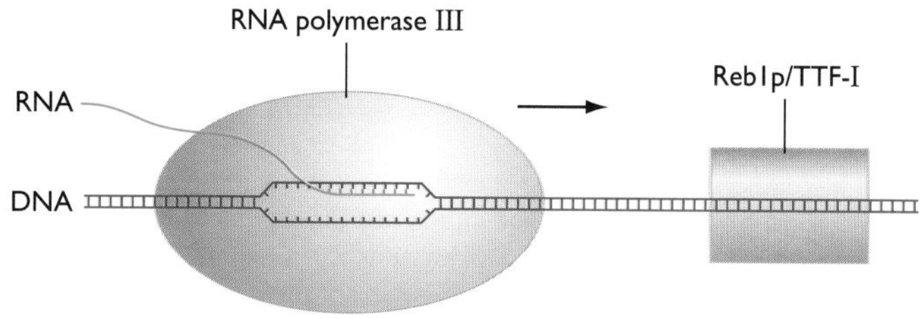

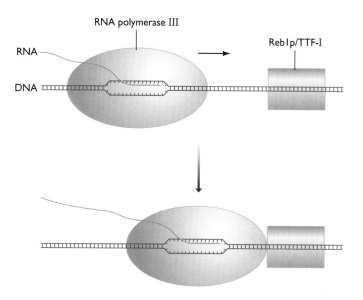

Figure 9.19 A possible scheme for termination of transcription by RNA polymerase III.

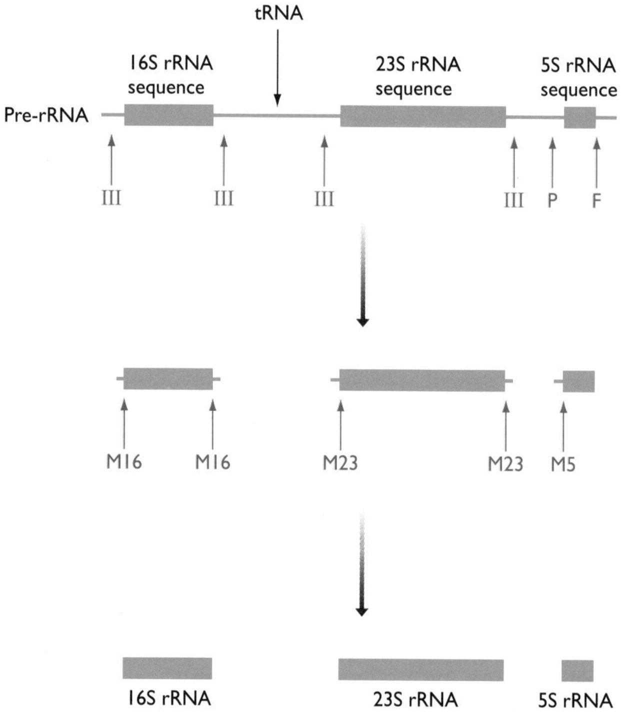

Figure 9.20 Pre-rRNA processing in *Escherichia coli.*

The pre-rRNA, containing copies of the 16S, 23S and 5S rRNAs, is cleaved by ribonucleases III, P and F and the resulting molecules trimmed to give the mature rRNAs by ribonucleases M16, M23 and M5. Eukaryotic pre-rRNAs, which contain the rRNA sequences in the order 18S–5.8S–28S, are processed in a similar fashion. Note that a tRNA is located between the 16S and 23S sequences in the *E. coli* pre-rRNA. This tRNA is processed as shown in *Figure 9.21*.

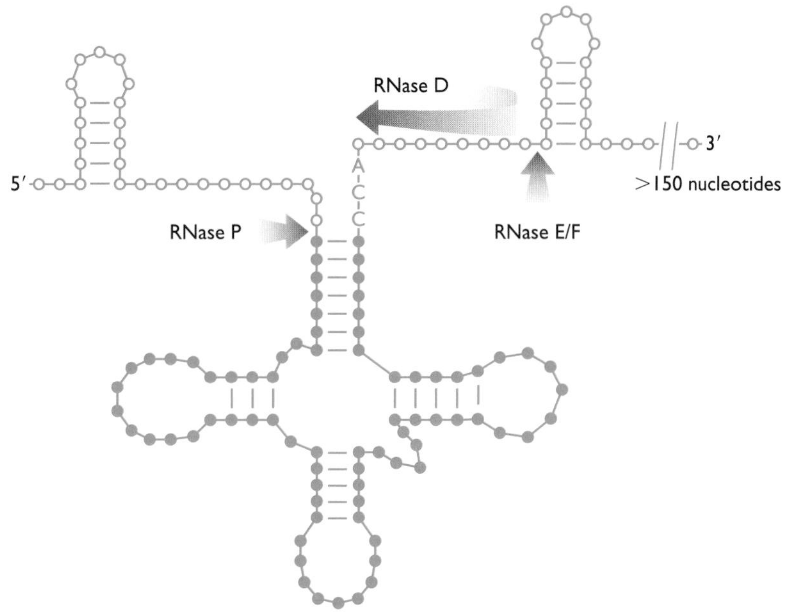

Figure 9.21 Processing of an *Escherichia coli* pre-tRNA.

The example shown results in synthesis of tRNA^{tyr}. The tRNA sequence in the primary transcript adopts its base-paired cloverleaf structure (see *Figure 10.2*, p. 233) and two additional hairpin structures form, one on either side of the tRNA. Processing begins with the cut by ribonuclease E or F, forming a new 3′-end just upstream of one of the hairpins. Ribonuclease D, which is an exonuclease, trims seven nucleotides from this new 3′-end and then pauses while ribonuclease P makes a cut at the start of the cloverleaf, forming the 5′-end of the mature mRNA. Ribonuclease D then removes two more nucleotides, creating the 3′-end of the mature molecule. With this tRNA the 3′-terminal CCA sequence is present in the RNA and not removed by ribonuclease D. With some other tRNAs this sequence has to be completely or partly added by tRNA nucleotidyltransferase. Abbreviation: RNase, ribonuclease. Based on Turner *et al.*, 1997.

end-trimming. Several nucleases are involved, such as **ribonuclease MRP** (Lygerou *et al.*, 1996).

In both bacteria and eukaryotes, tRNA genes occur singly and as multigene transcription units, or, in bacteria, as infiltrators within the rRNA transcription unit. These are also processed by a series of ribonucleases, as illustrated in *Figure 9.21*. All mature tRNAs must end with the trinucleotide 5′-CCA-3′. Some tRNAs have this sequence already; those that do not, or from which the 5′-CCA-3′ has been removed by the processing ribonucleases, have the motif added by **tRNA nucleotidyltransferase**.

9.3.3 Introns in eukaryotic pre-rRNA and pre-tRNA

Some eukaryotic pre-rRNAs and pre-tRNAs contain introns which must be spliced during the processing of these transcripts into mature RNAs. Neither type of intron is similar to the GU–AG and AU–AC introns of pre-mRNA.

Eukaryotic pre-rRNA introns are self-splicing

Introns are quite uncommon in eukaryotic pre-rRNAs but a few are known in microbial eukaryotes such as *Tetrahymena*. These introns are members of the Group I family (see *Table 9.3*, p. 212) and are also found in mitochondrial and chloroplast genomes, where they occur in pre-mRNA as well as pre-rRNA. A few isolated examples are known in bacteria, for instance in a tRNA gene of the cyanobacterium *Anabaena* and in the thymidylate synthase gene of the *E. coli* bacteriophage T4.

The splicing pathway for Group I introns is similar to that of pre-mRNA introns in that two transesterifications are involved. The first is induced not by a nucleotide within the intron but by a free nucleoside or nucleotide, any one of guanosine or guanosine mono-, di- or triphosphate (*Figure 9.22*). The 3′-OH of this cofactor attacks the phosphodiester bond at the 5′ splice site, cleaving it with transfer of the G to the 5′-end of the intron. The second transesterification involves the 3′-OH at the end of the exon, which attacks the phosphodiester bond at the 3′ splice site, causing cleavage, joining of the two exons, and release of the intron. The released intron is linear, rather than the lariat seen with pre-mRNA introns, but may undergo additional transesterifications leading to circular products as part of its degradation process.

The remarkable feature of the Group I splicing pathway is that it proceeds in the absence of proteins and hence is autocatalytic, the RNA itself possessing enzymatic activity. This was the first example of an RNA enzyme or **ribozyme** to be discovered, back in the early 1980s. Initially this caused quite a stir but it is now realized that, although uncommon, there are several examples of ribozymes (*Table 9.5*). They are thought to be relics from the **RNA world**, the early period of evolution when all biological reactions centered around RNA (see Section 14.1.1).

The self-splicing activity of Group I introns resides in the base-paired structure taken up by the RNA. This structure was first described in two-dimensional terms, by comparing the sequences of different Group I introns and working out a common base-paired arrangement that could be adopted by all versions. This resulted in a model comprising nine major base-paired regions (*Figure 9.23*; Burke *et al.*, 1987). More recently, the three-dimensional structure has been solved by X-ray crystallography (Cate *et al.*, 1996; Golden *et al.*, 1998). The ribozyme consists of a catalytic core made up of two domains, each one comprising two of the base-paired regions, with the splice sites brought into proximity by interactions between two other parts of the secondary structure. Although this RNA structure is sufficient for splicing, it is possible that with some introns the stability of the ribozyme is enhanced by noncatalytic protein factors that bind to it. This has long been suspected with the Group I introns in organelle genes as many of these contain an open reading frame coding for a protein called a **maturase** that appears to play a role in splicing.

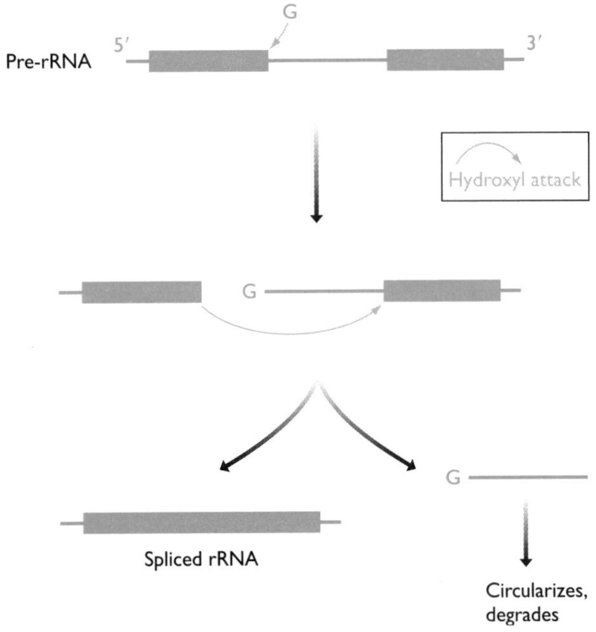

Figure 9.22 The splicing pathway for the *Tetrahymena* rRNA intron.

Eukaryotic tRNA introns are variable but all are spliced by the same mechanism

Introns in eukaryotic pre-tRNA are relatively short and usually found at the same position in the mature tRNA sequence, within the anticodon loop (see *Figure 10.2*, p. 233). Their sequences are variable and no common motifs can be identified. Unlike all other types, splicing of pre-tRNA introns does not involve transesterifications. Instead, the two splice sites are cut by ribonuclease action, leaving a cyclic phosphate structure attached to the 3′-end of the upstream exon, and a hydroxyl group at the 5′-end of the downstream exon. These ends are held in proximity by the natural base-pairing adopted by the tRNA sequence and are ligated by an RNA ligase.

9.4 PROCESSING OF PRE-RNA BY CHEMICAL MODIFICATION

The processing events that we have studied so far have been either chemical modifications that affect the ends of transcripts (capping, polyadenylation) or physical changes to the lengths of transcripts (splicing, cutting

Table 9.5 Ribozymes

Ribozyme	Description
Self-splicing introns	Some introns of Groups I, II and III splice themselves by an autocatalytic process
Ribonuclease P	The enzyme that creates the 5′-ends of bacterial tRNAs (see Section 9.3.2) consists of an RNA subunit and a protein subunit, with the catalytic activity residing in the RNA
Ribosomal RNA	The peptidyl transferase activity required for peptide bond formation during protein synthesis (Section 10.2.2) is associated with the 23S rRNA of the large subunit of the ribosome (see Research Briefing 10.2, p. 249)
$tRNA^{Phe}$	Undergoes self-catalyzed cleavage in the presence of divalent lead ions
Virus genomes	Replication of the RNA genomes of some viruses involves self-catalyzed cleavage of chains of newly-synthesized genomes linked head-to-tail. Examples are the plant viroids and virusoids and the animal hepatitis delta virus. These viruses form a diverse group with the self-cleaving activity specified by a variety of different base-paired structures, including a well-studied one that resembles a **hammerhead**

For further details see Scott and Klug (1996) and Nitta *et al.* (1998).

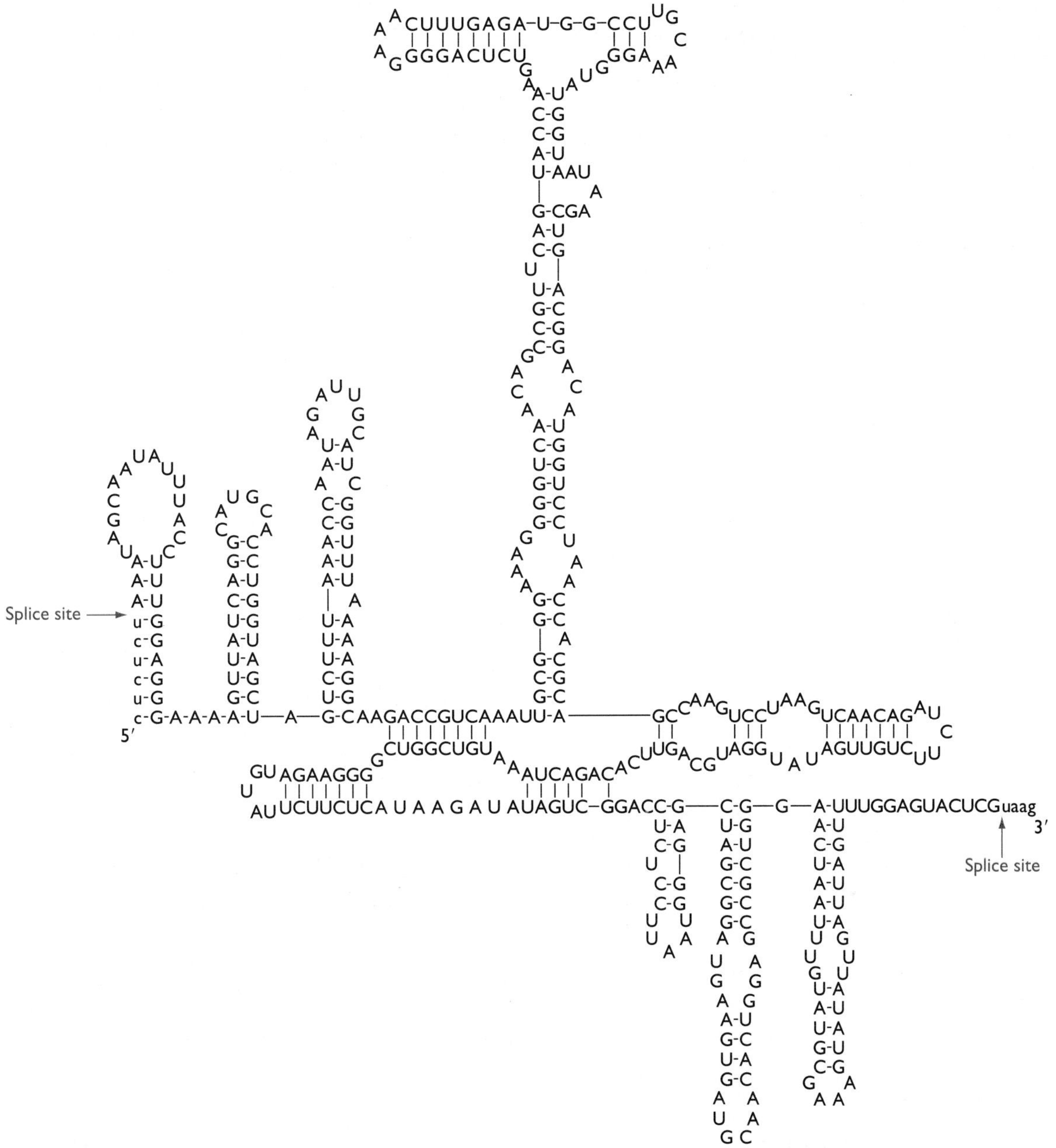

Figure 9.23 The base-paired structure of the *Tetrahymena* rRNA intron.

The intron is shown in capital letters with the exons in lower case. Additional interactions fold the intron into a three-dimensional structure that brings the two splice sites close together. Reprinted with permission from Burke *et al.* (1987) *Nucleic Acids Research*, **15**, 7217–7221, Oxford University Press.

events). The final type of processing that occurs with pre-RNAs is the chemical modification of nucleotides within the transcript. This occurs with pre-rRNAs and pre-tRNAs of both bacteria and eukaryotes and, to a much lesser extent, with pre-mRNAs of eukaryotes. Equivalent events in the archaea are poorly understood.

A broad spectrum of chemical changes have been identified with different pre-RNAs, over fifty different modifications known in total (*Table 9.6*). Most of these are carried out directly on an existing nucleotide within the transcript but two modified nucleotides, queosine and wyosine, are put in place by cutting out an entire nucleotide and replacing it with the modified version. Many of these modifications were first identified in tRNAs, within which approximately one in ten nucleotides becomes altered. These modifications are

Table 9.6 Examples of chemical modifications occurring with nucleotides in rRNA and tRNA

Modification	Details	Example
Methylation	Addition of one or more $-CH_3$ groups to the base or sugar	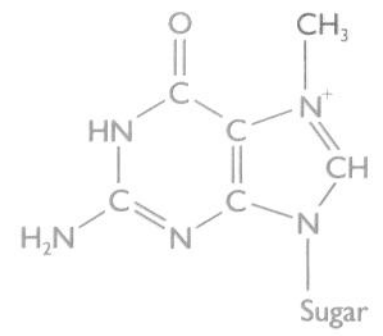 Methylation of guanosine gives 7-methylguanosine
Deamination	Removal of an amino ($-NH_2$) group from the base	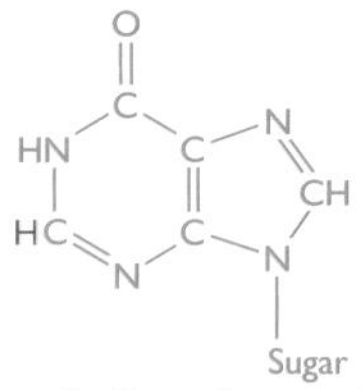 Deamination of guanosine gives inosine
Sulfur substitution	Replacement of oxygen with sulfur	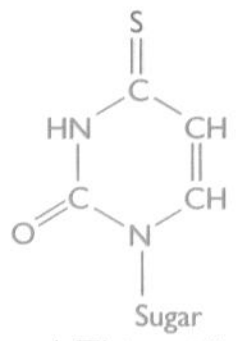 4-Thiouridine
Base isomerization	Changing the positions of atoms in the ring component of the base	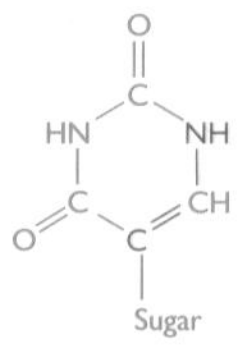 Isomerization of uridine gives pseudouridine
Double-bond saturation	Converting a double bond to a single bond	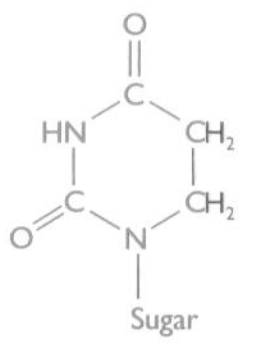 Double bond saturation converts uridine to dihydrouridine
Nucleotide replacement	Replacement of an existing nucleotide with a new one	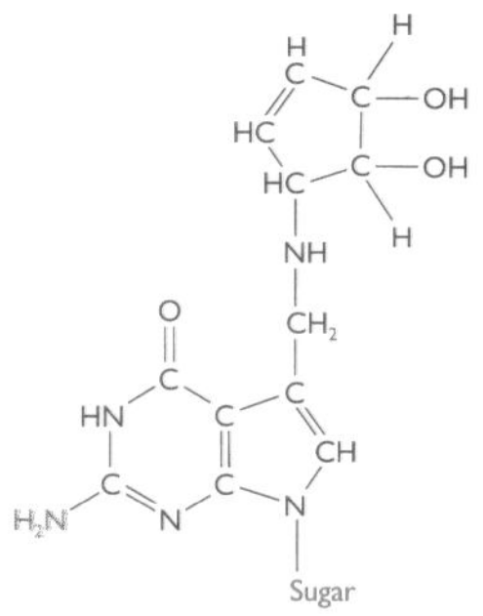 Queosine

Box 9.3: Other types of intron

There are eight different types of intron (see *Table 9.3*, p. 212). Four types are described in the text: the nuclear pre-mRNA introns of the GU–AG and AU–AC classes, the self-splicing Group I introns, and the introns in eukaryotic pre-tRNA genes. The details of the four other categories are as follows:

- **Group II introns** are found in the organelle genomes of fungi and plants and a few are known in prokaryotes. Group II introns take up a characteristic secondary structure and they are self-splicing, but they are distinct from Group I introns. The secondary structure is different and the splicing mechanism is more closely allied to that of pre-mRNA introns, the initial trans-esterification being promoted by the hydroxyl group of an internal adenosine nucleotide and the intron being converted into a lariat structure. These similarities have prompted suggestions that Group II and pre-mRNA introns may have a common evolutionary origin (see Section 14.3.2).
- **Group III introns** are also found in organelle genomes and they self-splice via a mechanism very similar to that of Group II introns, but Group III introns are smaller and have their own distinctive secondary structure. The resemblance with Group II introns again suggests an evolutionary relationship.
- **Twintrons** are composite structures made up of two or more Group II and/or Group III introns. The simplest twintrons consist of one intron embedded in another, but more complex ones contain multiple embedded introns. The individual introns that make up a twintron are usually spliced in a defined sequence.
- **Archaeal introns** are present in tRNA and rRNA genes. They are cleaved by a ribonuclease similar to the one involved in eukaryotic pre-tRNA splicing.

For more details concerning Group II and Group III introns and twintrons see Copertino and Hallick (1993). Archaeal introns are described by Lykke-Andersen *et al.* (1997).

thought to mediate the recognition of individual tRNAs by the enzymes that attach amino acids to these molecules (Section 10.1.1), and to increase the range of the interactions that can occur between tRNAs and codons during translation, enabling a single tRNA to recognize more than one codon (Section 10.1.2).

We know relatively little about how tRNA modifications are carried out, beyond the fact that there are a number of enzymes that catalyze these changes. How the enzymes are directed to the correct nucleotides on which they must act has not been explained. With rRNA and mRNA modifications we know rather less about the reasons for the chemical alterations but rather more about how the alterations are carried out. These issues are covered in the next two sections.

9.4.1 Chemical modification of pre-rRNAs

Ribosomal RNAs are modified in two ways: by addition of methyl groups to, mainly, the 2′-OH group on nucleotide sugars, and by conversion of uridine to pseudouridine (see *Table 9.6*). The same modifications occur at the same positions on all copies of an rRNA, and these modified positions are, to a certain extent, the same in different species. Some similarities in modification patterns are even seen when bacteria and eukaryotes are compared, although bacterial rRNAs are less heavily modified than eukaryotic ones. Functions for the modifications have not been identified, though most occur within those parts of rRNAs thought to be most critical to the activity of these molecules in ribosomes (Section 10.2.1). Modified nucleotides might, for example, be involved in rRNA-catalyzed reactions such as synthesis of peptide bonds.

It is not easy, simply by intuition, to imagine how specificity of rRNA modification can be ensured. Human pre-rRNA, for example, undergoes 106 methylations and 95 pseudouridinylations, each alteration at a specified position, with no obvious sequence similarities that can be inferred as target motifs for the modifying enzymes. Not surprisingly, progress in understanding rRNA modification was slow to begin with. The breakthrough came when it was shown that in eukaryotes the short RNAs called snoRNAs are involved in the modification process. These molecules are 70–100 nucleotides in length and located in the nucleolus, the region within the nucleus where rRNA processing takes place. The initial discovery was that by base-pairing to the relevant region snoRNAs pinpoint positions at which the pre-rRNA must be methylated. The base-pairing does not involve the entire length of the snoRNA, only a few nucleotides, but these nucleotides are always located immediately upstream of a conserved sequence called the D box (*Figure 9.24A*). The base pair involving the nucleotide that will be modified is five positions away from the D box. The hypothesis is that the D box is the recognition signal for the methylating enzyme, which is therefore directed towards the appropriate nucleotide (Bachellerie and Cavaillé, 1997).

After these initial discoveries with regard to methylation, it was shown that a different family of snoRNAs carries out the same guiding role in conversion of uridines to pseudouridines (Maden, 1997). These snoRNAs do not have D boxes but still have conserved motifs that could be recognized by the modifying enzyme, and each is able to form a specific base-paired interaction with its target site, specifying the nucleotide to be modified.

The implication is that there is a different snoRNA for each modified position in a pre-rRNA, except possibly for a few sites that are close enough together to be dealt with

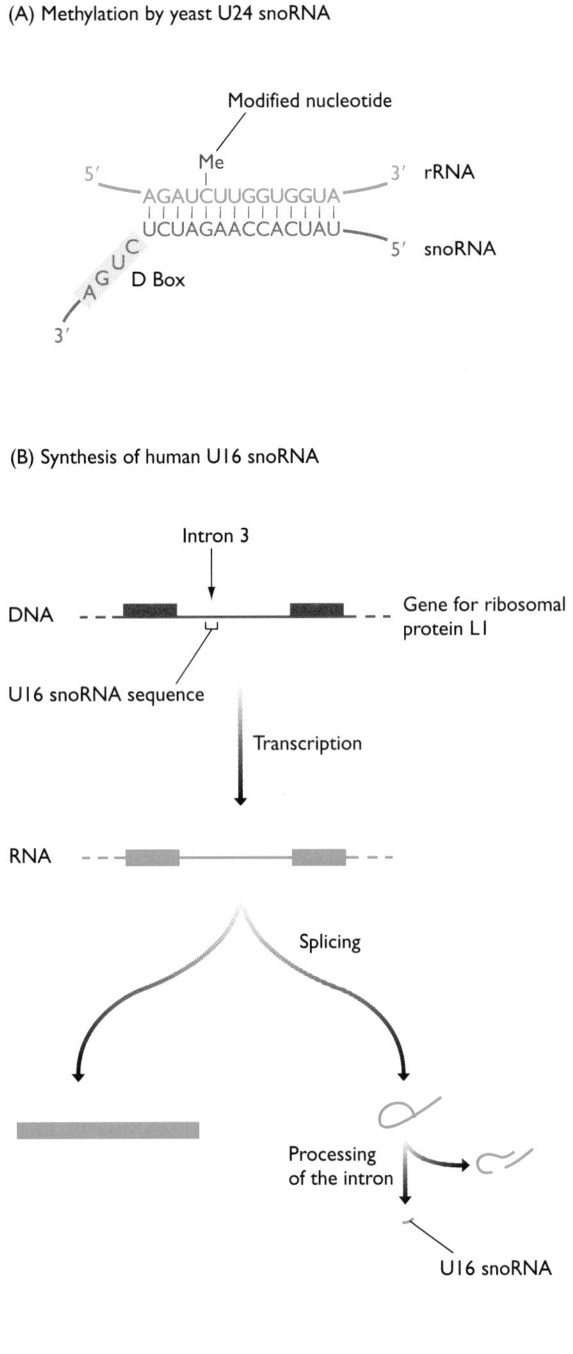

Figure 9.24 Methylation of rRNA by a snoRNA.

(A) This example shows methylation of the C at position 1436 in the *Saccharomyces cerevisiae* 25S rRNA (equivalent to the 28S rRNA of vertebrates), directed by U24 snoRNA. The D box of the snoRNA is highlighted. Modification always occurs at the base pair five positions away from the D box. Note that the interaction between rRNA and snoRNA involves an unusual G–U base pair, which is permissible between RNA polynucleotides (see also *Figure 10.9A*, p. 241). The diagram is based on Tollervey (1996). (B) Many snoRNAs are synthesized from intron RNA, as shown here for human U16 snoRNA which is specified by a sequence in intron 3 of the gene for ribosomal protein L1.

Box 9.4: Chemical modification without snoRNAs

One of the most surprising features of the snoRNA system is that it applies only to eukaryotic rRNAs. Bacterial rRNAs also undergo chemical modification, not to the same extent as eukaryotic rRNA but with the same specificity, and it might therefore be anticipated that bacteria would use a similar mechanism for carrying out the alterations. This is not the case; bacterial rRNA modifications, as well as tRNA modifications in both bacteria and eukaryotes, are carried out by enzymes that directly recognize the sequence and/or structure of the regions of RNA that contain the nucleotides to be modified. Often two or more nucleotides in the same region are modified at once. Presumably this relatively simple modification system evolved into the more complex snoRNA-directed process seen in eukaryotic cells, but exactly how this transition occurred is not known (Lafontaine and Tollervey, 1998).

by a single snoRNA. This means that there must be several hundred snoRNAs per cell. At one time this seemed unlikely because very few snoRNA genes could be located, but now it appears that only a fraction of all the snoRNAs are transcribed from these standard genes, most being specified by sequences within the introns of other genes and released by cutting up the intron after splicing (*Figure 9.24B*).

9.4.2 RNA editing

Because rRNAs and tRNAs are noncoding, chemical modifications to their nucleotides affect only the structural features of the molecules. With mRNAs the situation is very different, chemical modification having the potential to change the coding properties of the transcript, resulting in an equivalent alteration in the amino acid sequence of the protein that is specified. A notable example of **RNA editing** occurs with the human mRNA for apolipoprotein-B. The gene for this protein codes for a 4563 amino acid polypeptide, called apolipoprotein-B100, which is synthesized in liver cells and secreted into the bloodstream where it transports lipids from site to site within the body. A related protein, apolipoprotein-B48 is made by intestinal cells. This protein is only 2153 amino acids in length and is synthesized from an edited version of the mRNA for the full length protein (*Figure 9.25*). In intestinal cells this mRNA is modified by deamination of a cytosine, converting this into a uracil. This changes a CAA codon, specifying glutamine, into a UAA codon, which causes translation to stop, resulting in the truncated protein. The deamination is carried out by an RNA-binding enzyme that, in conjunction with a set of

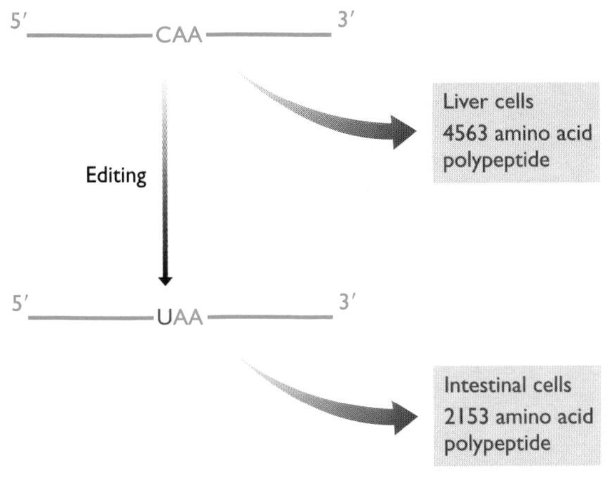

Figure 9.25 Editing of the human apolipoprotein-B mRNA.

Conversion of a C to a U creates a termination codon, resulting in a shortened form of apolipoprotein-B being synthesized in intestinal cells.

auxiliary protein factors, binds to a sequence immediately downstream of the modification position within the mRNA (Smith and Sowden, 1996).

Although not common, RNA editing occurs in a number of different organisms and includes a variety of different nucleotide changes (*Table 9.7* and Box 9.5). Deaminations of adenosine to inosine, carried out by enzymes called **double-stranded RNA adenosine deaminases (dsRADs)**, are particularly interesting. Some mRNAs are selectively edited at a limited number of positions, these positions apparently defined by double-stranded segments of the pre-mRNA, formed by base-pairing between the modification site and sequences from adjacent introns. This type of editing occurs, for example, during processing of mRNAs for mammalian glutamate receptors (Scott, 1997). Selective editing contrasts with the second type of modification carried out by dsRAD, in which the target molecules become extensively deaminated, over 50% of the adenosines in the RNA becoming converted to inosines. Hyperediting has so far been observed mainly, but not exclusively, with viral RNAs and is thought to occur by chance, these RNAs adopting base-paired structures that fortuitously act as substrates for dsRAD. It may, however, have physiological importance in the etiology of diseases caused by the edited viruses. This possibility is raised by the discovery that viral RNAs associated with persistent measles infections (as opposed to the more usual transient version of the disease) are hyperedited (Bass, 1997).

RNA editing appears, at first glance, to be a convenient way of obtaining two biochemical activities from a single gene, by providing a means of modifying the resulting protein in certain tissues. A more in-depth consideration of what is involved questions the convenience of the process. In order to obtain two versions of apolipoprotein-B from one gene, the human genome has to encode a gene for the modifying enzyme as well as additional genes for the protein factors involved in the modification process. In terms of 'convenience' the better choice would be to have two apolipoprotein-B genes. At present, the reason for RNA editing eludes us.

9.5 TURNOVER OF mRNAs

So far this chapter has concentrated on synthesis of RNAs. Their degradation is equally important, especially with regard to mRNAs whose presence or absence in the cell determines which proteins will be synthesized. Turnover of mRNAs is therefore a potentially important means of regulating genome expression.

The rate of turnover of an mRNA can be estimated by determining its half-life in the cell. The estimates show that there are considerable variations between and within organisms. Bacterial mRNAs are generally turned over very rapidly, their half-lives rarely being longer than a few minutes, a reflection of the rapid changes in protein synthesis patterns that can occur in an actively growing bacterium with a generation time of 20 minutes or so. Eukaryotic mRNAs are longer lived, with half-lives of, on average, 10–20 minutes for yeast and several hours for mammals. Within individual cells the variations are almost equally striking: some yeast mRNAs have half-lives of only one minute whereas for others the figure is more like 35 minutes (Tuite, 1996). These observations raise two questions: what are the processes for mRNA turnover, and how are these processes controlled?

Convincing answers are lacking for both questions. In bacteria, mRNAs are degraded by a multienzyme complex called the **degradosome** which includes endonucleases that cut the molecules at internal sites and exonucleases that sequentially remove nucleotides from

Table 9.7 Examples of RNA editing in mammals

Tissues	Target RNA	Change	Comments
Liver, intestine	Apolipoprotein-B mRNA	C→U	Converts a glutamic acid codon to a stop codon
Muscle	α-Galactosidase mRNA	U→A	Converts a phenylalanine codon into a tyrosine codon
Testis, tumors	Wilms tumor-1 mRNA	U→C	Converts a leucine codon into a proline codon
Tumors	Neurofibromatosis type-1 mRNA	C→U	Converts an arginine codon into a stop codon
Brain	Glutamate receptor mRNA	A→inosine	Multiple positions leading to various codon changes

Based on Smith and Sowden (1996) and Scott (1997). See also Box 9.5.

Box 9.5: More complex forms of RNA editing

The examples of RNA editing described in the text and in *Table 9.7* are relatively straightforward events that lead to nucleotide changes at a single or limited number of positions in selected mRNAs. More complex types of RNA editing are also known:

- **Pan-editing** involves the extensive insertion of nucleotides into abbreviated RNAs in order to produce functional molecules. It is particularly common in the mitochondria of trypanosomes, the parasitic protozoa that cause diseases such as sleeping sickness. Many of the RNAs transcribed in trypanosome mitochondria are specified by **cryptogenes**, sequences lacking many of the nucleotides present in the mature RNAs. The pre-RNAs transcribed from these cryptogenes are processed by multiple insertions of U nucleotides, at positions defined by short **guide RNAs**. These are short RNAs that can base-pair to the pre-RNA and which contain As at the positions where Us must be inserted (see figure opposite):

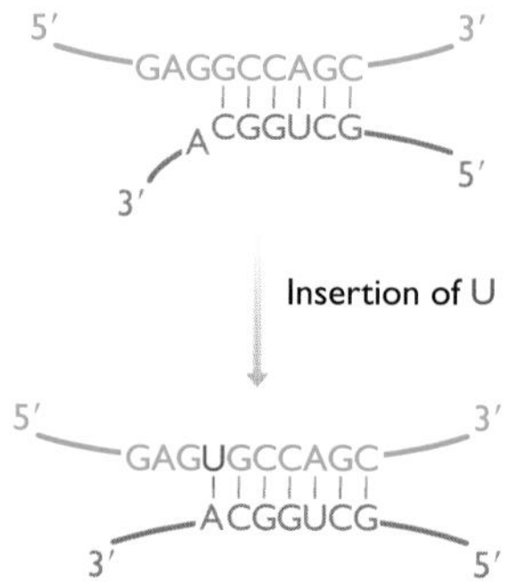

- Less extensive **insertional editing** occurs with some viral RNAs. For example, the paramyxovirus P gene gives rise to at least two different proteins because of the insertion of Gs at specific positions in the mRNA. These insertions are not specified by guide RNAs: instead they are added by the RNA polymerase as the mRNA is being synthesized.
- **Polyadenylation editing** is seen with many animal mitochondrial mRNAs. Five of the mRNAs transcribed from the human genome end with just a U or UA, rather than with one of the three termination codons (UAA, UAG in the human mitochondrial genetic code: see Section 10.1.2). Polyadenylation converts the terminal U or UA into UAAAA..., and so creates a termination codon. This is just one of several features that appear to have evolved in order to make vertebrate mitochondrial genomes as small as possible (see Box 10.2, p. 242).

the 5′- and 3′-ends (Py *et al.*, 1996). Stability is associated with an absence of endonuclease target sites and the presence of hairpin loops in the 3′ regions, the latter blocking the action of the exonucleases. This suggests that the stability of a bacterial mRNA is an inherent feature of its nucleotide sequence, which is a satisfyingly straightforward way of conferring turnover properties on an mRNA. To what extent, if at all, these turnover properties can be changed in order to regulate gene expression, is not known.

Among eukaryotes, most progress in understanding mRNA turnover has been made with yeast (Tuite, 1996). Two pathways exist:

- A **deadenylation-dependent pathway** (*Figure 9.26*), which is triggered by removal of the poly(A) tail, possibly by exonuclease cleavage or possibly by loss of the polyadenylate binding protein which stabilizes the tail (Section 9.2.2). Poly(A) tail removal is followed by cleavage of the 5′ cap and rapid exonuclease digestion from the 5′-end of the mRNA.
- A **deadenylation-independent pathway**, which is induced by the presence in the mRNA of a termination codon which prevents it being translated correctly. Cap cleavage and exonuclease degradation occurs even though the poly(A) tail is still in place.

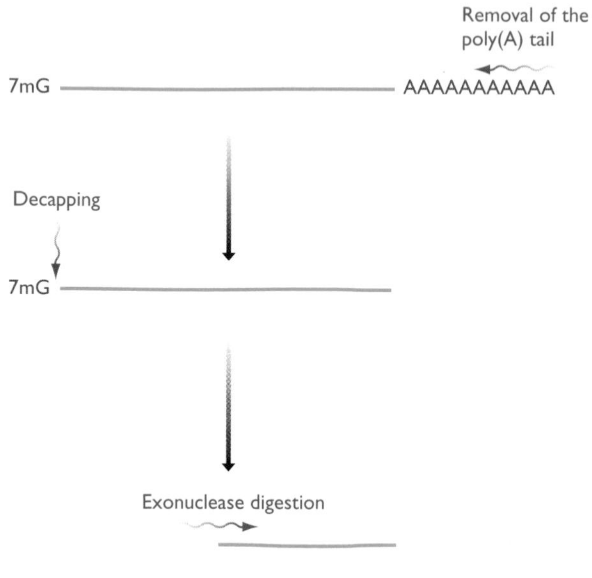

Figure 9.26 The deadenylation-dependent pathway for degradation of an mRNA.

Although this pathway is designed primarily to degrade mRNAs that have become altered by mutation or have been incorrectly spliced (Nagy and

Maquat, 1998), there is evidence that it is also involved in turnover of normal mRNAs.

Although equivalent pathways are poorly defined in higher eukaryotes, the available data suggest that loss of the poly(A) tail and removal of the cap are also preliminaries to mRNA degradation in these organisms. This raises the possibility that, as in bacteria, inherent features of the mRNA sequence, which may influence the stability of the poly(A) tail and cap, can confer a preset half-life on a particular mRNA. In mammalian mRNAs, internal sequences that appear to influence stability, possibly by forming a hairpin loop structure, have been identified. These may be binding sites for proteins that determine the stability of the mRNA and which, in turn, could respond to external stimuli.

REFERENCES

Alberts B, Bray D, Lewis J, Raff M, Roberts K and Watson JD (1994) *Molecular Biology of the Cell*, 3rd edn. Garland Publishing, New York.

Bachellerie J-P and Cavaillé J (1997) Guiding ribose methylation of rRNA. *Trends Biochem. Sci.*, **22**, 257–261.

Bass BL (1997) RNA editing and hypermutation by adenosine deamination. *Trends Biochem. Sci.*, **22**, 157–162.

Burke JM, Belfort M, Cech TR, Davies RW, Schweyen RJ, Shub DA, Szostak JW and Tabak HF (1987) Structural conventions for Group I introns. *Nucleic Acids Res.*, **15**, 7217–7221.

Cate JH, Gooding AR, Podell E, Zhou K, Golden BL, Kundrot CE, Cech TR and Doudna JA (1996) Crystal structure of a Group I ribozyme domain: principles of RNA packing. *Science*, **273**, 1678–1685.

Chabot B (1996) Directing alternative splicing: cast and scenarios. *Trends Genet.*, **12**, 472–478.

Colgan DF, Murthy KGK, Prives C and Manley JL (1996) Cell cycle regulation of poly(A) polymerase by phosphorylation. *Nature*, **384**, 282–285.

Copertino DW and Hallick RB (1993) Group II and Group III introns of twintrons: potential relationships with nuclear pre-mRNA introns. *Trends Biochem. Sci.*, **18**, 467–471.

Corden JL and Patturajan M (1997) A CTD function linking transcription to splicing. *Trends Biochem. Sci.*, **22**, 413–416.

Friedman DI, Imperiale MJ and Adhya SL (1987) RNA 3′ end formation in the control of gene expression. *Annu. Rev. Genet.*, **21**, 453–488.

Golden BL, Gooding AR, Podell ER and Cech TR (1998) A preorganized active site in the crystal structure of the *Tetrahymena* ribozyme. *Science*, **282**, 259–264.

Guo Z and Sherman F (1996) 3′-end forming signals of yeast mRNA. *Trends Biochem. Sci.*, **21**, 477–481.

Hirose Y and Manley JL (1998) RNA polymerase II is an essential mRNA polyadenylation factor. *Nature*, **395**, 93–96.

Jansa P, Mason SW, Hoffman-Rohrer U and Grummt I (1998) Cloning and functional characterization of PTRF, a novel protein which induces dissociation of paused ternary transcription complexes. *EMBO J.*, **17**, 2855–2864.

Lafontaine DLJ and Tollervey D (1998) Birth of the snoRNPs: the evolution of the modification-guide snoRNAs. *Trends Biochem. Sci.*, **23**, 383–388.

Landick R and Roberts JW (1996) The shrewd grasp of RNA polymerase. *Science*, **273**, 202–203.

Lygerou Z, Allmang C, Tollervey D and Séraphin B (1996) Accurate processing of a eukaryotic precursor ribosomal RNA by ribonuclease MRP *in vitro*. *Science*, **272**, 268–270.

Lykke-Andersen J, Aagaard C, Semionenkov M and Garrett RA (1997) Archaeal introns: splicing, intercellular mobility and evolution. *Trends Biochem. Sci.*, **22**, 326–331.

Maden BEH (1997) Eukaryotic ribosomal RNA: guides to 95 new angles. *Nature*, **389**, 129–131.

Manley JL and Takagaki Y (1996) The end of the message – another link between yeast and mammals. *Science*, **274**, 1481–1482.

McCracken S, Fong N, Yankulov K, Ballantyne S, Pan G, Greenblatt J, Patterson SD, Wickens M and Bentley DL (1997) The C-terminal domain of RNA polymerase II couples mRNA processing to transcription. *Nature*, **385**, 357–361.

Muto A, Ushida C and Himeno H (1998) A bacterial RNA that functions as both a tRNA and an mRNA. *Trends Biochem. Sci.*, **23**, 25–29.

Nagy E and Maquat LE (1998) A rule for termination-codon position within intron-containing genes: when nonsense affects RNA abundance. *Trends Biochem. Sci.*, **23**, 198–199.

Nigg EA (1997) Nucleocytoplasmic transport: signals, mechanisms and regulation. *Nature*, **386**, 779–787.

Nilsen TW (1996) A parallel spliceosome. *Science*, **273**, 1813.

Nitta I, Kamada Y, Noda H, Ueda T and Watanabe K (1998) Reconstitution of peptide bond formation with *Escherichia coli* 23S ribosomal RNA domains. *Science*, **281**, 666–669.

Nudler E, Avetissova E, Markovtsov V and Goldfarb A (1996) Transcription processivity: protein–DNA interactions holding together the elongation complex. *Science*, **273**, 211–217.

Nudler E, Gusarov I, Avetissova E, Kozlov M and Goldfarb A (1998) Spatial organization of transcription elongation complex in *Escherichia coli*. *Science*, **281**, 424–428.

Py B. Higgins CF, Krisch HM and Carpousis AJ (1996) A DEAD-box RNA helicase in the *Escherichia coli* RNA degradosome. *Nature*, **381**, 169–172.

Reeder RH and Lang WH (1997) Terminating transcription in eukaryotes: lessons learned from RNA polymerase I. *Trends Biochem. Sci.*, **22**, 473–477.

Reines D, Conaway JW and Conaway RC (1996) The RNA polymerase II general elongation factors. *Trends Biochem. Sci.*, **21**, 351–355.

Scott J (1997) RNA editing: message change for a fat controller. *Nature*, **387**, 242–243.

Scott WG and Klug A (1996) Ribozymes: structure and mechanism in RNA catalysis. *Trends Biochem. Sci.*, **21**, 220–224.

Smith HC and Sowden MP (1996) Base-modification mRNA editing through deamination – the good, the bad and the unregulated. *Trends Genet.*, **12**, 418–424.

Strachan T and Read AP (1996) *Human Molecular Genetics*. BIOS Scientific Publishers, Oxford.

Studitsky VM, Kassavetis GA, Gieduschek EP and Felsenfeld G (1997) Mechanism of transcription through the nucleosome by eukaryotic RNA polymerase. *Science*, **278**, 1960–1963.

Tarn W-Y and Steitz JA (1996) Highly diverged U4 and U6 small nuclear RNAs required for splicing rare AT-AC introns. *Science*, **273**, 1824–1832.

Tarn W-Y and Steitz JA (1997) Pre-mRNA splicing: the discovery of a new spliceosome doubles the challenge. *Trends Biochem. Sci.*, **22**, 132–137.

Tollervey D (1996) Small nucleolar RNAs guide ribosomal RNA methylation. *Science*, **273**, 1056–1057.

Tuite MF (1996) RNA processing: death by decapitation for mRNA. *Nature*, **382**, 577–579.

Turner PC, McLennan AG, Bates AD and White MRH (1997) *Instant Notes in Molecular Biology*. BIOS Scientific Publishers, Oxford.

Valcárcel J and Green MR (1996) The SR protein family: pleiotropic functions in pre-mRNA splicing. *Trends Biochem. Sci.*, **21**, 296–301.

Velculescu VE, Zhang L, Zhou W, Vogelstein J, Basrai MA, Bassett DE, Hieter P, Vogelstein B and Kinzler KW (1997) Characterization of the yeast transcriptome. *Cell*, **88**, 243–251.

von Hippel PH (1998) An integrated model of the transcription complex in elongation, termination and editing. *Science*, **281**, 660–665.

Weis K (1998) Importins and exportins: how to get in and out of the nucleus. *Trends Biochem. Sci.*, **23**, 185–189.

FURTHER READING

Beebee TJC and Burke J (1992) *Gene Structure and Transcription: In Focus*. IRL Press, Oxford. — *A good general description of the topic.*

Cech TR (1990) Self-splicing of group I introns. *Annu. Rev. Biochem.*, **59**, 543–568. — *Written by one of the discoverers of autocatalytic RNA.*

Henkin TM (1996) Control of transcription termination in prokaryotes. *Annu. Rev. Genet.*, **30**, 35–57. — *Detailed accounts of antitermination and attenuation.*

Maxwell ES and Fournier MJ (1995) The small nucleolar RNAs. *Annu. Rev. Biochem.*, **64**, 897–934.

Nagai K and Mattaj IW (eds) (1994) *RNA–Protein Interactions*. IRL Press, Oxford. — *Chapter 7 describes RNA–protein interactions in snRNPs.*

Richardson JP (1993) Transcription termination. *Crit. Rev. Biochem. Mol. Biol.*, **28**, 1–30.

Uptain SM, Kane CM and Chamberlin MJ (1997) Basic mechanisms of transcript elongation and its regulation. *Annu. Rev. Biochem.*, **66**, 117–172. — *A particularly useful source of information on regulation of transcription elongation in both bacteria and eukaryotes.*

Young RA (1991) RNA polymerase II. *Annu. Rev. Biochem.*, **60**, 689–715.

CHAPTER

10 Synthesis and Processing of the Proteome

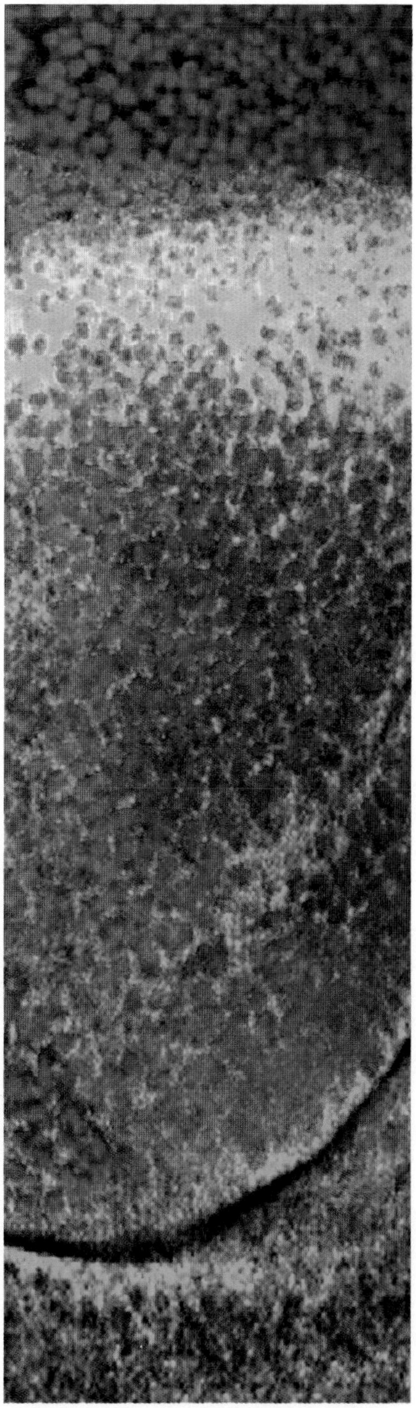

Contents

10.1 *The Role of tRNA in Protein Synthesis* 232
10.1.1 Aminoacylation: the attachment of amino acids to tRNAs 232
10.1.2 Codon–anticodon interactions: the attachment of tRNAs to mRNA 235
10.2 *The Role of the Ribosome in Protein Synthesis* 240
10.2.1 Ribosome structure 242
10.2.2 Translation in detail 243
10.3 *Post-translational Processing of Proteins* 248
10.3.1 Protein folding 250
10.3.2 Processing by proteolytic cleavage 253
10.3.3 Processing by chemical modification 255
10.3.4 Inteins 256
10.4 *Protein Turnover* 258
10.4.1 Degradation of ubiquitin-tagged proteins in the proteasome 259

Concepts

- *tRNAs form the link between the mRNA and the polypeptide being synthesized*
- *The genetic code specifies the amino acid sequence of the polypeptide*
- *Aminoacyl-tRNA synthetases attach amino acids to tRNAs which then base-pair to the appropriate codons*
- *Ribosomes coordinate and catalyze protein synthesis*
- *Bacteria and eukaryotes use different mechanisms for initiation of translation, but later stages of the process are similar in all organisms*
- *In order to become functional, all proteins must be folded, some must be cut into segments and chemically modified, and a few must have their inteins removed*
- *Protein turnover results in degradation of proteins whose functions are no longer needed*

THE END RESULT of genome expression is the proteome, the collection of functioning proteins in a living cell. The identity and relative abundances of the individual proteins in a proteome represents a balance between the synthesis of new proteins and the degradation of existing ones. The biochemical capabilities of the proteome can also be changed by chemical modification and other processing events. The combination of synthesis, degradation, and modification/processing enables the proteome to meet the changing requirements of the cell and to respond to external stimuli.

In this chapter we will study the synthesis and processing of the proteome. To understand its synthesis we will examine the role of tRNAs in decoding the genetic code and investigate the events, occurring at the ribosome, that result in polymerization of amino acids into polypeptides. The ribosomal events are sometimes looked on as the final stage in gene expression but the polypeptide that is initially synthesized is inactive until it has been folded, and may also have to undergo cutting and chemical modification before it becomes functional. We will study these protein processing events in Section 10.3. At the end of the chapter we will investigate how the cell degrades proteins that it no longer requires.

10.1 THE ROLE OF tRNA IN PROTEIN SYNTHESIS

Transfer RNAs play the central role in translation. They are the adaptor molecules, whose existence was predicted by Francis Crick in 1956 (Crick, 1990), which form the link between the mRNA and the polypeptide that is being synthesized. This is both a *physical* link, tRNAs binding to both the mRNA and the growing polypeptide, and an *informational* link, tRNAs ensuring that the polypeptide being synthesized has the amino acid sequence that is denoted, via the genetic code, by the sequence of nucleotides in the mRNA (*Figure 10.1*). To understand how tRNAs play this dual role we must examine **aminoacylation**, the process by which the correct amino acid is attached to each tRNA, and **codon–anticodon recognition**, the interaction between tRNA and mRNA.

10.1.1 Aminoacylation: the attachment of amino acids to tRNAs

Bacteria contain 30–45 different tRNAs and eukaryotes have up to 50. As only 20 amino acids are designated by the genetic code, this means that all organisms have at least some **isoaccepting tRNAs**, different tRNAs that are

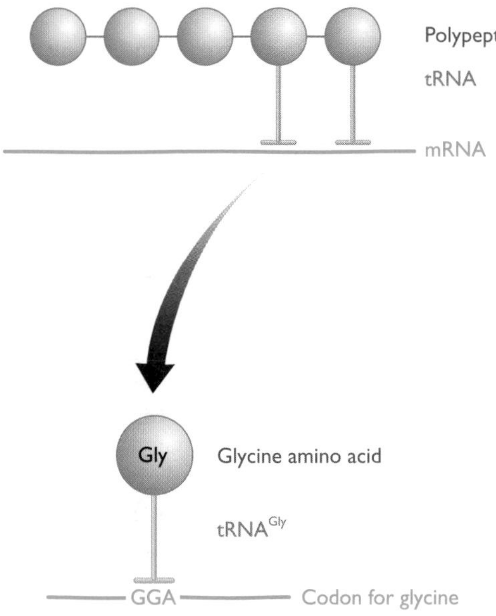

Figure 10.1 The adaptor role of tRNA in translation.

In the top drawing, the physical role of tRNA is shown, forming an attachment between the polypeptide and the mRNA. The lower drawing shows the informational link, the tRNA carrying the amino acid specified by the triplet codon to which it attaches.

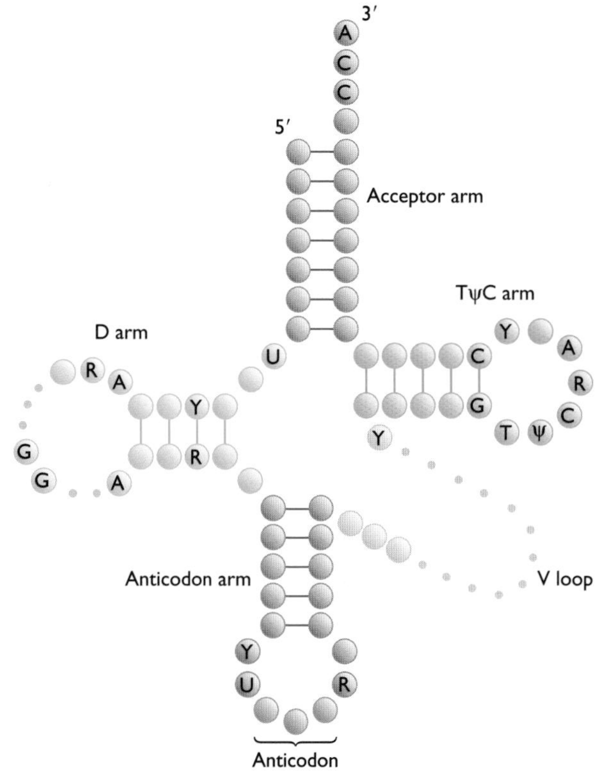

Figure 10.2 The cloverleaf structure of a tRNA.

The tRNA is drawn in the conventional cloverleaf structure with the different components labeled. Invariant nucleotides are indicated (A, C, G, T, U, Ψ, where Ψ = pseudouridine) as are semi-invariant ones. (Abbreviations: R, purine; Y, pyrimidine). Optional nucleotides not present in all tRNAs are shown as smaller dots. The standard numbering system places position 1 at the 5′-end and position 76 at the 3′-end; it includes some but not all of the optional nucleotides. The invariant and semi-invariant nucleotides are at positions 8, 11, 14, 15, 18, 19, 21, 24, 32, 33, 37, 48, 53, 54, 55, 56, 57, 58, 60, 61, 74, 75 and 76. The nucleotides of the anticodon are at positions 34, 35 and 36.

specific for the same amino acid. The terminology used when describing tRNAs is to indicate the amino acid specificity with a superscript suffix, using the numbers 1, 2, etc. to distinguish different isoacceptors: for example, two tRNAs specific for glycine would be written as $tRNA^{Gly1}$ and $tRNA^{Gly2}$.

All tRNAs have a similar structure

Transfer RNAs are the shortest functional RNAs known: the smallest are only 74 nucleotides in length, and the largest are rarely more than 90 nucleotides. Because of their small size, and because it is possible to purify individual tRNAs, they were among the first nucleic acids to be sequenced, way back in 1965 by Robert Holley's group at Cornell University, New York. The sequences revealed one unexpected feature, that as well as the standard RNA nucleotides (A, C, G and U), tRNAs contain a number of modified nucleotides, 5–10 in any particular tRNA with over 50 different modifications known altogether (Section 9.4).

Examination of the first tRNA sequence, for $tRNA^{Ala}$ of *Saccharomyces cerevisiae*, showed that the molecule could adopt various base-paired secondary structures. After more tRNAs had been sequenced, it became clear that one particular structure could be taken up by all of them. This is the **cloverleaf** (*Figure 10.2*), the features of which are:

- The **acceptor arm**, formed by 7 base pairs between the 5′ and 3′ ends of the molecule. The amino acid is attached to the extreme 3′-end of the tRNA, to the adenosine of the invariant CCA terminal sequence (Section 9.3.2).
- The **D arm**, named after the modified nucleoside dihydrouridine (see *Table 9.6*, p. 224), which is always present in this structure.
- The **anticodon arm**, containing the triplet of nucleotides called the **anticodon** which base-pair with the mRNA during translation.
- The **V loop**, which contains 3–5 nucleotides in Class 1 tRNAs or 13–21 nucleotides in Class 2 tRNAs.
- The **TΨC arm**, named after the sequence thymidine–pseudouridine–cytosine, which it always contains.

The cloverleaf structure can be formed by virtually all tRNAs, the main exceptions being the tRNAs used in vertebrate mitochondria, which are coded by the mitochondrial genome and which sometimes lack parts of the structure. An example is the human mitochondrial

$tRNA^{Ser}$, which has no D arm. As well as the conserved secondary structure, the identities of nucleotides at some positions are completely invariant (always the same nucleotide) or semi-invariant (always a purine or always a pyrimidine), and the positions of the modified nucleotides are almost always the same.

Many of the invariant nucleotide positions are important in the tertiary structure of tRNA. X-ray crystallography studies have shown that nucleotides in the D and TΨC loops form base pairs that fold the tRNA into a compact, L-shaped structure (*Figure 10.3*). Each arm of the L-shape is approximately 7 nm long and 2 nm in diameter, with the amino acid binding site at the end of one arm and the anticodon at the end of the other. The additional base-pairing means that the base-stacking (see p. 147) is almost continuous from one end of the tRNA to the other, providing stability to the structure.

Aminoacyl-tRNA synthetases attach amino acids to tRNAs

The aminoacylation of tRNAs – 'charging' in molecular biology jargon – is the function of the group of enzymes called **aminoacyl-tRNA synthetases**. The chemical reaction that results in aminoacylation occurs in two steps. An activated amino acid intermediate is first formed by reaction with ATP, and then the amino acid is transferred to the 3′-end of the tRNA, the link being formed between the -COOH group of the amino acid and the -OH group attached to either the 2′ or 3′ carbon on the sugar of the last nucleotide, which is always an A (*Figure 10.4*).

With few exceptions, organisms have 20 aminoacyl-tRNA synthetases, one for each amino acid. This means that groups of isoaccepting tRNAs are aminoacylated by a single enzyme. Although the basic chemical reaction is the same for each amino acid, the twenty aminoacyl-tRNA synthetases fall into two distinct groups, Class I and Class II, with several important differences between them (*Table 10.1*). In particular, Class I enzymes attach the amino acid to the 2′-OH group of the terminal nucleotide of the tRNA, whereas Class II enzymes attach the amino acid to the 3′-OH group (Arnez and Moras, 1997).

Aminoacylation must be carried out accurately: the correct amino acid must be attached to the correct tRNA if the rules of the genetic code are to be followed during protein synthesis. It appears that an aminoacyl-tRNA synthetase has high fidelity for its tRNA, due to an extensive interaction between the two, covering some 25 nm^2 of surface area and involving the acceptor arm and anticodon loop of the tRNA, as well as individual nucleotides in the D and TΨC arms (Arnez and Moras, 1997). The interaction between enzyme and amino acid is, of necessity, less extensive, amino acids being much smaller than tRNAs, and presents greater problems with regard to specificity as several pairs of amino acids are structurally similar. Errors do therefore occur, at a very low rate for most amino acids but possibly as frequently as one aminoacylation in 80 for difficult pairs such as isoleucine

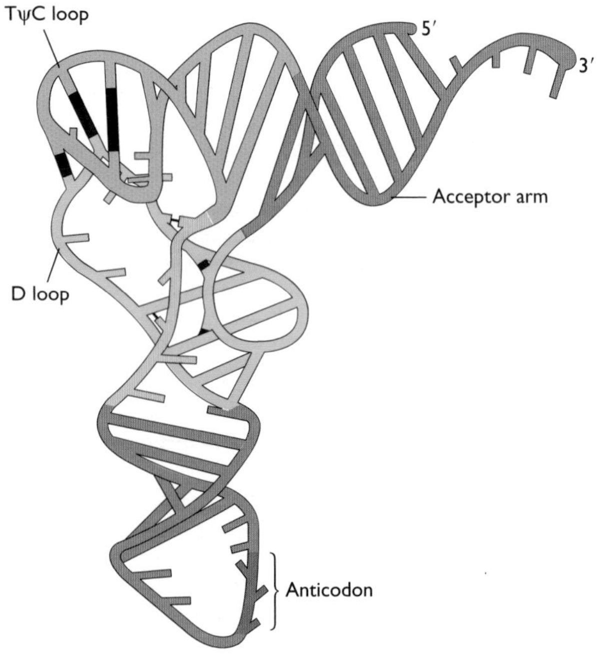

Figure 10.3 The three-dimensional structure of a tRNA.

Additional base pairs, shown in black and mainly between the D and TΨC loops, fold the cloverleaf structure shown in *Figure 10.2* into this L-shaped configuration. Depending on its sequence, the V loop might also form interactions with the D arm, as indicated by thin black lines. From Freifelder D, *Molecular Biology*, 2nd edn, 1986, Sudbury, MA: Jones and Bartlett Publishers. www.jbpub.com. Reprinted with permission.

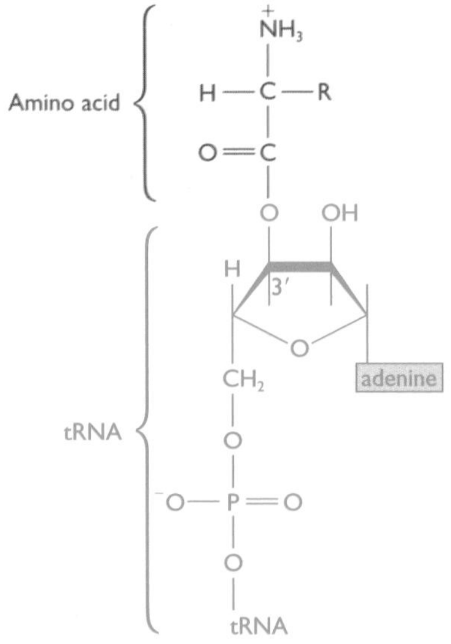

Figure 10.4 Aminoacylation of a tRNA.

The result of aminoacylation by a Class II aminoacyl-tRNA synthetase is shown, the amino acid being attached via its -COOH group to the 3′-OH of the terminal nucleotide of the tRNA. A Class I aminoacyl-tRNA synthetase attaches the amino acid to the 2′-OH group.

Table 10.1 Features of aminoacyl-tRNA synthetases

Feature	Class I enzymes	Class II enzymes
Structure of the enzyme active site	Parallel β-sheet	Antiparallel β-sheet
Interaction with the tRNA	Minor groove of the acceptor stem	Major groove of the acceptor stem
Orientation of the bound tRNA	V loop faces away from the enzyme	V loop faces the enzyme
Amino acid attachment	To the 2′-OH of the terminal nucleotide of the tRNA	To the 3′-OH of the terminal nucleotide of the tRNA

For more details see Arnez and Moras (1997).

and valine. Most errors are corrected by the aminoacyl-tRNA synthetase itself, by an editing process that appears to be distinct from aminoacylation as it involves different contacts with the tRNA (Hale *et al.*, 1997).

In most organisms, aminoacylation is carried out by the process just described, but a few unusual events have been documented. These include a number of instances where the aminoacyl-tRNA synthetase attaches the incorrect amino acid to a tRNA, this amino acid subsequently being transformed into the correct one by a second, separate chemical reaction. This was first discovered in the bacterium *Bacillus megaterium* for synthesis of glutamine-$tRNA^{Gln}$ (i.e. glutamine attached to its tRNA). This aminoacylation is carried out by the enzyme responsible for synthesis of glutamic acid-$tRNA^{Glu}$, and initially results in attachment of a glutamic acid to the $tRNA^{Gln}$ (*Figure 10.5A*). This glutamic acid is then converted to glutamine by transamidation by a second enzyme. The same process is used by other bacteria (though not *E. coli*) and by the archaea. Some archaea also use transamidation to synthesize asparagine-$tRNA^{Asn}$ from aspartic acid-$tRNA^{Asn}$ (Ibba *et al.*, 1997). There are two other examples where an attached amino acid is modified after aminoacylation, involving two systems that we will return to later in this chapter. The first is the conversion of methionine to *N*-formylmethionine (*Figure 10.5B*), producing the special tRNA used in initiation of bacterial translation (Section 10.2.2), and the second leads to synthesis of a tRNA charged with the unusual amino acid selenocysteine (*Figure 10.5C*; Section 10.1.2).

10.1.2 Codon–anticodon interactions: the attachment of tRNAs to mRNA

Aminoacylation represents the first level of specificity displayed by a tRNA. The second level is the specificity of the interaction between the anticodon of the tRNA and the mRNA being translated. This specificity ensures that protein synthesis follows the rules of the **genetic code**.

The genetic code specifies how an mRNA sequence is translated into a polypeptide

It was recognized in the 1950s that a triplet genetic code – one in which each codeword or **codon** comprises three nucleotides – is required to account for all 20 amino acids found in proteins. A two-letter code would have only $4^2 = 16$ codons, whereas a three-letter code would give $4^3 = 64$ codons. It was also assumed, as a working hypothesis, that mRNAs contain nonoverlapping series of codons that are colinear with the polypeptides they encode (*Figure 10.6*). Although experimental proof of these assumptions was difficult to obtain, all three turned out to be correct. There are qualifications, such as the lack of strict colinearity displayed by many eukaryotic genes because of the presence of introns, a complication that was not appreciated until introns were discovered in 1977, long after the main work on the genetic code was carried out. This work was completed in the mid-1960s when the meanings of all 64 codons were determined, partly by analysis of polypeptides resulting from translation of artificial mRNAs of known or predictable sequence in cell-free protein-synthesizing systems, and partly by determining which aminoacyl-tRNAs associate with which RNA sequences in a reconstituted ribosome assay (Research Briefing 10.1).

The 64 codons of the genetic code (*Figure 10.7*) fall into groups, the members of each group coding for the same amino acid. Only tryptophan and methionine have just a single codon each, all others are coded by two, three, four or six codons. This feature of the code is called **degeneracy**. The code has four punctuation codons: the initiation codon 5′-AUG-3′, which also specifies methionine (though with a few genes other codons such as 5′-GUG-3′ and 5′-UUG-3′ are used as the initiation codon), and three termination codons, 5′-UAG-3′, 5′-UAA-3′ and 5′-UGA-3′. The termination codons are sometimes called amber, opal and ochre, respectively, these being the whimsical names given to the original *E. coli* mutants whose analysis led to their discovery.

An important feature of the genetic code is that it is not universal. The code shown in *Figure 10.7* holds for the vast majority of genes in the vast majority of organisms, but three types of deviation have been discovered (*Table 10.2*):

- **Mitochondrial genetic codes** have variations that reduce the number of tRNAs that are needed. This unusual feature of mitochondrial genes was first discovered in 1979 by Frederick Sanger's group in Cambridge, UK, who found that several human mitochondrial genes contain 5′-UGA-3′ codons,

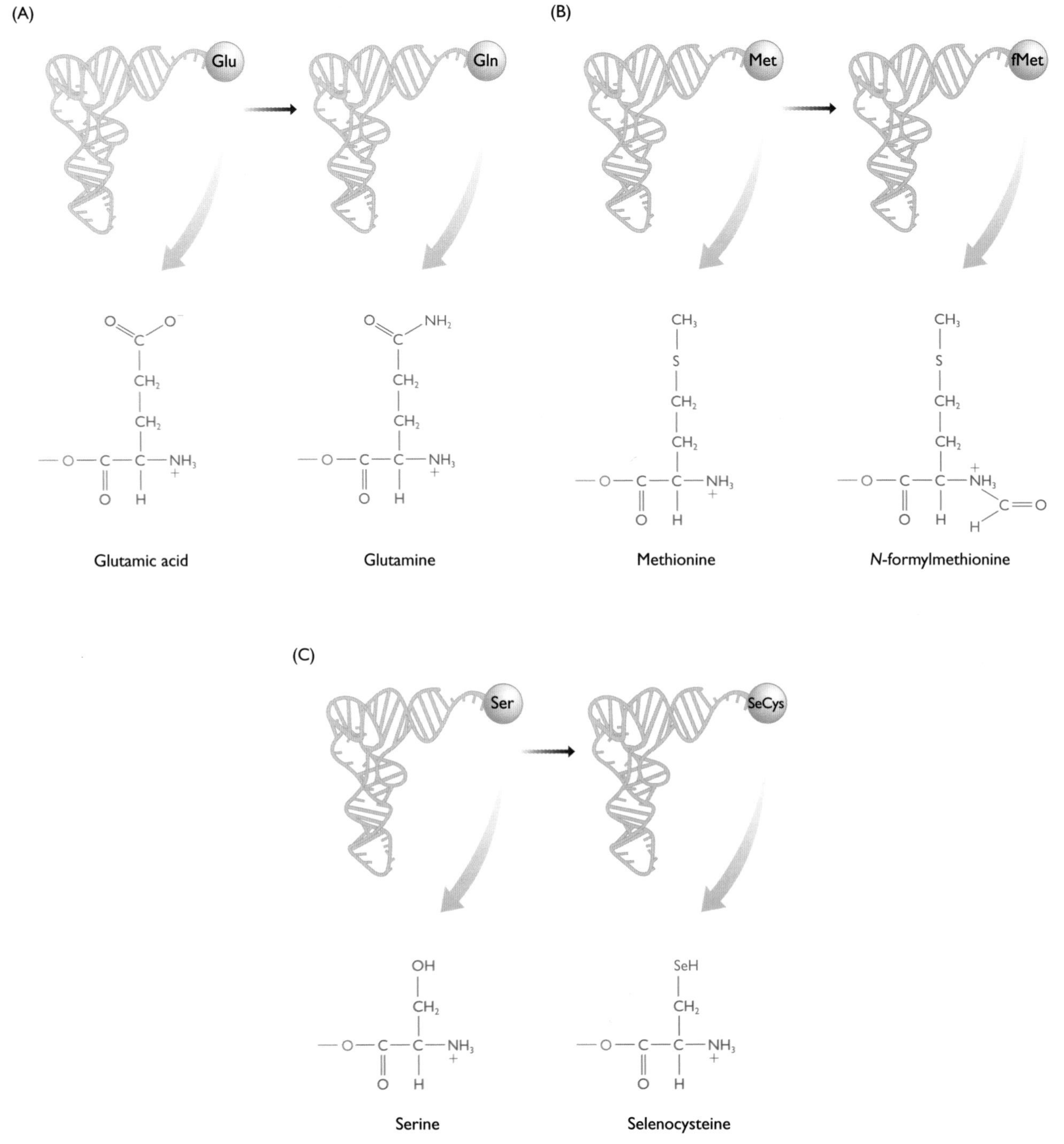

Figure 10.5 Unusual types of aminoacylation.

(A) In some bacteria, $tRNA^{Gln}$ is aminoacylated with glutamic acid which is then converted to glutamine by transamidation. (B) The special tRNA used in initiation of translation in *Escherichia coli* is aminoacylated with methionine which is then converted to *N*-formylmethionine. (C) $tRNA^{SeCys}$ in various organisms is initially aminoacylated with serine.

normally coding for termination, at internal positions where translation was not expected to stop. Comparisons with the amino acid sequences of the proteins coded by these genes showed that 5′-UGA-3′ is a tryptophan codon in human mitochondria, and that this is just one of four code deviations in this particular genetic system (see *Table 10.2*). Mitochondrial genes in other organisms also display code deviations, though at least one of these, the use of 5′-CGG-3′ as a tryptophan codon in plant mitochondria, is probably corrected by RNA editing (Section 9.4.2) before translation occurs (Covello and Gray,

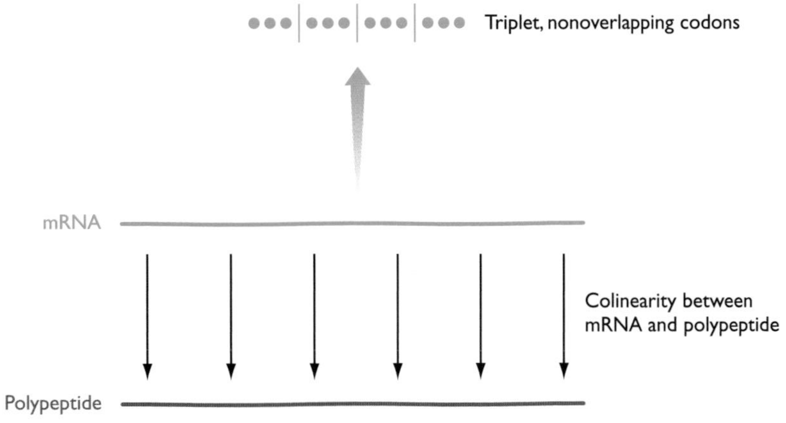

Figure 10.6 Early assumptions about the genetic code.

It was assumed that the genetic code is triplet, that codons do not overlap, and that there is a colinear relationship between the sequences of an mRNA and the polypeptide it encodes.

1989; Gualberto *et al.*, 1989). The purpose of these modified genetic codes appears to be to reduce the number of tRNAs needed in mitochondrial protein synthesis, by allowing 'superwobble' to occur, as described in the next section.

- **Nuclear genetic codes** of lower eukaryotes display a variety of modifications, often restricted to just a small group of organisms and frequently involving reassignment of the termination codons (see *Table 10.2*). Modifications are less common among prokaryotes but one example is known in *Mycoplasma* species. The reasons for these changes are not understood.
- **Context-dependent codon reassignment** occurs when the protein to be synthesized contains selenocysteine. This applies to many organisms, both prokaryotes and eukaryotes, including humans, because selenoproteins are widespread (*Table 10.3*). Selenocysteine (see *Figure 7.10*, p. 155) is coded by 5′-UGA-3′, which therefore has two meanings because it is still used as a termination codon in the organisms concerned. A special $tRNA^{SeCys}$ is aminoacylated initially with serine which is subsequently converted to selenocysteine (see *Figure 10.5C*). 5′-UGA-3′ codons that specify selenocysteine are distinguished from those which are true termination codons by the presence of a hairpin loop structure in the mRNA, positioned just downstream of the selenocysteine

UUU	phe	UCU	ser	UAU	tyr	UGU	cys
UUC	phe	UCC	ser	UAC	tyr	UGC	cys
UUA	leu	UCA	ser	UAA	stop	UGA	stop
UUG	leu	UCG	ser	UAG	stop	UGG	trp
CUU	leu	CCU	pro	CAU	his	CGU	arg
CUC	leu	CCC	pro	CAC	his	CGC	arg
CUA	leu	CCA	pro	CAA	gln	CGA	arg
CUG	leu	CCG	pro	CAG	gln	CGG	arg
AUU	ile	ACU	thr	AAU	asn	AGU	ser
AUC	ile	ACC	thr	AAC	asn	AGC	ser
AUA	ile	ACA	thr	AAA	lys	AGA	arg
AUG	met	ACG	thr	AAG	lys	AGG	arg
GUU	val	GCU	ala	GAU	asp	GGU	gly
GUC	val	GCC	ala	GAC	asp	GGC	gly
GUA	val	GCA	ala	GAA	glu	GGA	gly
GUG	val	GCG	ala	GAG	glu	GGG	gly

Figure 10.7 The genetic code.

See *Table 7.3*, p. 155, for the three-letter abbreviations of the amino acids.

RESEARCH BRIEFING 10.1

Elucidation of the genetic code

Assigning amino acids to triplet codons was the main objective of molecular biology during the first half of the 1960s.

Once it had been accepted that the genetic code is triplet and nonoverlapping, and that genes display colinearity with the proteins that they specify, attention turned to elucidation of the code with the objective of assigning amino acids to individual codons. Two types of experiment enabled the *Escherichia coli* code to be completely worked out during the years 1961–1966.

Cell-free translation of artificial RNAs

The first type of experiment was pioneered by Marshall Nirenberg and Heinrich Matthaei at the National Institutes of Health, Maryland, and made use of an extract, prepared from *E. coli* cells, that contained all the components needed to carry out translation except for mRNA. This **cell-free protein synthesizing system** therefore only made protein when RNA was added. If the sequence of the added RNA was known or could be predicted then the composition of the proteins that were made could be used to assign codons to amino acids.

The system was first used with the simplest artificial RNAs, those that contain just a single nucleotide, such as the **homopolymer** poly(U), whose sequence is 5′-UUU...UUU-3′. When added into the cell-free system poly(U) directed synthesis of polyphenylalanine, showing that the codon 5′-UUU-3′ codes for phenylalanine. Equivalent experiments enabled 5′-AAA-3′ to be assigned to lysine and 5′-CCC-3′ to proline; for unexplained reasons poly(G) gave no protein product.

Reliable methods for sequencing RNA were not developed until the late 1960s. This meant that when the cell-free experiments were performed it was not possible to determine the exact sequences of artificial **heteropolymers**, RNAs containing more than one nucleotide. These could still, however, be used in the cell-free system because their codon compositions could be deduced statistically from the identity and relative amounts of the nucleotides used in the reaction mixture from which each heteropolymer was made. Random heteropolymers of different compositions enabled amino acids to be assigned to over half the codons in the genetic code. A few more codons were assigned when the technique for synthesis of artificial RNAs was refined so that ordered heteropolymers could be synthesized. These RNAs are polymerized from dinucleotides such as 5′-GC-3′ and so have predictable sequences, the ordered heteropolymer poly(GC) having the sequence 5′-GCGC...GCGC-3′ and and therefore containing just two codons, 5′-GCG-3′ and 5′-CGC-3′.

The triplet binding assay

The genetic code could not be completed with the standard cell-free protein synthesizing system because it was simply not possible to devise random and ordered heteropolymers that enabled every codon to be assigned unambiguously. A new approach was therefore needed. This was the triplet binding assay, devised by Nirenberg and Philip Leder in 1964 and based on their discovery that a ribosome will attach to an RNA triplet if the appropriate aminoacyl-tRNA is also present. The code could therefore be completed, and all previously assigned codons checked, by synthesizing triplets of all possible sequences and testing each one individually with different aminoacyl-tRNAs. The final remaining ambiguity concerned the termination codons, which could not be directly identified by the triplet binding assay as it could not be proven that the inability of 5′-UAA-3′, 5′-UAG-3′ and 5′-UGA-3′ to bind any aminoacyl-tRNA was not due simply to deficiencies in this assay system. Confirmation that these are the termination codons was provided by Sydney Brenner and coworkers at Cambridge, UK, through genetic analysis of **suppressor mutations**.

References

Nirenberg MW and Matthaei H (1961) The dependence of cell-free protein synthesis in *E. coli* upon naturally occurring or synthetic polyribonucleic acids. *Proc. Natl Acad. Sci. USA*, **47**, 1588–1602.

Nirenberg MW and Leder P (1964) RNA codewords and protein synthesis. *Science*, **145**, 1399–1407.

Brenner S, Stretton AOW and Kaplan S (1965) Genetic code: the 'nonsense' triplets for chain termination and their suppression. *Nature*, **206**, 994–998.

codon in prokaryotes and in the 3′ untranslated region (i.e. the part of the mRNA after the termination codon) in eukaryotes. Recognition of the codon requires interaction between the hairpin and a special selenocysteine-specific translation elongation factor (Section 10.2.2; Low and Berry, 1996).

Codon–anticodon recognition is made flexible by wobble

In principle, codon–anticodon recognition is a straightforward process involving base-pairing between the anticodon of the tRNA and a codon in the mRNA (*Figure 10.8*). The specificity of aminoacylation ensures that the

Table 10.2 Deviations from the standard genetic code

Organism	Codon	Should code for	Actually codes for
Mitochondrial genomes			
Mammals	UGA	Stop	Trp
	AGA, AGG	Arg	Stop
	AUA	Ile	Met
Drosophila	UGA	Stop	Trp
	AGA	Arg	Ser
	AUA	Ile	Met
Saccharomyces cerevisiae	UGA	Stop	Trp
	CUN	Leu	Thr
	AUA	Ile	Met
Fungi	UGA	Stop	Trp
Maize	CGG	Arg	Trp
Nuclear and prokaryotic genomes			
Some protozoa	UAA, UAG	Stop	Gln
Candida cylindracea	CUG	Leu	Ser
Mycoplasma sp.	UGA	Stop	Trp
Context-dependent codon reassignments			
Various	UGA	Stop	Selenocysteine

Abbreviation: N, any nucleotide.

tRNA carries the amino acid denoted by the codon that it pairs with, and the ribosome controls the topology of the interaction in such a way that only a single triplet of nucleotides is available for pairing. Because base-paired polynucleotides are always antiparallel, and because the mRNA is read in the 5′→3′ direction, the first nucleotide of the codon pairs with nucleotide 36 of the tRNA, the second with nucleotide 35, and the third with nucleotide 34.

In practice, codon recognition is complicated by the possibility of **wobble**. This is another of the principles of gene expression originally proposed by Crick and subsequently shown to be correct. Because the anticodon is in a loop of RNA, the triplet of nucleotides is slightly curved (see *Figures 10.2* and *10.3*, pp. 233 and 234) and so cannot make an entirely uniform alignment with the anticodon. As a result, a nonstandard base pair can form between the

Table 10.3 Examples of proteins that contain selenocysteine

Enzyme	Organisms
Prokaryotic enzymes	
Formate dehydrogenase	*Clostridium thermoaceticum, Clostridium thermoautotrophicum, Enterobacter aerogenes, Escherichia coli, Methanococcus vaniellii*
Glycine reductase	*Clostridium purinolyticum, Clostridium sticklandii*
NiFeSe hydrogenase	*Desulfomicrobium baculatum, Methanococcus voltae*
Eukaryotic enzymes	
Glutathione peroxidase	Human, cow, rat, mouse
Selenoprotein P	Human, cow, rat
Selenoprotein W	Rat
Type 1 deiodinase	Human, rat, mouse, dog
Type 2 deiodinase	Frog
Type 3 deiodinase	Human, rat, frog

See Low and Berry (1996).

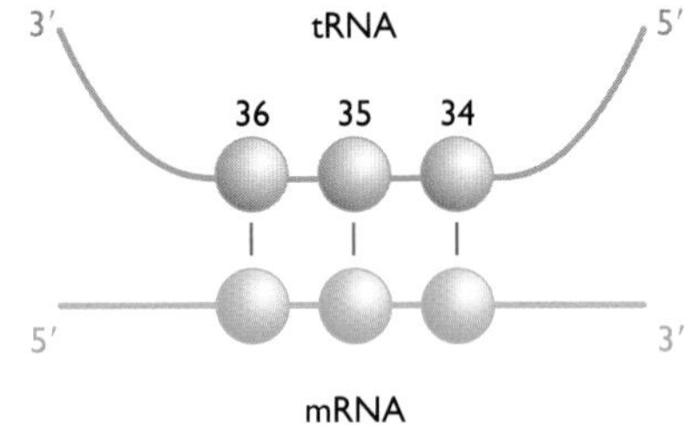

Figure 10.8 The interaction between a codon and an anticodon.

third nucleotide of the codon and the first nucleotide (number 34) of the anticodon. This is called 'wobble'. A variety of pairings are possible, especially if the nucleotide at position 34 is modified. The two main possibilities are:

- **G–U base-pairs** are permitted. This means that an anticodon with the sequence 3′-xxG-5′ can base-pair with both 5′-xxC-3′ and 5′-xxU-3′. Similarly, the anticodon 3′-xxU-5′ can base-pair with both 5′-xxA-3′ and 5′-xxG-3′. The consequence is that, rather than needing a different tRNA for each codon, the four members of a codon family (e.g. 5′-GCN-3′, all coding for alanine) can be decoded by just two tRNAs (*Figure 10.9A*).
- **Inosine**, abbreviated to I, is a modified version of guanosine that can base-pair with A, C and U. Inosine can only occur in the tRNA because the mRNA is not modified in this way. The triplet 3′-UAI-5′ is sometimes used as the anticodon in a $tRNA^{Ile}$ molecule because it pairs with 5′-AUA-3′, 5′-AUC-3′ and 5′-AUU-3′ (*Figure 10.9B*), which form the three-codon family for this amino acid in the standard genetic code.

Wobble reduces the number of tRNAs needed in a cell by enabling one tRNA to read two or possibly three codons. Hence bacteria can decode their mRNAs with as few as 30 tRNAs. Some mitochondria can get by with even fewer: human mitochondria, for example, use only 22 tRNAs. With these tRNAs the nucleotide in the wobble position of the anticodon is virtually redundant because it can base-pair with any nucleotide, enabling all four codons of a family to be recognized by the same tRNA. This phenomenon has been called **superwobble**. It appears to be part of a general drive by vertebrate mitochondrial genomes to be as compact as possible.

10.2 THE ROLE OF THE RIBOSOME IN PROTEIN SYNTHESIS

An *E. coli* cell contains approximately 20 000 ribosomes, distributed throughout its cytoplasm. The average human cell contains rather more (nobody has ever counted them all), some free in the cytoplasm and some attached to the outer surface of the endoplasmic reticulum, the membranous network of tubes and vesicles that permeates the cell. Originally, ribosomes were looked on as passive partners in protein synthesis, merely the structures on which translation occurs. This view has changed over the years and ribosomes are now considered to play two active roles in protein synthesis:

- Ribosomes *coordinate* protein synthesis by placing the mRNA, aminoacyl-tRNAs and associated protein factors in their correct positions relative to one another.
- Components of ribosomes *catalyze* at least some of the chemical reactions occurring during translation.

To understand how ribosomes play these roles we will first survey the structural features of ribosomes in bacteria and eukaryotes, and then examine the detailed mechanism for protein synthesis in these two types of organism.

Box 10.1: The origin and evolution of the genetic code

The evolution of the genetic code has provoked argument ever since DNA was established as the genetic material back in the 1950s. Many geneticists support the 'frozen accident' theory, which suggests that codons were randomly allocated to amino acids during the earliest stages of evolution, the code subsequently becoming 'frozen' because any changes would result in widespread disruption of the amino acid sequences of proteins. Various lines of evidence suggest that the code might have evolved in a less random manner (Cedergren and Miramontes, 1996). First, controversial experimental results suggest that at least some amino acids bind directly to RNAs containing the appropriate codons, this occurring in the absence of the tRNA that mediates the interaction in present-day cells. These results suggest that the code might have originated from natural interactions between RNA molecules and amino acids. Second, the deviations from the standard code listed in *Table 10.2* indicate that the same codon reallocations have occurred more than once. For example, three unrelated groups of protozoa use 5′-UAA-3′ and 5′-UAG-3′ as glutamine rather than termination codons, suggesting that these codons were reallocated in the same way on at least three separate occasions. Similarly, the use of 5′-UGA-3′ as a tryptophan codon must have evolved twice, once in mitochondria and once in the ancestors of *Mycoplasma*. If the relationship between codon and amino acid is entirely random, as suggested by the 'frozen accident' theory, then we would not expect to see the same codon reallocations recurring on different occasions. Unfortunately, ideas on the origin and evolution of the genetic code will always be speculative because of the difficulty in obtaining evidence relating to these issues.

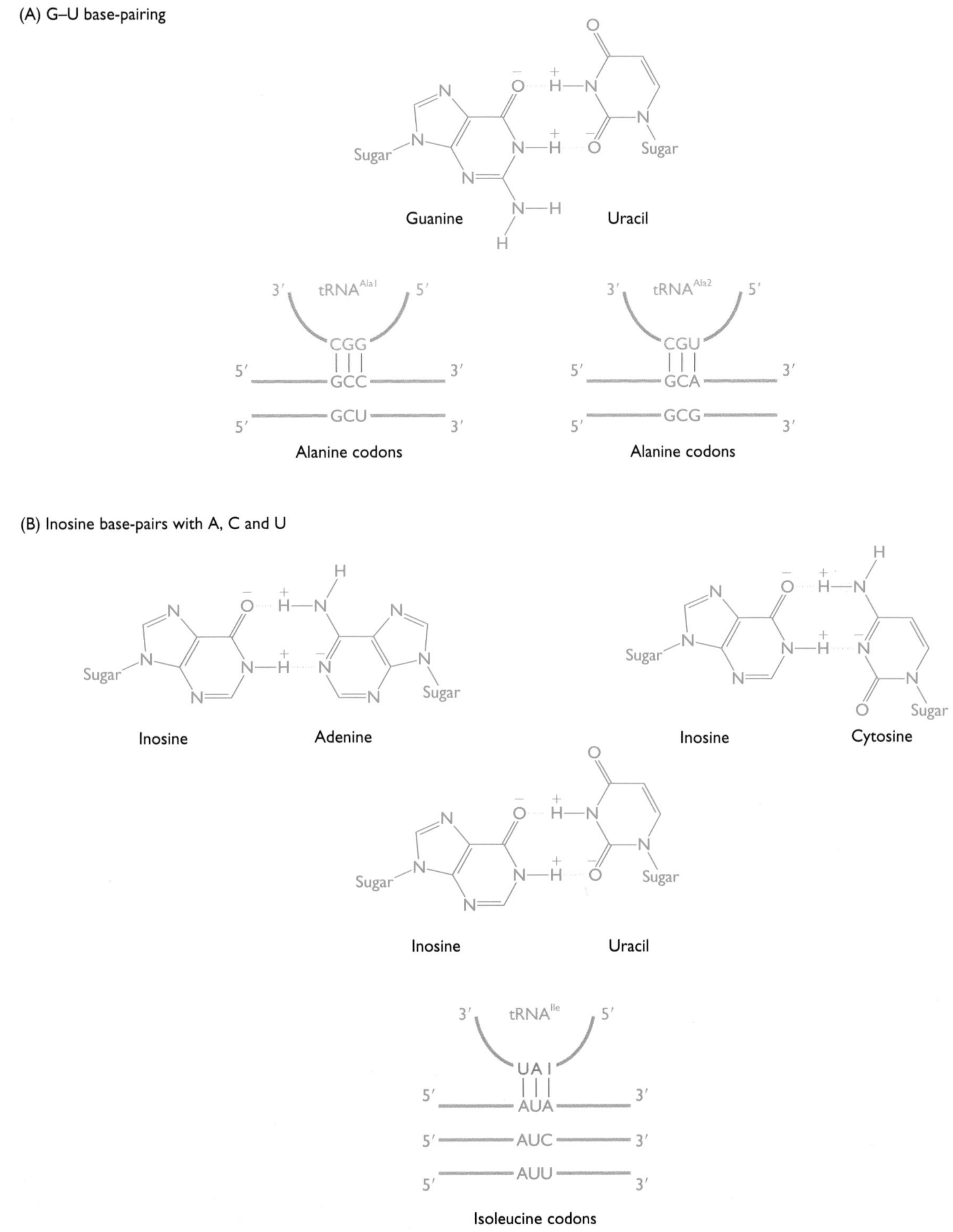

Figure 10.9 Two examples of wobble.

(A) Wobble involving a G–U base pair enables the four codon family for alanine to be decoded by just two tRNAs. Note that wobble involving G–U also enables accurate decoding of a four-codon family that specifies two amino acids. For example, the anticodon 3′-AAG-5′ can decode 5′-UUC-3′ and 5′-UUU-3′, both coding for phenylalanine (see *Figure 10.7*), and the anticodon 3′-AAU-5′ can decode the other two members of this family, 5′-UUA-3′ and 5′-UUG-3′, which code for leucine. (B) Inosine can base-pair with A, C or U, meaning that a single tRNA can decode all three codons for isoleucine. ...., Hydrogen bond; I, inosine.

Box 10.2: The compactness of vertebrate mitochondrial genomes

One consequence of superwobble is that the human mitochondrial genome, in common with those of other vertebrates, has reduced to a bare minimum the number of tRNA genes that it needs. This is one of several features that appear to be designed to enable these genomes to be as small as possible. Other features include:

- The absence of intergenic spacers between most pairs of genes (see *Figure 6.9A*, p. 126). In general, these genomes have very little wasted space: only 87 bp of the human mitochondrial DNA cannot be ascribed a function.
- The ribosomal RNA genes are extremely short, the large and small subunit rRNAs having sedimentation coefficients of only 16S and 12S, respectively. These rRNAs are substantially smaller than those found in bacterial ribosomes and the cytoplasm of eukaryotic cells (see *Figure 10.10*).
- As described in Box 9.5, p. 228, some genes are truncated and do not contain standard termination codons, the termination codon being completed by polyadenylation editing of the mRNA.

10.2.1 Ribosome structure

Our understanding of ribosome structure has gradually developed over the last 50 years as more and more powerful techniques have been applied to the problem. Originally called 'microsomes', ribosomes were first observed in the early decades of the 20th century, as tiny particles almost beyond the resolving power of light microscopy. In the 1940s and 1950s the first electron micrographs showed that bacterial ribosomes are oval-shaped with dimensions of 29 nm × 21 nm, rather smaller than eukaryotic ribosomes, the latter varying a little in size depending on species but averaging about 32 nm × 22 nm. In the mid-1950s the discovery that ribosomes are the sites of protein synthesis stimulated attempts to define the structures of these particles in greater detail.

Ultracentrifugation was used to measure the sizes of ribosomes and their components

The initial progress in understanding the detailed structure of the ribosome came not from observing them with the electron microscope but by analyzing their components by ultracentrifugation (Technical Note 6.1, p. 123). Intact ribosomes have sedimentation coefficients of 80S for eukaryotes and 70S for bacteria, and each can be broken down into smaller components (*Figure 10.10*):

- Each ribosome comprises two subunits. In eukaryotes these subunits are 60S and 40S, in bacteria they are 50S and 30S. Note that sedimentation coefficients are not additive because they depend on shape as well as mass: it is perfectly acceptable for the intact ribosome to have an S value less than the sum of its two subunits.
- The large subunit contains three rRNAs in eukaryotes (the 28S, 5.8S and 5S rRNAs) but only two in bacteria (23S and 5S rRNAs). In bacteria the equivalent of the eukaryotic 5.8S rRNA is contained within the 23S rRNA.
- The small subunit contains a single rRNA in both types of organism: an 18S rRNA in eukaryotes and a 16S rRNA in bacteria.
- Both subunits are associated with a variety of **ribosomal proteins**, the numbers detailed in *Figure 10.10*. The ribosomal proteins of the small subunit are called S1, S2, etc.; those of the large subunit are L1, L2, etc. There is just one of each protein per ribosome, except for L7 and L12, which are present as dimers.

Probing the fine structure of the ribosome

Once the basic composition of eukaryotic and bacterial ribosomes had been worked out, attention was focused on the way in which the various rRNAs and proteins fit together. Important information was provided by the first rRNA sequences, comparisons between these identifying conserved regions that can base-pair to form complex two-dimensional structures (*Figure 10.11*). This suggested that the rRNAs provide a scaffolding within the ribosome, to which the proteins are attached, an interpretation that understates the active role rRNAs play in protein synthesis but nonetheless was a useful foundation on which to base subsequent research.

Much of that subsequent research has concentrated on the *E. coli* ribosome, as it is smaller than the eukaryotic version and available in large amounts from extracts of bacteria grown to high density in cell cultures. A number of technical approaches have been applied to the *E. coli* ribosome, including:

- **Nuclease protection studies** (Section 6.1.1), which enable contacts between rRNAs and proteins to be identified.
- **Protein–protein crosslinking**, which identifies pairs or groups of proteins that are located close to one another in the ribosome.
- **Electron microscopy**, which has gradually become more sophisticated, enabling the overall structure of the ribosome to be resolved in greater detail. For example, innovations such as **immunoelectron microscopy**, in which ribosomes are labeled with antibodies specific for individual ribosomal proteins prior to examination, have been used to locate the positions of these proteins on the surface of the ribosome.
- **Site-directed hydroxyl radical probing** makes use of the ability of Fe(II) ions to generate hydroxyl radicals

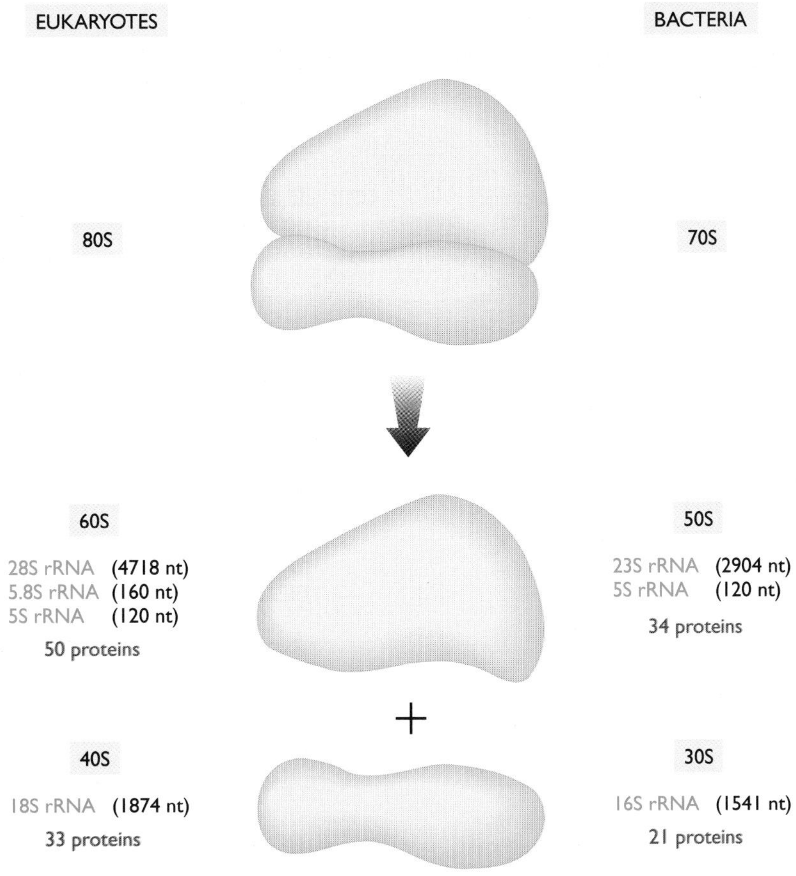

Figure 10.10 The composition of eukaryotic and bacterial ribosomes.

The details refer to a 'typical' eukaryotic ribosome and the *Escherichia coli* ribosome. Variations between different species mainly concern the numbers of ribosomal proteins. Abbreviation: nt, nucleotides.

which cleave RNA phosphodiester bonds located within 1 nm of the site of radical production. The technique has been used to determine the exact positioning of ribosomal protein S5 in the *E. coli* ribosome. Different amino acids within S5 were labeled with Fe(II) and hydroxyl radical generation induced in reconstituted ribosomes. The positions at which the 16S rRNA was cleaved were then used to infer the topology of the rRNA in the vicinity of S5 protein (*Figure 10.12*; Heilek and Noller, 1996).

These studies have gradually built up a detailed picture of the fine structure of the ribosome. Research is now being conducted in two main directions. First, X-ray crystallography and NMR studies are being used to work out the structures of individual ribosomal proteins and to identify the RNA-binding domain(s) for each one (Ramakrishnan and White, 1998). Second, details of the interactions between the ribosome, mRNA and aminoacyl-tRNAs are being investigated by high-resolution electron microscopy (e.g. Agrawal *et al.*, 1996) and hydroxyl-radical probing (e.g. Joseph *et al.*, 1997).

10.2.2 Translation in detail

Although ribosomal architecture is similar in bacteria and eukaryotes, there are distinctions in the way in which translation is carried out in the two types of organism. The most important of these differences occurs during the first stage of translation, when the ribosome is assembled on the mRNA at a position upstream of the initiation codon.

Initiation in bacteria requires an internal ribosome binding site

The main difference between initiation of translation in bacteria and eukaryotes is that in bacteria the translation initiation complex is built up directly over the initiation codon, the point at which protein synthesis will begin. Eukaryotes, as we will see in the next section, use a more indirect process for locating the initiation point.

When not actively participating in protein synthesis, ribosomes dissociate into their subunits which remain in the cytoplasm waiting to be selected for a new round of translation. In bacteria, the process initiates when a small

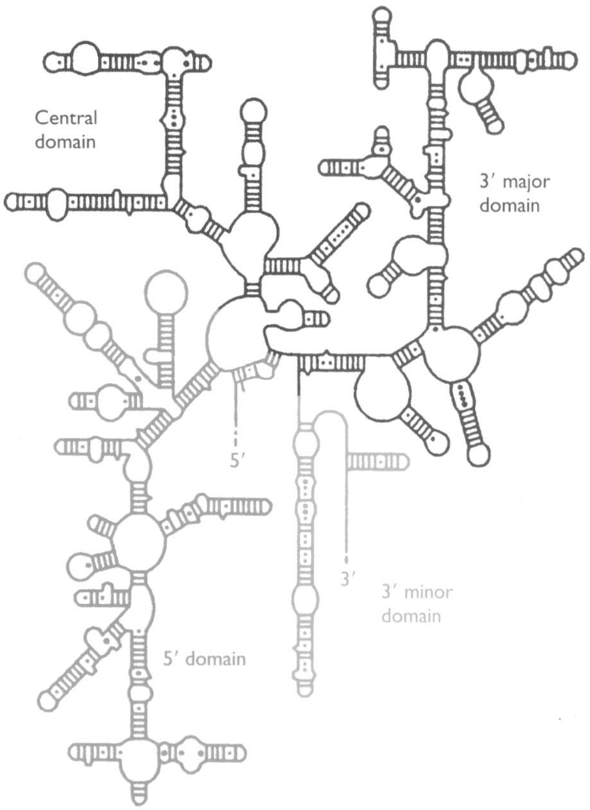

Figure 10.11 The base-paired structure of the *Escherichia coli* 16S rRNA.

In this representation, standard base pairs (G–C, A–U) are shown as bars and nonstandard base pairs (e.g. G–U) as dots.

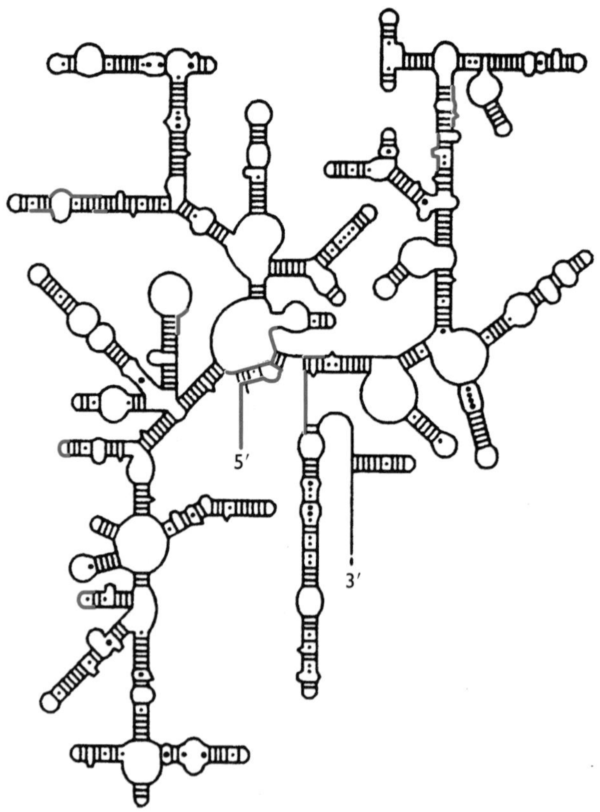

Figure 10.12 Positions within the *E. coli* 16S rRNA that form contacts with ribosomal protein S5.

The distribution of the contact positions (shown in red) for this single ribosomal protein emphasizes the extent to which the base-paired secondary structure of the rRNA is further folded within the three-dimensional structure of the ribosome. For details of the work that led to these results, see Heilek and Noller (1996).

subunit, in conjunction with the translation **initiation factor** IF-3, attaches to the **ribosome binding site** (also called the **Shine-Dalgarno sequence**). This is a short target site, consensus 5′-AGGAGGU-3′ in *E. coli* (*Table 10.4*), located about 3–10 nucleotides upstream of the initiation codon, the point at which translation will begin (*Figure 10.13*). The ribosome binding site is complementary to a region at the 3′-end of the 16S rRNA, the one present in the small subunit, and it is thought that base-pairing between the two is involved in the attachment of the small subunit to the mRNA.

Attachment to the ribosome binding site positions the small subunit of the ribosome over the initiation codon (*Figure 10.14*). This codon is usually 5′-AUG-3′, coding for methionine, though 5′-GUG-3′ and 5′-UUG-3′ are sometimes used. All three codons can be recognized by the same initiator tRNA, the last two by wobble. This initiator tRNA is the one that was aminoacylated with methionine and subsequently modified by conversion of the methionine to *N*-formylmethionine (see *Figure 10.5B*, p. 236). The modification attaches a formyl group, -COH, to the amino group which means that only the carboxyl

Table 10.4 Examples of ribosome binding sequences in *Escherichia coli*

Gene	Codes for	Ribosome binding sequence	Nucleotides to start codon
E. coli consensus	—	5′-AGGAGGU-3′	10
Lactose operon	Lactose utilization enzymes	5′-AGGA-3′	7
galE	Hexose-1-phosphate uridyltransferase	5′-GGAG-3′	6
rplJ	Ribosomal protein L10	5′-AGGAG-3′	8

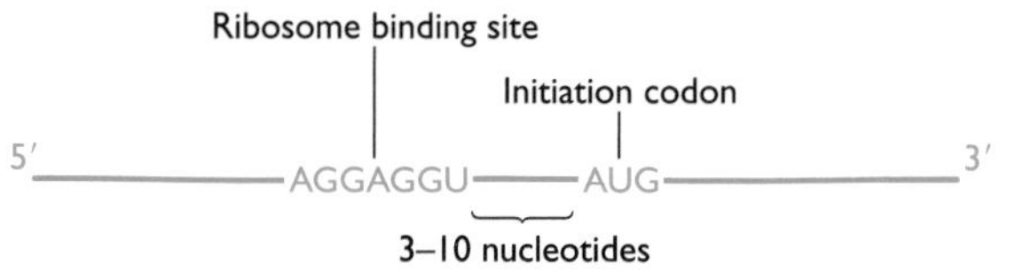

Figure 10.13 The ribosome binding site for bacterial translation.

In *Escherichia coli*, the ribosome binding site has the consensus sequence 5′-AGGAGGU-3′ and is located between 3 and 10 nucleotides upstream of the initiation codon.

group of the initiator methionine is free to participate in peptide bond formation. This ensures that polypeptide synthesis can take place only in the N→C direction. The initiator $tRNA_i^{Met}$ is brought to the small subunit of the ribosome by a second initiation factor, IF-2, along with a molecule of GTP, the latter acting as a source of energy for the final step of initiation. Note that the $tRNA_i^{Met}$ is only able to decode the initiation codon: it cannot enter the complete ribosome during the elongation phase of translation during which internal 5′-AUG-3′ codons are recognized by a different $tRNA^{Met}$ carrying an unmodified methionine.

Completion of the initiation phase occurs when IF-1 binds to and stabilizes the initiation complex, enabling the large subunit of the ribosome to attach. Attachment of the large subunit requires energy, generated by hydrolysis of the bound GTP, and results in release of the initiation factors.

Initiation in eukaryotes is mediated by the cap structure and poly(A) tail

Only a small number of eukaryotic mRNAs have internal ribosome binding sites. Instead, the small subunit of the ribosome makes its initial attachment at the 5′-end of the mRNA and then **scans** along the sequence until it locates the initiation codon. The details are as follows. The first step involves assembly of the **preinitiation complex**, which comprises the 40S subunit of the ribosome, the initiator $tRNA_i^{Met}$, the eukaryotic initiation factor eIF-2 (a trimer of three different proteins), and a molecule of GTP (*Figure 10.15A*). As in bacteria, the initiator tRNA is distinct from the normal $tRNA^{Met}$ that recognizes internal 5′-AUG-3′ codons but, unlike bacteria, it is aminoacylated with normal methionine, not the formylated version. After assembly, the preinitiation complex associates with the 5′-end of the mRNA. This step requires the **cap binding complex** (sometimes called eIF-4F), which comprises the initiation factors eIF-4A, eIF-4E and eIF-4G. One of these factors, eIF-4A, does not come into play just yet. The other two form an indirect attachment between the cap structure and the preinitiation complex, with eIF-4E bound to the cap, and eIF-4G forming a bridge between eIF-4E and another initiation factor, eIF-3 (see *Figure 10.15A*; Hentze, 1997). The result is that the preinitiation complex becomes attached to the 5′ region of the mRNA.

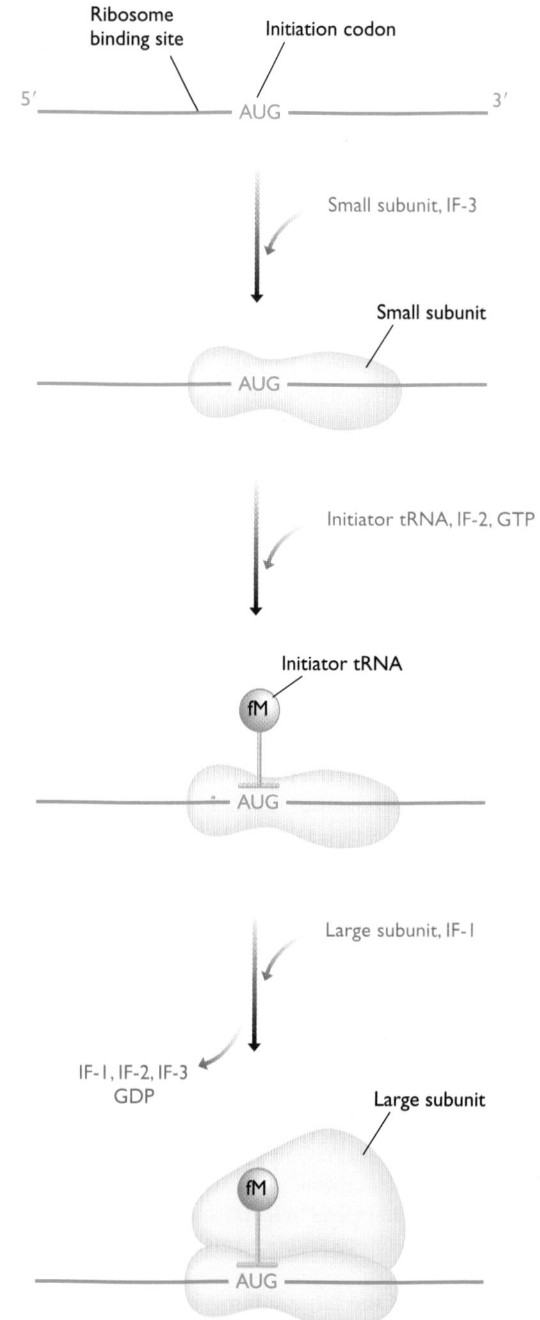

Figure 10.14 Initiation of translation in *E. coli*.

See the text for details. Note that the different components of the initiation complex are not drawn to scale. Abbreviation: fM, *N*-formylmethionine.

Attachment of the preinitiation complex to the mRNA is also influenced by the poly(A)-tail, at the distant 3′-end of the mRNA. This is thought to be mediated by the polyadenylate-binding protein (PADP), which is attached to the poly(A)-tail (Section 9.2.2). In yeast and plants it has been shown that PADP can form an association with eIF-4G, this association requiring that the mRNA bends back on itself. With artificially uncapped mRNAs, the PADP interaction is sufficient to load the preinitiation complex on to the 5′-end of the mRNA, but under normal circumstances the cap structure and poly(A)-tail probably work together (Preiss and Hentze, 1998). The poly(A)-tail could have an important regulatory role as the length of the tail appears to be correlated with the extent of initiation that occurs with a particular mRNA. The association between PADP and eIF-4G has not yet been documented in mammals, but probably occurs (Craig *et al.*, 1998).

After becoming attached to the 5′-end of the mRNA, the **initiation complex**, as it is now called, has to scan along the molecule and find the initiation codon. The leader regions of eukaryotic mRNAs can be several tens, or even hundreds, of nucleotides in length and often contain regions that form hairpins and other base-paired structures. These are probably removed by a combination of eIF-4A (the third component of the cap binding complex) and eIF-4B (which we have not met yet). eIF-4A, and possibly also eIF-4B, has a helicase activity and so is able to break intramolecular base-pairs in the mRNA, freeing the passage for the initiation complex (*Figure 10.15B*). The initiation codon, which is invariably 5′-AUG-3′ in eukaryotes, is recognizable because it is contained in a short consensus sequence, 5′-ACCAUGG-3′, referred to as the **Kozak consensus**. Once the initiation complex is positioned over the initiation codon, the large subunit of the ribosome attaches. As in bacteria, this requires hydrolysis of GTP and leads to release of the initiation factors. Two final initiation factors are involved at this stage: eIF-5, which aids release of the other factors, and eIF-6, which is associated with the unbound large subunit and prevents it from attaching to small subunits in the cytoplasm.

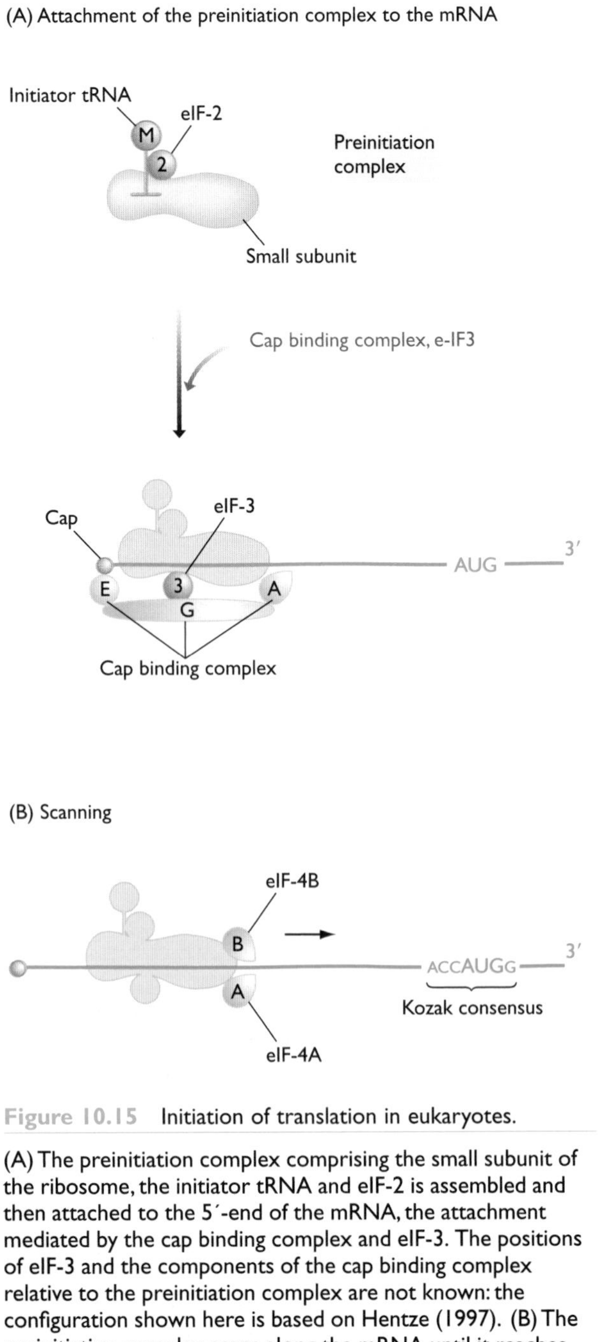

Figure 10.15 Initiation of translation in eukaryotes.

(A) The preinitiation complex comprising the small subunit of the ribosome, the initiator tRNA and eIF-2 is assembled and then attached to the 5′-end of the mRNA, the attachment mediated by the cap binding complex and eIF-3. The positions of eIF-3 and the components of the cap binding complex relative to the preinitiation complex are not known: the configuration shown here is based on Hentze (1997). (B) The preinitiation complex scans along the mRNA until it reaches the initiation codon, which is recognizable because it is located within the Kozak consensus sequence. Scanning is aided by eIF-4A and eIF-4B, which are thought to have helicase activities. It is probable that eIF-3 remains attached to the preinitiation complex during scanning, as shown here. It is not clear whether eIF-4E and eIF-4G also remain attached at this stage. Note that scanning is an energy-dependent process that requires hydrolysis of ATP. Abbreviation: M, methionine.

Elongation is similar in bacteria and eukaryotes

The main differences between translation in bacteria and eukaryotes occur during the initiation phase: the events after the large subunit of the ribosome becomes associated with the initiation complex are similar in both types of organism. We can therefore deal with them together, by looking at what happens in *E. coli* and referring to the distinctive features of eukaryotic translation at the appropriate places.

Attachment of the large subunit results in two sites at which aminoacyl-tRNAs can bind. The first of these, the **P** or **peptidyl site**, is already occupied by the initiator $tRNA_i^{Met}$, charged with *N*-formylmethionine or methionine, and base-paired with the initiation codon. The second site, the **A** or **acceptor site**, covers the second codon in the open reading frame (*Figure 10.16*). These sites are thought to be located in the cavity between the large and small subunits of the ribosome, the codon–anticodon interaction associated with the small subunit and the aminoacyl end of the tRNA with the large subunit. The A site becomes filled with the appropriate aminoacyl-tRNA, which in *E. coli* is brought into position by the **elongation factor** EF-Tu. This factor is an example of a G protein, meaning that it binds a molecule of GTP which it can hydrolyze to release energy. In eukaryotes the equivalent factor is called eEF-1, which is a complex of four subunits, eEF-1α, eEF-1β, eEF-1γ and eEF-1δ. The first of these exists in at least two forms, eEF-1α1 and eEF-1α2, which

Box 10.3: Initiation of eukaryotic translation without scanning

The scanning system for initiation of translation does not apply to every eukaryotic mRNA. This was first recognized with the picornaviruses, a group of eukaryotic viruses (i.e. those that infect eukaryotic cells) with RNA genomes which include the human poliovirus and rhinovirus, the latter responsible for the common cold. Transcripts from these viruses are not capped but instead have **internal ribosome entry sites (IRES)**, similar in function to the ribosome binding site of bacteria, though the sequences of IRESs and their positions relative to the initiation codon are more variable than the bacterial versions (Mountford and Smith, 1995). The presence of IRESs on their transcripts means that picornaviruses can block protein synthesis in the host cell, by inactivating the cap binding complex, without affecting translation of their own transcripts, although this is not a normal part of the infection strategy of all picornaviruses.

Remarkably, no virus proteins are required for recognition of an IRES by a host ribosome. In other words, the normal eukaryotic cell possesses proteins and/or other factors that enable it to initiate translation by the IRES method. Because of their variability, IRESs are difficult to identify by inspection of DNA sequences, but it is becoming clear that a few nuclear gene transcripts possess them and that these are translated, at least under some circumstances, not by scanning but via their IRES. Examples are the mRNAs for the mammalian immunoglobulin heavy chain binding protein and the *Drosophila* Antennapedia protein (Section 11.3.3). The presence of an IRES on these mRNAs may enable them to be translated during periods of the cell cycle, such as mitosis, when cap-mediated protein synthesis is repressed.

are very similar proteins that probably have equivalent functions in different tissues (Hafezparast and Fisher, 1998).

When the aminoacyl-tRNA has entered the A site, a peptide bond is formed between the two amino acids. This involves a **peptidyl transferase** enzyme which releases the amino acid from the initiator $tRNA_i^{Met}$ and then forms a peptide bond between this amino acid and the one attached to the second tRNA. In *E. coli*, the peptidyl transferase activity resides in the 23S rRNA of the large subunit, and so is an example of a ribozyme (Section 9.3.3; Nitta *et al.*, 1998). The reaction is energy-dependent and requires hydrolysis of the GTP attached to EF-Tu (eEF-1 in eukaryotes). This inactivates EF-Tu, which is ejected from the ribosome and regenerated by EF-Ts. A eukaryotic equivalent of EF-Ts has not been identified, and it is possible that one of the subunits of eEF-1 has the regenerative activity.

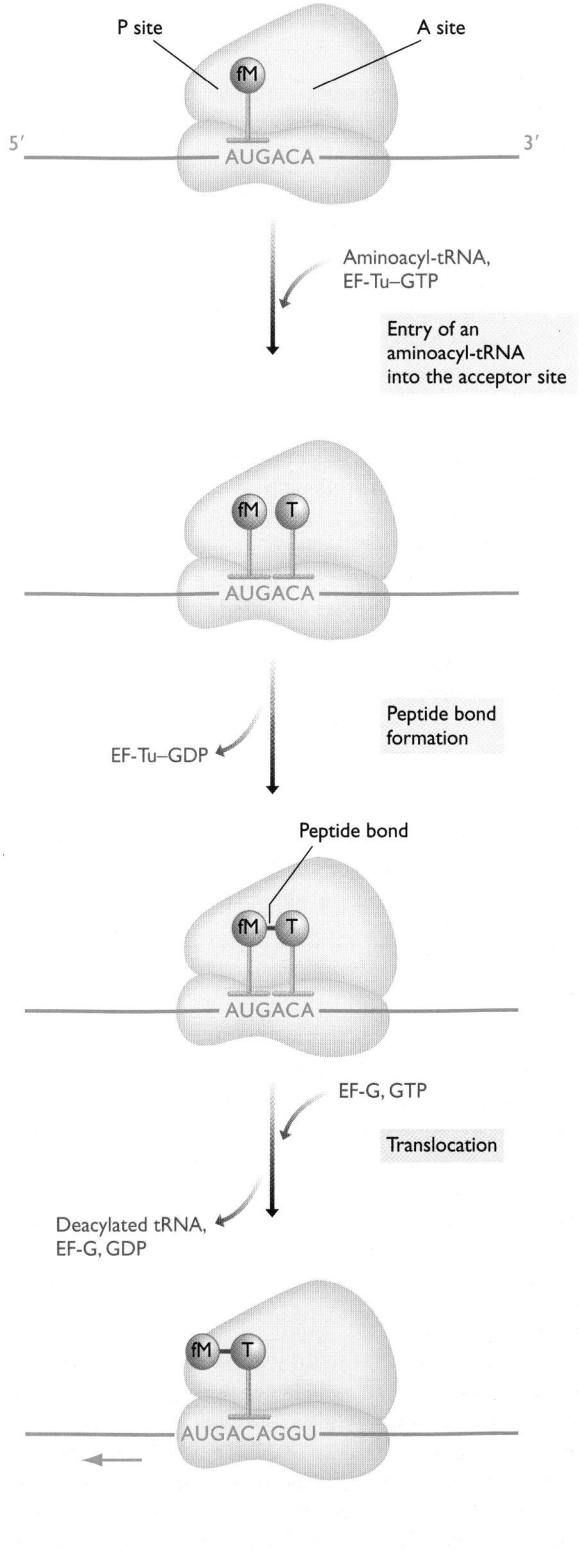

Figure 10.16 Elongation of translation.

The diagram shows the events occurring during a single elongation cycle in *E. coli*. See the text for details regarding eukaryotic translation. Abbreviations: fM, *N*-formylmethionine; T, threonine.

Box 10.4: Regulation of translation initiation

The initiation of translation is an important control point in protein synthesis, at which two levels of regulation are exerted:

- **Global regulation** involves a general alteration in the amount of protein synthesis occurring, with all mRNAs being affected to a similar extent. In eukaryotes this is commonly achieved by phosphorylation of eIF-2 which results in repression of translation initiation by preventing eIF-2 from binding the molecule of GTP that it needs before it can transport the initiator tRNA to the small subunit of the ribosome. Phosphorylation of eIF-2 occurs during heat shock, when the overall level of protein synthesis is decreased, and is used by some viruses as a way of switching off protein synthesis in the host cell.
- **Transcript-specific regulation** involves mechanisms that act on a single transcript or a small group of transcripts coding for related proteins. The most frequently cited example involves the operons for the ribosomal protein genes of *Escherichia coli*. The leader region of the mRNA transcribed from each operon contains a sequence that acts as a binding site for one of the proteins coded by the operon. When this protein is synthesized it can either attach to its position on the ribosomal RNA, or it can bind to the leader region of the mRNA. The rRNA attachment is favored and occurs if there are free rRNAs in the cell. Once the rRNAs have all been assembled into ribosomes the ribosomal protein binds to its mRNA, blocking translation initiation and hence switching off further synthesis of the ribosomal proteins coded by that particular mRNA. Similar events involving other mRNAs ensure that synthesis of each ribosomal protein is coordinated with the amount of free rRNA in the cell.

An example of transcript-specific regulation in mammals occurs with the mRNA for ferritin, an iron-storage protein. In the absence of iron, ferritin synthesis is inhibited by proteins that bind to sequences called iron-response elements located in the leader region of the ferritin mRNA. The bound proteins block the ribosome as it attempts to scan along the mRNA in search of the initiation codon. When iron is present, the binding proteins detach and the mRNA is translated. Interestingly, the mRNA for a related protein, the transferrin receptor involved in the uptake of iron, also has iron-response elements, but in this case detachment of the binding proteins in the presence of iron results not in translation of the mRNA but in its degradation. This is logical because when iron is present in the cell there is less requirement for transferrin receptor activity as there is less need to import iron from outside.

Now the dipeptide corresponding to the first two codons in the open reading frame is attached to the tRNA in the A site. The next step is **translocation**, during which three things happen at once (see *Figure 10.16*):

- The ribosome moves along three nucleotides, so the next codon enters the A site.
- The dipeptide-tRNA in the A site moves to the P site.
- The deacylated tRNA in the P site moves to a third position, the **E** or **exit site**, in bacteria or, in eukaryotes, is simply ejected from the ribosome.

Translocation requires hydrolysis of a molecule of GTP and is mediated by EF-G in bacteria and eEF-2 in eukaryotes. It results in the A site becoming vacant, allowing a new aminoacyl-tRNA to enter. The elongation cycle is now repeated, and continues until the end of the open reading frame is reached.

Termination requires release factors

Protein synthesis ends when one of the three termination codons is reached. The A site is now entered not by a tRNA but by a protein **release factor** (*Figure 10.17*). Bacteria have three of these: RF-1 which recognizes the termination codons 5′-UAA-3′ and 5′-UAG-3′, RF-2 which recognizes 5′-UAA-3′ and 5′-UGA-3′, and RF-3 which plays a supporting role. Eukaryotes have just one release factor, eRF. The process is energy-independent in bacteria but requires hydrolysis of GTP in eukaryotes. Termination results in release of the completed polypeptide from the tRNA in the P site, and dissociation of the translation complex. The ribosome subunits enter the cytoplasmic pool where they remain until used again in another round of translation.

10.3 POST-TRANSLATIONAL PROCESSING OF PROTEINS

Translation is not the end of the gene expression pathway. The polypeptide that emerges from the ribosome is inactive, and before taking on its functional role in the cell must undergo at least the first of the following four types of post-translational processing (*Figure 10.18*):

- **Protein folding**. The polypeptide is inactive until it is folded into its correct tertiary structure.

RESEARCH

10.2

BRIEFING

Peptidyl transferase is a ribozyme

A ribosome-associated protein that has the peptidyl transferase activity needed to synthesize peptide bonds during translation has never been isolated. The reason for this lack of success is now known: the enzyme activity is specified by part of the 23S rRNA.

When the base-paired structures of rRNAs (see *Figure 10.11*, p. 244) were first determined in the early 1980s, the possibility that an RNA molecule can have enzymatic activity was unheard of – the breakthrough discoveries with regards to ribozymes not being made until the period 1982–86. Ribosomal RNAs were therefore initially assigned purely structural roles in the ribosome, their base-paired conformations being looked on as scaffolds to which the important components of the ribosome, the proteins, were attached. Problems with this interpretation began to arise in the late 1980s when difficulties were encountered in identifying the protein or proteins responsible for the central catalytic activity of the ribosome, the formation of peptide bonds. By now the existence of ribozymes had been established and molecular biologists began to take seriously the possibility that rRNAs might have an enzymatic role in protein synthesis.

Peptide bond formation by intact 23S rRNA

The first experimental evidence that the peptidyl transferase activity of the ribosome is ribozymal came from studies of purified 23S rRNA from *Escherichia coli*. It was shown that these rRNA preparations can catalyze the formation of a peptide bond between the amino acids attached to two $tRNA^{Phe}$ molecules. One of these tRNAs was aminoacylated with normal phenylalanine and the second with *N*-acetylphenylalanine. In the presence of 23S rRNA the dipeptide *N*-acetylphenylalanine-phenylalanine was synthesized.

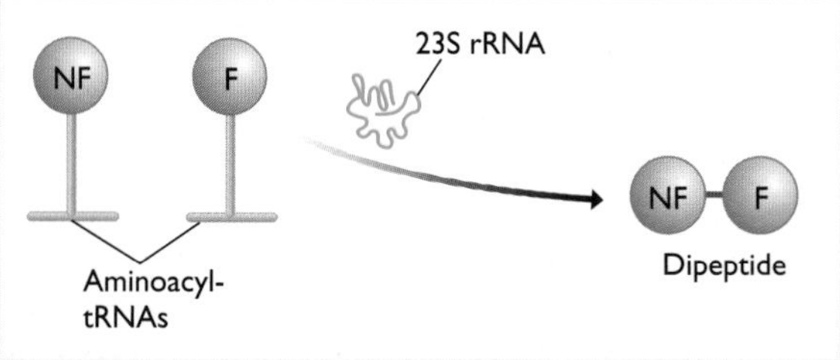

Synthetic 23S rRNA fragments

Although strongly suggestive, the experiments with intact 23S rRNA were not definite proof that the peptidyl transferase activity is ribozymal. The problem is that the 23S rRNA molecules used in this experiment were purified from *E. coli* ribosomes and so might contain some residual protein, these contaminating proteins possibly possessing the catalytic activity ascribed to the rRNA. To counter this problem a second set of experiments was carried out, not with purified rRNA, but with RNA synthesized in the test tube. This RNA had never been in contact with ribosomal proteins, or any other kinds of protein, so any activity it possessed could not be ascribed to contamination. When tested in the peptide bond assay the synthetic RNA was indeed shown to have peptidyl transferase activity, conclusive evidence that the 23S rRNA is a ribozyme.

The use of synthetic RNA enabled more sophisticated questions to be asked about the peptidyl transferase activity, as the 23S rRNA could be synthesized in segments, each segment corresponding to one of the six domains of the intact molecule. This meant that the location of the peptidyl transferase activity within the rRNA could be examined. Peptide bond assays were performed with a complete set of domain fragments and with a series of mixtures lacking individual fragments. When domains I to IV or VI were removed from the mixture peptide bond formation still occurred, albeit with slightly reduced efficiency compared to the control. However, when the domain V fragment was omitted the activity was almost completely absent.

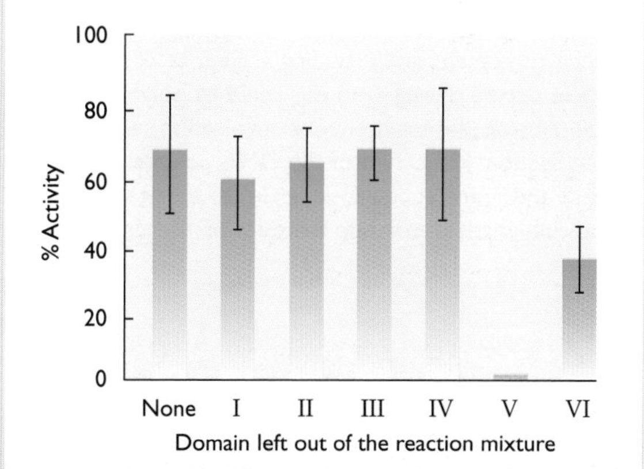

Reprinted with permission from Nitta *et al. Science*, **281**, 666–669. Copyright 1998 American Association for the Advancement of Science.

Recognition of the ribozymal activity of 23S rRNA is important not only for research concerning the function of the ribosome in protein synthesis. It also has a bearing on ideas regarding the earliest evolution of biochemical systems, especially those concerning the 'RNA world', which propose that RNA evolved before DNA and at one time carried genetic information and possessed the enzymatic activities needed to convert that information into protein (Section 14.1.1).

References

Nitta I, Ueda T and Watanabe K (1998) Possible involvement of *Escherichia coli* 23S ribosomal RNA in peptide bond formation. *RNA*, **4**, 257–267.

Nitta I, Kamada Y, Noda H, Ueda T and Watanabe K (1998) Reconstitution of peptide bond formation with *Escherichia coli* 23S ribosomal RNA domains. *Science*, **281**, 666–669.

Box 10.5: Frameshifting

Frameshifting occurs when a ribosome pauses in the middle of an mRNA, moves back one nucleotide or, less frequently, forward one nucleotide, and then continues translation (Farabaugh, 1996). The result is that the codons that are read after the pause are not contiguous with the preceding set of codons; they lie in a different reading frame:

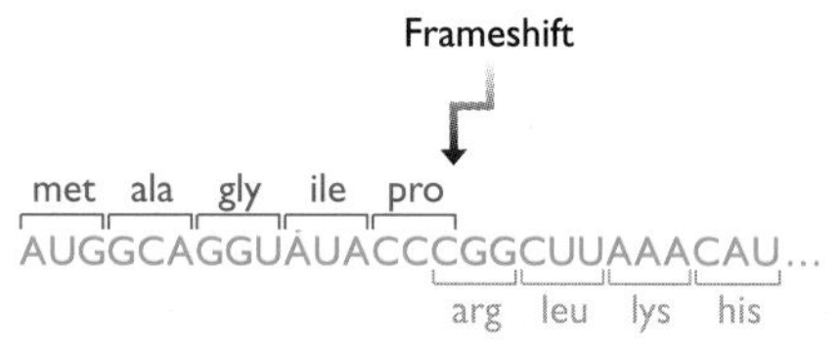

Spontaneous frameshifts occur randomly and are deleterious as the polypeptide synthesized after the frameshift has the incorrect amino acid sequence. But not all frameshifts are spontaneous, a few mRNAs utilizing programmed frameshifting to induce the ribosome to change frame at a specific point within the transcript. Programmed frameshifting occurs in all types of organism from bacteria through to humans and including a number of viruses. An example occurs during synthesis of DNA polymerase III of *Escherichia coli*, the main enzyme involved in replication of DNA (Section 12.2). Two of the DNA polymerase III subunits, γ and τ, are coded by a single gene, *dnaX*. Subunit τ is the full-length translation product of the *dnaX* mRNA and subunit γ is a shortened version. Synthesis of γ involves a frameshift in the middle of the *dnaX* mRNA, the ribosome encountering a termination codon immediately after the frameshift and so producing the truncated γ version of the translation product. It is thought that the frameshift is induced by three features of the *dnaX* mRNA:

- A hairpin loop, located immediately after the frameshift position, which stalls the ribosome.
- A sequence similar to a ribosome-binding site immediately upstream of the frameshift position, which is thought to base-pair with the 16S rRNA (as does an authentic ribosome-binding site), again causing the ribosome to stall.
- The codon 5′-AAG-3′ at the frameshift position. The presence of a modified nucleotide at the wobble position of the $tRNA^{Lys}$ that decodes 5′-AAG-3′ means that the codon–anticodon interaction is relatively weak at this position, enabling the frameshift to occur.

A similar phenomenon, translational **slippage**, enables a single ribosome to translate an mRNA that contains copies of two or more genes. This means that, for example, a single ribosome can synthesize each of the five proteins coded by the mRNA transcribed from the tryptophan operon of *E. coli* (see *Figure 6.14B*, p. 133). When the ribosome reaches the end of one series of codons it releases the protein it has just made, slips to the next initiation codon, and begins synthesizing the next protein.

- **Proteolytic cleavage**. Some proteins are processed by cutting events carried out by enzymes called **proteases**. These cutting events may remove segments from one or both ends of the polypeptide, resulting in a shortened form of the protein, or they may cut the polypeptide into a number of different segments, each one of which is active.
- **Chemical modification**. Individual amino acids in the polypeptide might be modified by attachment of new chemical groups.
- **Intein splicing**. Inteins are intervening sequences in some proteins, similar in a way to introns in mRNAs. They have to be removed and the **exteins** ligated in order for the protein to become active.

Often these different types of processing occur together, the polypeptide being cut and modified at the same time that it is folded. The cutting, modification and/or splicing events may be necessary for the polypeptide to take up its correct three-dimensional conformation, because this is dependent in part on the relative positioning of the various chemical groups along the molecule. Alternatively, a cutting event or a chemical modification might occur after the protein has been folded, possibly as part of a regulatory mechanism that converts a folded but inactive protein into an active form.

10.3.1 Protein folding

Protein folding was introduced in Chapter 7 when we examined the four levels of protein structure (primary, secondary, tertiary and quaternary) and learnt that all of the information that a polypeptide needs to adopt its correct tertiary structure is contained within its amino acid sequence (Section 7.2.1). This is one of the central principles of molecular biology. We must therefore examine its experimental basis and consider how the information contained in the amino acid sequence is utilized during the folding processes for newly-translated polypeptides.

Not all proteins fold spontaneously in the test tube

The notion that the amino acid sequence contains all the information needed to fold the polypeptide into its correct tertiary structure derives from experiments carried out with ribonuclease in the 1960s (Anfinsen, 1973). Ribonuclease is a small protein, just 124 amino acids in length, containing four disulfide bridges and with a ter-

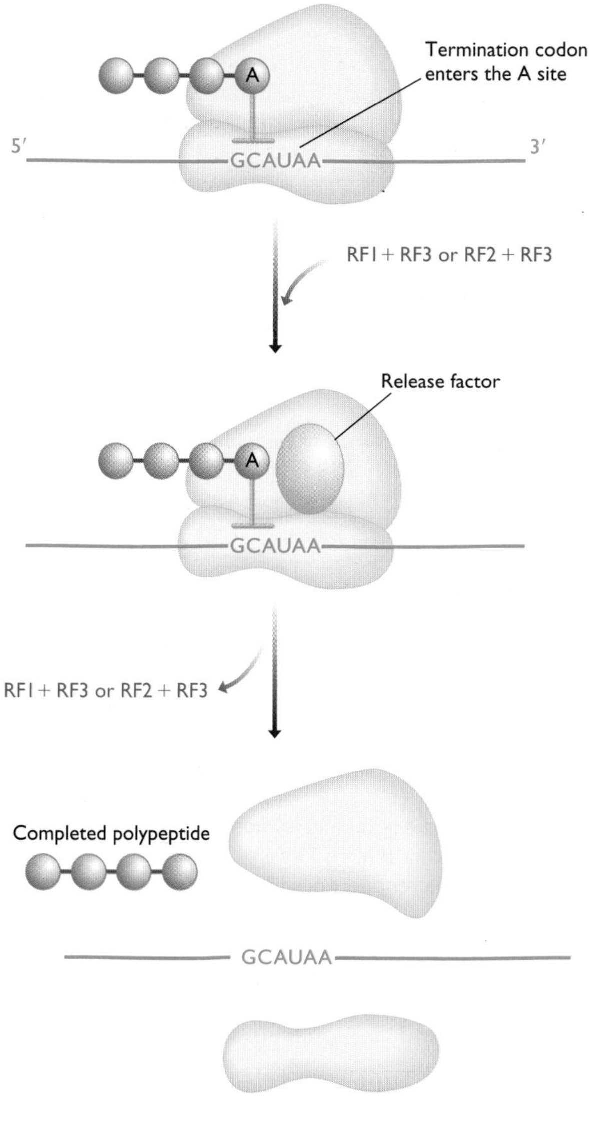

Figure 10.17 Termination of translation.

Termination in *E. coli* is illustrated. For differences in eukaryotes, see the text. 'A' indicates alanine.

tiary structure that is made up predominantly of β-sheet, with very little α-helix. Studies of its folding were carried out with ribonuclease that had been purified from cow pancreas and resuspended in buffer. Addition of urea, a compound that disrupts hydrogen bonding, resulted in a decrease in the activity of the enzyme (measured by testing its ability to cut RNA) and an increase in the viscosity of the solution (*Figure 10.19*), indicating that the protein was being **denatured** by unfolding into an unstructured polypeptide chain. The critical observation was that when the urea was removed by dialysis, the viscosity decreased and the enzyme activity reappeared. The conclusion is that the protein refolds spontaneously when the denaturant (in this case, urea) is removed. In these initial experiments the four disulfide bonds remained intact as these are not disrupted by urea, but the same result occurs when the urea treatment is combined with addition of a reducing agent that breaks the disulfide bonds: the activity is still regained on renaturation. This shows that the disulfide bonds are not critical to the protein's ability to refold, they merely stabilize the tertiary structure once it has been adopted.

More detailed study of the spontaneous folding pathways for ribonuclease and other small proteins has led to the following general, two-step description of the process (Hartl, 1996):

1. The secondary structural motifs along the polypeptide chain form within a few milliseconds of the denaturant being removed. This step is accompanied by the protein collapsing into a compact, but not folded, organization with its hydrophobic groups on the inside, shielded from water.
2. During the next few seconds or minutes, the secondary structural motifs interact with one another and the tertiary structure gradually takes shape, often via a series of intermediate conformations that are seen every time the protein folds. In other words, the protein follows a defined folding pathway. It may, however, make the wrong choice at certain steps in the pathway and temporarily adopt incorrect structures which unfold, presumably due to their lack of stability.

For several years it was more or less assumed that all proteins would fold spontaneously in the test tube, but

Box 10.6: Translation in the archaea

In most respects, translation in the archaea more closely resembles the equivalent events in the eukaryotic cytoplasm rather than in bacteria. The one apparent exception is that the archaeal ribosome, at 70S, is comparable in size to the bacterial ribosome and, like bacterial ribosomes, contains 23S, 16S and 5S rRNAs. This apparent similarity is illusory because the archaeal rRNAs form base-paired secondary structures that are significantly different from the equivalent bacterial structures. These structures are also different from the eukaryotic versions, but the ribosomal proteins that attach to the rRNAs are homologs of the eukaryotic proteins. Archaeal mRNAs are capped and polyadenylated and translation initiation is thought to involve a scanning process similar to that described for eukaryotic mRNAs. Archaeal tRNAs display a few unique features, including the absence of thymine in the so-called TΨC arm of the cloverleaf and the presence at various positions of modified nucleotides not seen in either bacteria or eukaryotes. The methionine carried by the initiator tRNA is not *N*-formylated and the initiation and elongation factors resemble the eukaryotic molecules.

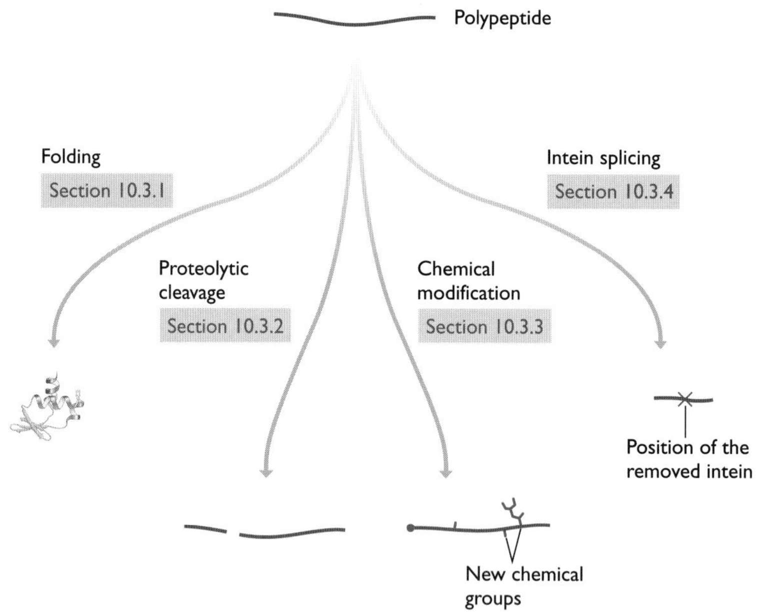

Figure 10.18 Schematic representation of the four types of post-translational processing event. Not all events occur in all organisms: see the text for details.

experiments have shown that only smaller proteins, with less complex structures, possess this ability. Two factors seem to prevent larger proteins from folding spontaneously. The first of these is their tendency to form insoluble aggregates when the denaturant is removed: the polypeptides may collapse into interlocked networks when they attempt to protect their hydrophobic groups from water, in step 1 of the general folding pathway. Experimentally, this can be avoided by using a low dilution of the protein, but this is not an option that the cell can take to prevent its unfolded proteins from aggregating. The second factor that prevents folding is that a large protein tends to get stuck in nonproductive side branches of its folding pathway, taking on an intermediate form that is incorrectly folded but which is too stable to unfold to any significant extent. Concerns have also been raised about the relevance of test tube folding (carried out with ribonuclease), to the folding of proteins in the cell, because a cellular protein might begin to fold before it has been fully synthesized. If the initial folding occurs when only part of the polypeptide is available, then there might be an increased possibility of incorrect branches of the folding pathway being followed. These various considerations prompted research into folding in living cells.

In cells, folding is aided by molecular chaperones

Most of our current understanding of protein folding in the cell is founded on the discovery of proteins that help other proteins to fold. These are called **molecular chaperones** and have been studied in most detail in *E. coli*, but it is clear that both eukaryotes and archaea possess equivalent proteins although some of the details of the way they work are different (Hartl, 1996).

The molecular chaperones in *E. coli* can be divided into two groups:

- **The Hsp70 chaperones**, which include the proteins called Hsp70 (coded by the *dnaK* gene and sometimes called DnaK protein), Hsp40 (coded by *dnaJ*) and GrpE.
- **The chaperonins**, the main version of which in *E. coli* is the GroEL/GroES complex.

Molecular chaperones do not specify the tertiary structure of a protein, they merely help the protein find that correct structure. The two types of chaperone do this in different ways. The Hsp70 family bind to hydrophobic regions of proteins, including proteins that are still being translated (*Figure 10.20A*). They prevent protein aggregation by holding the protein in an open conformation until it is completely synthesized and ready to fold. The Hsp70 chaperones are also involved in other processes that require shielding of hydrophobic regions in proteins, such as transport through membranes and disaggregation of proteins that have been damaged by heat stress.

The chaperonins work in a quite different way. GroEL and GroES form a multisubunit structure that looks like a hollowed-out bullet with a central cavity (*Figure 10.20B*; Xu *et al.*, 1997). A single unfolded protein enters the cavity and emerges folded. The mechanism is not known but it is postulated that GroEL/GroES acts as a cage that prevents the unfolded protein aggregating with other pro-

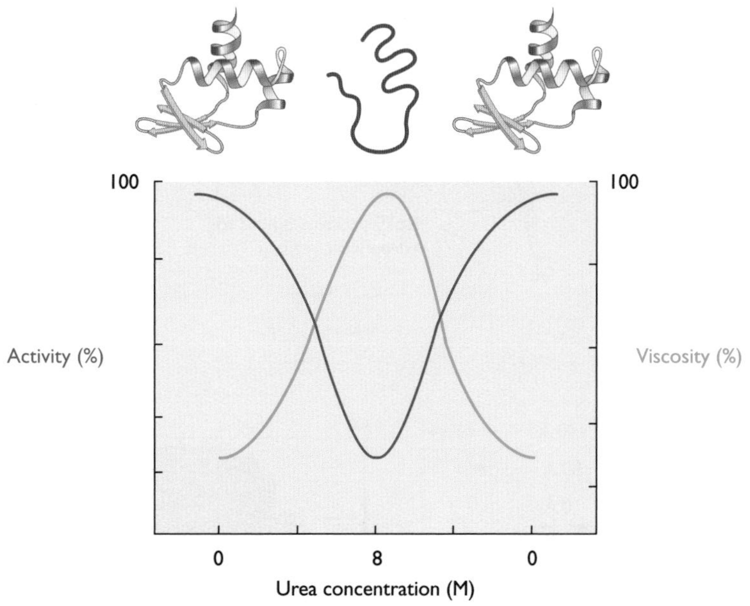

Figure 10.19 Denaturation and spontaneous renaturation of a small protein.

As the urea concentration increases to 8 M the protein becomes denatured by unfolding: its activity decreases and the viscosity of the solution increases. When the urea is removed by dialysis this small protein re-adopts its folded conformation. The activity of the protein increases back to the original level and the viscosity of the solution decreases.

teins, and that the inside surface of the cavity changes from hydrophobic to hydrophilic in such a way as to promote the burial of hydrophobic amino acids within the protein. This is not the only hypothesis: other researchers hold that the cavity unfolds proteins that have folded incorrectly, passing these unfolded proteins back to the cytoplasm so they can have a second attempt at adopting their correct tertiary structure (Hartl, 1996).

Eukaryotic proteins equivalent to both the Hsp70 family of chaperones and the GroEL/GroES chaperonins have been found, but it seems that in eukaryotes protein folding makes less use of chaperonins and depends more on the action of the Hsp70 proteins. This might be because of general differences between bacterial and eukaryotic proteins. Most bacterial proteins are relatively small (less than 300 amino acids), and all the secondary structural units interact to form a single tertiary structural unit or **folding domain**. Proteins of this type must be folded after translation has been completed, because all of the protein contributes to the single folding domain. Eukaryotic proteins, on the other hand, tend to be longer and usually comprise two or more folding domains, each one confined to a single segment of the amino acid sequence. Individual domains can fold before later domains have been synthesized: folding could therefore occur co-translationally. The chaperonin system is suitable for post-translational folding and so is used extensively by bacteria, whereas in eukaryotes there is more emphasis on Hsp70-type chaperones, as these can mediate co-translational folding (Netzer and Hartl, 1997).

10.3.2 Processing by proteolytic cleavage

Proteolytic cleavage has two functions in post-translational processing of proteins (*Figure 10.21*):

- It is used to remove short pieces from the N- and/or C-terminal regions of polypeptides, leaving a single, shortened molecule that folds into the active protein.
- It is used to cut **polyproteins** into segments, each one of which is an active protein.

These events are relatively common in eukaryotes but less frequent in bacteria.

Cleavage of the ends of polypeptides

Processing by cleavage is common with secreted polypeptides whose biochemical activities might be deleterious to the cell producing the protein. An example is provided by melittin, the most abundant protein in bee venom and the one responsible for causing cell lysis after injection of the bee-sting into the person or animal being stung. Melittin lyses cells in bees as well as animals and so must initially be synthesized as an inactive precursor. This precursor, promelittin, has 22 additional amino acids at its N terminus. The pre-sequence is removed by an extracellular protease that cuts it at eleven positions, releasing the active venom protein. The protease does not cleave within the active sequence because its mode of action is to release dipeptides with the sequence X–Y, where X is alanine, aspartic acid or glutamic acid, and Y

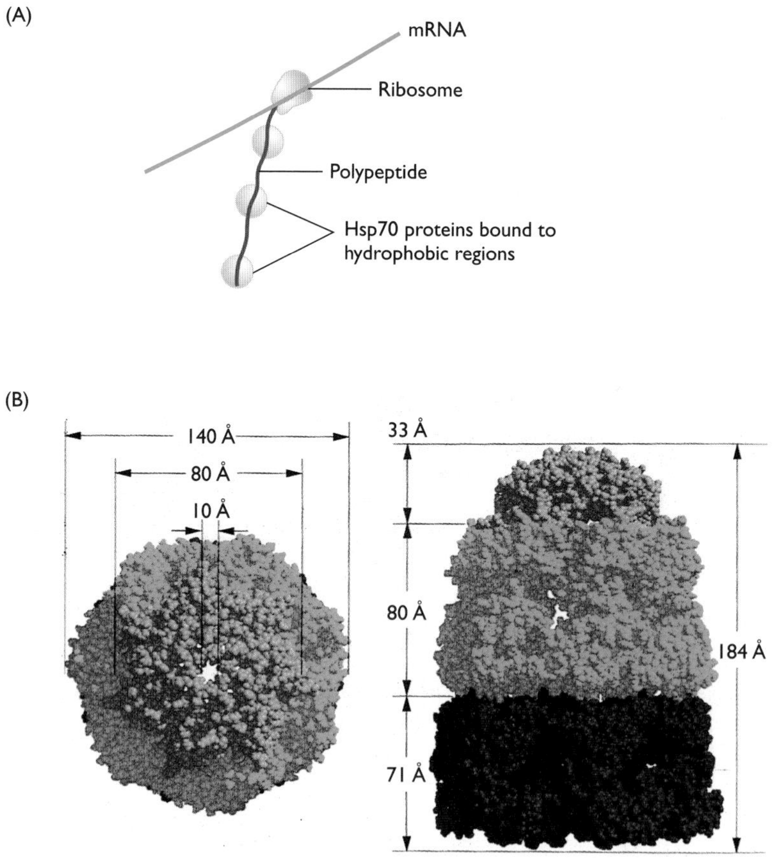

Figure 10.20 Molecular chaperones of *Escherichia coli*.

(A) Hsp70 chaperones bind to hydrophobic regions in unfolded polypeptides, including those that are still being translated, and hold the protein in an open conformation until it is ready to be folded. (B) The structure of the GroEL/GroES chaperonin. On the left is a view from the top and on the right a view from the side. 1 Å is equal to 0.1 nm. The GroES part of the structure is made up of seven identical protein subunits and is shown in gold. The GroEL components consist of 14 identical proteins arranged into two rings, each containing seven subunits: these two rings are shown in red and green. The main entrance into the central cavity is through the bottom of the structure shown on the right. Reprinted with permission from Xu *et al.*, *Nature*, **388**, 741–750. Copyright 1997 Macmillan Magazines Limited. Original image kindly supplied by Dr Zhaohui Xu, Department of Biological Chemistry, The University of Michigan.

is alanine or proline; these motifs do not occur in the active sequence (*Figure 10.22A*).

A similar type of processing occurs with insulin, the protein made in the Islets of Langerhans in the vertebrate pancreas and responsible for controlling blood sugar levels. Insulin is synthesized as preproinsulin, which is 105 amino acids in length (*Figure 10.22B*). The processing pathway involves the removal of the first 24 amino acids to give proinsulin, followed by two additional cuts which excise a central segment, leaving two active parts of the protein, the A and B chains, which link together by formation of two disulfide bonds to form mature insulin. The first segment to be removed, the 24 amino acids from the N terminus, is a **signal peptide**, a highly hydrophobic stretch of amino acids that attaches the precursor protein to a membrane prior to transport across that membrane and out of the cell. Signal peptides are commonly found on proteins that bind to and/or cross membranes, in both eukaryotes and prokaryotes.

Proteolytic processing of polyproteins

In the examples shown in *Figure 10.22*, proteolytic processing results in a single mature protein. This is not always the case. Some proteins are initially synthesized as polyproteins, long polypeptides that contain a series of mature proteins linked together in head-to-tail fashion. Cleavage of the polyprotein releases the individual proteins, which may have very different functions from one another.

Polyproteins are not uncommon in eukaryotes. Several types of eukaryotic virus use them as a way of reducing the sizes of their genomes, a single polyprotein gene with

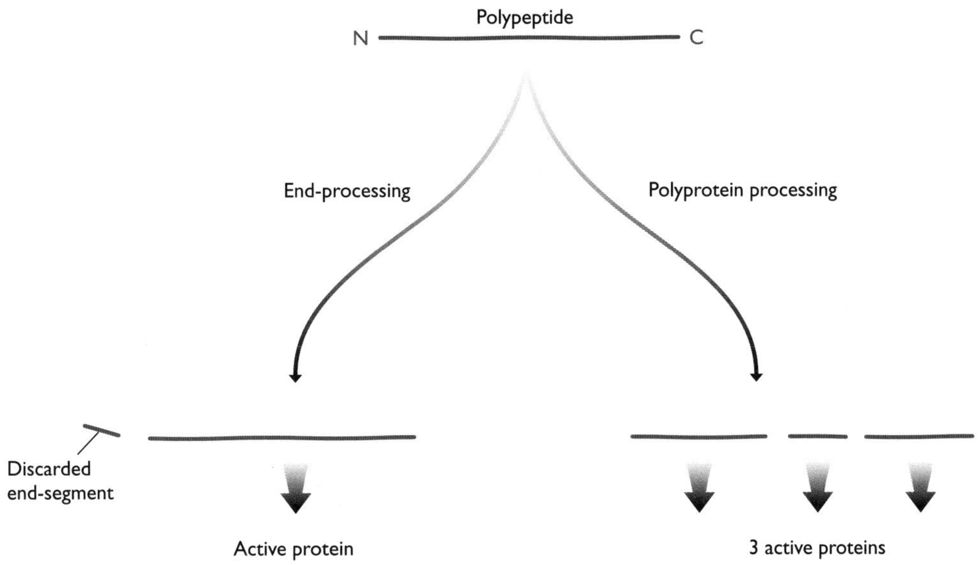

Figure 10.21 Protein processing by proteolytic cleavage.

On the left, the protein is processed by removal of the N-terminal segment. C-terminal processing also occurs with some proteins. On the right, a polyprotein is processed to give three different proteins. Note that not all proteins undergo proteolytic cleavage.

one promoter and one terminator taking up less space than a series of individual genes. Polyproteins are also involved in the synthesis of peptide hormones in vertebrates. For example, the polyprotein called proopiomelanocortin, made in the pituitary gland, contains at least ten different peptide hormones. These are released by proteolytic cleavage of the polyprotein (*Figure 10.23*), but not all can be produced at once because of overlaps between individual peptide sequences. Instead, the exact cleavage pattern is different in different cells.

10.3.3 Processing by chemical modification

The genome has the capacity to code for 21 different amino acids, the 20 specified by the standard genetic code, and selenocysteine, inserted in polypeptides by the context-dependent reading of a 5′-UGA-3′ codon (Section 10.1.2). This repertoire is increased dramatically by post-translational chemical modification of proteins, which results in a vast array of different amino acid types. The simpler types of modification occur in all organisms; the more complex ones, especially glycosylation, are rare in bacteria.

The simplest types of chemical modification involve addition of a small chemical group (e.g. an acetyl, methyl or phosphate group; *Table 10.5*) to an amino acid side chain, or to the amino or carboxyl groups of the terminal amino acids in a polypeptide (for an example, see Bradshaw *et al.*, 1998). Over 150 different modified amino acids have been documented in different proteins, with each modification carried out in a highly specific manner, the same amino acids being modified in the same way in every copy of the protein. This is illustrated in *Figure 10.24* for histone H3. This example reminds us that chemical modification often plays an important role in determining the precise biochemical activity of the target protein: we saw in Section 8.1.1 how acetylation of H3 and other histones has an important influence on chromatin fine structure and hence on gene expression. Other types of chemical modification have important regulatory roles, an example being phosphorylation, which is used to activate many proteins involved in signal transduction (Section 11.1.2).

More complex types of modification include **glycosylation**, the attachment of large carbohydrate side chains to polypeptides (Drickamer and Taylor, 1998). Two general types of glycosylation occur (*Figure 10.25*):

- **O-linked glycosylation** is the attachment of a sugar side chain via the hydroxyl group of a serine or threonine amino acid.
- **N-linked glycosylation** involves attachment through the amino group on the side chain of asparagine.

Glycosylation can result in attachment to the protein of grand structures comprising branched networks of 10–20 sugar units of various types. These side chains target proteins to particular sites in cells and determine the stability of proteins circulating in the bloodstream. Other types of large-scale modification involve attachment of long-chain

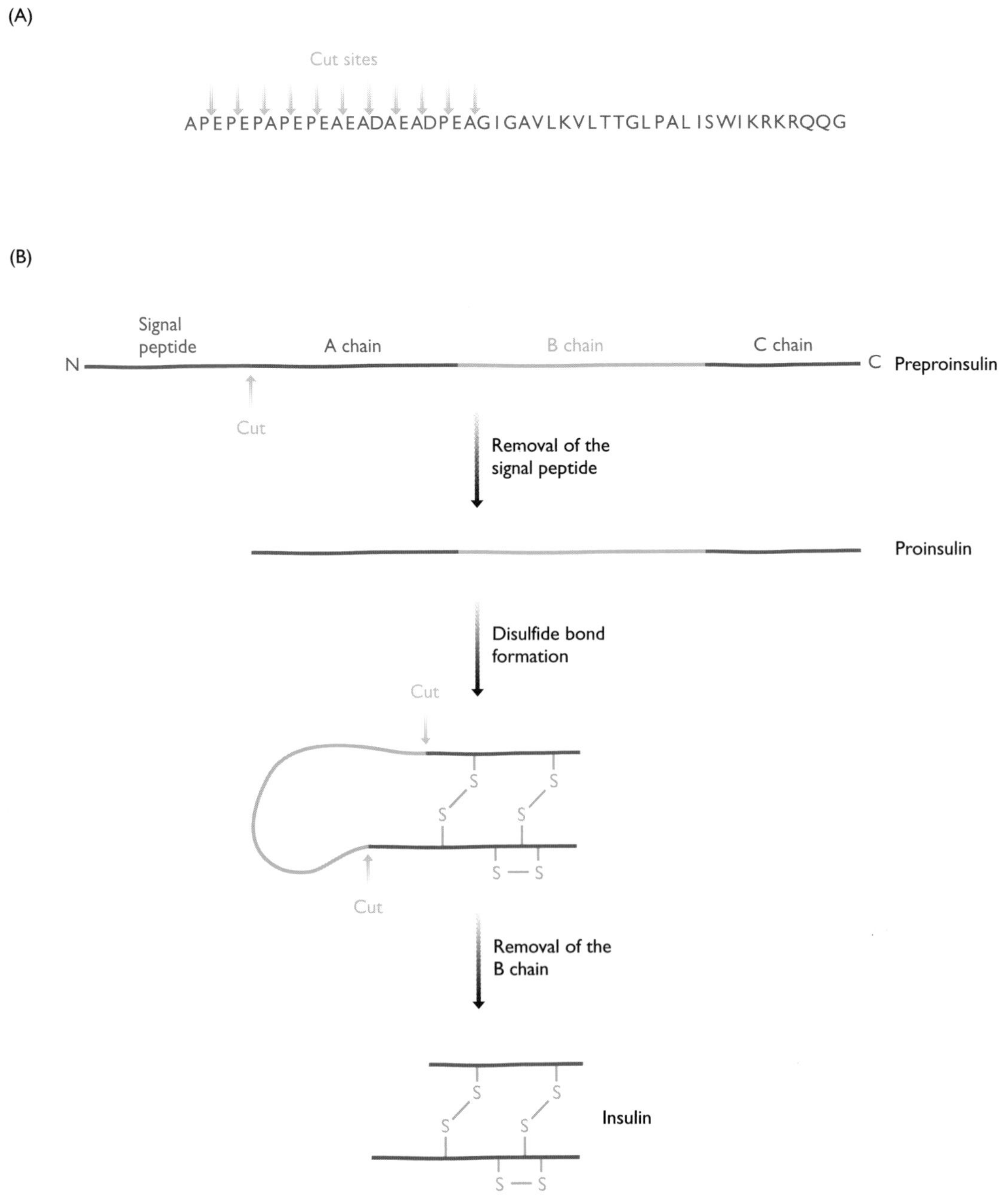

Figure 10.22 Post-translational processing by proteolytic cleavage.

(A) Processing of promelittin, the bee-sting venom. Arrows indicate the cut sites. For the one-letter abbreviations of the amino acids see *Table 7.3*, p. 155. (B) Processing of preproinsulin.

lipids, often to serine or cysteine amino acids. This process is called **acylation** and occurs with many proteins that become associated with membranes.

10.3.4 Inteins

The final type of post-translational processing that we must consider is intein splicing, a protein version of the more extensive intron splicing that occurs with pre-RNAs. Inteins are internal segments of proteins that are removed soon after translation, the two external segments or exteins becoming linked together (*Figure 10.26*). The first intein was discovered in 1990, in *Saccharomyces cerevisiae*, and there have only been 20 confirmed identifications so far, although there are almost 100 'hypothetical inteins' – ones that look like inteins because of their

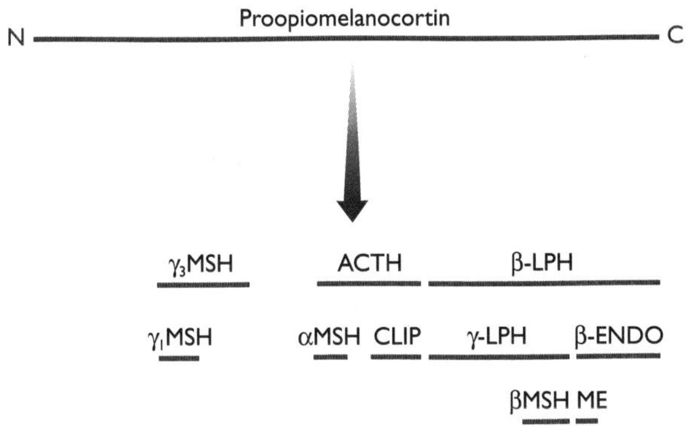

Figure 10.23 Processing of the proopiomelanocortin polyprotein.

Abbreviations: ACTH, adrenocorticotropic hormone; CLIP, corticotropin-like intermediate lobe protein; ENDO, endorphin; LPH, lipotrophin; ME, met-encephalin; MSH, melanotropin.

sequence features but whose splicing ability has not yet been demonstrated. Despite their scarcity, they are widespread. Most are known in the archaea, where there are a few cases of more than one intein in a single gene, but there are also examples in bacteria and lower eukaryotes. The eukaryotic inteins are present in both nuclear and organelle genes.

Most inteins are between 300 and 600 amino acids in length. Like pre-mRNA introns (Section 9.2.3), the sequences at the splice junctions of inteins are similar in all the known examples. The first amino acid is usually cysteine or, less frequently, serine, and the last two are almost always histidine followed by asparagine. The first amino acid of the downstream extein is cysteine, serine or threonine. A few other amino acids within the intein sequence are also conserved. It is assumed that these conserved amino acids are involved in the splicing process, which is self-catalyzed by the intein itself, but the detailed mechanism has not yet been worked out (Cooper and Stevens, 1995).

Two interesting features of inteins have recently come to light. The first of these was discovered when the structures of two inteins were determined by X-ray crystallography (Duan *et al.*, 1997; Klabunde *et al.*, 1998). These structures are similar in some respects to that of a *Drosophila* protein called Hedgehog, which is involved in development of the segmentation pattern of the fly embryo. Hedgehog is an autoprocessing protein that cuts itself in two, and the structural similarity with inteins lies in the part of the Hedgehog protein that catalyzes its

Table 10.5 Examples of post-translational chemical modifications

Modification	Amino acids that are modified	Examples of proteins
Addition of small chemical groups		
Acetylation	Lysine	Histones
Methylation	Lysine	Histones
Phosphorylation	Serine	Some proteins involved in signal transduction
Hydroxylation	Proline, lysine	Collagen
N-formylation	N-terminal glycine	Melittin
Addition of sugar side chains		
O-linked glycosylation	Serine, threonine	Many membrane proteins and secreted proteins
N-linked glycosylation	Asparagine	Many membrane proteins and secreted proteins
Addition of lipid side chains		
Acylation	Serine, threonine, cysteine	Many membrane proteins
N-myristoylation	N-terminal glycine	Some protein kinases involved in signal transduction

See Section 11.1.2 for more information on the role of chemical modification during signal transduction.

Figure 10.24 Post-translational chemical modification of calf histone H3.

The first 30 amino acids of this 135-amino-acid protein are listed, using the one-letter abbreviations (see *Table 7.3*, p. 155). Four modifications occur, two methylations and two acetylations. For the role of acetylation of histones in determining chromatin structure see Section 8.1.1.

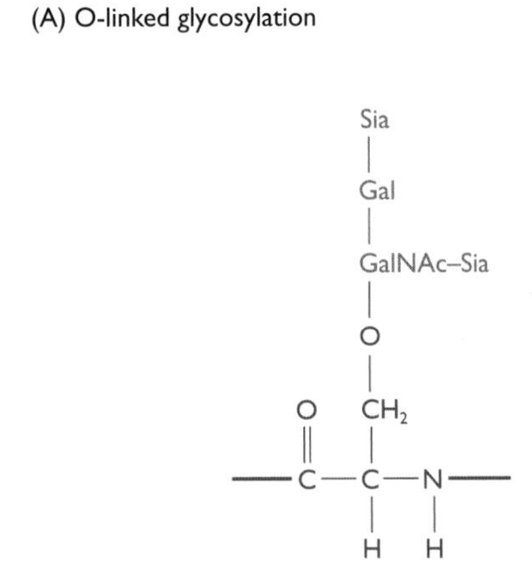

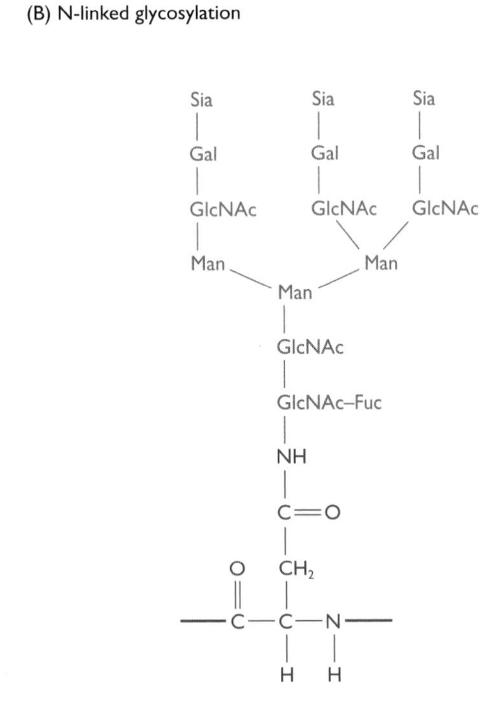

Figure 10.25 Glycosylation.

(A) O-linked glycosylation. The structure shown is found in a number of glycoproteins. It is drawn here attached to a serine amino acid but can also be linked to a threonine. (B) N-linked glycosylation usually results in larger sugar structures than seen with O-linked glycosylation. The drawing shows a typical example of a complex glycan attached to an asparagine amino acid. Abbreviations: Fuc, fucose; Gal, galactose; GalNAc, N-acetylgalactosamine; GlcNAc, N-acetylglucosamine; Man, mannose; Sia, sialic acid.

self-cleavage. Possibly the same protein structure has evolved twice, or possibly inteins and Hedgehog shared a common link at some stage in the evolutionary past.

The second interesting feature is that with some inteins the excised segment is a sequence-specific endonuclease.

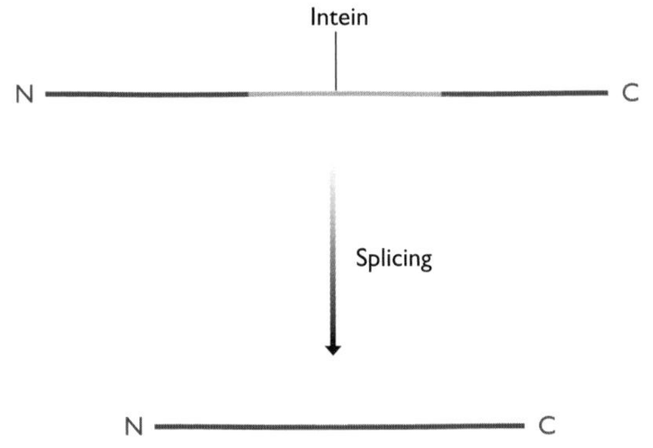

Figure 10.26 Intein splicing.

The intein cuts DNA at the sequence corresponding to its insertion site in a gene coding for an intein-free version of the protein that it is derived from (*Figure 10.27*). If the cell also contains a gene coding for the intein-containing protein, then the DNA sequence for the intein is able to jump into the cut site, converting the intein-minus gene into an intein-plus version, a process called **intein homing**. The same type of event occurs with some Group I introns (Section 9.3.3), which code for proteins that direct **intron homing**. It is possible that transfer of inteins and Group I introns might also occur between cells or even between species (Cooper and Stevens, 1995). This is thought to be a mechanism by which **selfish DNA** is able to propagate (see Box 14.5, p. 386).

10.4 PROTEIN TURNOVER

The protein synthesis and processing events that we have studied so far in this chapter result in new, active proteins that take up their place in the cell's proteome. These proteins either replace existing ones that have reached the end of their working lives or provide new protein functions in response to the changing requirements of the cell. The concept that the proteome of a cell can change over time requires not only *de novo* protein synthesis but also the removal of proteins whose functions are no longer required. This removal must be highly selective so that only the correct proteins are degraded, and must also be rapid in order to account for the abrupt changes that occur under certain conditions, for example during key transitions in the cell cycle (Hunt, 1997).

For many years, protein degradation was an unfashionable subject and it was not until the 1990s that real progress was made in understanding how specific proteolysis events are linked with processes such as the cell cycle and differentiation. Even now, our knowledge centers largely on descriptions of protein breakdown pathways and less on the regulation of the pathways and the mechanisms used to target specific proteins. There also appear to be a number of different types of breakdown

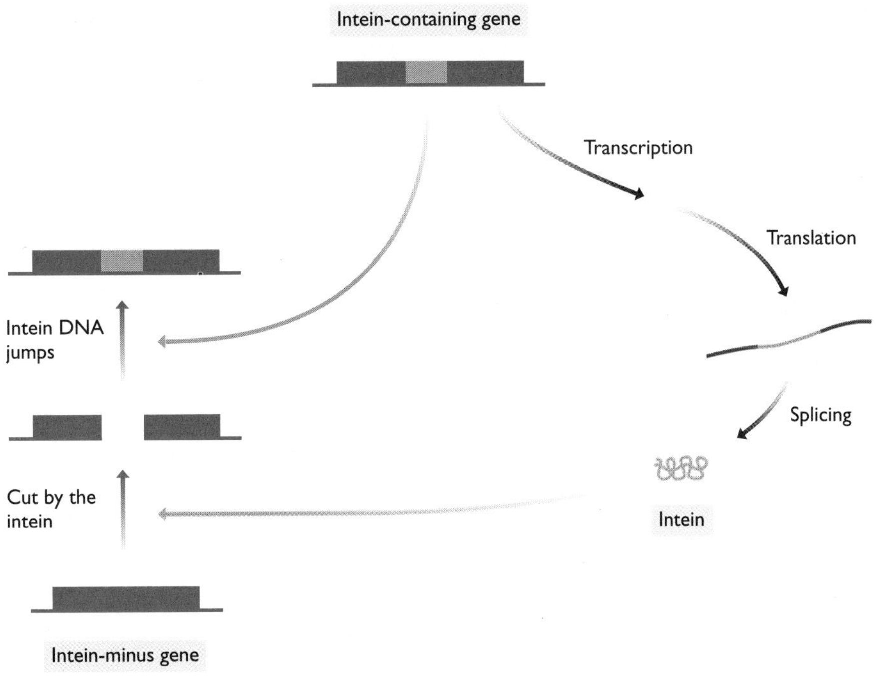

Figure 10.27 Intein homing.

The cell is heterozygous for the intein-containing gene, possessing one allele with the intein and one allele without the intein. After protein splicing, the intein cuts the intein-minus gene at the appropriate place, allowing a copy of the intein DNA sequence to jump into this gene, converting it into the intein-plus version.

pathway whose interconnectivities have not yet been traced. This is particularly true in bacteria, which seem to have a range of proteases that work together in controlled turnover of proteins. In eukaryotes, on the other hand, most breakdown involves a single system, that involving **ubiquitin** and the **proteasome**.

10.4.1 Degradation of ubiquitin-tagged proteins in the proteasome

A link between ubiquitin and protein degradation was first established in 1975 when it was shown that this abundant, 76 amino acid protein is involved in energy-dependent proteolysis reactions in rabbit cells (Varshavsky, 1997). Subsequent research identified a series of enzymes that attach ubiquitin molecules, singly or in chains, to lysine amino acids in proteins that are targeted for breakdown. Whether or not a protein becomes ubiquitylated depends on the presence or absence within it of amino acid motifs that act as degradation signals. These signals have not been completely characterized but there are thought to be at least ten different types in *Saccharomyces cerevisiae*, including:

- **The N-degron**, a sequence element present at the N terminus of a protein.
- **PEST sequences**, internal sequences that are rich in proline (P), glutamic acid (E), serine (S) and threonine (T).

These sequences are permanent features of the proteins that contain them and so cannot be straightforward 'degradation signals': if they were then these proteins would be broken down as soon as they are synthesized. Instead they must determine susceptibility to degradation and hence the general stability of a protein in the cell. How this might be linked to the controlled breakdown of selected proteins at specific times, for instance during the cell cycle, is not yet clear.

The second component of the ubiquitin-dependent degradation pathway is the proteasome, the structure within which ubiquitylated proteins are broken down. In eukaryotes, the proteasome is a large, multisubunit structure with a sedimentation coefficient of 26S, comprising a hollow cylinder of 20S and two 'caps' of 19S (Groll *et al.*, 1997). Archaea also have proteasomes of about the same size but these are less complex as they are composed of multiple copies of just two proteins: eukaryotic proteasomes contain 14 different types of protein subunit. The entrance into the cavity within the proteasome is narrow and a protein must be unfolded before it can get in. This probably occurs through an energy-dependent process which may involve structures similar to chaperonins

(Section 10.3.1) but with unfolding rather than folding activity (Lupas *et al.*, 1997). After unfolding, the protein can enter the proteasome within which it is cleaved into short peptides 4–10 amino acids in length. These are released back into the cytoplasm where they are broken down into single amino acids which can be reutilized in protein synthesis.

REFERENCES

Agrawal RK, Penczek P, Grassucci RA, Li Y, Leith A, Nierhaus KH and Frank J (1996) Direct visualization of the A-, P-, and E-site transfer RNAs in the *Escherichia coli* ribosome. *Science*, **271**, 1000–1002.

Anfinsen CB (1973) Principles that govern the folding of protein chains. *Science*, **181**, 223–230.

Arnez JG and Moras D (1997) Structural and functional considerations of the aminoacylation reaction. *Trends Biochem. Sci.*, **22**, 211–216.

Bradshaw RA, Brickey WW and Walker KW (1998) N-terminal processing: the methionine aminopeptidase and N^{α}-acetyl transferase families. *Trends Biochem. Sci.*, **23**, 263–267.

Cedergren R and Miramontes P (1996) The puzzling origin of the genetic code. *Trends Biochem. Sci.*, **21**, 199–200.

Cooper AA and Stevens TH (1995) Protein splicing: self-splicing of genetically mobile elements at the protein level. *Trends Biochem. Sci.*, **20**, 352–357.

Covello PS and Gray MW (1989) RNA editing in plant mitochondria. *Nature*, **341**, 662–666.

Craig AWB, Haghighat A, Yu ATK and Sonenberg N (1998) Interaction of polyadenylate-binding protein with the eIF4G homologue PAIP enhances transcription. *Nature*, **392**, 520–523.

Crick FHC (1990) *What Mad Pursuit: A Personal View of Scientific Discovery.* Penguin Books, London.

Drickamer K and Taylor ME (1998) Evolving views of protein glycosylation. *Trends Biochem. Sci.* **23**, 321–324.

Duan XQ, Gimble FS and Quiocho FA (1997) Crystal structure of PI-SceI, a homing endonuclease with protein splicing activity. *Cell*, **89**, 555–564.

Farabaugh PJ (1996) Programmed translational frameshifting. *Annu. Rev. Genet.*, **30**, 507–528.

Groll M, Ditzel L, Löwe J, Stock D, Bochtler M, Bartunik HD and Huber R (1997) Structure of 20S proteasome from yeast at 2.4 Å resolution. *Nature*, **386**, 463–471.

Gualberto JM, Lamattinia L, Bonnard G, Weil J-H and Grienenberger J-M (1989) RNA editing in wheat mitochondria results in the conservation of protein sequences. *Nature*, **341**, 660–662.

Hafezparast M and Fisher E (1998) Wasted by an elongation factor. *Trends Genet.*, **14**, 215–217.

Hale SP, Auld DS, Schmidt E and Schimmel P (1997) Discrete determinants in transfer RNA for editing and aminoacylation. *Science*, **276**, 1250–1252.

Hartl FU (1996) Molecular chaperones in cellular protein folding. *Nature*, **381**, 571–580.

Heilek GM and Noller HF (1996) Site-directed hydroxyl radical probing of the rRNA neighbourhood of ribosomal protein S5. *Science*, **272**, 1659–1662.

Hentze MW (1997) eIF4G: a multipurpose ribosome adapter? *Science*, **275**, 500–501; **275**, 1553.

Hunt T (1997) Extinction is forever. *Trends Biochem. Sci.*, **22**, 371.

Ibba M, Curnow AW and Söll D (1997) Aminoacyl-tRNA synthesis: divergent routes to a common goal. *Trends Biochem. Sci.*, **22**, 39–42.

Joseph S, Weiser B and Noller HF (1997) Mapping the inside of the ribosome with an RNA helical ruler. *Science*, **278**, 1093–1098.

Klabunde T, Sharma S, Telenti A, Jacobs WR and Sacchettini JC (1998) Crystal structure of *gyrA* intein from *Mycobacterium xenopi* reveals structural basis of protein splicing. *Nature Struct. Biol.*, **5**, 31–36.

Low SC and Berry MJ (1996) Knowing when not to stop: selenocysteine incorporation in eukaryotes. *Trends Biochem. Sci.*, **21**, 203–208.

Lupas A, Flanaghan JM, Tamaru T and Baumeister W (1997) Self-compartmentalizing proteases. *Trends Biochem. Sci.*, **22**, 399–404.

Mountford PS and Smith AG (1995) Internal ribosome entry sites and dicistronic RNAs in mammalian transgenesis. *Trends Genet.*, **11**, 179–184.

Netzer WJ and Hartl FU (1997) Recombination of protein domains facilitated by co-translational folding in eukaryotes. *Nature*, **388**, 343–349.

Nitta I, Kamada Y, Noda H, Ueda T and Watanabe K (1998) Reconstitution of peptide bond formation with *Escherichia coli* 23S ribosomal RNA domains. *Science*, **281**, 666–669.

Preiss T and Hentze MW (1998) Dual function of the messenger RNA cap structure in poly(A)-tail-promoted translation in yeast. *Nature*, **392**, 516–520.

Ramakrishnan V and White SW (1998) Ribosomal protein structures: insights into the architecture, machinery and evolution of the ribosome. *Trends Biochem. Sci.*, **23**, 208–212.

Varshavsky A (1997) The ubiquitin system. *Trends Biochem. Sci.*, **22**, 383–387.

Xu Z, Horwich AL and Sigler PB (1997) The crystal structure of the asymmetric GroEL-GroES-$(ADP)_7$ chaperonin complex. *Nature*, **388**, 741–750.

FURTHER READING

Arnstein HRV and Cox RA (1992) *Protein Biosynthesis: In Focus*. IRL Press, Oxford. — *Excellent source of general information on translation.*

Green R and Noller HF (1997) Ribosomes and translation. *Annu. Rev. Biochem.*, **66**, 679–716. — *Detailed description of the* Escherichia coli *ribosome.*

Kozak M (1992) Regulation of translation in eukaryotic systems. *Annu. Rev. Cell Biol.*, **8**, 197–225.

McCarthy JEG (1998) Posttranscriptional control of gene expression in yeast. *Microbiol. Mol. Biol. Rev.*, **62**, 1492–1553. — *Detailed review of translation and its control in yeast.*

Merrick WC (1992) Mechanism and regulation of eukaryotic protein synthesis. *Microbiol. Rev.*, **56**, 291–315.

Nagai K and Mattaj IW (eds) (1994) *RNA–Protein Interactions*. IRL Press, Oxford. — *Chapters 3 and 4 are on aminoacyl-tRNA synthetases and ribosomes.*

CHAPTER

11 Regulation of Genome Activity

Contents

11.1 Transient Changes in Genome Activity 266
11.1.1 Signal transmission by import of the extracellular signaling compound 267
11.1.2 Signal transmission mediated by cell surface receptors 270
11.2 Permanent and Semipermanent Changes in Genome Activity 277
11.2.1 Genome rearrangements 278
11.2.2 Changes in chromatin structure 280
11.2.3 Genome regulation by feedback loops 281
11.3 Regulation of Genome Activity During Development 282
11.3.1 Sporulation in *Bacillus* 282
11.3.2 Vulval development in *Caenorhabditis elegans* 285
11.3.3 Development in *Drosophila melanogaster* 287

Concepts

- *Transient changes in genome activity enable the cell to respond to external stimuli*
- *Transient changes occur in response to external signaling compounds that enter the cell or act via cell surface receptors*
- *Permanent changes in genome activity underlie differentiation*
- *Permanent changes in genome activity can be brought about by DNA rearrangements, changes in chromatin structure, and feedback loops*
- *During developmental processes, coordination between genomes in different cells can be achieved by intercellular signaling*
- *Gradients of signaling compounds can convey positional information to cells*
- *Genes that control the body plan of the fruit fly are similar to genes that carry out equivalent functions in vertebrates including humans*

WE HAVE FOLLOWED the gene expression pathway from initiation of transcription to synthesis of functioning protein. This expression pathway, repeated many times in parallel and combined with the turnover of existing proteins, is the means by which the genome specifies the content of the proteome, which in turn defines the biochemical signature of the cell. In no organism is this biochemical signature entirely constant. Even the simplest unicellular organisms are able to alter their proteomes to take account of changes in the environment, so that their biochemical capabilities are continually in tune with the available nutrient supply and the prevailing physical and chemical conditions. Cells in multicellular organisms are equally responsive to changes in the extracellular environment, the only difference being that the major stimuli are hormones and growth factors rather than nutrients. The resulting *transient* changes in genome activity enable the proteome to be remodeled continually to satisfy the demands that the outside world places on the cell (*Figure 11.1*). Other changes in genome activity are *permanent* or at least *semipermanent*, and result in the cell's biochemical signature becoming altered in a way that is not readily reversible. These changes lead to cellular **differentiation**, the adoption by the cell of a specialized physiological role. Many unicellular organisms are able to differentiate, an example being the production of spore cells by bacteria such as *Bacillus*, but we more frequently associate differentiation with multicellular organisms, most of which are composed of different types of specialized cell (over 250 types in humans) organized into tissues and organs. Assembly of these complex multicellular structures, and of the organism as a whole, requires coordination of the activities of genomes in different cells. This coordination involves both transient and permanent changes, and must continue over a long period of time during the **development** of the organism.

Within the expression pathways for individual genes there are many steps at which regulation can be exerted (*Table 11.1*) and examples of the biological roles of different mechanisms were provided at the appropriate places in Chapters 8–10. The objective of this chapter is not to reiterate these gene-specific control systems, but to explain how the activity of the genome as a whole is regulated. In doing this we should bear in mind that the biosphere is so diverse, and the numbers of genes in individual genomes so large, that it is reasonable to assume that any mechanism that could have evolved to regulate gene expression is likely to have done so. It is therefore no surprise that we can nominate examples of regulation for every point in the gene expression pathway. But are all these control points of equal importance in regulating the activity of the genome as a whole? Our current perception is that they are not. Our understanding may be imperfect, based as it is on investigation of

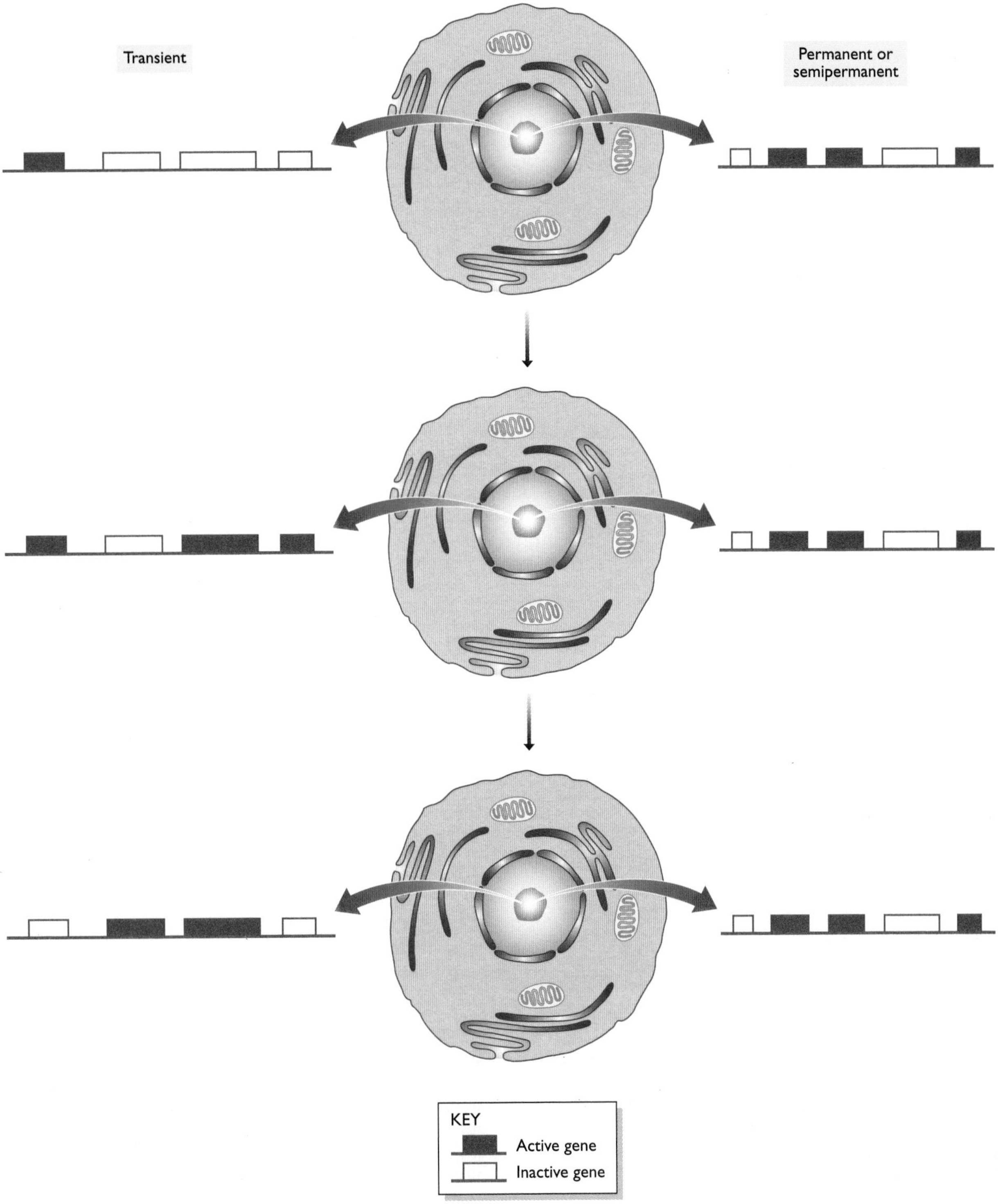

Figure 11.1 Two ways in which genome activity is regulated.

The set of genes on the left are subject to transient regulation and are switched on and off in response to changes in the extracellular environment. The cluster of genes on the right have undergone a permanent or semipermanent change in their expression pattern, resulting in the same three genes being expressed continuously.

Table 11.1 Examples of steps in the gene expression pathway at which regulation can be exerted

Step	Example of regulation	Cross-reference
Transcription		
Gene accessibility	Locus control regions determine chromatin structure in areas that contain genes	Section 8.1.1
	Nucleosome positioning controls access of RNA polymerase and transcription factors to the promoter region	Section 8.1.1
Initiation of transcription	Productive initiation is influenced by transcription factors and other control systems	Section 8.3
Synthesis of RNA	Prokaryotes use antitermination and attenuation to control the amount and nature of individual transcripts	Section 9.2.1
Eukaryotic mRNA processing		
Capping	Some animals use capping as a means of regulating protein synthesis during egg maturation	
Polyadenylation	Translation of *bicoid* mRNA in *Drosophila* eggs is activated after fertilization by extension of the poly(A) tail	Section 11.3.3
Splicing	Alternative splice site selection controls sex determination in *Drosophila*	Section 9.2.3
Chemical modification	RNA editing of apolipoprotein-B mRNA results in liver- and intestine-specific versions of this protein	Section 9.4.2
mRNA turnover	Iron controls turnover of transferrin receptor mRNA	Box 10.4, p. 248
Protein synthesis and processing		
Initiation of translation	Phosphorylation of eIF-2 results in a general reduction in translation initiation in eukaryotes	Box 10.4, p. 248
	Ribosomal proteins in bacteria control their own synthesis by modulating ribosome attachment to their mRNAs	Box 10.4, p. 248
	In some eukaryotes, iron controls ribosome scanning on ferritin mRNAs	Box 10.4, p. 248
Protein synthesis	Frameshifting enables two DNA polymerase III subunits to be translated by the *E. coli dnaX* gene	Box 10.5, p. 250
Cutting events	Alternative cleavage pathways for polyproteins result in tissue-specific protein products	Section 10.3.2
Chemical modification	Many proteins involved in signal transduction are activated by phosphorylation	Section 11.1.2

just a limited number of genes in a few organisms, but it appears that the critical controls over genome expression, the decisions about which genes are switched on and which are switched off, are exerted at the level of transcription initiation. For most genes, control that is exerted at later steps serves to modulate expression but does not act as the primary determinant of whether the gene is on or off (see *Figure 8.13*, p. 185). Most, but not all, of what we will discuss in this chapter therefore concerns control of genome activity by mechanisms that specify which genes are transcribed and which are silent. We will address two issues: the way in which transient and permanent changes in genome activity are brought about, and the way in which these changes are linked in time and space to result in developmental pathways.

11.1 TRANSIENT CHANGES IN GENOME ACTIVITY

Transient changes in genome activity occur predominantly in response to external stimuli. For unicellular organisms, the most important external stimuli relate to nutrient availability, these cells living in variable environments in which the identities and relative amounts of the nutrients change over time. The genomes of unicellular organisms therefore include genes for uptake and utilization of a range of nutrients, and changes in nutrient availability are shadowed by changes in genome activity, so that at any one time only those genes needed to utilize the available nutrients are expressed. Most cells in multicellular organisms live in less variable environments, but an environment whose maintenance requires coordination between the activities of different cells. For these cells, the major external stimuli are therefore hormones, growth factors and related compounds that convey signals within the organism and stimulate coordinated changes in genome activity.

To exert an effect on genome activity, the nutrient, hormone, growth factor or other extracellular compound that represents the external stimulus must influence events within the cell. There are two ways in which it can do this (*Figure 11.2*):

- **Directly**, by acting as a signaling compound that is transported across the cell membrane and into the cell.
- **Indirectly**, by binding to a cell surface receptor which transmits a signal into the cell.

Signal transmission is one of the major research areas in cell biology (Alberts *et al.*, 1994) with attention focused in particular on its relevance to the abnormal biochemical activities that underlie cancer. Many examples of direct and indirect signal transmission have been discovered, some of general importance in a variety of organisms and others restricted to just a few species. In the first part of this chapter we will survey the field.

11.1.1 Signal transmission by import of the extracellular signaling compound

In the direct method of signal transmission the extracellular compound that represents the external stimulus crosses the cell membrane and enters the cell. After import into the cell, there are three ways in which the signaling compound can influence genome activity:

- By acting as a transcription factor.
- By directly influencing the activity of a transcription factor.
- By indirectly influencing the activity of a transcription factor.

Examples of each of these three types of imported signaling compound are described below.

Lactoferrin is an extracellular signaling compound which acts as a transcription factor

If the extracellular signaling compound that is imported into the cell is a protein with suitable DNA-binding properties then it could function as a transcription factor, directly activating or repressing target genes. This might appear to be an attractively straightforward way of regulating genome activity, especially as it would provide a convenient system for coordinating changes in the gene expression patterns of groups of cells in a multicellular organism, but it is not a common mechanism. The reason for this is not clear but probably relates, at least partly, to the difficulty in designing a protein that combines the hydrophobic properties needed for effective transport across a membrane with the hydrophilic properties needed for migration through the aqueous cytoplasm to the protein's site of action in the nucleus.

The one clear example of a signaling protein that acts as a transcription factor is provided by lactoferrin, a mammalian protein found mainly in milk and to a lesser extent in the bloodstream. The specific function of lactoferrin has been difficult to pin down, but it seems to play a role in the body's defenses against microbial attack. As its name suggests, lactoferrin is able to bind iron, and it is thought that at least part of its protective role arises from its ability to reduce free iron levels in milk, thereby starving invading microbes of this essential cofactor. Lactoferrin might therefore appear to be an unlikely candidate for a transcription factor, but it has been known since the early 1980s that the protein is multitalented and, among other things, can bind to DNA. This property was linked to a second function of lactoferrin – stimulation of the blood cells involved in the immune response – when in 1992 it was shown that the protein is taken up by immune cells, enters their nuclei, and attaches to the DNA (Garre *et al.*, 1992). Subsequently the DNA-binding was shown to be sequence-specific and to result in gene activation, showing that lactoferrin is a true transcription factor (He and Furmanski, 1995).

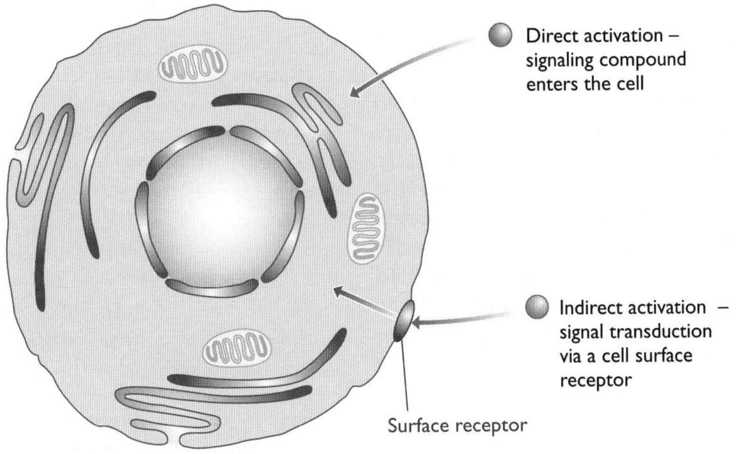

Figure 11.2 Two ways in which an extracellular signaling compound can influence events occurring within a cell.

Some imported signaling compounds directly influence the activity of pre-existing transcription factors

Although few imported signaling compounds are able themselves to act as transcription factors, many have the ability to directly influence the activity of transcription factors already present in the cell. We encountered one example of this type of regulation in Section 8.3.1 when we studied the lactose operon of *E. coli*. This operon responds to extracellular levels of lactose, the latter acting as a signaling molecule which enters the cell and, after conversion to its isomer allolactose, influences the DNA-binding properties of the lactose repressor and hence determines whether or not the lactose operon is transcribed (see *Figure 8.15*, p. 187). Many other bacterial operons coding for genes involved in sugar utilization are controlled in this way.

Direct interaction with transcription factors is also a common means of regulating genome activity in order to ensure that the metal ion content of the cell is at an appropriate level. Cells need metal ions such as copper and zinc as cofactors in biochemical reactions, but these metals are toxic if they accumulate in the cell above a certain amount. Their uptake therefore has to be carefully controlled so that the cell contains sufficient metal ions when the environment is lacking in metal compounds, but does not over-accumulate metal ions when the environmental concentrations are high. The strategies used are illustrated by the copper control system of *Saccharomyces cerevisiae*. This yeast has two copper-dependent transcription factors, Mac1p and Ace1p. Both of these factors bind copper ions, the binding inducing a conformational change that activates the factor and enables it to stimulate expression of its target genes (*Figure 11.3*). For Mac1p these target genes code for copper-uptake proteins, whereas for Ace1p they are genes coding for proteins such as superoxide dismutase that are involved in copper detoxification. The balance between activation of the two transcription factors ensures that the copper content of the cell remains within acceptable levels (Pena *et al*., 1998; Winge *et al*., 1998).

Transcription factor activation is also the main way in which **steroid hormones** such as progesterone, estrogen and glucocorticoid hormone coordinate genome activities in the cells of higher eukaryotes (Tsai and O'Malley, 1994). Steroid hormones are hydrophobic and so easily penetrate the cell membrane. Once inside the cell, each hormone binds to a specific **steroid receptor** protein, which is usually located in the cytoplasm (*Figure 11.4*). After binding, the activated receptor migrates into the nucleus, where it attaches to **hormone response elements** upstream of target genes. The typical hormone response element is a 15 bp sequence comprising a 6 bp inverted palindrome separated by a 3 bp spacer to which the steroid receptor binds via a special version of the zinc finger (see *Figure 7.23*, p. 166). Each receptor activates some 50–100 genes, and hence induces a large-scale change in the biochemical properties of the cell.

All steroid receptors are structurally similar, not just with regards to their DNA-binding domain but also in other parts of their protein structures (*Figure 11.5*). Recognition of these similarities has led to identification of a number of putative or orphan steroid receptors whose hormonal partners and cellular functions are not yet known. The structural similarities have also shown that a second set of receptor proteins, the **nuclear receptor superfamily**, belong to the same general class as steroid receptors although the hormones that they work with are not themselves steroids. As their name suggests, these receptors are located in the nucleus rather than the cytoplasm. They include the receptors for vitamin D_3, whose roles include control of bone development, and thyroxine, which stimulates the tadpole to frog metamorphosis.

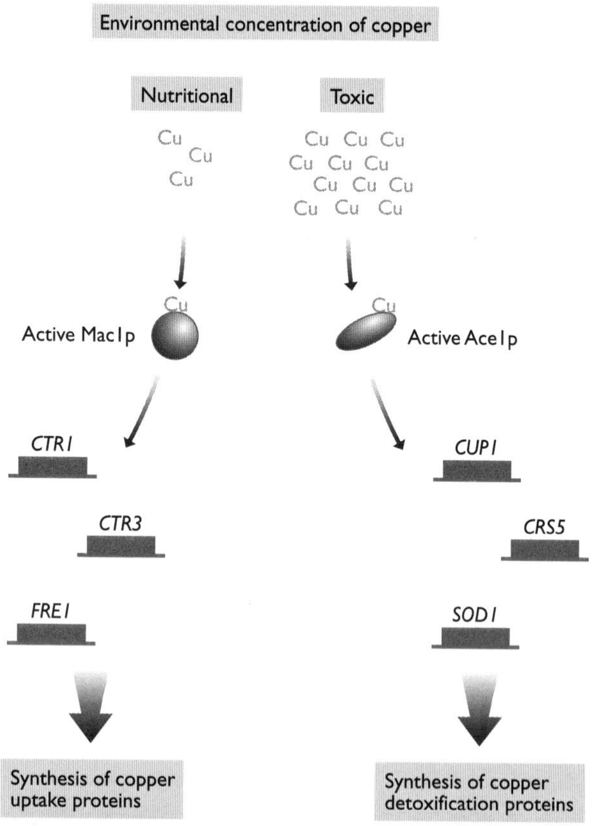

Figure 11.3 Copper-regulated gene expression in *Saccharomyces cerevisiae*.

Yeast cells require low amounts of copper because a few of its enzymes (e.g. cytochrome *c* oxidase and tyrosinase) are copper-containing metalloproteins, but too much copper is toxic for the cell. When copper levels are low, the Mac1p transcription factor is activated by copper binding and switches on expression of genes for copper uptake. When the copper levels are too high, a second transcription factor, Ace1p, is activated, switching on expression of a different set of genes, these coding for proteins involved in copper detoxification.

Some imported signaling compounds influence genome activity indirectly

There are a number of cases where the signaling molecule does not interact directly with a transcription factor, but

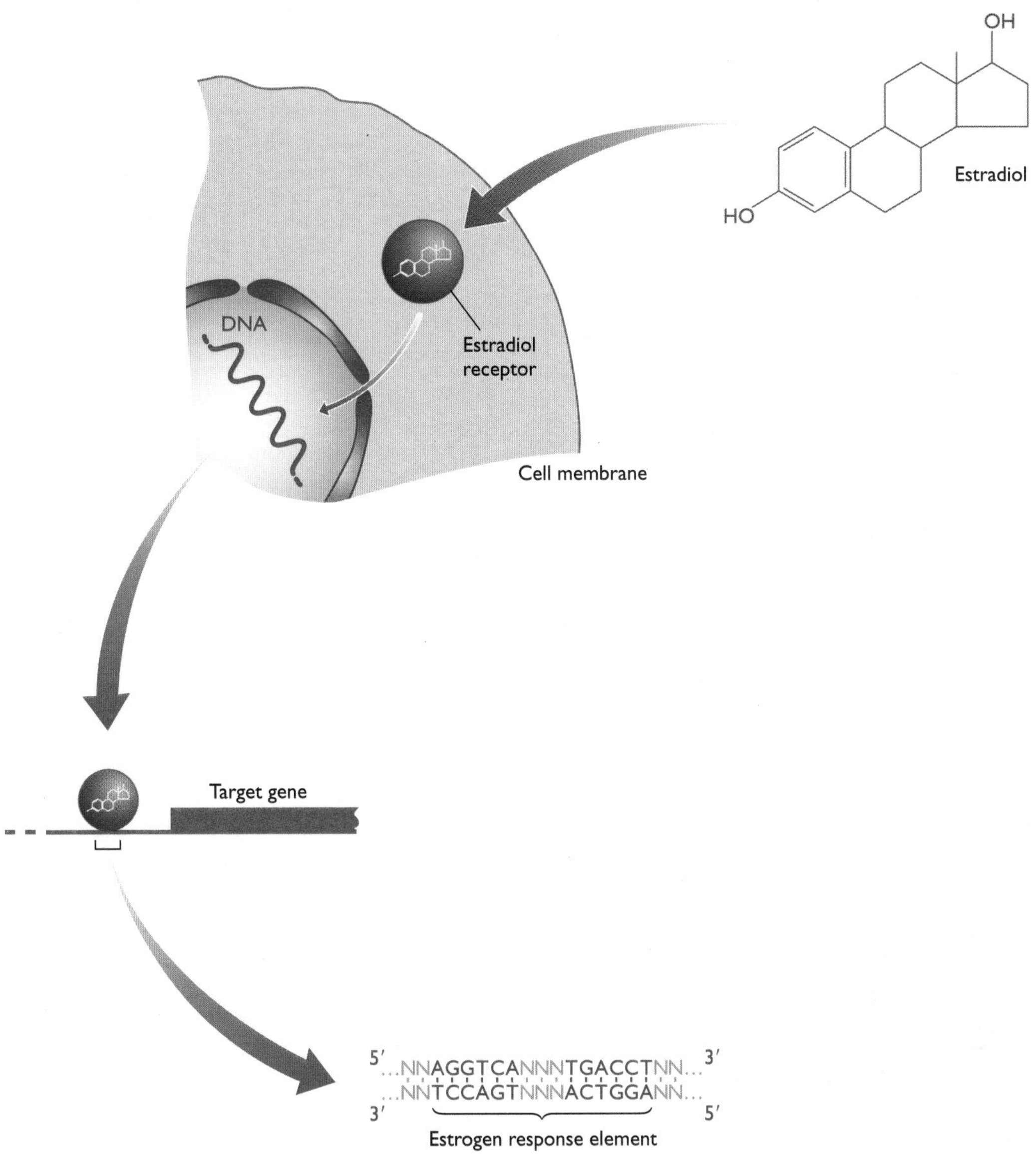

Figure 11.4 Gene activation by a steroid hormone.

Estradiol is one of the estrogen steroid hormones. After entering the cell, estradiol attaches to its receptor protein and the complex enters the nucleus where it binds to the 15-bp estrogen response element (abbreviation: N, any nucleotide), which is located upstream of those genes activated by estradiol and other estrogens. Other steroid hormone receptors recognize other response elements: for example, glucocorticoid hormones target the sequence 5′-AGAACANNNTGTTCT-3′. Note that this sequence, and that of the estrogen response element, is an inverted palindrome. The response element for vitamin D_3, which is a steroid derivative that activates transcription via a nuclear receptor (see the text), has the sequence 5′-AGGTCANNNAGGTCA-3′, which is a direct repeat rather than an inverted palindrome.

instead influences genome activity in an indirect manner. An example is provided by the **catabolite repression** system of bacteria. This is the means by which extracellular and intracellular glucose levels dictate whether or not operons for utilization of other sugars are switched on when those alternative sugars are present in the medium.

This phenomenon was discovered by Jacques Monod in 1941, who showed that if *E. coli* or *Bacillus subtilis* is provided with a mixture of sugars, then one will be metabolized first, the bacteria turning to the second sugar only when the first is used up. Monod used a French word to describe this: **diauxie** (Brock, 1990). One combination of sugars that elicits a diauxic response is glucose plus lactose, glucose being used before the lactose (*Figure 11.6A*). When the details of the lactose operon were worked out some 20 years later (Section 8.3.1) it became clear that the diauxie between glucose and lactose must involve a mechanism whereby the presence of

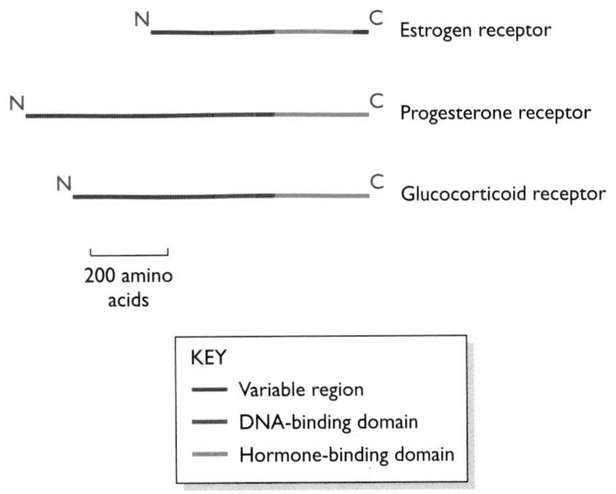

Figure 11.5 All steroid hormone receptor proteins have similar structures.

Three receptor proteins are compared. Each one is shown as an unfolded polypeptide with the two main functional domains aligned. The DNA-binding domain is very similar in all steroid receptors, displaying 50–90% amino acid sequence identity. The hormone-binding domain is less well conserved, with 20–60% sequence identity between different receptors. The region between the N terminus and the DNA-binding domain is the part of the protein that is involved in activation of RNA polymerase II, but this region displays little sequence similarity in different receptors.

glucose can override the normal inductive effect that lactose has on its operon. In the presence of lactose plus glucose the lactose operon is switched off, even though some of the lactose in the mixture is converted into allolactose which binds to the lactose repressor, so that under normal circumstances the operon would be transcribed (*Figure 11.6B*).

The explanation for the diauxic response is that glucose acts as a signaling compound that represses expression of the lactose operon, as well as other sugar utilization operons, through an indirect influence on the **catabolite activator protein**. This protein binds to various sites in the bacterial genome and activates transcription initiation at downstream promoters. Productive initiation of transcription at these promoters is dependent on the presence of the bound protein: if the protein is absent then the genes it controls are not transcribed.

Glucose does not itself interact with the catabolite activator protein. Instead, glucose controls the level in the cell of the modified nucleotide **cyclic AMP** (**cAMP**; *Figure 11.6C*). It does this by inhibiting the activity of **adenylate cyclase**, the enzyme that synthesizes cAMP from ATP. This means that if glucose levels are high, the cAMP content of the cell is low. The catabolite activator protein can bind to its target sites only in the presence of cAMP, so when glucose is present the protein remains detached and the operons it controls are switched off. In the specific case of diauxie involving glucose plus lactose, the indirect effect of glucose on the catabolite activator protein means that the lactose operon remains unactivated, even though the lactose repressor is not bound, and so the glucose in the medium is used up first. When the glucose is gone the cAMP levels rise and the catabolite activator protein binds to its target sites, including the site upstream of the lactose operon, and transcription of the lactose genes is activated.

11.1.2 Signal transmission mediated by cell surface receptors

Many extracellular signaling compounds are unable to enter the cell because they are too hydrophilic to penetrate the lipid membrane and the cell lacks a specific transport mechanism for their uptake. In order to influence genome activity these signaling compounds must bind to cell surface receptors that carry their signals across the cell membrane. These receptors are proteins that span the membrane, with a site for binding the signaling compound on the outer surface. Binding of the signaling compound results in a conformational change in the receptor, inducing a biochemical event within the cell, often phosphorylation of an intracellular protein. This event forms the first step in the intracellular stage of the signal transduction pathway (*Figure 11.7*). Several types of cell surface receptor have been discovered (*Table 11.2*) and the intracellular events that they initiate are diverse, with many variations on each theme, not all of these specifically involved in regulating genome activity. Three examples will help us gain an appreciation of the complexity of the system.

The simplest signal transduction pathways involve direct activation of transcription factors

With some signal transduction systems, stimulation of the cell surface receptor by attachment of the extracellular signaling compound results in direct activation of a transcription factor. This is the simplest system by which an extracellular signal can be transduced into a genomic response.

The direct system is used by many cytokines, such as interleukins and interferons, these being extracellular signaling polypeptides that control cell growth and division. Binding of these polypeptides to their cell surface receptors results in activation of a type of transcription factor called a **STAT** (signal transducer and activator of transcription). Activation is by phosphorylation of a single tyrosine amino acid at a position near to the C terminus of the STAT polypeptide. If the cell surface receptor is a member of the tyrosine kinase family (see *Table 11.2*) then it is able to activate the STAT directly (*Figure 11.8A*). If it is a tyrosine kinase associated receptor then it does not itself have the ability to phosphorylate a STAT, or any other intracellular protein, but acts through intermediaries called **Janus kinases** (**JAKs**). Binding of the signaling molecule to a tyrosine kinase associated receptor causes a change in the conformation of the receptor, often by

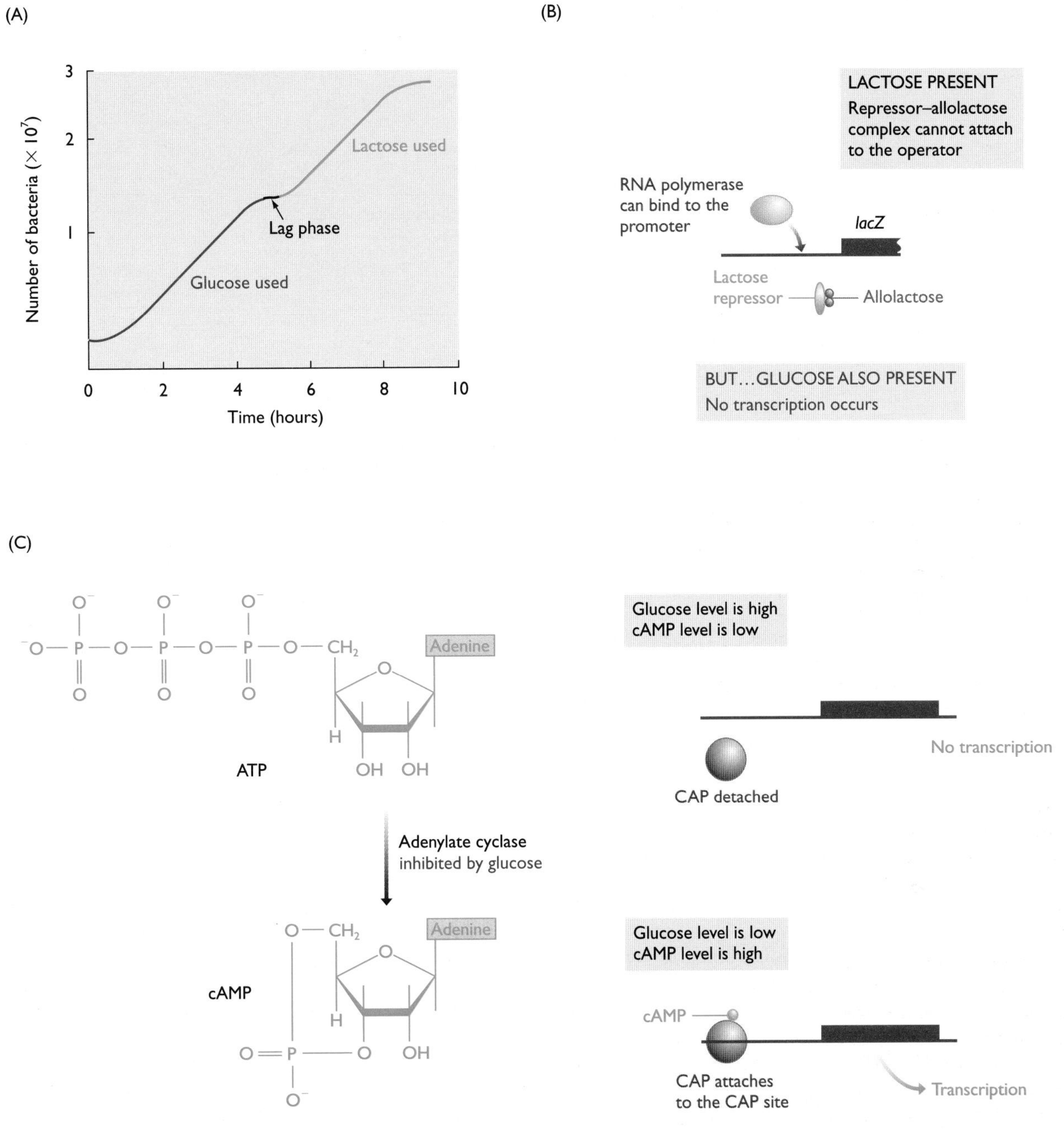

Figure 11.6 Catabolite repression.

(A) A typical diauxic growth curve, as seen when *Escherichia coli* is grown in a medium containing a mixture of glucose and lactose. During the first few hours the bacteria divide exponentially, using the glucose as the carbon and energy source. When the glucose is used up there is a brief lag period while the *lac* genes are switched on before the bacteria return to exponential growth, now using up the lactose. (B) Glucose overrides the lactose repressor. If lactose is present then the repressor detaches from the operator and the lactose operon should be transcribed, but it remains silent if glucose is also present. Refer to *Figure 8.15B*, p. 187 for details of how the lactose repressor controls expression of the lactose operon. (C) Glucose exerts its effect on the lactose operon and other target genes by controlling the activity of adenylate cyclase and hence regulating the amount of cAMP in the cell. The catabolite activator protein (CAP) can attach to its DNA-binding site only in the presence of cAMP. If glucose is present the cAMP level is low, so CAP does not bind to the DNA and does not activate the RNA polymerase. Once the glucose has been used up the cAMP level rises, allowing CAP to bind to the DNA and activate transcription of the lactose operon and its other target genes.

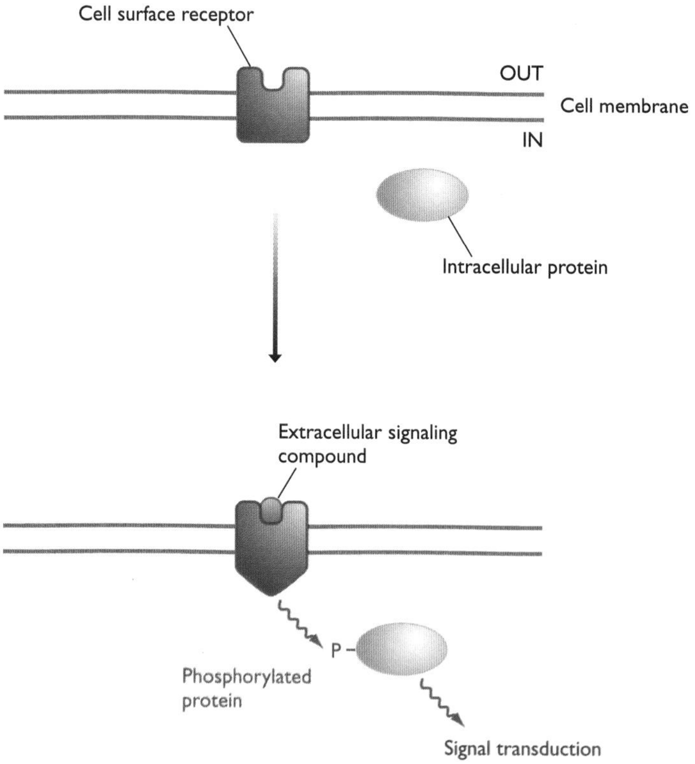

Figure 11.7 The role of a cell surface receptor in signal transduction.

Binding of the extracellular signaling compound to the outer surface of the receptor protein causes a conformational change that results in activation of an intracellular protein, for example by phosphorylation. The events occurring 'downstream' of this initial protein activation are diverse, as described in the text. 'P' indicates a phosphate group, PO_3^{2-}.

inducing dimerization. This causes a JAK that is associated with the receptor to phosphorylate itself, this auto-activation being followed by phosphorylation of the STAT by the JAK (*Figure 11.8B*).

Seven STATs have so far been identified in mammals (Darnell, 1997). Three of these – STATs 2, 4 and 6 – are specific for just one or two extracellular cytokines, but the others are broad spectrum and can be activated by several different interleukins and interferons. Discrimination is provided by receptors on the surface of cells: a particular receptor binds just one type of cytokine, and most cells have only one or a few types of receptor. Different cells

Table 11.2 Cell surface receptor proteins involved in signal transmission into eukaryotic cells

Receptor type	**Description**	**Signals**
G-protein-coupled receptors	Activate intracellular G-proteins, which bind GTP and control biochemical activities by conversion of this GTP to GDP with the release of energy	Diverse: epinephrine, peptides (e.g. glucagon), protein hormones, odorants, light
Tyrosine kinases	Activate intracellular proteins by tyrosine phosphorylation	Hormones (e.g. insulin), various growth factors
Tyrosine kinase associated	Similar to tyrosine kinase receptors but activate intracellular proteins indirectly (e.g. see description of STATs in the text)	Hormones, growth factors
Serine–threonine kinases	Activate intracellular proteins by serine and/or threonine phosphorylation	Hormones, growth factors
Ion channels	Control intracellular activities by regulating the movement of ions and other small molecules into and out of cells	Chemical stimuli (e.g. glutamate), electrical charges

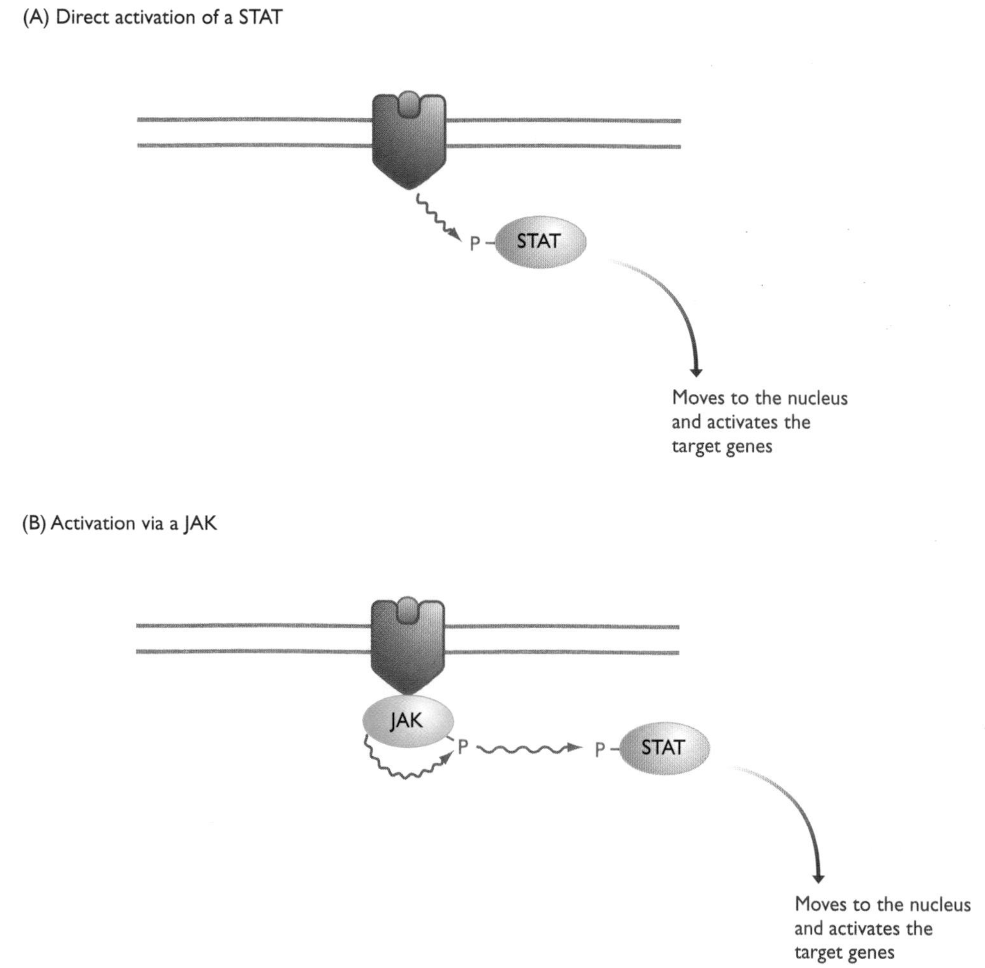

Figure 11.8 Signal transduction involving STATs.

(A) If the receptor is a member of the tyrosine kinase family then it can activate the STAT directly. (B) If the receptor is a tyrosine kinase associated type then it acts via a JAK, which autophosphorylates when the extracellular signal binds and then activates the STAT. Note that activation of the JAK usually involves dimerization, the extracellular signal inducing two subunits to associate, resulting in the version of the JAK with phosphorylation activity. Dimerization is also central to activation of STATs, phosphorylation resulting in two STATs, not necessarily of the same type, forming a dimer which is the active version of the transcription factor. 'P' indicates a phosphate group, PO_3^{2-}.

therefore respond in different ways to the presence of particular cytokines, even though the internal signaling process involves only a limited number of STATs.

The consensus sequence of the DNA-binding sites for STATs has been defined as 5′-TTCCGGGAA-3′, largely by studies using purified STATs and oligonucleotides of known sequence. The DNA-binding domain of the STAT protein is made up of three loops emerging from a barrel-shaped β-sheet structure (Becker *et al.*, 1998). This is an unusual type of DNA-binding domain that has not been identified in precisely the same form in any other type of protein, though it has similarities with the DNA-binding domains of the NK-κB and Rel transcription factors. These similarities refer only to the tertiary structures of the domains as the proteins as a whole have very little amino acid sequence identity. Many target genes are activated by STATs but the overall genomic response is modulated by other proteins which interact with STATs and influence which genes are switched on under a particular set of circumstances. Complexity is entirely expected because the cellular processes that STATs mediate – growth and division – are themselves complex and we anticipate that changes in these processes will require extensive remodeling of the proteome and hence large scale changes in genome activity.

More complex signal transduction pathways have many steps between receptor and genome

The simplicity of the system whereby the cell surface receptor activates a transcription factor, directly or through a JAK associated with the receptor, contrasts with the more prevalent forms of signal transduction, in

RESEARCH 11.1 BRIEFING

Unraveling a signal transduction pathway

A typical set of experiments for studying the functions of proteins involved in a signal transduction pathway is described.

One of the most important extracellular signaling compounds is transforming growth factor-β (TGF-β), a family of some 30 related polypeptides that control processes such as cell division and differentiation in vertebrates. The cell surface receptors for TGF-β are serine–threonine kinases (see *Table 11.2*) which activate a variety of target proteins within the cell. Part of the signal transduction process initiated by TGF-β binding involves a set of proteins called the **SMAD family**, the name being an abbreviation of 'SMA/MAD related', referring to proteins in *Drosophila melanogaster* and *Caenorhabditis elegans*, respectively, which were the original members of the family to be isolated.

SMAD signaling in vertebrates

Initially, five SMADs were discovered in vertebrate cells. Four of these – Smad1, Smad2, Smad3 and Smad5 – are associated with specific types of serine–threonine receptor and hence respond to different members of the TGF-β family. Binding of the extracellular signal induces a receptor to phosphorylate its SMAD, which then associates with Smad4, moves to the nucleus and, via interactions with DNA-binding proteins, activates a set of target genes. The SMADs are therefore the only intermediates between the receptor at the start of the pathway and the transcription factors at the end.

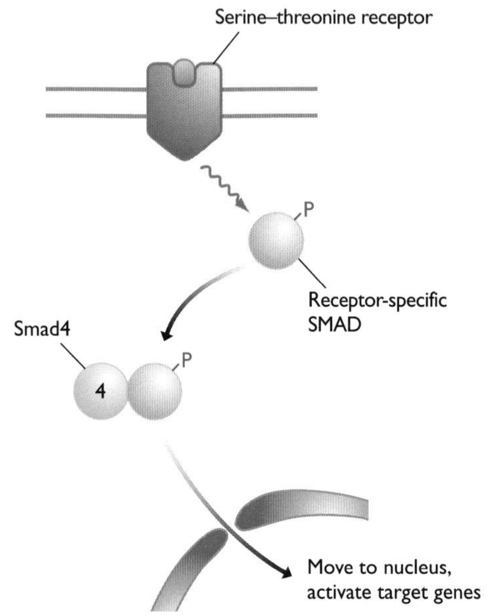

This interpretation of the SMAD pathway has been complicated by the discovery of additional SMADs that do not fit into this scheme. It is clear from their amino acid sequences that Smad6 and Smad7 do not respond directly to attachment of the extracellular signals to the receptor protein, because they lack the C-terminal polypeptide sequence that is phosphorylated by the receptor. In this respect, the new SMADs resemble Smad4, suggesting that they have a general rather than signal-specific role in TGF-β signal transduction.

The first step in understanding the functions of Smad6 and Smad7 was to see what effect over-expression of these proteins has on TGF-β signal transduction. Over-expression was achieved by attaching the appropriate SMAD gene to a strong promoter and then using cloning techniques to introduce the gene into cultured cells. It was found that genes normally switched on by TGF-β became nonresponsive to the extracellular signal in cells over-expressing Smad6 or Smad7. This is a clear indication that Smad6 and Smad7 have inhibitory effects on the pathway.

Inhibition could arise by Smad6 or Smad7 binding to one or more of the other SMADs, removing these from the pathway and hence blocking signal transduction. Studies of cell extracts suggested that this was not the case, because the Smad6 and Smad7 proteins were associated not with other SMADs but with the intracellular parts of the cell-surface receptors. This leads to the hypothesis that Smad6 and Smad7 inhibit signal transduction by preventing the activated receptors from phosphorylating the other SMADs.

The new model that emerges is of a more flexible TGF-β pathway than was originally envisaged, signal transduction now being modifiable by the inhibitory effects of Smad6 and Smad7, these proteins presumably responding to as-yet unidentified intracellular signals in order to modulate the effects of TGF-β binding in an appropriate way.

References

Whitman M (1998) Feedback from inhibitory SMADs. *Nature*, **389**, 549–551.

Nakao A, Afrakhte M, Morén A, *et al.* (1998) Identification of Smad7, a TGFβ-inducible antagonist of TGF-β signalling. *Nature*, **389**, 631–635.

Imamura T, Takase M, Nishihara A, Oeda E, Hanai J, Kawabata M and Miyazono K (1998) Smad6 inhibits signalling by the TGF-β superfamily. *Nature*, **389**, 622–626.

ten Dijke P and Heldin C-H (1999) An anchor for activation. *Nature*, **397**, 109–111.

which the receptor represents just the first in a series of steps that lead eventually to activation or deactivation of one or more transcription factors. A number of these **cascade** pathways have been delineated in different organisms. In mammals the important ones are:

- The **MAP** (mitogen activated protein) **kinase** system (*Figure 11.9*), which responds primarily to mitogens, extracellular signaling compounds with similar effects to cytokines but specifically stimulating cell division (Robinson and Cobb, 1997). Binding of the signaling compound causes the internal parts of the mitogen receptor to become phosphorylated. Phosphorylation stimulates attachment to the receptor, on the internal side of the membrane, of various cytoplasmic proteins, one of which is Raf, a protein kinase that is activated when it becomes membrane bound. Raf initiates a cascade of phosphorylation reactions. It phosphorylates Mek, activating this protein so that it, in turn, phosphorylates the MAP kinase. The activated MAP kinase now moves into the nucleus where it activates, again by phosphorylation, a series of transcription factors. The MAP kinase also phosphorylates another protein kinase, this one called Rsk, which phosphorylates and activates a second set of transcription factors. Additional flexibility is provided by the possibility of replacing one or more of the proteins in the MAP kinase pathway with related proteins, ones with slightly different specificities and so activating a different suite of transcription factors. The MAP kinase pathway is used by vertebrate cells and equivalent pathways, using intermediates similar to those identified in mammals, are known in other organisms (see Section 11.3.2 for an example).
- The **Ras** system is centered around the Ras proteins, three of which are known in mammalian cells (H-, K- and N-Ras), and similar proteins such as Rac and Rho. These proteins are involved in regulation of cell growth and differentiation and, as with many proteins in this category, when dysfunctional they can give rise to cancer. The Ras family proteins are not limited to mammals, examples being known in other eukaryotes such as the fruit fly. Ras proteins are intermediates in signal transduction pathways that initiate with autophosphorylation of a tyrosine kinase receptor in response to an extracellular signal. The phosphorylated version of the receptor forms protein– protein complexes with **GNRPs** (**guanine nucleotide releasing proteins**) and **GAPs** (**GTPase activating proteins**) which activate and inactivate Ras, respectively (*Figure 11.10*; Schlessinger, 1993). The extracellular signals can therefore switch Ras-mediated signal transduction on or off, the choice between the two depending on the nature of the signal and the relative amounts of active GNRPs and GAPs in the cell. When activated, Ras stimulates Raf activity, so in effect Ras provides a second entry point into the MAP kinase pathway, though this is unlikely to be the only function of Ras and it probably also activates proteins involved in signal transduction by second messengers (as described in the next section).
- The **SAP** (stress activated protein) **kinase** system is induced by stress-related signals such as UV radiation and growth factors associated with inflammation. The pathway has not been described in detail but is very similar to the MAP kinase system though targeting a different set of transcription factors.

Signal transduction via second messengers

Some signal transduction cascades do not involve the direct transfer of the external signal to the genome but instead utilize a more indirect means of influencing transcription factor activity. The indirectness is provided by **second messengers**, which are less specific, internal signaling compounds that transduce the signal from a cell surface receptor in several directions so that a variety of cellular activities, not just transcription, respond to the one signal.

The second messenger system is analogous in some respects to the way in which glucose modulates the catabolite activator protein by influencing cAMP levels in bacteria cells (see *Figure 11.6*). Cyclic nucleotides are also important second messengers in eukaryotic cells. Some cell surface receptors have guanidylate cyclase activity, and so convert GTP to cGMP, but most receptors in this family work indirectly by influencing the activity of cytoplasmic cyclases and decyclases. The cellular levels of cGMP and cAMP control the activities of various other enzymes. An example is protein kinase A, which is stimulated by cAMP. One of the activities of protein kinase A is to phosphorylate and hence activate a transcription factor called **CREB**. This is one of several proteins that influence activity of a variety of genes by interacting with a second factor, p300/CBP, which is able to modify histone proteins and so affect nucleosome positioning and chromatin structure (Sections 8.1.1 and 8.3.2).

As well as being activated indirectly by cAMP, p300/CBP also responds to another second messenger, calcium (Chawla *et al.*, 1998). The calcium ion concentration in the cell is substantially greater than that outside it, so proteins that open calcium channels in the cell membrane allow Ca^{2+} ions to enter. This can be induced by extracellular signals that activate tyrosine kinase receptors which in turn activate phospholipases that cleave phosphatidylinositol-4,5-bisphosphate (PtdIns(4,5)P_2), a component of the inner cell membrane, into inositol-1,4,5-trisphosphate (Ins(1,4,5)P_3) and 1,2-diacylglycerol (DAG). The first of these two products opens calcium channels (*Figure 11.11*). Ins(1,4,5)P_3 and DAG are themselves second messengers that can initiate other signal transduction cascades (Spiegel *et al.*, 1996; Toker and Cantley, 1997). Both the calcium and the lipid-induced cascades involve activation of transcription factors, but only indirectly: the primary targets are proteins. Calcium, for example, binds to and activates the protein called calmodulin, which regulates a variety of enzyme types including protein kinases, ATPases, phosphatases and nucleotide cyclases.

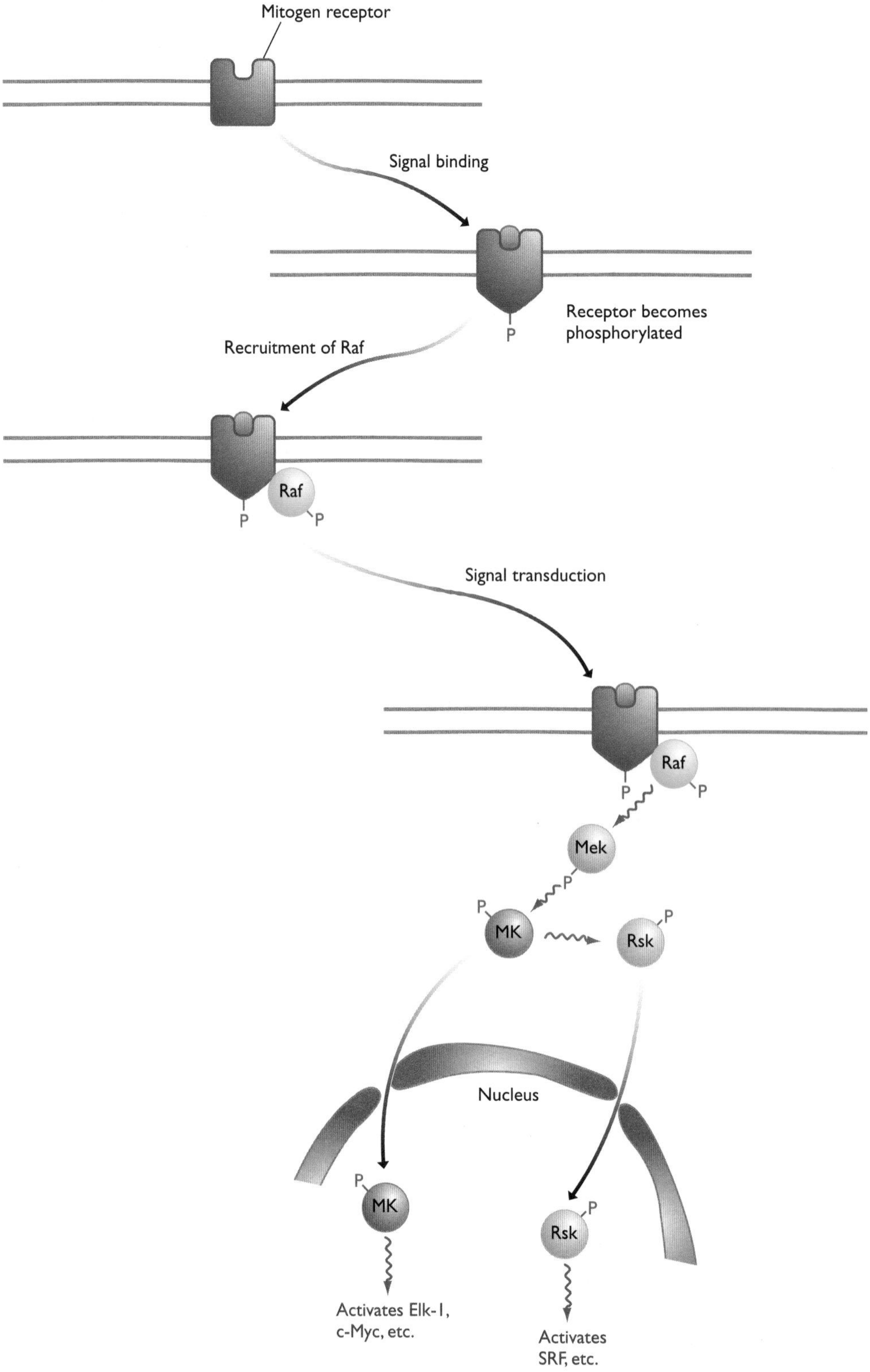

Figure 11.9 Signal transduction by the MAP kinase pathway.

See the text for details. 'MK' is the MAP kinase and 'P' indicates a phosphate group, PO_3^{2-}. Elk-1, c-Myc and SRF (serum response factor) are examples of transcription factors activated at the end of the pathway.

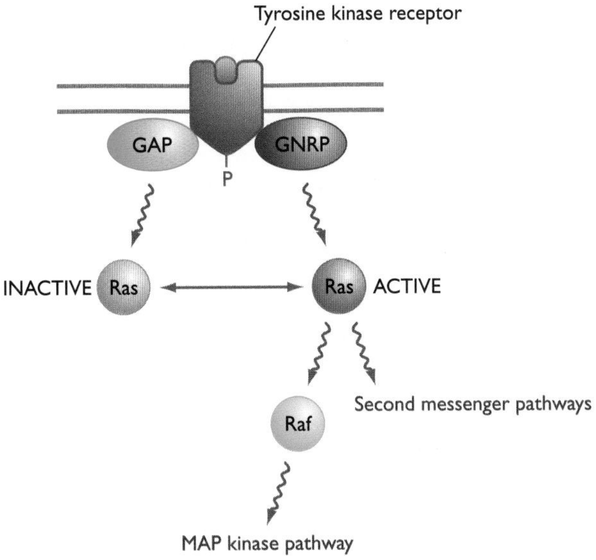

Figure 11.10 The Ras signal transduction system.

See the text for details. Abbreviations: GAP, GTPase activating protein; GNRP, guanine nucleotide releasing protein. 'P' indicates a phosphate group, PO_3^{2-}.

11.2 PERMANENT AND SEMI-PERMANENT CHANGES IN GENOME ACTIVITY

Transient changes in genome activity are, by definition, readily reversible, the gene expression pattern reverting to its original state when the external stimulus is removed or replaced by a contradictory stimulus. In contrast, the permanent and semipermanent changes in genome activity that underlie cellular differentiation must persist for long periods, and ideally should be maintained even when the stimulus that originally induced them has disappeared. We therefore anticipate that the regulatory mechanisms bringing about these longer term changes will involve systems in addition to the modulation of transcription factor activity. This expectation is correct: long-term changes in genome activity occur by a variety of means and many of these do not directly involve transcription factors. We will look at three mechanisms:

- Changes resulting from physical rearrangement of the genome.
- Changes due to chromatin structure.
- Changes maintained by feedback loops.

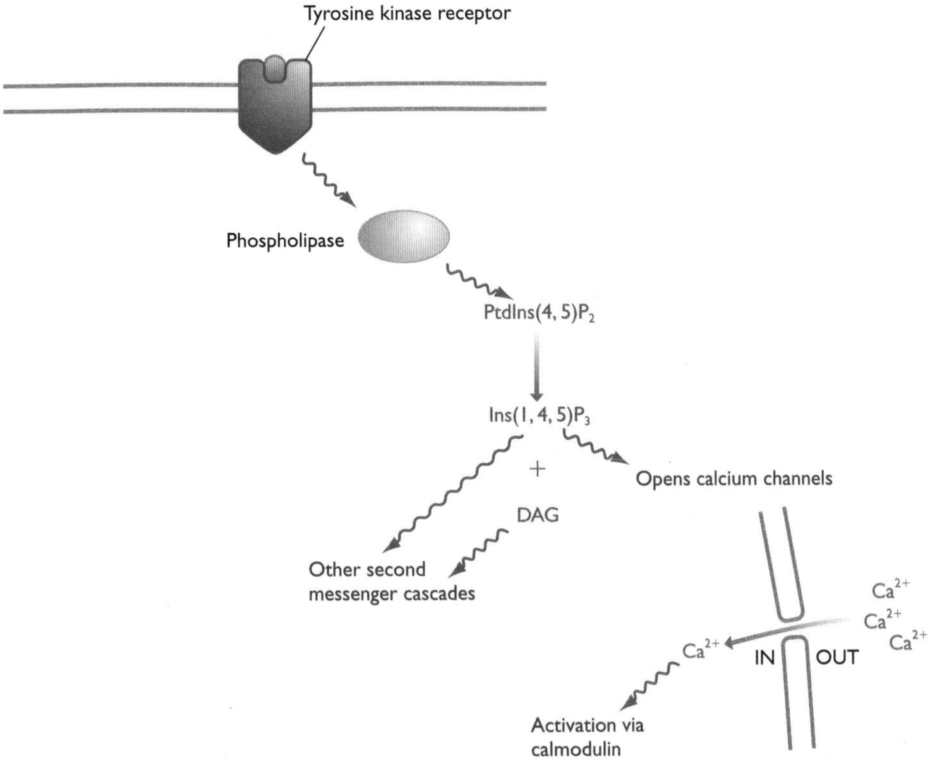

Figure 11.11 Induction of the calcium second messenger system.

See the text for details. Abbreviations: DAG, 1,2-diacylglycerol; Ins(1,4,5)P_3, inositol-1,4,5-trisphosphate; PtdIns(4,5)P_2, phosphatidylinositol-4,5-bisphosphate.

11.2.1 Genome rearrangements

Changing the physical structure of the genome is an obvious, although drastic, way of bringing about a permanent change in the pattern of gene expression. It is not a frequent mechanism for genome regulation, but several important examples are known.

Yeast mating types are determined by gene conversion events

Mating type is the equivalent of sex in yeasts and other eukaryotic microorganisms. Because these organisms reproduce mainly by vegetative cell division there is the possibility that a population, being derived from just one or a few ancestral cells, will be largely or completely composed of a single mating type and so will not be able to reproduce sexually. To avoid this problem, cells are able to change sex by the process called **mating type switching**. In *Saccharomyces cerevisiae* and some other species this switching involves a genome rearrangement called **gene conversion**.

The two *S. cerevisiae* mating types are called a and α. The mating type is specified by the *MAT* gene, located on chromosome III. This gene has two alleles, *MATa* and *MATα*, a haploid yeast cell displaying the mating type corresponding to whichever allele it possesses. Elsewhere on chromosome III there are two additional *MAT*-like genes, called *HMLα* and *HMRa* (*Figure 11.12*). These have the same sequences as *MATα* and *MATa* respectively, but neither gene is expressed because upstream of each one is a silencer that represses initiation of their transcription. These two genes are called 'silent mating type cassettes'.

Mating type switching is initiated by the HO endonuclease, which makes a double-stranded cut at a 24-bp sequence located within the *MAT* gene. This enables a gene conversion event to take place. We will examine the details of gene conversion in Section 13.2; all that concerns us at the moment is that one of free 3′-ends produced by the endonuclease can be extended by DNA synthesis, using one of the two silent cassettes as the template. The newly synthesized DNA subsequently replaces the DNA currently at the *MAT* locus. The silent cassette chosen as the template is usually the one that is different to the allele originally at *MAT* (Haber, 1998), so replacement with the newly synthesized strand converts the *MAT* gene from *MATa* to *MATα* or *vice versa* (see *Figure 11.12*). This results in mating type switching.

The *MAT* gene codes for a regulatory protein that interacts with a transcription factor, MCM1, determining which set of genes this factor activates. The *MATa* and *MATα* gene products have different effects on MCM1, and so specify different allele-specific gene expression patterns. These expression patterns are maintained in a semipermanent fashion until another *MAT* gene conversion occurs.

Genome rearrangements are responsible for immunoglobulin and T-cell receptor diversities

In vertebrates there are two striking examples of the use of DNA rearrangements to achieve permanent changes in genome activity. These two examples, which are very similar to one another, are responsible for the generation of immunoglobulin and T-cell receptor diversities.

Immunoglobulins and T-cell receptors are proteins that

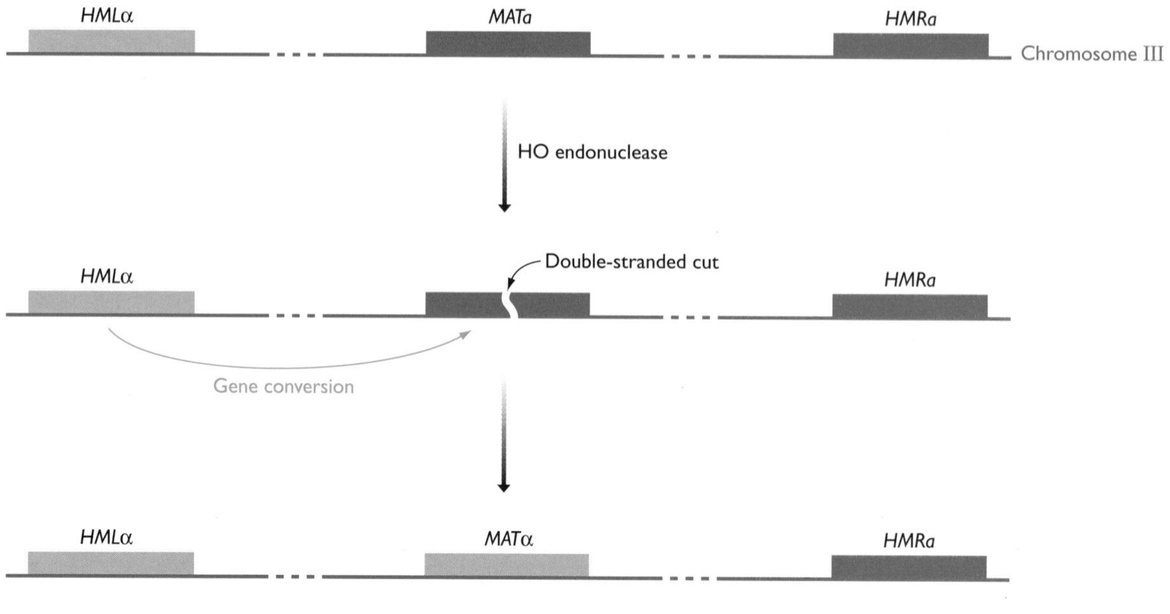

Figure 11.12 Mating type switching in yeast.

In this example, the cell begins as mating type a. The HO endonuclease cuts the *MATa* locus, initiating gene conversion by the *HMLα* locus. The result is that the mating type switches to type α. For details of the molecular basis of gene conversion, see Section 13.2.

are synthesized by B and T lymphocytes respectively. Both types of protein become attached to the outer surfaces of their cells, and immunoglobulins are also released into the bloodstream. The proteins help to protect the body against invasion by bacteria, viruses and other unwanted substances by binding to these **antigens**, as they are called. For every antigen there is an immunoglobulin protein and a T-cell receptor protein that recognizes and binds to that antigen and to no other. This means that humans must make approximately 10^8 different immunoglobulin and T-cell receptor proteins. But there are only 10^5 genes in the human genome, so where do all these proteins come from?

To understand the answer we will look at the structure of a typical immunoglobulin protein. Each immunoglobulin is a tetramer of four polypeptides linked by disulfide bonds (*Figure 11.13*). There are two long 'heavy' chains and two short 'light' chains. When the sequences of different heavy chains are compared it becomes clear that the variability between them lies mainly in the N-terminal regions of these polypeptides, the C-terminal parts being very similar, or 'constant', in all heavy chains. The same is true for the light chains, except that two families, κ and λ, can be distinguished, differing in the sequences of their constant regions.

In the human genome there are no complete genes for the immunoglobulin heavy and light polypeptides. Instead, these proteins are specified by gene segments, three segments coding for each light chain and four for each heavy chain. For example, on chromosome 14 there are 11 C_H gene segments coding for slightly different versions of the constant region of the heavy chain, and similar gene segments for the κ and λ constant regions on chromosomes 2 and 22, respectively (*Table 11.3*). Upstream of each of these sets of gene segments, stretching over several megabases, there is an array of additional gene segments, coding for different parts of the variable regions of the heavy and light chains. The C_H gene segments, for example, are preceded by 86 V_H gene segments, 30 D_H gene segments and 9 J_H gene segments, these coding for different versions of the V (variable), D (diverse) and J (joining) components of the variable part of the heavy chain (*Figure 11.14A*). A similar arrangement is seen with the light chain loci, the only difference being that the light chains do not have D segments.

As a B lymphocyte develops, the immunoglobulin loci in its genome undergo rearrangements. Within the heavy chain loci, these rearrangements link one of the V_H gene segments with one of D_H gene segments, and then links this V–D combination with a J_H gene segment, and finally attaches the resulting sequence to a C_H gene segment (*Figure 11.14B*). The end result is a complete heavy chain gene, but one that is specific for just that one lymphocyte. A similar series of DNA rearrangements results in the lymphocyte's light chain gene, and transcription of the two genes produces one of the 10^8 immunoglobulins that the human body needs.

Diversity of T-cell receptors is based on similar rearrangements which link V, D, J and C gene segments in different combinations to produce cell-specific genes. We met two small components of this system – the Tβ gene segments V28 and V29-1, in Chapter 1 in the 50-kb segment of the human genome (see *Figure 1.5*, p. 6) with which we began our exploration of genomes.

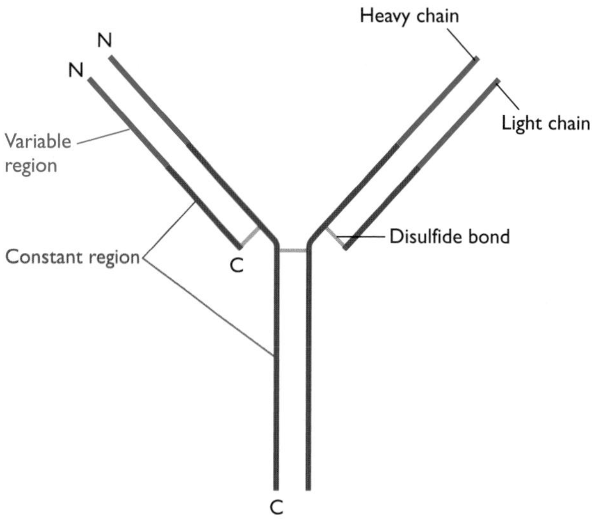

Figure 11.13 Immunoglobulin structure.

Each immunoglobulin protein is made up of two heavy and two light chains linked by disulfide bonds. Each heavy chain is 446 amino acids in length and consists of a variable region (shown in red) spanning amino acids 1–108 followed by a constant region. Each light chain is 214 amino acids, again with an N-terminal variable region of 108 amino acids. Additional disulfide bonds form between different parts of individual chains: these and other interactions fold the protein into a more complex three-dimensional structure.

Table 11.3 Immunoglobulin gene segments in the human genome

			Number of gene segments			
Component	**Locus**	**Chromosome**	**V**	**D**	**J**	**C**
Heavy chain	*IGH*	14	86	30	9	11
Light κ chain	*IGK*	2	76	0	5	1
Light λ chain	*IGL*	22	52	0	7	7

It is not known if all the gene segments are functional. For further details see Strachan and Read (1996).

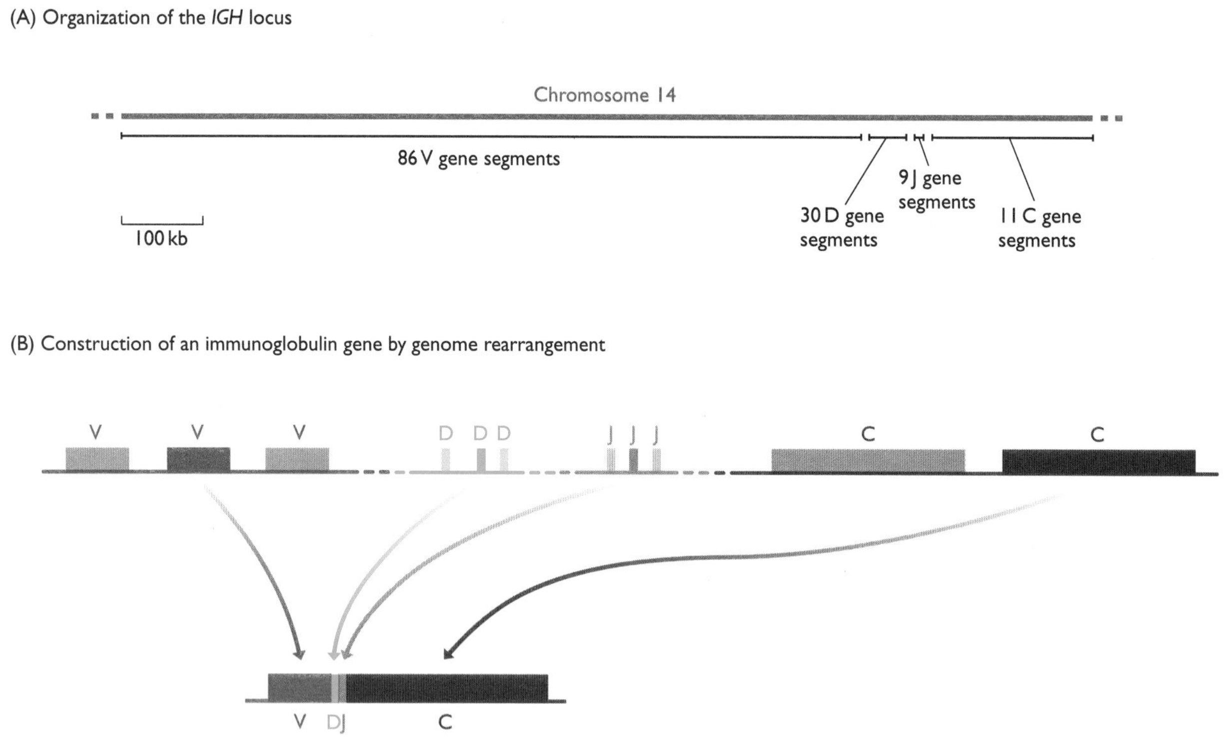

Figure 11.14 Immunoglobulin gene segments and construction of a functional gene.

(A) Organization of the human *IGH* locus on chromosome 14, containing gene segments for the immunoglobulin heavy chain. For a more detailed map of this region see Strachan and Read (1996). (B) A functional heavy chain gene is constructed by genome rearrangements which link V, D, J and C gene segments.

11.2.2 Changes in chromatin structure

Some of the effects that chromatin structure can have on gene expression were described in Section 8.1. These range from the modulation of transcription initiation at an individual promoter by nucleosome positioning, through to the silencing of large segments of DNA locked up in higher order chromatin structure. The latter is an important means of bringing about long-term changes in genome activity and is implicated in a number of important regulatory events. One of these concerns the yeast mating type loci that we looked at earlier in this section, the silencing of the *HMLα* and *HMRa* cassettes resulting mainly from these loci being buried in inaccessible chromatin due to the influence of their upstream silencer sequences, which are typical locus control regions (Section 8.1.1; Haber, 1998).

Two other chromatin silencing examples merit attention. The first of these we will meet again later in the chapter when we look at development processes in the fruit fly. It concerns the *Polycomb* gene family. The proteins coded by these genes bind to DNA sequences called Polycomb-response elements and induce formation of heterochromatin, the condensed form of chromatin that prevents transcription of the genes that it contains (*Figure 11.15*). The heterochromatin nucleates around the attached Polycomb protein and then propagates along the DNA for tens of kilobases (Pirrotta, 1997). The regions that become silenced contain homeotic genes which, as we will see in Section 11.3.3, specify the development of the individual body parts of the fly. As only one body part must be specified at a particular position it is important that a cell expresses only the correct homeotic gene. This is ensured by the action of Polycomb, which permanently silences the homeotic genes that must be switched off. An important point is that the heterochromatin induced by Polycomb is heritable: after division the two new cells retain the heterochromatin established in the parent cell. This type of regulation over genome activity is therefore permanent not only in a single cell, but also in a cell lineage.

The second example of chromatin silencing concerns **X inactivation**. This occurs because females have two X chromosomes whereas males have just one. If both of the female X chromosomes were active then proteins coded by genes on the X chromosome might be synthesized at twice the rate in females compared to males. To avoid this undesirable state of affairs, one of the female X chromosomes is inactivated and is seen in the nucleus as a condensed structure called the **Barr body**, which is comprised entirely of heterochromatin. Inactivation occurs early in embryo development and is controlled by

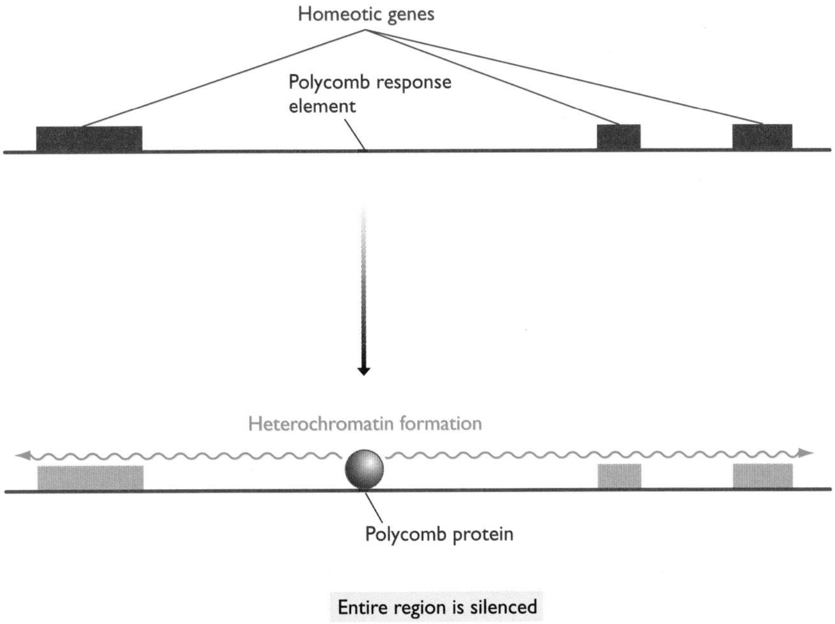

Figure 11.15 Polycomb silences regions of the *Drosophila* genome by initiating heterochromatin formation.

the X inactivation center, a discrete region present on each X chromosome. In each cell undergoing X inactivation, the inactivation center on one of the X chromosomes initiates the formation of heterochromatin. As with the Polycomb-induced system, heterochromatin formation spreads out from the nucleation center, but in this case continues for megabases in both directions until the entire chromosome is affected, with the exception of a few short segments containing small clusters of genes that remain active. The process takes several days to complete. The exact mechanism is not understood but it involves, though is not entirely dependent upon, each of the following:

- A gene called *Xist*, located in the inactivation center, which is transcribed into a long noncoding RNA that coats the chromosome as heterochromatin is formed.
- Replacement of histone H2A, one of the members of the core octamer of the nucleosome (Section 6.1.1), with a special histone, macroH2A1 (Costanzi and Pehrson, 1998).
- Deacetylation of histone H4 (Jeppesen and Turner, 1993), as usually occurs in heterochromatin (Section 8.1.1).
- Hypermethylation of certain DNA sequences (see Box 8.1, p. 177), although this appears to occur after the inactive state has been set up.

As with Polycomb-induced heterochromatin formation, X inactivation is heritable and is displayed by all cells descended from the initial one within which the inactivation took place.

11.2.3 Genome regulation by feedback loops

The final mechanism that we will consider for bringing about long-term changes in genome activity involves the use of a feedback loop. In this system a regulatory protein activates its own transcription so that once its gene has been switched on it is expressed continuously (*Figure 11.16*). A number of examples of this type of feedback regulation are known, including:

- ***The MyoD transcription factor***, which is involved in muscle development, one of the best understood examples of cellular differentiation in vertebrates. A cell becomes committed to becoming a muscle cell when it begins to express the *myoD* gene. The product of this gene is a transcription factor that activates a number of other genes coding for muscle-specific proteins, such as actin and myosin, and is also indirectly responsible for one of the key features of muscle cells, the absence of a normal cell cycle, these cells being stopped in the G1 phase (Section 12.4). The MyoD protein also binds upstream of *myoD*, ensuring that its own gene is continually expressed. The result of this positive feedback loop is that the cell continues to synthesize the muscle-specific proteins and remains a muscle cell. The differentiated state is heritable because cell division is accompanied by transmission of MyoD to the daughter cells, ensuring that these are muscle cells too.
- ***Deformed of Drosophila*** is one of several proteins coded by homeotic selector genes and responsible for specifying segment identity in the fruit fly (Section 11.3.3). The Deformed (Dfd) protein is responsible for

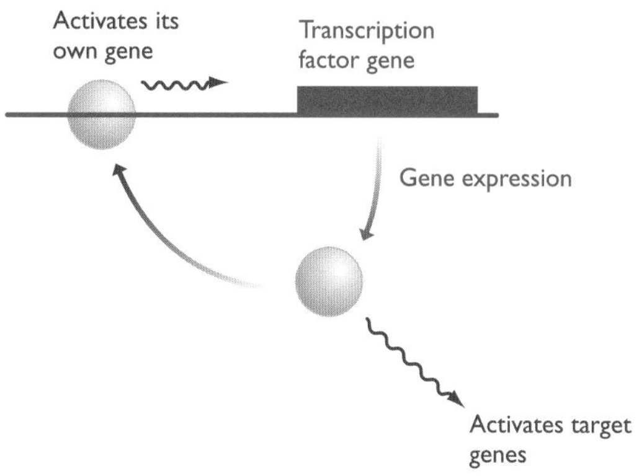

Figure 11.16 Feedback regulation of gene expression.

the identity of the head segments. To perform this function, Dfd must be continually expressed in the relevant cells. This is achieved by a feedback system, Dfd binding to an enhancer located upstream of the *Dfd* gene (Regulski *et al.*, 1991). At least some homeotic selector genes of vertebrates also make use of feedback autoregulation (Popperl *et al.*, 1995).

11.3 REGULATION OF GENOME ACTIVITY DURING DEVELOPMENT

The developmental pathway of a multicellular eukaryote begins with a fertilized egg cell and ends with an adult form of the organism. In between lies a complex series of genetic, cellular and physiological events that must occur in the correct order, in the correct cells, and at the appropriate times if the pathway is to reach a successful culmination. With humans, this developmental pathway results in an adult containing 10^{13} cells differentiated into approximately 250 specialized types, the activity of each individual cell coordinated with that of every other cell. Developmental processes of such complexity might appear intractable, even to the powerful investigative tools of modern molecular biology, but remarkably good progress towards understanding them has been made in recent years. The research that has underpinned this progress has been designed around three guiding principles:

1. That it should be possible to describe and comprehend the genetic and biochemical events that underlie differentiation of individual cell types. This in turn means that an understanding of how specialized tissues and even complex body parts are constructed should be within reach.
2. That the signaling processes that coordinate events in different cells are amenable to study. We saw in Section 11.1 that a start is being made to describing these systems at the molecular level.
3. That there are similarities and parallels between developmental processes in different organisms, reflecting common evolutionary origins. This means that information relevant to human development can be obtained from studies of model organisms chosen for the relative simplicity of their developmental pathways.

Developmental biology encompasses areas of genetics, molecular biology, cell biology, physiology and biochemistry. We are concerned only with the role of the genome in development and so will not attempt a wide ranging overview of developmental research in all its guises. Instead we will concentrate on three model systems, of increasing complexity, in order to investigate the types of change in genome activity that occur during development.

11.3.1 Sporulation in *Bacillus*

The first developmental pathway that we will examine is formation of spores by the bacterium *Bacillus subtilis* (Grossman, 1995; Errington, 1996; Stragier and Losick, 1996). Strictly speaking, this is not a developmental pathway, merely a type of cellular differentiation, but the process illustrates two of the fundamental issues that have to be addressed when genuine development in multicellular organisms is studied. These issues are how a series of changes in genome activity over time are controlled, and how signaling establishes coordination between events occurring in different cells. The advantages of *Bacillus* as a model system are that it is easy to grow in the laboratory and is amenable to study by genetic and molecular biological techniques such as analysis of mutants and sequencing of genes.

Sporulation involves coordinated activities in two distinct cell types

Bacillus is one of several genera of bacteria that produce endospores in response to unfavorable environmental conditions. These spores are highly resistant to physical and chemical abuse and can survive for decades or even centuries: the possibility of infection with anthrax spores produced by *Bacillus anthracis* is taken very seriously by archaeologists excavating sites containing human and animal remains. Resistance is due to the specialized nature of the spore coat, which is impermeable to many chemicals, and to biochemical changes which retard the decay of DNA and other polymers and enable the spore to survive a prolonged period of dormancy.

In the laboratory, sporulation is usually induced by nutrient starvation. This causes the bacteria to abandon their normal, vegetative mode of cell division involving synthesis of a cross-wall, or septum, in the center of the cell. Instead the cells construct an unusual septum, one that is thinner than normal, at one end of the cell (*Figure 11.17*). This produces two cellular compartments, the smaller of which is called the prespore and the larger the

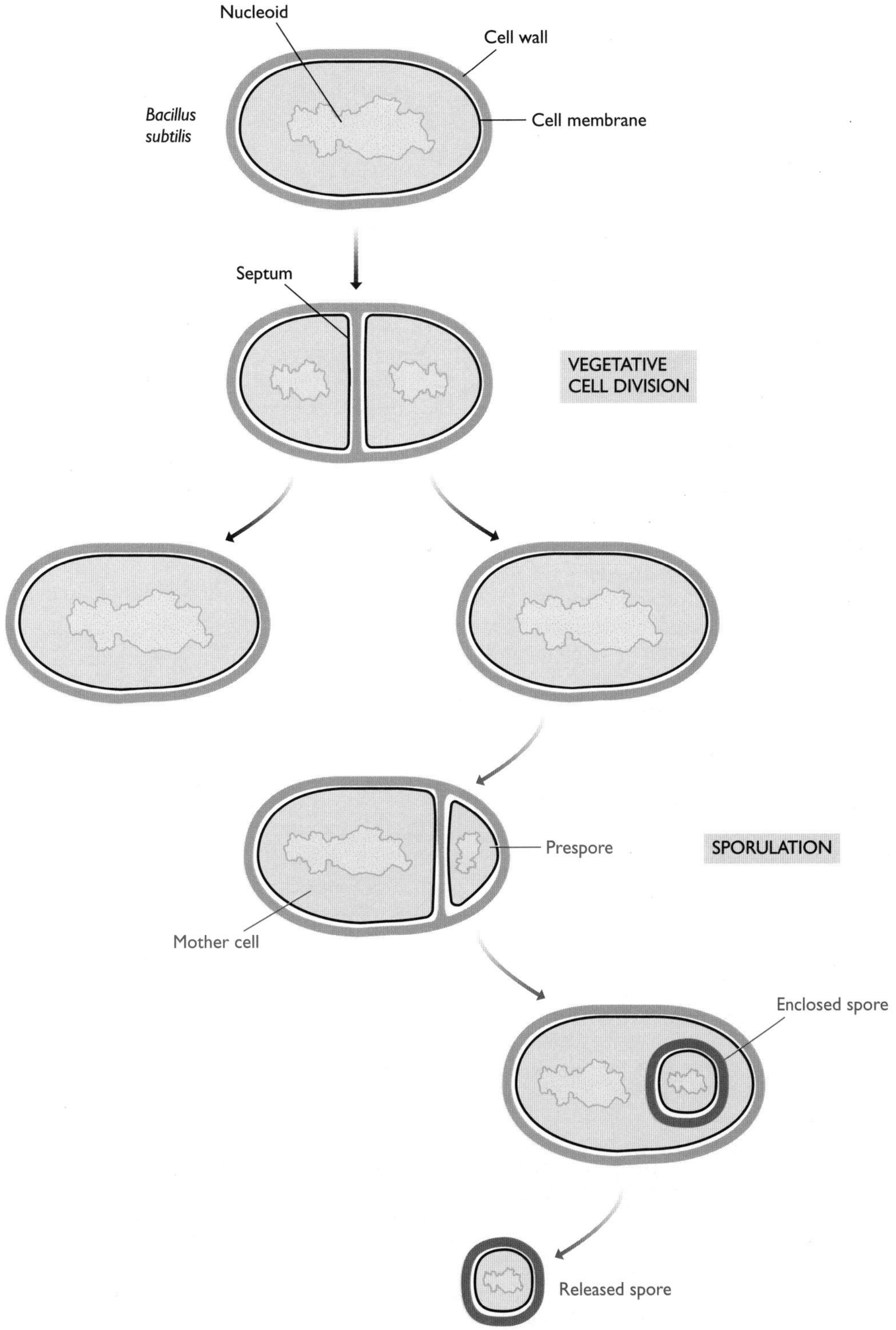

Figure 11.17 Sporulation in *Bacillus subtilis*.

The top part of the diagram shows the normal, vegetative mode of cell division, involving formation of a septum across the center of the bacterium and resulting in two identical daughter cells. The lower part of the diagram shows sporulation, in which the septum forms near one end of the cell, leading to a mother cell and prespore of different sizes. Eventually the mother cell completely engulfs the prespore. At the end of the process, the mature, resistant spore is released.

mother cell. As sporulation proceeds the prespore becomes entirely engulfed by the mother cell. By now the two cells are committed to different, but coordinated differentiation pathways, the prespore undergoing the biochemical changes that enable it to become dormant, and the mother cell constructing the resistant coat around the spore and eventually dying.

Special σ subunits control genome activity during sporulation

Changes in genome activity during sporulation are controlled largely by the synthesis of special σ subunits that change the promoter specificity of the *Bacillus* RNA polymerase. Recall that the σ subunit is the part of the RNA polymerase that recognizes the bacterial promoter sequence, and that replacement of one σ subunit with another, with a different DNA-binding specificity, can result in a different set of genes being transcribed (Section 8.3.1). We have seen how this simple control system is used by *E. coli* in response to heat stress (see *Figure 8.14*, p. 186). It is also the key to the changes in genome activity that occur during sporulation.

The standard *B. subtilis* σ subunits are called σ^A and σ^H. These subunits are synthesized in vegetative cells and enable the RNA polymerase to recognize promoters for all the genes it needs to transcribe in order to maintain normal growth and cell division. In the prespore and mother cell these subunits are replaced by σ^F and σ^E respectively, which recognize different promoter sequences and so result in large scale changes in gene expression patterns. The master switch from vegetative growth to spore formation is provided by a protein called SpoOA, which is present in vegetative cells but in an inactive form. This protein is activated by phosphorylation; the protein kinases that phosphorylate it respond to various extracellular signals that indicate the presence of environmental stresses such as lack of nutrients. Activated SpoOA is a transcription factor that modulates the expression of various genes transcribed by the vegetative RNA polymerase and hence recognized by the regular σ^A and σ^H subunits. The genes that are switched on include those for σ^F and σ^E, resulting in the switch to prespore and mother cell differentiation (*Figure 11.18*).

Initially, both σ^F and σ^E are present in each of the two differentiating cells. This is not exactly what is wanted because σ^F is the prespore-specific subunit and so should be active only in this cell, and σ^E is mother spore-specific. A means is therefore needed of activating or inactivating the appropriate subunit in the correct cell. This is thought to be achieved as follows (*Figure 11.19*; Errington, 1996):

- **σ^F is activated by release from a complex with a second protein, SpoIIAB.** This is controlled by a third protein SpoIIAA which, when unphosphorylated, can also attach to SpoIIAB and prevent the latter from binding to σ^F. If SpoIIAA is unphosphorylated then σ^F is released and is active; when SpoIIAA is phosphorylated then σ^F remains bound to SpoIIAB and so is inactive. In the mother cell, SpoIIAB phosphorylates SpoIIAA and so keeps σ^F in its bound, inactive state. But in the prespore, SpoIIAB's attempts to phosphorylate SpoIIAA are antagonized by yet another protein, SpoIIE, and so σ^F is released and becomes active. SpoIIE's ability to antagonize SpoIIAB in the prespore but not the mother cell derives from the fact that SpoIIE molecules are bound to the membrane on the surface of the septum. Because the prespore is much smaller than the mother cell, but the septum surface area is similar in both, the concentration of SpoIIE is greater in the prespore, and this enables it to antagonize SpoIIAB.
- **σ^E is activated by proteolytic cleavage of a precursor protein.** The protease that carries out this cleavage is the SpoIIGA protein, which spans the septum between the prespore and mother cell. The protease domain, which is on the mother cell side of the septum, is activated by binding of SpoIIR to a receptor domain on the prespore side. It is a typical receptor-mediated signal transduction system (Section 11.1.2). SpoIIR is one of the genes whose promoter is recognized specifically by σ^F, so activation of the protease, and conversion of pre-σ^E to active σ^E, occurs once σ^F-directed transcription is underway in the prespore.

Activation of σ^F and σ^E is just the beginning of the story. In the prespore, about one hour after its activation, σ^F responds to an unknown signal (possibly from the mother cell) which results in a slight change in genome activity in the spore. This includes transcription of a gene for another σ subunit, σ^G, which recognizes promoters

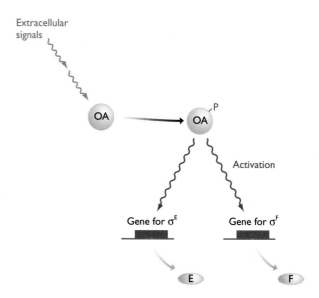

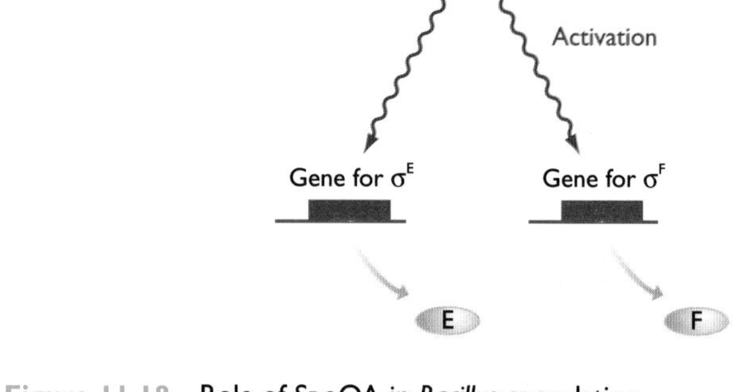

Figure 11.18 Role of SpoOA in *Bacillus* sporulation.

SpoOA is phosphorylated in response to extracellular signals derived from environmental stresses. It is a transcription factor with roles including activation of the genes for the σ^F and σ^E RNA polymerase subunits. Abbreviations: E, σ^E subunit; F, σ^F subunit; OA, SpoOA. 'P' indicates a phosphate group, PO_3^{2-}.

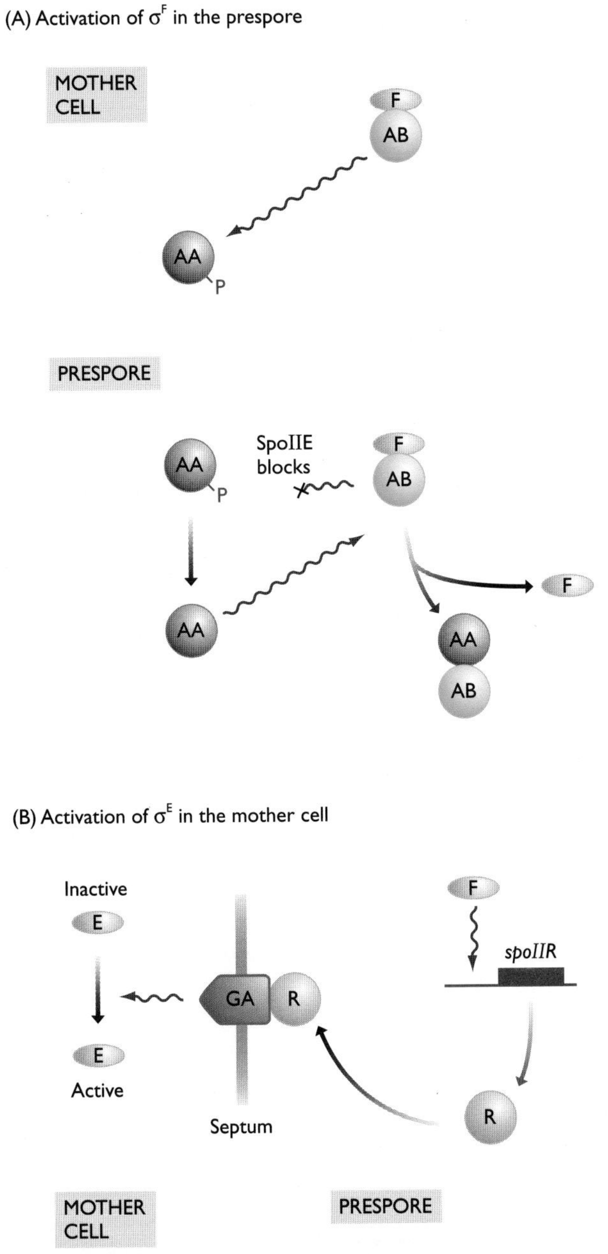

Figure 11.19 Activation of the prespore- and mother cell-specific σ subunits during *Bacillus* sporulation.

(A) Activation of σ^F in the prespore occurs by release from its complex with SpoAB, which is indirectly influenced by the concentration of membrane-bound SpoIIE. (B) In the mother cell, σ^E is activated by proteolytic cleavage by SpoIIGA, which responds to the presence in the prespore of the σ^F-dependent protein SpoIIR. See the text for more details. Abbreviations: AA, SpoIIAA; AB, SpoIIAB; E, σ^E subunit; F, σ^F subunit; GA, SpoIIGA; R, SpoIIR.

upstream of genes whose products are required during the later stages of spore differentiation. One of these proteins is SpoIVB, which activates another septum-bound protease, SpoIVF (*Figure 11.20*). This protease then activates a second mother cell σ subunit, σ^K, which is coded by a σ^E-transcribed gene but retained in the mother cell in inactive form until the signal for its activation is received from the prespore. σ^K directs transcription of the genes whose products are needed during the later stages of the mother cell differentiation pathway.

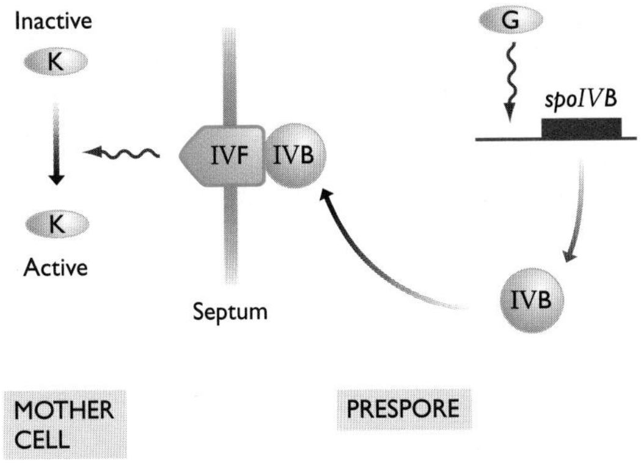

Figure 11.20 Activation of σ^K during *Bacillus* sporulation.

See the text for details. Note that the scheme is very similar to the procedure used to activate σ^E (see *Figure 11.19B*). Abbreviations: G, σ^G subunit; K, σ^K subunit; IVB, SpoIVB; IVF, SpoIVF.

To summarize the key features of *Bacillus* sporulation:

- The master protein, SpoOA, responds to external stimuli to determine if and when the switch to sporulation should occur.
- A cascade of σ subunits in prespore and mother cell brings about time-dependent changes in genome activity in the two cells.
- Cell–cell signaling ensures that the events occurring in prespore and mother cell are coordinated.

11.3.2 *Vulval development in* Caenorhabditis elegans

Bacillus subtilis is a unicellular organism and although sporulation involves the coordinated differentiation of two cell types it can hardly be looked on as comparable to the developmental processes occurring in multicellular organisms. Sporulation provides pointers to the general ways in which genome activity might be regulated during the development of a multicellular organism, but it does not indicate what specific events to expect. We therefore need to examine development in a simple multicellular eukaryote.

C. elegans *is a model for multicellular eukaryotic development*

Research with *Caenorhabditis elegans* (*Figure 11.21*) was initiated by Sydney Brenner in the 1960s with the aim of utilizing this microscopic nematode worm as a simple model for multicellular eukaryotic development. *C. elegans* is easy to grow in the laboratory and has a short generation time, measured in days but still convenient for

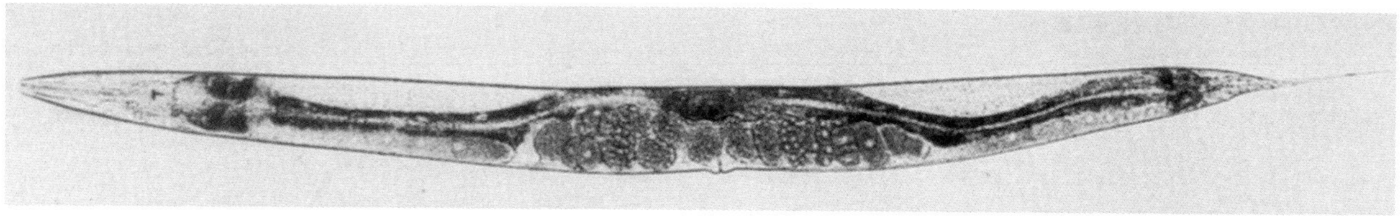

Figure 11.21 The nematode worm *Caenorhabditis elegans*.

The micrograph shows an adult hermaphrodite worm, approximately 1 mm in length. The vulva is the small projection located on the underside of the animal about halfway along. Egg cells can be seen inside the worm's body in the region either side of the vulva. Reprinted with permiss3ion from Kendrew J (1994) *The Encyclopaedia of Molecular Biology*, p. 127, Blackwell Science, Oxford.

genetic analysis. The worm is transparent at all stages of its life cycle, so internal examination is possible without killing the animal. This is an important point as it has enabled researchers to follow the entire developmental process of the worm at the cellular level. Every cell division in the pathway from fertilized egg to adult worm has been charted, and every point at which a cell adopts a specialized role has been identified. In addition, the complete connectivity of the 302 cells that comprise the nervous system of the worm has been mapped.

The genome of *C. elegans* is relatively small, just 100 Mb (see *Table 1.1*, p. 9), and the entire sequence is known. (*C. Elegans* Sequencing Consortium, 1998) This means that any mutation with an interesting effect can be mapped to a gene whose sequence is available, enabling molecular biological experiments to be devised with the aim of defining the function of the gene and determining how this function underlies the physiological changes seen in the mutant. This is an excellent basis for understanding the links between genome activity and developmental pathways and with this organism the goal of a complete genetic description of development is attainable in the not too distant future.

Determination of cell fate during development of the C. elegans vulva

A critical feature that underpins the usefulness of *C. elegans* as a tool for research is the fact that its development is more or less invariant: the pattern of cell division and differentiation is virtually the same in every individual. This appears to be due in large part to cell–cell signaling, which induces each cell to follow its appropriate differentiation pathway. To illustrate this we will look at development of the *C. elegans* vulva.

Most *C. elegans* worms are hermaphrodites, meaning that they have both male and female sex organs. The vulva is part of the female sex apparatus, being the tube through which sperm enter and fertilized eggs are laid. The adult vulva comprises 22 cells which are the progeny of three ancestral cells originally located in a row on the undersurface of the developing worm (*Figure 11.22*). Each of these ancestral cells becomes committed to the differentiation pathway that leads to production of vulval cells. The central cell, called P6.p, adopts the 'primary vulval cell fate' and divides to produce eight new cells, and the other two cells – P5.p and P7.p – take on the 'secondary vulval cell fate' and divide into seven cells each. These 22

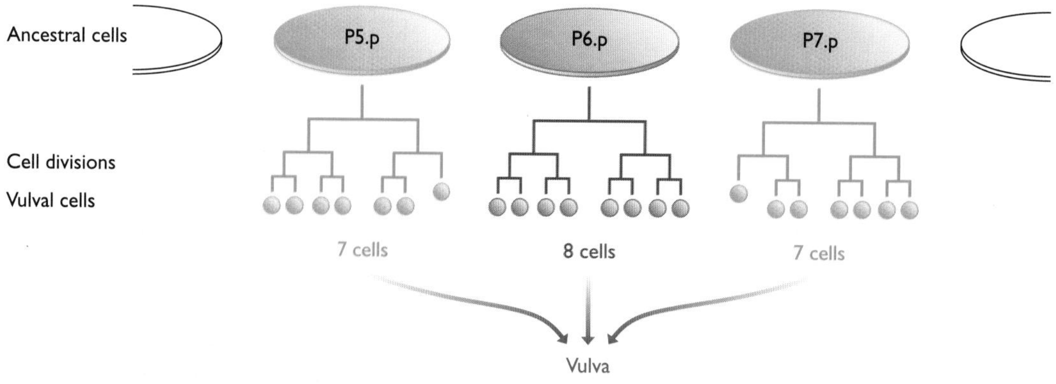

Figure 11.22 Cell divisions resulting in production of the vulval cells of *C. elegans*.

Three ancestral cells divide in a programmed manner to produce 22 progeny cells which reorganize their positions relative to one another to construct the vulva.

cells then reorganize their positions to construct the vulva.

A critical aspect of vulval development is that it must occur in the correct position relative to the gonad, the structure containing the egg cells. If the vulva develops in the wrong place then the gonad will not receive sperm and the egg cells will never be fertilized. The positional information needed by the vulval progenitor cells is provided by a cell within the gonad called the anchor cell (*Figure 11.23*). The importance of the anchor cell is demonstrated by experiments in which it is artificially destroyed in the embryonic worm: in the absence of the anchor cell, vulvas do not develop. The implication is that the anchor cell secretes an extracellular signaling compound that induces P5.p, P6.p and P7.p to differentiate. This signaling compound is the protein called LIN-3, coded by the *lin-3* gene.

Why does P6.p adopt the primary cell fate whereas P5.p and P7.p take on secondary cell fates? There are two possibilities. The first is that LIN-3 forms a concentration gradient so it has a different effect on P6.p, the cell which is closest to it, than it has on the more distant P5.p and and P7.p (as shown in *Figure 11.23*). Evidence in favor of this idea comes from studies showing that isolated cells adopt the secondary fate when exposed to low levels of LIN-3 (Katz *et al.*, 1995). Alternatively, the signal that commits P5.p and P7.p to their secondary fates might not come directly from the anchor cell but via P6.p in the form of a different extracellular signaling compound whose synthesis by P6.p is switched on by LIN-3 activation (Kornfeld, 1997). This hypothesis is supported by the abnormal features displayed by certain mutants in which more than three cells become committed to vulval development. With these mutants there is more than one primary cell, but each one is invariably surrounded by two

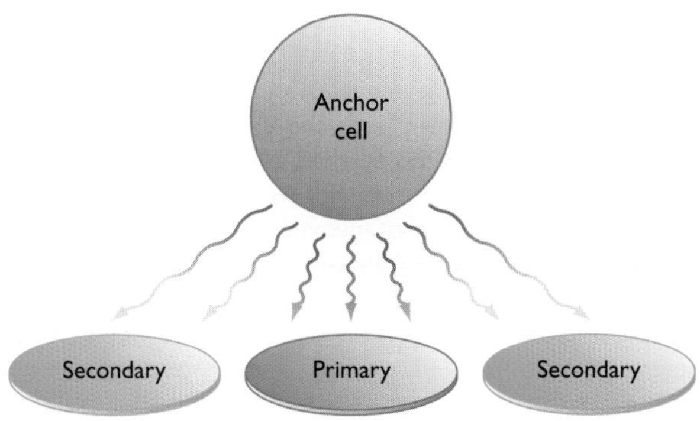

Figure 11.23 The postulated role of the anchor cell in determining cell fate during *C. elegans* vulva development.

It is thought that release of the signaling compound LIN-3 by the anchor cell commits P6.p (shown in blue), the cell closest to the anchor cell, to the primary vulval cell fate. P5.p and P7.p (shown in green) are further away from the anchor cell and so are exposed to a lower concentration of LIN-3 and become secondary vulval cells. As described in the text, there is evidence that commitment of the secondary cells to their fates is also influenced by signals from the primary vulval cell.

secondary cells, suggesting that in the living worm adoption of the secondary cell fate is dependent on the presence of an adjacent primary cell.

There are other instructive features of vulval development in *C. elegans*. The first is that the signaling process that commits P6.p to its primary cell fate has many similarities with the MAP kinase signal transduction system of vertebrates (see *Figure 11.9*, p. 276). The cell surface receptor for LIN-3 is a protein kinase called LET-23 which, when activated by binding LIN-3, initiates a series of intracellular reactions that leads to activation of a MAP kinase-like protein which in turn activates a variety of transcription factors (Sternberg and Han, 1998). Unfortunately the target genes that become switched on have not yet been delineated in either the primary or secondary vulval progenitor cells, but the system is open to study.

A second noteworthy feature is that as well as the activation signal provided by the anchor cell in the form of LIN-3, the vulval progenitor cells are also subject to the deactivating effects of a second signaling compound, secreted by the hypodermal cell, a multinuclear sheath that surrounds most of the worm's body. This repressive signal is overcome by the positive signals that induce P5.p, P6.p and P7.p to differentiate, but prevents the unwanted differentiation of three adjacent cells, P3.p, P4.p and P8.p, each of which can become committed to vulval development if the repressive signal malfunctions, for example in a mutant worm.

In summary, the general concepts to emerge from vulval development in *C. elegans* are:

- In a multicellular organism, positional information is important: the correct structure must develop at the appropriate place.
- The commitment to differentiation of a small number of progenitor cells can lead to construction of a multicelled structure.
- Cell–cell signaling can utilize a concentration gradient to induce different responses in cells at different positions relative to the signaling cell.
- Cells might be subject to competitive signaling, where one signal tells it to do one thing and a second signal tells it to do the opposite.

11.3.3 Development in *Drosophila melanogaster*

The last organism whose development we will study is *Drosophila melanogaster*. The experimental history of the fruit fly dates back to 1910 when Thomas Hunt Morgan first used this organism as a model system in genetic research. For Morgan the advantages of *Drosophila* were its small size, enabling large numbers to be studied in a single experiment, its minimal nutritional requirements (the flies like bananas), and the presence in natural populations of occasional variants with easily recognized genetic characteristics such as unusual eye colors. Morgan was not aware that other advantages are a small

RESEARCH
11.2
BRIEFING

A *Drosophila* mutant with an extended lifespan

Genetic studies with *Drosophila* have identified a single gene that has a major influence on the lifespan of the fly.

Research into aging has been hindered by an absence of hard facts about the biological basis of the aging process. Faced with an information vacuum of this kind, the standard suggestion of a geneticist is 'Let's isolate some mutants'. This is because examination of mutants that display altered phenotypes has proven in the past to be a good starting point for research into 'difficult' areas of biology, as illustrated by the stimulus that the *Drosophila* homeotic mutants have provided to the understanding of development in higher eukaryotes.

Aging mutants of Caenorhabditis elegans

Progress has already been made in studying the aging process in *C. elegans*, mutants with extended lifespans having been isolated and their gene products identified. One general theme that has emerged is that, at least in *C. elegans*, aging is linked with stress tolerance, several of the *C. elegans* mutants displaying an enhanced ability to withstand heat shock and other stresses. This was an interesting observation but unfortunately suggested that *C. elegans* might not be a good model for studying aging in higher eukaryotes, because this species has the unusual ability of being able to enter a semidormant state, called the dauer stage, in response to stress. Possibly the link between stress tolerance and extended lifespan is due in some way to the dauer stage, and hence not applicable to eukaryotes in general. These considerations stimulated efforts to obtain aging mutants of *Drosophila*.

The Drosophila methuselah *gene*

If the objective of a mutational project is to obtain the sequence of a mutated gene, in order to gain some idea of its function, then the question of how the mutants will be generated has to be carefully considered. Simply inducing small-scale mutations by treatment with X-rays or mutagenic chemicals (Section 13.1.1) is unsuitable because the precise nature of the resulting mutations are unpredictable, which means that it is impossible to design a hybridization probe to isolate the mutated genes from a clone library. For this reason, mutants are often generated by **transposon tagging**, by inducing a transposable element (Section 6.3.2) to move to new positions in the genome under study. If the transposon jumps into a gene it will disrupt it, resulting in a mutant whose inactivated gene can be isolated by probing a library for clones containing the transposon.

Transposon tagging in *Drosophila* routinely makes use of the **P element**, a 3-kb DNA transposon whose movement can be induced by crossing flies that carry inactive P elements with flies that possess the gene for the transposase needed to activate the P elements. This procedure was used in the search for aging mutants. After mutagenesis, a single mutant with an extended lifespan was identified. Rather than living for just 57 days, the average lifespan for normal fruit flies, this mutant, called *methuselah*, survived for an average of 77 days.

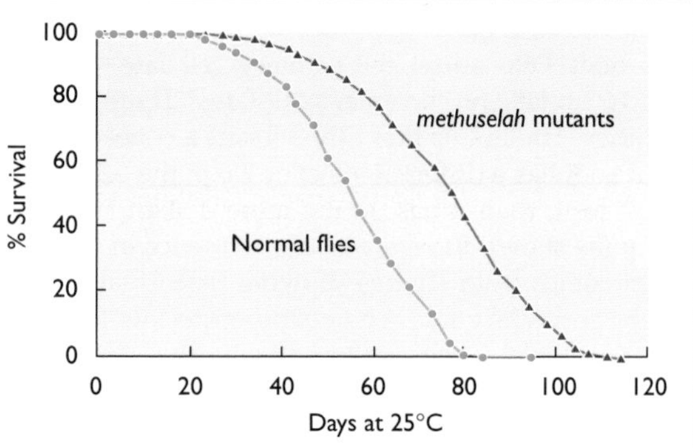

Reprinted from Lin *et al*., *Science*, **282**, 943–946. Copyright 1998 American Association for the Advancement of Science.

Southern hybridization (Technical Note 2.1, p. 19) showed that *methuselah* contains a single mutated gene coding for a 514-amino-acid protein with the characteristic features of a G-protein-coupled cell surface receptor (*Table 11.2*, p. 272). In functional terms, this is the most diverse type of cell surface receptor, different versions responding to a variety of extracellular signals. The sequence of the segment of the *methuselah* receptor that binds the extracellular signal is quite unusual, and provides no clues as to the nature of the stimulus that this protein responds to.

As with the *C. elegans* mutants, *methuselah* combines extended lifespan with enhanced stress tolerance, mutant flies showing significantly better resistances to high temperature, starvation and oxidative stress. Females are more tolerant than males, an observation that might be explained by the fact that female fruit flies are bigger than the males. The link between aging and stress in *Drosophila* indicates that the *C. elegans* mutants are not irrelevant to the study of aging in higher eukaryotes, and suggest that the key to future progress in aging research lies with understanding the signal transduction pathways involved in stress responses.

Reference

Lin Y-J, Seroude L and Benzer S (1998) Extended life-span and stress resistance in the *Drosophila* mutant *methuselah*. *Science*, **282**, 943–946.

genome (140 Mb; see *Table 1.1*, p. 9) and the fact that gene isolation is aided by the presence in a fly's salivary glands of 'giant' chromosomes made up of multiple copies of the same DNA molecule laid side by side, displaying banding patterns that can be correlated with the physical map of each chromosome to pinpoint the positions of desired genes. But Morgan did foresee that *Drosophila* might become an important organism for developmental research, a topic that he was as interested in as we are today.

The major contribution that *Drosophila* has made to our understanding of development has been through the insights it has provided into how an undifferentiated embryo acquires positional information that eventually results in the construction of complex body parts at the correct places in the adult organism. Although in some respects *Drosophila* is quite unusual in its embryonic organization (as we will see in the next section), it has turned out that the genetic mechanisms that the fly uses to specify its body plan are similar to those used by other organisms, including humans. Knowledge gained with *Drosophila* has therefore directed research into areas of human development that for a long time were thought to be inaccessible. To explore this story we must start first with the events occurring in the developing fruit fly embryo.

Maternal genes establish protein gradients in the Drosophila embryo

The unusual feature of the early *Drosophila* embryo is that it is not made up of lots of cells, as in most organisms, but instead is a single **syncytium** comprising a mass of cytoplasm and multiple nuclei (*Figure 11.24*). This structure persists until successive rounds of nuclear division have produced some 1500 nuclei: only then do individual, uninuclear cells start to appear around the outside of the syncytium, producing the structure called the blastoderm. Before the blastoderm stage has been reached the positional information has begun to be established.

Initially the positional information that the embryo needs is a definition of which end is the front (anterior) and which the back (posterior), as well as similar information relating to up (dorsal) and down (ventral). This information is provided by concentration gradients of proteins that become established in the syncytium. The bulk of these proteins are not synthesized from genes in the embryo, but are translated from mRNAs injected into the embryo by the mother; these are called **maternal-effect genes**. To see how they work we will examine the synthesis of Bicoid, one of the four proteins involved in determining the anterior–posterior axis.

The *bicoid* gene is transcribed in the maternal nurse cells, which are in contact with the egg cells, and the mRNA is injected into the anterior end of the unfertilized egg. This position is defined by the orientation of the egg cell in the egg chamber. The *bicoid* mRNA remains in the anterior region of the egg cell, attached by its 3′ untranslated region to the cell's cytoskeleton. It is not translated immediately, probably because its poly(A) tail is too short. This is inferred because translation, which occurs after fertilization of the egg, is preceded by extension of the poly(A) tail through the combined efforts of the Cortex, Grauzone and Staufen proteins, all of which are synthesized from genes in the egg. Bicoid protein then diffuses through the syncytium, setting up a concentration gradient, highest at the anterior end and lowest at the posterior end (*Figure 11.25*).

Three other maternal-effect gene products are also involved in setting up the anterior–posterior gradient. These are the Hunchback, Nanos and Caudal proteins. All are injected as mRNAs into the anterior region of the unfertilized egg. The *nanos* mRNA is transported to the posterior part of the egg and attached to the cytoskeleton while it awaits translation. The *hunchback* and *caudal* mRNAs become distributed evenly through the cytoplasm, but their proteins subsequently form gradients through the action of Bicoid and Nanos:

- Bicoid activates the *hunchback* gene in the embryonic nuclei and represses translation of the maternal *caudal* mRNA, increasing the concentration of the Hunchback protein in the anterior region and decreasing that of Caudal.
- Nanos represses translation of *hunchback* mRNA, contributing further to the anterior–posterior gradient of the Hunchback protein.

The net result is a gradient of Bicoid and Hunchback, greater at the anterior end, and of Nanos and Caudal, greater at the posterior end (see *Figure 11.25*). The gradient is supplemented with Torso protein, another maternal-effect gene product, which accumulates at the extreme anterior and posterior ends. Similar events result in a dorsal to ventral gradient, predominantly of the protein called Dorsal.

A cascade of gene expression converts positional information into a segmentation pattern

The body plan of the adult fly, as well as that of the larva, is built from a series of segments, each with a different structural role. This is clearest in the thorax, which has three segments each carrying one pair of legs, and the abdomen, which is made up of eight segments, but is also true for the head, even though in the head the segmented structure is less visible (*Figure 11.26*). The objective of embryo development is therefore production of a young larva with the correct segmentation pattern.

The gradients established in the embryo by the maternal-effect gene products are the first stage in formation of the segmentation pattern. These gradients provide the interior of the embryo with a basic amount of positional information, each point in the syncytium now having its own unique chemical signature defined by the relative amounts of the various maternal-effect gene products. This positional information is made more precise by expression of the **gap genes**.

Three of the anterior–posterior gradient proteins – Bicoid, Hunchback and Caudal – are transcription factors. They activate transcription of gap genes in the nuclei that

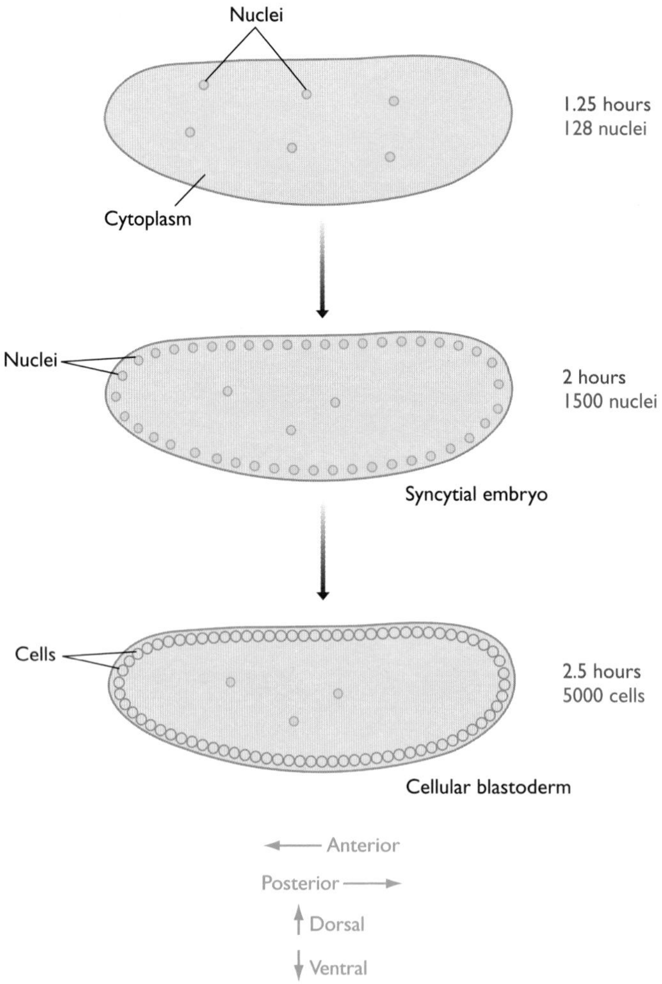

Figure 11.24 Early development of the *Drosophila* embryo.

To begin with, the embryo is a single syncytium containing a gradually increasing number of nuclei. These nuclei migrate to the periphery of the embryo after about 2 hours, and within another 30 minutes cells begin to be constructed. The embryo is approximately 500 μm in length and 170 μm in diameter.

now line the inside of the embryo (see *Figure 11.24*). The identities of the gap genes expressed in a particular nucleus depend on the relative concentrations of the gradient proteins and hence on the position of the nucleus along the anterior–posterior axis. Some gaps genes are activated directly by Bicoid, Hunchback and Caudal, examples being *buttonhead, empty spiracles* and *orthodenticle* which are activated by Bicoid. Other gap genes are switched on indirectly, as is the case with *hucklebein* and *tailless* which are dependent on transcription factors activated by Torso. There are also repressive effects (e.g. Bicoid represses *knirps*) and the gap gene products regulate their own expression in various ways. This complex interplay results in the positional information in the embryo, now carried by the relative concentrations of the gap gene products, becoming more detailed (*Figure 11.27*).

The next set of genes to be activated, the **pair-rule genes**, establish the basic segmentation pattern. Transcription of these genes responds to the relative concen-

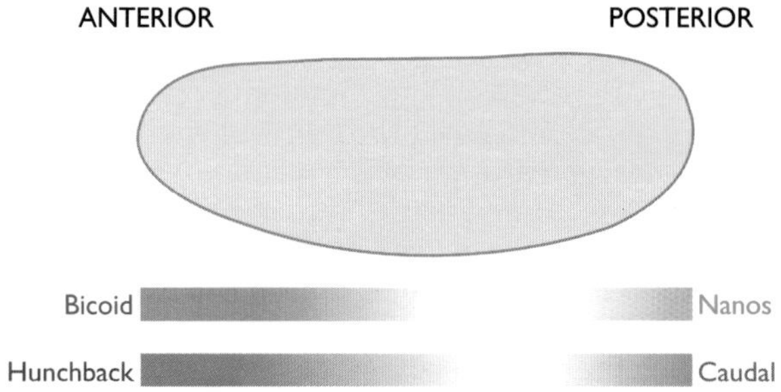

Figure 11.25 Establishment of the anterior–posterior axis in a *Drosophila* embryo.

The anterior–posterior axis is established by gradients of Bicoid, Nanos, Caudal and Hunchback proteins, as described in the text. In this diagram, the concentration gradients are indicated by the colored bars under the outline of the embryo.

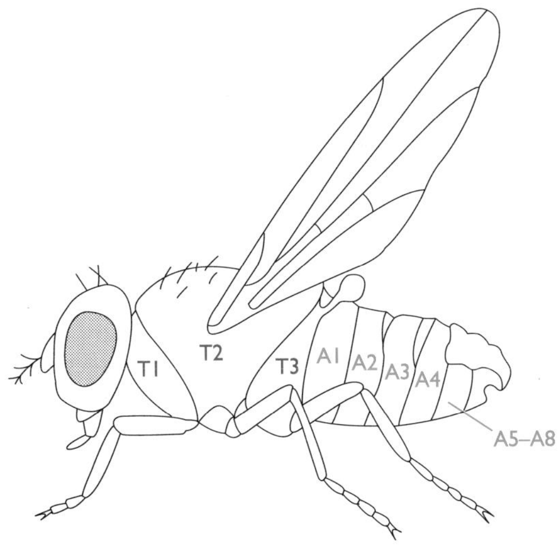

Figure 11.26 The segmentation pattern of the adult *Drosophila melanogaster.*

Note that the head is also segmented, but the pattern is not easily discernable from the morphology of the adult fly. Reprinted with permission from Lewis EB, *Nature*, **276**, 565. Copyright 1978 Macmillan Magazines Limited.

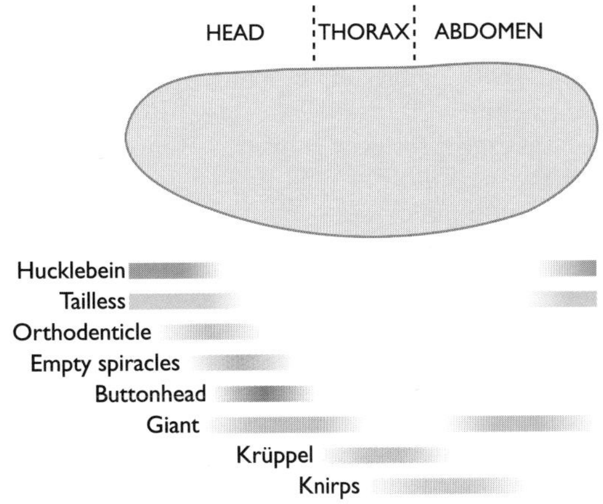

Figure 11.27 The role of the gap gene products in conferring positional information during embryo development in *D. melanogaster.*

As in *Figure 11.25*, the concentration gradient of each gap gene product is denoted by the colored bars. The parts of the embryo that give rise to the head, thorax and abdomen regions of the adult fly are indicated. Based on Twyman (1998).

trations of the gap gene products and occurs in nuclei that have become enclosed in cells. The pair-rule gene products therefore do not diffuse through the syncytium but remain localized within the cells that express them. The result is that the embryo can now be looked on as comprising a series of stripes, each stripe consisting of a set of cells expressing a particular pair-rule gene. In a further round of gene activation the **segment polarity** genes become switched on, providing greater definition to the stripes by setting the sizes and precise locations of what will eventually be the segments of the larval fly. Gradually we have converted the imprecise positional information of the maternal-effect gradients into a sharply defined segmentation pattern.

Segment identity is determined by the homeotic selector genes

The pair rule and segment polarity genes establish the segmentation pattern of the embryo but do not determine the identities of the individual segments. This is the job of the **homeotic selector genes**, which were first discovered by virtue of the extravagant effects that mutations in these genes have on the appearance of the adult fly. The *antennapedia* mutation, for example, transforms the head segment that usually produces an antenna into one that makes a leg, so the mutant fly has a pair of legs where its antennae should be. The early geneticists were fascinated by these monstrous, **homeotic** mutants and many were collected during the first few decades of the 20th century.

Genetic mapping of homeotic mutations has revealed that the selector genes are clustered in two groups on chromosome 3. These clusters are called the Antennapedia complex (ANT-C), which contains genes involved in determination of the head and thorax segments, and the Bithorax complex (BX-C), containing genes for abdomen segments (*Figure 11.28*). Some additional, nonselector development genes, such as *bicoid*, are also located in ANT-C. One interesting feature of the ANT-C and BX-C clusters, which is still not understood, is that the order of genes corresponds with the order of the segments in the

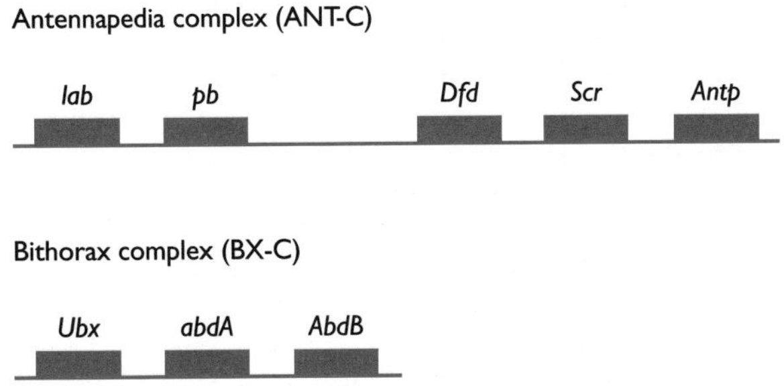

Figure 11.28 The Antennapedia and Bithorax gene complexes of *D. melanogaster.*

Both complexes are located on the fruit fly chromosome 3, ANT-C upstream of BX-C. The genes are usually drawn in the order shown, though this means that they are transcribed from right to left. The diagram does not reflect the actual lengths of the genes. The full gene names are as follows: *lab, labial palps*; *pb, proboscipedia*; *Dfd, Deformed*; *Scr, Sex combs reduced*; *Antp, Antennapedia*; *Ubx, Ultrabithorax*; *abdA, abdominal A*; *AbdB, Abdominal B.* In ANT-C, the nonselector genes *zerknüllt* and *bicoid* occur between *pb* and *Dfd*, and *fushi tarazu* lies between *Scr* and *Antp*.

fly, the first gene in ANT-C being *labial palps* which controls the most anterior segment of the fly, and the last gene in BX-C being *abdomenB*, which specifies the most posterior segment.

The correct selector gene is expressed in each segment because the activation of each one is responsive to the positional information represented by the distributions of the gap and pair-rule gene products. The selector gene products are themselves transcription factors, each containing a homeodomain version of the helix-turn-helix DNA-binding structure (Section 7.4.2). Each transcription factor, possibly in conjunction with a coactivator such as Extradenticle, switches on the set of genes needed to initiate development of the specified segment. Maintenance of the differentiated state is ensured partly by the repressive effect that each selector gene product has on expression of the other selector genes, and partly by the work of Polycomb which, as we saw in Section 11.2.2, constructs inactive chromatin over the selector genes that are not expressed in a particular cell (Pirrotta, 1997).

Homeotic selector genes are universal features of higher eukaryotic development

The homeodomains of the various *Drosophila* selector genes are strikingly similar. This observation led researchers in the 1980s to search for other homeotic genes by using the homeodomain as a probe in hybridization experiments. First, the *Drosophila* genome was searched, resulting in isolation of several previously unknown homeodomain-containing genes. These have turned out not to be selector genes, but other types of gene coding for transcription factors involved in development. Examples include the pair rule genes *even-skipped* and *fushi tarazu*, and the segment polarity gene *engrailed*.

The real excitement came when the genomes of other organisms were probed and it was realized that homeodomains are present in genes in a wide variety of animals, including humans. Even more unexpected was the discovery that some of the homeodomain genes in these other organisms are homeotic selector genes organized

Box 11.1: The genetic basis of flower development

Developmental processes in plants are, in most respects, very different from those of fruit flies and other animals, but at the genetic level there are certain similarities, sufficient for the knowledge gained about *Drosophila* development to be of value in aiding interpretation of similar research carried out with plants. In particular, the recognition that a limited number of homeotic selector genes control the *Drosophila* body plan has led to a model for plant development which postulates that the structure of the flower is determined by a small number of homeotic genes.

All flowers are constructed along similar lines, made up of four concentric whorls, each comprising a different floral organ. The outer whorl, number 1, contains sepals, which are modified leaves that envelop and protect the bud during its early development. The next whorl, number 2, contains the distinctive petals, and within these are whorls 3 (stamens, the male reproductive organs) and 4 (carpels, female reproductive organs).

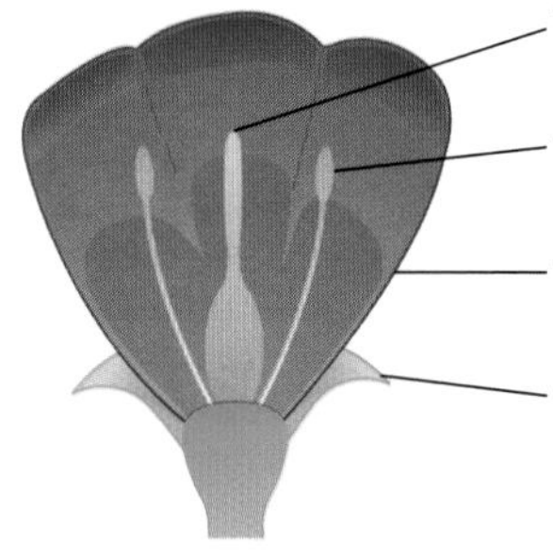

Most of the research on plant development has been carried out with *Antirrhinum*, the snapdragon, and *Arabidopsis thaliana*, a small vetch that has been adopted as a model species, partly because it has a genome of only 100 Mb (see *Table 1.1*, p. 9), one of the smallest known among flowering plants. Although these plants do not appear to contain homeodomain proteins they do have genes which, when mutated, lead to homeotic changes in the floral architecture, such as replacement of sepals by carpels. Analysis of these mutants has led to the 'ABC model', which states that there are three types of homeotic genes – A, B and C – controlling flower development:

- Whorl 1 is specified by A-type genes: examples in *Arabidopsis* are *apetala1* and *apetala2*.
- Whorl 2 is specified by A genes acting in concert with B genes, examples of the latter including *apetala3* and *pistillata*.
- Whorl 3 is specified by B plus C genes, the latter including *agamous*.
- Whorl 4 is specified by C genes acting on their own.

As anticipated from the work with *Drosophila*, the A, B and C homeotic gene products are transcription factors. Most of them contain the same DNA-binding domain, the **MADS box**. Other components of the developmental system include at least one master gene, called *floricaula* in *Antirrhinum* and *leafy* in *Arabidopsis*, which controls the switch from vegetative to reproductive growth, hence initiating flower development, and also has a role in establishing the pattern of homeotic gene expression (Ma, 1998; Parcy *et al.*, 1998). In *Arabidopsis* there is also a gene, called *curly leaf*, whose product acts like Polycomb of *Drosophila* (Section 11.2.2), maintaining the differentiated state of each cell by repressing those homeotic genes that are inactive in a particular whorl (Goodrich *et al.*, 1997).

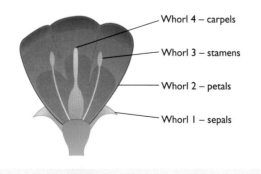

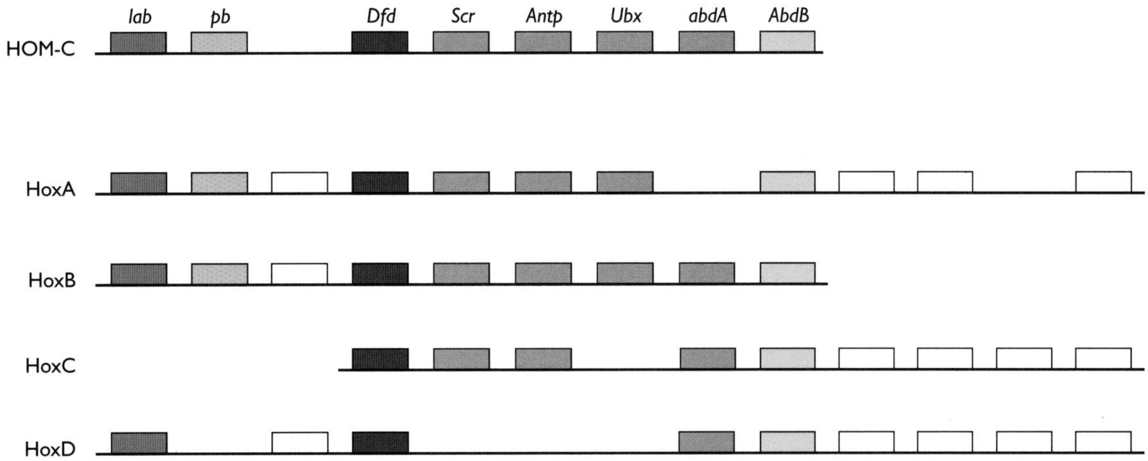

Figure 11.29 Comparison between the *Drosophila* HOM-C gene complex and the four Hox clusters of vertebrates.

Genes that code for proteins with related structures and functions are indicated by the colors. For more details on the evolution of the Hox clusters, see Section 14.2.1. The diagram does not reflect the actual lengths of the genes. Gene names are given in the legend to *Figure 11.28*.

into clusters similar to ANT-C and BX-C, and that these genes have equivalent functions to the fruit fly versions, specifying construction of the body plan.

We now look on the ANT-C and BX-C clusters of selector genes in *Drosophila* as two parts of a single complex, usually referred to as the homeotic gene complex or HOM-C. In vertebrates there are four homeotic gene clusters called HoxA to HoxD. When these four clusters are aligned with one another and with HOM-C (*Figure 11.29*), similarities are seen between the genes at equivalent positions, such that the evolutionary history of the homeotic selector gene clusters can be traced from insects through to humans (see Section 14.2.1). The genes in the vertebrate clusters specify the development of body structures and, as in *Drosophila*, the order of genes reflects the order of these structures in the adult body plan. This is clearly seen with the mouse HoxB cluster, which controls development of the nervous system (*Figure 11.30*).

The remarkable conclusion is that, at this fundamental level, developmental processes in fruit flies and other 'simple' eukaryotes are similar to the processes occurring in human and other 'complex' organisms. The discovery that studies of fruit flies are directly relevant to human development opens up vast vistas of future research possibilities.

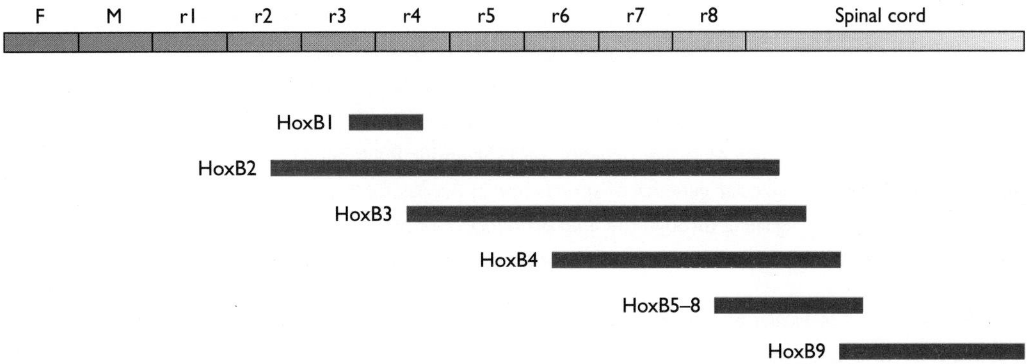

Figure 11.30 Specification of the mouse nervous system by selector genes of the HoxB cluster.

The nervous system is shown schematically and the positions specified by the individual HoxB genes (HoxB1 to HoxB9) indicated by the red bars. The components of the nervous system are: F, forebrain; M, midbrain; r1–r8, rhombomeres 1–8; followed by the spinal cord. Rhombomeres are segments of the hindbrain seen during development.

REFERENCES

Alberts B, Bray D, Lewis J, Raff M, Roberts K and Watson JD (1994) *Molecular Biology of the Cell*, 3rd edn. Garland Publishing, New York.

Becker S, Groner B and Müller CW (1998) Three-dimensional structure of the Stat3β homodimer bound to DNA. *Nature*, **394**, 145–151.

Brock TD (1990) *The Emergence of Bacterial Genetics*. Cold Spring Harbor Laboratory Press, Cold Spring Harbor, New York.

***C. elegans* Sequencing Consortium** (1998) Genome sequence of the nematode *C. elegans*: a platform for investigating biology. *Science*, **282**, 2012–2018.

Chawla S, Hardingham GE, Quinn DR and Bading H (1998) CBP: a signal-regulated transcriptional coactivator controlled by nuclear calcium and CaM kinase IV. *Science*, **281**, 1505–1509.

Costanzi C and Pehrson JR (1998) Histone macroH2A1 is concentrated in the inactive X chromosome of female mammals. *Nature*, **393**, 599–601.

Darnell JE (1997) STATs and gene regulation. *Science*, **277**, 1630–1635.

Errington J (1996) Determination of cell fate in *Bacillus subtilis*. *Trends Genet.*, **12**, 31–34.

Garre C, Bianchiscarra G, Sirito M, Musso M and Ravazzolo R (1992) Lactoferrin binding sites and nuclear localization in K562(S) cells. *J. Cell Physiol.*, **153**, 477–482.

Goodrich J, Puangsomlee P, Martin M, Long D, Meyerowitz EM and Coupland G (1997) A Polycomb-group gene regulates homeotic gene expression in *Arabidopsis*. *Nature*, **386**, 44–51.

Grossman AD (1995) Genetic networks controlling the initiation of sporulation and the development of genetic competence in *Bacillus subtilis*. *Annu. Rev. Genet.*, **29**, 477–508.

Haber JE (1998) A locus control region regulates yeast recombination. *Trends Genet.*, **14**, 317–321.

He J and Furmanski P (1995) Sequence specificity and transcriptional activation in the binding of lactoferrin to DNA. *Nature*, **373**, 721–724.

Jeppesen P and Turner BM (1993) The inactive X chromosome in female mammals is distinguished by a lack of histone H4 acetylation, a cytogenetic marker for gene expression. *Cell*, **74**, 281–289.

Katz WS, Hill RJ, Clandinin TR and Sternberg PW (1995) Different levels of the *C. elegans* growth factor LIN-3 promote distinct vulval precursor fates. *Cell*, **82**, 297–307.

Kornfeld K (1997) Vulval development in *Caenorhabditis elegans*. *Trends Genet.*, **13**, 55–61.

Ma H (1998) To be, or not to be, a flower – control of floral meristem identity. *Trends Genet.*, **14**, 26–32.

Parcy F, Nilsson O, Busch MA, Lee I and Weigel D (1998) A genetic framework for floral patterning. *Nature*, **395**, 561–566.

Pena MMO, Koch KA and Thiele DJ (1998) Dynamic regulation of copper uptake and detoxification genes in *Saccharomyces cerevisiae*. *Mol. Cell. Biochem.*, **18**, 2514–2523.

Pirrotta V (1997) Chromatin-silencing mechanisms in *Drosophila* maintain patterns of gene expression. *Trends Genet.*, **13**, 314–318.

Popperl H, Bienz M, Studer M, Chan SK, Aparicio S, Brenner S, Mann RS and Krumlauf R (1995) Segmental expression of HoxB-1 is controlled by a highly conserved autoregulatory loop dependent upon exd/pbx. *Cell*, **81**, 1031–1042.

Regulski M, Dessain S, McGinnis N and McGinnis W (1991) High affinity binding sites for the Deformed protein are required for the function of an autoregulatory enhancer of the *deformed* gene. *Genes Devel.*, **5**, 278–286.

Robinson MJ and Cobb MH (1997) Mitogen-activated kinase pathways. *Curr. Opin. Cell Biol.*, **9**, 180–186.

Schlessinger J (1993) How receptor tyrosine kinases activate Ras. *Trends Biochem. Sci.*, **18**, 273–275.

Spiegel S, Foster D and Kolesnick R (1996) Signal transduction through lipid second messengers. *Curr. Opin. Cell Biol.*, **8**, 159–167.

Sternberg PW and Han M (1998) Genetics of RAS signaling in *C. elegans*. *Trends Genet.*, **14**, 466–472.

Strachan T and Read AP (1996) *Human Molecular Genetics*. BIOS Scientific Publishers, Oxford.

Stragier P and Losick R (1996) Molecular genetics of sporulation in *Bacillus subtilis*. *Annu. Rev. Genet.*, **30**, 297–341.

Toker A and Cantley LC (1997) Signaling through the lipid products of phosphoinositide-3-OH kinase. *Nature*, **387**, 673–676.

Tsai M-J and O'Malley BW (1994) Molecular mechanisms of action of steroid/thyroid receptor superfamily members. *Annu. Rev. Biochem.*, **63**, 451–486.

Twyman RM (1998) *Advanced Molecular Biology: A Concise Reference*. BIOS Scientific Publishers, Oxford.

Winge DR, Jensen LT and Srinivasan C (1998) Metal ion regulation of gene expression in yeast. *Curr. Opin. Chem. Biol.*, **2**, 216–221.

FURTHER READING

Ballabio A and Willard HF (1992) Mammalian X-chromosome inactivation and the XIST gene. *Curr. Opin. Genet. Devel.*, **2**, 439–448.

Gehring WJ, Affolter M and Bürglin T (1994) Homeodomain proteins. *Annu. Rev. Biochem.*, **63**, 487–526. — *Details of the* Drosophila *homeodomain proteins with emphasis on the DNA–protein interactions.*

Karin M and Hunter T (1995) Transcriptional control by protein phosphorylation: signal transmission from the cell surface to the nucleus. *Curr. Biol.*, **5**, 747–757.

Maconochie M, Nonchev S, Morrison A and Krumlauf R (1996) Paralogous *Hox* genes: function and regulation. *Annu. Rev. Genet.*, **30**, 529–556. — *Describes homeotic selector genes in vertebrates.*

Maruta H and Burgess AW (1994) Regulation of the Ras signaling network. *Bioessays*, **16**, 489–496.

Massagué J (1998) TGF-β signal transduction. *Annu. Rev. Biochem.*, **67**, 753–791. — *Comprehensive description of the SMAD signaling pathway, as introduced in Research Briefing 11.1.*

Stark GR, Kerr IM, Williams BRG, Silverman RH and Schreiber RD (1998) How cells respond to interferons. *Annu. Rev. Biochem.*, **67**, 227–264. — *Includes detailed information on JAKs and STATs.*

PART

3 How Genomes Replicate and Evolve

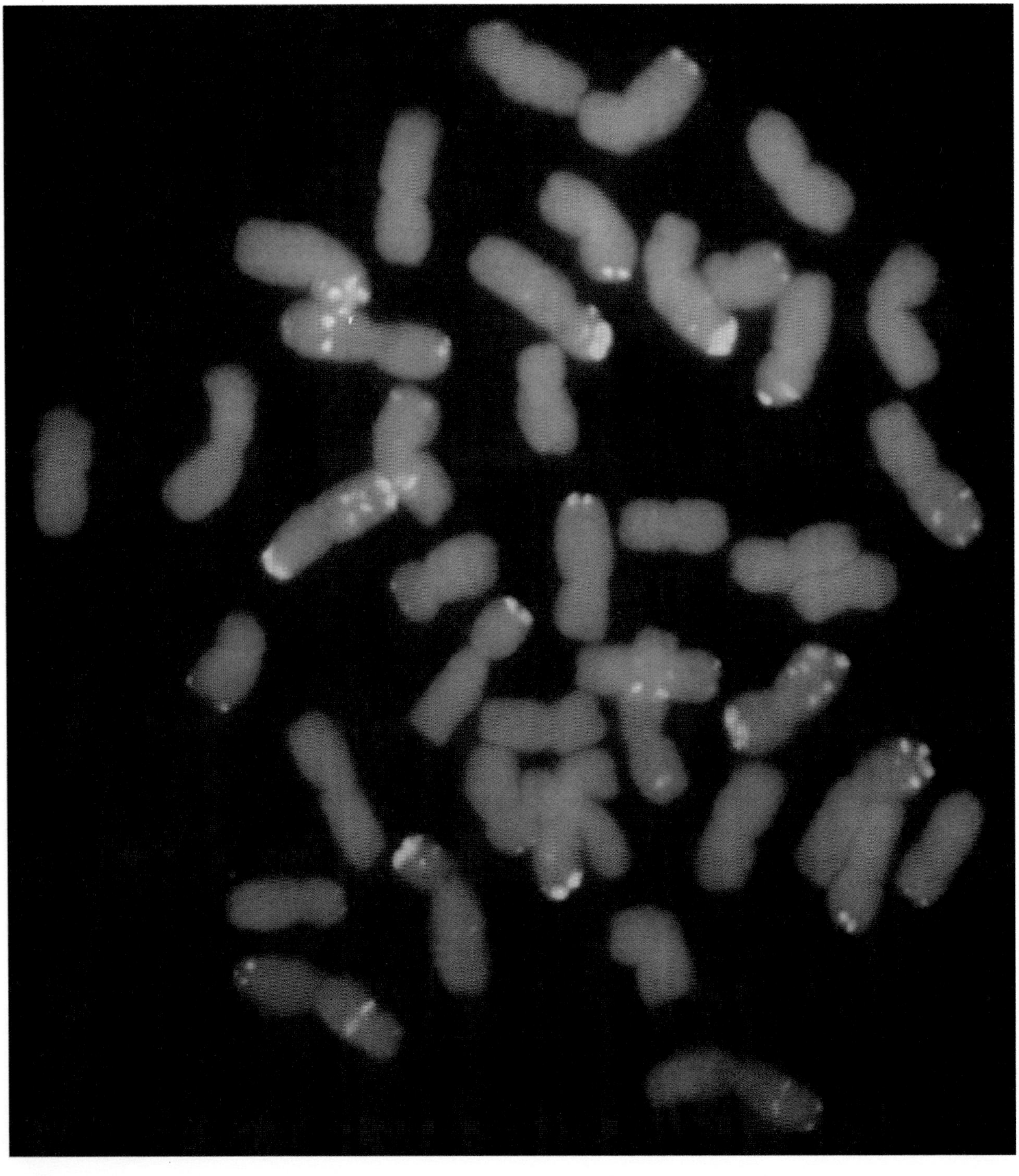

Chapter 12
Genome Replication

Chapter 13
The Molecular Basis of Genome Evolution

Chapter 14
Patterns of Genome Evolution

Chapter 15
Molecular Phylogenetics

Opposite page: A wheat genome incorporating two *Aegilops umbellulata*-origin chromosomes which carry high molecular weight glutenin genes giving high bread-making quality. The yellow-green probe is a repetitive DNA sequence isolated from rye enabling the identification of many of the different chromosomes.

Acknowledgement: Reproduced from A Castilho, TE Miller and JS Heslop-Harrison (1996) Physical mapping of sub-centromeric and intercalary translocation breakpoints in a set of wheat-*Aegilops umbellulata* recombinant lines using *in situ* hybridization, *Theor. Appl. Genet.*, **93**, 816–825. Reproduced with kind permission from Springer–Verlag and Professor JS Heslop-Harrison.

The image was supplied by Professor JS Heslop-Harrison.

CHAPTER

12 Genome Replication

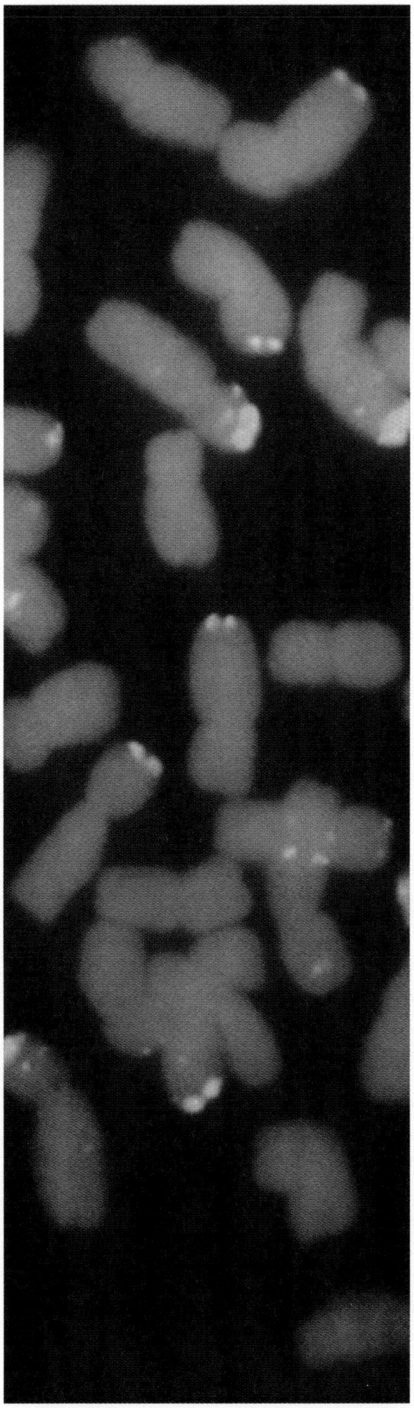

Contents

12.1 The Issues Relevant to Genome Replication 300
12.2 The Topological Problem 301
12.2.1 Experimental proof of the Watson–Crick scheme for DNA replication 301
12.2.2 DNA topoisomerases provide a solution to the topological problem 305
12.3 The Replication Process 306
12.3.1 Initiation of genome replication 307
12.3.2 The elongation phase of replication 311
12.3.3 Termination of replication 319
12.3.4 Maintaining the ends of a linear DNA molecule 322
12.4 Regulation of Eukaryotic Genome Replication 323
12.4.1 Coordination of genome replication and cell division 323
12.4.2 Control within S phase 326

Concepts

- *Genome replication occurs every time the cell divides*
- *The double helix structure indicated a possible mode of DNA replication, which was subsequently confirmed by the Meselson–Stahl experiment*
- *The topological problem is solved by DNA topoisomerases which relieve torsional stress ahead of the replication fork*
- *Origins of replication are well defined in bacteria and yeast but less so in eukaryotes*
- *During replication, the leading strand is copied continuously but the lagging strand is copied discontinuously*
- *Strand synthesis requires a primer*
- *An array of enzymes and proteins are involved in the events at the replication fork*
- *Termination of replication is poorly understood*
- *The ends of linear DNA molecules will shorten unless special measures are taken to repair them: in eukaryotic chromosomes this problem is solved by telomere synthesis*
- *Cell cycle checkpoints ensure that DNA replication is coordinated with cell division*

THE PRIMARY FUNCTION of a genome is to specify the biochemical signature of the cell in which it resides. In Part 2 we saw that the genome achieves this objective by the coordinated expression of genes and groups of genes, resulting in maintenance of a proteome whose individual protein components carry out and regulate the cell's biochemical activities. In order to continue carrying out this function, the genome must replicate each time that the cell divides. This means that the entire DNA content of the cell must be copied at the appropriate period in the cell cycle, and the resulting DNA molecules distributed to the daughter cells so that each one receives a complete copy of the genome. This elaborate process, which spans the interface between molecular biology, biochemistry and cell biology, is the subject of this chapter.

Normally we think of replication as producing two identical copies of the genome, this being a requirement if the daughter cells are to have the same biochemical capabilities as the parent cell. In a general sense, genome replication does result in identical copies, but over time the genome undergoes change, nucleotide sequence alterations accumulating as a result of mutations, occasional errors in replication, and sequence rearrangements caused by recombination and related events. These sequence alterations, which are described in Chapter 13, enable the genome to evolve. This is **molecular evolution**, which underlies the evolution of organisms, though in a complex manner that is not yet fully understood. In Chapter 14 we will explore how molecular evolution has resulted in the vast variety of genomes present in the different organisms alive today, and in Chapter 15 we will see how the techniques of **molecular phylogenetics** can be used to make comparisons between the sequences of genes and of entire genomes, enabling the evolutionary relationships between organisms to be inferred.

12.1 THE ISSUES RELEVANT TO GENOME REPLICATION

Genome replication has been studied since Watson and Crick first discovered the double helix structure of DNA back in 1953. In the years since then research has been driven by three related but distinct issues:

1. ***The topological problem*** was the primary concern in the years from 1953 to 1958. The problem arose from the need to unwind the double helix in order to make copies of its two polynucleotides (see *Figure 12.1*). The issue assumed center stage in the mid-1950s because it was the main stumbling block to acceptance of the double helix as the correct structure for DNA, but moved into the background in 1958 when Matthew Meselson and Franklin Stahl demonstrated that, despite the perceived difficulties, DNA replication in *Escherichia coli* occurs by the method predicted

by the double helix structure. This enabled research into genome replication to move forward, even though the topological problem itself was not solved until the early 1980s when the mode of action of **DNA topoisomerases** was understood (Section 12.2.2).

2. ***The replication process*** has been intensively studied since 1958. During the 1960s the enzymes and proteins involved in replication in *E. coli* were identified and their functions delineated, and in the following years similar progress was made in understanding the details of eukaryotic DNA replication. This work is ongoing with research today centered on topics such as the initiation of replication and the precise modes of action of the proteins active at the replication fork.
3. ***The regulation of genome replication***, particularly in the context of the cell cycle, has become the predominant area of research in recent years. This work has shown that initiation is the key control point in genome replication and has begun to explain how replication is synchronized with the cell cycle so that daughter genomes are available when the cell divides.

Our study of genome replication will deal with each of these three topics in the order in which they are listed above.

12.2 THE TOPOLOGICAL PROBLEM

In their *Nature* paper announcing the discovery of the double helix structure of DNA, Watson and Crick (1953a) made one of the most famous statements in molecular biology:

> It has not escaped our notice that the specific pairing we have postulated immediately suggests a possible copying mechanism for the genetic material.

The pairing process that they refer to is one in which each strand of the double helix acts as a template for synthesis of a second, complementary strand, the end result being that both of the two daughter double helices are identical to the parent molecule (*Figure 12.1*). The scheme is almost implicit in the double helix structure, but it presents problems, as admitted by Watson and Crick in a second *Nature* paper published just a month after the report of the structure. This paper (Watson and Crick, 1953b) describes the postulated replication process in more detail, but points out the difficulty that arises from the need to unwind the double helix. The most trivial of these difficulties is the possibility of the daughter molecules getting tangled up. More critical is the rotation that would accompany the unwinding: with one turn occurring for every 10 bp of the double helix, complete replication of the DNA molecule in human chromosome 1, which is 250 Mb in length, would require 25 million rotations of the chromosomal DNA. It is difficult to imagine how this could occur within the constrained volume of the nucleus, but the

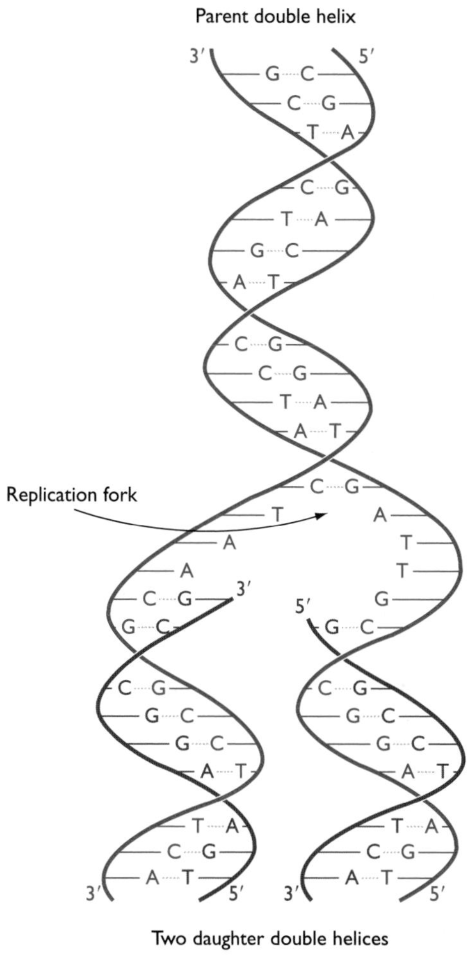

Figure 12.1 DNA replication, as predicted by Watson and Crick.

The polynucleotides of the parent double helix are shown in red. Both act as templates for synthesis of new strands of DNA, shown in blue. The sequences of these new strands are determined by base-pairing with the template molecules. Note that the two polynucleotides of the parent helix cannot simply be pulled apart: the helix has to be unwound in some way.

unwinding of a linear chromosomal DNA is not physically impossible. In contrast, a circular double-stranded molecule, for example a bacterial or bacteriophage genome, having no free ends, would not be able to rotate in the required manner and so, apparently, could not be replicated by the Watson–Crick scheme. Finding an answer to this dilemma was a major preoccupation of molecular biology during the 1950s.

12.2.1 Experimental proof of the Watson–Crick scheme for DNA replication

The topological problem was considered so serious by some molecular biologists, notably Max Delbrück, that there was initially some resistance to accepting the double helix as the correct structure of DNA (Holmes, 1998). The difficulty relates to the **plectonemic** nature of the double helix, this being the topological arrangement that prevents the two strands of a coil being separated without unwinding, and would therefore be resolved if the double helix was in fact **paranemic**, because this would mean that the two strands could be separated simply by moving each one sideways without unwinding the molecule. It was suggested that the double helix could be converted into a paranemic structure by supercoiling (see *Figure 6.11*, p. 129) in the direction opposite to the turn of the helix itself, or that within a DNA molecule the right-handed helix proposed by Watson and Crick might be 'balanced' by equal lengths of a left-handed helical structure. The possibility that double-stranded DNA was not a helix at all, but a side-by-side ribbon structure, was also briefly considered, this idea surprisingly being revived in the late 1970s (e.g. Rodley *et al.*, 1976) and receiving a rather ascerbic response from Crick and his colleagues (Crick *et al.*, 1979). Each of these structural solutions to the topological problem were individually rejected for one reason or another, most of them because they required alterations to the double helix structure, alterations that were not compatible with the X-ray diffraction results and other experimental data pertaining to DNA structure.

The first real progress towards a solution of the topological problem came in 1954 when Delbrück proposed a 'breakage-and-reunion' model for separating the strands of the double helix (Holmes, 1998). In this model, the strands are separated not by unwinding the helix with accompanying rotation of the molecule, but by breaking one of the strands, passing the second strand though the gap, and rejoining the first strand. This scheme is in fact very close to the correct solution to the topological problem, being one of the ways in which DNA topoisomerases work (see *Figure 12.4A*), but unfortunately Delbrück over-complicated the issue by attempting to combine breakage-and-reunion with the actual replication process. This led him to a model for DNA replication which results in each polynucleotide in the daughter molecule being made up partly of parental DNA and partly of newly-synthesized DNA (*Figure 12.2A*). This **dispersive** mode of replication contrasts with the **semiconservative** system proposed by Watson and Crick (*Figure 12.2B*). A third possibility is that replication is fully **conservative**, one of the daughter double helices being made entirely from newly-synthesized DNA and the other comprising the two parental strands (*Figure 12.2C*). Models for conservative replication are difficult to devise, but it can be envisaged that this type of replication might be accomplished without unwinding the parent helix.

The Meselson–Stahl experiment

Delbrück's breakage-and-reunion model was important because it stimulated experiments designed to test between the three modes of DNA replication illustrated in *Figure 12.2*. Radioactive isotopes had recently been introduced into molecular biology so DNA labeling (Technical Note 3.3, p. 46) was attempted in order to

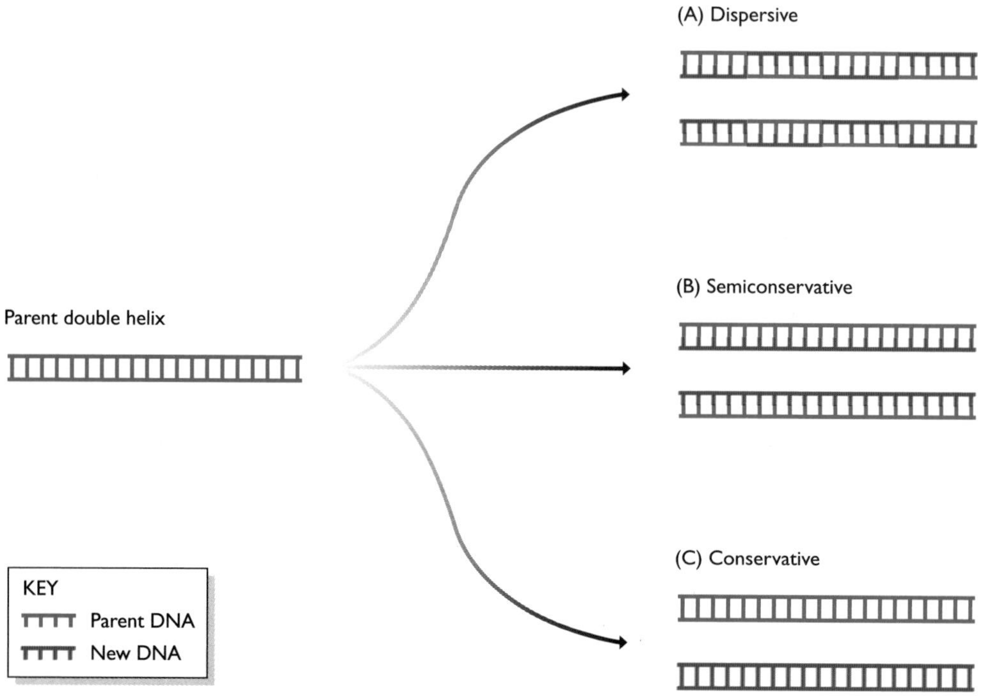

Figure 12.2 Three possible schemes for DNA replication.

For the sake of clarity, the DNA molecules are drawn as ladders rather than helices.

Box 12.1: Variations on the semiconservative theme

No exceptions to the semiconservative mode of DNA replication are known but there are several variations on this basic theme. DNA copying via a replication fork, as shown in *Figure 12.1*, is the predominant system, being used by chromosomal DNA molecules in eukaryotes and by the circular genomes of prokaryotes. Some smaller circular molecules, such as the human mitochondrial genome (Section 6.1.2) use a slightly different process called **displacement replication**, which involves continuous copying of one strand of the helix, the second strand being displaced and subsequently copied after synthesis of the first daughter genome has been completed (see figure A).

The advantage of displacement replication as performed by human mitochondrial DNA is not clear. In contrast, the special type of displacement process called **rolling circle replication** is an efficient mechanism for the rapid synthesis of multiple copies of a circular genome (Novick, 1998). Rolling circle replication, which is used by λ and various other bacteriophages, initiates at a nick which is made in one of the parent polynucleotides. The free 3′-end that results is extended, displacing the 5′-end of the polynucleotide. Continued DNA synthesis 'rolls off' a complete copy of the genome, and further synthesis eventually results in a series of genomes linked head-to-tail (see Figure B). These genomes are single-stranded and linear, but can easily be converted to double-stranded, circular molecules by complementary strand synthesis followed by cleavage at the junction points between genomes and circularization of the resulting segments.

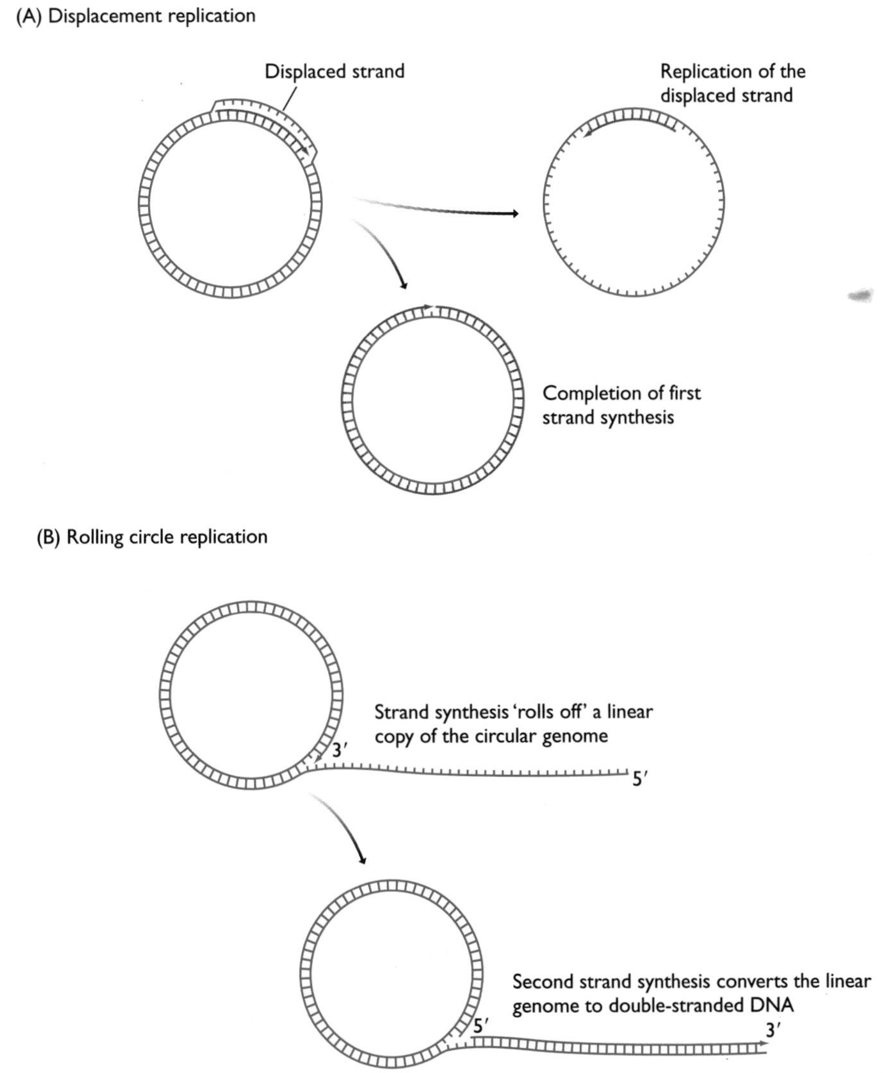

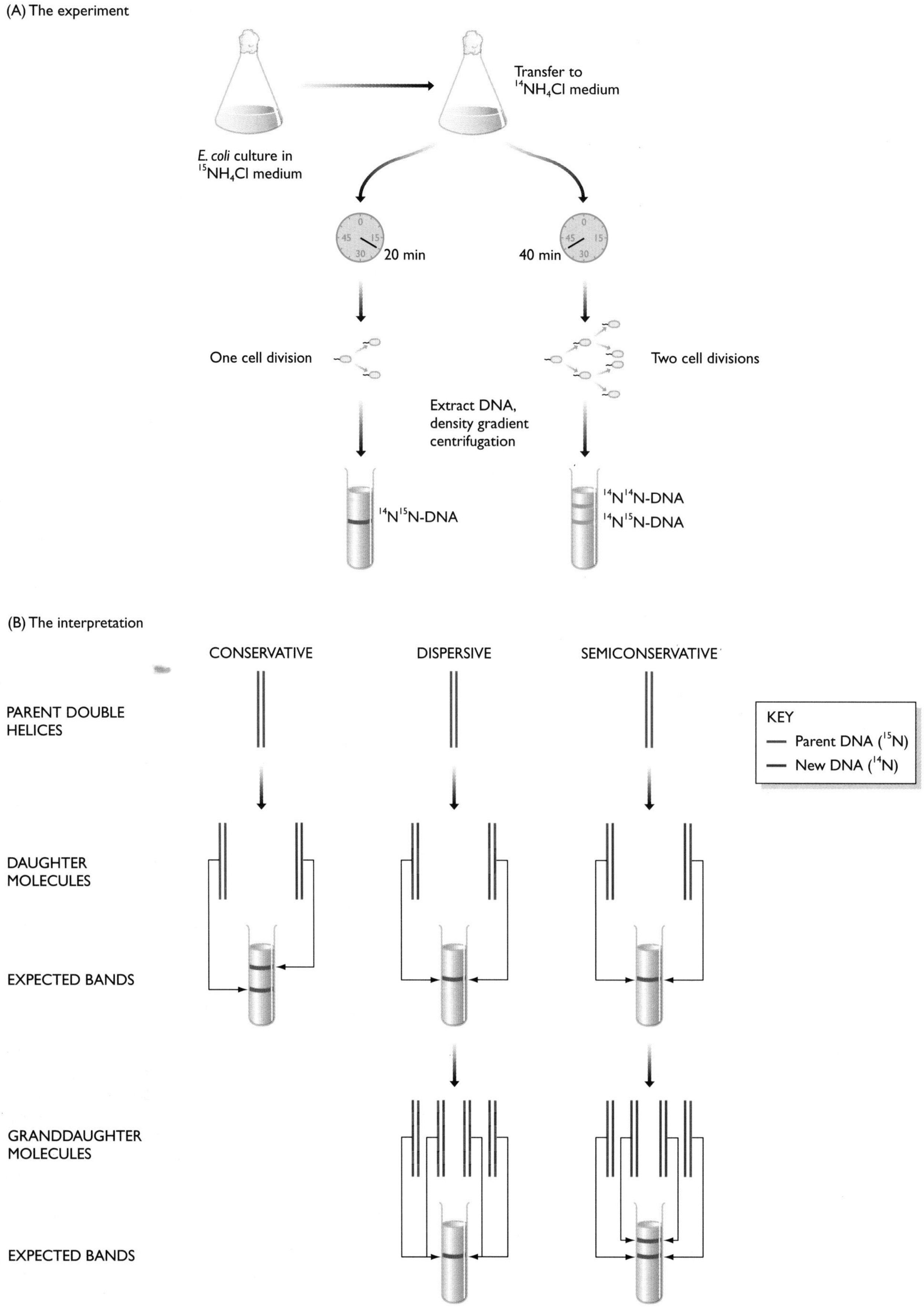
(A) The experiment
Transfer to
$^{14}NH_4Cl$ medium
E. coli culture in
$^{15}NH_4Cl$ medium
20 min
40 min
One cell division
Two cell divisions
Extract DNA,
density gradient
centrifugation
$^{14}N^{15}N$-DNA
$^{14}N^{14}N$-DNA
$^{14}N^{15}N$-DNA
(B) The interpretation
CONSERVATIVE
DISPERSIVE
SEMICONSERVATIVE
PARENT DOUBLE
HELICES
KEY
Parent DNA (^{15}N)
New DNA (^{14}N)
DAUGHTER
MOLECULES
EXPECTED BANDS
GRANDDAUGHTER
MOLECULES
EXPECTED BANDS

distinguish newly synthesized DNA from the parental polynucleotides. Note that each mode of replication predicts a different distribution of newly synthesized DNA, and hence of radioactive label, in the double helices resulting after two or more rounds of replication. Analysis of the radioactive contents of these molecules should therefore determine which replication scheme operates in living cells. Unfortunately, it proved impossible to obtain a clearcut result, largely because of the difficulty in measuring the precise amount of radioactivity in the DNA molecules, the analysis being complicated by the rapid decay of the ^{32}P isotope that was used as the label.

The breakthrough was eventually made by Matthew Meselson and Franklin Stahl who, in 1958, carried out the required experiment not with a radioactive label but with ^{15}N, the nonradioactive 'heavy' isotope of nitrogen. Now it was possible to analyze the replicated double helices by density gradient centrifugation (Technical Note 6.1, p. 123), because a DNA molecule labeled with ^{15}N has a higher buoyant density than an unlabeled molecule. Meselson and Stahl (1958) started with a culture of *E. coli* cells that had been grown with $^{15}NH_4Cl$ and whose DNA molecules therefore contained heavy nitrogen. The cells were transferred to normal medium and samples taken after 20 minutes and 40 minutes, corresponding to one and two cell divisions. DNA was extracted from each sample and the molecules examined by density gradient centrifugation (*Figure 12.3A*). After one round of DNA replication, the daughter molecules synthesized in the presence of normal nitrogen formed a single band in the density gradient, indicating that each double helix was made up of equal amounts of newly synthesized and parental DNA. This result immediately enabled the conservative mode of replication to be discounted, as this predicts that there will be two bands after one round of replication (*Figure 12.3B*), but did not provide a distinction between Delbrück's dispersive model and the semiconservative process favored by Watson and Crick. The distinction was, however, possible when the DNA molecules resulting from two rounds of replication were examined. Now the density gradient revealed two bands of DNA, the first corresponding to a hybrid composed of equal parts of newly synthesized and old DNA, and the second corresponding to molecules made up entirely of new DNA. This result agrees with the semiconservative scheme but is incompatible with dispersive replication, the latter predicting that after two rounds of replication all molecules will be hybrids.

12.2.2 DNA topoisomerases provide a solution to the topological problem

The Meselson–Stahl experiment proved that DNA replication in living cells follows the semiconservative scheme proposed by Watson and Crick, and hence indicated that the cell must have a solution to the topological problem. This solution was not understood by molecular biologists until some 25 years later when the activities of the groups of enzymes called DNA topoisomerases began to be characterized.

DNA topoisomerases are enzymes that carry out breakage-and-reunion reactions similar but not identical to that envisaged by Delbrück. Three types of DNA topoisomerases are recognized (*Table 12.1*):

- **Type IA topoisomerases** introduce a break in one polynucleotide and pass the second polynucleotide through the gap that is formed (*Figure 12.4A*). The two ends of the broken strand are then religated (Lima *et al.*, 1994). This mode of action results in the linking number (the number of times one strand crosses the other in a circular molecule) to be changed by one.
- **Type 1B topoisomerases** act in a similar way to the Type 1A enzymes, though the detailed mechanism is different (Redinbo *et al.*, 1998; Stewart *et al.*, 1998). Type 1A and 1B topoisomerases are distinct from one another and probably evolved separately.
- **Type II topoisomerases** break both strands of the double helix, creating a 'gate' through which a second segment of the helix is passed (*Figure 12.4B*; Berger *et al.*, 1996; Cabral *et al.*, 1997). This changes the linking number by two.

DNA topoisomerases do not themselves *unwind* the double helix. Instead they solve the topological problem

Figure 12.3 The Meselson–Stahl experiment.

(A) The experiment carried out by Meselson and Stahl involved growing a culture of *E. coli* bacteria in a medium containing $^{15}NH_4Cl$ (ammonium chloride labeled with the heavy isotope of nitrogen). Cells were then transferred to normal medium (containing $^{14}NH_4Cl$) and samples taken after 20 minutes (one cell division) and 40 minutes (two cell divisions). DNA was extracted from each sample and the molecules analyzed by density gradient centrifugation. After 20 minutes all the DNA contained similar amounts of ^{14}N and ^{15}N, but after 40 minutes two bands were seen, one corresponding to hybrid ^{14}N-^{15}N-DNA, and the other to DNA molecules made entirely of ^{14}N. (B) The predicted outcome of the experiment is shown for each of the three possible modes of DNA replication. The banding pattern seen after 20 minutes enables conservative replication to be discounted because this scheme predicts that after one round of replication there will be two different types of double helix, one containing just ^{15}N and the other containing just ^{14}N. The single ^{14}N-^{15}N-DNA band that was actually seen after 20 minutes is compatible with both dispersive and semiconservative replication, but the two bands seen after 40 minutes are consistent only with semiconservative replication. Dispersive replication continues to give hybrid ^{14}N-^{15}N molecules after two rounds of replication, whereas the granddaughter molecules produced at this stage by semiconservative replication include two that are made entirely of ^{14}N-DNA.

Table 12.1 DNA topoisomerases

Type	Substrate	Examples
Type IA	Single-stranded DNA	*E. coli* topoisomerases I and III; yeast and human topoisomerase III; archaeal reverse gyrase
Type IB	Single-stranded DNA	Eukaryotic topoisomerase I
Type II	Double-stranded DNA	*E. coli* topoisomerases II (DNA gyrase) and IV; eukaryotic topoisomerases II and IV

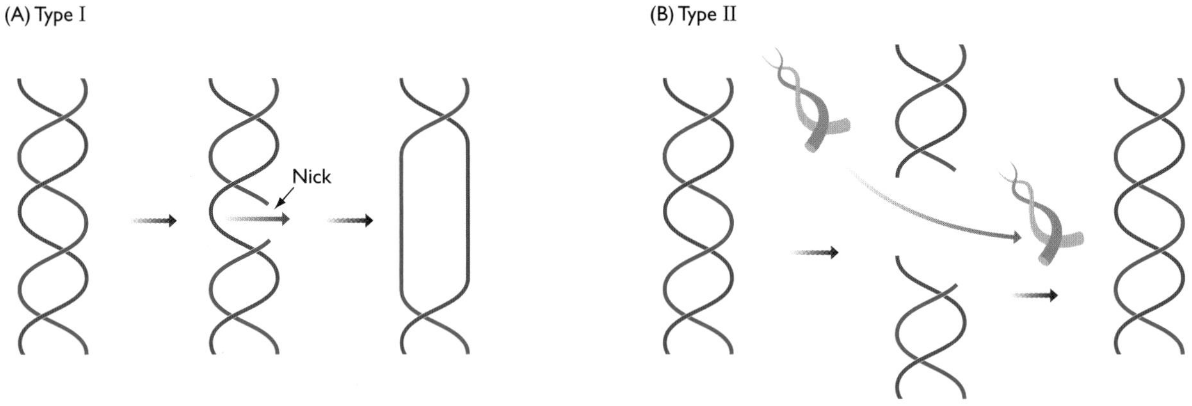

Figure 12.4 The mode of action of Type I and Type II DNA topoisomerases.

(A) A type I topoisomerase makes a nick in one strand of a DNA molecule, passes the intact strand through the nick, and reseals the gap. (B) A type II topoisomerase makes a double-stranded break in the double helix, creating a gate through which a second segment of the helix is passed.

by relieving the torsional stress introduced into the molecule by the progression of the replication fork. This enables the helix to be 'unzipped' with the two strands pulled apart sideways without the molecule having to rotate (*Figure 12.5*).

12.3 THE REPLICATION PROCESS

As with many processes in molecular biology, we conventionally look on replication as being made up of three phases, these being initiation, elongation and termination:

- **Initiation** (Section 12.3.1) involves recognition of the position(s) on a DNA molecule where replication will begin.
- **Elongation** (Section 12.3.2) concerns the events occurring at the replication fork, where the parent polynucleotides are copied.
- **Termination** (Section 12.3.3), which in general is only vaguely understood, occurs when the parent molecule has been completely replicated.

As well as these three stages in replication, one additional topic demands attention. This relates to a limitation in the replication process that, if uncorrected, would lead to linear, double-stranded DNA molecules getting shorter each time they are replicated (see *Figure 12.23*). The solution to this problem, which concerns the structure and synthesis of the telomeres at the ends of chromosomes (Section 6.1.1), will be described in Section 12.3.4.

Box 12.2: The diverse functions of DNA topoisomerases

In recent years it has become increasingly clear that DNA topoisomerases are important enzymes that have a number of functions in living cells. As well as relieving the torsional strain that builds up in the double helix during replication, topoisomerases play similar roles during transcription, recombination and other processes that result in over- or underwinding of the helix. In eukaryotes, topoisomerases form a major part of the nuclear matrix, the scaffold-like network that permeates the nucleus, and have roles in maintaining chromatin structure and unlinking DNA molecules during chromosome division. Most topoisomerases are only able to relax DNA, but prokaryotic Type II enzymes such as the bacterial DNA gyrase and the archaeal reverse gyrase are able to introduce supercoils into DNA molecules.

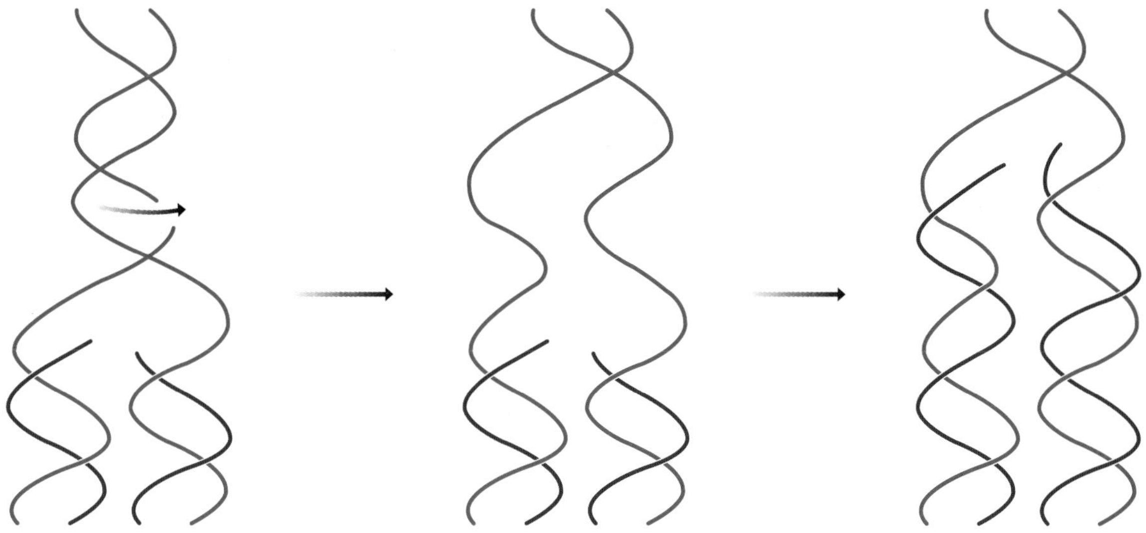

Figure 12.5 Unzipping the double helix.

During replication, the double helix is 'unzipped' as a result of the action of DNA topoisomerases. The replication fork is therefore able to proceed along the molecule without the helix having to rotate.

12.3.1 Initiation of genome replication

Initiation of replication is not a random process and always begins at the same position or positions on a DNA molecule, these points being called the **origins of replication**. Once initiated, two replication forks emerge from the genome and progress in opposite directions along the DNA: replication is therefore bidirectional (*Figure 12.6*). A circular bacterial genome has a single origin of replication, meaning that several thousand kb of DNA are copied by each replication fork. This situation differs from that seen with eukaryotic chromosomes, which have multiple origins and whose replication forks progress for shorter distances. The yeast *Saccharomyces cerevisiae*, for example, has about 300 origins, corresponding to one per 40 kb of DNA, and humans have some 20 000 origins, or one for every 150 kb.

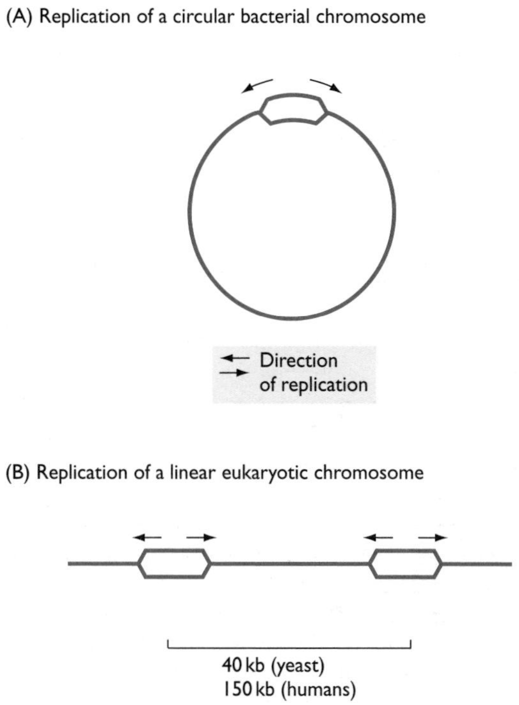

Figure 12.6 Bidirectional DNA replication of (A) a circular bacterial chromosome and (B) a linear eukaryotic chromosome.

Initiation at the E. coli origin of replication

We know substantially more about initiation of replication in bacteria than in eukaryotes. The *E. coli* origin of replication is referred to as *oriC*. By transferring segments of DNA from the *oriC* region into plasmids that lack their own origins it has been estimated that the *E. coli* origin of replication spans approximately 245 bp of DNA. Sequence analysis of this segment shows that it contains two short repeat motifs, one of nine nucleotides and the other of 13 nucleotides (*Figure 12.7A*). The nine-nucleotide repeat, five copies of which are dispersed throughout *oriC*, is the binding site for a protein called

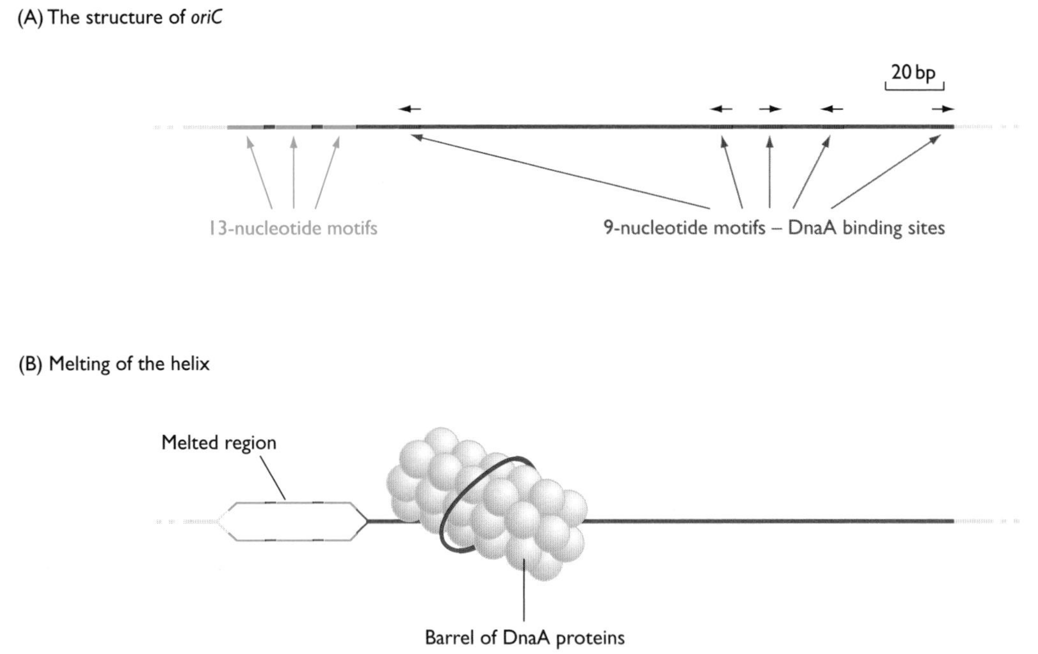

Figure 12.7 The *E. coli* origin of replication.

(A) The *E. coli* origin of replication is called *oriC* and is approximately 245 bp in length. It contains three copies of a 13-nucleotide repeat motif, consensus sequence 5′-GATCTNTTNTTTT-3′ where 'N' is any nucleotide, and five copies of a 9-nucleotide repeat, consensus 5′-TT$^{A}_{T}$T$^{A}_{C}$CA$^{A}_{C}$A-3′. The 13-nucleotide sequences form a tandem array of direct repeats at one end of *oriC*. The 9-nucleotide sequences are distributed through *oriC* with different units in the direct and inverted repeat configurations. Three of these repeats – numbers one, three and five when counted from the left hand end of *oriC* as drawn here – are regarded as major sites for DnaA attachment; the other two repeats are minor sites. The overall structure of the origin is similar in all bacteria and the sequences of the repeats do not vary greatly. (B) Model for the attachment of DnaA proteins to *oriC*, resulting in melting of the helix within the AT-rich 13-nucleotide sequences.

DnaA. With five copies of the binding sequence it might be imagined that five copies of DnaA attach to the origin, but in fact bound DnaA proteins cooperate with unbound molecules until some 30 are associated with the origin. Attachment occurs only when the DNA is negatively supercoiled, as is the normal state of affairs for the *E. coli* chromosome (Section 6.2.1).

The result of DnaA binding is that the double helix opens up ('melts') within the tandem array of three AT-rich, 13-nucleotide repeats located at one end of the *oriC* sequence (*Figure 12.7B*). The exact mechanism is unknown but DnaA does not appear to possess the enzymatic activity needed to break base pairs, and it is therefore assumed that the helix is melted by torsional stresses introduced by attachment of the DnaA proteins. An attractive model imagines the DnaA proteins forming a barrel-like structure around which the helix is wound. Melting the helix is promoted by HU, the most abundant of the DNA packaging proteins of *E. coli* (Section 6.2.1).

Melting of the helix initiates a series of events that culminates in the start of the elongation phase of replication. The first step is attachment of a complex of two proteins, DnaBC, forming the **prepriming complex**. DnaC has a transitory role and is released from the complex soon after it is formed, its function probably being simply to aid the attachment of DnaB. The latter is a **helicase**, an enzyme which can break base pairs and whose mode of action we will examine in Section 12.3.2. DnaB begins to increase the single-stranded region within the origin, enabling the enzymes involved in the elongation phase of DNA replication to attach. This represents the end of the initiation phase of replication in *E. coli* as now the replication forks start to progress away from the origin and DNA copying begins.

Origins of replication in yeast have also been clearly defined

The technique used to delineate the *E. coli oriC* sequence, involving transfer of DNA segments into a nonreplicating plasmid, has also proved valuable in identifying origins of replication in the yeast *Saccharomyces cerevisiae*. Origins identified in this way are called **autonomously replicating sequences** or **ARSs**. A typical yeast ARS is shorter than the *E. coli* origin, usually less than 200 bp in length, but like the *E. coli* origin contains discrete segments with different functional roles, these 'subdomains' having similar sequences in different ARSs (*Figure 12.8A*). Four subdomains are recognized, two of which – subdomains A and B1 – make up the **origin recognition sequence**, a stretch of some 40 bp in total that is the binding site for the **origin recognition complex** (**ORC**), a set of six proteins that attach to the ARS (*Figure 12.8B*). ORCs have been described as initiator proteins for yeast replication but this interpretation is probably not strictly correct because ORCs appear to remain attached to yeast origins throughout the cell cycle (Bell and Stillman, 1992; Diffley and Cocker, 1992). Instead they have a key role in the regulation of DNA replication, acting as mediators between replication origins and the regulatory signals that coordinate the initiation of DNA replication with the cell cycle (Section 12.4; Stillman, 1996).

We must therefore look elsewhere in yeast ARSs for sequences with functions equivalent to *oriC* of *E. coli*. This leads us to the two other conserved sequences in the typical yeast ARS, subdomains B2 and B3 (see *Figure 12.8A*). Our current understanding suggests that these two subdomains function in a manner similar to the *E. coli* origin. Subdomain B2 appears to correspond with the 13-nucleotide repeat array of the *E. coli* origin, being the position at which the two strands of the helix are first separated. This melting is induced by torsional stress introduced by attachment of a DNA binding protein, ARS binding factor 1 (ABF1), which attaches to subdomain B3 (see *Figure 12.8B*). As in *E. coli*, melting of the helix within a yeast replication origin is followed by attachment of the helicase and other replication enzymes to the DNA, completing the initiation process and enabling the replication forks to begin their progress along the DNA, as described in Section 12.3.2.

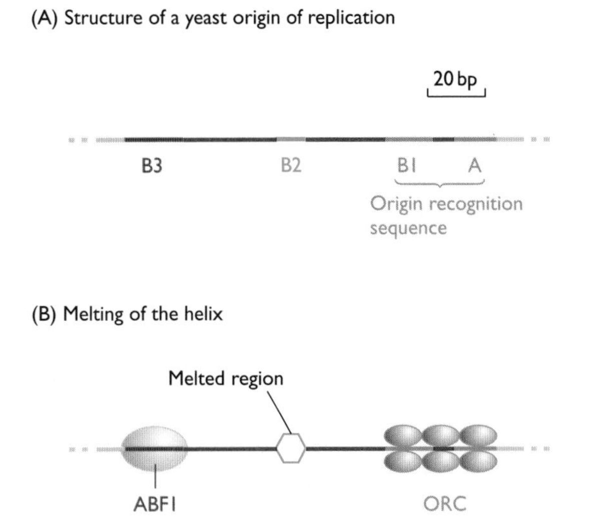

Figure 12.8 Structure of a yeast replication origin.

(A) Structure of ARS1, a typical autonomously replicating sequence that acts as an origin of replication in *Saccharomyces cerevisiae*. The relative positions of the functional sequences A, B1, B2 and B3 are shown. For more details see Bielinsky and Gerbi (1998). (B) Melting of the helix occurs within subdomain B2, induced by attachment of the ARS binding protein 1 (ABF1) to subdomain B3. The proteins of the origin replication complex (ORC) are permanently attached to subdomains A and B1.

Replication origins in higher eukaryotes have been less easy to identify

Attempts to identify replication origins in humans and other higher eukaryotes have, until recently, been less successful. **Initiation regions** (parts of the chromosomal DNA where replication initiates) were delineated by various biochemical methods, for example by allowing replication to initiate in the presence of labeled nucleotides, then arresting the process, purifying the newly-synthesized DNA and determining the positions of these nascent strands in the genome. These experiments suggested that there are specific regions in mammalian chromosomes where replication begins, but some researchers were doubtful whether these regions contained replication origins equivalent to yeast ARSs. One alternative hypothesis was that replication is initiated by protein structures that have specific positions in the nucleus, the chromosome initiation regions simply being those DNA segments located close to these protein structures in the three dimensional organization of the nucleus.

Doubts about mammalian replication origins were fueled by the failure of mammalian initiation regions to confer replicative ability on replication-deficient plasmids, though these experiments were not considered conclusive because it was recognized that a mammalian origin might be too long to be cloned in a plasmid or might function only when activated by distant sites on chromosomal DNA. The breakthrough eventually came when an 8 kb segment of a human initiation region was transferred to the monkey genome, where it still directed replication despite being removed from any hypothetical initiator structure in the human nucleus (Aladjem *et al.*, 1998). Analysis of this transferred initiation region showed that there are primary sites within the region where initiation occurs at high frequency, surrounded by secondary sites, spanning the entire 8 kb region, at which replication initiates with lower frequency. The presence of discrete functional domains within the initiation region could also be demonstrated by examining the effects of deletions of parts of the region on the efficiency of replication initiation (Research Briefing 12.1).

The demonstration that the human genome contains replication origins equivalent to those in yeast raises the question of whether mammals possess an equivalent of the yeast ORC. The answer appears to be yes, as several genes whose protein products have similar sequences to the yeast ORC proteins have been identified in higher eukaryotes and some of these have been shown to be able to replace the equivalent yeast protein in the yeast ORC (Carpenter *et al.*, 1996). These results indicate that initiation of replication in yeast is a good model for events occurring in mammals, a conclusion that is very relevant to studies of the control of replication initiation, as we will see in Section 12.4.

RESEARCH **12.1** BRIEFING

Identification of mammalian origins of replication

Recent research has shown that initiation of DNA synthesis in mammalian cells is associated with discrete replication origins that might be similar in many respects to those of yeast and lower eukaryotes.

Technical problems have hindered attempts to identify replication origins in mammalian DNA molecules and although it has been known for some time that replication initiates in particular regions of the genome, it has not proved possible to identify specific DNA sequences that act as replication origins. The main problem has been that mammalian DNA fragments thought to contain replication origins do not usually confer replicative ability on nonreplicating plasmids, possibly because the resulting plasmids lack some feature of chromosomal DNA structure needed to 'fire' a replication origin. The plasmid approach that has been so useful in characterizing bacterial and yeast origins has therefore not been informative with regard to mammalian replication, and it has been necessary to use alternative techniques to make direct studies of chromosomal DNA.

One of these alternative techniques has been to use biochemical methods to identify regions of mammalian genomes where new DNA synthesis originates during early S phase, the part of the cell cycle when DNA replication occurs (Section 12.4.1). These studies have shown that an 8 kb segment of human chromosome 11, spanning the β-globin gene cluster (see *Figure 6.6*, p. 123), contains one of the many initiation regions in the human genome, but it is not clear from the biochemical analysis whether or not DNA synthesis begins at specific sequences within the β-globin locus, or nonspecifically at various points within this region.

Pinpointing initiation sites for DNA synthesis in the β-globin region

A combination of pulse-labeling and PCR has been used to pinpoint the positions within the β-globin region where DNA synthesis is initiated. Pulse-labeling involves exposing cells to a limited amount of label so that biochemical reactions occurring during a brief, defined period of time can be studied. In this example, cells entering S phase are pulse-labeled with nucleotides containing the unusual base, 5-bromouracil (Section 13.1.1). These nucleotides, which have a higher molecular weight than A nucleotides, are incorporated into newly synthesized DNA in place of As enabling these new strands to be purified by density gradient centrifugation (Technical Note 6.1, p. 123).

After purification, the shortest labeled strands are examined by PCRs directed at different segments of the β-globin initiation region. If DNA synthesis begins randomly within this region then there will be similar amounts of template DNA for each of these PCRs and each one will give a similar amount of product. This is not what is seen: instead, some PCRs give abundant product and others give very little or none at all. The conclusion is that two, or possibly four short segments of the initiation region act as specific origins of DNA synthesis.

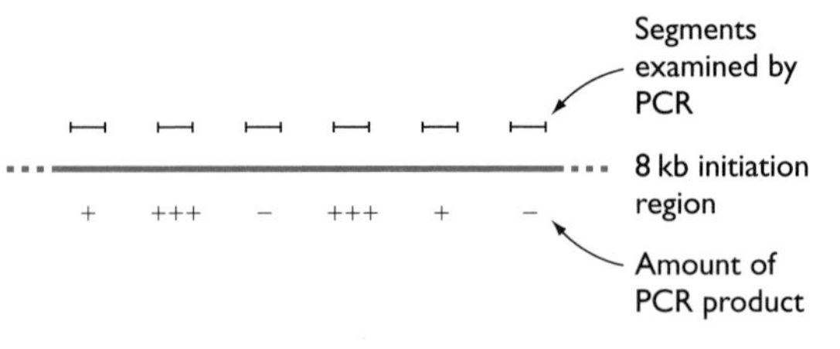

Are mammalian origins similar to those of yeast?

We have seen that the replication origins of yeast comprise discrete domains that play different roles during the initiation of DNA synthesis (see *Figure 12.8*). To determine whether mammalian origins have an equivalent domain-like structure, the 8 kb human β-globin region, as well as versions in which different segments had been deleted, were transferred into the monkey genome in such a way that in each resulting cell line the human DNA was located at the same position in the monkey chromosomes and was therefore subject to the same chromatin 'environment'. Any differences in the patterns of DNA synthesis within the transferred human DNA segments would therefore be due to the effects of the deletions rather than the position of the transferred segment in the monkey genome. These studies enabled the human initiation region to be divided into a central core segment that is essential for initiation of DNA synthesis, plus two flanking regions, either, but not both, of which can be deleted without affecting initiation. These results suggest that mammalian origins of replication are made up of cooperating domains and so might be similar in structure to yeast origins. Further studies are needed to determine exactly how close these similarities are.

References

Aladjem MI, Rodewald LW, Kolman JL and Wahl GM (1998) Genetic dissection of a mammalian replicator in the human β-globin locus. *Science*, **281**, 1005-1009.

12.3.2 The elongation phase of replication

Once replication has been initiated, the replication forks progress along the DNA and carry out the central activity of DNA replication, the synthesis of new strands of DNA complementary to the parent polynucleotides. The template-dependent synthesis of DNA is illustrated in *Figure 12.9* and has been studied in great detail by X-ray crystallographic analyses of DNA polymerases in the act of adding a new nucleotide to the 3′-end of a polynucleotide (Steitz, 1998).

At the chemical level, DNA replication and transcription are very similar (compare *Figure 12.9* with *Figure 9.6*, p. 202). This similarity should not mislead us into making an extensive analogy between transcription and replication. The mechanics of the two processes are quite different, replication being complicated by two factors that do not apply to transcription:

- During DNA replication both strands of the double helix must be copied. This is an important complication because, as noted in Section 7.1, DNA polymerase enzymes are only able to synthesize DNA in the 5′→3′ direction. This means that one strand of the parent double helix, called the **leading strand**, can be copied in a continuous manner, but replication of the **lagging strand** has to be carried out in a discontinuous fashion, as a series of short segments that must be ligated together to produce the intact daughter strand (*Figure 12.10*).
- The second complication arises because template-dependent DNA polymerases cannot initiate DNA synthesis on a molecule that is entirely single-stranded: there must be a short double-stranded region to provide a 3′ end onto which the enzyme can add new nucleotides. This means that **primers** are needed, one to initiate complementary strand synthesis on the leading polynucleotide, and one for every segment of discontinuous DNA synthesized on the lagging strand (*Figure 12.10*).

Before dealing with these two complications we will first examine the DNA polymerase enzymes themselves.

The DNA polymerases of bacteria and eukaryotes

The principal chemical reaction catalyzed by a DNA polymerase is the 5′→3′ synthesis of a DNA polynucleotide, as shown in *Figure 12.9*. As noted in Box 4.2, p. 66, some DNA polymerases combine this function with at least one exonuclease activity, which means that these enzymes can degrade polynucleotides as well as synthesize them (*Figure 12.11*):

- A **3′→5′ exonuclease** is possessed by many bacterial and eukaryotic template-dependent DNA polymerases (*Table 12.2*). This activity enables the enzyme to remove nucleotides from the 3′-end of the strand that it has just synthesized. It is looked on as a **proofreading** activity whose function is to correct the occasional base-pairing error that might occur during strand synthesis (see Section 13.1.1).
- A **5′→3′ exonuclease** activity is less common but possessed by some polymerases whose function in replication requires that they must be able to remove at least part of a polynucleotide that is already attached to the template strand that the polymerase is copying. This activity is utilized during the process that joins together the discontinuous DNA fragments synthesized on the lagging strand during bacterial DNA replication (see *Figure 12.17*).

The search for DNA polymerases began in the mid-1950s, as soon as it was realized that DNA synthesis was the key to replication of genes. It was thought that bacteria would probably have just a single DNA polymerase, and when the enzyme now called **DNA polymerase I** was isolated by Arthur Kornberg in 1957 there was a widespread assumption that this was the main replicating enzyme. The discovery that inactivation of the *E. coli polA* gene, coding for DNA polymerase I, was not lethal, (as cells were still able to replicate their genomes) therefore came as something of a surprise, especially when the same result was obtained with inactivation of *polB*, coding for a second enzyme, **DNA polymerase II.** It was not until 1972 that the main replicating polymerase of *E. coli*, **DNA polymerase III**, was eventually isolated. In fact both DNA polymerases I and III are involved in DNA replication, but the role of DNA polymerase I is less significant, as we will see in the next section.

The properties of the three *E. coli* DNA polymerases are described in *Table 12.2A*. DNA polymerases I and II are single polypeptides but DNA polymerase III, befitting its role as the main replicating enzyme, is multisubunit with a molecular mass of approximately 900 kDa. The three main subunits, which form the core enzyme, are called α, ε and θ, with the polymerase activity specified by the α subunit and the 3′→5′ exonuclease by ε. The function of θ is not clear: it may have a purely structural role in bringing together the other two core subunits and in assembling the various accessory subunits. The latter include τ and γ, both coded by the same gene with synthesis of γ involving a frameshift (see Box 10.5, p. 250), β, which acts as a 'sliding clamp' and holds the polymerase complex tightly to the template, δ, δ′, χ and ψ.

Eukaryotes have five DNA polymerases, called α, β, γ, δ and ε in mammals (*Table 12.2B*). The nomenclature is unfortunate as it tempts confusion with the identically-named subunits of *E. coli* DNA polymerase III. The main replicating enzyme is **DNA polymerase δ**, which has two subunits (three according to some researchers) and works in conjunction with an accessory protein called the **proliferating cell nuclear antigen (PCNA)**, the latter being the functional equivalent of the β subunit of *E. coli* DNA polymerase III, holding the enzyme tightly to the template. **DNA polymerase α** also has an important function in DNA synthesis, being the enzyme that primes eukaryotic replication (see *Figure 12.12B*) and **DNA polymerase γ**, although coded by a nuclear gene, is responsible for replicating the mitochondrial genome.

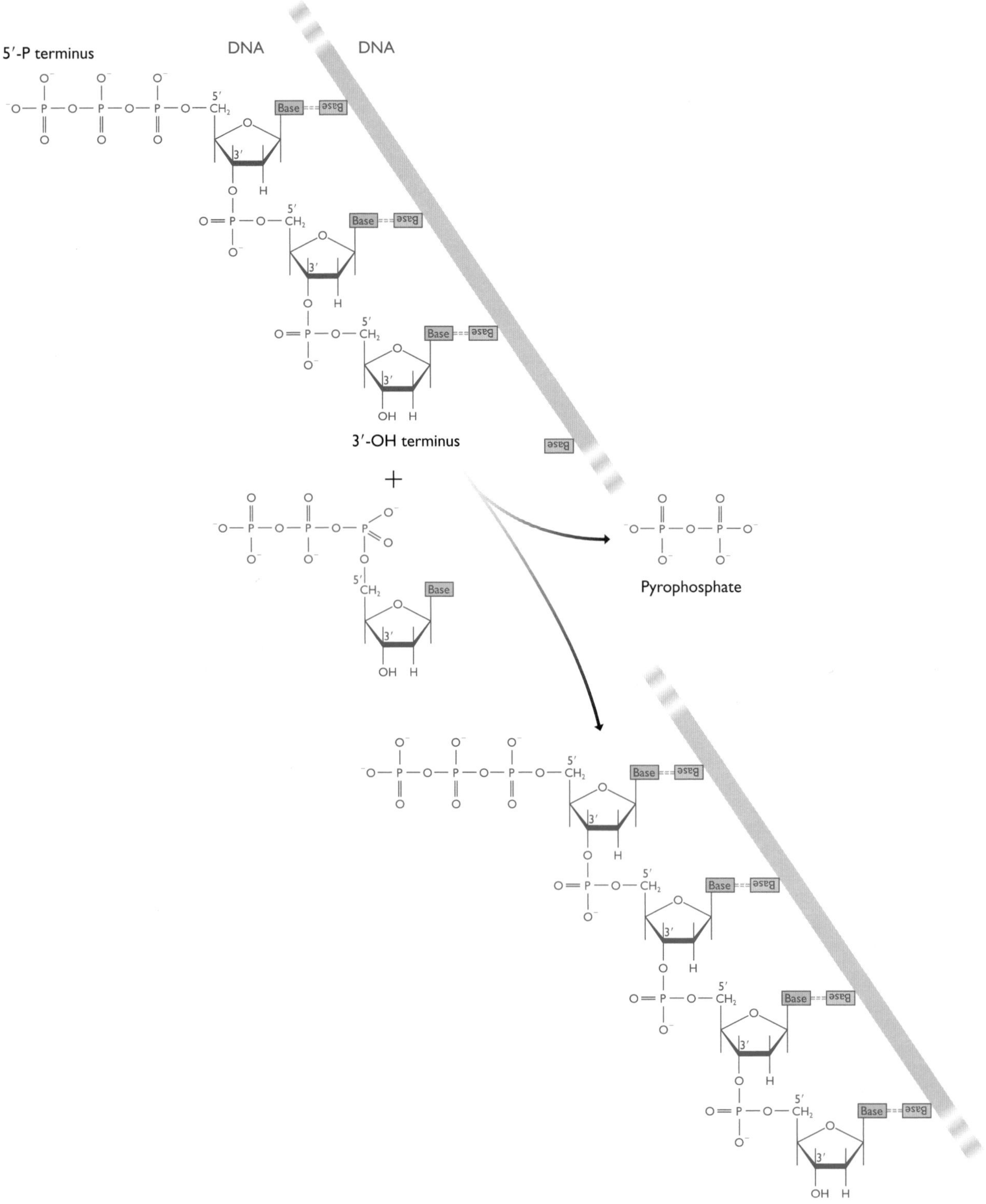

Figure 12.9 Template-dependent synthesis of DNA.

Compare this reaction with template-dependent synthesis of RNA, shown in *Figure 9.6*, p. 202.

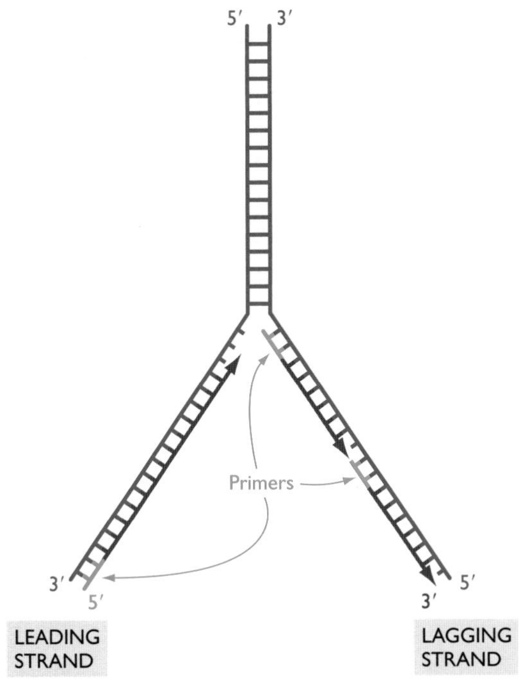

Figure 12.10 Complications with DNA replication.

Two complications have to be solved when double-stranded DNA is replicated. First, only the leading strand can be continuously replicated by 5′→3′ DNA synthesis: replication of the lagging strand has to be carried out discontinuously. Second, initiation of DNA synthesis requires a primer. This is true both of cellular DNA synthesis, as shown here, and DNA synthesis reactions that are carried out in the test tube: recall that both PCR (Technical Note 2.2, p. 20) and chain termination DNA sequencing (see *Figure 4.2, p. 63*) require primers.

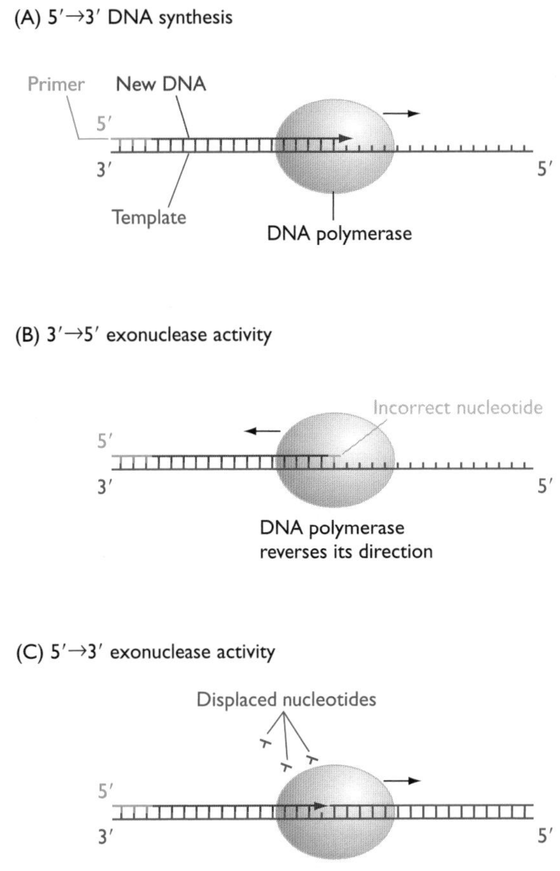

Figure 12.11 The DNA synthesis and exonuclease activities of DNA polymerases.

All DNA polymerases can make DNA but not all of them have exonuclease activities: see *Table 12.2* for details.

Table 12.2 DNA polymerases of bacteria and eukaryotes

Enzyme	Subunits	Exonuclease activities 3′→5′	Exonuclease activities 5′→3′	Function
A. Bacterial DNA polymerases				
DNA polymerase I	1	Yes	Yes	DNA repair, replication
DNA polymerase II	1	Yes	No	DNA repair
DNA polymerase III	At least 10	Yes	No	Main replicating enzyme
B. Eukaryotic DNA polymerases				
DNA polymerase α	4	No	No	Priming during replication
DNA polymerase β	1	No	No	DNA repair
DNA polymerase γ	2	Yes	No	Mitochondrial DNA replication
DNA polymerase δ	2 or 3	Yes	No	Main replicative enzyme
DNA polymerase ε	At least 1	Yes	No	DNA replication, precise function unknown

The functions of the DNA polymerases involved in repair are described in Section 13.1.4.

Discontinuous strand synthesis and the priming problem

The limitation that DNA polymerases can synthesize polynucleotides only in the 5′→3′ direction means that the lagging strand of the parent molecule must be copied in a discontinuous fashion, as shown in *Figure 12.10*. The implication of this model – that the initial products of lagging strand replication are short segments of polynucleotide – was confirmed in 1969 when **Okazaki fragments**, as these segments are now called, were first isolated from *E. coli* (Okazaki and Okazaki, 1969). In bacteria, Okazaki fragments are 1000–2000 nucleotides in length, but in eukaryotes the equivalent fragments appear to be much shorter, perhaps less than 200 nucleotides, an interesting observation that might indicate that each round of discontinuous synthesis replicates the DNA associated with a single nucleosome (140 and 150 bp wound around the core particle plus 50–70 bp of linker DNA: Section 6.1.1).

The second difficulty illustrated in *Figure 12.10* is the need for a primer to initiate synthesis of each new polynucleotide. It is not known for certain why DNA polymerases cannot begin synthesis on an entirely single-stranded template, but it may relate to the proofreading activity of these enzymes, which is essential for the accuracy of replication. As described in Section 13.1.1, for proofreading to work the 3′→5′ exonuclease activity of a DNA polymerase must be more effective than the 5′→3′polymerase activity when the nucleotide at the 3′-terminus of the polynucleotide being synthesized is not base-paired to the template. This implies that the polymerase can extend a polynucleotide efficiently only if its 3′-nucleotide is base-paired, which in turn could be the reason why an entirely single-stranded template, which by definition lacks a base-paired 3′-nucleotide, cannot be used by a DNA polymerase.

Whatever the reason, priming is a necessity in DNA replication. Obtaining the primer is not a problem: although DNA polymerases cannot deal with an entirely single-stranded template, RNA polymerases have no difficulty in this respect, so the primers for DNA replication are made of RNA. In bacteria, primers are synthesized by **primase**, a special RNA polymerase unrelated to the transcribing enzyme, with each primer about 5 nucleotides in length. Once the primer has been completed, strand synthesis is continued by DNA polymerase III (*Figure 12.12A*). In eukaryotes the situation is slightly more complex, because the primase is an integral part of DNA polymerase α, specified by the smallest subunit of this enzyme (Foiani *et al.*, 1997). This polymerase synthesizes RNA primers of 10 nucleotides or so, and then extends the primers by adding about 30 nucleotides of DNA (often with some ribonucleotides mixed in) before the main replicative enzyme, DNA polymerase δ, takes over (*Figure 12.12B*).

Priming needs to occur just once on the leading strand, within the replication origin, because once primed the leading strand copy is synthesized continuously until replication is completed. On the lagging strand, priming is a repeated process that must occur every time a new Okazaki fragment is initiated. In *E. coli*, which makes Okazaki fragments of 1000–2000 nucleotides in length, approximately 4000 priming events are needed every time the genome is replicated. In eukaryotes the Okazaki fragments are much shorter and priming is a highly repetitive event.

Events at the bacterial replication fork

Now that we have considered the complications introduced by discontinuous strand synthesis and the priming problem we can move on to study the combination of events occurring at the replication fork during the elongation phase of DNA replication.

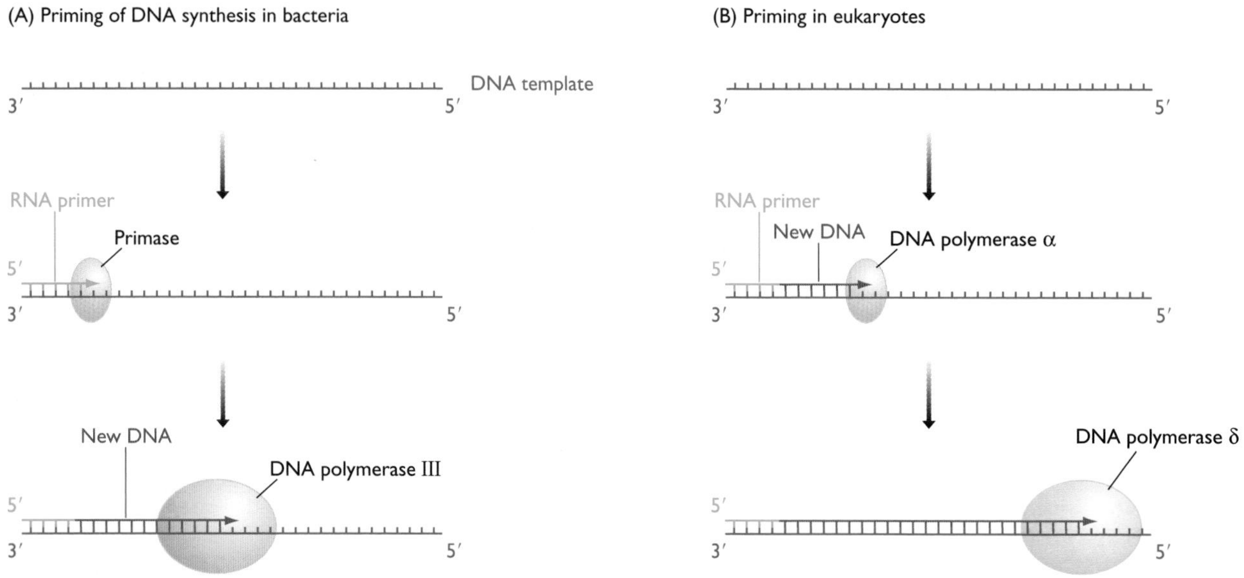

Figure 12.12 Priming of DNA synthesis in (A) bacteria and (B) eukaryotes.

In Section 12.3.1 we identified attachment of the DnaB helicase, followed by extension of the melted region of the replication origin, as representing the end of the initiation phase of replication in *E. coli*. To a large extent, the division between initiation and elongation is artificial as the two processes run seamlessly one into the other. After the helicase has bound to the origin, forming the prepriming complex, the primase is recruited, resulting in the **primosome** which initiates replication of the leading strand. It does this by synthesizing the RNA primer that DNA polymerase III needs in order to begin copying the template.

DnaB is the main helicase involved in DNA replication in *E. coli*, but it is by no means the only helicase that this bacterium possesses: in fact there were twelve at the last count (Lohman and Bjornson, 1996), the size of the collection reflecting the fact that DNA unwinding is required not only during replication but also during diverse processes such as transcription, recombination and DNA repair. The mode of action of a typical helicase has not been precisely defined, but it is thought that these enzymes bind to single-stranded rather than double-stranded DNA, and migrate along the polynucleotide in either the 5′→3′ or 3′→5′ direction, depending on the specificity of the helicase. Breakage of base pairs in advance of the helicase requires energy generated by hydrolysis of ATP. According to this model, a single DnaB helicase could migrate along the lagging strand (DnaB is a 5′→3′ helicase) unzipping the helix and generating the replication fork, the torsional stress generated by the unwinding activity being relieved by DNA topoisomerase action (*Figure 12.13*). This model is probably a good approximation of what actually happens, although it does not provide a function for the two other *E. coli* helicases thought to be involved in DNA replication. Both of these, PriA and Rep, are 3′→5′ helicases and so could conceivably complement DnaB activity by migrating along the leading strand, but they may have lesser roles. The involvement of Rep in DNA replication might in fact be limited to participation in the rolling circle process used by λ and a few other *E. coli* bacteriophages (see Box 12.1).

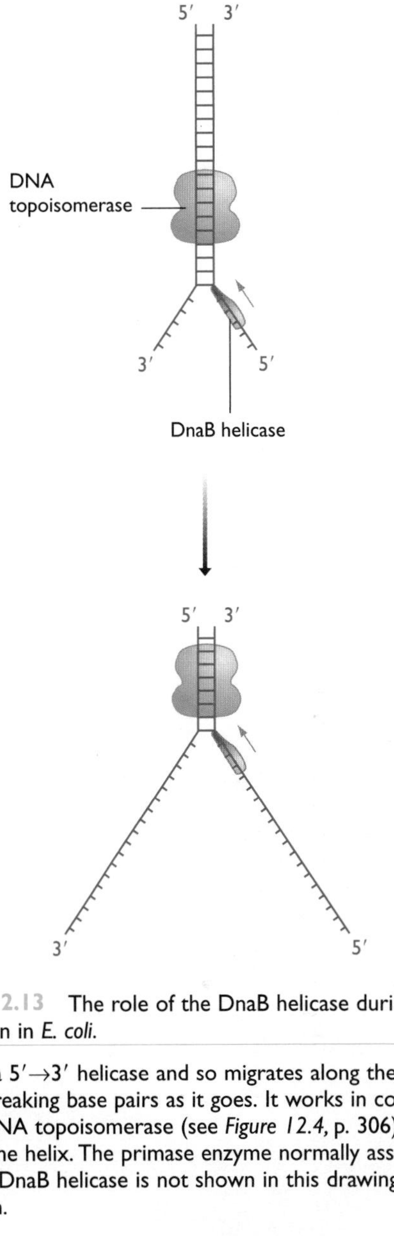

Figure 12.13 The role of the DnaB helicase during DNA replication in *E. coli*.

DnaB is a 5′→3′ helicase and so migrates along the lagging strand, breaking base pairs as it goes. It works in conjunction with a DNA topoisomerase (see *Figure 12.4*, p. 306) to unwind the helix. The primase enzyme normally associated with the DnaB helicase is not shown in this drawing, to avoid confusion.

Single-stranded DNA is naturally 'sticky' and the two separated polynucleotides produced by helicase action would immediately reform base pairs after the enzyme has passed if allowed to. To prevent this, **single-strand binding proteins (SSBs)**, which lack enzymatic activity, attach to the polynucleotides and prevent them from re-associating (*Figure 12.14A*). The *E. coli* SSB is made up of of four identical subunits and probably works in a similar way to the major eukaryotic SSB, called RPA, by enclosing the polynucleotide in a channel formed by a series of SSBs attached side by side on the strand (*Figure 12.14B*; Bochkarev *et al*., 1997). As with helicases, SSBs have diverse roles in different processes involving DNA unwinding.

After 1000–2000 nucleotides of the leading strand have been replicated, the first round of discontinuous strand synthesis on the lagging strand can begin. The primase, which is still associated with the DnaB helicase in the primosome, makes an RNA primer which is then extended by DNA polymerase III (*Figure 12.15*). This is the same DNA polymerase III complex that is synthesizing the leading strand copy, the complex comprising, in effect, two copies of the polymerase. In fact it is not two complete enzymes because there is only a single copy of the **γ-complex**, containing subunit γ in association with δ, δ′, χ and ψ. The main function of the γ-complex is to interact with the β subunit (the 'sliding clamp') and hence control the attachment and removal of the enzyme from the template, a function that is required primarily during lagging strand replication when the enzyme has to attach and detach repeatedly at the start and end of each Okazaki fragment. Some models of the DNA polymerase III complex place the two enzymes in opposite

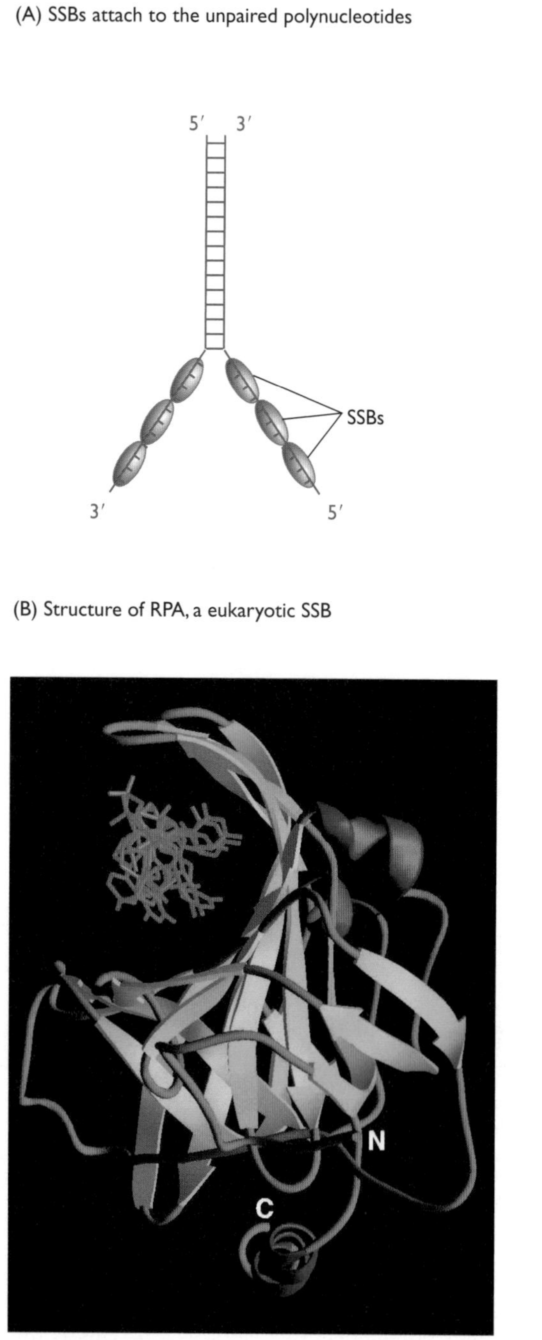

Figure 12.14 The role of single-strand binding proteins (SSBs) during DNA replication.

(A) SSBs attach to the unpaired polynucleotides produced by helicase action and prevent the strands from base-pairing with one another. (B) Structure of the eukaryotic SSB called RPA. The protein contains a β-sheet structure that forms a channel in which the DNA (shown in orange, viewed from the end) is bound. Reproduced with permission from Bochkarev *et al.*, *Nature* **385**, 176–181. Copyright 1997 Macmillan Magazines Limited. Image supplied courtesy of Dr Lori Frappier, Department of Medical Genetics and Microbiology at the University of Toronto.

orientations to reflect the different directions in which DNA synthesis occurs, towards the replication fork on the leading strand and away from it on the lagging strand. It is more likely, however, that the pair of enzymes face the same direction and the lagging strand forms a loop, so that DNA synthesis can proceed in parallel as the polymerase complex moves forward in pace with the progress of the replication fork (*Figure 12.16*).

The combination of the DNA polymerase III dimer and the primosome, migrating along the parent DNA and carrying out most of the replicative functions, is called the **replisome**. After its passage, the replication process must be completed by joining up the individual Okazaki fragments. This is not a trivial event because one member of each pair of adjacent Okazaki fragments still has its RNA primer attached at the point where ligation should take

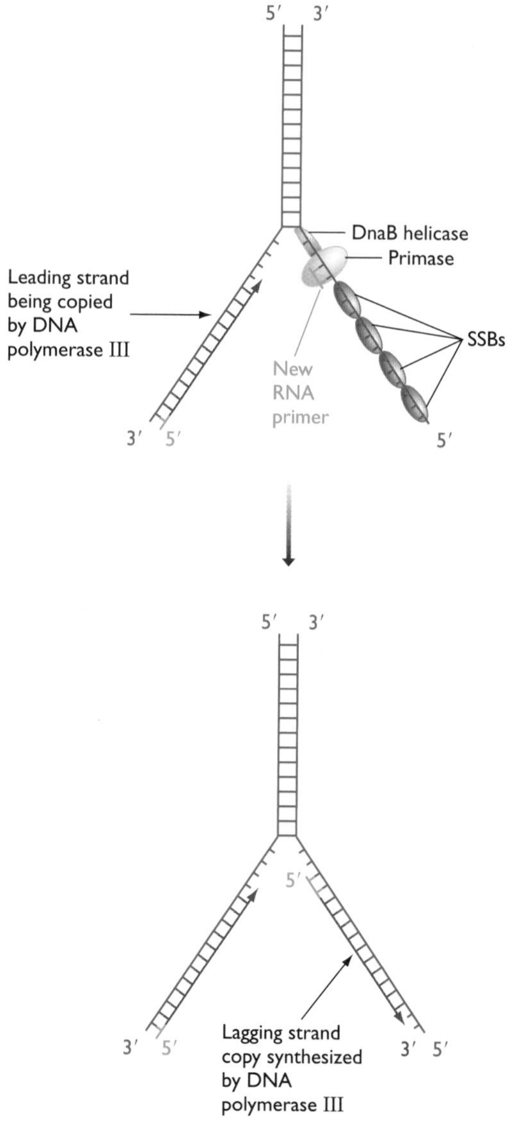

Figure 12.15 Priming and synthesis of the lagging strand copy during DNA replication in *E. coli*.

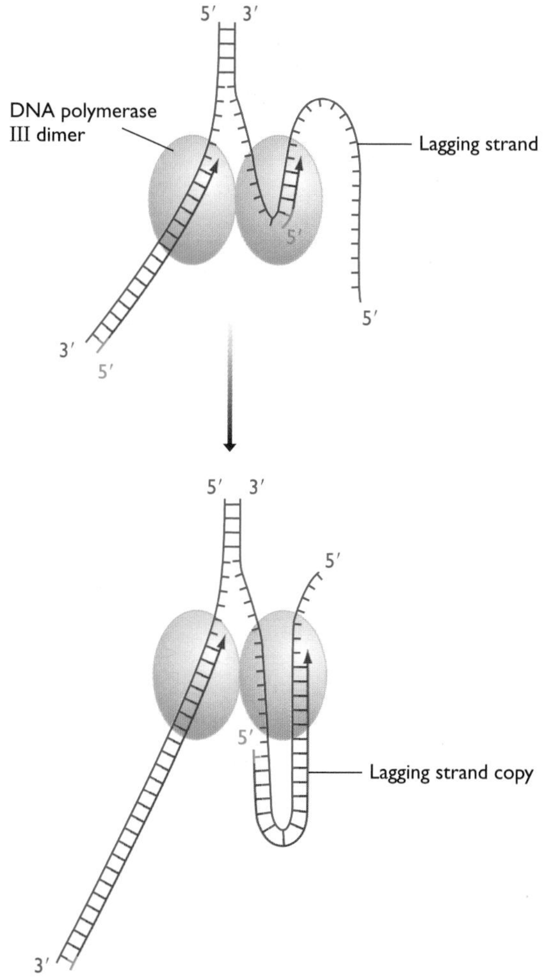

Figure 12.16 A model for parallel synthesis of the leading and lagging strand copies by a dimer of DNA polymerase III enzymes.

It is thought that the lagging strand loops through its copy of the DNA polymerase III enzyme, in the manner shown, so that both the leading and lagging strands can be copied as the dimer moves along the molecule being replicated. The two components of the DNA polymerase III dimer are not identical because there is only one copy of the γ-complex.

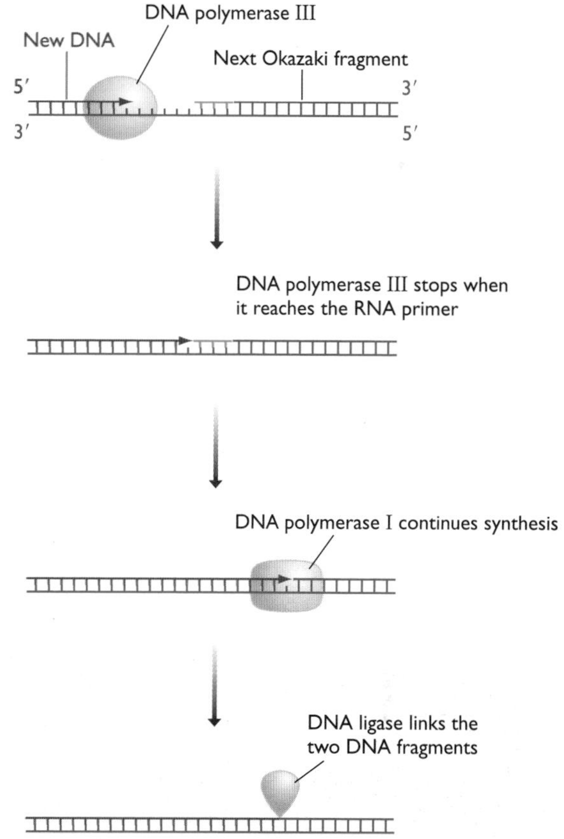

Figure 12.17 The series of events involved in joining up adjacent Okazaki fragments during DNA replication in *E. coli*.

DNA polymerase III lacks a 5′→3′ exonuclease activity and so stops making DNA when it reaches the RNA primer of the next Okazaki fragment. At this point DNA synthesis is continued by DNA polymerase I, which does have a 5′→3′ exonuclease activity, and which works in conjunction with RNase H to remove the RNA primer and replace it with DNA. DNA polymerase I usually also replaces some of the DNA from the Okazaki fragment before detaching from the template. This leaves a single missing phosphodiester bond, which is synthesized by DNA ligase, completing this step in the replication process.

place (*Figure 12.17*). *Table 12.2* shows us that this primer cannot be removed by DNA polymerase III, because this enzyme lacks the required 5′→3′ exonuclease activity. At this point, DNA polymerase III releases the lagging strand and its place is taken by DNA polymerase I, which does have a 5′→3′ exonuclease and so removes the primer, and usually the start of the DNA component of the Okazaki fragment as well, extending the adjacent fragment into the region of the template that is exposed. The two Okazaki fragments now abut, with the terminal regions of both composed entirely of DNA. All that remains is for the missing phosphodiester bond to be put in place by a **DNA ligase**, linking the two fragments and completing replication of this region of the lagging strand.

The eukaryotic replication fork: variations on the bacterial theme

The overall process of DNA replication is similar in bacteria and eukaryotes, although the details are different. The progress of the replication fork in eukaryotes is maintained by helicase activity, though which of the several eukaryotic helicases that have been identified are primarily responsible for DNA unwinding during replication has not been established. The separated polynucleotides are prevented from reattaching by single-strand binding proteins, the main one of these in eukaryotes being **replication protein A (RPA)**.

Box 12.3: Genome replication in the archaea

Much of what is known about DNA replication in archaea has been deduced by searching DNA sequences for genes coding for proteins with similarities to those known to be involved in replication in bacteria and/or eukaryotes. We therefore have information about the proteins that might be involved in archaeal DNA replication but little idea about their mode of action.

Origins of replication have not been identified in any archaeal genome, indicating that the origins, presuming that they exist, do not contain the characteristic sequences found in bacteria. The sequences of most of the proteins involved in the elongation phase of replication, as predicted from their genes, are similar to the equivalent eukaryotic versions. In particular, archaea have proteins that appear to be homologs of the eukaryotic RFC and PCNA. The archaeal DNA polymerase is interesting because the subunit that specifies the DNA synthesis activity is similar to the equivalent subunit of the eukaryotic DNA polymerase δ, whereas the proofreading function is conferred by a protein that appears to be a homolog of subunit ε of *E. coli* DNA polymerase III.

We begin to encounter unique features of the eukaryotic replication process when we examine the method used to prime DNA synthesis. As described on p. 311, the eukaryotic DNA polymerase α has an intrinsic primase activity and so puts in place the RNA primers both at the start of the copy of the leading strand and at the beginning of each Okazaki fragment. However, DNA polymerase α is not capable of lengthy DNA synthesis, presumably because it lacks the stabilizing effect of a sliding clamp equivalent to the β subunit of *E. coli* DNA polymerase III or the PCNA accessory protein that aids the eukaryotic DNA polymerase δ. This means that although DNA polymerase α can both synthesize an RNA primer and extend this primer with about 30 nucleotides of DNA, it must then be replaced by the main replicative enzyme, DNA polymerase δ (see *Figure 12.12B*, p. 314).

The DNA polymerase enzymes that copy the leading and lagging strands in eukaryotes do not associate into a dimeric complex equivalent to the one formed by DNA polymerase III during replication in *E. coli*: instead, the two copies of the polymerase remain separate. The function performed by the γ-complex of the *E. coli* polymerase – controlling attachment and detachment of the enzyme from the lagging strand – appears to be carried out by a multisubunit accessory protein called **replication factor C (RFC)**.

As in *E. coli*, completion of lagging strand synthesis requires removal of the RNA primer from each Okazaki fragment. There appears to be no eukaryotic DNA polymerase with the 5′→3′ exonuclease needed for this purpose and the process is therefore very different to that described for bacterial cells. The central player is the 'flap endonuclease' **FEN1** (previously called MF1), which associates with the DNA polymerase δ complex, at the 3′-end of one Okazaki fragment, in order to degrade the primer from the 5′-end of the adjacent fragment. Understanding exactly how this occurs is complicated by the inability of FEN1 to initiate primer degradation because it is unable to remove the ribonucleotide at the extreme 5′-end of the primer, because this ribonucleotide carries a 5′-triphosphate group which blocks FEN1 activity (*Figure 12.18*). Two alternative models have been proposed to circumvent this problem (Waga and Stillman, 1998):

- The first possibility is that a helicase breaks the base pairs holding the primer to the template strand, enabling the primer to be pushed aside by DNA polymerase δ as it extends the adjacent Okazaki fragment into the region thus exposed (*Figure 12.19A*). The flap that results can be cut off by FEN1, whose endonuclease activity can cleave the phosphodiester bond at the branch point where the displaced region attaches to the part of the fragment that is still base-paired.
- Alternatively, most of the RNA component of the primer could be removed by RNase H, which can degrade the RNA part of a base-paired RNA–DNA hybrid, but cannot cleave the phosphodiester bond between the last ribonucleotide and the first deoxyribonucleotide. However, this ribonucleotide will carry a 5′-monophosphate rather than triphosphate and so can be removed by FEN1 (*Figure 12.19B*).

Both schemes are made more attractive by the possibility that not just the RNA primer but also all of the DNA originally synthesized by DNA polymerase α is removed. This is because DNA polymerase α has no 3′→5′ proofreading activity (see *Table 12.2*, p. 313) and therefore synthesizes DNA in a relatively error-prone manner. To prevent these errors from becoming permanent features of the daughter double helix this region of DNA might be degraded and resynthesized by DNA polymerase δ, which does have a proofreading activity and so makes a very accurate copy of the template. At present this possibility remains speculative.

The final difference between replication in bacteria and eukaryotes is that in eukaryotes there is no replisome. Instead, the enzymes and proteins involved in replication form sizeable structures within the nucleus, each containing hundreds or thousands of individual replication complexes. These structures are immobile because of attachments with the nuclear matrix, so DNA molecules are threaded through the complexes as they are replicated. The structures are referred to as **replication factories** (*Figure 12.20*) and may in fact also be features of the replication process in at least some bacteria (Lemon and Grossman, 1998).

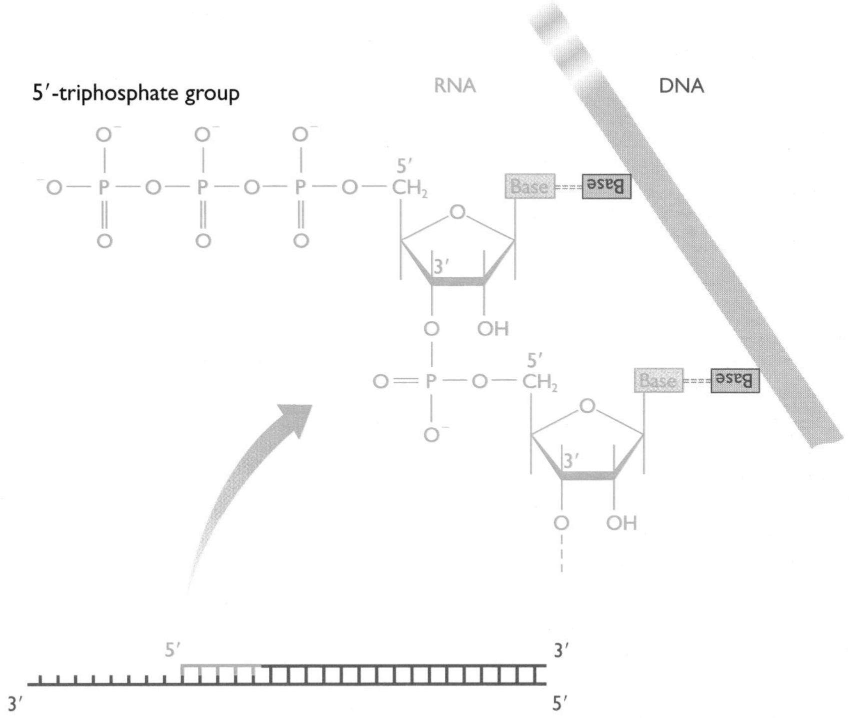

Figure 12.18 The 'flap endonuclease' FEN1 cannot initiate primer degradation because its activity is blocked by the triphosphate group present at the 5′-end of the primer.

12.3.3 Termination of replication

Replication forks proceed along linear genomes, or around circular ones, generally unimpeded except when a region that is being transcribed is encountered. DNA synthesis occurs at approximately five times the rate of RNA synthesis, so the replication complex can easily overtake an RNA polymerase, but this probably does not happen: instead it is thought that the replication fork pauses behind the RNA polymerase, proceeding only when the transcript has been completed (Deshpande and Newlon, 1996).

Eventually the replication fork reaches the end of the molecule or meets a second replication fork moving in the opposite direction. What happens next is one of the least well understood aspects of DNA replication.

Replication of the *E. coli* genome terminates within a defined region

Bacterial genomes are replicated bidirectionally from a single point (see *Figure 12.6*, p. 307), which means that the two replication forks should meet at a position diametrically opposite the origin of replication on the genome map. However, if one fork is delayed, possibly because it has to replicate extensive regions where transcription is occurring, then it might be possible for the other fork to overshoot the halfway point and continue replication on the 'other side' of the genome (*Figure 12.21*). It is not immediately apparent why this should be undesirable, the daughter molecules presumably being unaffected, but it is not allowed to happen because of the presence of **terminator sequences**. Seven of these have been identified in the *E. coli* genome (*Figure 12.22A*), each one acting as the recognition site for a sequence-specific DNA-binding protein called **Tus**.

The mode of action of Tus is quite unusual. When bound to a terminator sequence, a Tus protein allows a replication fork to pass if the fork is moving in one direction, but blocks progress if the fork is moving the opposite way round the genome. The directionality is set by the orientation of the Tus protein on the double helix. When approached from one direction, Tus blocks the passage of the DnaB helicase, which is responsible for progression of the replication fork, because the helicase is faced with a 'wall' of β-strands which it is unable to penetrate. But when approaching from the other direction, DnaB is able to disrupt the structure of the Tus protein, probably because of the effect that helix unwinding has on Tus, and so is able to pass by (*Figure 12.22B*; Kamada *et al.*, 1996).

The orientation of the termination sequences, and hence of the bound Tus proteins, in the *E. coli* genome is such that both replication forks become trapped within a relatively short region at the opposite side of the genome to the origin (see *Figure 12.22A*). This ensures that

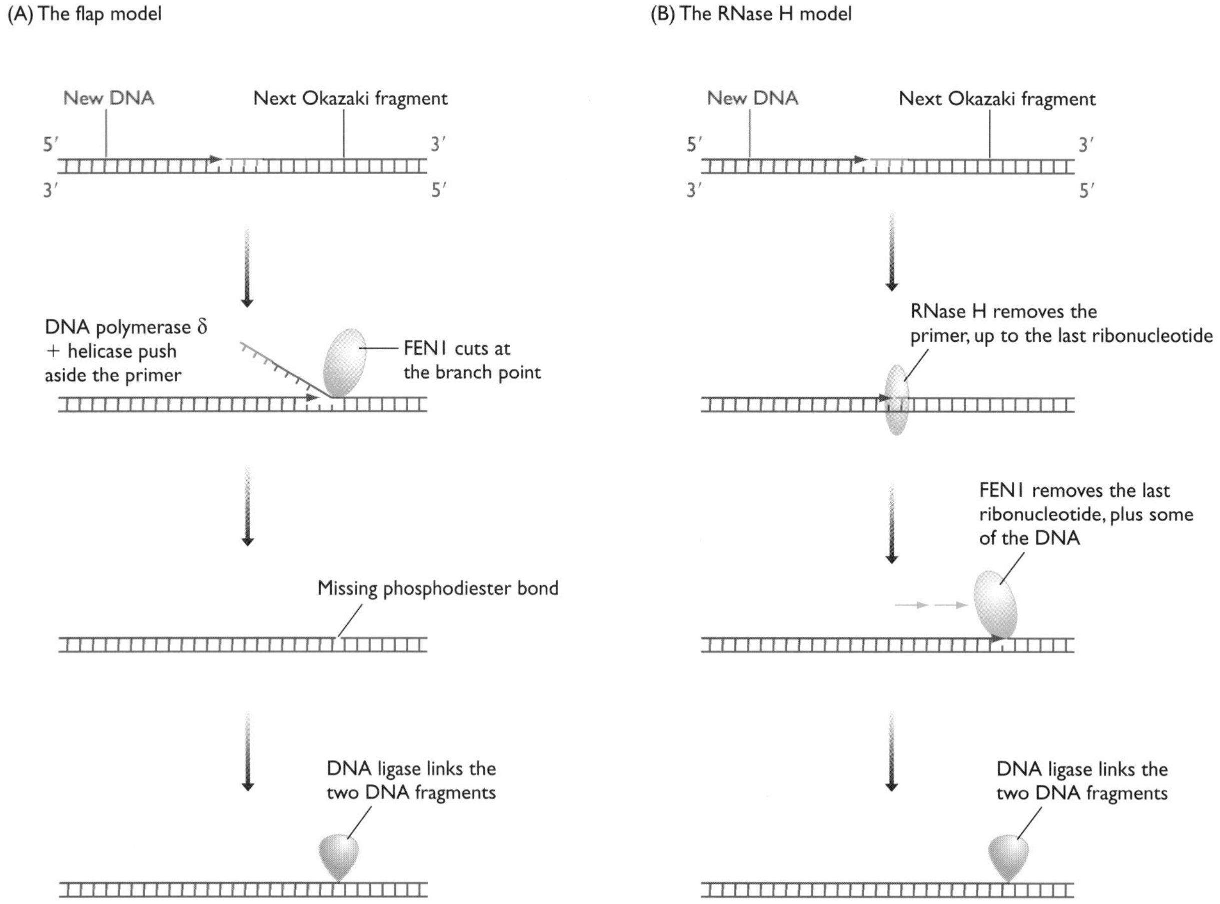

Figure 12.19 Two models for completion of lagging strand replication in eukaryotes.

See the text for details. The new DNA (blue strand) is synthesized by DNA polymerase δ but this enzyme is not shown in order to increase the clarity of the diagrams.

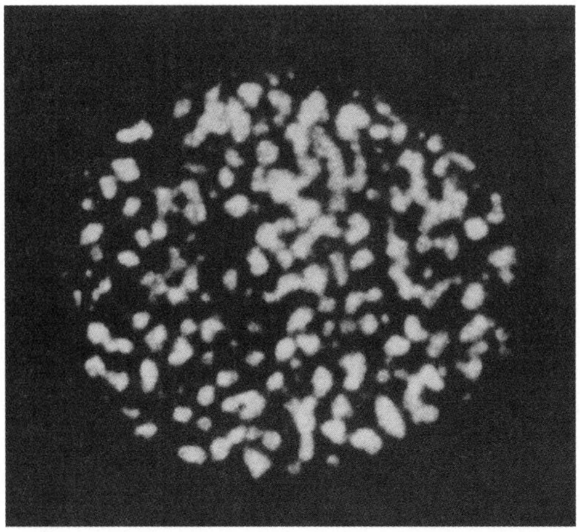

Figure 12.20 Replication factories in a eukaryotic nucleus.

Equivalent **transcription factories** are responsible for RNA synthesis. Reproduced with permission from Nakamura H *et al.* (1986) *Exp. Cell Res.*, **165**, 291–297, Academic Press, Inc., Orlando, FL.

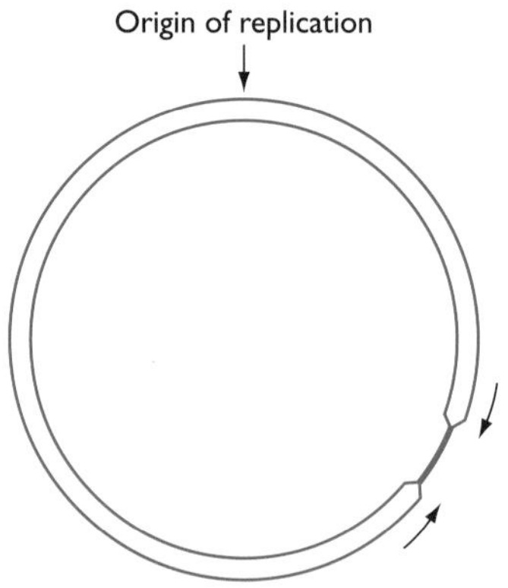

Figure 12.21 A situation that is not allowed to occur during replication of the circular *E. coli* genome.

One of the replication forks has proceeded some distance past the halfway point. This does not happen during *E. coli* DNA replication because of the action of the Tus proteins (see *Figure 12.22B*).

termination always occurs at or near the same position. Exactly what happens when the two replication forks meet is unknown, but the event is followed by disassembly of the replisomes, either spontaneously or in a controlled fashion. The result is two interlinked daughter molecules, which are separated by topoisomerase IV.

Little is known about termination of replication in eukaryotes

No sequences equivalent to bacterial terminators are known in eukaryotes and proteins similar to Tus have not been identified. Quite possibly, replication forks meet at random positions and termination simply involves ligation of the ends of the new polynucleotides. We do know that the replication complexes do not break down because these factories are permanent features of the nucleus (see *Figure 12.20*).

Rather than concentrating on the molecular events occurring when replication forks meet, attention has been focused more on the difficult question of how the daughter DNA molecules produced in a eukaryotic nucleus do not become impossibly tangled up. Although DNA topoisomerases have the ability to untangle DNA molecules it is generally assumed that tangling is kept to a minimum so that extensive breakage-and-reunion reactions, as catalyzed by topoisomerases (see *Figure 12.4*, p. 306), can be avoided. Various models have been proposed to solve this problem. One of these (Cook, 1998; Wei *et al.*, 1998) suggests that a eukaryotic genome is not randomly packed into the nucleus, but is ordered around the replication factories, which appear to be present in only

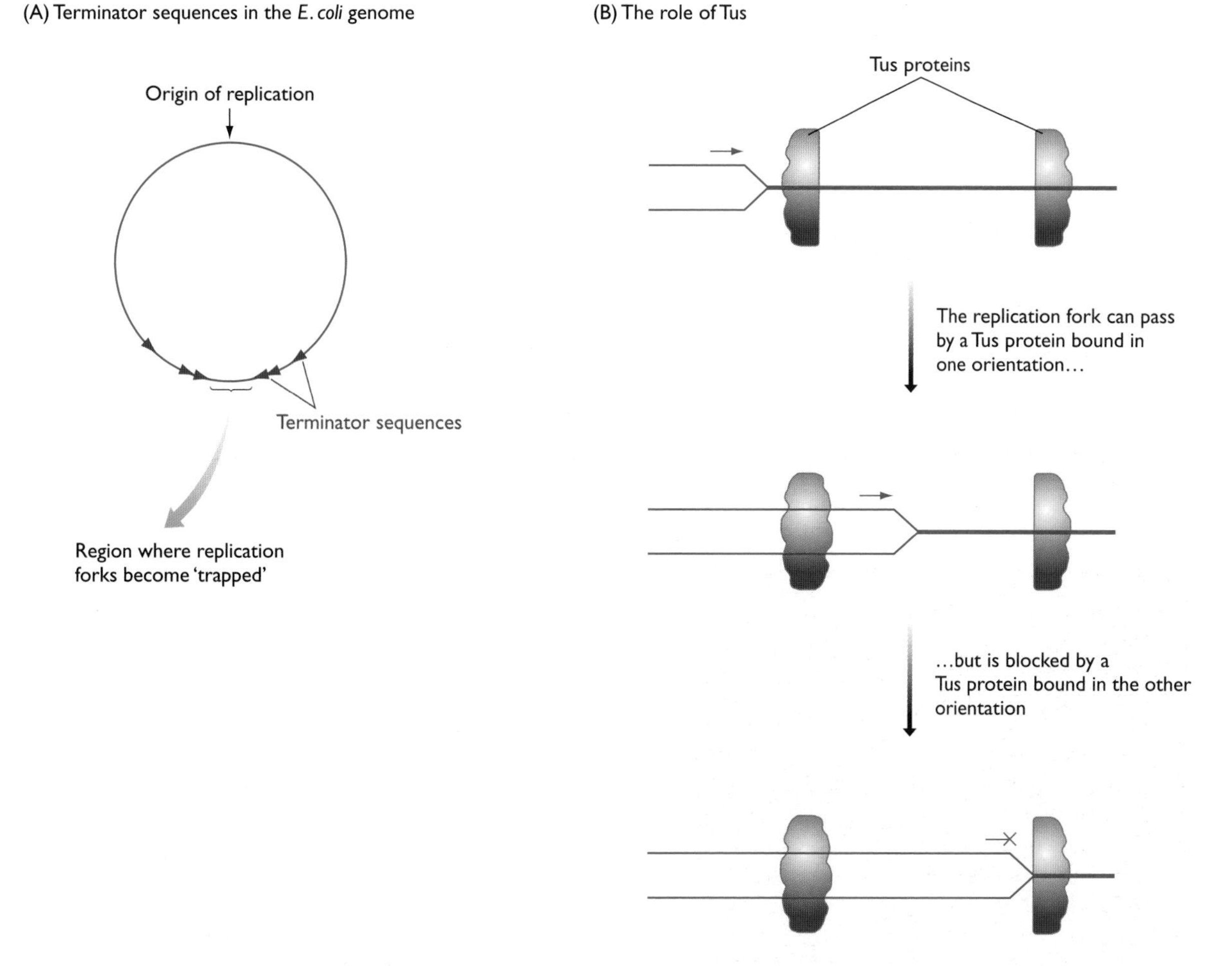

Figure 12.22 The role of terminator sequences during DNA replication in *E. coli*.

(A) The positions of the six terminator sequences on the *E. coli* genome are shown, with the arrowheads indicating the direction that each terminator sequence can be passed by a replication fork. (B) Bound Tus proteins allow a replication fork to pass when the fork approaches from one direction but not when it approaches from the other direction. The diagram shows a replication fork passing by the lefthand Tus, because the DnaB helicase that is moving the fork forward can disrupt the Tus when it approaches it in this direction. The fork is then blocked by the second Tus, because this one has its impenetrable wall of β-strands facing towards the fork.

Box 12.4: Telomeres in *Drosophila*

The text describes the reverse transcriptase reaction carried out by the protein component of telomerase. When the amino acid sequences of telomerase and other reverse transcriptases are compared similarities are seen with the enzymes coded by the non-LTR retroelements called retroposons (Section 6.3.2; Eickbush, 1997). This is a fascinating observation when taken in conjunction with the unusual structure of the telomeres of *Drosophila*. These telomeres are not made up of the short repeated sequences seen in most other organisms, but instead consist of tandem arrays of much longer repeats, 6 or 10 kb in length. These repeats are full-length copies of two typical retroposons, related to LINE-1 of humans, called *HeT-A* and *TART* (Pardue *et al.*, 1996). It is not known how these telomeres are maintained, but it is conceivable that the process is analogous to that carried out by telomerase, with a template RNA obtained by transcription of the telomeric retroposons being copied by the reverse transcriptase coded by the *TART* sequences (*HeT-A* does not have a reverse transcriptase gene).

The unusual structure of the *Drosophila* telomere could simply be a quirk of nature, but the attractive possibility that the telomeres of other organisms are degraded retroposons, as suggested by the similarities between telomerase and retroposon reverse transcriptases, cannot be discounted.

limited numbers. It is envisaged that each factory replicates a single region of the DNA, maintaining the daughter molecules in an ordered arrangement that avoids their entanglement.

12.3.4 Maintaining the ends of a linear DNA molecule

There is one final problem that we must consider before leaving the replication process. This concerns the steps that have to be taken to prevent the ends of a linear, double-stranded molecule from gradually getting shorter during successive rounds of DNA replication. There are two ways in which this shortening might occur:

- The extreme 3′-end of the lagging strand might not be copied because the final Okazaki fragment cannot be primed, the natural position for the priming site being beyond the end of the template (*Figure 12.23*). The absence of this Okazaki fragment means that the lagging strand copy is shorter than it should be. If the copy remains this length then when it acts as a parental polynucleotide in the next round of replication the resulting daughter molecule will be shorter than its grandparent.
- If the primer for the last Okazaki fragment is placed at the extreme 3′-end of the lagging strand, then shortening will still occur, though to a lesser extent, because this terminal RNA primer cannot be converted into DNA by the standard processes for primer removal, as illustrated in *Figure 12.17*, p. 317, for bacteria and *Figure 12.19*, p. 320, for eukaryotes. This is because these methods for primer replacement require extension of the 3′-end of an adjacent Okazaki fragment, which cannot exist at the very end of the molecule.

Once this problem had been recognized, attention was directed at the telomeres, the unusual DNA sequences at the ends of eukaryotic chromosomes. We noted in Section 6.3.1 that telomeric DNA is made up of a type of minisatellite sequence, being comprised of multiple copies of a short repeat motif, 5′-TTAGGG-3′ in most higher eukaryotes, a thousand or more copies of this sequence occurring in tandem repeats at each end of every chromosome. The solution to the end-shortening problem lies with the way in which this telomeric DNA is synthesized.

Telomeric DNA is synthesized by the telomerase enzyme

Most of the telomeric DNA is copied in the normal fashion during DNA replication but this is not the only way in which it can be synthesized. To compensate for the limitations of the replication process, telomeres can be extended by an independent mechanism catalyzed by the enzyme called **telomerase**. This is an unusual enzyme in that it consists of both protein and RNA. In the human enzyme this RNA is 450 nucleotides in length and contains near its 5′-end the sequence 5′-CUAACCCUAAC-3′, whose central region is the reverse complement of the human telomere repeat sequence 5′-TTAGGG-3′ (Feng *et al.*, 1995). This enables telomerase to extend the telomeric DNA at the 3′-end of a polynucleotide by the copying mechanism shown in *Figure 12.24*, in which the telomerase RNA is used as a template for each extension step, the DNA synthesis being carried out by the protein component of the enzyme, which is a reverse transcriptase (Lingner *et al.*, 1997). The correctness of this model is indicated by comparisons between telomere repeat sequences and the telomerase RNAs of other species (*Table 12.3*): in all organisms that have been looked at, the telomerase RNA contains a sequence that enables it to make copies of the repeat motif present at the organism's telomeres. An interesting feature is that in all organisms the strand synthesized by telomerase has a preponderance of G nucleotides and is therefore referred to as the G-rich strand.

Telomerase can only synthesize this G-rich strand. It is not clear how the other polynucleotide – the C-rich strand – is extended, but it is presumed that when the G-rich strand is long enough, DNA polymerase α attaches at its end and initiates synthesis of complementary DNA in the

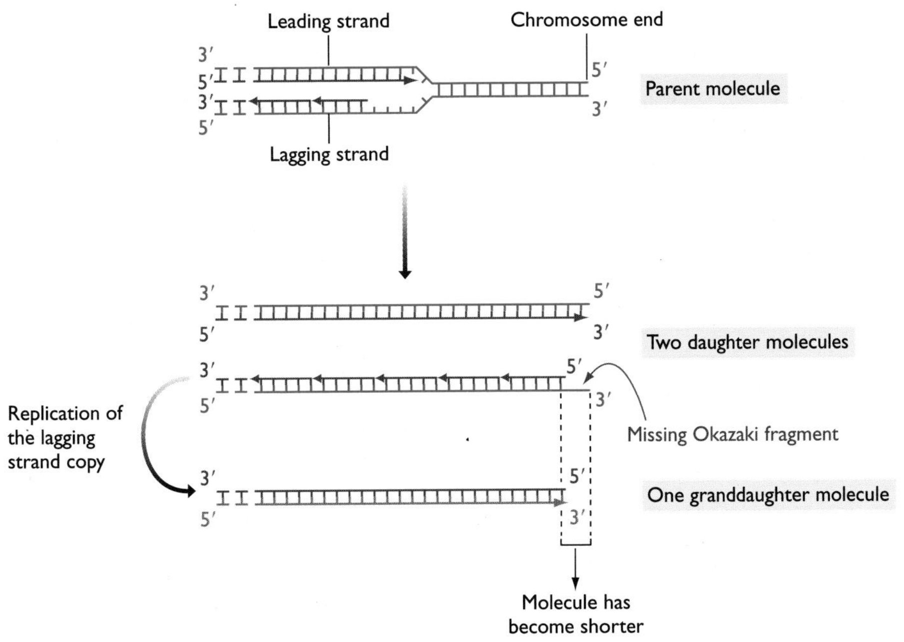

Figure 12.23 One of the reasons why linear DNA molecules could become shorter after DNA replication.

The parent molecule is replicated in the normal way. A complete copy is made of its leading strand, but the lagging strand copy is incomplete because the last Okazaki fragment is not made. This is because primers for Okazaki fragments are synthesized at positions approximately 200 bp apart on the lagging strand. If one Okazaki fragment begins at a position less than 200 bp from the 3′-end of the lagging strand then there will not be room for another priming site, so the remaining segment of the lagging strand is not copied. The resulting daughter molecule therefore has a 3′-overhang and, when replicated, gives rise to a granddaughter molecule that is shorter than the original parent.

normal way (*Figure 12.25*). The paradox is, because this requires the use of a new RNA primer, the C-rich strand is still shorter than the G-rich one. The important point is, however, that the overall length of the chromosomal DNA has not been reduced in comparison with the parent molecule whose daughter we have been considering.

What prevents telomerase from over-extending telomeres? If its activity is not controlled in some way then there is the possibility that chromosomes would continue to grow, which presumably would be equally as deleterious as end-shortening. Telomere length is thought to be regulated by **telomere binding proteins (TBPs)**, such as TRF1 in humans (Smith and de Lange, 1997). Experiments with yeasts and human cells have shown that overproduction of TBPs leads to telomere shortening, whereas mutations that prevent TBPs from binding to the DNA result in the telomeres becoming longer than normal. These results suggest that telomerase activity becomes repressed when a certain number of TBPs have bound to the telomere (Shore, 1997).

12.4 REGULATION OF EUKARYOTIC GENOME REPLICATION

Genome replication in eukaryotic cells is regulated at two levels:

1. Replication is coordinated with the cell cycle so that two copies of the genome are available when the cell divides.
2. The replication process itself can be arrested under certain circumstances, for example if the DNA is damaged and must be repaired before copying can be completed.

We will complete this chapter by looking at these regulatory mechanisms.

12.4.1 Coordination of genome replication and cell division

The concept of a **cell cycle** emerged from the light microscopy studies carried out by the early cell biologists. Their observations showed that dividing cells pass through repeated cycles of mitosis (see *Figure 2.9*, p. 25), when nuclear and cell division occurs, and interphase, the latter being a less dramatic period when few dynamic changes can be detected with the light microscope. It was understood that chromosomes divide during interphase, so when DNA was identified as the genetic material interphase took on a new importance as the period when genome replication must take place. This led to a

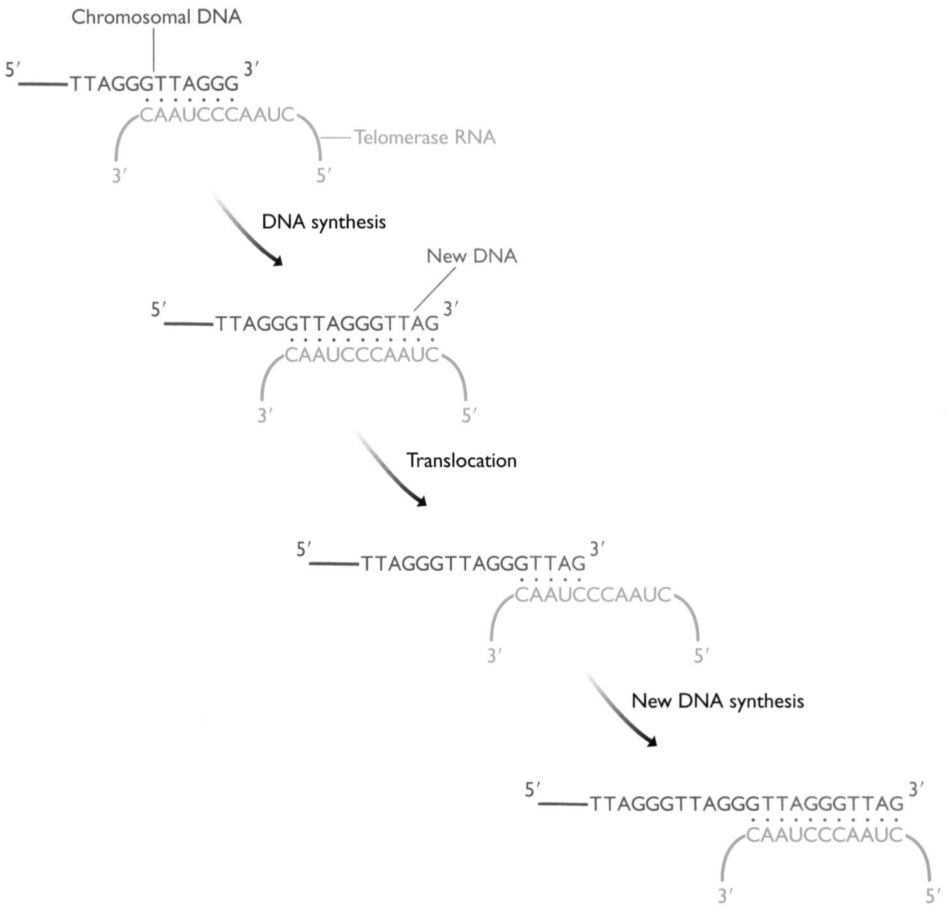

Figure 12.24 Extension of the end of a human chromosome by telomerase.

The 3′-end of a human chromosomal DNA molecule is shown. The sequence comprises repeats of the human telomere motif 5′-TTAGGG-3′. The telomerase RNA base-pairs to the end of the DNA molecule and extends the polynucleotide a short distance. The telomerase RNA then translocates to a new base-pairing position slightly further along the DNA polynucleotide and extends the molecule by a few more nucleotides. The process can be repeated until the chromosome end has been extended by a sufficient amount.

reinterpretation of the cell cycle as a four-stage process (*Figure 12.26*), comprising:

- **Mitosis** or **M phase**, the period when the nucleus and cell divide.
- **Gap 1** or **G1 phase**, an interval when transcription, translation and other general cellular activities occur.
- **Synthesis** or **S phase**, when the genome is replicated.
- **Gap 2** or **G2 phase**, a second interval period.

It is clearly important that the S and M phases be coordinated so that the genome is completely replicated, but replicated only once, before mitosis occurs. The periods

Table 12.3 Sequences of telomere and telomerase RNAs in various organisms

Species	Telomere repeat sequence	Telomerase RNA template sequence
Human	5′-TTAGGG-3′	5′-CUAACCCUAAC-3′
Oxytricha	5′-TTTTGGGG-3′	5′-CAAAACCCCAAAACC-3′
Tetrahymena	5′-TTGGGG-3′	5′-CAACCCCAA-3′

Oxytricha and *Tetrahymena* are protozoans which are particularly useful for telomere studies because at certain developmental stages their chromosomes break into small fragments, all of which have telomeres: they therefore have many telomeres per cell (Greider, 1996).

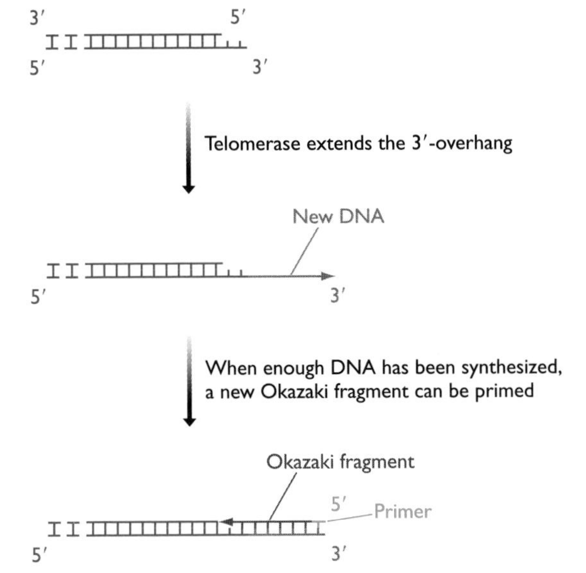

Figure 12.25 Completion of the extension process at the end of a chromosome.

It is believed that after telomerase has extended the 3′-end a sufficient amount, as shown in *Figure 12.24*, DNA polymerase α primes and synthesizes a new Okazaki fragment, converting the 3′-extension into a completely double-stranded end.

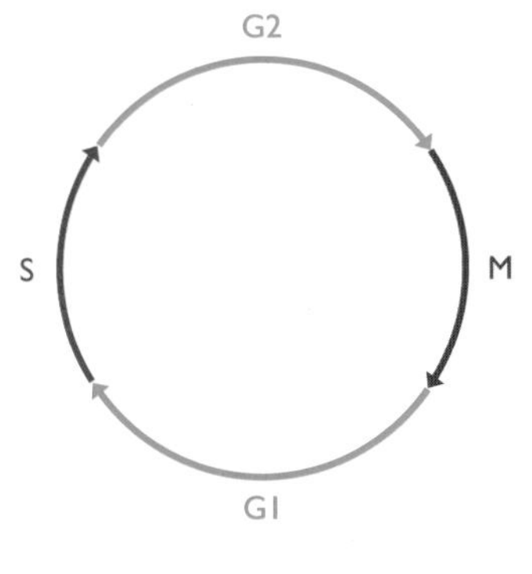

Figure 12.26 The cell cycle.

The lengths of the individual phases vary in different cells. Abbreviations: G1 and G2, gap phases; M, mitosis; S, synthesis phase.

immediately before entry into S and M phases are looked on as key **cell cycle checkpoints**, and it is at one of these two points that the cycle becomes arrested if critical genes involved in cell cycle control are mutated, or if the cell undergoes trauma such as extensive DNA damage. Attempts to understand how genome replication is coordinated with mitosis have therefore concentrated on these two checkpoints, especially the pre-S checkpoint, the period immediately before replication.

Establishment of the pre-replication complex enables genome replication to commence

Studies primarily with *Saccharomyces cerevisiae* have led to a model for controlling the timing of S phase which postulates that genome replication requires construction of **pre-replication complexes (pre-RCs)** at origins of replication, these pre-RCs being converted to **post-RCs** as replication proceeds, the latter unable to initiate replication and so preventing the cell from re-copying its genome before mitosis has occurred (Stillman, 1996). The ORC, the complex of six proteins that is assembled on to domains A and B1 of a yeast ARS (see *Figure 12.8B*, p. 309), was an early contender for the pre-RC but is probably not a central component because ORCs are present at origins of replication at all stages of the cell cycle. Instead, the ORC is looked on as the 'landing pad' on which the pre-RC is constructed.

Two types of protein have been implicated as components of the pre-RC. The first is Cdc6p, which was originally identified in yeast and subsequently shown to have homologs in higher eukaryotes. Yeast Cdc6p is synthesized at the end of G2, as the cell enters mitosis, and becomes associated with chromatin in early G1 before disappearing at the end of G1, when replication begins (*Figure 12.27*). The involvement of the protein in the pre-RC is suggested by experiments in which the gene coding for Cdc6p is repressed, which results in an absence of pre-RCs, and other experiments in which Cdc6p is overproduced, which leads to multiple genome replications in the absence of mitosis. There is also biochemical evidence for a direct interaction between Cdc6p and yeast ORCs.

A second component of the pre-RC is thought to be the group of proteins called **replication licensing factors (RLFs)**. As with Cdc6p, the first examples of these proteins were identified in yeast (the MCM family of proteins) with homologs in higher eukaryotes discovered at a later date. RLFs become bound to chromatin towards the end of M phase and remain in place until the start of S phase, after which they are gradually removed from the DNA as it is replicated. Their attachment to chromatin appears to involve an interaction of some kind with Cdc6p, but the nature of this interaction is not clearly understood, nor is it known whether RLFs associate directly with ORCs or form a more general coating to the DNA.

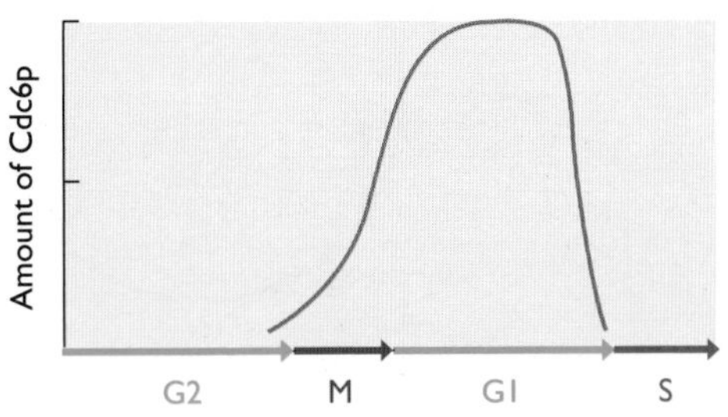

Figure 12.27 Graph showing the amount of Cdc6p in the nucleus at different stages of the cell cycle.

Regulation of pre-RC assembly

Identification of the components of the pre-RC takes us some distance towards understanding how genome replication is initiated, but still leaves open the question of how replication is coordinated with other events in the cell cycle. Cell cycle control is a complex process, mediated largely by protein kinases which phosphorylate and activate enzymes and other proteins that have specific functions during the cell cycle. The same protein kinases are present in the nucleus throughout the cell cycle, so they must themselves be subject to control. This control is exerted partly by proteins called **cyclins**, whose abundance varies at different stages of the cell cycle, partly by other protein kinases that activate the cyclin-dependent kinases, and partly by inhibitory proteins. Even before we start looking for regulators of pre-RC assembly we can anticipate that the control system will be convoluted.

A number of cyclins have been linked with activation of DNA replication and prevention of pre-RC reassembly after replication has been completed (Stillman, 1996). These include the mitotic cyclins, whose main function has been looked on as activating mitosis but which also repress DNA replication. When the effects of these cyclins are blocked by, for example, overproduction of proteins that inhibit their activity, the cell is not only incapable of entering M phase but it also undergoes repeated DNA replication. There are also more specific S phase cyclins, such as Clb5p and Clb6p in *S. cerevisiae*, inactivation of which delays or prevents DNA replication, and other mitotic cyclins that are active during G2 phase and prevent the assembly of pre-RCs in the period after genome replication and before cell division (*Figure 12.28*).

In addition to these cyclin-dependent control systems, DNA replication is also regulated by a cyclin-independent protein kinase, Cdc7p-Dbf4p, found in organisms as diverse as yeasts and mammals. The proteins activated by this kinase have not been identified, separate lines of evidence suggesting that both RLFs and ORCs are targeted. Whatever the mechanism, Cdc7p-Dbf4p activity is a prerequisite for replication, the cyclin-dependent processes on their own being insufficient to push the cell into S phase.

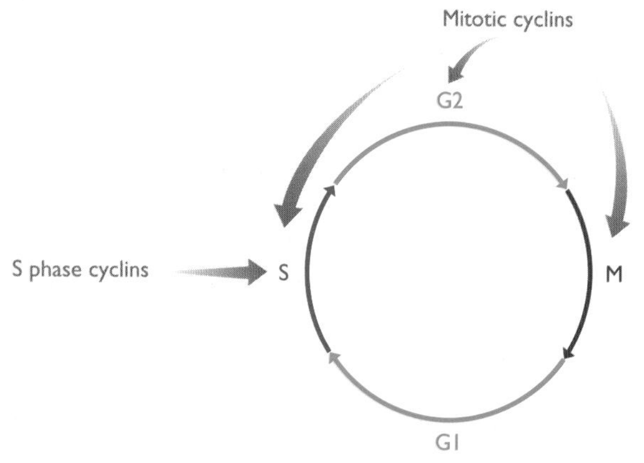

Figure 12.28 Cell cycle control points for cyclins involved in regulation of DNA replication.

See the text for details.

12.4.2 Control within S phase

Regulation of the G1–S transition can be looked on as the major control process affecting genome regulation, but it is not the only one. The specific events occurring during S phase are also subject to regulation.

Early and late replication origins

Initiation of replication does not occur at the same time at all replication origins, nor is 'origin firing' an entirely random process. Some parts of the genome are replicated early in S phase and some later, the pattern of replication being consistent from cell division to cell division (Fangman and Brewer, 1992).

The general pattern is that actively transcribed genes are replicated early in S phase and nontranscribed regions of the genome later on, with the centromere usually being the very last part of a chromosome to be replicated. Early-firing origins are therefore tissue-specific and reflect the pattern of gene expression occurring in a particular cell.

Understanding what determines the firing time of a replication origin is proving to be quite difficult. It is not simply the sequence of the origin because transfer of a DNA segment from its normal position to another site in the same or a different chromosome can result in a change in the firing pattern of origins contained in that segment. This positional effect may be linked with chromatin organization and hence influenced by structures such as locus control regions (Section 8.1.1) which control DNA packaging. The position of the origin in the nucleus may also be important as origins that become active at similar periods within S phase appear to be clustered together, at least in mammals.

Checkpoints within S phase

The final aspect of the regulation of DNA replication that we will study is the function of the checkpoints that exist within S phase. These were first identified when it was shown that one of the responses of yeast cells to DNA damage is a slowing down and possibly a complete halting of the DNA replication process (Paulovich and Hartwell, 1995). This is linked with the activation of genes whose products are involved in DNA repair (Elledge, 1996).

As with entry into S phase, cyclin-dependent kinases are implicated in the regulation of S phase checkpoints. These may respond to damage-detection signals from proteins associated with the replication fork. DNA polymerase ε, which has not been assigned a precise function during DNA synthesis (Section 12.3.2), is a particularly strong candidate in this regard because mutant yeast cells that have abnormal DNA polymerase ε enzymes do not respond to DNA damage in the same way as normal cells. Other replication fork proteins, including the single-strand binding protein RPA and the accessory protein RFC, have also been assigned roles in damage detection (Waga and Stillman, 1998). The regulatory enzymes that respond to these signals probably act by repressing the firing of origins of replication that are usually activated at later stages in S phase (Santocanale and Diffley, 1998).

REFERENCES

Aladjem MI, Rodewald LW, Kolman JL and Wahl GM (1998) Genetic dissection of a mammalian replicator in the human β-globin locus. *Science*, **281**, 1005–1009.

Bell SP and Stillman B (1992) ATP-dependent recognition of eukaryotic origins of DNA replication by a multiprotein complex. *Nature*, **357**, 128–134.

Berger JM, Gamblin SJ, Harrison SC and Wang JC (1996) Structure and mechanism of DNA topoisomerase II. *Nature*, **379**, 225–232; **380**, 179.

Bielinsky A-K and Gerbi SA (1998) Discrete start sites for DNA synthesis in the yeast *ARS1* origin. *Science*, **279**, 95–98.

Bochkarev A, Pfuetzner RA, Edwards AM and Frappier L (1997) Structure of the single-stranded-DNA-binding domain of replication protein A bound to DNA. *Nature*, **385**, 176–181.

Cabral JHM, Jackson AP, Smith CV, Shikotra N, Maxwell A and Liddington RC (1997) Crystal structure of the breakage-reunion domain of DNA gyrase. *Nature*, **388**, 903–906.

Carpenter PB, Mueller PR and Dunphy WG (1996) Role for a *Xenopus* Orc2-related protein in controlling DNA replication. *Nature*, **379**, 357–360.

Cook P (1998) Duplicating a tangled genome. *Science*, **281**, 1466–1467.

Crick FHC, Wang JC and Bauer WR (1979) Is DNA really a double helix? *J. Mol. Biol.*, **129**, 449–461.

Deshpande AM and Newlon CS (1996) DNA replication fork pause sites dependent on transcription. *Science*, **272**, 1030–1033.

Diffley JFX and Cocker JH (1992) Protein–DNA interactions at a yeast replication origin. *Nature*, **357**, 169–172.

Eickbush TH (1997) Telomerase and retrotransposons: which came first? *Science*, **277**, 911–912.

Elledge SJ (1996) Cell cycle checkpoints: preventing an identity crisis. *Science*, **274**, 1664–1672.

Fangman WL and Brewer BJ (1992) A question of time – replication origins of eukaryotic chromosomes. *Cell*, **71**, 363–366.

Feng J, Funk WD, Wang S-S, *et al.* (1995) The RNA component of human telomerase. *Science*, **269**, 1236–1241.

Foiani M, Lucchini G and Plevani P (1997) The DNA polymerase α–primase complex couples DNA replication, cell-cycle progression and DNA-damage response. *Trends Biochem. Sci.*, **22**, 424–427.

Greider CW (1996) Telomere length regulation. *Annu. Rev. Biochem.*, **65**, 337–365.

Holmes FL (1998) The DNA replication problem, 1953–1958. *Trends Biochem. Sci.*, **23**, 117–120.

Kamada K, Horiuchi T, Ohsumi K, Shimamoto N and Morikawa K (1996) Structure of a replicator-terminator protein complexed with DNA. *Nature*, **383**, 598–603.

Lemon KP and Grossman AD (1998) Localization of bacterial DNA polymerase: evidence for a factory model of replication. *Science*, **282**, 1516–1519.

Lima CD, Wang JC and Mondragón A (1994) Three-dimensional structure of the 67K N-terminal fragment of *E. coli* DNA topoisomerase I. *Nature*, **367**, 138–146.

Lingner J, Hughes TR, Shevchenko A, Mann M, Lundblad V and Cech TR (1997) Reverse transcriptase motifs in the catalytic subunit of telomerase. *Science*, **276**, 561–567.

Lohman TM and Bjornson KP (1996) Mechanisms of helicase-catalyzed DNA unwinding. *Annu. Rev. Biochem.*, **65**, 169–214.

Meselson M and Stahl F (1958) The replication of DNA in *Escherichia coli*. *Proc. Natl Acad. Sci. USA*, **44**, 671–682.

Novick RP (1998) Contrasting lifestyles of rolling-circle phages and plasmids. *Trends Biochem. Sci.*, **23**, 434–438.

Okazaki T and Okazaki R (1969) Mechanisms of DNA chain growth. *Proc. Natl Acad. Sci. USA*, **64**, 1242–1248.

Pardue ML, Danilevskaya ON, Lowenhaupt K, Slot F and Traverse KL (1996) *Drosophila* telomeres: new views on chromosome evolution. *Trends Genet.*, **12**, 48–52.

Paulovich AG and Hartwell LH (1995) A checkpoint regulates the rate of progression through S phase in *S. cerevisiae* in response to DNA damage. *Cell*, **82**, 841–847.

Redinbo MR, Stewart L, Kuhn P, Champoux JJ and Hol WGJ (1998) Crystal structures of human topoisomerase I in covalent and noncovalent complexes with DNA. *Science*, **279**, 1504–1513.

Rodley GA, Scobie RS, Bates RHT and Lewitt RM (1976) A possible conformation for double-stranded polynucleotides. *Proc. Natl Acad. Sci. USA*, **73**, 2959–2963.

Santocanale C and Diffley JFX (1998) A Mec1- and Rad53-dependent checkpoint controls late-firing origins of DNA replication. *Nature*, **395**, 615–618.

Shore D (1997) Telomerase and telomere-binding proteins: controlling the endgame. *Trends Biochem. Sci.*, **22**, 233–235.

Smith S and de Lange T (1997) TRF1, a mammalian telomeric protein. *Trends Genet.*, **13**, 21–26.

Steitz TA (1998) A mechanism for all polymerases. *Nature*, **391**, 231–232.

Stewart L, Redinbo MR, Qiu X, Hol WGJ and Champoux JJ (1998) A model for the mechanism of human topoisomerase I. *Science*, **279**, 1534–1541.

Stillman B (1996) Cell cycle control of DNA replication. *Science*, **274**, 1659–1664.

Waga S and Stillman B (1998) The DNA replication fork in eukaryotic cells. *Annu. Rev. Biochem.*, **67**, 721–751.

Watson JD and Crick FHC (1953a) Molecular structure of nucleic acids: a structure for deoxyribose nucleic acid. *Nature*, **171**, 737–738.

Watson JD and Crick FHC (1953b) Genetical implications of the structure of deoxyribonucleic acid. *Nature*, **171**, 964–967.

Wei X, Samarabandu J, Devdhar RS, Siegel AJ, Acharya R and Berezney R (1998) Segregation of transcription and replication sites into higher order domains. *Science*, **281**, 1502–1505.

FURTHER READING

Adams RLP (1991) *DNA Replication: In Focus*. IRL Press, Oxford. — *A concise introduction to the subject.*

DePamphilis ML (ed.) (1996) *DNA Replication in Eukaryotic Cells*. Cold Spring Harbor Laboratory Press, Cold Spring Harbor, NY. — *A more advanced and detailed treatment.*

Judson HF (1979) *The Eighth Day of Creation: Makers of the Revolution in Biology*. Penguin Books, London. — *Includes an account of the topological problems and the Meselson–Stahl experiment.*

Kornberg A (1989) *For the Love of Enzymes: The Odyssey of a Biochemist*. Harvard University Press, Boston. — *A fascinating autobiography by the discoverer of DNA polymerase.*

Kornberg A and Baker T (1991) *DNA Replication*, 2nd edn. W.H. Freeman, New York. — *The best detailed textbook on this subject.*

Wang JC (1996) DNA topoisomerases. *Annu. Rev. Biochem.*, **65**, 635–692.

CHAPTER

13 The Molecular Basis of Genome Evolution

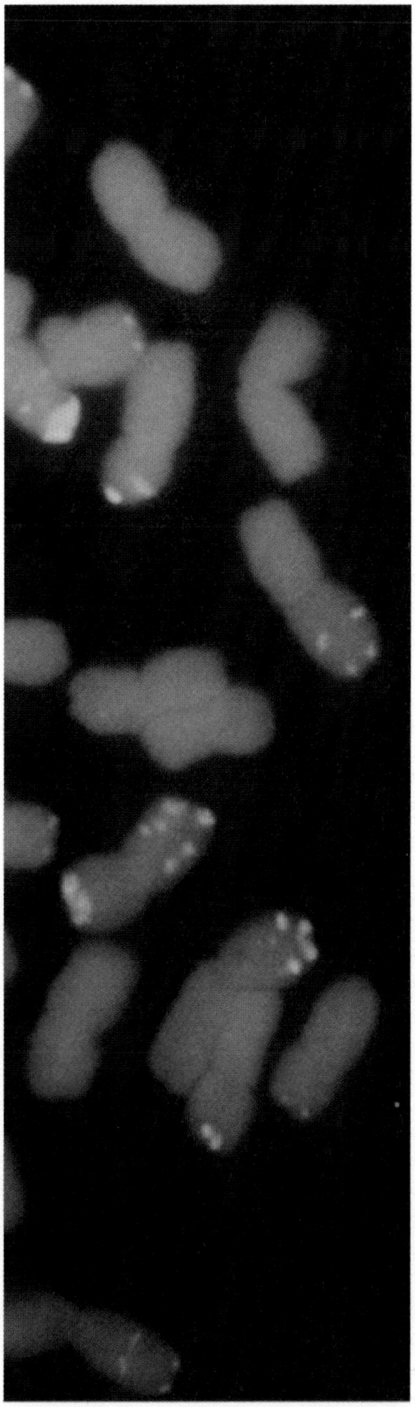

Contents

13.1	*Mutations*		**331**
	13.1.1	The causes of mutations	331
	13.1.2	The effects of mutations	340
	13.1.3	Hypermutation and the possibility of programmed mutations	345
	13.1.4	DNA repair	346
13.2	*Recombination*		**353**
	13.2.1	Homologous recombination	353
	13.2.2	Site-specific recombination	357
	13.2.3	Transposition	359

Concepts

- *Mutation and recombination are the key processes responsible for genome evolution*
- *A mutation is a change in the nucleotide sequence of the genome, caused by a replication error or by a mutagen*
- *Replication errors are infrequent but can lead to point mutations; replication slippage can occur when short repetitive sequences are copied*
- *A variety of chemical and physical mutagens cause damage to DNA molecules*
- *Mutations have various effects on genome function*
- *Some cells can modify their repair processes to induce hypermutation and bacteria may be able to carry out programmed mutations*
- *Most mutations can be corrected by DNA repair processes which include excision procedures for removing damaged nucleotides and mismatch repair systems for correcting replication errors*
- *Recombination results in restructuring of the genome*
- *Models for recombination involve the formation of heteroduplexes containing one or more Holliday structures*
- *Site-specific recombination is responsible for insertion of the λ genome into* E. coli *DNA*
- *Transposition of DNA transposons and retroelements involves recombination events*

GENOMES ARE DYNAMIC entities that evolve over time due to the cumulative effects of small-scale sequence alterations caused by **mutation** and larger scale rearrangements arising from **recombination**. In Chapter 14 we will survey the patterns of genome evolution that are believed to have occurred in the past and hence resulted in the different genome organizations that we see today, and in Chapter 15 we will examine how our knowledge of these patterns can be used to infer the evolutionary relationships between organisms. Before dealing with these issues we must first, in this chapter, gain an understanding of the molecular basis of genome evolution by studying the events involved in mutation and recombination.

Mutation and recombination can both be defined as *processes that result in changes to a genome*, but they are unrelated and we must make a clear distinction between them:

- A **mutation** (Section 13.1) is a change in the nucleotide sequence of a short region of a genome (*Figure 13.1A*). Many mutations are **point mutations** that replace one nucleotide with another, others involve **insertion** or **deletion** of one or a few nucleotides. Mutations result either from errors in DNA replication or from the damaging effects of **mutagens** such as chemicals and radiation that react with DNA and change the structures of individual nucleotides. All cells possess **DNA repair** enzymes that attempt to minimize the number of mutations that occur (Section 13.1.4). These enzymes work in two ways. Some are post-replicative and check newly-synthesized DNA for errors, correcting those that they find; others are pre-replicative and search the DNA for nucleotides with unusual structures, these being replaced before replication occurs. A possible definition of mutation is therefore *a deficiency in DNA repair*.
- **Recombination** (Section 13.2) results in a restructuring of part of a genome, for example by exchange of segments of homologous chromosomes during meiosis or by transposition of a mobile element from one position to another within a chromosome or between chromosomes (*Figure 13.1B*). Various other events that we have studied, including construction of immunoglobulin genes (see *Figure 11.14*, p. 280) and mating type switching in yeast (see *Figure 11.12*, p. 278), are also the results of recombination. Recombination is a cellular process which, like other cellular processes involving DNA (e.g. transcription and

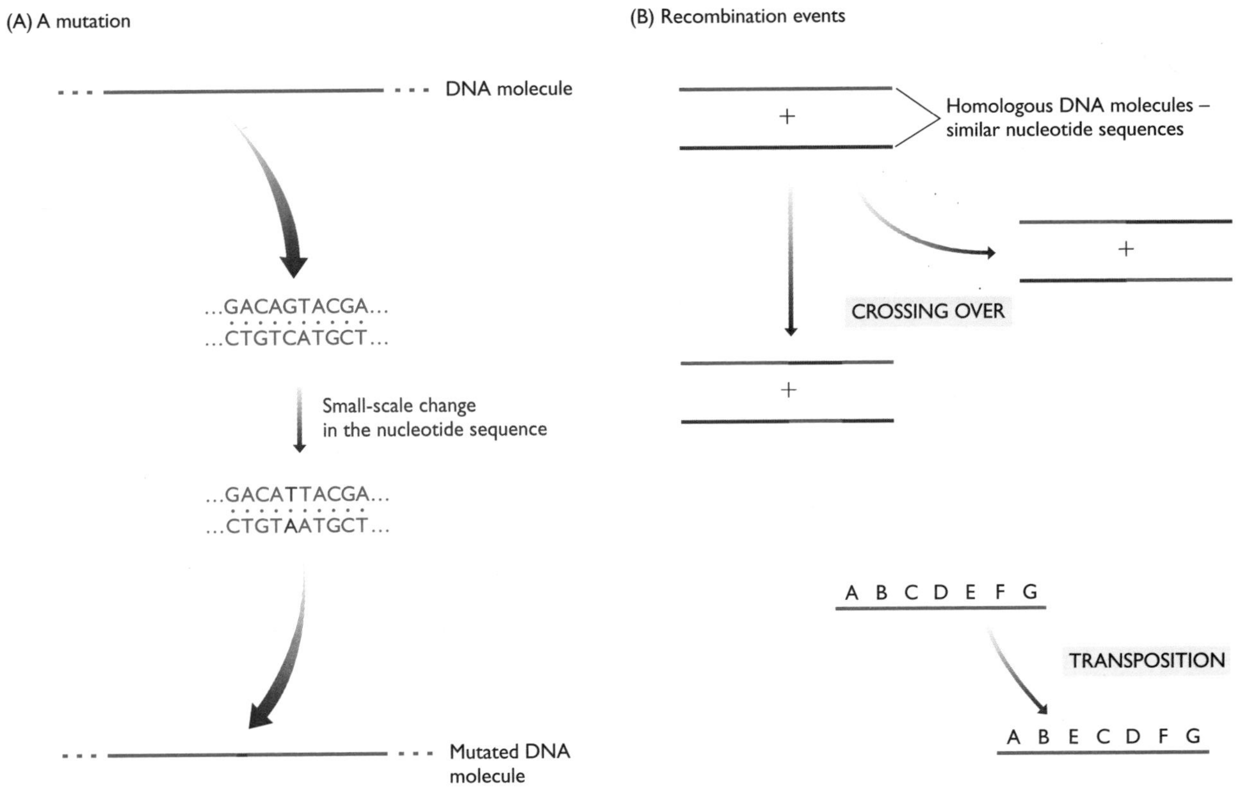

Figure 13.1 Mutation and recombination.

(A) A mutation is a small-scale change in the nucleotide sequence of a DNA molecule. A point mutation is shown but there are several other types of mutations, as described in the text. (B) Recombination events include exchange of segments of DNA molecules, as occurs during meiosis (see *Figure 2.10*, p. 26) and the movement of a segment from one position in a DNA molecule to another, for example by transposition (Section 13.2.3).

replication), is carried out and regulated by enzymes and other proteins.

Both mutation and recombination can have important effects on the cell in which they occur. A mutation in a key gene may cause the cell to die if it results in the protein coded by this gene being defective (Section 13.1.2) and some recombination events lead to changes in the biochemical capabilities of the cell, examples being those involved in immunoglobulin gene construction and yeast mating type switching. Other mutations and recombination events have a less significant impact on the phenotype of the cell and many have none at all. As we will see in Chapter 14, all mutations and recombination events that are not lethal have the potential to contribute to the evolution of the genome but for this to happen they must be inherited when the organism reproduces. With a single-celled organism such as a bacterium or yeast, all genome alterations that are not lethal or reversible are inherited by daughter cells and become permanent features of the lineage that descends for the original cell in which the alteration occurred. In a multicellular organism, only those events that occur in germ cells are relevant to genome evolution. Changes to the genomes of somatic cells are unimportant in an evolutionary sense, but they will have biological relevance if they result in a deleterious phenotype that affects the health of the organism.

13.1 MUTATIONS

With mutations, the issues that we have to consider are: how they arise; what effects they have on the genome and on the organism in which the genome resides; whether it is possible for a cell to increase its mutation rate and induce programmed mutations under certain circumstances; and how mutations are repaired.

13.1.1 The causes of mutations

Mutations arise in two ways:

- Some mutations are **spontaneous** errors in replication that evade the proofreading function of the DNA polymerases that synthesize new polynucleotides at the replication fork (Section 12.3.2). These mutations are called **mismatches** because they are

Box 13.1: Terminology for describing point mutations

Point mutations are also called simple mutations or single-site mutations. They are sometimes described as **substitution mutations** but this risks confusion because to an evolutionary geneticist 'substitution' occurs only when a mutation becomes fixed in a population (see Box 15.5, p. 408), so every individual displays it, as opposed to when the mutation first appears in a single organism.

Point mutations are divided into two categories:

- **Transitions** are purine-to-purine or pyrimidine-to-pyrimidine changes: A→G, G→A, C→T or T→C.
- **Transversions** are purine-to-pyrimidine or pyrimidine-to-purine changes: A→C, A→T, G→C, G→T, C→A, C→G, T→A or T→G.

positions where the nucleotide that is inserted into the daughter polynucleotide does not match, by base-pairing, the nucleotide at the corresponding position in the template DNA (*Figure 13.2A*). If the mismatch is retained in the daughter double helix then *one* of the granddaughter molecules produced during the next round of DNA replication will carry a permanent, double-stranded version of the mutation.

- Other mutations arise because a mutagen has reacted with the parent DNA, causing a structural change that affects the base-pairing capability of the altered nucleotide. Usually this alteration affects only one strand of the parent double helix, so only one of the daughter molecules carries the mutation, but two of the granddaughter molecules produced during the next round of replication will have it (*Figure 13.2B*).

Errors in replication are a source of point mutations

When considered purely as a chemical reaction, complementary base-pairing is not particularly accurate and if it was possible to copy a DNA template in the test tube, without the aid of any enzymes, then the resulting polynucleotide would probably have point mutations at 5–10 positions out of every hundred. This represents an error rate of 5–10%, which would be completely unacceptable during genome replication. The template-dependent DNA polymerases that carry out DNA replication must therefore increase the accuracy of the process by several orders of magnitude. This improvement is brought about in two ways:

- The DNA polymerase operates a nucleotide selection process that dramatically increases the accuracy of template-dependent DNA synthesis (*Figure 13.3A*). This selection process probably acts at three different stages during the polymerization reaction, discrimination against an incorrect nucleotide occurring when the nucleotide is first bound to the DNA polymerase, when it is shifted to the active site of the enzyme, and when it is attached to the 3′-end of the polynucleotide that is being synthesized.
- The accuracy of DNA synthesis is increased still further if the DNA polymerase possesses a 3′→5′ exonuclease activity and so is able to remove an incorrect nucleotide that evades the base selection process and becomes attached to the 3′-end of the new polynucleotide (see *Figure 12.11B*, p. 313). This is called **proofreading** (Section 12.3.2), but the name is a misnomer because the process is not an active checking mechanism. Instead, each step in the synthesis of a polynucleotide should be viewed as a competition between the polymerase and exonuclease functions of the enzyme, the polymerase usually winning because it is more active than the exonuclease, at least when the 3′-terminal nucleotide is base-paired to the template. But the polymerase activity is less efficient if the terminal nucleotide is not base-paired, the resulting pause in polymerization allowing the exonuclease activity to predominate so the incorrect nucleotide is removed (see *Figure 13.3B*).

Escherichia coli is able to synthesize DNA with an error rate of only 1 per 10^7 nucleotide additions. Interestingly, these errors are not evenly distributed between the two daughter molecules, the product of lagging strand replication being prone to about 20 times as many errors as the leading strand replicant. This asymmetry might indicate that DNA polymerase I, which is involved only in lagging strand replication (Section 12.3.2), has a less effective base selection and proofreading capability compared with DNA polymerase III, the main replicating enzyme (Francino and Ochman, 1997).

Not all of the errors that occur during DNA synthesis can be blamed on the polymerase enzymes: sometimes an error occurs even though the enzyme adds the 'correct' nucleotide, the one that base-pairs with the template. This is because each nucleotide base can occur as either of two alternative **tautomers**, structural isomers that are in dynamic equilibrium. For example, thymine exists as two tautomers, the *keto* and *enol* forms, with individual molecules occasionally undergoing a shift from one tautomer to the other. The equilibrium is biased very much towards the *keto* form but every now and then the *enol* version of thymine occurs in the template DNA at the precise time that the replication fork is moving past. This will lead to an 'error', because *enol*-thymine base-pairs with G rather than A (*Figure 13.4*). The same problem can occur with adenine, the rare *imino* tautomer of this base preferentially forming a pair with C, and guanine, *enol*-guanine pairing with thymine. After replication, the rare tautomer will inevitably revert to its more common form, leading to a mismatch in the daughter double helix.

As stated above, the error rate for DNA synthesis in *E. coli* is 1 in 10^7. The overall error rate for replication of the *E. coli* genome is only 1 in 10^{10} to 1 in 10^{11}, the

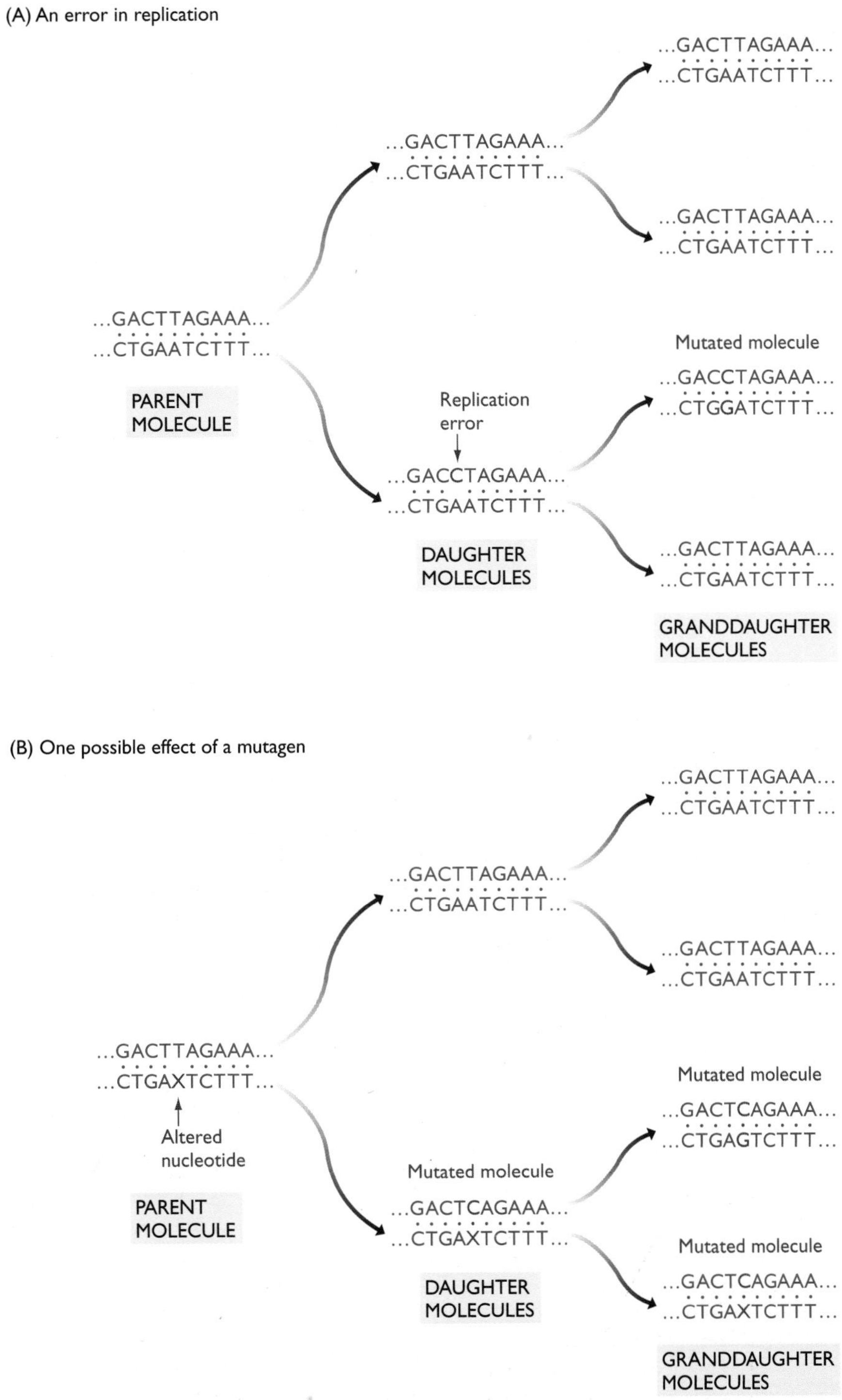

Figure 13.2 Examples of mutations.

(A) An error in replication leads to a mismatch in one of the daughter double helices, in this case a T to C change because one of the As in the template DNA was miscopied. When the mismatched molecule is itself replicated it gives one double helix with the correct sequence and one with a mutated sequence. (B) A mutagen has altered the structure of an A in the lower strand of the parent molecule, giving nucleotide X, which does not base-pair with the T in the other strand so, in effect, a mismatch has been created. When the parent molecule is replicated, X base-pairs with C, giving a mutated daughter molecule. When this daughter molecule is replicated, both granddaughters inherit the mutation.

TECHNICAL
13.1
NOTES

Mutation detection

Rapid procedures for detecting mutations in DNA molecules.

Many genetic diseases are caused by point mutations that result in modification or inactivation of a gene product. Methods for detecting these mutations are important in two contexts. First, when a new gene responsible for a genetic disease is first identified it is usually necessary to examine many versions of that gene, from different individuals, to identify the mutation or mutations responsible for the disease state. Second, when a disease-causing mutation has been characterized, high-throughput methods are needed so that clinicians can screen many DNA samples in order to identify individuals who have the mutation and are at risk of developing the disease or passing it on to their children.

Any mutation can be identified by DNA sequencing but sequencing is relatively slow and would be inappropriate for screening a large number of samples. DNA chip technology (Technical Note 2.3, p. 23) could also be employed, but this is not yet a widely available option. For these reasons, a number of 'low technology' methods have been devised. These can be divided into two categories: **mutation scanning** techniques, which require no prior information about the position of a mutation, and **mutation screening** techniques, which determine whether a specific mutation is present.

Most *scanning* techniques involve analysis of the heteroduplex formed between a single strand of the DNA being examined and the complementary strand of a control DNA that has the unmutated sequence:

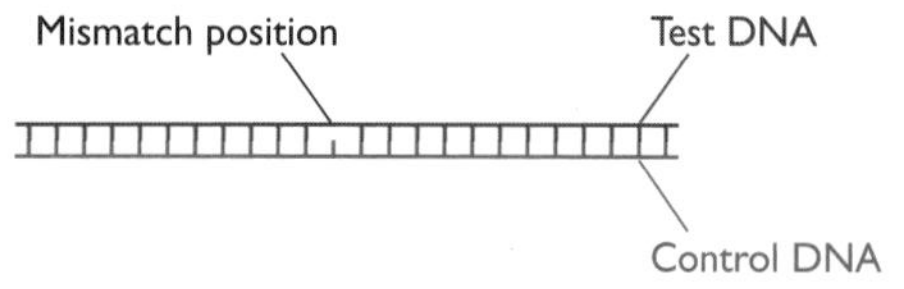

If the test DNA contains a mutation then there will be a single mismatched position in the heteroduplex, where a base pair has not formed. Various techniques can be used for detecting whether this mismatch is present or not (Cotton, 1997):

- **Electrophoresis** or **high-performance liquid chromatography** (**HPLC**) can detect the mismatch by identifying the difference in the mobility of the mismatched hybrid, compared with a fully base-paired one, in a polyacrylamide gel or HPLC column. This approach determines if a mismatch is present but does not provide information on where in the test DNA the mutation is located.
- **Cleavage** of the heteroduplex at the mismatch position followed by gel electrophoresis will locate the position of a mismatch. If the heteroduplex stays intact then no mismatch is present; if it is cleaved then it contains a mismatch, the position of the mutation in the test DNA being indicated by the sizes of the cleavage products. Cleavage is carried out by treatment with enzymes or chemicals that cut at single-stranded regions of mainly double-stranded DNA, or with a single-strand-specific ribonuclease such as S1 (see *Figure* 5.7, p. 94) if the hybrid has been formed between the control DNA and an RNA version of the test DNA.

Most *screening* methods for detection of specific mutations make use of the ability of oligonucleotides to distinguish between target DNAs whose sequences differ at just one nucleotide position (see *Figure* 2.7, p. 22). In **allele-specific oligonucleotide** (**ASO**) hybridization the DNA samples are screened by probing with an oligonucleotide that hybridizes only to the mutant sequence:

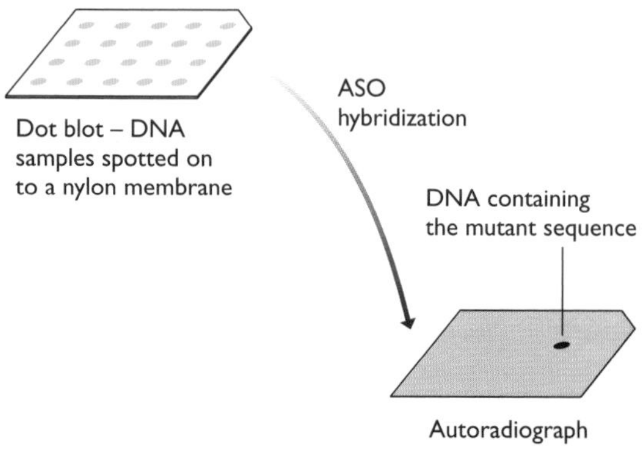

This is an efficient procedure but it is unnecessarily long-winded. The DNA samples are usually obtained by PCR of clinical isolates so a more rapid alternative is to use the diagnostic oligonucleotide as one of the PCR primers, so that presence or absence of the mutation in the test DNA is indicated by the synthesis or otherwise of a PCR product.

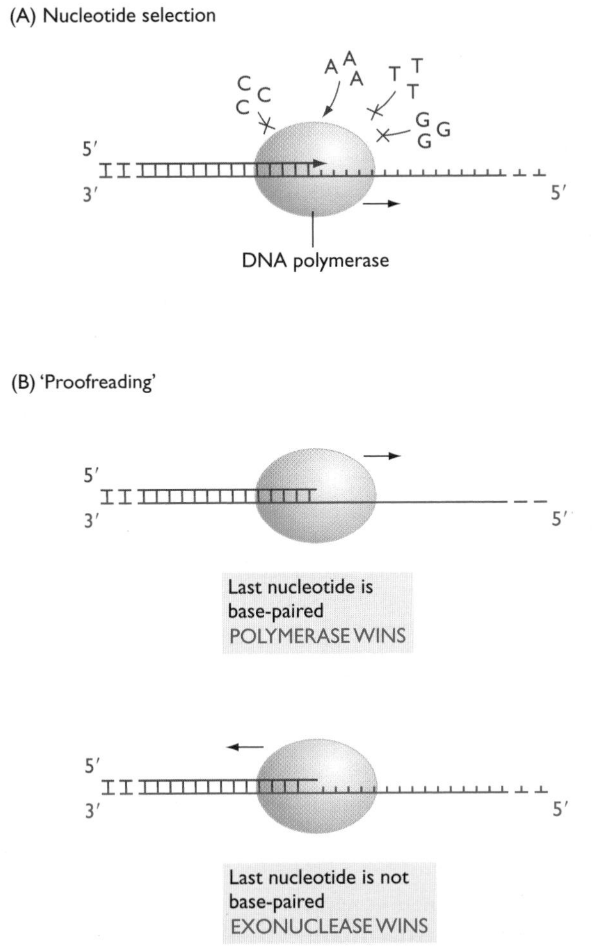

Figure 13.3 Mechanisms for ensuring the accuracy of DNA replication.

(A) The DNA polymerase actively selects the correct nucleotide to insert at each position. (B) Those errors that occur can be corrected by 'proofreading' if the polymerase has a 3′→5′ exonuclease activity. If the last nucleotide that was inserted is base-paired to the template then the polymerase activity predominates, but if the last nucleotide is not base-paired then the exonuclease activity is favored.

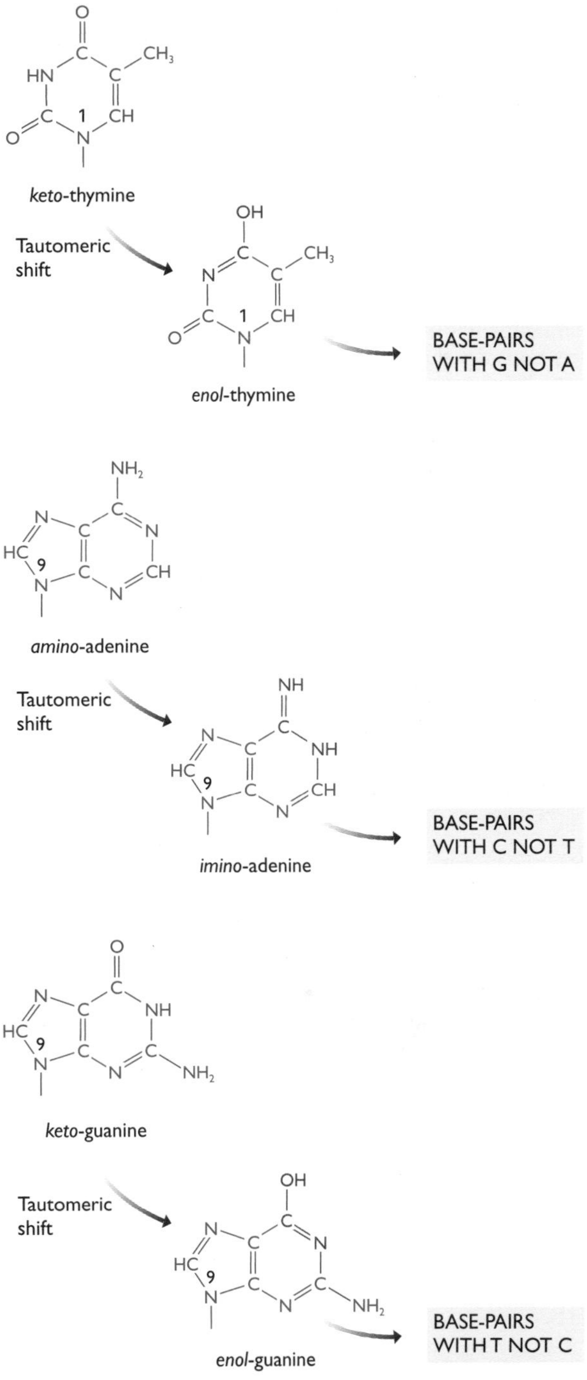

Figure 13.4 The effects of tautomerism on base-pairing.

In each of these three examples, the two tautomeric forms of the base have different pairing properties. Cytosine also has *amino* and *imino* tautomers but both pair with G.

improvement compared to the polymerase error rate being due to the mismatch repair system (Section 13.1.4) that scans newly replicated DNA for positions where the bases are unpaired and hence corrects the few mistakes that the replication enzymes makes. The implication is that only one uncorrected replication error occurs every 1000 times that the *E. coli* genome is copied.

Replication errors can also lead to insertion and deletion mutations

Not all errors in replication are point mutations. Aberrant replication can also result in small numbers of extra nucleotides being inserted into the polynucleotide being synthesized, or some nucleotides in the template not being copied. Insertions and deletions are often called **frameshift** mutations because when one occurs within a coding region it can result in a shift in the reading frame used for translation of the protein specified by the gene (see *Figure 13.12*, p. 342). However, it is inaccurate to use 'frameshift' to describe all insertions and deletions because they can occur anywhere, not just in genes, and not all insertions or deletions in coding regions result in frameshifts: an insertion or deletion of three nucleotides, or multiples of three, simply adds or removes codons or parts of adjacent codons without affecting the reading frame.

Insertion and deletion mutations can affect all parts of the genome but are particularly prevalent when the template DNA contains short repeated sequences, such as are found in microsatellites (Section 6.3.1). This is because repeated sequences can induce **replication slippage**, in which the template strand and its copy shift their relative positions so that part of the template is either copied twice or missed out. The result is that the new polynucleotide has a larger or smaller number, respectively, of the repeat units (*Figure 13.5*). This is the main reason why microsatellite sequences are so variable, replication slippage occasionally generating a new length variant, adding to the collection of alleles already present in the population (Section 15.3.2).

Replication slippage is probably also responsible for the **trinucleotide repeat expansion diseases** that have been discovered in humans in recent years (Ashley and Warren, 1995). Each of these neurodegenerative diseases is due to a relatively short series of trinucleotide repeats becoming elongated to two or more times its normal length. For example, the human *Hdh* gene contains the sequence 5′-CAG-3′ repeated between 10 and 35 times in tandem, coding for a series of glutamines in the protein product. In Huntington's disease this repeat expands to a copy number of 36–121, increasing the length of the polyglutamine tract and resulting in a dysfunctional protein. Several other human diseases are also due to expansions of polyglutamine codons (*Table 13.1*). Some diseases associated with mental retardation result from trinucleotide expansions in the leader region of a gene, giving a **fragile site**, a position where the chromosome is likely to break (Sutherland *et al.*, 1998) and expansions involving intron and trailer regions are also known.

How triplet expansions are generated is not precisely understood. The size of the insertion is much greater than occurs with normal replication slippage, such as is seen with microsatellite sequences, and once the expansion reaches a certain length it appears to become susceptible to further expansion in subsequent rounds of replication, leading to the disease becoming increasingly severe in succeeding generations. The possibility that expansion involves formation of hairpin loops in the DNA has been raised, based on the observation that only a limited number of trinucleotide sequences are known to undergo expansion, and all of these sequences are GC-rich and so might form stable secondary structures. Studies of similar triplet expansions in yeast have shown that these are more prevalent when the *RAD27* gene is inactivated (Freudenreich *et al.*, 1998), an interesting observation as *RAD27* is the yeast version of the mammalian gene for

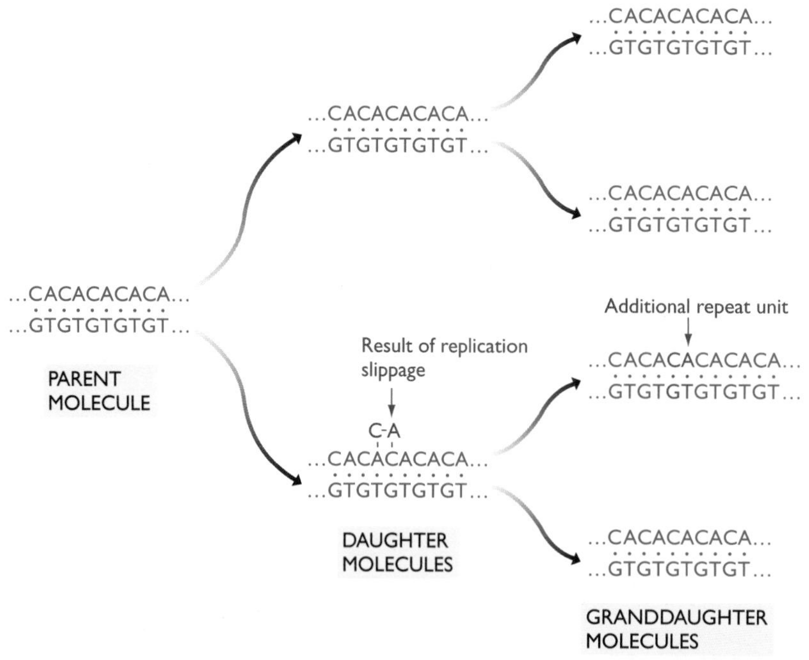

Figure 13.5 Replication slippage.

The diagram shows replication of a 5-unit CA repeat microsatellite. Slippage has occurred during replication of the parent molecule, inserting an additional repeat unit into the newly synthesized polynucleotide of one of the daughter molecules. When this daughter molecule replicates it gives a granddaughter molecule whose microsatellite array is one unit longer than that of the original parent.

Table 13.1 Examples of human trinucleotide repeat expansions

Locus	Repeat sequence: Normal	Repeat sequence: Mutated	Associated disease
Polyglutamine expansions (all in coding regions of genes)			
Hdh	$(CAG)_{10-35}$	$(CAG)_{36-121}$	Huntington's disease
AR	$(CAG)_{11-33}$	$(CAG)_{38-66}$	Spinal and bulbar muscular atrophy
B37	$(CAG)_{7-25}$	$(CAG)_{49-75}$	Dentatoribral-pallidoluysian atrophy
MJD1	$(CAG)_{12-37}$	$(CAG)_{61-84}$	Machado–Joseph disease
SCA1	$(CAG)_{6-39}$	$(CAG)_{41-81}$	Spinocerebellar ataxia type 1
Fragile site expansions (probably all in the untranslated leader regions of genes)			
FRAXA	$(CGG)_{6-52}$	$(CGG)_{60-1000}$	Fragile X syndrome
FRAXE	$(GCC)_{7-35}$	$(GCC)_{130-750}$	Fragile XE mental retardation
FRAXF	$(GCC)_{6-29}$	$(GCC)_{300-1000}$	None
FRA11B	$(CGG)_{11}$	$(CGG)_{80-1000}$	Predisposed to Jacobsen syndrome
FRA16A	$(CCG)_{16-49}$	$(CCG)_{1000-1900}$	None
Other expansions (positions described below)			
DMPK	$(CTG)_{5-37}$	$(CTG)_{50-3000}$	Myotonic dystrophy
FRDA	$(GAA)_{40-60}$	$(GAA)_{>200}$	Friedreich's ataxia

For more details see Ashley and Warren (1995). The DMPK and FRDA expansions are in the trailer and intron regions of their genes, respectively, and are thought to affect RNA processing (Ashley and Warren, 1995; Campuzano *et al.*, 1996). There are also a few disease-causing mutations that involve expansions of longer sequences, such as progressive myoclonus epilepsy caused by a $(CCCCGCCCCGCG)_{2-3}$ to $(CCCCGCCCCGCG)_{>12}$ expansion in the promoter region of the *EPM1* locus (Mandel, 1997).

FEN1, the protein involved in processing of Okazaki fragments (Section 12.3.2). This might indicate that a trinucleotide repeat expansion is caused by an aberration in lagging strand synthesis.

Mutations are also caused by chemical and physical mutagens

Many chemicals that occur naturally in the environment have mutagenic properties and these have been supplemented in recent years with other chemical mutagens that result from human industrial activity. Physical agents such as radiation are also mutagenic. Most organisms are exposed to greater or lesser amounts of these various mutagens and their genomes suffer damage as a result.

The definition of the term 'mutagen' is *a chemical or physical agent that causes mutations*. This definition is important because it distinguishes mutagens from other types of environmental agent that cause damage to cells in ways other than by causing mutations (*Table 13.2*). There are overlaps between these categories (for example, some mutagens are also carcinogens) but each type of agent has a distinct biological effect. The definition of mutagen also makes a distinction between true mutagens and other agents that damage DNA without causing mutations, for example by causing breaks in DNA molecules. This type of damage may block replication and cause the cell to die, but it is not a mutation in the strict sense of the term and the causative agents are therefore not mutagens.

Mutagens cause mutations in three different ways:

- Some act as **base analogs** and are mistakenly used as substrates when new DNA is synthesized at the replication fork.
- Some react directly with DNA, causing structural changes that lead to miscopying of the template strand when the DNA is replicated. These structural changes are diverse, as we will see when we look at individual mutagens.
- Some mutagens act indirectly on DNA. They do not themselves affect DNA structure, but instead cause the cell to synthesize chemicals such as peroxides that have the direct mutagenic effect.

The range of mutagens is so vast that it is difficult to devise an all-embracing classification. We will therefore restrict our study to the most common types. For chemical mutagens these are as follows:

- **Base analogs** are purine and pyrimidine bases that are similar enough to the standard bases to be

Table 13.2 Categories of environmental agent that cause damage to living cells

Agent	Effect on living cells
Carcinogen	Causes cancer – the neoplastic transformation of eukaryotic cells
Clastogen	Causes fragmentation of chromosomes
Mutagen	Causes mutations
Oncogen	Induces tumor formation
Teratogen	Results in developmental abnormalities

Based on Twyman (1998).

incorporated into nucleotides when these are synthesized by the cell. The resulting unusual nucleotides can then be used as substrates for DNA synthesis during genome replication. For example, **5-bromouracil** (**5-bU**; *Figure 13.6A*) has the same base-pairing properties as thymine and nucleotides containing this base can be added to the daughter polynucleotide at positions opposite A's in the template. The mutagenic effect arises because the equilibrium between the two tautomers of 5-bU is shifted more towards the rarer *enol* form than is the case with thymine. This means that during the next round of replication there is a relatively high chance of the polymerase encountering *enol*-5bU, which (like *enol*-thymine) pairs with G rather than A (*Figure 13.6B*). This results in a point mutation (*Figure 13.6C*). **2-Aminopurine** acts in a similar way: it is an analog of adenine with an *amino*-tautomer that pairs with thymine and an *imino*-tautomer that pairs with cytosine, the *imino* form being less uncommon than *imino*-adenine and hence inducing T to C transitions during DNA replication.

- **Deaminating agents** also cause point mutations. A

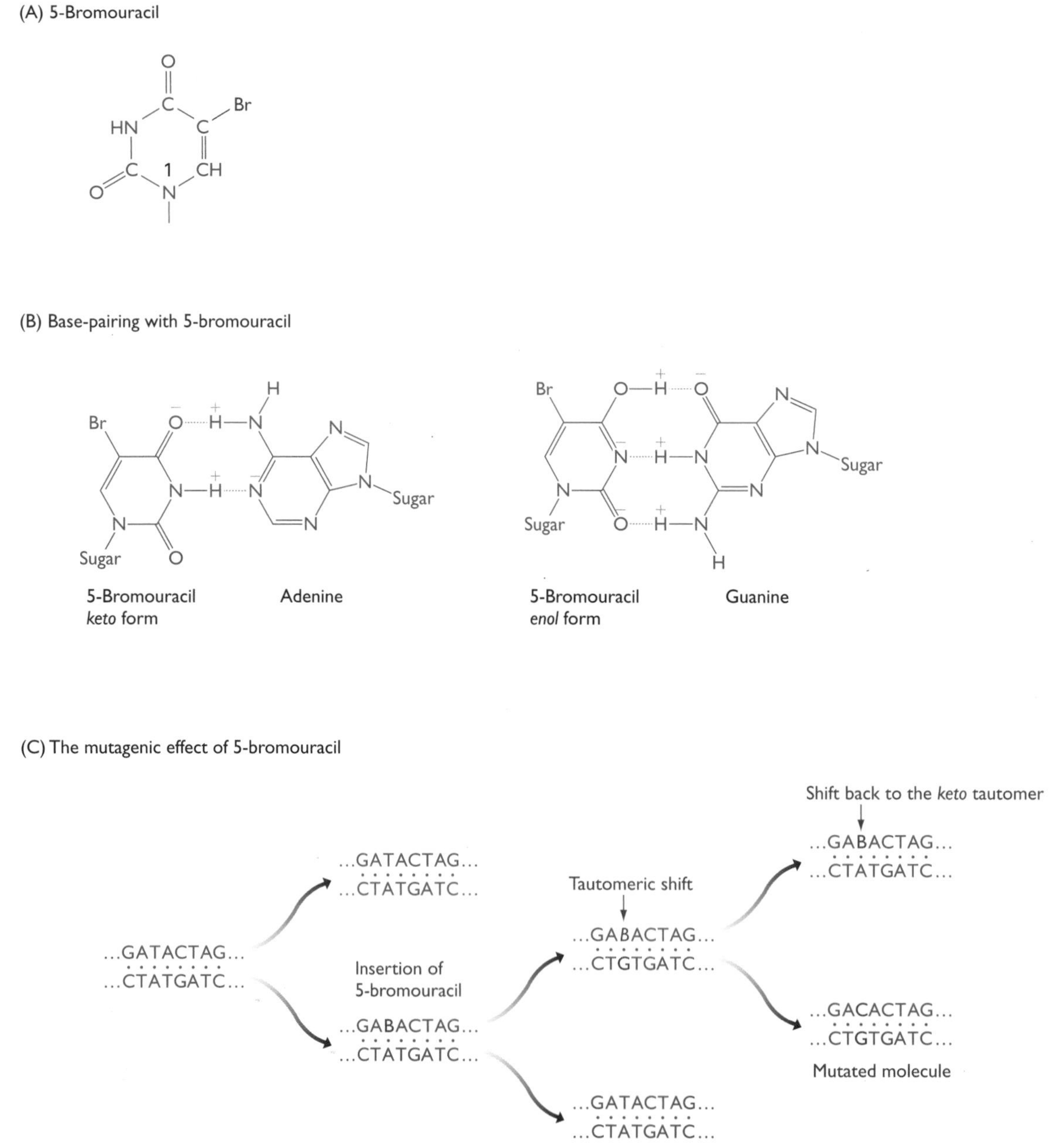

Figure 13.6 5-Bromouracil and its mutagenic effect.

See the text for details.

certain amount of base deamination (removal of an amino group) occurs spontaneously in genomic DNA molecules, with the rate being increased by chemicals such as nitrous acid, which deaminates adenine, cytosine and guanine (thymine has no amino group and so cannot be deaminated), and sodium bisulfite, which acts only on cytosine. Deamination of guanine is not mutagenic because the resulting base, xanthine, blocks replication when it appears in the template polynucleotide. Deamination of adenine gives hypoxanthine (*Figure 13.7*), which pairs with C rather than T, and deamination of cytosine gives uracil, which pairs with A rather than G. Deaminations of these two bases therefore result in point mutations when the template strand is copied.

- **Alkylating agents** are a third type of mutagen that can give rise to point mutations. Chemicals such as **ethylmethane sulfonate** (**EMS**) and dimethylnitrosamine add alkyl groups to nucleotides in DNA molecules, as do methylating agents such as methyl halides which are present in the atmosphere, and the products of nitrite metabolism. The effect that alkylation has depends on the position at which the nucleotide is modified and the type of alkyl group that is added. Methylations, for example, often result in modified nucleotides with altered base-pairing properties and so lead to point mutations. Other alkylations block replication by forming crosslinks between the two strands of a DNA molecule, or by adding large alkyl groups that prevent progress of the replication complex.
- **Intercalating agents** are usually associated with insertion mutations. The best known mutagen of this type is **ethidium bromide**, which fluoresces when exposed to UV radiation and so is used to reveal the positions of DNA bands after agarose gel electrophoresis (see Technical Note 3.2, p. 43). Ethidium bromide and other intercalating agents are flat molecules that can slip in between base pairs in the double helix, slightly unwinding the helix and hence increasing the distance between adjacent base pairs (*Figure 13.8*).

The most important types of physical mutagen are:

- **UV radiation** of 260 nm induces dimerization of adjacent pyrimidine bases, especially if these are both thymines (*Figure 13.9A*), resulting in a **cyclobutyl dimer**. Other pyrimidine combinations also form dimers, the order of frequency being 5′-CT-3′ > 5′-TC-3′ > 5′-CC-3′. Purine dimers are much less common. UV-induced dimerization usually results in a deletion mutation when the modified strand is copied. Another type of UV-induced **photoproduct** is the **(6–4) lesion** in which carbons number 4 and 6 of adjacent pyrimidines become covalently linked (*Figure 13.9B*).
- **Ionizing radiation** has various effects on DNA depending on the type of radiation and its intensity. Point, insertion and/or deletion mutations might arise, as well as more severe forms of DNA damage that prevent subsequent replication of the genome. Some types of ionizing radiation act directly on DNA, others act indirectly by stimulating the formation of reactive molecules such as peroxides in the cell.

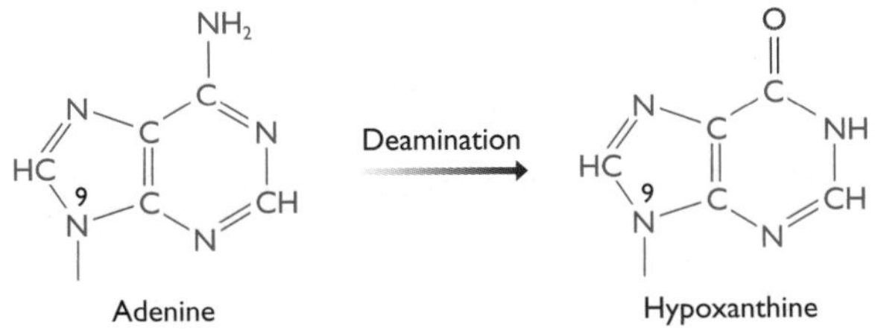

Figure 13.7 Hypoxanthine is a deaminated version of adenine.

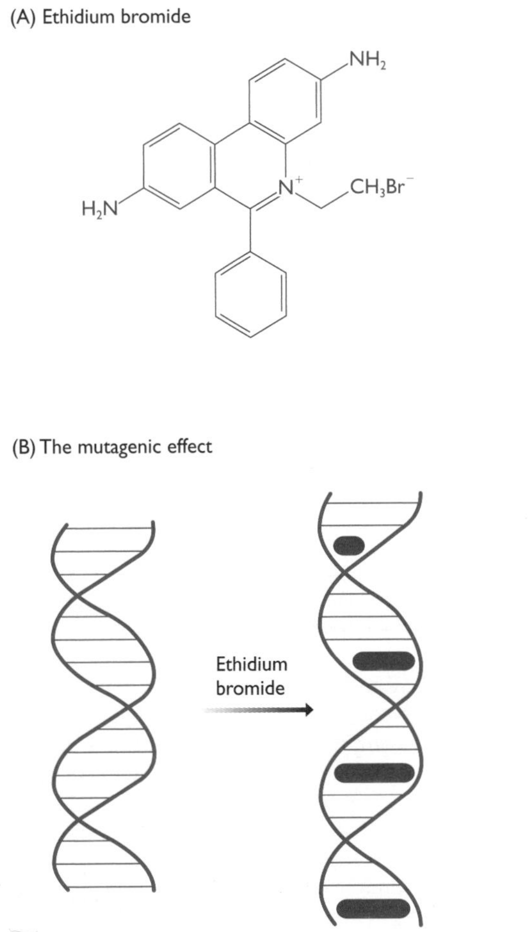

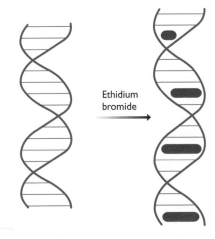

Figure 13.8 The mutagenic effect of ethidium bromide.

(A) Ethidium bromide is a flat, plate-like molecule that is able to slot in between the base pairs of the double helix. (B) Ethidium bromide molecules are seen intercalated into the helix: the molecules are viewed sideways on. Note that intercalation results in the distance between adjacent base pairs being increased.

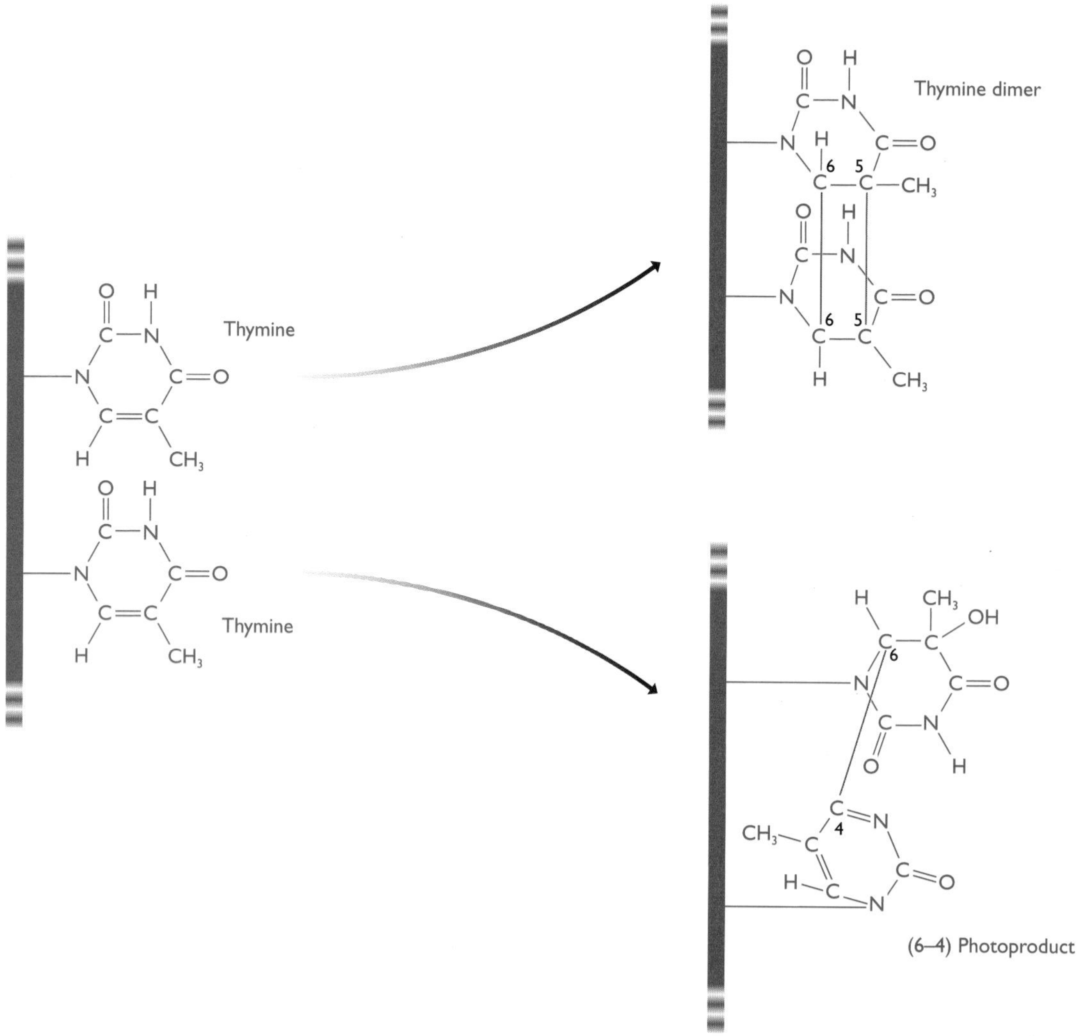

Figure 13.9 Photoproducts induced by UV irradiation.

A segment of a polynucleotide containing two adjacent thymine bases is shown. (A) A thymine dimer contains two UV-induced covalent bonds, one linking the carbons at position 6 and the other linking the carbons at position 5. (B) The (6–4) lesion involves formation of a covalent bond between carbons 4 and 6 of the adjacent nucleotides.

- **Heat** stimulates the water-induced cleavage of the β-*N*-glycosidic bond that attaches the base to the sugar component of the nucleotide (*Figure 13.10A*). This occurs more frequently with purines rather than pyrimidines and results in an **AP** (apurinic/apyrimidinic) or **baseless site**. The sugar–phosphate that is left is unstable and rapidly degrades, leaving a gap if the DNA molecule is double-stranded (*Figure 13.10B*). This reaction is not normally mutagenic because cells have effective systems for repairing nicks (Section 13.1.4), which is reassuring when one considers that 10 000 AP sites are generated in each human cell per day. Gaps do, however, lead to mutations under certain circumstances, for example in *E. coli* when the SOS response is activated, when gaps are filled with A's regardless of the identity of the nucleotide in the other strand (Section 13.1.3).

13.1.2 The effects of mutations

When considering the effects of mutations we must make a distinction between the *direct* effect that a mutation has on the functioning of a genome and its *indirect* effect on the phenotype of the organism in which it occurs. The direct effect is relatively easy to assess because we can use our understanding of gene structure and expression to predict the impact that a mutation will have on genome function. The indirect effects are more complex because these relate to the phenotype of the mutated organism which, as described in Section 5.2.2, is often difficult to correlate with the activities of individual genes.

The effects of mutations on genomes

Many mutations result in nucleotide sequence changes that have no effect on the functioning of the genome.

(A) Heat-induced hydrolysis of a β-*N*-glycosidic bond

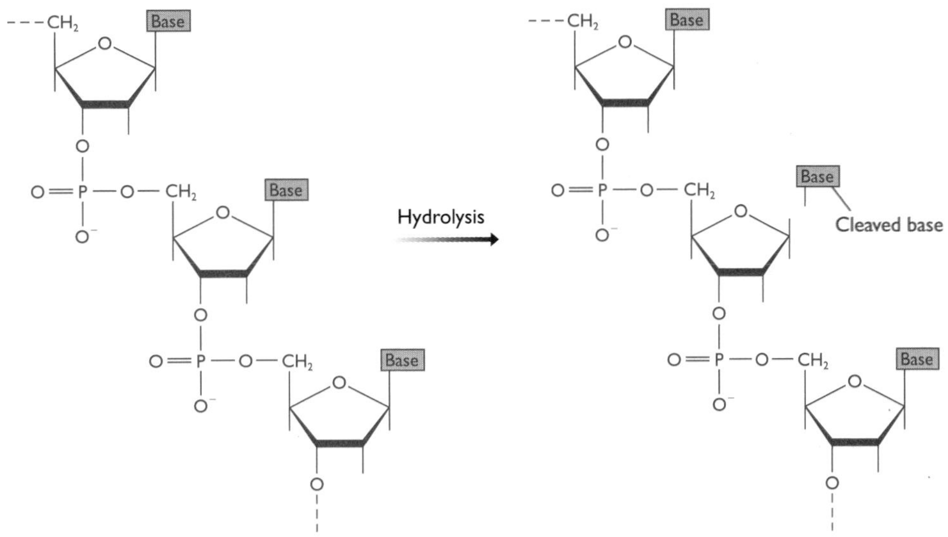

(B) The effect of hydrolysis on double-stranded DNA

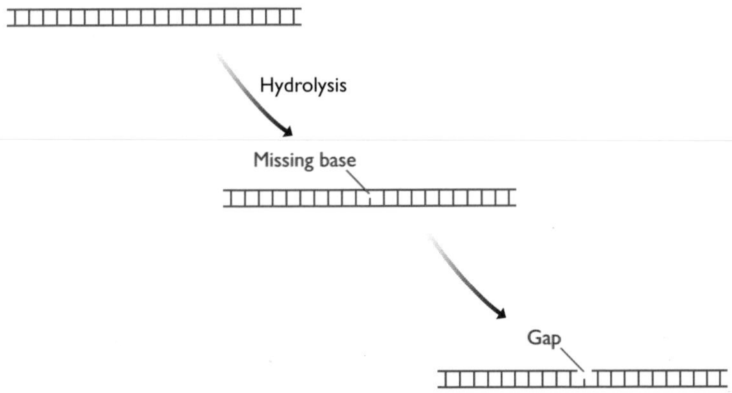

Figure 13.10 The mutagenic effect of heat.

(A) Heat induces hydrolysis of β-*N*-glycosidic bonds resulting in a baseless site in a polynucleotide. (B) Schematic representation of the effect of heat-induced hydrolysis on a double-stranded DNA molecule. The baseless site is unstable and degrades leaving a gap in one strand.

These **silent mutations** include virtually all of those that occur in extragenic DNA and in the noncoding components of genes and gene-related sequences. In other words, some 97% of the human genome (see Box 6.4, p. 135) can be mutated without significant effect.

Mutations in the coding regions of genes are much more important. First, we will look at point mutations that change the sequence of a triplet codon. A mutation of this type will have one of four effects (*Figure 13.11*):

- It may result in a **synonymous** change, the new codon specifying the same amino acid as the unmutated codon. A synonymous change is therefore a silent mutation because it has no effect on the coding function of the genome: the mutated gene codes for exactly the same protein as the unmutated gene.
- It may result in a **nonsynonymous** change, the mutation altering the codon so that it specifies a different amino acid. The protein coded by the mutated gene therefore has a single amino acid change, which often has no significant effect on the biological activity of the protein: most proteins can tolerate at least a few amino acid changes without noticeable effect on their ability to function in the cell, but changes to

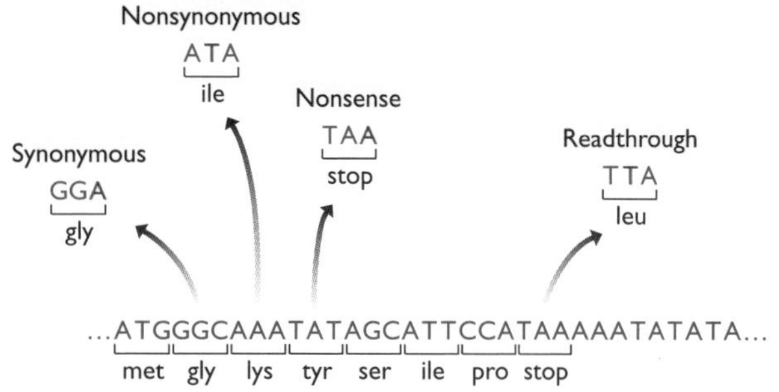

Figure 13.11 Effects of point mutations on the coding region of a gene.

Four different effects of point mutations are shown, as described in the text. The readthrough mutation results in the gene being extended beyond the end of the sequence shown here, the leucine codon created by the mutation being followed by AAA = lys, TAT = tyr and ATA = ile. See *Figure 10.7*, p. 237 for the genetic code.

some amino acids, such as those at the active site of an enzyme, have a greater impact. A nonsynonymous change is also called a **missense** mutation.

- The mutation may convert a codon that specifies an amino acid into a termination codon. This is a **nonsense** mutation and it results in a shortened protein because translation of the mRNA stops at this new termination codon rather than proceeding to the correct termination codon which is further downstream. The effect that this has on the protein activity depends on how much of the polypeptide is lost: usually the effect is drastic and the protein is nonfunctional.
- The mutation could convert a termination codon into one specifying an amino acid, resulting in **readthrough** of the stop signal so the protein is extended by an additional series of amino acids at its C-terminus. Most proteins can tolerate short extensions without an effect on function, but longer extensions might interfere with folding of the protein and so result in reduced activity.

Deletion and insertion mutations also have distinct effects on the coding capabilities of genes (*Figure 13.12*). If the number of deleted or inserted nucleotides is three or a multiple of three then one or more codons are removed or added, the resulting loss or gain of amino acids having varying effects on the function of the encoded protein. Deletions or insertions of this type are often inconsequential but will have an impact if, for example, amino acids involved in an enzyme's active site are lost, or if an insertion disrupts an important secondary structure in the protein. On the other hand, if the number of deleted or inserted nucleotides is not three or a multiple of three then a **frameshift** results, all of the codons downstream of the mutation being taken from a different reading frame from that used in the unmutated gene. This usually has a significant effect on the protein function, because a greater or lesser part of the mutated polypeptide has a completely different sequence to the normal polypeptide.

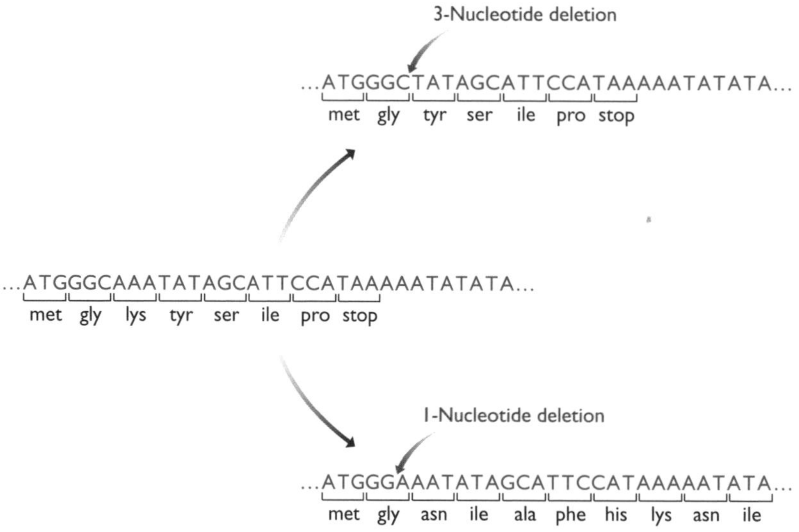

Figure 13.12 Deletion mutations.

In the top sequence three nucleotides comprising a single codon are deleted. This shortens the resulting protein product by one amino acid but does not affect the rest of its sequence. In the lower section, a single nucleotide is deleted. This results in a frameshift so all the codons downstream of the deletion are changed, including the termination codon which is now read through. See *Figure 10.7*, p. 237, for the genetic code. Note that if a three-nucleotide deletion removes parts of adjacent nucleotides then the result is more complicated than shown here. Consider, for example, deletion of the trinucleotide GCA from the sequence ...ATGGGCAAATAT... coding for Met–Gly–Lys–Tyr. The new sequence is ...ATGGAATAT..., coding for Met–Glu–Tyr. Two amino acids have been replaced by a single, different one.

It is less easy to make generalizations about the effects of mutations that occur outside of the coding regions of the genome. Any protein binding site is susceptible to point, insertion or deletion mutations that change the identity or relative positioning of nucleotides involved in the DNA–protein interaction. These mutations therefore have the potential to inactivate promoters or regulatory sequences, with predictable consequences for gene expression (*Figure 13.13*; Sections 8.2 and 8.3). Origins of replication could conceivably be made nonfunctional by mutations that change, delete or disrupt sequences recognized by the relevant binding proteins (Section 12.3.1) but these possibilities are not well-documented. There is also little information about the potential impact on gene expression of mutations that affect nucleosome positioning (Section 8.1.1).

One area that has been better researched concerns mutations that occur in introns or at intron–exon boundaries. In these regions, single point mutations will be important if they change nucleotides involved in the RNA–protein and RNA–RNA interactions that occur during splicing of different types of intron (Sections 9.2.3 and 9.3.3). For example, mutation of either the G or T in the DNA copy of the 5′ splice site of a GU–AG intron, or of the A or G at the 3′ splice site, will disrupt splicing because the correct intron–exon boundary will no longer be recognized. This may mean that the intron is not removed from the pre-mRNA, but it is more likely that a cryptic splice site (see p. 215) will be used as an alternative. It is also possible for a mutation within an intron or an exon to create a new cryptic site that is preferred over a genuine splice site that is not itself mutated. Both types of event have the same result, relocation of the active splice site, leading to aberrant splicing. This might delete or add new amino acids into the resulting protein, or lead to a frameshift. Several versions of the blood disease β-thalassemia are caused by mutations that lead to cryptic splice site selection during processing of β-globin transcripts.

The effects of mutations on multicellular organisms

Now we turn to the indirect effects that mutations have on organisms, beginning with multicellular, diploid eukaryotes such as humans. The first issue to consider is the relative importance of the same mutation in a somatic cell compared with a germ cell. Because somatic cells do not pass copies of their genomes to the next generation, a somatic cell mutation is important only for the organism in which it occurs: it has no potential evolutionary impact. In fact, most somatic cell mutations have no significant effect, even if they result in cell death, because there are many other identical cells in the same tissue and the loss of one cell is immaterial. An exception is when a mutation causes a somatic cell to malfunction in a way that is harmful to the organism, for instance by inducing tumor formation or other cancerous activity.

Mutations in germ cells are more important because they can be transmitted to members of the next generation and will then be present in all the cells of any individual who inherits the mutation. Most mutations, including all silent ones as well as many in coding regions, will still not change the phenotype of the organism in any significant way. Those that do have an effect can be divided into two categories:

- **Loss-of-function** is the normal result of a mutation that reduces or abolishes a protein activity. Most loss-of-function mutations are recessive, because in a heterozygote the second chromosome copy carries an unmutated version of the gene coding for a fully functional protein whose presence compensates for the effect of the mutation (*Figure 13.14*). There are some exceptions where a loss-of-function mutation is dominant, one example being **haploinsufficiency**, where the organism is unable to tolerate the approximately 50% reduction in protein activity suffered by the heterozygote. In humans this is the explanation of a few genetic diseases, including Marfan syndrome which results from a mutation in the gene for the connective tissue protein called fibrillin.

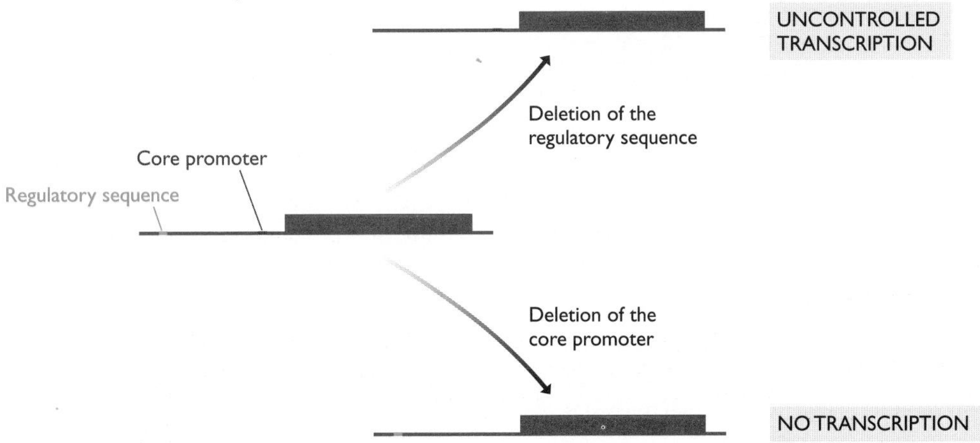

Figure 13.13 Two possible effects of deletion mutations in the region upstream of a gene.

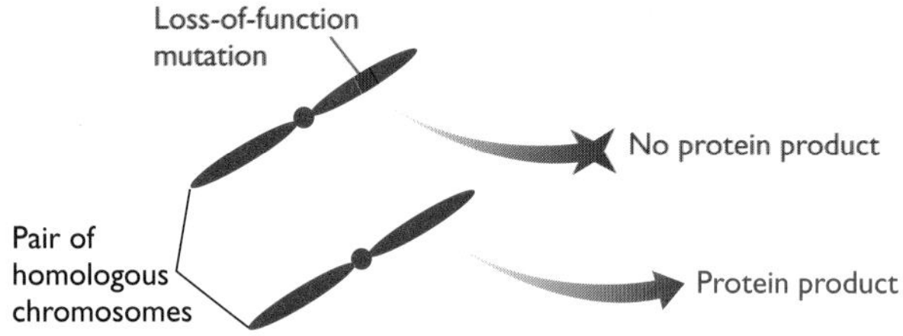

Figure 13.14 A loss-of-function mutation is usually recessive because a functional version of the gene is present on the second chromosome copy.

- **Gain-of-function** mutations are much less common. The mutation must be one that confers an abnormal activity on a protein. Many gain-of-function mutations are in regulatory sequences rather than coding regions, and can therefore have a number of consequences. For example, a mutation might lead to one or more genes being expressed in the wrong tissues, these tissues gaining functions that they normally lack. Alternatively the mutation could lead to over-expression of one or more genes involved in control of the cell cycle, thus leading to uncontrolled cell division and hence to cancer. Because of their nature, gain-of-function mutations are usually dominant.

There are added complications when considering the effects of mutations on the phenotypes of multicellular organisms. Not all mutations have an immediate effect on the organism: some are **delayed-onset** and only confer an altered phenotype later in the individual's life. Others display **nonpenetrance** in some individuals, never being expressed even though the individual has a dominant mutation or is a homozygous recessive. With humans, these factors complicate attempts to map disease-causing mutations by pedigree analysis (Section 2.3.2), because they introduce uncertainties regarding which members of a pedigree carry a mutant allele.

The effects of mutations on microorganisms

Mutations in microbes such as bacteria and yeast can also be described as loss-of-function or gain-of-function, but with microorganisms this is neither the normal nor most useful classification scheme. Instead, a more detailed description of the phenotype is usually attempted, based on the growth properties of mutated cells in various culture media. This enables most mutations to be assigned to one of four categories:

- **Auxotrophs** are cells that will only grow when provided with a nutrient not required by the unmutated organism. For example, *E. coli* normally makes its own tryptophan, courtesy of the enzymes coded by the five genes in the tryptophan operon (*Figure 6.14B*, p. 133). If one of these genes is mutated in such a way that its protein product is inactivated, then the cell is no longer able to make tryptophan and so is a tryptophan auxotroph. It cannot survive on a medium that lacks tryptophan, being able to grow only when this amino acid is provided as a nutrient (*Figure 13.15*). Unmutated bacteria, which do not require extra supplements in their growth media, are called **prototrophs**.

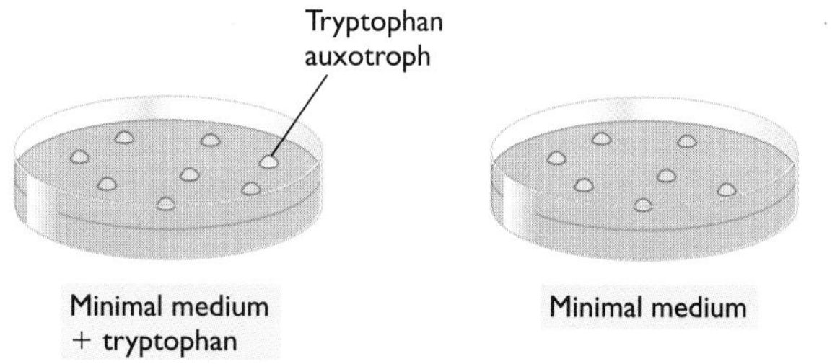

Figure 13.15 A tryptophan auxotrophic mutant.

Two Petri-dish cultures are shown. Both contain **minimal medium**, which contains just the basic nutritional requirements for bacterial growth (nitrogen, carbon and energy sources, plus some salts). The medium on the left is supplemented with tryptophan but the medium on the right is not. Unmutated bacteria, plus tryptophan auxotrophs, can grow on the plate on the left, the auxotrophs growing because the medium supplies the tryptophan that they cannot make themselves. Tryptophan auxotrophs cannot grow on the plate on the right, because this does not contain tryptophan. To identify a tryptophan auxotroph, colonies are first grown on the minimal medium + tryptophan plate and then transferred to the minimal medium plate by **replica plating**. In this procedure, a sterile felt pad is pressed onto the colonies on the minimal medium + tryptophan plate, carefully removed, and then pressed onto the surface of the minimal medium plate, transferring a few bacteria from one plate to the other. After incubation, colonies appear on the minimal medium plate in the same relative positions as on the plate containing tryptophan, except for the tryptophan auxotrophs which do not grow. These colonies can therefore be identified and samples of the tryptophan auxotrophic bacteria recovered from the minimal medium + tryptophan plate.

- **Conditional-lethal** mutants are unable to withstand certain growth conditions: under **permissive conditions** they appear to be entirely normal but when transferred to **restrictive conditions** the mutant phenotype is seen. **Temperature-sensitive** mutants are typical examples of conditional-lethals. Temperature-sensitive mutants behave like wild-type cells at low temperatures but exhibit their mutant phenotype when the temperature is raised above a certain threshold value, which is different for each mutant. Usually this is because the mutation reduces the stability of a protein, so the protein becomes unfolded and hence inactive when the temperature is raised.
- **Inhibitor-resistant** mutants are able to resist the toxic effects of an antibiotic or other type of inhibitor. There are various molecular explanations for this type of mutant. In some cases the mutation changes the structure of the protein that is targeted by the inhibitor, so the latter can no longer bind to the protein and interfere with its function. This is the basis of streptomycin-resistance in *E. coli*, which results

from a change in the structure of ribosomal protein S12. Another possibility is that the mutation changes the properties of a protein responsible for transporting the inhibitor into the cell, this often being the way in which resistance to toxic metals is acquired.

- **Regulatory** mutants have defects in promoters and other regulatory sequences. This category includes **constitutive** mutants, which continually express genes that are normally switched on and off under different conditions. For example, a mutation in the operator sequence of the lactose operon (Section 8.3.1) can prevent the repressor from binding and so results in the lactose operon being expressed all the time, even when lactose is absent and the genes should be switched off (*Figure 13.16*).

In addition to these four categories, many mutations are lethal and so result in death of the mutant cell, and others have no effect. The latter are less common in microorganisms than in higher eukaryotes, because most microbial genomes are relatively compact with little noncoding DNA. Mutations can also be **leaky**, meaning that a less extreme form of the mutant phenotype is expressed. For example, a leaky version of the tryptophan auxotroph illustrated in *Figure 13.15* would grow slowly on minimal medium, rather than not growing at all.

13.1.3 Hypermutation and the possibility of programmed mutations

Is it possible for cells to utilize mutations in a positive fashion, either by increasing the rate at which mutations appear in their genomes, or by directing mutations towards specific genes? Both types of event might appear, at first glance, to go against the accepted wisdom that mutations occur randomly but, as we shall see, **hypermutation** and **programmed mutations** are possible without contravening this dogma.

Hypermutation occurs when a cell causes the rate at which mutations occur in its genome to increase. This might appear to be an illogical thing to do as it is difficult to imagine situations where an increased mutation rate would be beneficial, and with the best studied example of hypermutation we do in fact know rather more about the process itself than about the reasons why it occurs. This is the **SOS response** of *E. coli*, which is induced when the genome of the bacterium suffers extensive damage, typically as a result of exposure to UV radiation or chemical mutagens. The SOS response enables the cell to replicate its DNA even though the template polynucleotides contain AP sites and/or cyclobutyl dimers and other photoproducts that would normally block or at least delay the replication complex. This requires construction of a **mutasome**, comprising several copies of the RecA protein and of the UmuD$'_2$C complex, the latter a trimer made up of two UmuD′ proteins and one copy of UmuC (Goodman, 1998). The RecA proteins coat the DNA in the region adjacent to the damage position and the UmuD$'_2$C complexes bind to the attached RecA proteins. Somehow this enables DNA polymerase III to proceed past the damaged site and continue replicating the DNA.

The SOS response is primarily looked on as the last best chance that the bacterium has to replicate its DNA and hence survive under adverse conditions. However, the price of survival is an increased mutation rate because the mutasome does not repair damage, it simply allows a damaged region of a polynucleotide to be replicated. When it encounters a damaged position in the template DNA the polymerase selects a nucleotide more or less at random, though with some preference for placing an A opposite an AP site: in effect the error rate of the replication process becomes increased. It has been suggested that this increased mutation rate is the purpose of the SOS response, mutation for some reason or other being an advantageous response to DNA damage, but this idea remains controversial (Walker, 1995).

Less controversial is the way in which vertebrates, including humans, are able to increase the mutation rate at one specific gene in one type of cell. This phenomenon takes us back to the way in which immunoglobulin diversity is generated, which we have already touched upon in Section 11.2.1 when we examined the genome rearrangements that result in joining of the V, D, J and H segments of the immunoglobulin heavy and light genes (see *Figure 11.14*, p. 280). Additional diversity is produced by hypermutation of the V gene segments, after assembly of the intact immunoglobulin gene (*Figure 13.17*), the mutation rate for these segments being 6–7 orders of magnitude greater than the background mutation rate experienced by the rest of the genome (Shannon and Weigert, 1998). This enhanced mutation rate appears to result from the unusual behavior of the mismatch repair system which normally corrects replication errors. At all other positions within the genome, the mismatch repair system corrects

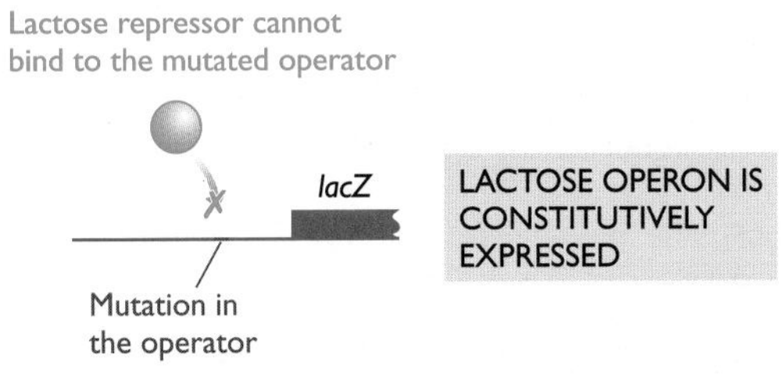

Figure 13.16 Effect of a constitutive mutation in the lactose operator.

The operator sequence has been altered by a mutation and the lactose repressor can no longer bind to it. The result is that the lactose operon is transcribed all the time, even when lactose is absent from the medium. This is not the only way in which a constitutive *lac* mutant can arise. For example, the mutation could be in the gene coding for the lactose repressor, changing the tertiary structure of the repressor protein so that its DNA-binding motif is disrupted and it can no longer recognize the operator sequence, even when the latter is unmutated. See *Figure 8.15*, p. 187, for more details about the lactose repressor and its regulatory effect on expression of the lactose operon.

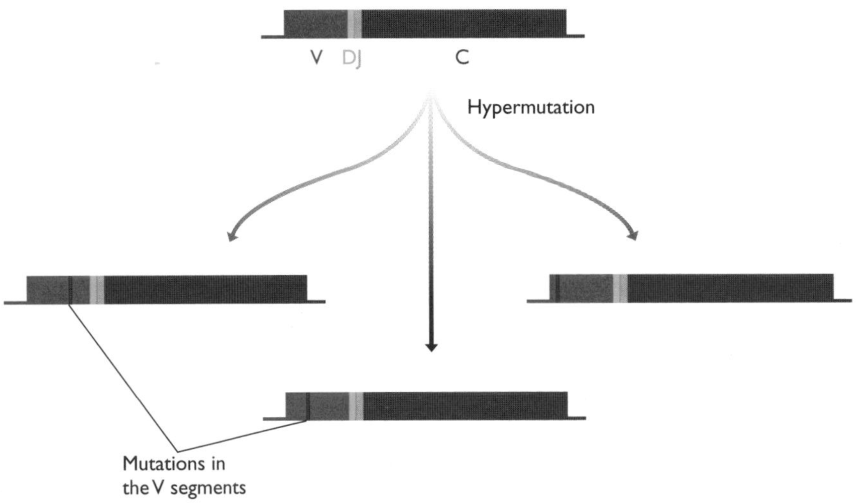

Figure 13.17 Hypermutation of the V gene segment of an intact immunoglobulin gene.

See *Figure 11.14*, p. 280, for a description of the events leading to assembly of an immunoglobulin gene.

errors of replication by searching for mismatches and replacing the nucleotide in the daughter strand, this being the strand that has just been synthesized and so contains the error (see Section 13.1.4). At V gene segments, the repair system changes the nucleotide in the parent strand, and so stabilizes the mutation rather than correcting it (Cascalho *et al.*, 1998). The mechanism by which this is achieved has not yet been described.

An apparent increase in mutation rate arising from modifications to the normal DNA repair process does not contradict the dogma regarding the randomness of mutations. Where problems have arisen is with reports, dating back to 1988 (Cairns *et al.*, 1988), suggesting that *E. coli* is able to direct mutations towards genes whose mutation would be advantageous under the environmental conditions that the bacterium is encountering. The original experiments involved a strain of *E. coli* that has a frameshift mutation in the lactose operon, inactivating the proteins needed for utilization of this sugar (Research Briefing 13.1). The bacteria were spread on an agar medium in which the only carbon source was lactose. This meant that a cell could grow and divide only if a second mutation occurred in the lactose operon, restoring the correct reading frame and therefore allowing the lactose enzymes to be synthesized. Mutations with this effect appeared to occur significantly more frequently than expected, and at a rate that was greater than mutations in other parts of the genomes of these *E. coli* cells.

These experiments suggested that bacteria can programme mutations according to the selective pressures that they are placed under. In other words, the environment can directly affect the phenotype of the organism, as suggested by Lamarck, rather than operating through the random processes postulated by Darwin. With the implications being so radical it is not surprising that the experiments have been debated at length with numerous attempts to discover flaws in their design or alternative explanations for the results. At present, the possibility that the enhanced mutation rate is not programmed towards specific genes but occurs throughout the genome is being re-examined (Bridges, 1997), and models based on gene amplification rather than selective mutation are being tested (Andersson *et al.*, 1998).

13.1.4 DNA repair

In view of the thousands of damage events that genomes suffer every day, coupled with the errors that occur when the genome replicates, it is essential that cells possess efficient repair systems. Without these repair systems a genome would not be able to maintain its essential cellular functions for more that a few hours before key genes became inactivated by DNA damage. Similarly, cell lineages would accumulate replication errors at such a rate that their genomes would become dysfunctional after a few cell divisions.

Most cells possess five different categories of DNA repair system:

- **Direct repair systems**, as the name suggests, act directly on damaged nucleotides, converting each one back to its original structure.
- **Base excision repair** involves removal of a damaged nucleotide base, excision of a short piece of the polynucleotide around the AP site thus created, and resynthesis with a DNA polymerase.
- **Nucleotide excision repair** is similar to base excision repair but is not preceded by removal of a damaged base and can act on more substantially damaged areas of DNA.
- **Mismatch repair** corrects errors of replication, again by excising a stretch of single-stranded DNA containing the offending nucleotide and then repairing the resulting gap.

RESEARCH

13.1

BRIEFING

Adaptive mutations?

In 1988 startling results were published suggesting that under some circumstances *Escherichia coli* bacteria are able to mutate in a directed way that enables cells to adapt to an environmental stress.

The randomness of mutations is an important concept in biology because it is a requirement of the Darwinian view of evolution, which holds that changes in the characteristics of an organism occur by chance and are not influenced by the environment in which the organism is placed. Beneficial changes are positively selected and harmful ones are negatively selected (see Box 15.5, p. 408). In contrast, the Lamarckian theory of evolution, which biologists rejected well over a century ago, states that organisms acquire changes that enable them to adapt to their environment. The Darwinian view requires that mutations occur at random, whereas Lamarckian evolution demands that adaptive mutations occur in response to the environment.

Random mutations in E. coli

The randomness of mutations in bacteria was first demonstrated by Luria and Delbrück in 1943. They grew a series of *E. coli* cultures in different flasks and then added T1 bacteriophages to each one. Most of the bacteria were killed by the phage, but a few T1-resistant mutants were able to survive. These were identified by plating samples from each culture, soon after T1 infection, onto an agar medium. If mutations leading to T1 resistance occurred randomly in the cultures before the bacteriophages were added, each culture would contain different numbers of resistant mutants, the numbers depending on how early during the growth period the first mutant cells arose. Those that arose early would divide many times to give rise to a large number of resistant progeny in the culture at the end of the growth period, whereas those that arose later would give rise to just a few progeny. Some cultures would therefore contain many T1-resistant cells and others would contain just a few. Alternatively, if resistant bacteria arose by adaptive mutation only when the T1 phage were added, then all cultures would have similar numbers of mutants (see figure opposite).

Luria and Delbrück found that each of their cultures contained a different number of T1-resistant bacteria; thus, they concluded that mutations occur randomly and not in response to T1 phage.

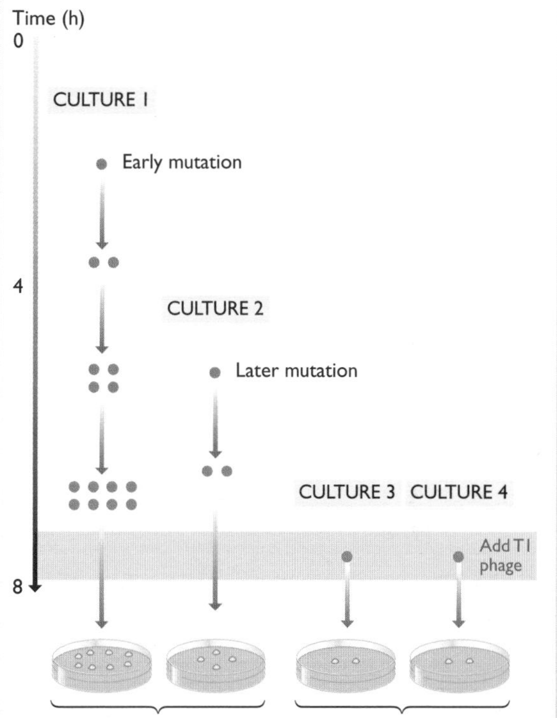

Adaptive mutations in E. coli

The possibility that Luria and Delbrück's conclusion might not be universally true for *E. coli* mutations was first suggested by studies of an *E. coli* strain that carries a nonsense mutation in its *lacZ* gene. The presence of the termination codon in *lacZ* means that these cells are unable to synthesize functional β-galactosidase enzymes and so cannot use lactose as a carbon and energy source – they are therefore lactose auxotrophs. This is not necessarily a permanent situation because a cell could undergo a mutation that converts the termination codon back into one specifying an amino acid. These new mutants would be able to make β-galactosidase and use any lactose that is available. According to Luria and Delbrück's results, such mutations should occur at random and should not be influenced by the presence of lactose in the medium. The results of Cairns *et al.* (1988) showed that when the lactose auxotrophs were plated onto a minimal medium containing lactose as the only sugar – circumstances that require that the bacteria must mutate into lactose prototrophs in order to survive – then the number of lactose prototrophs that arose was significantly higher than that expected if mutations occurred randomly. In other words, some cells mutated adaptively and acquired the specific DNA sequence change needed to withstand the selective pressure.

Since 1988, a number of examples of what appear to be adaptive mutations have been published, but the notion that bacteria, and possibly other organisms, can program mutations in response to environmental stress is by no means accepted by the scientific community. It is quite possible that these mutations will eventually be disproved or shown to have an orthodox basis. However, until this happens we are left with the tantalizing possibility that even at this very fundamental level our knowledge about genomes might be far from complete.

Reference

Cairns J, Overbaugh J and Miller S (1988) The origin of mutants. *Nature*, **335**, 142–145.

Box 13.2: DNA repair and human disease

The importance of DNA repair is emphasized by the number and severity of inherited human diseases which have been linked with defects in one of the repair processes. One of the best characterized of these is xeroderma pigmentosum, which results from a mutation in any one of several genes for proteins involved in nucleotide excision repair. The disease symptoms include hypersensitivity to UV radiation, so patients suffer more mutations on exposure to sunlight, this often leading to skin cancer (Lehmann, 1995). Two other diseases, Cockayne syndrome and trichothiodystrophy, are also caused by defects in nucleotide excision repair, but these are more complex disorders which, although not involving cancer, usually include problems with both the skin and nervous system.

A few diseases have been linked with defects in the transcription-coupled component of nucleotide excison repair. These include breast and ovarian cancers, the BRCA1 gene that confers susceptibility to these cancers coding for a protein that has been implicated, at least indirectly, with transcription-coupled repair (Gowen *et al.*, 1998). A deficiency in transcription-coupled repair has also been identified in humans suffering from the cancer susceptibility syndrome called HNPCC (hereditary nonpolyposis colorectal cancer; Mellon *et al.*, 1996) though this disease was originally identified as a defect in mismatch repair (Kolodner, 1995). Other diseases that seem to involve a breakdown in some aspect of DNA repair, but whose direct causes have not been uncovered, are ataxia telangiectasia, whose symptoms include sensitivity to ionizing radiation, Bloom's syndrome, which is probably due to inactivation of DNA ligase genes, and Fanconi's anemia, which confers sensitivity to chemicals that cause crosslinks in DNA.

- **Recombination repair** is used to mend double-strand breaks.

In this section we will look at the first four types of repair system, leaving recombination repair until Box 13.4, this last system being easier to understand after we have dealt with the more general principles of recombination.

Direct repair systems fill in nicks and correct some types of nucleotide modification

Relatively few forms of DNA damage can be repaired without excision of nucleotides. Those that can be repaired by direct methods are as follows:

- **Nicks** can be repaired by a DNA ligase if all that has happened is that a phosphodiester bond has been broken, without damage to the 5′-phosphate and 3′-hydroxyl groups of the nucleotides either side of the nick (*Figure 13.18*). This is often the case with nicks resulting from the effects of ionizing radiation.

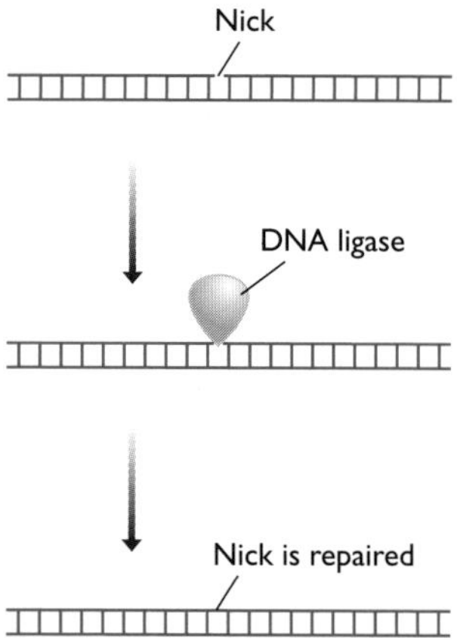

Figure 13.18 Repair of a nick by DNA ligase.

- Some forms of **alkylation** damage are directly reversible by enzymes that transfer the alkyl group from the nucleotide to their own polypeptide chains. Enzymes capable of doing this are known in many different organisms and include the **Ada enzyme** of *E. coli*, which is involved in an adaptive process that this bacterium is able to activate in response to DNA damage. Ada removes alkyl groups attached to the oxygen groups at positions 4 and 6 of thymine and guanine, respectively, and can also repair phosphodiester bonds that have become methylated. Other alkylation repair enzymes have more restricted specificities, an example being human **MGMT** (O^6-methylguanine-DNA methyltransferase) which, as its name suggests, only removes alkyl groups from position 6 of guanine.
- **Cyclobutyl dimers** are repaired by a light-dependent direct system called **photoreactivation**. In *E. coli*, the process involves the enzyme called **DNA photolyase** (more correctly named deoxyribodipyrimidine photolyase). When stimulated by light with a wavelength between 300 and 500 nm the enzyme binds to cyclobutyl dimers and converts them back to the original monomeric nucleotides. Photoreactivation is a widespread but not universal type of repair: it is known in many but not all bacteria and also in quite a few eukaryotes, including some vertebrates, but is absent in humans. A similar type of photoreactivation involves the **(6–4) photoproduct photolyase** and results in repair of (6–4) lesions. Neither *E. coli* nor humans have this enzyme but it is possessed by a variety of other organisms.

Base excision repairs many types of damaged nucleotide

Base excision is the least complex of the various repair systems that involve removal of a damaged nucleotide followed by resynthesis of DNA to span the resulting gap. It is used to repair many modified nucleotides that have

Box 13.3: Cell cycle checkpoints for monitoring DNA damage

In a multicellular organism, the death of a single somatic cell as a result of DNA damage is usually less dangerous than allowing that cell to replicate its mutated DNA and possibly give rise to a tumor or other cancerous growth. Eukaryotic cells therefore monitor their genomes for damage, principally at the checkpoints immediately before the entry into the S and M phases of the cell cycle (Section 12.4.1; Russell, 1998). These checkpoints ensure that a damaged genome is not replicated, which would lead to mutations being perpetuated in the cell lineage, and prevent problems with the distribution of chromosomes to daughter cells, which might occur if one or more chromosomes has extensive DNA damage. A cell that fails to pass one or other of the checkpoint tests might undergo cell cycle arrest, permanently or until its DNA is repaired, or it might be forced into programmed cell death or **apoptosis** (Chernova et *al.*, 1995; Enoch and Norbury, 1995).

In mammals, a central player in induction of cell cycle arrest and apoptosis is the protein called p53. This is classified as a tumor-suppressor protein, because when this protein is defective, cells with damaged genomes can avoid the cell cycle checkpoints and possibly proliferate into a cancer. p53 is a sequence-specific DNA-binding protein that activates a number of genes thought to be directly responsible for arrest and apoptosis, and also represses expression of others that must be switched off to facilitate these processes. A second protein that might play a regulatory role at the checkpoints is the product of the human *ATM* gene which, when defective, gives rise to ataxia telangiectasia, one of the diseases associated with a deficiency in DNA repair (see Box 13.2).

suffered relatively minor damage to their bases. The process is initiated by a **DNA glycolyase** which cleaves the β-*N*-glycosidic bond between a damaged base and the sugar component of the nucleotide (*Figure 13.19A*). Each DNA glycolyase has a limited specificity, the specificities of the glycolyases possessed by a cell determining the range of damaged nucleotides that can be repaired by the base excision pathway. Most organisms are able to deal with deaminated bases such as uracil (deaminated cytosine) and hypoxanthine (deaminated adenine), oxidation products such as 5-hydroxycytosine and thymine glycol, and methylated bases such as 7-methylguanine and 2-methylcytosine (Seeberg *et al.*, 1995). Other DNA glycolyases remove normal bases as part of the mismatch repair system.

DNA glycolyase removes a damaged base by 'flipping' the structure to a position outside of the helix and then detaching it from the polynucleotide (Kunkel and Wilson, 1996; Roberts and Cheng, 1998). This creates an AP or baseless site (see *Figure 13.10*, p. 341) which is converted into a single nucleotide gap in the second step of the repair pathway (*Figure 13.19B*). This step can be carried out in a variety of ways. The standard method makes use of an **AP endonuclease**, such as exonuclease III or endonuclease IV of *E. coli*, which cuts the phosphodiester bond on the 5′ side of the AP site. Some AP endonucleases can also remove the sugar from the AP site, this being all that remains of the damaged nucleotide, but others lack this ability and so work in conjunction with a separate **phosphodiesterase**. An alternative pathway for converting the AP site into a gap utilizes the endonuclease activity possessed by some DNA glycolyases, which can make a cut at the 3′ side of the AP site, probably at the same time that the damaged base is removed, followed again by removal of the sugar by a phosphodiesterase.

The single nucleotide gap is filled by a DNA polymerase, using the undamaged base in the other strand of the DNA molecule to ensure that the correct nucleotide is inserted. In *E. coli* the gap is filled by DNA polymerase I and in mammals by DNA polymerase β (see *Table 12.2*, p. 313; Sobol *et al.*, 1996). Yeast seems to be unusual in that it uses its main DNA replicating enzyme, DNA polymerase δ, for this purpose (Seeberg *et al.*, 1995). After gap filling, the final phosphodiester bond is put in place by a DNA ligase.

Nucleotide excision repair is used to correct more extensive types of damage

Nucleotide excision repair has a much broader specificity than the base excision system and is able to deal with more extreme forms of damage such as intrastrand crosslinks and bases that have become modified by attachment of large chemical groups. It is also able to correct cyclobutyl dimers by a **dark repair** process, providing those organisms that do not have the photoreactivation system with a means of repairing these dimers.

In nucleotide excision repair, a segment of single-stranded DNA containing the damaged nucleotide(s) is excised and replaced with new DNA. The process is therefore similar to base excision repair except that it is not preceded by selective base removal and a longer stretch of polynucleotide is excised. The best studied example of nucleotide excision repair is the **short patch** process of *E. coli*, so called because the region of polynucleotide that is excised and subsequently 'patched' is relatively short, usually 12 nucleotides in length.

Short patch repair is initiated by a multienzyme complex called the **UvrABC endonuclease**, sometimes also referred to as the 'excinuclease'. In the first stage of the process a trimer comprising two UvrA proteins and one copy of UvrB attaches to the DNA at the damaged site. How the site is recognized is not known but the broad specificity of the process indicates that individual types of damage are not directly detected and that the complex must search for a more general attribute of DNA damage such as distortion of the double helix. UvrA may be the part of the complex most involved in damage location because it dissociates once the site has been found and plays no further part in the repair process. Departure of

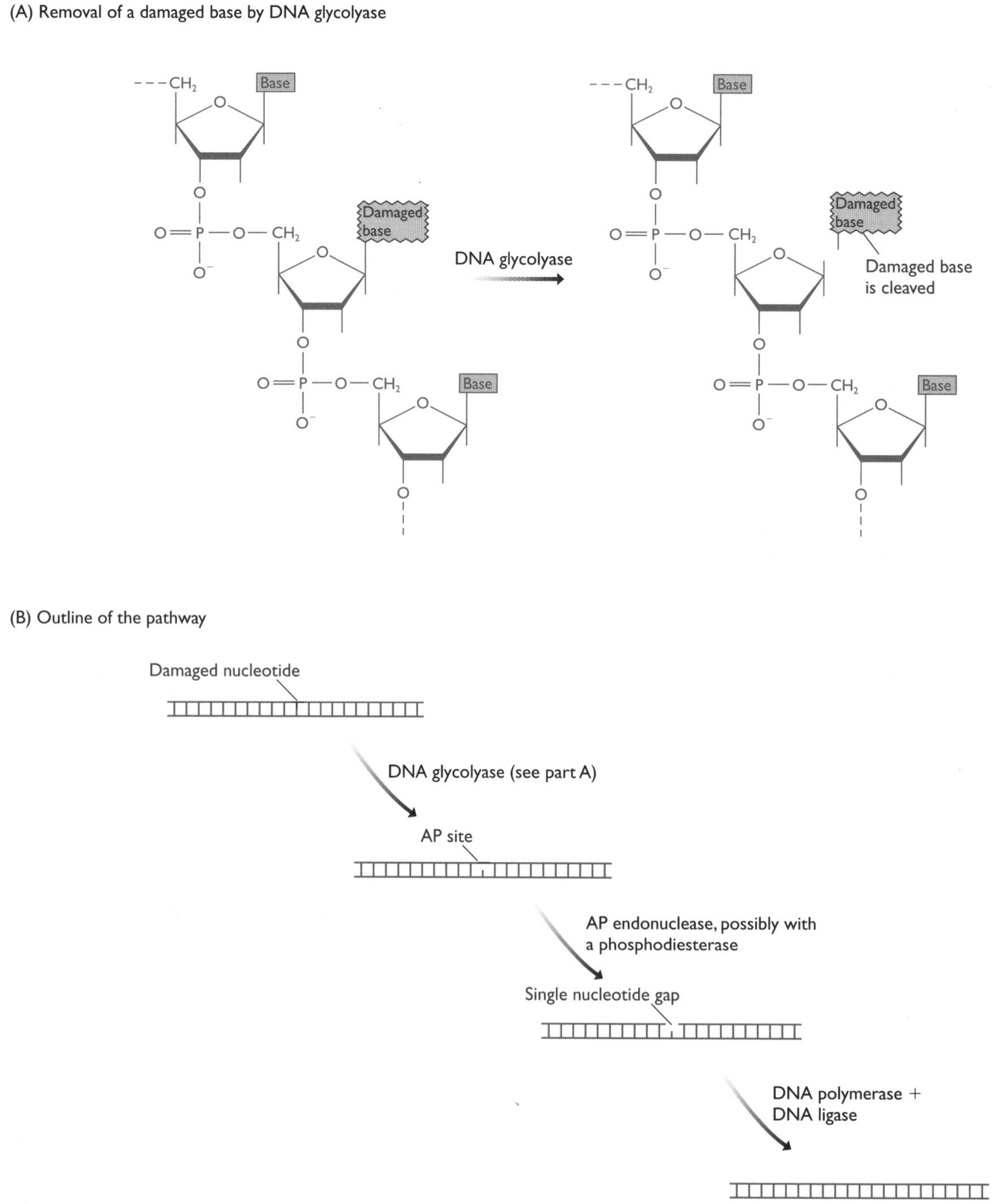

Figure 13.19 Base excision repair.

(A) Excision of a damaged nucleotide by a DNA glycolyase. (B) Schematic representation of the base excision repair pathway. Alternative versions of the pathway are described in the text.

UvrA allows UvrC to bind (*Figure 13.20*), forming a UvrBC dimer that cuts the polynucleotide either side of the damaged site. The first cut is made by UvrB at the fifth phosphodiester bond downstream of the damaged nucleotide, and the second cut is made by UvrC at the eighth phosphodiester bond upstream, resulting in the 12 nucleotide excision, though there is some variability, especially in the position of the UvrB cut site. The excised segment is then removed, usually as an intact oligonucleotide, by DNA helicase II which presumably detaches the segment by breaking the base pairs holding it to the second strand. UvrC also detaches at this stage, but UvrB remains in place and bridges the gap produced by the excision, possibly to prevent the single-stranded region

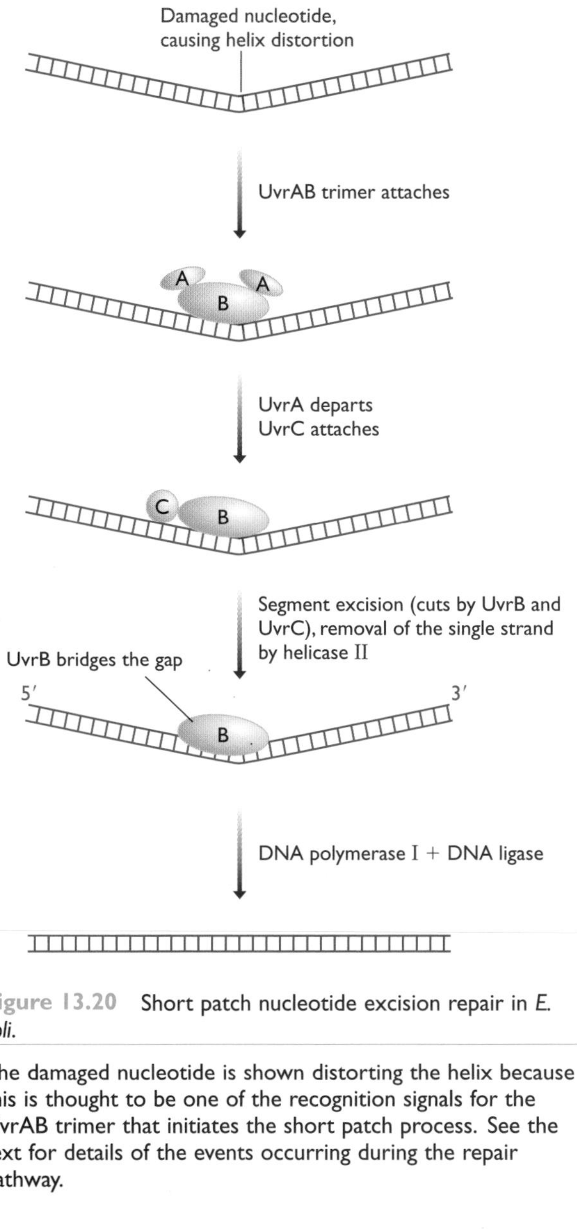

Figure 13.20 Short patch nucleotide excision repair in *E. coli.*

The damaged nucleotide is shown distorting the helix because this is thought to be one of the recognition signals for the UvrAB trimer that initiates the short patch process. See the text for details of the events occurring during the repair pathway.

that has been exposed from base-pairing with itself, possibly to prevent this strand from becoming damaged, or possibly to direct the DNA polymerase to the site that needs to be repaired. As in base excision repair, the gap is filled by DNA polymerase I and the last phosphodiester bond is synthesized by DNA ligase.

E. coli also has a **long patch** nucleotide excision repair system that involves Uvr proteins but differs in that the piece of DNA that is excised can be anything up to 2 kb in length. Long patch repair has been less well studied and the process is not understood in detail, but it is presumed to work on more extensive forms of damage, possibly regions where groups of nucleotides, rather than just single bases, have become modified. The eukaryotic nucleotide excision repair process is also called 'long patch' but results in replacement of only 24–29 nucleotides of DNA. In fact, there is no 'short patch' system in eukaryotes and the name is used to distinguish the process from base excision repair. The system is more complex than in *E. coli* and the relevant enzymes do not seem to be homologs of the Uvr proteins. In humans at least 16 proteins are involved, with the downstream cut being made at the same position as in *E. coli* – the fifth phosphodiester bond – but with a more distant upstream cut, resulting in the longer excision. Both cuts are made by endonucleases that attack single-stranded DNA specifically at its junction with a double-stranded region, indicating that before the cuts are made the DNA around the damage site has been melted, presumably by a helicase (*Figure 13.21*). This activity is provided at least in part by TFIIH, one of the components of the RNA polymerase II initiation complex (see *Table 8.3*, p. 184). At first it was assumed that TFIIH simply has a dual role in the cell, functioning separately in both transcription and repair, but now it is thought that there is a more direct link between the two processes (Lehmann, 1995; Svejstrup *et al.*, 1996). This view is supported by the discovery of **transcription-coupled repair**, which results in the template strands of genes being repaired more quickly than other parts of the eukaryotic genome, an observation that is entirely logical as these template strands contain the genome's biological information and maintaining their integrity is the highest priority for the repair systems.

Mismatch repair: correcting errors of replication

Each of the three repair systems that we have looked at so far recognize and act upon DNA damage caused by mutagens. This means that they search for abnormal chemical structures such as modified nucleotides, cyclobutyl dimers and intrastrand crosslinks. They cannot correct mismatches resulting from errors in replication because the mismatched nucleotide is not abnormal in any way, it is simply an A, C, G or T that has been inserted at the wrong position. As these nucleotides look exactly like any other nucleotide the mismatch repair system that corrects replication errors has to detect not the mismatched nucleotide itself but the absence of base-pairing between the parent and daughter strands. Once it has found a mismatch, the repair system excises part of the daughter polynucleotide and fills in the gap, in a manner similar to what we have already seen with base and nucleotide excision repair.

The scheme described above leaves one important question unanswered. The repair must be made in the daughter polynucleotide because it is in this newly synthesized strand that the error has occurred: the parent polynucleotide has the correct sequence. How does the repair process know which strand is which? In *E. coli* the answer is that the daughter strand is, at this stage, under-methylated and therefore can be distinguished from the parent polynucleotide, which has a full complement of methyl groups. *E. coli* DNA is methylated because of the activities of the **DNA adenine methylase** (**Dam**), which converts adenines to 6-methyladenines in the sequence 5′-GATC-3′, and the **DNA cytosine methylase**

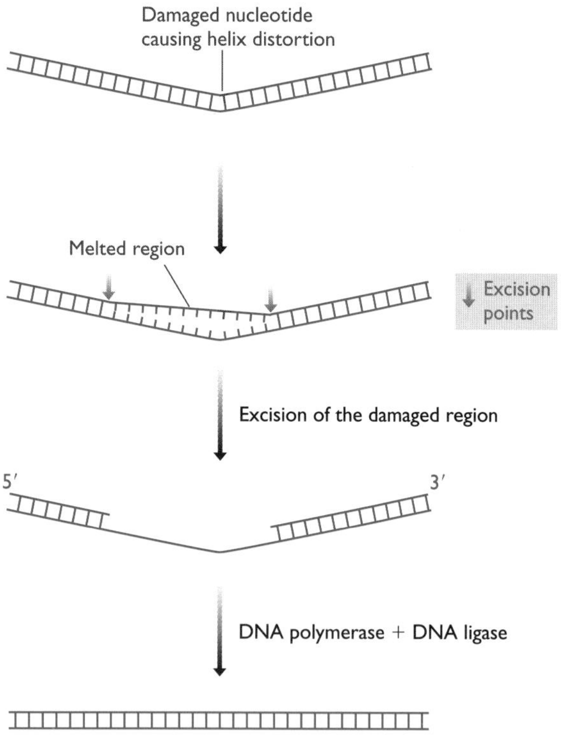

Figure 13.21 Outline of the events involved in nucleotide excision repair in eukaryotes.

The endonucleases that remove the damaged region make cuts specifically at the junction between single-stranded and double-stranded regions of a DNA molecule. The DNA is therefore thought to melt either side of the damaged nucleotide, as shown in the diagram, possibly due to the helicase activity of TFIIH.

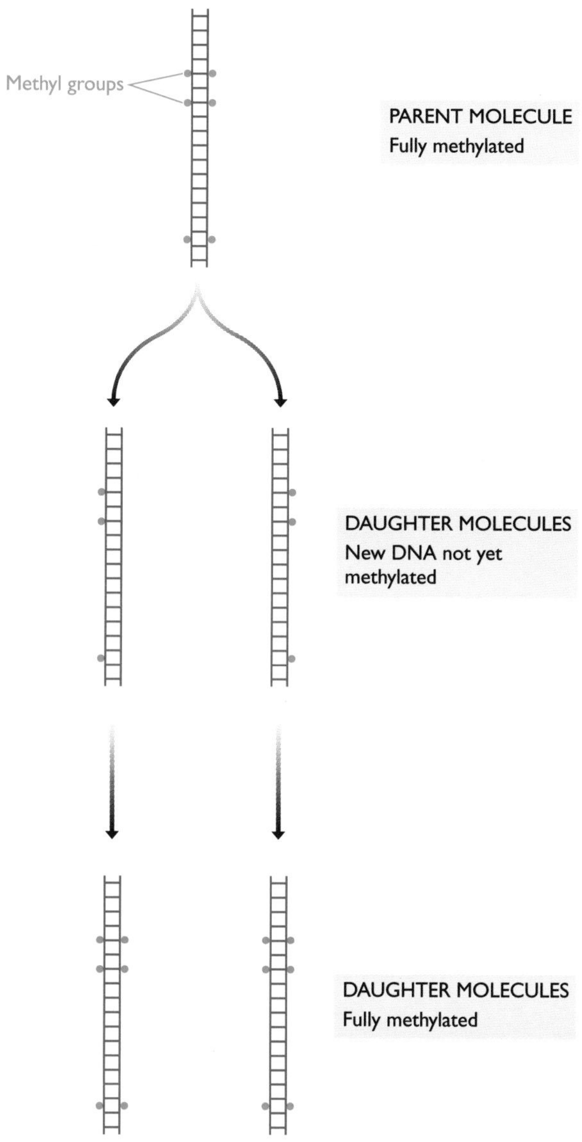

Figure 13.22 Methylation of newly-synthesized DNA in *E. coli* does not occur immediately after replication, providing a window of opportunity for the mismatch repair proteins to recognize the daughter strands and correct replication errors.

(**Dcm**), which converts cytosines to 5-methylcytosines in 5′-CCAGG-3′ and 5′-CCTGG-3′. These methylations are not mutagenic, the modified nucleotides having the same base-pairing properties as the unmodified versions. There is delay between DNA replication and methylation of the daughter strand, and it is during this window of opportunity that the repair system scans the DNA for mismatches and makes the required corrections in the undermethylated, daughter strand (*Figure 13.22*).

E. coli has at least three mismatch repair systems, called 'long patch', 'short patch and 'very short patch', the names indicating the relative lengths of the excised and resynthesized segments. The long patch system replaces up to a kb or more of DNA and requires the MutH, MutL and MutS proteins, as well as the DNA helicase II that we met during nucleotide excision repair. MutS recognizes the mismatch and MutH distinguishes the two strands by binding to unmethylated 5′-GATC-3′ sequences (*Figure 13.23*). The role of MutL is unclear but it might coordinate the activities of the two other proteins so that MutH binds to 5′-GATC-3′ sequences only in the vicinity of mismatch sites recognized by MutS. After binding, MutH cuts the phosphodiester bond immediately upstream of the G in the methylation sequence and DNA helicase II detaches the single strand. There does not appear to be an enzyme that cuts the strand downstream of the mismatch; instead the detached single-stranded region is degraded by an exonuclease that follows the helicase and continues beyond the mismatch site. The gap is then filled in by DNA polymerase I and DNA ligase.

Similar events are thought to occur during short and very short mismatch repair, the difference being the specificities of the proteins that recognize the mismatch. The short patch system, which results in excision of a segment

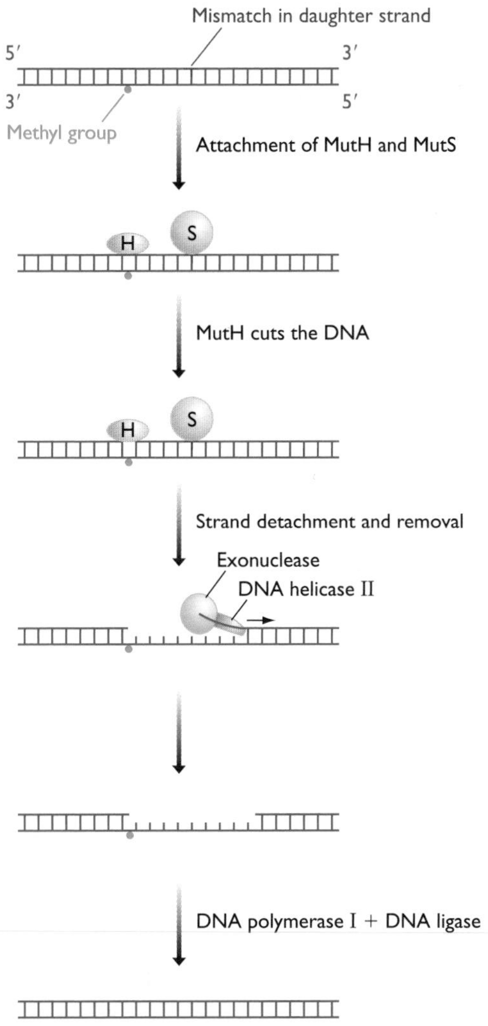

Figure 13.23 Long patch mismatch repair in *E. coli*. See the text for details.

less than 10 nucleotides in length, begins when MutY recognizes an A–G or A–C mismatch, and the very short repair system corrects G–T mismatches which are recognized by the Vsr endonuclease.

Eukaryotes have homologs of the *E. coli* Mut proteins and their mismatch repair processes probably work in a similar way. The one difference is that methylation might not be the method used to distinguish between the parent and daughter polynucleotides (Modrich and Lahue, 1996). Methylation has been implicated in mismatch repair in mammalian cells, but the DNA of some eukaryotes, including fruit flies and yeast, is not extensively methylated (see Box 8.1, p. 177); it is thought that these organisms must therefore use a different method. Possibilities include an association between the repair enzymes and the replication complex, so that repair is coupled with DNA synthesis, or use of single-strand binding proteins that mark the parent strand.

13.2 RECOMBINATION

Without recombination, genomes would be relatively static structures, undergoing very little change. Over a long period of time the gradual accumulation of mutations would result in small scale alterations in the nucleotide sequence of the genome, but more extensive restructuring, which is the role of recombination, would not occur. The evolutionary potential of the genome would be severely restricted.

Recombination was first recognized as the process responsible for crossing-over and exchange of DNA segments between homologous chromosomes during meiosis of eukaryotic cells (see *Figure 2.10*, p. 26), and subsequently implicated in the integration of transferred DNA into bacterial genomes after conjugation, transduction or transformation (Section 2.3.2). The biological importance of these processes stimulated the first attempts to describe the molecular events involved in recombination and led to the Holliday model (Holliday, 1964), with which we will begin our study of recombination.

13.2.1 Homologous recombination

The Holliday model refers to a type of recombination called **general** or **homologous recombination**. This is the most important version of recombination in nature, being responsible for meiotic crossing-over and integration of transferred DNA into bacterial genomes.

The Holliday model for homologous recombination

The Holliday model describes recombination between two homologous double-stranded molecules, ones with identical or nearly identical sequences, but is equally applicable to two different molecules that share a limited region of homology, or a single molecule that recombines with itself because it contains two separate regions that are homologous with one another.

The central feature of the model is formation of a **heteroduplex** resulting from the exchange of polynucleotide segments between the two homologous molecules (*Figure 13.24*). The heteroduplex is initially stabilized by base-pairing between each transferred strand and the intact polynucleotide of the recipient molecule, this base-pairing being possible because of the sequence similarity between the two molecules. Subsequently the gaps are sealed by DNA ligase, giving a **Holliday structure**. This structure is dynamic, **branch migration** resulting in exchange of longer segments of DNA being possible if the two helices rotate in the same direction.

Separation, or **resolution**, of the Holliday structure back into individual double-stranded molecules occurs by cleavage across the branchpoint. This is the key to the entire process because the cut can be made in either of two orientations, as becomes apparent when the three dimensional configuration or **chi form** of the Holliday structure is examined (see *Figure 13.24*). These two cuts have very different results. If the cut is made left–right across the chi form as drawn in *Figure 13.24*, then all that happens is that a short segment of polynucleotide, corresponding to the distance migrated by the branch of the Holliday structure, is transferred between the two molecules. On the other hand, an up–down cut results in **reciprocal strand exchange**, double-stranded DNA being transferred between the two molecules so that the end of one molecule is exchanged for the end of the other molecule. This is the DNA transfer seen in crossing-over.

So far we have ignored one aspect of the Holliday model. This is the way in which the two double-stranded molecules interact at the beginning of the process to produce the heteroduplex. In the original scheme, the two molecules lined up with one another and single-stranded nicks appeared at equivalent positions in each helix. This produced free single-stranded ends that could be exchanged, resulting in the heteroduplex (*Figure 13.25A*). This feature of the model was criticized as no mechanism could be proposed for ensuring that the nicks occurred at precisely the same position on each molecule. The Meselson–Radding modification (Meselson and Radding, 1975) proposes a more satisfactory scheme whereby a single-stranded nick occurs in just one of the double helices, the free end that is produced 'invading' the unbroken double helix at the homologous position and displacing one of its strands, forming a **D-loop** (*Figure*

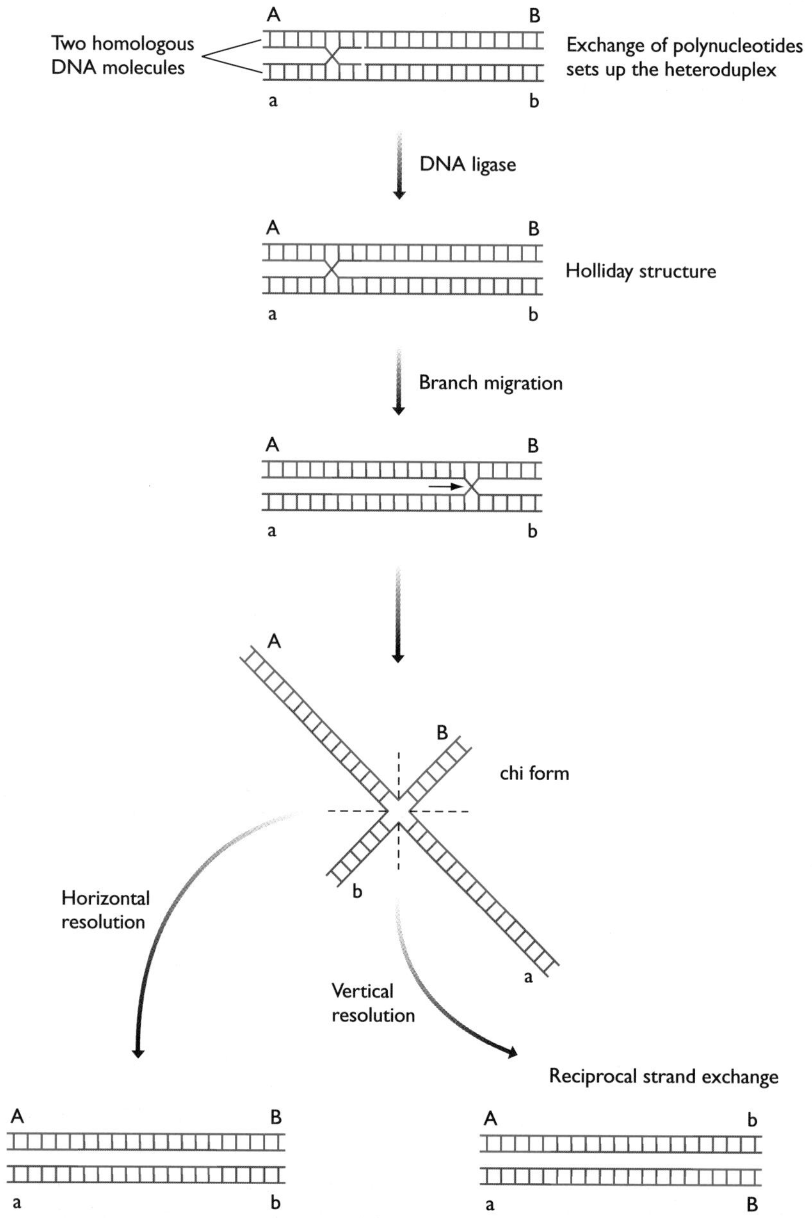

Figure 13.24 The Holliday model for homologous recombination.

13.25B). Subsequent cleavage of the displaced strand at the junction between its single-stranded and base-paired regions produces the heteroduplex.

Proteins involved in homologous recombination in E. coli

The Holliday model and Meselson–Radding modification refer to homologous recombination in all organisms but, as with many areas of molecular biology, the initial progress in understanding how the process is carried out in the cell was made with *E. coli*. The specific recombination system that has been studied has the circular *E. coli* genome as one partner and a linear chromosome fragment as the second partner, this being the situation that occurs during conjugation, transduction or transformation of bacterial cells (Section 2.3.2).

Mutational studies have identified a number of *E. coli* genes which, when inactivated, give rise to defects in homologous recombination, indicating that their protein products are involved in the process in some way. Three distinct recombination systems have been described, these being the RecBCD, RecE and RecF pathways, with RecBCD apparently being the most important in the bacterium (Camerini-Otero and Hsieh, 1995). In this pathway, recombination is initiated by the **RecBCD enzyme**, which has both nuclease and helicase activities. Its precise mode of action is uncertain: in the simplest model the enzyme binds to one end of the linear molecule and unwinds it until it reaches the first copy of the eight-nucleotide consensus sequence 5′-GCTGGTGG-3′, rather confusingly called the **chi site**, which occurs once every 6 kb in *E. coli* DNA (Blattner *et al.*, 1997). The nuclease activity of the enzyme then makes the single-stranded nick at a position approximately 56 nucleotides to the 3′ side of the chi site (*Figure 13.26*). Alternative proposals

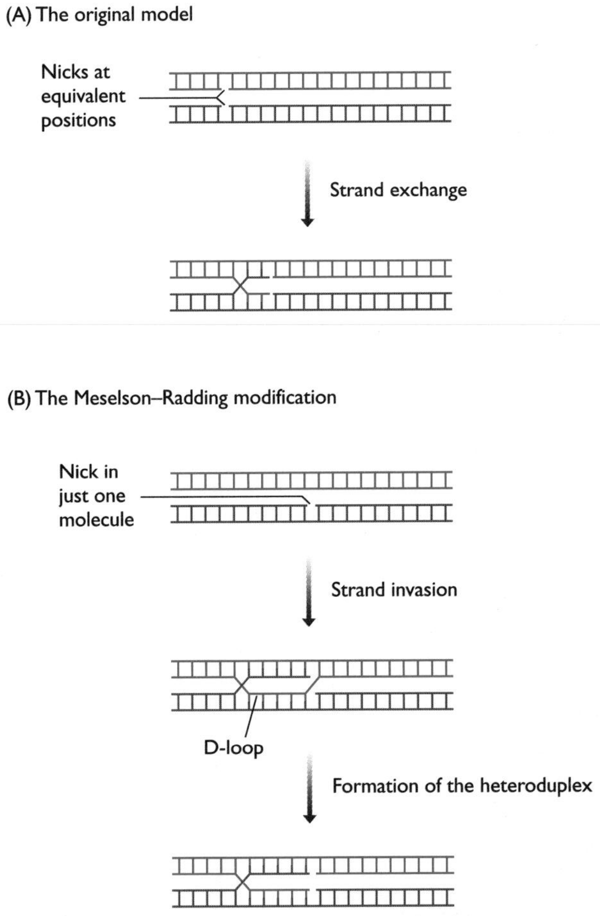

Figure 13.25 Two schemes for initiation of homologous recombination.

(A) Initiation as described by the original model for homologous recombination. (B) The Meselson–Radding modification which proposes a more plausible series of events for formation of the heteroduplex.

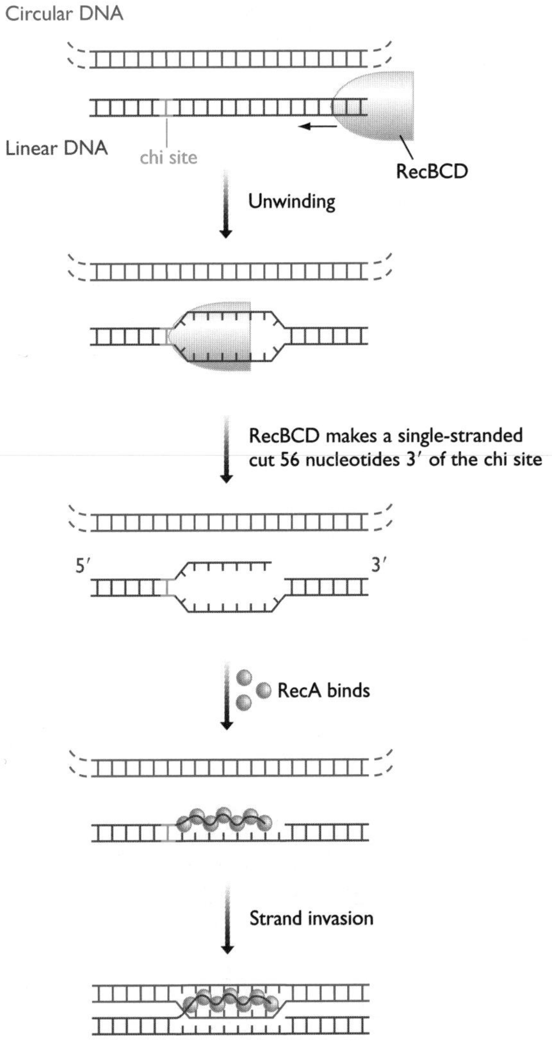

Figure 13.26 The RecBCD pathway for homologous recombination in *E. coli*.

The events leading to formation of the heteroduplex are shown. In the bottom structure the RecA-coated DNA filament has formed a triplex structure, which is thought to be an intermediate in the process. See the text for details.

have the RecBCD enzyme making nicks as it progresses along the linear DNA, this activity being inhibited when the chi site is reached, the last of these progressive nicks being equivalent to the single nick envisaged in the first model (Eggleston and West, 1996).

Whatever the precise mechanism, the RecBCD enzyme produces the free single-stranded end which, according to the Meselson–Radding modification, invades the intact partner, in this case the circular *E. coli* genome. This stage is mediated by the RecA protein, a single-stranded DNA binding protein that forms a protein-coated DNA filament that is able to invade the intact double helix and set up the D-loop (see *Figure 13.26*). An intermediate in formation of the D-loop is probably a **triplex** structure, a three-stranded DNA helix in which the invading polynucleotide lies within the major groove of the intact helix and forms hydrogen bonds with the base pairs it encounters (Camerini-Otero and Hsieh, 1995).

Branch migration is catalyzed by the RuvA and RuvB proteins, both of which attach to the branch point of the Holliday structure. X-ray crystallographic studies suggest that four copies of RuvA bind directly to the branch, forming a core to which two RuvB rings, each consisting of eight proteins, attach, one to either side (*Figure 13.27*; Rafferty *et al.*, 1996). The resulting structure might act as a 'molecular motor' rotating the helices in the required manner so that the branch point moves. The RecG protein also has a role in branch migration but it is not clear if this is in conjunction with RuvAB, or as part of an alternative mechanism (Eggleston and West, 1996).

Branch migration does not appear to be a random process, but instead stops preferentially at the sequence $5'\text{-}\frac{A}{T}TT\frac{G}{C}\text{-}3'$. This sequence occurs frequently in the *E. coli* genome so presumably migration does not halt at the first instance of the motif that is reached. The RuvAB complex then detaches, to be replaced by two RuvC proteins (see *Figure 13.27*) which carry out the cleavage that resolves the Holliday structure. The cuts are made between the T and $\frac{G}{C}$ components of the recognition sequence.

The double-strand break model for recombination in yeast

Although the Holliday model for homologous recombination, either in its original form or as modified by Meselson and Radding, explains most of the results of recombination in all organisms, it has a few inadequacies that have prompted the development of alternative schemes. In particular, the Holliday model was thought to be unable to explain **gene conversion**, a phenomenon first described in yeasts and fungi but now known to occur with many eukaryotes. In yeast, fusion of a pair of gametes results in a zygote that gives rise to an ascus containing four haploid spores whose genotypes can be individually determined. If the gametes have different alleles at a particular locus then under normal circumstances two of the spores will display one genotype and two will display the other genotype, but sometimes this expected 2:2 segregation pattern is replaced by an unexpected 3:1 ratio (*Figure 13.28*). This is called gene conversion because the ratio can only be explained by one of the alleles 'converting' from one type to the other, presumably by recombination during the meiosis that occurs after the gametes have fused.

The double-strand break model provides an opportunity for gene conversion to take place during the recombination process. It initiates not with a single-strand nick, as in the Holliday scheme, but with a double-strand cut that breaks one of the partners in the recombination into two pieces (*Figure 13.29*). This might appear to be a drastic move to make but it has been shown that the protein responsible for the cut is a Type II DNA topoisomerase (Section 12.2.2) which forms covalent linkages with the two pieces of DNA and hence prevents them completely drifting apart. After the double-stranded cut, one strand in each half of the molecule is trimmed back by a $5' \rightarrow 3'$ exonuclease, so each end now has a $3'$ overhang of approximately 500 nucleotides. One of these invades the homologous DNA molecule in a manner similar to that envisaged by the Meselson–Radding scheme, setting up a

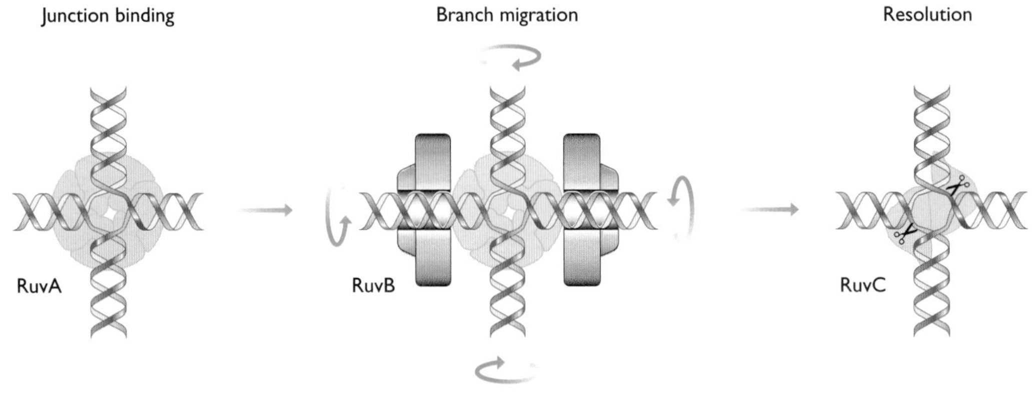

Figure 13.27 The role of the Ruv proteins in homologous recombination in *E. coli*.

Branch migration is induced by a structure comprising four copies of RuvA bound to the Holliday junction with an RuvB ring on either side. After RuvBC has detached, two RuvC proteins bind to the junction, the orientation of their attachment determining the direction of the cuts which resolve the structure. Reprinted with permission from Rafferty JB *et al.* (1996) *Science*, **274**, 415–421. Copyright 1996 American Association for the Advancement of Science.

Box 13.4: Double-strand break repair

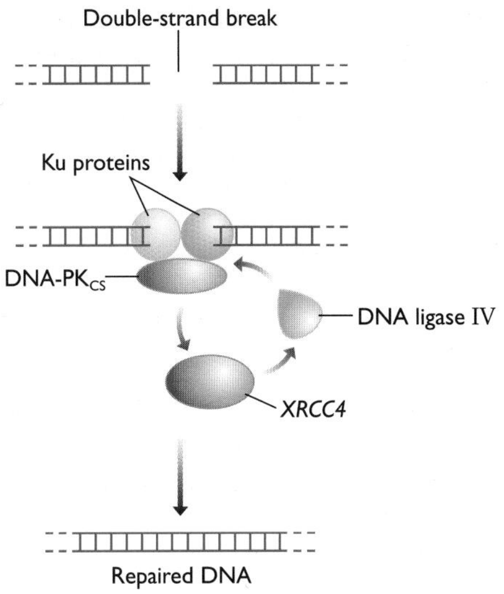

Not to be confused with the double-strand break model for recombination is the double-strand break repair system that has been studied in mammalian cells and which appears to work in a similar way in yeast. Double-strand breaks are generated by exposure to ionizing radiation and some chemical mutagens and are also made by the cell, in a controlled fashion, during the genome rerrangements that join together immunoglobulin gene segments and T-cell receptor gene segments in B and T lymphocytes (Section 11.2.1).

Progress in understanding the break repair system has been stimulated by studies of mutant human cell lines, which have resulted in the identification of four sets of genes involved in the process (Critchlow and Jackson, 1998). These genes specify a multicomponent protein complex that directs a DNA ligase to the break. The complex includes a protein called Ku, made up of two nonidentical subunits, which binds the DNA ends either side of the break. Ku binds to the DNA in association with a protein kinase called DNA-PK$_{CS}$ which appears to activate a third protein, *XRCC4*, which interacts with the mammalian DNA ligase IV, directing this repair protein to the double-strand break.

The repair process is called **nonhomologous end-joining** (**NHEG**), the name indicating that there is no need for there to be any homology between the two molecules whose ends are being joined. It is looked on as a type of recombination, because as well as repairing breaks it can be used to join together molecules or fragments that were not previously joined, producing new combinations. A version of the NHEG system is probably used during construction of immunoglobulin and T-cell receptor genes, but the details are likely to be different because these programmed rearrangements of the genome involve intermediate structures, such as DNA hairpin loops, not seen during the repair of DNA breaks resulting from damage.

Holliday junction that can migrate along the heteroduplex if the invading strand is extended by a DNA polymerase. To complete the heteroduplex the other broken strand, the one not involved in the Holliday junction, is also extended. Note that both DNA syntheses involve extension of strands from the partner that suffered the double-stranded cut, using as templates the equivalent regions of the uncut partner. This is the basis of the gene conversion because it means that the polynucleotide segments removed from the cut partner by the exonuclease have been replaced with copies of the DNA from the uncut partner.

The resulting heteroduplex has a pair of Holliday structures that can be resolved in a number of ways, some resulting in gene conversion and others giving a standard reciprocal strand exchange. An example leading to gene conversion is shown in *Figure 13.29*.

The double-strand break model has been sufficiently well characterized in yeast for there to be little doubt that it occurs, at least in a form approximating to that shown in *Figure 13.29*. Some of the proteins involved in recombination in yeast are very similar to their counterparts in *E. coli* – eukaryotic RAD51, for example, has sequence similarity with RecA and is believed to work in the same way (Baumann and West, 1998) – prompting the suggestion that recombination in all organisms follows the double-strand break system. As yet there is little evidence to support this idea, especially for the larger chromosomes of higher eukaryotes, and many geneticists are resistant to the suggestion that vertebrate DNA undergoes frequent double-strand breaks during meiosis.

13.2.2 Site-specific recombination

A region of extensive homology is not an essential prerequisite for recombination: the process can also be initiated between two DNA molecules with only very short sequences in common. This is called **site-specific recombination** and it has been extensively studied because of the part that it plays during the infection cycle of bacteriophage λ.

Integration of λ DNA into the *E. coli* genome involves site-specific recombination

After injecting its DNA into an *E. coli* cell, bacteriophage λ can follow either of two infection pathways. One of these, the **lytic pathway**, results in the rapid synthesis of λ coat proteins, combined with replication of the λ genome, leading to death of the bacterium and release of new phage particles within about 20 minutes of the initial infection. In contrast, if the phage follows the **lysogenic**

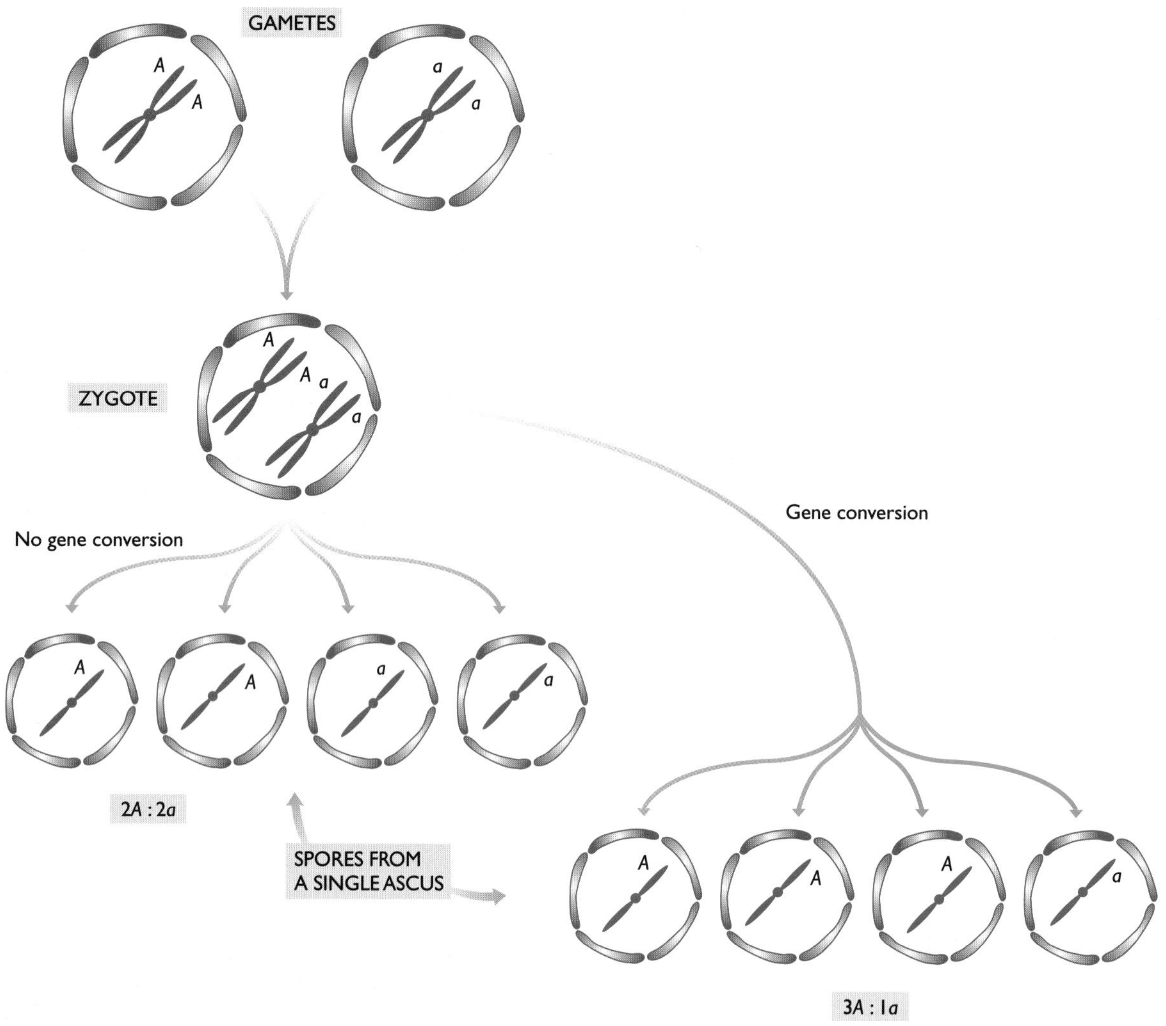

Figure 13.28 Gene conversion.

One gamete contains allele *A* and the other contains allele *a*. These fuse to produce a zygote that gives rise to four haploid spores, all contained in a single ascus. Normally, two of the spores will have allele *A* and two will have allele *a*, but if gene conversion occurs the ratio will be changed, possibly to 3*A*:1*a* as shown here.

pathway, new phage particles do not immediately appear. The bacterium divides as normal, possibly for many cell divisions, with the phage in a quiescent form called the **prophage**. Eventually, possibly as the result of DNA damage or some other triggering stimulus, the phage becomes active again, replicating its genome, directing synthesis of coat proteins, and bursting from the cell.

During the lysogenic phase the λ genome becomes integrated into the *E. coli* chromosome. It is therefore replicated whenever the *E. coli* DNA is copied, and so is passed on to daughter cells just as though it was a standard part of the bacterium's genome. Integration occurs by site-specific recombination between the *att* sites, one on the λ genome and one on the *E. coli* chromosome, which have at their center an identical 15bp sequence (*Figure 13.30*). Because this is recombination between two circular molecules the result is that one bigger circle is formed, in other words the λ DNA becomes integrated into the bacterial genome. A second site-specific recombination between the two *att* sites, now both contained in the same molecule, reverses the original process and releases the λ DNA, which can now return to the lytic mode of infection and direct synthesis of new phage particles.

The recombination event is catalyzed by a specialized Type I topoisomerase (Section 12.2.2) called **integrase** (Kwon *et al.*, 1997), a member of a diverse family of **recombinases** present in bacteria, archaea and yeast. The enzyme makes a staggered double-stranded cut at equivalent positions in the λ and bacterial *att* sites. The two short single-stranded overhangs are then exchanged between the DNA molecules, producing a Holliday junction which migrates a few base pairs along the

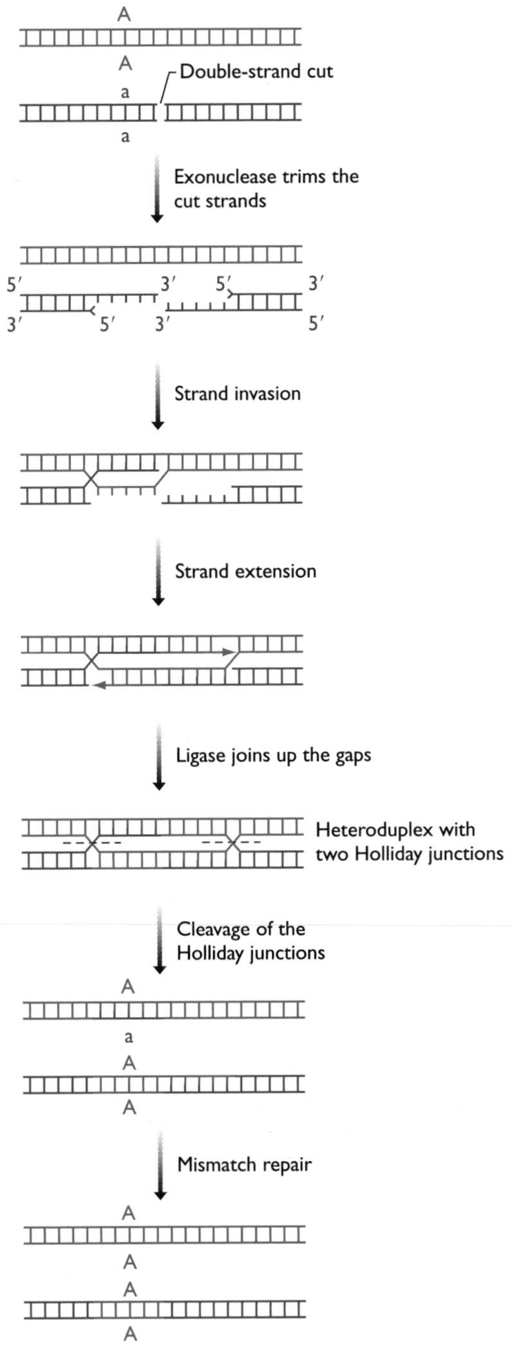

Figure 13.29 The double-strand break model for recombination in yeast.

This model explains how gene conversion can occur. See the text for details.

Box 13.5: The RecE and RecF recombination pathways of *E. coli*

The RecBCD, RecE and RecF pathways involve similar mechanisms and share several of the same proteins: RecA is involved in each pathway, RecF is a component of both the RecE and RecF pathways, and other proteins such as RecJ, RecO and RecQ are common to two or more of the systems. The precise functions of the different pathways are not entirely clear. In normal *E. coli* cells most recombination takes place via RecBCD, but if this pathway is inactivated by mutation the RecE system is able to take over, to be replaced by RecF when RecE is inactivated. These results suggests that the functions of the three pathways overlap, but there are dissimilarities between them. For example, only the RecBCD pathway initiates recombination at the chi sites scattered around the *E. coli* genome, and only RecF is able to induce recombination between a pair of plasmids.

heteroduplex before being cleaved. This cleavage, providing it is made in the appropriate orientation, resolves the Holliday structure in such a way that the λ DNA becomes inserted into the *E. coli* genome. A similar process underlies excision, which is also carried out by integrase, but in conjunction with a second protein, 'excisionase', coded by the λ *xis* gene. If integrase could carry out excision on its own then it would probably excise the λ DNA as soon as it had integrated it.

13.2.3 Transposition

Transposition is not a type of recombination but a process that utilizes recombination, the end result being the transfer of a segment of DNA from one position in the genome to another. A characteristic feature of transposition is that the transferred segment is flanked by a pair of short direct repeats (*Figure 13.31*) which, as we will see, are formed during the transposition process.

In Section 6.3.2 we surveyed the various types of transposable element known in eukaryotes and prokaryotes and discovered that these could be broadly divided into three categories based on their transposition mechanism (*Figure 13.32*):

- DNA transposons that transpose replicatively, the original transposon remaining in place and a new copy appearing elsewhere in the genome.
- DNA transposons that transpose conservatively, the original transposon moving to a new site by a cut-and-paste process.
- Retroelements, all of which transpose via an RNA intermediate.

We will now examine the recombination events that are responsible for each of these three types of transposition.

Replicative and conservative transposition of DNA transposons

A number of models for replicative and conservation transposition have been proposed down through the years but most are modifications of a scheme originally

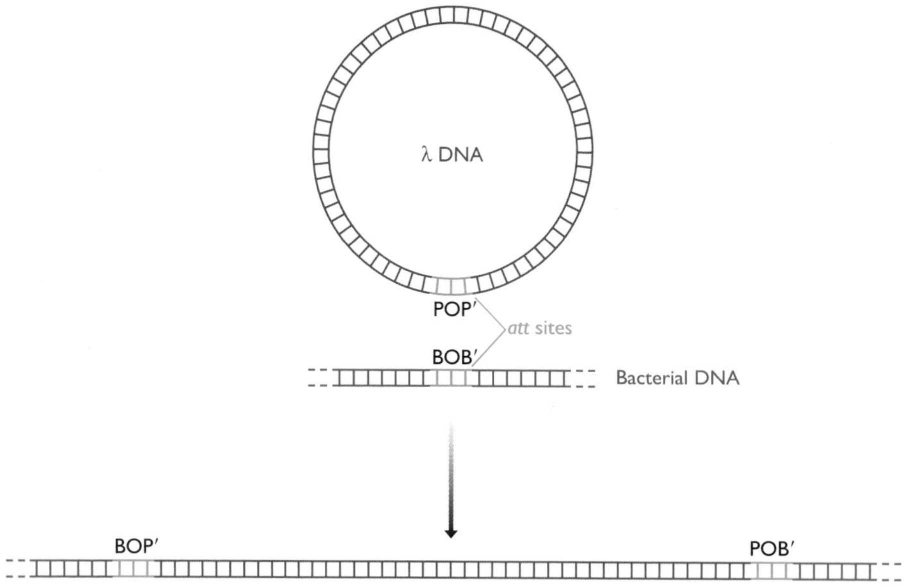

Figure 13.30 Integration of the bacteriophage λ genome into *E. coli* chromosomal DNA.

Both λ and *E. coli* DNA have a copy of the *att* site, each one comprising an identical central sequence called 'O' and flanking sequences P and P′ (for the phage *att* site) or B and B′ (bacterial *att* site). Recombination between the O regions integrates the λ genome into the bacterial DNA.

outlined by Shapiro (1979). According to this model, the replicative transposition of a bacterial element such as a Tn3–type transposon or a transposable phage (Section 6.3.2) is initiated by one or more endonucleases that make single-stranded cuts either side of the transposon and in the target site where the new copy of the element will be inserted (*Figure 13.33*). At the target site the two cuts are separated by a few base pairs, so that the cleaved double-stranded molecule has short 5′-overhangs.

Ligation of these 5′-overhangs to the free 3′-ends either side of the transposon produces a hybrid molecule in which the original two DNAs – the one containing the transposon and the one containing the target site – are linked together by the transposable element flanked by a pair of structures resembling replication forks. DNA synthesis at these replication forks copies the transposable element and converts the initial hybrid into a **cointegrate**, in which the two original DNAs are still linked. Homologous recombination between the two copies of the transposon uncouples the cointegrate, separating the original DNA molecule, with its copy of the transposon still in place, from the target molecule, which now contains a copy of the transposon. Replicative transposition has therefore occurred.

A modification of the process just described changes the mode of transposition from replicative to conservative (see *Figure 13.33*). Rather than carrying out DNA synthesis, the hybrid structure is converted back to two

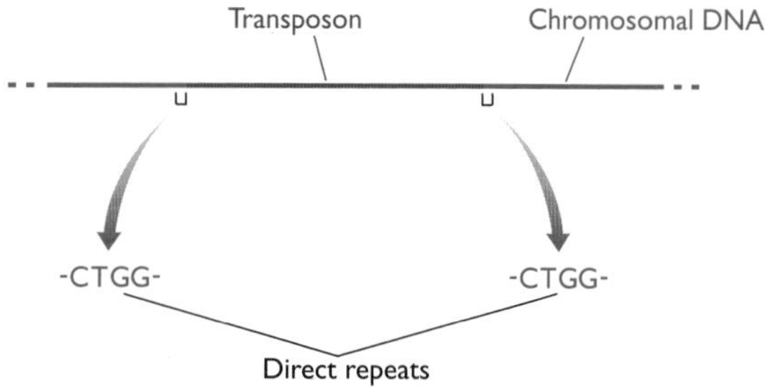

Figure 13.31 Integrated transposable elements are flanked by short direct repeat sequences.

This particular transposon is flanked by the tetranucleotide repeat 5′-CTGG-3′. Other transposons have different direct repeat sequences.

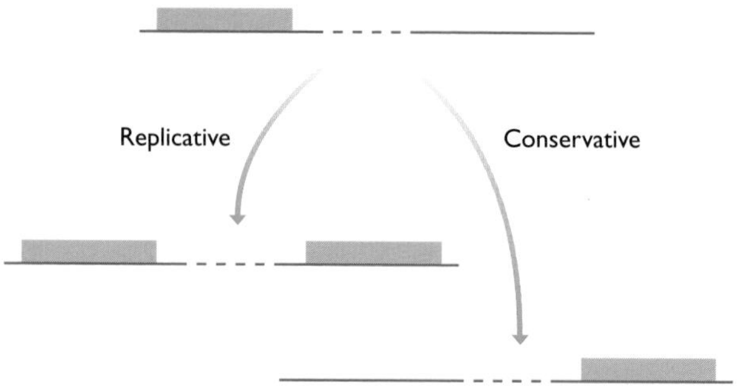

Figure 13.32 Replicative and conservative transposition.

DNA transposons use either the replicative or conservative pathway (some can use both). Retroelements transpose replicatively via an RNA intermediate.

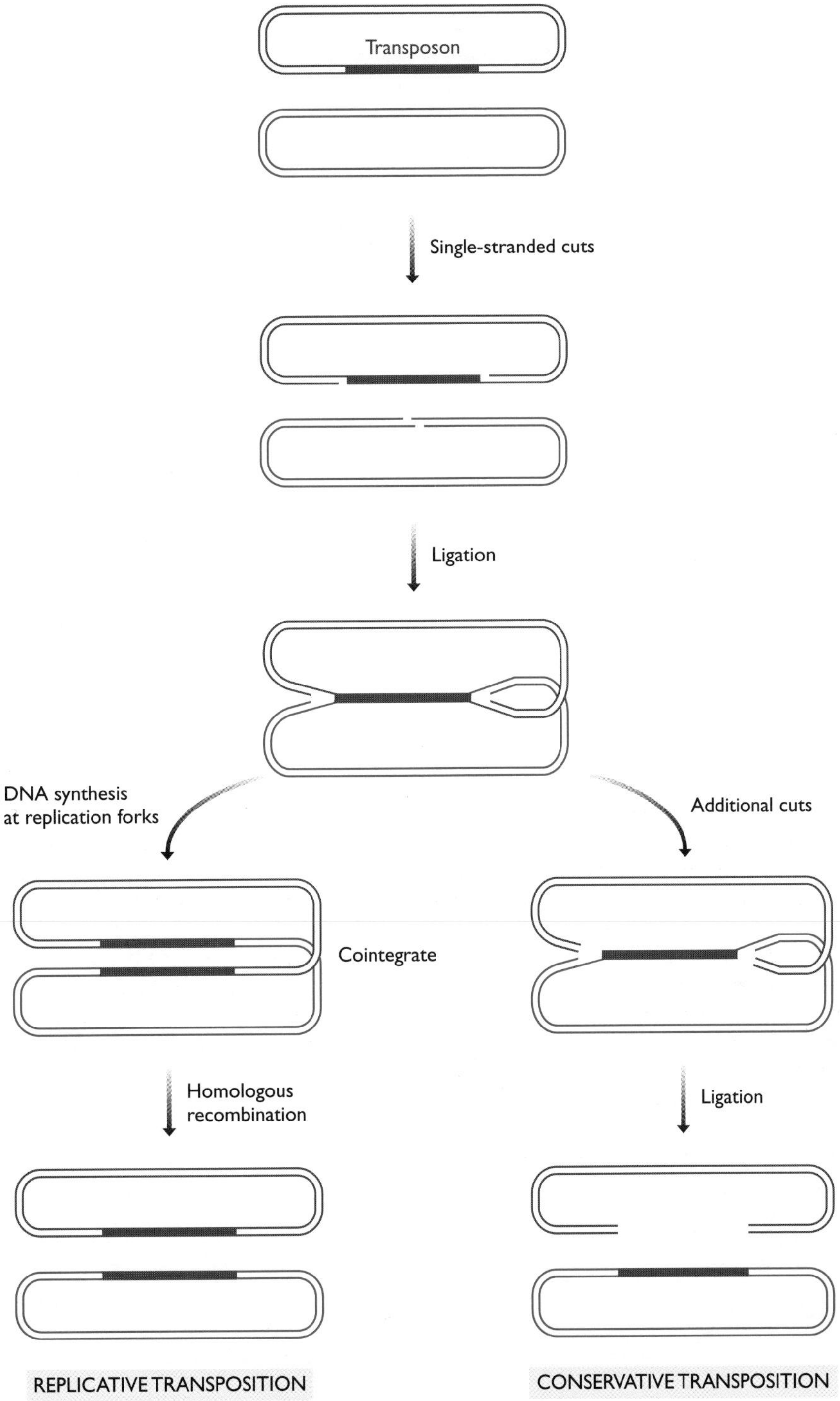

Figure 13.33 A model for the process resulting in replicative and conservative transposition. See the text for details.

separate DNA molecules simply by making additional single-stranded nicks either side of the transposon. This cuts the transposon out of its original molecule, leaving it 'pasted' into the target DNA. If this occurs in a bacterium then the gap at the site vacated by the transposon cannot be repaired and this molecule, which is usually a plasmid, is degraded. In a eukaryote the gap can be repaired by double-strand break recombination.

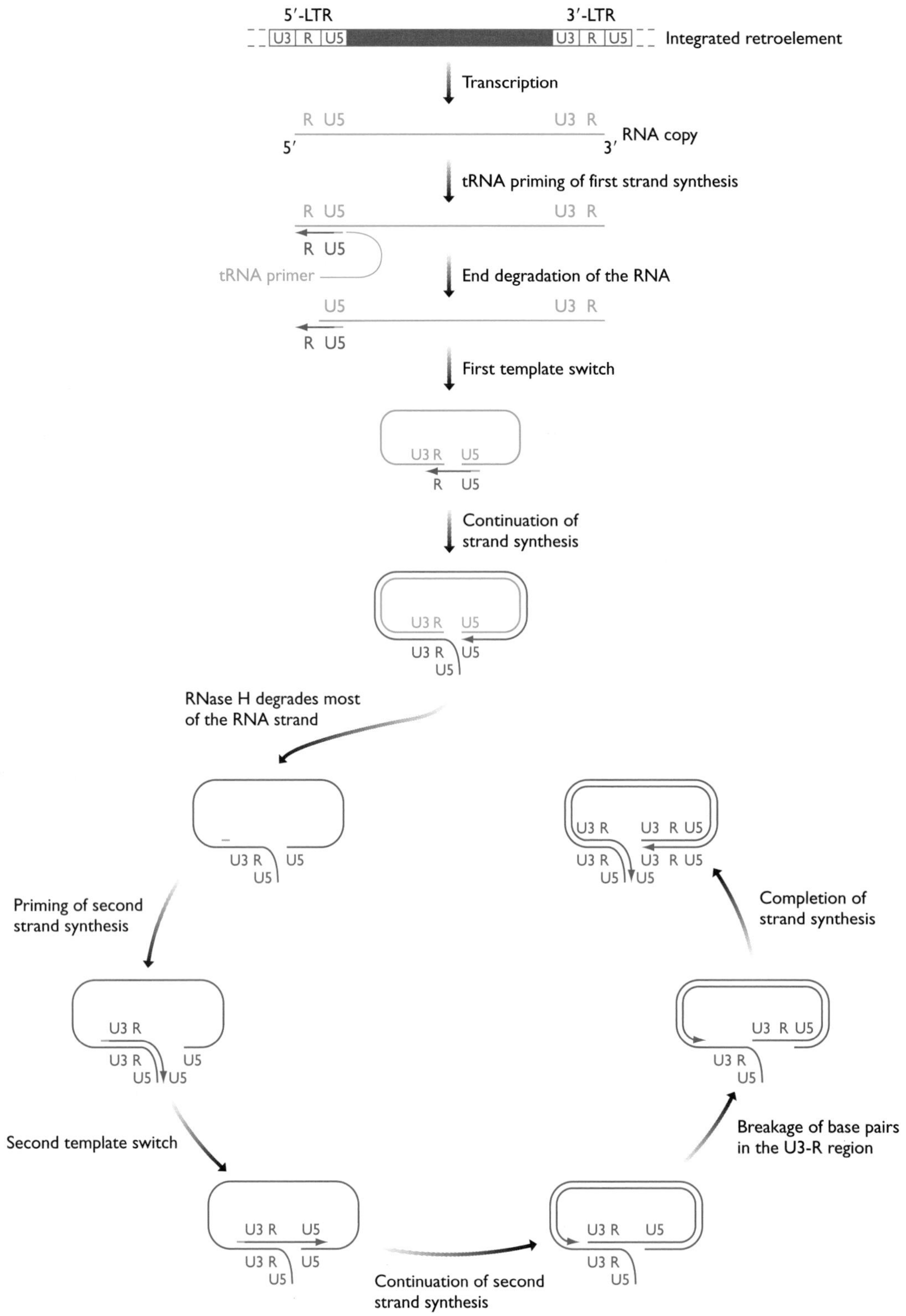

Figure 13.34 Transposition of a retroelement – Part 1.

This diagram shows how an integrated retroelement is copied into a free, double-stranded DNA version. The first step is synthesis of an RNA copy which is then converted to double-stranded DNA by a series of events that involves two template switches, as described in the text.

Transposition of retroelements

From the human perspective, the most important retroelements are the retroviruses, which include the human immunodeficiency viruses which cause AIDS and various other virulent types. Most of what we know about retrotransposition refers specifically to retroviruses though it is believed that other retroelements, such as retrotransposons of the *Ty1/copia* and *Ty3/gypsy* families, transpose by very similar mechanisms.

The first step in retrotransposition is synthesis of an RNA copy of the inserted retroelement (*Figure 13.34*). The long terminal repeat (LTR) at the 5′-end of the element contains a TATA sequence which acts as a promoter for transcription by RNA polymerase II (Section 8.2.2), and some retroelements also have enhancer sequences (Section 8.3.1) that are thought to regulate the amount of transcription that occurs. Transcription continues through the entire length of the element, up to a polyadenylation sequence (Section 9.2.2) in the 3′-LTR. The transcript now acts as the template for RNA-dependent DNA synthesis, catalyzed by a reverse transcriptase enzyme coded by the *pol* gene of the retroelement (see *Figure 6.19*, p. 138). Because this is synthesis of DNA a primer is required (Section 12.3.2) and, as during DNA replication, the primer is made of RNA rather than DNA. During DNA replication, the primer is synthesized *de novo* by a polymerase enzyme (see *Figure 12.12*, p. 314), but retroelements do not code for RNA polymerases so cannot make primers in this way. Instead they use one of the cell's tRNA molecules as a primer, which one depending on the retroelement: the *Ty1/copia* family of elements always use $tRNA^{Met}$ but other retroelements use different tRNAs.

The tRNA primer anneals to a site within the 5′-LTR (see *Figure 13.34*) At first glance this appears to be a strange location for the priming site because it means that DNA synthesis is directed away from the central region of the retroelement and so results in only a short copy of part of the 5′-LTR. In fact, when the DNA copy has been extended to the end of the LTR a part of the RNA is degraded and the DNA overhang that is produced re-anneals to the 3′-LTR of the retroelement which, being a long terminal *repeat*, has the same sequence as the 5′-LTR and so can base-pair with the DNA copy. DNA synthesis now continues along the transcript, eventually displacing the tRNA primer. Note that the result is a DNA copy of the entire template, including the priming site: the template switching is, in effect, the strategy that the retroelement uses to solve the 'end-shortening' problem, the same problem that chromosomal DNAs address through telomere synthesis (Section 12.3.4).

Completion of synthesis of the first DNA strand results in a DNA–RNA hybrid. The RNA is partially degraded by an RNase H enzyme, coded by another part of the *pol* gene. The RNA that is not degraded, usually just a single fragment attached to a short polypurine sequence adjacent to the 3′-LTR, primes synthesis of the second DNA strand, again by reverse transcriptase, which is able to act as both an RNA- and DNA-dependent DNA polymerase. As with the first DNA strand, second strand synthesis initially results in a DNA copy of just the LTR, but a second template switch, to the other end of the molecule, enables the DNA copy to be extended until it is full length. This creates a template for further extension of the first DNA strand, so that the resulting, double-stranded DNA is a complete copy of the internal region of the retroelement plus the two LTRs.

All that remains is to insert the new copy of the retroelement into the genome. It was originally thought that insertion occurred randomly, but it now appears that although no particular sequence is used as a target site, integration occurs preferentially at certain positions (Devine and Boeke, 1996). Insertion involves removal of two nucleotides from the 3′-ends of the double-stranded retroelement by the integrase enzyme (coded by yet another part of *pol*). The integrase also makes a staggered cut in the genomic DNA so both the retroelement and the integration site now have 5′-overhangs (*Figure 13.35*). These overhangs might not have complementary

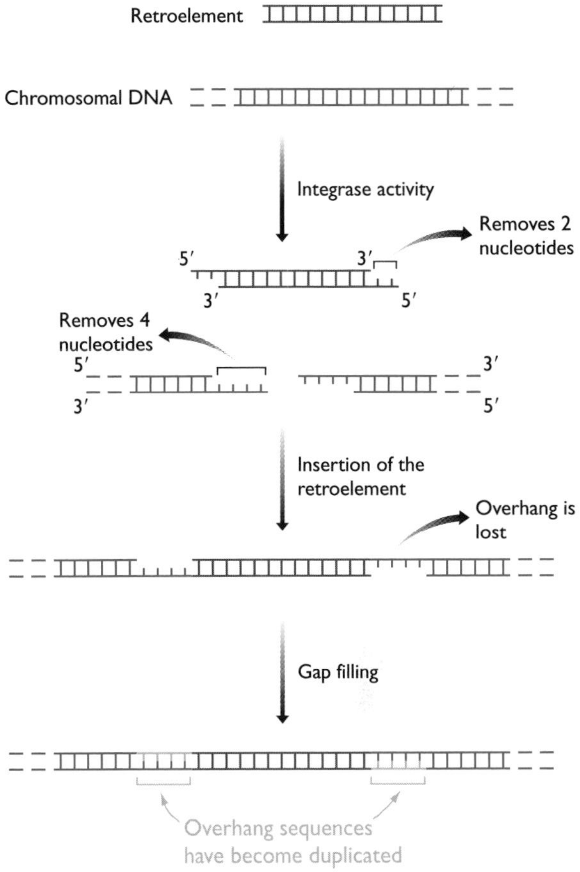

Figure 13.35 Transposition of a retroelement – Part 2.

Integration of the retroelement into the genome results in a 4-nucleotide direct repeat either side of the inserted sequence.

sequences but still appear to interact in some way so that the retroelement becomes inserted into the genomic DNA. The interaction results in loss of the retroelement overhangs and filling in of the gaps that are left, which means that the integration site becomes duplicated into a pair of direct repeats, one at either end of the inserted retroelement.

REFERENCES

Andersson DI, Slechta ES and Roth JR (1998) Evidence that gene amplification underlies adaptive mutability of the bacterial *lac* operon. *Science*, **282**, 1133–1135.

Ashley CT and Warren ST (1995) Trinucleotide repeat expansion and human disease. *Annu. Rev. Genet.*, **29**, 703–728.

Baumann P and West SC (1998) Role of the human RAD51 protein in homologous recombination and double-stranded-break repair. *Trends Biochem. Sci.*, **23**, 247–251.

Blattner FR, Plunkett G, Bloch CA, *et al.* (1997) The complete genome sequence of *Escherichia coli* K-12. *Science*, **277**, 1453–1462.

Bridges BA (1997) Hypermutation under stress. *Nature*, **387**, 557–558.

Cairns J, Overbaugh J and Miller S (1988) The origin of mutants. *Nature*, **335**, 142–145.

Camerini-Otero RD and Hsieh P (1995) Homologous recombination proteins in prokaryotes and eukaryotes. *Annu. Rev. Genet.*, **29**, 509–552.

Campuzano V, Montermini L, Moltò MD, *et al.* (1996) Friedreich's ataxia: autosomal recessive disease caused by an intronic GAA triplet repeat expansion. *Science*, **271**, 1423–1427.

Cascalho M, Wong J, Steinberg C and Wabl M (1998) Mismatch repair co-opted by hypermutation. *Science*, **279**, 1207–1210.

Chernova OB, Chernov AV, Agarwal ML, Taylor WR and Stark GR (1995) The role of p53 in regulating genomic stability when DNA and RNA synthesis are inhibited. *Trends Biochem. Sci.*, **20**, 431–434.

Cotton RGH (1997) Slowly but surely towards better scanning for mutations. *Trends Genet.*, **13**, 43–46.

Critchlow SE and Jackson SP (1998) DNA end-joining: from yeast to man. *Trends Biochem. Sci.*, **23**, 394–398.

Devine SE and Boeke JD (1996) Integration of the yeast retrotransposon *Ty1* is targeted to regions upstream of genes transcribed by RNA polymerase III. *Genes Devel.*, **10**, 620–633.

Eggleston AK and West SC (1996) Exchanging partners in *E. coli*. *Trends Genet.*, **12**, 20–26.

Enoch T and Norbury C (1995) Cellular responses to DNA damage: cell-cycle checkpoints, apoptosis and the roles of p53 and ATM. *Trends Biochem. Sci.*, **20**, 426–430.

Francino MP and Ochman H (1997) Strand asymmetries in DNA evolution. *Trends Genet.*, **13**, 240–245.

Freudenreich CH, Kantrow SM and Zakian VA (1998) Expansion and length-dependent fragility of CTG repeats in yeast. *Science*, **279**, 853–856.

Goodman MF (1998) Purposeful mutations. *Nature*, **395**, 221–223.

Gowen LC, Avrutskaya AV, Latour AM, Koller BH and Leadon SA (1998) BRCA1 required for transcription-coupled repair of oxidative DNA damage. *Science*, **281**, 1009–1012.

Holliday R (1964) A mechanism for gene conversion in fungi. *Genet. Res.*, **5**, 282–304.

Kolodner RD (1995) Mismatch repair: mechanisms and relationship to cancer susceptibility. *Trends Biochem. Sci.*, **20**, 397–401.

Kunkel TA and Wilson SH (1996) Push and pull of base flipping. *Nature*, **384**, 25–26.

Kwon HJ, Tirumalai R, Landy A and Ellenberger T (1997) Flexibility in DNA recombination: structure of the lambda integrase catalytic core. *Science*, **276**, 126–131.

Lehmann AR (1995) Nucleotide excision repair and the link with transcription. *Trends Biochem. Sci.*, **20**, 402–405.

Mandel J-L (1997) Breaking the rule of three. *Nature*, **386**, 767–769.

Mellon I, Rajpal DK, Koi M, Boland CR and Champe GN (1996) Transcription-coupled repair deficiency and mutations in human mismatch repair genes. *Science*, **272**, 557–560.

Meselson M and Radding CM (1975) A general model for genetic recombination. *Proc. Natl. Acad. Sci. USA*, **72**, 358–361.

Modrich P and Lahue R (1996) Mismatch repair in replication fidelity, genetic recombination, and cancer biology. *Annu. Rev. Biochem.*, **65**, 101–133.

Rafferty JB, Sedelnikova SE, Hargreaves D, Artymiuk PJ, Baker PJ, Sharples GJ, Mahdi AA, Lloyd RG and Rice DW (1996) Crystal structure of DNA recombination protein RuvA and a model for its binding to the Holliday junction. *Science*, **274**, 415–421.

Roberts RJ and Cheng X (1998) Base flipping. *Annu. Rev. Biochem.*, **67**, 181–198.

Russell P (1998) Checkpoints on the road to mitosis. *Trends Biochem. Sci.*, **23**, 399–402.

Seeberg E, Eide L and Bjørås M (1995) The base excision repair pathway. *Trends Biochem. Sci.*, **20**, 391–397.

Shannon M and Weigert M (1998) Fixing mismatches. *Science*, **279**, 1159–1160.

Shapiro JA (1979) Molecular model for the transposition and replication of bacteriophage Mu and other transposable elements. *Proc. Natl Acad. Sci. USA*, **76**, 1933–1937.

Sobol RW, Horton JK, Kühn R, Gu H, Singhal RK, Prasad R, Rajewsky K and Wilson SH (1996) Requirement of mammalian DNA polymerase-β in base-excision repair. *Nature*, **379**, 183–186.

Sutherland GR, Baker E and Richards RI (1998) Fragile sites still breaking. *Trends Genet.*, **14**, 501–506.

Svejstrup JQ, Vichi P and Egly J-M (1996) The multiple roles of transcription factor/repair factor TFIIH. *Trends Biochem. Sci.*, **21**, 346–350.

Twyman RM (1998) *Advanced Molecular Biology: A Concise Reference.* BIOS Scientific Publishers, Oxford.

Walker GC (1995) SOS-regulated proteins in translesion DNA synthesis and mutagenesis. *Trends Biochem. Sci.*, **20**, 416–420.

FURTHER READING

Cooper DN and Krawczak M (1993) *Human Gene Mutation.* BIOS Scientific Publishers, Oxford. — *An account of the effects of mutation on the human genome.*

Friedberg EC (1996) Relationships between DNA repair and transcription. *Annu. Rev. Biochem.*, **65**, 15–42. — *A detailed description of the coupling of repair with transcription.*

Friedberg EC, Walker GC and Siede W (1995) *DNA Repair and Mutagenesis.* American Society for Microbiology Press, Washington, DC. — *Particularly useful for its description of the* E. coli *SOS response.*

Humphries S and Malcolm S (1994) *From Genotype to Phenotype.* BIOS Scientific Publishers, Oxford. — *Emphasis on the biological effects of mutation.*

Sancar A (1996) DNA excision repair. *Annu. Rev. Biochem.*, **65**, 43–81.

Sancar A (1995) DNA repair in humans. *Annu. Rev. Genet.*, **29**, 69–105.

Shinagawa H and Iwasaki H (1996) Processing the Holliday junction in homologous recombination. *Trends Biochem. Sci.*, **21**, 107–111. — *An illuminating description of the central event in recombination.*

West SC (1997) Processing of recombination intermediates by the RuvABC proteins. *Annu. Rev. Genet.*, **31**, 213–244. — *Comprehensive information on branch migration and resolution of Holliday junctions.*

Wood RD (1996) DNA repair in eukaryotes. *Annu. Rev. Biochem.*, **65**, 135–167.

CHAPTER

14 Patterns of Genome Evolution

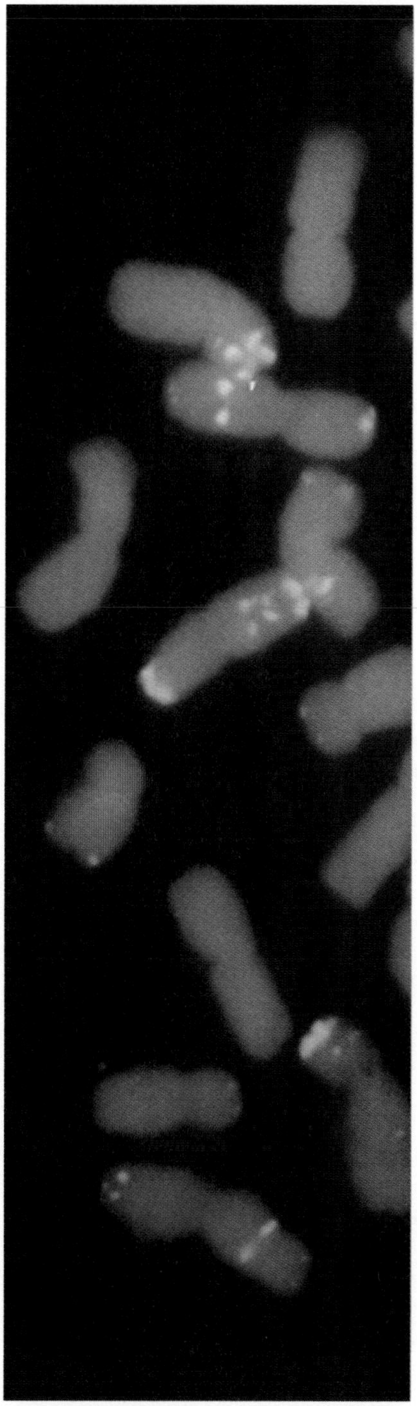

Contents

14.1 Genomes: The First Ten Billion Years **368**
14.1.1 The origins of genomes 369
14.2 Acquisition of New Genes **372**
14.2.1 Acquisition of new genes by gene duplication 373
14.2.2 Acquisition of new genes from other species 383
14.3 Noncoding DNA and Genome Evolution **384**
14.3.1 Transposable elements and genome evolution 384
14.3.2 The origins of introns 385
14.4 The Human Genome: The Last Five Million Years **387**

Concepts

- *The first biochemical systems were based on RNA*
- *The first DNA genomes appeared after RNA molecules had evolved protein-coding properties*
- *Genomes acquire new genes by genome and gene duplications*
- *After duplication, a new gene copy might degrade into a pseudogene or might evolve a new function*
- *The sequences of some gene copies stay the same because of concerted evolution*
- *Genome evolution also involves the rearrangement of domains in existing genes and, in some cases, domain shuffling*
- *Some organisms have probably acquired genes from other species*
- *Transposable elements have had an important influence on genome evolution, by acting as sites for homologous recombination and by moving to new locations in the genome*
- *Introns may have evolved early or they may have evolved late*
- *The human genome is 97% identical to that of the chimpanzee*

We learn very little about the evolutionary history of genomes simply by studying mutation and recombination events in living cells. Instead we must combine our understanding of these events with comparisons between the genomes of different organisms in order to infer the patterns of genome evolution that have occurred in the past. Clearly, this approach is imprecise and uncertain but, as we will see, it is based on a surprisingly large amount of hard data and we can be reasonably confident that, at least in outline, the picture that emerges is not too far divorced from the truth.

In this chapter we will explore the evolution of genomes from the very origins of biochemical systems through to the present day. We will look at ideas regarding the **RNA world**, prior to the appearance of the first DNA molecules, and then examine how DNA genomes have gradually become more complex. Finally, in Section 14.4 we will compare the human genome with the genomes of other primates in order to identify the evolutionary changes that have occurred during the last five million years and which must, somehow, make us what we are.

14.1 GENOMES: THE FIRST TEN BILLION YEARS

Cosmologists believe that the universe began some 14 billion years ago with the gigantic 'primordial fireball' called the Big Bang. Mathematical models suggest that after about four billion years galaxies began to fragment from the clouds of gas emitted by the Big Bang, and that within our own galaxy the solar nebula condensed to form the sun and its planets about 4.6 billion years ago (*Figure 14.1*). The early Earth was covered with water and it was in this vast planetary ocean that the first biochemical systems appeared, cellular life being well established by the time land masses began to appear, some 3.5 billion years ago. But cellular life was a relatively late stage in biochemical evolution, being preceded by self-replicating polynucleotides that were the progenitors of the first genomes. We must begin our study of genome evolution with these precellular systems.

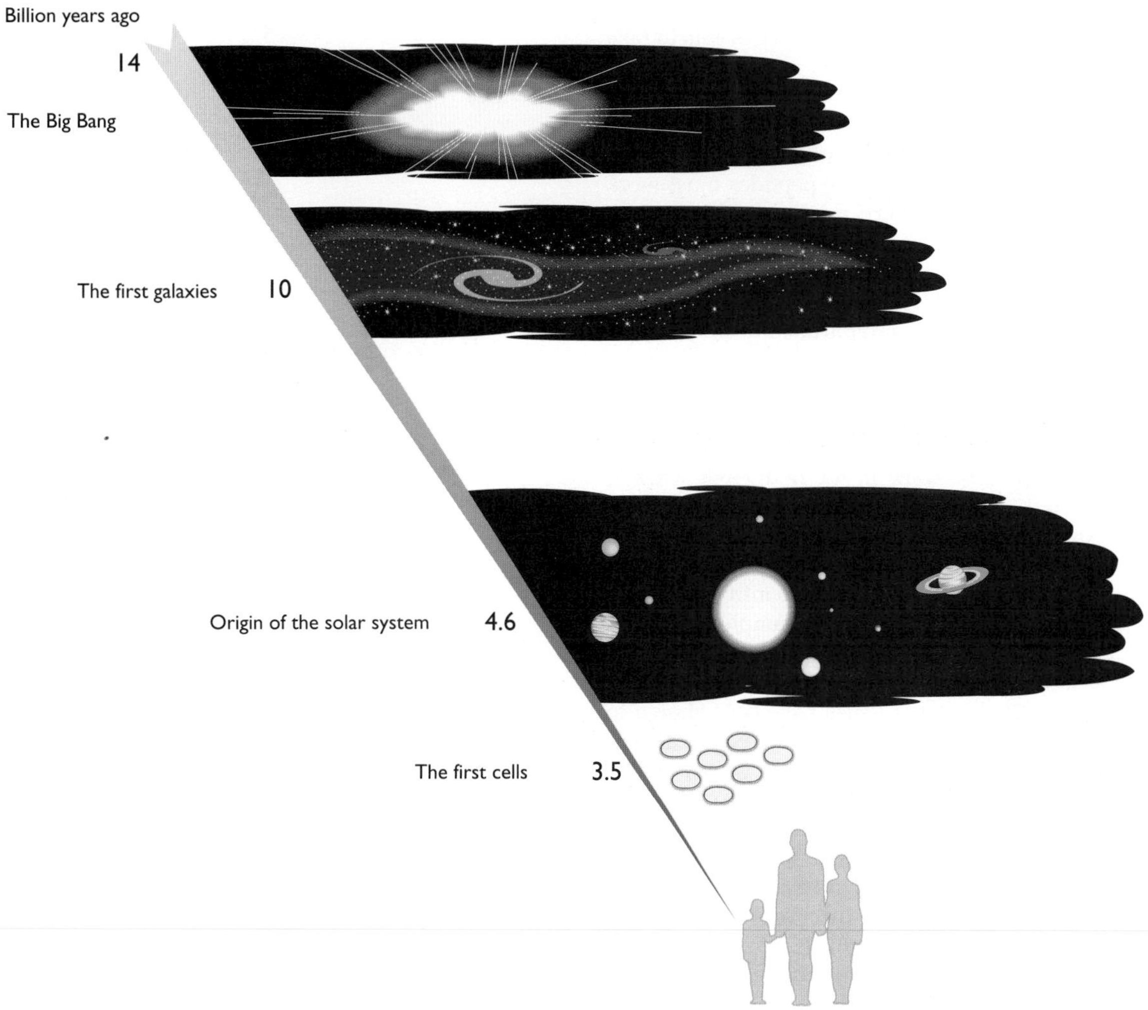

Figure 14.1 The origins of the universe, galaxies, solar system and cellular life.

14.1.1 The origins of genomes

The first oceans are thought to have had a similar salt composition to those of today but the Earth's atmosphere, and hence the dissolved gases in the oceans, was very different. The oxygen content of the atmosphere did not become significant until after photosynthesis had evolved, and to begin with the most abundant gases were probably methane and ammonia. Experiments attempting to recreate the conditions in the ancient atmosphere have shown that electrical discharges in a methane–ammonia mixture result in chemical synthesis of a range of amino acids, including alanine, glycine, valine and several of the others found in proteins (Miller, 1953). Hydrogen cyanide and formaldehyde are also formed, these participating in additional reactions to give other amino acids, as well as purines, pyrimidines and, in less abundance, sugars. At least some of the building blocks of biomolecules could therefore have accumulated in the ancient chemosphere.

The first biochemical systems were centered on RNA

Polymerization of the building blocks into biomolecules might have occurred in the oceans or could have been promoted by the repeated condensation and drying of droplets of water in clouds (Woese, 1979). Alternatively, polymerization might have taken place on solid surfaces, perhaps making use of monomers immobilized on clay particles (Wächtershäuser, 1988). The precise mechanism need not concern us; what is important is that it is possible to envisage purely geochemical processes that lead to synthesis of polymeric biomolecules similar to the ones found in living systems. It is the next steps that we must worry about. We have to go from a random collection of biomolecules to an ordered assemblage that displays at least some of the biochemical properties that we associate with life. These steps have never been reproduced experimentally and our ideas are therefore based mainly on speculation tempered by a certain amount of computer simulation. One problem is that the speculations are unconstrained because the global ocean could have contained as many as 10^{10} biomolecules per liter and we can allow a billion years for the necessary events to take place. This means that even the most improbable scenarios cannot be dismissed out of hand and a way through the resulting maze has been difficult to find.

Progress was initially stalled by the apparent requirement that polynucleotides and polypeptides must work in harness in order to produce a self-reproducing biochemical system. This is because proteins are required to catalyze biochemical reactions but cannot carry out their own self-replication. Polynucleotides can specify the synthesis of proteins and self-replicate, but they can do neither without the aid of proteins. It appeared that the biochemical system would have to spring fully formed from the random collection of biomolecules because any intermediate stage could not be perpetuated. The major breakthrough therefore came in the mid 1980s when it was discovered that RNA can have catalytic activity. Those ribozymes that are known today carry out three types of biochemical reaction: self-cleavage, as displayed by the self-splicing Group I, II and III introns (Section 9.3.3), cleavage of other RNAs (for example, RNase P; *Table 9.5*, p. 222 and Section 9.3.2) and synthesis of peptide bonds (Section 10.2.2 and Research Briefing 10.2). In the test tube, synthetic RNA molecules have been shown to carry out other biologically relevant reactions such as synthesis of ribonucleotides (Unrau and Bartel, 1998), synthesis of RNA molecules (Ekland and Bartel, 1996) and transfer of an RNA-bound amino acid to a second amino acid forming a dipeptide, in a manner analogous to the role of tRNA in protein synthesis (Section 10.1.1; Lohse and Szostak, 1996). The discovery of these catalytic properties solved the polynucleotide–polypeptide dilemma by showing that the first biochemical systems could have been centered entirely on RNA.

Ideas about the RNA world have taken shape in recent years (Robertson and Ellington, 1998). We now envisage that RNA molecules initially replicated in a slow and haphazard fashion simply by acting as templates for binding of complementary nucleotides which polymerized spontaneously (*Figure 14.2*). This process would have been very inaccurate so a variety of RNA sequences would have been generated, eventually leading to one or more with nascent ribozyme properties that were able to direct their own, more accurate self-replication. It is possible that a form of natural selection operated so that the most efficient replicating systems began to predominate, as has been shown to occur in experimental systems. A greater accuracy in replication would have enabled RNAs to increase in length without losing their sequence specificity, providing the potential for more sophisticated catalytic properties, possibly culminating in structures as complex as present-day Group I introns (see *Figure 9.23*, p. 223) and ribosomal RNAs (see *Figure 10.11*, p. 244).

To call these RNAs 'genomes' is a little fanciful, but the term **protogenome** has attractions as a descriptor for molecules that are self-replicating and able to direct simple biochemical reactions. These reactions might have included energy metabolism, based, as today, on the release of free energy by hydrolysis of the phosphate–phosphate bonds in the ribonucleotides ATP and GTP, and the reactions might have become compartmentalized within lipid membranes, forming the first cell-like structures. There are difficulties in envisaging how long-chain unbranched lipids could form by chemical or ribozyme-catalyzed reactions, but once present in sufficient quantities they would spontaneously assemble into membranes, possibly encapsulating one or more protogenomes and providing the RNAs with an enclosed environment in which more controlled biochemical reactions could be carried out.

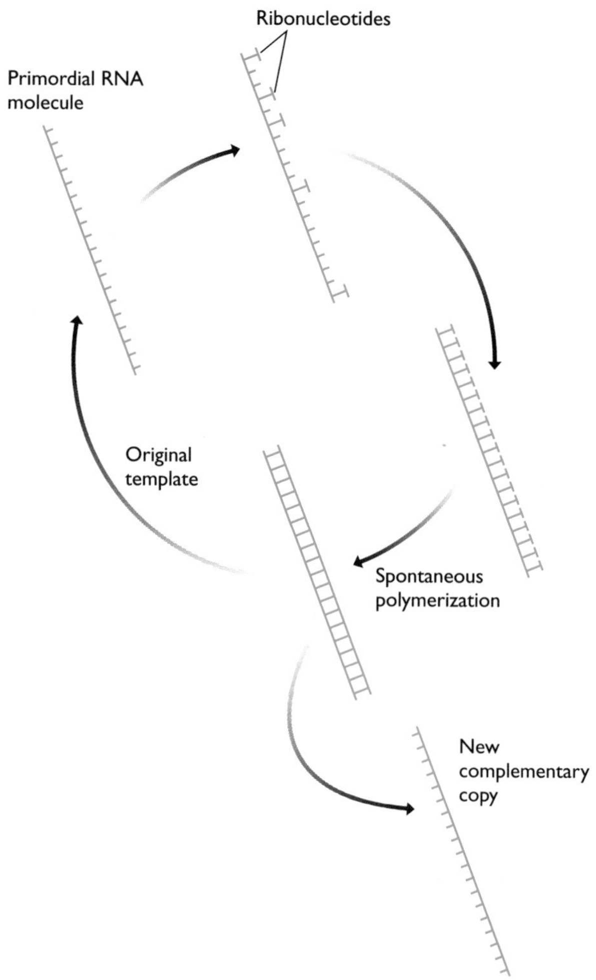

Figure 14.2 Copying of RNA molecules in the early RNA world.

Before the evolution of RNA polymerases, ribonucleotides that became associated with an RNA template would have had to polymerize spontaneously. This process would have been inaccurate and many RNA sequences would have been generated.

The first DNA genomes

How did the RNA world develop into the DNA world? The first major change was probably the development of

Box 14.1: How unique is life?

If the experimental simulations and computer models are correct then it is likely that the initial stages in biochemical evolution occurred many times in parallel in the oceans or atmosphere of the early Earth. It is therefore quite possible that 'life' arose on more than one occasion, even though all present-day organisms appear to derive from a single origin. This single origin is indicated by the remarkable similarity between the basic molecular biological and biochemical mechanisms in bacterial, archaeal and eukaryotic cells. To take just one example, there is no obvious biological or chemical reason why any particular triplet of nucleotides should code for any particular amino acid, but the genetic code, although not universal, is virtually the same in all organisms that have been studied. If these organisms derived from more than one origin then we would anticipate two or more very different codes.

If multiple origins are possible, but modern life is derived from just one, then at what stage did this particular biochemical system begin to predominate? The question cannot be answered precisely, but the most likely scenario is that the predominant system was the first to develop the means to synthesize protein enzymes and therefore probably also the first to adopt a DNA genome. The greater catalytic potential and more accurate replication conferred by protein enzymes and DNA genomes would have given these cells a significant advantage compared with those still containing RNA protogenomes. The DNA-RNA-protein cells would have multiplied more rapidly, enabling them to out-compete the RNA cells for nutrients which, before long, would have included the RNA cells themselves.

Are life forms based on informational molecules other than DNA and RNA possible? Orgel (1998) has reviewed the possibility that RNA was preceded by some other informational molecule at the very earliest period of biochemical evolution and concluded that a pyranosyl version of RNA, in which the sugar takes on a slightly different structure, could form a stable double helix and in some respects might be a better choice than normal RNA for an early protogenome. The same is true of **peptide nucleic acid (PNA)**, a polynucleotide analog in which the sugar–phosphate backbone is replaced by amide bonds. PNAs have been synthesized in the test tube and have been shown to form base pairs with normal polynucleotides. However, there are no indications that either pyranosyl RNA or PNA were more likely than RNA to have evolved in the prebiotic soup.

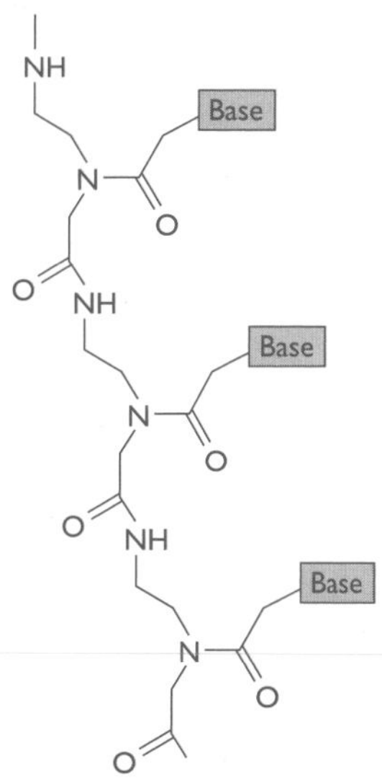

Peptide nucleic acid

protein enzymes, which supplemented and eventually replaced most of the catalytic activities of ribozymes. There are several unanswered questions relating to this stage of biochemical evolution, including the reason why the transition from RNA to protein occurred in the first place. Originally, it was assumed that the 20 amino acids in polypeptides provided proteins with greater chemical variability than the four ribonucleotides in RNA, enabling protein enzymes to catalyze a broader range of biochemical reactions, but this explanation has become less attractive as more and more ribozyme-catalyzed reactions have been demonstrated in the test tube. A more recent suggestion is that protein catalysis is more efficient because of the inherent flexibility of folded polypeptides compared with the greater rigidity of base-paired RNAs (Csermely, 1997).

The transition to protein catalysis demanded a radical shift in the function of the RNA protogenomes. Rather than being directly responsible for the biochemical reactions occurring in the early cell-like structures, the protogenomes became coding molecules whose main function was to specify the construction of the catalytic proteins. Whether the ribozymes themselves became coding molecules, or coding molecules were synthesized by the ribozymes is not known, though the most persuasive theories about the origins of translation and the genetic code suggest that the latter alternative is more likely to be correct (*Figure 14.3*; Szathmáry, 1993). Whatever the mechanism, the result was the paradoxical situation whereby the RNA protogenomes had abandoned their roles as enzymes, which they were good at, and taken on a coding function for which they were less well suited because of the relative instability of the RNA phosphodiester bond, resulting from the indirect effect of the 2′-OH group (Section 7.1.1). A transfer of the coding function to the more stable DNA seems almost inevitable and would not have been difficult to achieve, reduction of ribonucleotides giving deoxyribonucleotides which could then

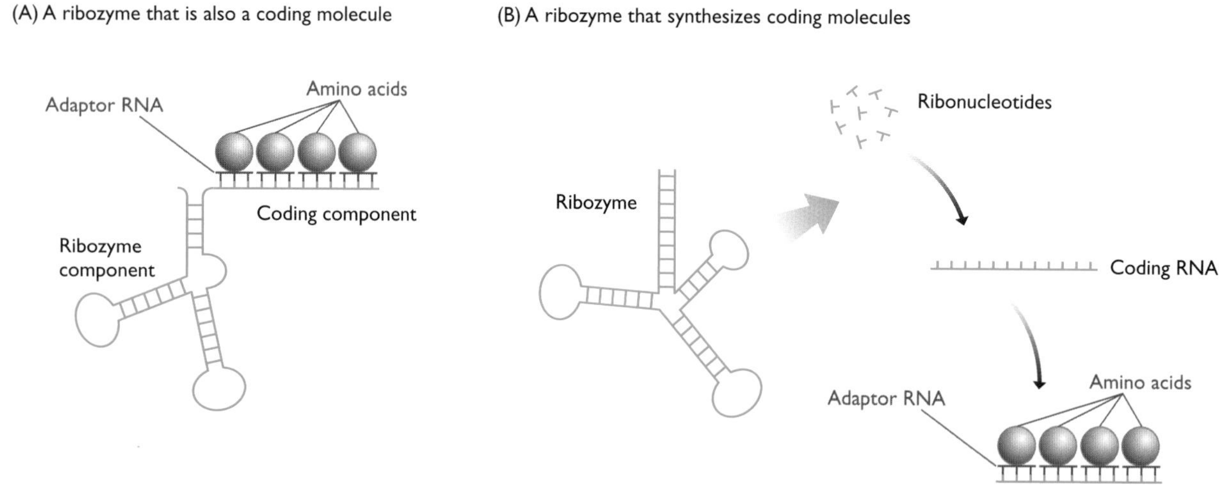

Figure 14.3 Two scenarios for the evolution of the first coding RNA.

A ribozyme could have evolved to have a dual catalytic and coding function (A), or a ribozyme could have synthesized a coding molecule (B). In both examples, the amino acids are shown attaching to the coding molecule via small adaptor RNAs, the presumed progenitors of today's tRNAs.

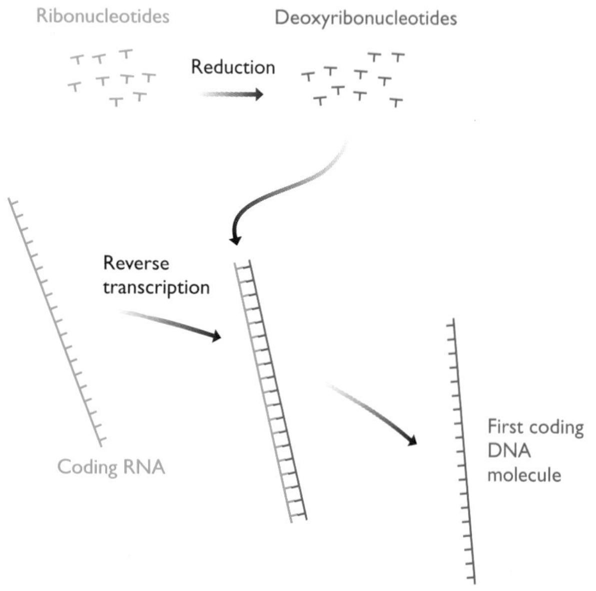

Figure 14.4 Conversion of a coding RNA molecule into the progenitor of the first DNA genome.

be polymerized into copies of the RNA protogenomes by a reverse transcriptase-catalyzed reaction (*Figure 14.4*). The replacement of uracil with its methylated derivative thymine probably conferred even more stability on the DNA polynucleotide, and the adoption of double-stranded DNA as the coding molecule was almost certainly prompted by the possibility of repairing DNA damage by copying the partner strand (Section 13.1.4).

According to this scenario, the first DNA genomes comprised many separate molecules, each specifying a single protein and each therefore equivalent to a single gene. The linking together of these genes into the first chromosomes, which could have occurred either before or after the transition to DNA, would have improved the efficiency of gene distribution during cell division, as it is easier to organize the equal distribution of a few large chromosomes than many separate genes. As with most stages in early genome evolution, several different mechanisms by which genes might have become linked have been proposed (Szathmáry and Maynard Smith, 1993).

14.2 ACQUISITION OF NEW GENES

Although the very old fossil record is difficult to interpret, there is reasonably convincing evidence that by 3.5 billion years ago biochemical systems had evolved into cells similar in appearance to modern bacteria. We cannot tell from the fossils what kinds of genomes these first real cells had, but from the preceding section we can infer that they were made of double-stranded DNA and consisted of a small number of chromosomes, possibly just one, each containing many linked genes.

If we follow the fossil record forward in time we see the first evidence for eukaryotic cells – structures resembling single-celled algae – about 1.4 billion years ago (*Figure 14.5*), and the first multicellular algae by 0.9 billion years ago. Multicellular animals appear around 640 million years ago, although there are enigmatic burrows suggesting that animals lived earlier than this. The Cambrian Revolution, when invertebrate life proliferated into many novel forms, occurred 530 million years ago and ended with the disappearance of many of the novel forms in a mass extinction 500 million years ago. Since then, evolution has continued apace and with increasing diversifica-

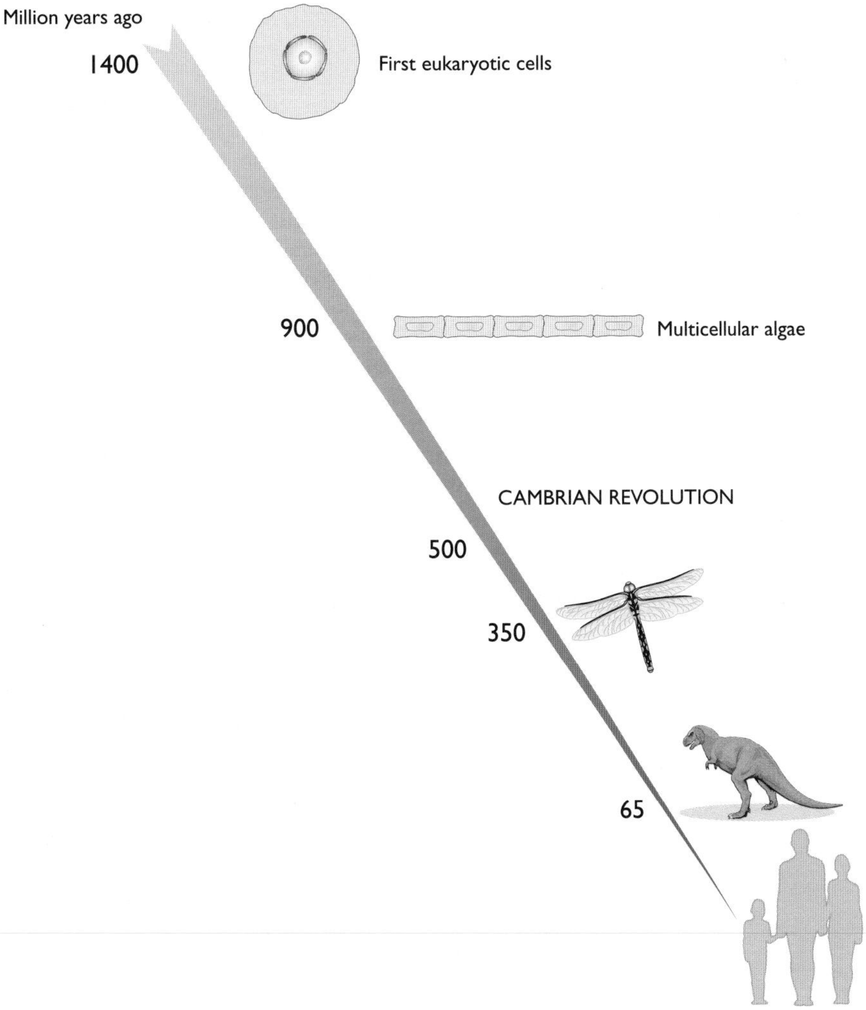

Figure 14.5 The evolution of life.

tion: the first terrestrial insects, animals and plants were established by 350 million years ago, the dinosaurs had been and gone by the end of the Cretaceous, 65 million years ago, and the first hominoids appeared a mere 4.5 million years ago.

Morphological evolution was accompanied by genome evolution. It is dangerous to equate evolution with 'progress' but it is undeniable that as we move up the evolutionary tree we see increasingly complex genomes. One indication of this complexity is gene number, which varies from less than 1000 in some bacteria to 80 000 or possibly more in vertebrates such as humans. However, this increase in gene number has not occurred in a gradual fashion: instead there seem to have been two sudden bursts when gene numbers increased dramatically (Bird, 1995). The first of these expansions occurred when eukaryotes appeared about 1.4 billion years ago, accompanied by an increase from the few thousand genes typical of prokaryotes to the 10 000+ seen in the simplest eukaryotes. The second expansion is associated with the first vertebrates, which became established soon after the end of the Cambrian with each protovertebrate probably having at least 50 000 genes, this being the minimum number for any modern vertebrate, including the most 'primitive' types.

There are two ways in which new genes could be acquired by a genome:

- By duplicating some or all of the existing genes in the genome (Section 14.2.1).
- By acquiring genes from other species (Section 14.2.2).

Both events have been important in genome evolution, as we will see in the next two sections.

14.2.1 Acquisition of new genes by gene duplication

The duplication of existing genes is almost certainly the most important process for generation of new genes during genome evolution. There are several ways in which it could occur:

- By duplication of the entire genome.
- By duplication of a single chromosome or part of a chromosome.
- By duplication of a single gene or group of genes.

The second of these possibilities can probably be discounted as a major cause of gene number expansions based on our knowledge of the effects of chromosome duplications on modern organisms. Duplication of individual human chromosomes, resulting in a cell that contains three copies of one chromosome and two copies of all the others (the condition called **trisomy**), is either lethal or results in genetic disease such as Down syndrome, and similar effects have been observed in artificially generated trisomic mutants of *Drosophila*. Probably, the resulting increase in copy numbers for some genes results in an imbalance of the gene products and disruption of the cellular biochemistry (Ohno, 1970). The other two ways of generating new genes – whole genome duplication and duplication of a single or small number of genes – have probably been much more important.

Whole genome duplications can result in sudden expansions in gene number

The most rapid means of increasing gene number is by duplicating the entire genome. This can occur if an error during meiosis leads to the gametes produced by an organism being diploid rather than haploid (*Figure 14.6*). If two diploid gametes fuse then the result will be a type of **autopolyploid**, in this case a tetraploid cell whose nucleus contains four copies of each chromosome.

Autopolyploidy, as with other types of polyploidy (see p. 383), is not uncommon among plants. Autopolyploids are often viable because each chromosome still has a homologous partner and so can form a bivalent during meiosis. This allows an autopolyploid to reproduce successfully, but generally prevents interbreeding with the original organism from which it was derived. This is because a cross between, for example, a tetraploid and diploid would give a triploid offspring which would not itself be able to reproduce because one full set of its chromosomes would lack homologous partners (*Figure 14.7*). Autopolyploidy is therefore a mechanism by which speciation can occur, a pair of species usually being defined as two organisms that are unable to interbreed. The

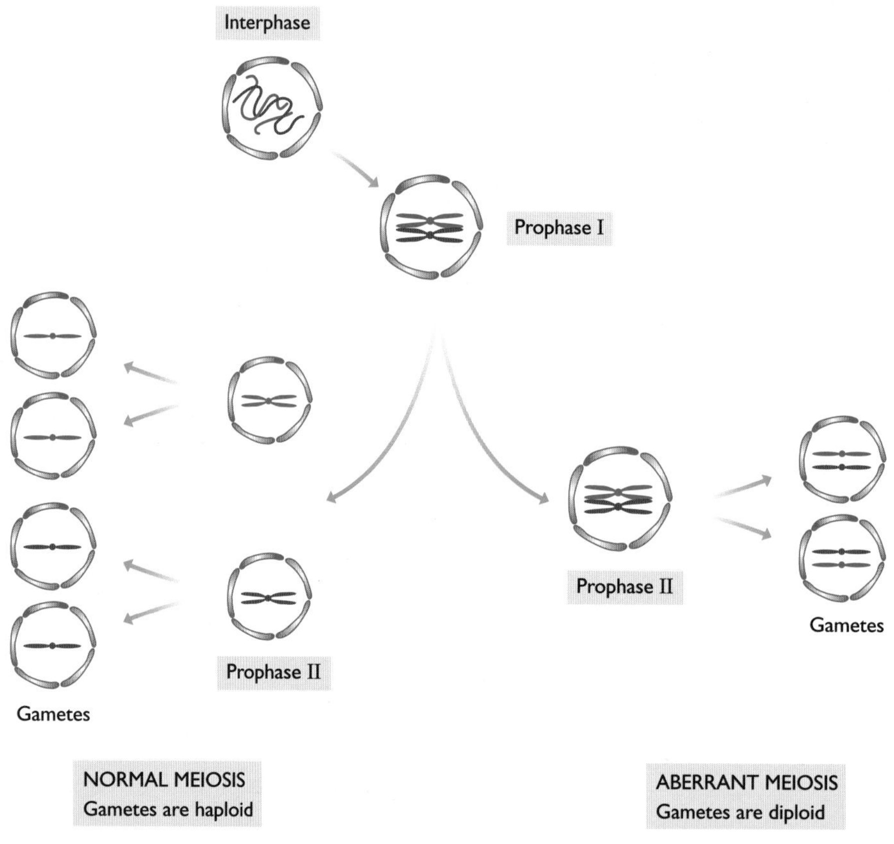

Figure 14.6 The basis of autopolyploidization.

The normal events occurring during meiosis are shown, in abbreviated form, on the left (compare with *Figure 2.10*, p. 26). On the right, an aberration has occurred between prophase I and prophase II and the pairs of homologous chromosomes have not separated into different nuclei. The resulting gametes will be diploid rather than haploid.

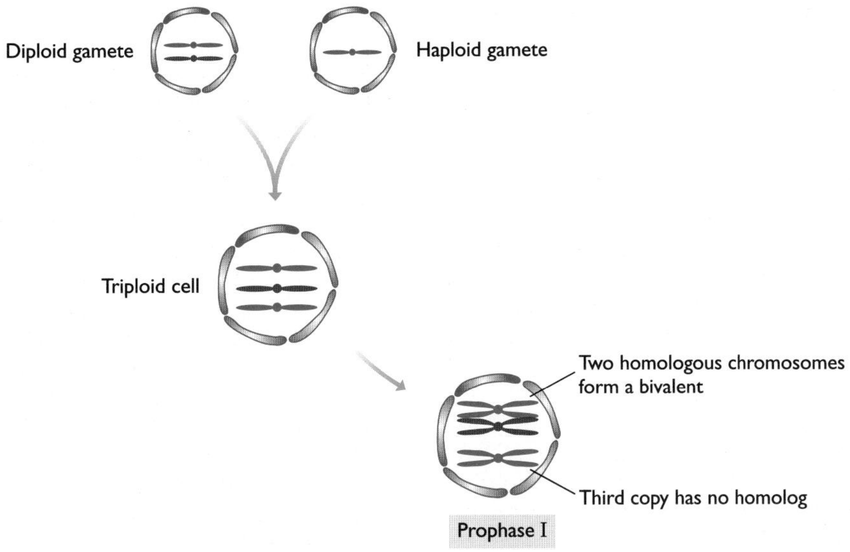

Figure 14.7 Autopolyploids cannot interbreed successfully with their parents.

Fusion of the diploid gamete produced by the aberrant meiosis shown in *Figure 14.6* with a haploid gamete produced by the normal meiosis leads to a triploid nucleus, one that has three copies of each homologous chromosome. During prophase I of the next meiosis, two of these homologous chromosomes will form a bivalent but the third will have no partner. This has a disruptive effect on the segregation of chromosomes during anaphase (see *Figure 2.10*, p. 26) and usually prevents meiosis from reaching a successful conclusion. This means that gametes are not produced and the triploid organism is sterile. Note that the bivalent could have formed between any two of the three homologous chromosomes, not just between the pair shown in the diagram.

generation of new plant species by autopolyploidy has in fact been observed, notably by Hugo de Vries, one of the rediscoverers of Mendel's experiments. During his work with evening primroses, *Oenothera lamarckiana*, de Vries isolated a tetraploid version of this normally diploid plant, which he named *Oenothera gigas*. Autopolyploidy among animals is less common, especially in those with two distinct sexes, possibly because of problems that arise if a nucleus possesses more than one pair of sex chromosomes.

Autopolyploidy does not lead directly to gene expansion because the initial product is an organism that simply has extra copies of every gene, rather than any new genes. It does, however, provide the potential for gene expansion because the extra genes are not essential to the functioning of the cell and so can undergo mutational change without harming the viability of the organism. With many genes, the resulting nucleotide sequence changes will be deleterious and the end result will be an inactive pseudogene, but occasionally the mutations will lead to a new gene function that is useful to the cell. This aspect of genome evolution is more clearly illustrated by considering duplications of single genes rather than entire genomes, so we will postpone a full discussion of it until the next section.

Are there are any indications of genome duplication in the evolutionary histories of present day genomes? The answer is yes, in particular with the yeast *Saccharomyces cerevisiae*, whose genome appears to be the product of a duplication that took place approximately 100 million years ago (Research Briefing 14.1; Wolfe and Shields, 1997). Some molecular biologists argue that the same is true of the genomes of vertebrates (Spring, 1997).

Duplications of genes and groups of genes have occurred frequently in the past

DNA sequencing has revealed that multigene families are common components of all genomes (Section 6.1.1). By comparing the sequences of individual members of a family, using the techniques to be described in Chapter 15, it is usually possible to trace the individual gene duplications involved in evolution of the family from a single progenitor gene that existed in an ancestral genome (*Figure 14.8*; Henikoff *et al.*, 1997). There are several mechanisms by which these gene duplications could have occurred:

- **Unequal crossing-over** is a recombination event that is initiated by similar nucleotide sequences that are not at identical places in a pair of homologous chromosomes. As shown in *Figure 14.9A*, the result of unequal crossover can be duplication of a segment of DNA in one of the recombination products.
- **Unequal sister chromatid exchange** occurs by the same mechanism as unequal crossover, but involves a pair of chromatids from a single chromosome (see *Figure 14.9B*).
- **DNA amplification** is sometimes used in this context to describe gene duplication in bacteria (Romero and Palacios, 1997). In bacteria and other haploid organisms it is thought that duplications can arise by unequal recombination between the two daughter

RESEARCH
14.1
BRIEFING

An ancient duplication of the yeast genome

Examination of the *Saccharomyces cerevisiae* DNA sequence indicates that this genome was completely duplicated approximately 100 million years ago.

Duplication of individual genes has been recognized for some time as having played an important role in genome evolution, largely because the results of gene duplication – multigene families – are common in present-day genomes. The evolutionary importance of whole genome duplication has been more controversial because the traces of this type of event are less easy to discern when modern genomes are examined. A complete genome sequence is a prerequisite for a rigorous search for evidence of whole genome duplication. The most comprehensive study so far has been made with *Saccharomyces cerevisiae*.

Indicators of an ancient genome duplication

From what we understand about the way in which genomes change over time, we might anticipate that evidence for whole genome duplication would be quite difficult to obtain. Many of the extra gene copies resulting from genome duplication would be expected to decay into pseudogenes and no longer be visible in the DNA sequence. Those genes that are retained, because their duplicated function is useful to the organism or because they have evolved new functions, should be identifiable, but it would be impossible to distinguish if they have arisen by genome duplication or simply by duplication of individual genes. For a genome duplication to be signaled it would be necessary to find duplicated *sets* of genes, with the same order of genes in both sets:

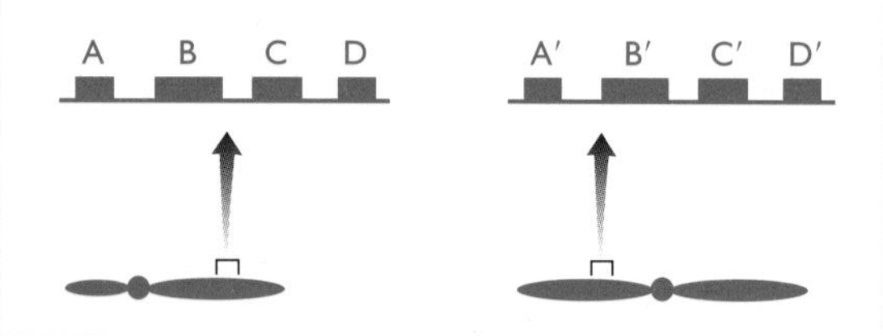

To what extent these duplicated sets are still visible in the genome will depend on how frequently past recombination events have moved genes to new positions.

Evidence for a yeast genome duplication

To search the yeast DNA sequence for evidence of a past genome duplication, homology analysis (Section 5.2.1) was carried out with every yeast gene tested against every other yeast gene. The objective was to identify pairs of genes that might have arisen by gene or genome duplication. To be considered as a possible pair, two genes had to display at least 25% identity when the predicted amino acid sequences of their protein products were compared. When the homology analysis was complete the positions on the genome map of each member of every gene pair was noted, to see if duplicated sets of genes, indicative of a past genome duplication, could be discerned. Fifty-five duplicate sets were identified, comprising 376 pairs of genes, with each of these sets containing at least three genes in the same order.

The next step was to compare the positions of different duplicated sets. Now evidence for a genome duplication started to emerge because many of the duplicated sets were in the same relative positions on different chromosomes. These chromosome pairs contained lengthy segments in which duplicated genes occurred in the same order, interspersed here and there with unique genes.

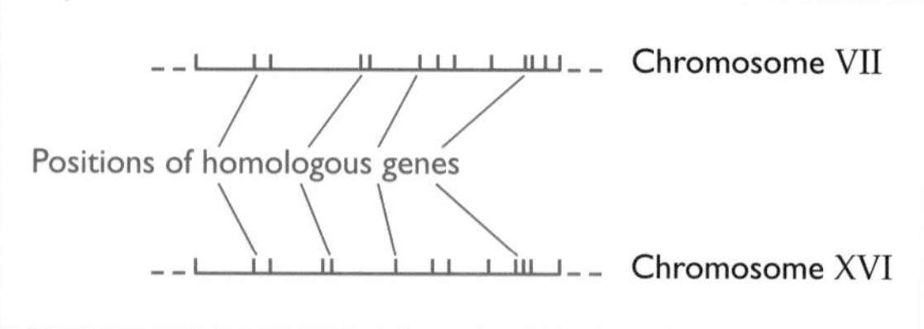

Altogether, these duplicated regions covered half the genome, good evidence that a genome duplication had occurred. To assign a date to this duplication, pairs of genes were compared by phylogenetic analysis and their degree of difference assessed with a molecular clock (Section 15.2.2), the results indicating that the event took place about 100 million years ago.

Reference

Wolfe KH and Shields DC (1997) Molecular evidence for an ancient duplication of the entire yeast genome. *Nature*, **387**, 708–713.

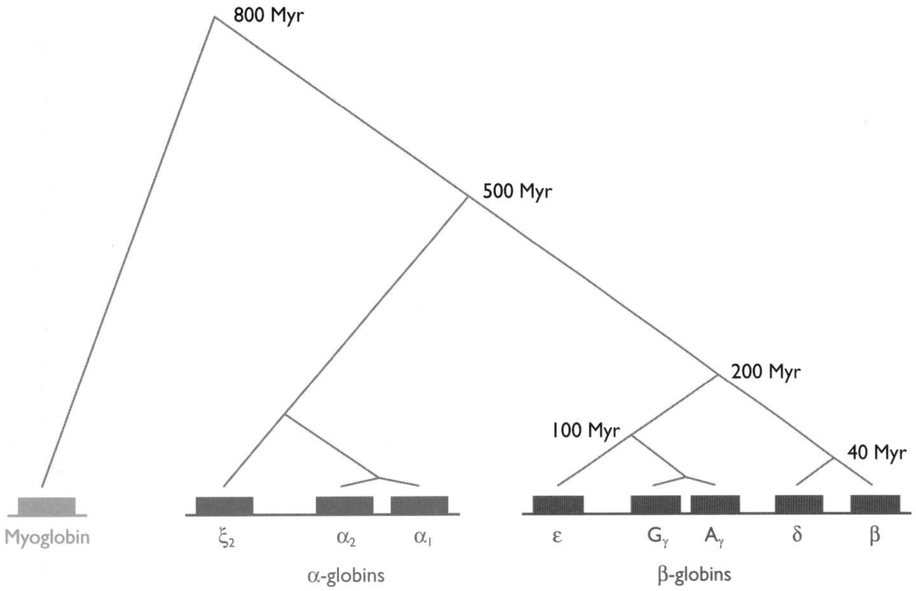

Figure 14.8 Gene duplications during the evolution of the human globin gene families.

Comparisons of their nucleotide sequences enables the evolutionary relationships between the globin genes to be deduced, using the molecular phylogenetics techniques described in Chapter 15. The dates of key duplications are shown. The initial split was between an ancestral gene that gave rise, in one lineage, to the modern gene for the muscle oxygen-binding protein, myoglobin, and, in the other lineage, to the globin genes. This duplication is estimated to have occurred approximately 800 million years ago. The proto-α and proto-β lineages split by a duplication that occurred 500 million years ago and the duplications within the α and β families took place during the last 200 million years. Note that each set of genes is now on a different chromosome: the myoglobin gene is on chromosome 22, the α-globin genes on chromosome 16, and the β-globin genes on chromosome 11. See *Figure 6.6*, p. 123 for more details about the globin genes. Based on Strachan and Read (1996). Abbreviation: Myr, million years.

DNA molecules in a replication bubble (*Figure 14.9C*).

- **Replication slippage** (see *Figure 13.5*, p. 336) could result in gene duplication if the genes are relatively short, this process being more commonly associated with the duplication of very short sequences such as the repeat units in microsatellites.

The initial result of gene duplication is two identical genes. As mentioned above with regards to genome duplication, selective constraints will ensure that one of these genes retains its original nucleotide sequence, or something very similar to it, so that it can continue to provide the protein function that was originally supplied by the single gene copy before the duplication took place. The second copy is probably not subject to the same selective pressures and so can accumulate mutations at random. The evidence is that the vast majority of new genes that arise by duplication acquire deleterious mutations that inactivate them so they become pseudogenes. From the sequences of the pseudogenes in the α- and β-globin gene families (Section 6.1.1), it appears that the commonest inactivating mutations are frameshifts and nonsense mutations that occur within the coding region of the gene, with mutations of the initiation codon and TATA box being less frequent.

Occasionally, the mutations that accumulate within a gene copy do not lead to inactivation of the gene, but instead result in a new gene function that is useful to the organism. We have already seen that gene duplication in the globin gene families led to the evolution of new globin proteins that are used by the organism at different stages in its development (see *Figure 6.6*, p. 123). We also noted (p. 124) that all the globin genes, both the α- and β-types, are related and hence form a gene superfamily that originated with a single ancestral globin gene that split to give the proto-α and proto-β globins about 500 million years ago (see *Figure 14.8*). Further back, about 800 million years ago, this ancestral globin gene itself arose by gene duplication, its sister duplicate evolving to give the modern gene for myoglobin, a muscle protein whose main function, like that of the globins, is the storage of oxygen (Doolittle, 1987). We observe similar patterns of evolution when we compare the sequences of other genes. The trypsin and chymotrypsin genes, for example, are related by a common ancestor approximately 1500 million years ago (Barker and Dayhoff, 1980). Both now code for proteases involved in protein breakdown in the vertebrate digestive tract, trypsin cutting other proteins at arginine and lysine amino acids and chymotrypsin cutting at phenylalanines, tryptophans and tyrosines. Genome evolution has therefore produced two

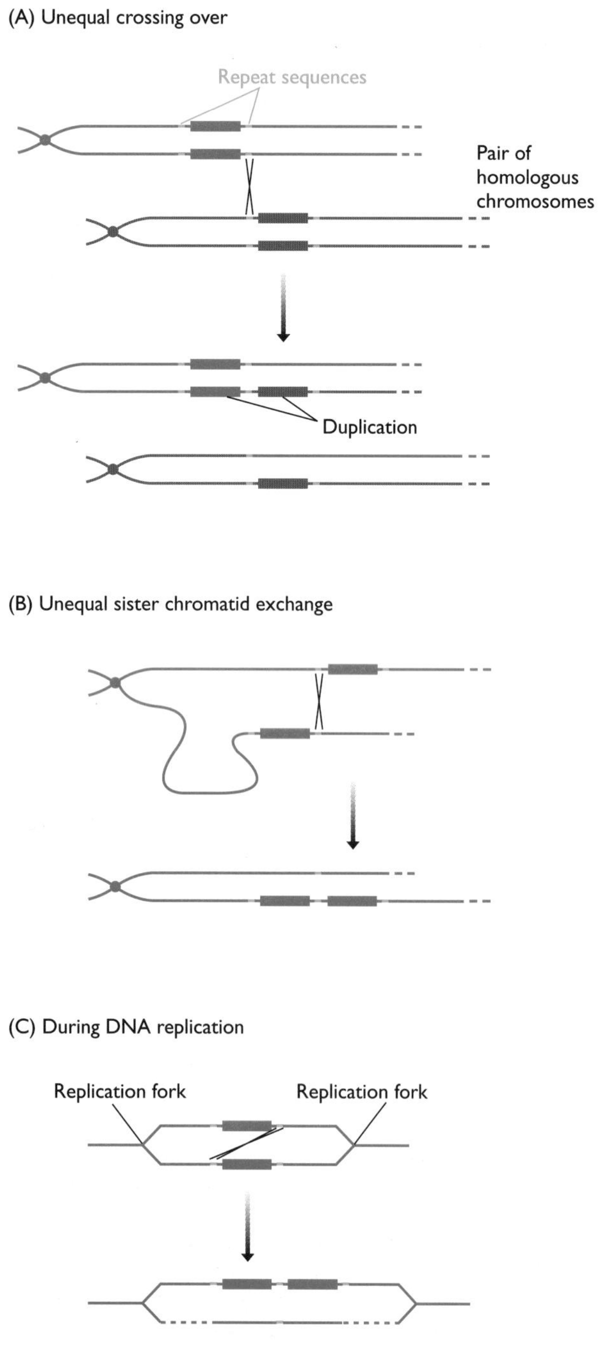

Figure 14.9 Models for gene duplication by (A) unequal crossing-over between homologous chromosomes, (B) unequal sister chromatid exchange, and (C) during replication of a bacterial genome.

In each case, recombination occurs between two different copies of a short repeat sequence, shown in green, leading to duplication of the sequence between the repeats. Unequal crossing-over and unequal sister chromatid exchange are essentially the same except that the first involves chromatids from a pair of homologous chromosomes and the second involves chromatids from a single chromosome. In (C), recombination occurs between two daughter double helices newly synthesized by DNA replication.

Box 14.2: Gene duplication and genetic redundancy

In the text we adopt the conventional scenario that states that after a duplication one of the two gene copies can accumulate mutations which either result in inactivation of that gene copy or lead to a new gene function. Implicit in this scenario is the assumption that the cell only needs one gene to perform the original function, so one of the copies can change. This interpretation of gene evolution is complicated by **genetic redundancy**, which occurs when two genes in a single genome perform the same function (Brookfield, 1997). Many examples of genetic redundancy have been discovered in the course of projects using gene inactivation to identify functions of newly sequenced genes (Section 5.2.2). On frequent occasions gene inactivation does not result in any change of phenotype because a second gene exists which is able to take over the function of the one that has been inactivated.

Genetic redundancy raises interesting evolutionary questions. If the functional gene is satisfying the biochemical requirement then there would appear to be no selective pressure on the redundant gene. Without this selective pressure we would anticipate that the sequence of the redundant gene would change because loss of its function would not result in any immediate disadvantage to the cell. In fact, three different situations have been identified which would lead to maintainance of the redundant gene copy (Nowak *et al.*, 1997), the simplest of these being if the redundant gene also has a second function, not satisfied by the first gene, so that its sequence is maintained by a different form of selective pressure.

complementary protein functions where originally there was just one.

The most striking example of gene evolution by duplication is provided by the homeotic selector genes, the key developmental genes responsible for specification of the body plans of animals. As described in Section 11.3.3, *Drosophila* has a single cluster of homeotic selector genes, called HOM-C, which consists of eight genes each containing a homeodomain sequence coding for a DNA-binding motif in the protein product (see *Figure 11.29*, p. 293). These eight genes, as well as other homeodomain genes in *Drosophila*, are believed to have arisen by a series of gene duplications that began with an ancestral gene that existed about 1000 million years ago. The functions of the modern genes, each specifying the identity of a different segment of the fruit fly, gives us a tantalizing glimpse of how gene duplication and sequence divergence could, in this case, have been the underlying processes responsible for increasing the morphological complexity of the series of organisms in the *Drosophila* evolutionary tree.

RESEARCH 14.2 BRIEFING

Directed evolution in the design of better genes

Experimental techniques based on our understanding of genome evolution are being used to obtain novel genes with improved functional properties.

Many proteins and enzymes are used by the biotechnology industry in applications as diverse as clinical diagnosis and biological washing powders. A naturally occurring enzyme can be used but often a **recombinant protein** is obtained by transferring the relevant gene into *Escherichia coli*, *Saccharomyces cerevisiae* or some other organism in which it can be expressed at a high level. Sometimes the nucleotide sequence of the gene is changed by site-directed mutagenesis (Technical Note 5.2, p. 101) so that the amino acid sequence of the resulting protein is altered in a predetermined fashion. This is done in order to modify the enzyme's properties so it is better suited for the use to which it will be put. For example, the enzyme's activity might be increased, its tertiary structure might be changed to make it more stable, or it might be engineered so that it can function in a nonaqueous environment.

The problem with site-directed mutagenesis is that it presupposes a fairly detailed knowledge of how the enzyme functions so that the appropriate mutations can be introduced. Often this detailed understanding is lacking and many different mutations have to be checked before a useful one is found. The procedure of **directed evolution** offers a radically different alternative to the standard methodology for identifying useful mutations by mimicking in the test tube the evolutionary events that normally occur over millions of years.

Directed evolution by DNA shuffling

There are several different types of directed evolution but one of the most powerful approaches is **DNA shuffling**. This procedure is carried out with a mixture of cloned genes, each one from a different organism and hence coding for a slightly different version of the protein of interest. The genes are randomly broken into short DNA fragments and a self-primed PCR carried out. As the name suggests, this is a PCR without the primers, the individual fragments of DNA themselves acting as primers after forming partial hybrids with complementary strands, as shown in the figure opposite.

If this technique is carried out with a set of fragments from a single gene then the self-priming PCR simply recreates the original DNA sequence. However, when a mixture of genes is used the result is a new set of shuffled genes, each shuffled gene being made up of segments of the original set. The shuffled genes code for novel versions of the protein being studied. Hopefully, one or more of these new versions will have enhanced properties.

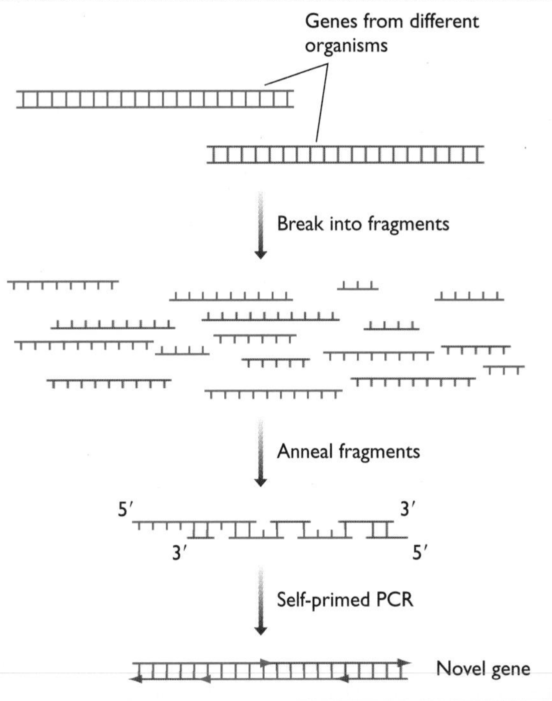

DNA shuffling in action

A striking illustration of the potential of DNA shuffling has been provided by a project in which genes coding for cephalosporinase enzymes from four bacteria – *Citrobacter freundii*, *Enterobacter cloacae*, *Klebsiella pneumonia* and *Yersinia enterocolitica* – were shuffled. Cephalosporinases are enzymes that confer resistance to the antibiotic moxalactam. Bacteria that possess a cephalosporinase gene are usually able to withstand up to 0.75 $\mu g\,ml^{-1}$ moxalactam in the growth medium. After shuffling, a number of genes coding for more effective cephalosporinases were obtained, the most active of these conferring resistance to 200 $\mu g\,ml^{-1}$ moxalactam, 250 times the original value. Such a dramatic increase tells us that it is indeed possible to alter the activity of an enzyme in a significant way by directed evolution.

Reference

Crameri A, Raillard S-A, Bermudez E and Stemmer WPC (1998) DNA shuffling of a family of genes from diverse species accelerates directed evolution. *Nature*, **391**, 288–291.

Vertebrates have four HOX gene clusters (see *Figure 11.29*, p. 293), each a recognizable copy of the *Drosophila* cluster with sequence similarities between genes in equivalent positions. The extra clusters may have been produced by duplication of the original cluster, or may be an indication of past genome duplications. Not all of the vertebrate HOX genes have been ascribed functions, but we believe that the additional versions possessed by vertebrates relate to the added complexity of the vertebrate body plan. Two observations support this conclusion. First, the amphioxus, an invertebrate that displays some primitive vertebrate features, has two HOX clusters (Brooke *et al.*, 1998), which is what we might expect for a primitive 'protovertebrate', and ray-finned fishes, probably the most diverse group of vertebrates with a vast range of different variations on the basic body plan, have seven HOX clusters (Amores *et al.*, 1998).

The sequences of some duplicated genes do not diverge

We should not become so enthusiastic about the interesting results of gene divergence that we forget that some multigene families are made up of genes with identical or near-identical sequences. The prime examples are the rRNA genes, whose copy numbers range from two in *Mycoplasma genitalium* to 500+ in *Xenopus laevis* (Section 6.1.1), with all of the copies having virtually the same sequence. These multiple copies of identical genes presumably reflect the need for rapid synthesis of the gene product at certain stages of the cell cycle. With these gene families there must be a mechanism that prevents the individual copies from accumulating mutations and hence diverging away from the functional sequence.

One possibility would be to continually duplicate the functional genes (*Figure 14.10*), but this does not appear to be the explanation because it implies that the genome should contain many rRNA pseudogenes, which is not the case. Intuition also forces us to reject this explanation as many of the pseudogenes would still be transcribed, resulting in synthesis of multiple copies of defective rRNA molecules, which we would expect to interfere with the functional rRNA – for example by binding ribosomal proteins and preventing these from being used to construct ribosomes.

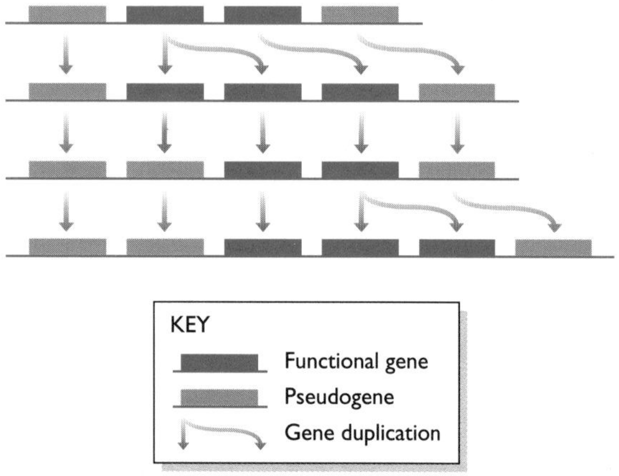

Figure 14.10 One possible way of maintaining the number of functional genes in a multigene family.

New gene duplications occur to compensate for the decay of functional genes into pseudogenes.

Instead of repeated gene duplication, the members of the rRNA gene families appear to undergo **concerted evolution**, so their sequences not only stay the same, they also evolve in parallel (Brown *et al.*, 1972). If one copy of the family acquires an advantageous mutation then it is possible for that mutation to spread throughout the family until all members possess it. The most likely way in which this is achieved is by gene conversion which, as described in Section 13.2.1, can result in the sequence of one copy of a gene being replaced with all or part of the sequence of a second copy. Multiple gene conversion events could therefore maintain identity among the sequences of the individual members of a multigene family.

Genome evolution also involves rearrangement of existing genes

As well as the generation of new genes by duplication followed by mutation, novel protein functions can also be produced by rearranging existing genes. This is possible because most proteins are made up of structural domains (Section 7.2.1), each comprising a segment of the polypeptide chain and hence encoded by a contiguous series of nucleotides (*Figure 14.11*). There are two ways in which rearrangement of domain-encoding gene segments can result in novel protein functions.

- **Domain duplication** occurs when the gene segment coding for a structural domain is duplicated by unequal crossing-over, replication slippage or one of the other methods that we have considered for duplication of DNA sequences (*Figure 14.12A*). Duplication results in the structural domain being repeated in the protein, which might itself be advantageous, for example, by making the protein product more stable. The duplicated domain might also change over time as its coding sequence becomes mutated, leading to a modified structure that might provide the protein with new activity. Note that domain duplication causes the gene to become longer. Gene elongation appears to be a general consequence of genome evolution, the genes of higher eukaryotes being longer, on average, than those of lower organisms.
- **Domain shuffling** occurs when segments coding for structural domains from completely different genes are joined together to form a new coding sequence that specifies a hybrid or mosaic protein, one that would have a novel combination of structural features and might provide the cell with an entirely new biochemical function (*Figure 14.12B*).

Implicit in these models of domain duplication and shuffling is the need for the relevant gene segments to be

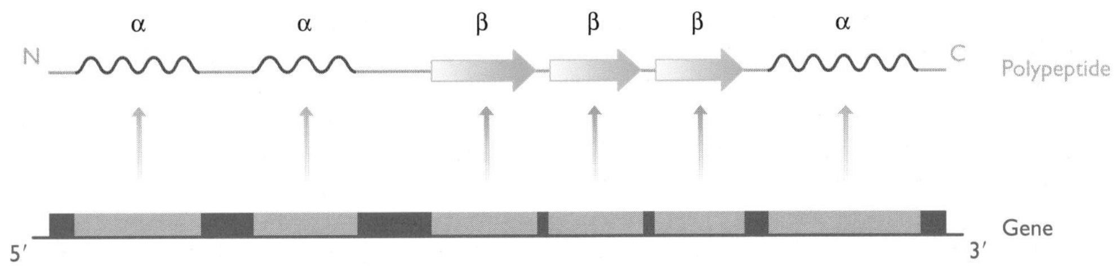

Figure 14.11 Structural domains are individual units in a polypeptide chain coded by a contiguous series of nucleotides.

In this simplified example, each secondary structure in the polypeptide is looked on as an individual structural domain. In reality, most structural domains comprise two or more secondary structural units.

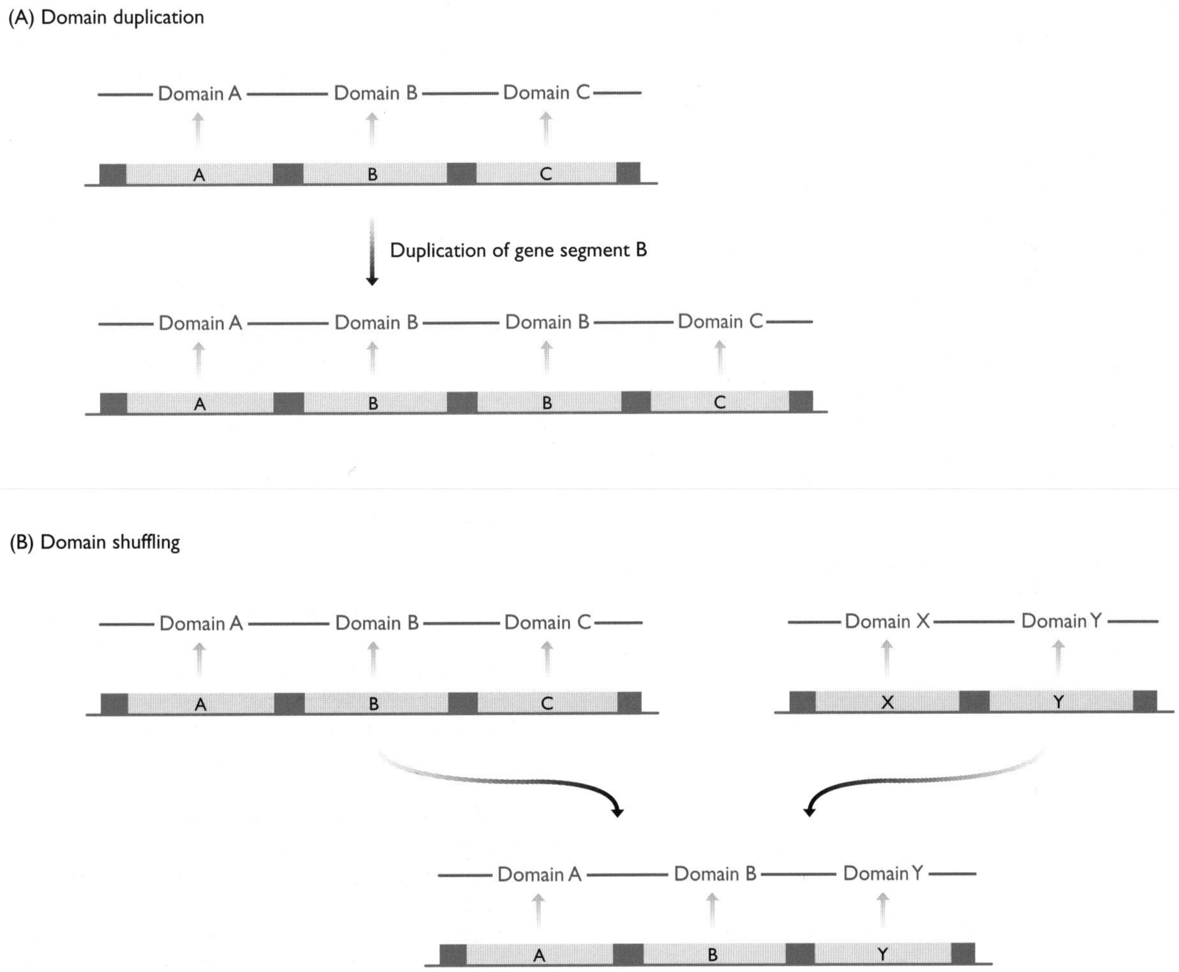

Figure 14.12 Creating new genes by (A) domain duplication and (B) domain shuffling.

separated so they can themselves be rearranged and shuffled. This requirement has led to the attractive suggestion that exons might code for structural domains. With some proteins, duplication or shuffling of exons does seem to have resulted in the structures seen today. An example is provided by the α2 Type I collagen gene of vertebrates, which codes for one of the three polypeptide chains of collagen. Each of the three collagen polypeptides has a highly repetitive sequence made up of repeats of the tripeptide glycine-X-Y, where X is usually proline and Y is usually hydroxyproline (*Figure 14.13*). The α2 Type I gene, which codes for 338 of these repeats, is split into 52 exons,

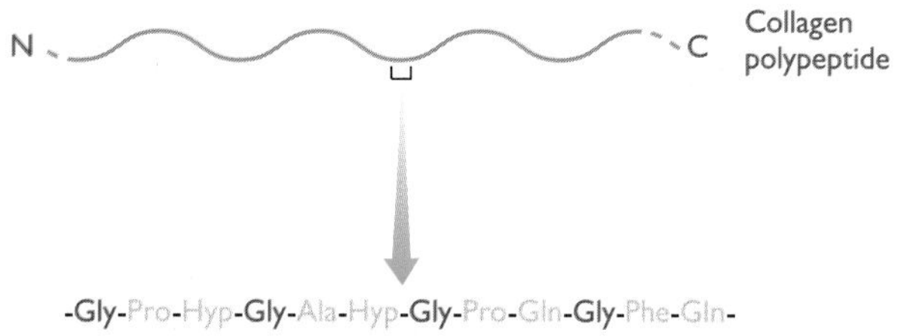

Figure 14.13 The α2 Type I collagen polypeptide has a repetitive sequence described as Gly-X-Y.

Every third amino acid is glycine, X is often proline and Y is often hydroxyproline (Hyp). See *Table 7.3*, p. 155 for other amino acid abbreviations. Hydroxyproline is a post-translationally modified version of proline (Section 10.3.3). The collagen polypeptide has a helical conformation, but one that is more extended than the standard α-helix.

42 of which cover the part of the gene coding for the glycine-X-Y repeats. Within this region, each exon encodes a set of complete tripeptide repeats. The number of repeats per exon varies but is either 15 (5 exons), 18 (23 exons), 33 (5 exons), 36 (8 exons) or 54 (1 exon). Clearly this gene could have evolved by duplication of exons leading to repetition of the structural domains.

Domain shuffling is illustrated by tissue plasminogen activator (TPA), a protein found in the blood of vertebrates which is involved in the blood clotting response. The TPA gene has four exons, each coding for a different structural domain (*Figure 14.14*). The upstream exon codes for a 'finger' module that enables the TPA protein to bind to fibrin, a fibrous protein found in blood clots which activates TPA. This exon appears to be derived from a second fibrin-binding protein, fibronectin, and is absent from the gene for a related protein, urokinase, which is not activated by fibrin. The second TPA exon specifies a growth-factor domain which has apparently been obtained from the gene for epidermal growth factor and which may enable TPA to stimulate cell proliferation. The last two exons code for 'kringle' structures which TPA uses to bind to fibrin clots; these kringle exons come from the plasminogen gene (Li and Graur, 1991).

Type I collagen and TPA provide elegant examples of gene evolution, but, unfortunately the clear links that they display between structural domains and exons are exceptional and rarely seen with other genes. Many other genes appear to have evolved by duplication and shuffling of segments, but in these the structural domains are coded by segments of genes that do not coincide with individual exons or even groups of exons. Domain duplication and shuffling still occur, but presumably in a less precise manner with many of the rearranged genes having no useful function. Despite being haphazard the process clearly works, as indicated by, among other examples, the number of proteins that share the same DNA-binding motifs (Section 7.4.2). Several of these motifs probably evolved *de novo* on more than one occasion, but it is clear that in many cases the nucleotide

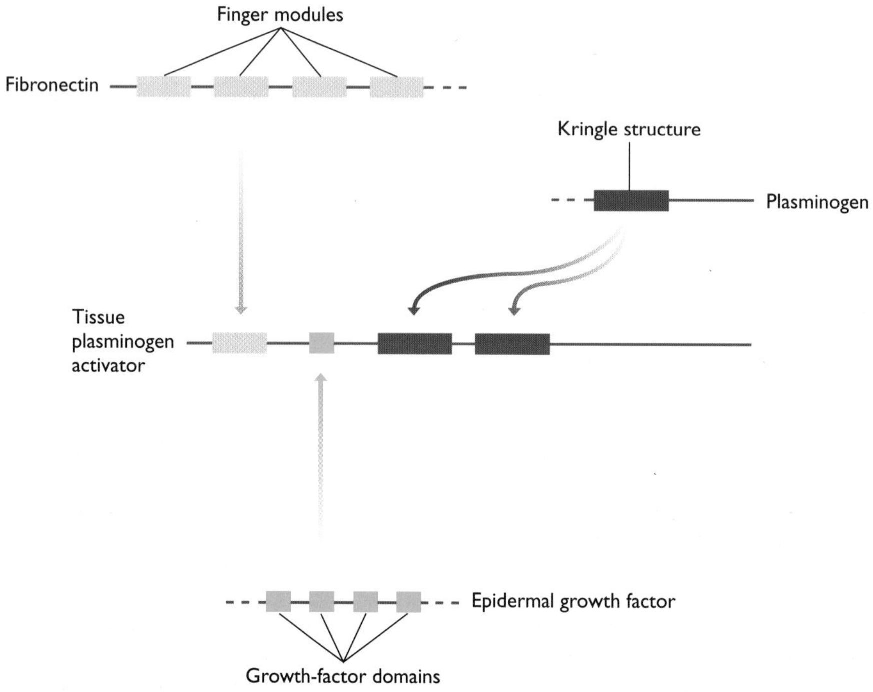

Figure 14.14 The modular structure of the tissue plasminogen activator protein.

See the text for details.

sequence coding for the motif has been transferred to a variety of different genes.

14.2.2 Acquisition o new genes from other species

The second possible way in which a genome can acquire new genes is to obtain them from another species. In bacteria, there are several mechanisms by which genes can be transferred between species but it is difficult to be sure how important these various processes have been in shaping genomes. Conjugation (Section 2.3.2), for example, enables plasmids to move between species and frequently results in the acquisition of new gene functions by the recipients. On a day-to-day basis, plasmid transfer is important because it is the means by which genes for resistance to antibiotics such as chloramphenicol, kanamycin and streptomycin spread through bacterial populations and across species barriers, but its evolutionary relevance is questionable. It is true that the genes transferred by conjugation can become integrated into the recipient bacterium's genome, but usually the process is reversible because the genes are carried by composite transposons (see *Figure 6.21B*, p. 140), so the process probably does not result in a substantial amount of permanent change to the genome.

A second process for DNA transfer between bacteria, transformation (Section 2.3.2), is more likely to have had an influence on genome evolution. Only a few bacteria, notably members of the *Bacillus*, *Pseudomonas* and *Streptococcus* genera, have efficient mechanisms for the uptake of DNA from the surrounding environment, but over evolutionary timescales all species have probably acquired at least a few genes through this route, and in some species transformation might have been a major factor in genome evolution. From analysis of completely sequenced prokaryotic genomes it is becoming increasingly clear that there has been widespread exchange of genes between species – for example, between the hyperthermophilic bacterium *Aquifex aeolicus* and archaeal counterparts such as *Methanococcus jannaschii* (Aravind *et al.*, 1998), and between *Helicobacter pylori* and various distantly related bacteria (Doolittle, 1997). Conjugation between these species is unlikely so the gene transfers are thought to have occurred by transformation.

In plants, new genes can be acquired by polyploidization. We have already seen how autopolyploidization can result in genome duplication in plants (see *Figure 14.6*, p. 374). **Allopolyploidy**, which results from interbreeding between two different species, is also common and, like autopolyploidy, can result in a viable hybrid. Usually, the two species that form the allopolyploid are closely related and have many genes in common, but each parent will possess a few novel genes or at least distinctive alleles of shared genes. For example, the bread wheats, *Triticum aestivum*, are hexaploids that arose by allopolyploidization between cultivated emmer wheat, which is a tetraploid, and a diploid wild grass, *Aegilops squarrosa*. The wild grass nucleus contained novel alleles for the high-molecular-weight glutenin genes which, when combined with the glutenin alleles already present in emmer wheat, resulted in the superior breadmaking properties displayed by the hexaploid wheats. Allopolyploidization can therefore be looked on as a combination of genome duplication and interspecies gene transfer.

Box 14.3: The origins of the eukaryotic cell

We have already examined the endosymbiont theory (Section 6.1.2) and seen that there is persuasive evidence for the view that chloroplasts and mitochondria are derived from bacteria that were engulfed by the protoeukaryotic cell and subsequently formed a stable symbiosis resulting in the eukaryotic cells that we see today. The nucleus of the protoeukaryote is thought to have been derived from the archaea because when the sequences of eukaryotic nuclear genes are compared with those of various types of prokaryote, the greatest similarities are seen with the group of archaea called the eocytes. One theory is that the protoeukaryote itself arose by endosymbiosis, possibly the engulfment of an eocyte by a cell whose biochemistry was still directed by RNA protogenomes. Another proposal is that the symbiosis was between the eocyte and a Gram-negative bacterium (Gupta and Golding, 1996), though the basis of this proposal – apparent similarities between some eukaryotic nuclear genes and equivalent genes in bacteria such as *Escherichia coli* – is not accepted by all evolutionary biologists (Roger and Brown, 1996).

Among animals, the species barriers appear to be less easy to cross and it is difficult to find evidence for **horizontal gene transfer** of any kind. Many eukaryotic genes have features associated with archaeal or bacterial sequences, but this is thought to reflect the endosymbiotic origins of eukaryotic cells (see Box 14.3) rather than recent gene acquisitions, although it has been suggested that incorporation into eukaryotic genomes of bacterial genes ingested with foodstuffs is more prevalent than might be imagined (Doolittle, 1998). Most proposals for gene transfer between animal species center on retroviruses and transposable elements. Transfer of retroviruses between animal species is well documented, as is their ability to carry animal genes between individuals of the same species, suggesting that they might be possible mediators of horizontal gene transfer. The same could be true of transposable elements such as P elements, which are known to spread from one *Drosophila* species to another, and *mariner* (Section 6.3.2), which has also been shown to transfer between *Drosophila* species and may also have crossed from other species into humans (Robertson *et al.*, 1996; Hartl *et al.*, 1997).

14.3 NONCODING DNA AND GENOME EVOLUTION

So far we have concentrated our attention on the evolution of the coding component of the genome. As coding DNA makes up only 3% of the human genome (see Box 6.4, p. 135) our view of genome evolution would be very incomplete if we did not devote some time to considering noncoding DNA. The problem is that in many respects there is little that can be said about the evolution of noncoding DNA. We envisage that duplications and other rearrangements have occurred through recombination and replication slippage, and that sequences have diverged through accumulation of mutations unfettered by the restraining selective forces acting on functional regions of the genome. We recognize that some parts of the coding DNA, for example the regulatory regions upstream of genes, have important functions, but as far as most of the noncoding DNA is concerned, all we can say is that it evolves in an apparently random fashion.

This randomness does not apply to all components of the noncoding DNA. In particular, transposable elements and introns have interesting evolutionary histories and are of general importance in genome evolution, as described in the following two sections.

Box 14.4: The origin of a microsatellite

There is no mystery about the origins of microsatellite repeat sequences (Section 6.3.1). A dimeric microsatellite, consisting of two repeat units in tandem array, can easily arise by chance mutational events. Replication slippage (Section 13.1.1) can then increase the copy number to three, four and more.

The birth of a microsatellite within a globin pseudogene of primates has been documented by comparing nucleotide sequences of various species (Messier *et al.*, 1996). In the orangutan, gibbon and other species some distance from humans, the sequence 5′-ATGT-GTGT-3′ occurs at the relevant position. In the gorilla, bonobo, chimpanzee and human a single point mutation has changed this sequence to 5′-ATGTATGT-3′, with subsequent expansion of the 5′-ATGT-3′ repeat motif.

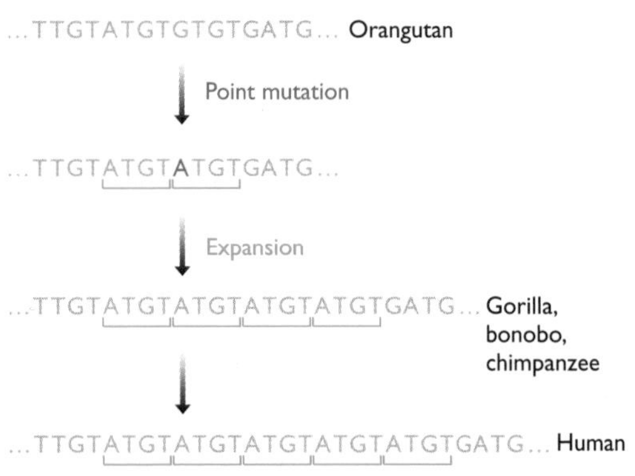

14.3.1 Transposable elements and genome evolution

Transposable elements have a number of effects on evolution of the genome as a whole. The most significant of these is the ability of transposons to initiate recombination events leading to genome rearrangements. This has nothing to do with the transposable activity of these elements, it simply relates to the fact that different copies of the same element have similar sequences and therefore can initiate recombination between two parts of the same chromosome or between different chromosomes (*Figure 14.15*). In many cases, the resulting rearrangement will be harmful because important genes will be deleted, but some instances where the result has been beneficial have been documented. For example, recombination between a pair of LINE-1 elements (Section 6.3.2) approximately 35 million years ago is thought to have caused the β-globin gene duplication that resulted in the Gγ and Aγ members of this gene family (see *Figure 14.8*, p. 377; Maeda and Smithies, 1986).

Transposition can also result in altered gene expression patterns. For example, if a transposon moves into a new site immediately upstream of a gene then it might disrupt the efficiency with which DNA-binding proteins attached to upstream regulatory sequences activate transcription of the gene (*Figure 14.16*). Transcription of the gene might also be influenced by the presence of promoters and/or enhancers within the transposon, so the gene becomes subject to an entirely new regulatory regime (McDonald, 1995). An interesting example of transposon-directed gene expression occurs with the mouse gene *Slp* which codes for a protein involved in the immune response, the tissue specificity of *Slp* being conferred by an enhancer located within an adjacent retrotransposon (Stavenhagen

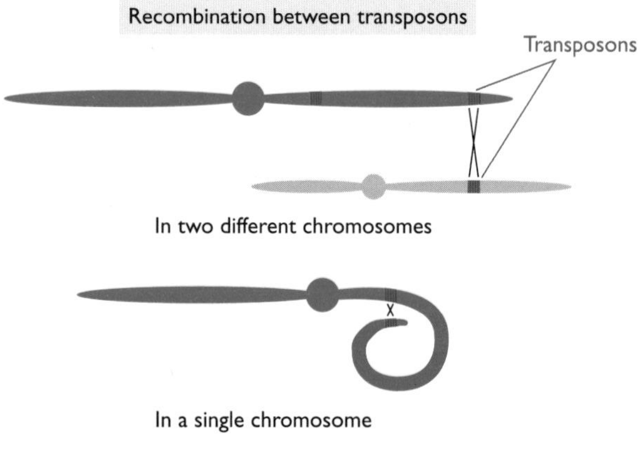

Figure 14.15 Transposons can initiate recombination events between chromosomes or between different sites in the same chromosome.

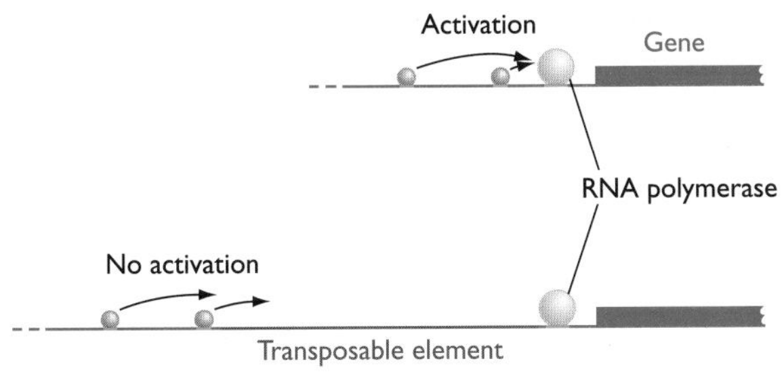

Figure 14.16 Insertion of a transposon into the region upstream of a gene could affect the ability of DNA-binding proteins to activate transcription.

and Robins, 1988). There are also examples where insertion of a transposon into a gene has resulted in an altered splicing pattern (Purugganan and Wessler, 1992).

In evolutionary terms, therefore, transposons have potentially beneficial effects for the cell that harbors them. In the short term, transposition is often deleterious because insertion of a transposon into the coding region of a gene usually results in inactivation of that gene. For this reason, in most cells, active transposons are repressed by methylation of their DNA sequences, this methylation inhibiting transcription of the genes contained within the transposon and preventing transposition (see Box 8.1, p. 177).

14.3.2 The origins of introns

Ever since introns were first discovered in the 1970s their origins have been debated. There are few controversies surrounding the Group I, II and III types (see *Table 9.3*, p. 212) as it is generally accepted that all these self-splicing introns evolved in the RNA world and have survived ever since without undergoing a great deal of change. The problems surround the origins of the GU-AG introns, the ones that are found in large numbers in eukaryotic nuclear genomes.

'Introns early' and 'introns late': two competing hypotheses

A number of proposals for the origins of GU-AG introns have been put forward but the debate is generally looked on as being between two opposing hypotheses:

- **'Introns early'** states that introns are very ancient and are gradually being lost from eukaryotic genomes.
- **'Introns late'** states that introns evolved relatively recently and are gradually accumulating in eukaryotic genomes.

There are several different models for each hypothesis. For 'introns early' the most persuasive model is the one also called the 'exon theory of genes' (Gilbert, 1987) which holds that introns were formed when the first DNA genomes were constructed, soon after the end of the RNA world. These genomes would have contained many short genes, each derived from a single coding RNA molecule and each specifying a very small polypeptide, perhaps just a single structural domain. These polypeptides would probably have had to associate together into larger, multidomain proteins in order to produce enzymes with specific and efficient catalytic mechanisms (*Figure 14.17*). To aid the synthesis of a multidomain enzyme it would have been beneficial for its individual polypeptides to become linked into a single protein, such as we see today. This was achieved by splicing together the transcripts of the relevant minigenes, a process that was aided by rearranging the genome so that groups of minigenes specifying the different parts of individual multidomain proteins were positioned next to each other. In other words, the minigenes became exons and the DNA sequences between them became introns.

According to the exon theory of genes and other 'introns early' hypotheses, all genomes originally possessed introns. But we know that bacterial genomes do not have GU-AG introns, so if these hypotheses are correct then we must assume that for some reason introns became lost from the ancestral bacterial genome at an early stage in its evolution. This is a stumbling block because it is difficult to envisage how a large number of introns could be lost from a genome without risking the disruption of many gene functions. If an intron is

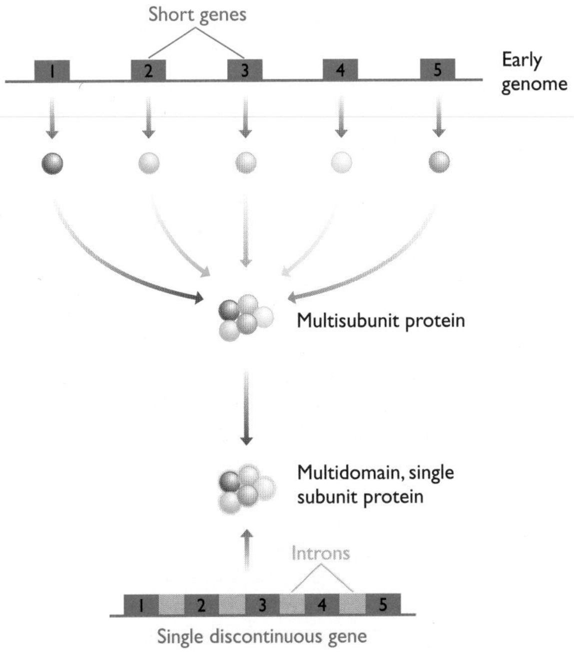

Figure 14.17 The 'exon theory of genes'.

The short genes of the first genomes probably coded for single-domain polypeptides that would have had to associate together to form a multisubunit protein to produce an effective enzyme. Later the synthesis of this enzyme could have been made more efficient by linking the short genes together into one discontinuous gene coding for a multidomain, single subunit protein.

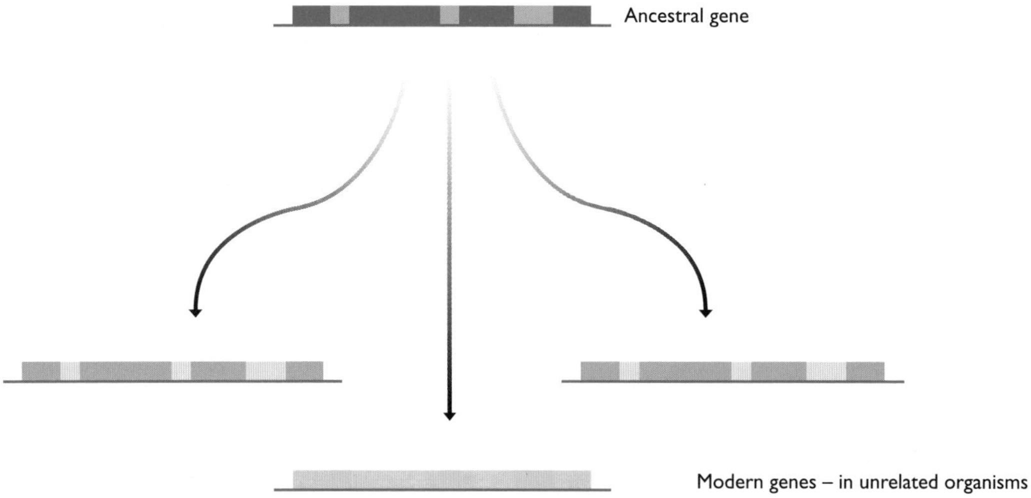

Figure 14.18 One prediction of the 'introns early' hypothesis is that the positions of introns in homologous genes should be similar in unrelated organisms, because all these genes are descended from an ancestral intron-containing gene.

removed from a gene with any imprecision then a part of the coding region will be lost or a frameshift mutation will occur, both of which would be expected to inactivate the gene. The 'introns late' hypothesis avoids this problem by proposing that to begin with no genes had introns, these structures invading the early eukaryotic nuclear genome and subsequently proliferating into the numbers seen today. The similarities between the splicing pathways for GU-AG and Group II introns (Section 9.3.3) suggest that the invaders that gave rise to GU-AG introns might well have been Group II sequences that escaped from organelle genomes. However, the similarity between GU-AG and Group II introns does not prove the 'introns late' view, because it is equally possible to devise an 'introns early' model, different to the exon theory of genes, in which Group II sequences gave rise to GU-AG introns, but at a very early stage in genome evolution.

The current evidence disproves neither hypothesis

One of the reasons why the debate regarding the origin of GU-AG introns has continued for over 20 years is because evidence in support of either hypothesis has been difficult to obtain and is often ambiguous. One prediction of 'introns early' is that there should be a close similarity between the positions of introns in homologous genes from unrelated organisms, because all these genes are descended from an ancestral intron-containing gene (*Figure 14.18*). Early support for 'introns early' came when this was shown to be the case for four introns in animal and plant genes for triosephosphate isomerase (Gilbert *et al.*, 1986), but when a larger number of species were examined the positions of the introns in this gene became less easy to interpret: it appeared that introns had been lost in some lineages but gained in others. This scenario fits both 'introns early' and 'introns late' as both allow for the loss, gain or repositioning of introns by recombination events occurring in individual lineages. When many genes in many organisms are examined the general picture that emerges is that intron numbers have gradually increased during the evolution of animal genomes, this being put forward as evidence for 'introns late' (Palmer and Logsdon, 1991), despite the fact that animal mitochondrial genomes do not contain Group II introns that could supplement the existing nuclear

Box 14.5: The role of noncoding DNA

The presence of extensive amounts of noncoding DNA in eukaryotic genomes (see Box 6.4, p. 135) is a puzzle for molecular evolutionists. Why is this apparently superfluous DNA tolerated?

One possibility is that the noncoding DNA has a function that has not yet been identified and, as such, must be maintained because without it the cell would be inviable. Plausible functions are not as difficult to identify as might be imagined. At several places in the preceding chapters the importance of chromatin structure has been stressed, including the attachment of chromatin to structures within the nucleus. Possibly, the noncoding component of genomes is involved in these aspects of genome organization (Cavalier-Smith, 1978). Alternatively, noncoding DNA might have a broad-ranging control function that so far has eluded discovery by molecular biologists (Zuckerkandl, 1976).

A second possibility is that noncoding DNA has no function but is tolerated by the genome because there is no selective pressure to get rid of it. Possession of noncoding DNA is therefore neither an advantage nor a disadvantage and so it is simply propagated along with the coding DNA. The noncoding DNA could simply be 'junk' or could be parasitic 'selfish DNA' (Orgel and Crick, 1980).

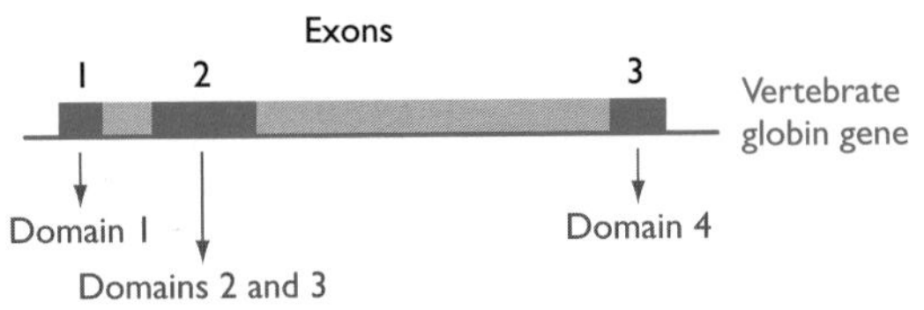

Figure 14.19 A vertebrate globin gene showing the relationship between the three exons and the four domains of the globin protein.

introns by repeated invasions. Intron numbers must therefore have increased by recombination events, which is possible with both hypotheses.

An alternative approach has been to try to correlate exons with protein structural domains, as the 'introns early ' hypothesis predicts that such a link should be evident, even allowing for the fuzzying effects of evolution since the primitive minigenes were assembled into the first real genes. Again, the first evidence to be obtained supported 'introns early'. A study of vertebrate globin proteins concluded that each of these comprises four structural domains, the first corresponding to exon 1 of the globin gene, the second and third to exon 2, and the fourth to exon 4 (*Figure 14.19*; Go, 1981). The prediction that there should be globin genes with another intron that splits the second and third domains was found to be correct when the leghemoglobin gene of soybean was shown to have an intron in exactly the expected position (Jensen *et al.*, 1981). Unfortunately, as more globin genes were sequenced more introns were discovered, over ten in all. The positions of the majority of these do not correspond to junctions between domains.

The globin genes therefore conform with the general principle that emerged from our discussion of domain shuffling (Section 14.2.1): that in most cases there are no clear links between gene exons and protein structural domains. But is our definition of 'structural domain' correct? Structural domains within a protein may not simply correspond with groups of secondary structures such as α-helices and β-sheets. A more subtle interpretation might be that a structural domain is a polypeptide segment whose amino acids are less than a certain distance apart in the protein's tertiary structure. It has been suggested that when this definition is adopted there is a better correlation between structural domain and exon (de Souza *et al.*, 1996).

14.4 THE HUMAN GENOME: THE LAST FIVE MILLION YEARS

Although the evolutionary history of humans is controversial, it is generally accepted that our closest relative among the primates is the chimpanzee and that the most recent ancestor that we share with the chimps lived about 4.6–5.0 million years ago (Takahata, 1995). Since the split the human lineage has embraced two genera – *Australopithecus* and *Homo* – and a number of species, not all of

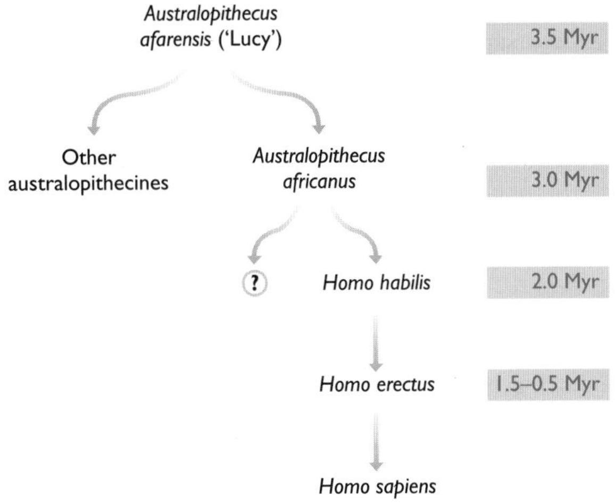

Figure 14.20 One possible scheme for the evolution of modern humans from australopithecene ancestors.

There are many controversies in this area of research and several different hypotheses have been proposed for the evolutionary relationships between different fossils. Abbreviation: Myr, million years.

which were on the direct line of descent to *Homo sapiens* (*Figure 14.20*). The result is us, a novel species in possession of what are, at least to our eyes, important biological attributes that make us very different from all other animals. So how different are we from the chimpanzees?

As far as our genomes are concerned the answer is 'about 1.5%', this being the extent of the nucleotide sequence dissimilarity between humans and chimpanzees. Within the coding DNA the difference is less than 1.5%, with many genes having identical sequences in the two genomes, but even in the noncoding regions the dissimilarity is rarely more than 3%. Only a few clear differences have been discovered:

- Several recent gene duplications have occurred, resulting in gene copies that can be described as human-specific or chimpanzee-specific, as they are present only in one or the other genome. However, as far as gene functions are concerned these new genes are not significant because they have not yet had time to accumulate mutations to any great extent and so, in effect, are simply second copies of the genes from which they were derived.
- Some components of the noncoding DNA in the two genomes have diverged extensively, illustrating how quickly repetitive DNA can evolve. For example, the alphoid DNA sequences present at human centromeres (Section 6.3.1) are quite different from the equivalent sequences in chimpanzee and gorilla chromosomes (Archidiacono *et al.*, 1995). The human genome also contains novel versions of the *Alu* element (Section 6.3.2; Zietkiewicz *et al.*, 1994).
- Human and chimpanzee genomes have undergone a few rearrangements, as revealed when the chromosome banding patterns are compared. The most

dramatic difference is that human chromosome 2 is two separate chromosomes in chimpanzees (*Figure 14.21*), so chimpanzees, as well as other apes, have 24 pairs of chromosomes whereas humans have just 23 pairs. Four other chromosomes – human numbers 5, 6, 9 and 12 – also have visible differences to their chimpanzee counterparts, though the other 18 chromosomes appear to be very similar if not identical (Yunis and Prakash, 1982).

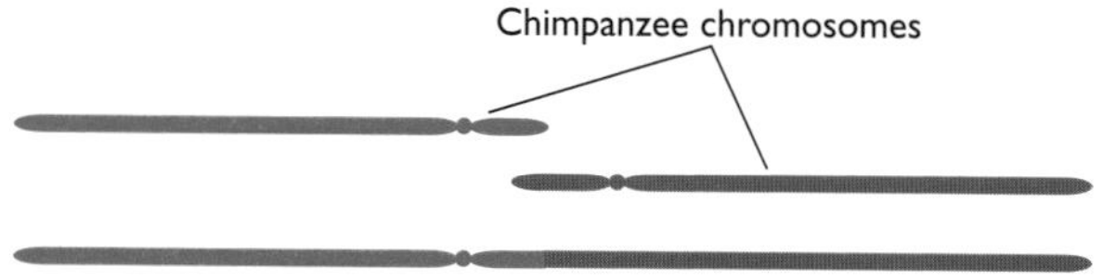

Figure 14.21 Human chromosome 2 is the product of a fusion between two chimpanzee chromosomes.

For more details about the banding patterns of these chromosomes, from which the fusion is deduced, see Strachan and Read (1996).

Some of these differences are significant as far as genome evolution is concerned but none of them reveal anything about the basis of the special biological attributes possessed by humans. This question – what makes us different from chimpanzees and other apes – is perplexing molecular biologists who have been frustrated by the absence of a sequencing project for any of the ape genomes (Gibbons, 1998). But this is only part of the problem because many of the key differences between humans and apes are likely to lie with subtle changes in the expression patterns of genes involved in developmental processes and in specification of interconnections within the nervous system, the latter being an aspect of gene function that we know virtually nothing about. What makes us human is probably not the human genome itself, but the way in which the genome functions.

REFERENCES

Amores A, Force A, Yan Y-L, *et al.* (1998) Zebrafish *hox* clusters and vertebrate genome evolution. *Science*, **282**, 1711–1714.

Aravind L, Tatisov RL, Wolf YI, Walker DR and Koonin EV (1998) Evidence for massive gene exchange between archaeal and bacterial hyperthermophiles. *Trends Genet.*, **14**, 442–444.

Archidiacono N, Antonacci R, Marzella R, Finelli P, Lonoce A and Rocchi M (1995) Comparative mapping of human alphoid sequences in great apes using fluoresence *in situ* hybridization. *Genomics*, **25**, 477–484.

Barker WC and Dayhoff MO (1980) Evolutionary and functional relationships of homologous physiological mechanisms. *BioScience*, **30**, 593–600.

Bird AP (1995) Gene number, noise reduction and biological complexity. *Trends Genet.*, **11**, 94–100.

Brooke NM, Garcia-Fernàndez J and Holland PWH (1998) The ParaHox gene cluster is an evolutionary sister of the Hox gene cluster. *Nature*, **392**, 920–922.

Brookfield JFY (1997) Genetic redundancy. *Adv. Genet.*, **36**, 137–155.

Brown DD, Wensink PC and Jordan E (1972) A comparison of the ribosomal DNAs of *Xenopus laevis* and *Xenopus mulleri*: evolution of tandem genes. *J. Mol. Biol.*, **63**, 57–73.

Cavalier-Smith T (1978) Nuclear volume control by nucleoskeletal DNA, selection for cell volume and cell growth rate and the solution to the DNA C-value paradox. *J. Cell Sci.*, **34**, 247–278.

Csermely P (1997) Proteins, RNAs and chaperones in enzyme evolution: a folding perspective. *Trends Biochem. Sci.*, **22**, 147–149.

de Souza SJ, Long M, Schoenbach L, Roy SW and Gilbert W (1996) Intron positions correlate with module boundaries in ancient proteins. *Proc. Natl Acad. Sci. USA*, **93**, 14632–14636.

Doolittle RF (1987) The evolution of the vertebrate plasma proteins. *Biol. Bull.*, **172**, 269–283.

Doolittle RF (1997) A bug with excess gastric avidity. *Nature*, **388**, 515–516.

Doolittle WF (1998) You are what you eat: a gene transfer ratchet could account for bacterial genes in eukaryotic nuclear genomes. *Trends Genet.*, **14**, 307–311.

Ekland EH and Bartel DP (1996) RNA-catalysed RNA polymerization using nucleoside triphosphates. *Nature*, **382**, 373–376.

Gibbons A (1998) Which of our genes makes us human? *Science*, **281**, 1432–1434.

Gilbert W (1987) The exon theory of genes. *Cold Spring Harbor Symp. Quant. Biol.*, **52**, 901–905.

Gilbert W, Marchionni M and McKnight G (1986) On the antiquity of introns. *Cell*, **46**, 151–153.

Go M (1981) Correlation of DNA exonic regions with protein structural units in hemoglobin. *Nature*, **291**, 90–92.

Gupta RS and Golding GB (1996) The origin of the eukaryotic cell. *Trends Biochem. Sci.*, **21**, 166–171.

Hartl DL, Lohe AR and Lozovskaya ER (1997) Modern thoughts on an ancyent *marinere*: function, evolution, regulation. *Annu. Rev. Genet.*, **31**, 337–358.

Henikoff S, Greene EA, Pietrokovski S, Bork P, Attwood TK and Hood L (1997) Gene families: the taxonomy of protein paralogs and chimeras. *Science*, **278**, 609–614.

Jensen EQ, Paludan K, Hyldig-Nielsen JJ, Jorgensen P and Markere KA (1981) The structure of a chromosomal leghemoglobin gene from soybean. *Nature*, **291**, 677–679.

Li W-H and Graur D (1991) *Fundamentals of Molecular Evolution*. Sinauer, Sunderland, MA.

Lohse PA and Szostak JW (1996) Ribozyme-catalysed amino-acid transfer reactions. *Nature*, **381**, 442–444.

Maeda N and Smithies O (1986) The evolution of multigene families: human haptoglobin genes. *Annu. Rev. Genet.*, **20**, 81–108.

McDonald JF (1995) Transposable elements: possible catalysts of organismic evolution. *Trends Ecol. Evol.*, **10**, 123–126.

Messier W, Li S-H and Stewart C-B (1996) The birth of microsatellites. *Nature*, **381**, 483.

Miller SL (1953) A production of amino acids under possible primitive Earth conditions. *Science*, **117**, 528–529.

Nowak MA, Boerlijst MC, Cooke J and Maynard Smith J (1997) Evolution of genetic redundancy. *Nature*, **388**, 167–170.

Ohno S (1970) *Evolution by Gene Duplication*. George Allen and Unwin, London.

Orgel LE (1998) The origin of life – a review of facts and speculations. *Trends Biochem. Sci.*, **23**, 491–495.

Orgel LE and Crick FHC (1980) Selfish DNA: the ultimate parasite. *Nature*, **284**, 604–607.

Palmer JD and Logsdon JM (1991) The recent origin of introns. *Curr. Opin. Genet. Dev.*, **1**, 470–477.

Purugganan MD and Wessler S (1992) The splicing of transposable elements and its role in intron evolution. *Genetica*, **86**, 295–303.

Robertson HM, Zumpano KL, Lohe AR and Hartl DL (1996) Reconstructing the ancient mariners of humans. *Nature Genet.*, **12**, 360–361.

Robertson MP and Ellington AD (1998) How to make a nucleotide. *Nature*, **395**, 223–225.

Roger AJ and Brown JR (1996) A chimeric origin for eukaryotes re-examined. *Trends Biochem. Sci.*, **21**, 370–371.

Romero D and Palacios R (1997) Gene amplification and genomic plasticity in prokaryotes. *Annu. Rev. Genet.*, **31**, 91–111.

Spring J (1997) Vertebrate evolution by interspecific hybridization – are we polyploid? *FEBS Lett.*, **400**, 2–8.

Stavenhagen JB and Robins DM (1988) An ancient provirus has imposed androgen regulation on the adjacent mouse sex limited protein gene. *Cell*, **55**, 247–254.

Strachan T and Read AP (1996) *Human Molecular Genetics*. BIOS Scientific Publishers, Oxford.

Szathmáry E (1993) Coding coenzyme handles: a hypothesis for the origin of the genetic code. *Proc. Natl Acad. Sci. USA*, **90**, 9916–9920.

Szathmáry E and Maynard Smith J (1993) The origin of chromosomes. II. Molecular mechanisms. *J. Theoret. Biol.*, **164**, 447–454.

Takahata N (1995) A genetic perspective on the origin and history of humans. *Annu. Rev. Ecol. System.*, **26**, 343–372.

Unrau PJ and Bartel DP (1998) RNA-catalysed nucleotide synthesis. *Nature*, **395**, 260–263.

Wächtershäuser G (1988) Before enzymes and templates: theory of surface metabolism. *Microbiol. Rev.*, **52**, 452–484.

Woese CR (1979) A proposal concerning the origin of life on the planet Earth. *J. Mol. Evol.*, **13**, 95–101.

Wolfe KH and Shields DC (1997) Molecular evidence for an ancient duplication of the entire yeast genome. *Nature*, **387**, 708–713.

Yunis JJ and Prakash O (1982) The origin of Man: a chromosomal pictorial legacy. *Science*, **215**, 1525–1530.

Zietkiewicz E, Richer C, Makalowski W, Jurka J and Labuda D (1994) A young *Alu* subfamily amplified independently in human and African great apes lineages. *Nucleic Acids Res.*, **22**, 5608–5612.

Zuckerkandl E (1976) Gene control in eukaryotes and c-value paradox: 'excess' DNA as an impediment to transcription of coding sequences. *J. Mol. Evol.*, **9**, 73–104.

FURTHER READING

Futuyama DJ (1998) *Evolutionary Biology*, 3rd edn. Sinauer, Sunderland, MA. — *An accessible description of evolutionary biology.*

Jackson M, Strachan T and Dover GA (1996) *Human Genome Evolution*. BIOS Scientific Publishers, Oxford. — *An advanced treatment of the subject.*

Li W-H (1997) *Molecular Evolution*. Sinauer, Sunderland, MA. — *Detailed descriptions of many of the topics covered in this chapter.*

Maconochie M, Nonchev S, Morrison A and Krumlauf R (1996) Paralogous *Hox* genes: function and regulation. *Annu. Rev. Genet.*, **30**, 529–556.

Maynard Smith J and Szathmáry E (1995) *The Major Transitions in Evolution*. WH Freeman, Oxford. — *A remarkable book that begins with the origin of life and ends with the evolution of human language.*

CHAPTER

15 Molecular Phylogenetics

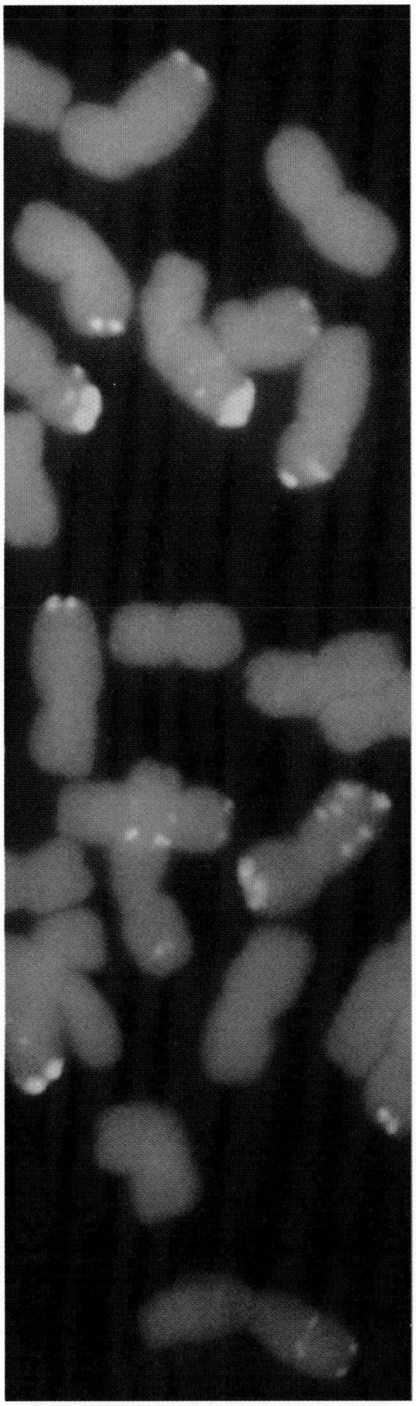

Contents

15.1 The Origins of Molecular Phylogenetics **392**

15.2 The Reconstruction of DNA-based Phylogenetic Trees **394**

15.2.1 The key features of DNA-based phylogenetic trees 394

15.2.2 Tree reconstruction 397

15.3 The Applications of Molecular Phylogenetics **402**

15.3.1 Examples of the use of phylogenetic trees 402

15.3.2 Molecular phylogenetics as a tool in the study of human prehistory 406

Concepts

- *Molecular phylogenetics enables the evolutionary relationships between DNA sequences to be deduced*
- *Gene trees are not the same as species trees*
- *The reconstruction of phylogenetic trees involves alignment of homologous DNA sequences followed by any one of several approaches to tree building*
- *Molecular clocks can be used to estimate the time of divergence of ancestral sequences, but there is no universal molecular clock so calibration is important*
- *Molecular phylogenetics has been applied to questions such as the relationships between humans and other primates, the time of the earliest life on Earth, the origins of AIDS and the relationships between prions of different species*
- *Molecular phylogenetics and related analyses are being used to study the origin of modern humans and the patterns of past human migrations*

IF GENOMES EVOLVE by the gradual accumulation of mutations, then the amount of nucleotide sequence difference between a pair of genomes should indicate how recently those two genomes shared a common ancestor. Two genomes that diverged in the recent past would be expected to have fewer differences than a pair of genomes whose common ancestor is more ancient. This means that by comparing three or more genomes with each other it should be possible to work out the evolutionary relationships between them. These are the objectives of **molecular phylogenetics**.

15.1 THE ORIGINS OF MOLECULAR PHYLOGENETICS

Molecular phylogenetics predates DNA sequencing by several decades. It is derived from the traditional method for classifying organisms according to their similarities and differences, as first practised in a comprehensive fashion by Linnaeus in the 18th century. Linnaeus was a systematicist not an evolutionist, his objective being to place all known organisms into a logically ordered classification which he believed would reveal the great plan used by the Creator – the *Systema Naturae*. However, he unwittingly laid the framework for later evolutionary schemes by dividing organisms into a hierarchical series of taxonomic categories, starting with kingdom and progressing down through phylum, class, order, family and genus to species. The naturalists of the 18th and early 19th centuries likened this hierarchy to a 'tree of life' (*Figure 15.1*), an analogy that was adopted by Darwin (1859) in *The Origin of Species* as a means of describing the interconnected evolutionary histories of living organisms. The classificatory scheme devised by Linnaeus therefore became reinterpreted as a **phylogeny** indicating not just the similarities between species but also their evolutionary relationships.

Whether the objective is to construct a classification or to infer a phylogeny, the data used in the analysis are obtained by examining variable characters in the organisms being compared. Originally, these characters were morphological features, but molecular data were introduced at a surprisingly early stage. In 1904 Nuttall used immunological tests to deduce relationships between a variety of animals, one of his objectives being to place humans in their correct evolutionary position relative to

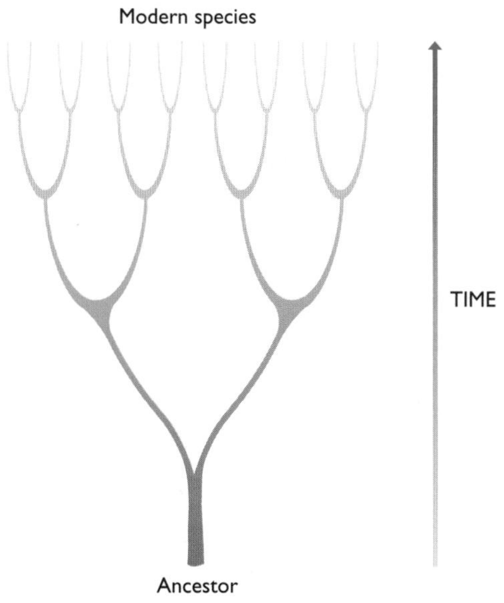

Figure 15.1 The tree of life.

An ancestral species is at the bottom of the 'trunk' of the tree. As time passes, new species evolve from earlier ones so the tree repeatedly branches until we reach the present time when there are many species descended from the ancestor.

other primates, an issue that we will return to in Section 15.3.1. Despite the success of Nuttall's work, the molecular approach was not widely adopted until the late 1950s, this delay being due largely to technical limitations, but also partly because classification and phylogenetics had to undergo their own evolutionary changes before the value of molecular data could be fully appreciated. These changes came about with the introduction of **phenetics** and **cladistics** (Box 15.1) which, although very different methods, both place emphasis on the need for large datasets that can be analyzed by rigorous mathematical procedures. The difficulty in meeting these requirements when morphological characters are used prompted a gradual shift towards proteins and DNA, which offer three advantages compared with other types of data:

- Many molecular characters can be scored at once: in effect, every nucleotide position in a DNA sequence is a character with four variable **character states**, A, C, G and T.
- Molecular character states are unambiguous: A, C, G and T are easily recognizable and one cannot be confused for another. Morphological character states, on the other hand, often 'overlap' and are less easy to distinguish.
- Molecular data are easily converted to numerical form and hence are amenable to mathematical and statistical analysis.

The sequences of protein and DNA molecules provide the most detailed and unambiguous data for molecular phylogenetics, but protein sequencing did not become routine until the late 1960s and rapid DNA sequencing

Box 15.1: Phenetics and cladistics

Phenetics, when first introduced (Michener and Sokal, 1957), challenged the prevailing view that classifications should be based on comparisons between a limited number of characters that taxonomists believed to be important for one reason or another. Pheneticists argued that classifications should encompass as many variable characters as possible, these characters being scored numerically and analyzed by rigorous mathematical methods.

Cladistics (Hennig, 1966) also emphasizes the need for large datasets but differs from phenetics in that it does not give equal weight to all characters. The argument is that in order to infer the branching order in a phylogeny it is necessary to distinguish those characters that provide a good indication of evolutionary relationships from other characters that might be misleading. This might appear to take us back to the pre-phenetic approach but cladistics is much less subjective: rather than making assumptions about which characters are 'important', cladistics demands that the evolutionary relevance of individual characters be defined. In particular, errors in the branching pattern within a phylogeny are minimized by recognizing two types of anomalous data.

- **Convergent evolution** or **homoplasy** occurs when the same character state evolves in two separate lineages. For example, both birds and bats possess wings, but bats are more closely related to wingless mammals than they are to birds. The character state 'possession of wings' is therefore misleading in the context of vertebrate phylogeny.
- **Ancestral character states** must be distinguished from **derived character states**. An ancestral (or **plesiomorphic**) character state is one possessed by a remote common ancestor of a group of organisms, an example being five toes in vertebrates. A derived (or **apomorphic**) character state is one that evolved from the ancestral state in a more recent common ancestor, and so is seen in only a subset of the species in the group being studied. Among vertebrates, the possession of a single toe, as displayed by modern horses, is a derived character state. If we did not realize this then we might conclude that humans are more closely related to lizards, which like us have five toes, rather than to horses.

Phenetics and cladistics have had an uneasy relationship over the last 40 years. Most of today's evolutionary biologists favor cladistics, even though a strictly cladistic approach throws up some apparently counterintuitive results, a notable example being the conclusion that the birds should not have their own class (Aves) but be included among the reptiles.

was not developed until ten years after that. Early studies, therefore, depended largely on indirect assessments of DNA or protein variations, using one of three methods:

- **Immunological data**, such as those obtained by Nuttall (1904), involve measurements of the amount of cross-reactivity seen when an antibody specific for a protein from one organism is mixed with the same protein from a different organism. Remember that in Section 11.2.1 we learned that antibodies are immunoglobulin proteins that help to protect the body against invasion by bacteria, viruses and other unwanted substances by binding to these 'antigens'. Proteins also act as antigens so if human β-globin, for example, is injected into a rabbit then the rabbit makes an antibody that binds specifically to this protein. The antibody will also cross-react with β-globins from other vertebrates, because these β-globins have similar structures to the human version. The degree of cross-reactivity depends on how similar the β-globin being tested is to the human protein, providing the similarity data used in the phylogenetic analysis.
- **Protein electrophoresis** is used to compare the electrophoretic properties, and hence degree of similarity, of proteins from different organisms. This technique has proved useful for comparing closely related species and variations between members of a single species.
- **DNA–DNA hybridization data** are obtained by hybridizing DNA samples from the two organisms being compared. The DNA samples are denatured and mixed together so that hybrid molecules form. The stability of these hybrid molecules depends on the degree of similarity between the nucleotide sequences of the two DNAs, and is measured by determining the melting temperature (see *Figure 2.7*, p. 22), a stable hybrid having a higher melting temperature than a less stable one. The melting temperatures obtained with DNAs from different pairs of organisms provide the data used in the phylogenetic analysis.

By the end of the 1960s these indirect methods were supplemented with an increasing number of protein sequence studies (e.g. Fitch and Margoliash, 1967) and during the 1980s DNA-based phylogenetics began to be carried out on a large scale. Protein sequences are still used today in some contexts, but DNA has now become by far the predominant molecule. This is mainly because DNA yields more phylogenetic information than protein: the nucleotide sequences of a pair of homologous genes has a higher information content than the amino sequences of the corresponding proteins, because mutations that result in nonsynonymous changes affect only the DNA sequence (*Figure 15.2*), and entirely novel information can be obtained by DNA sequence analysis because variabilities in both the coding and noncoding regions of the genome can be examined. The ease with

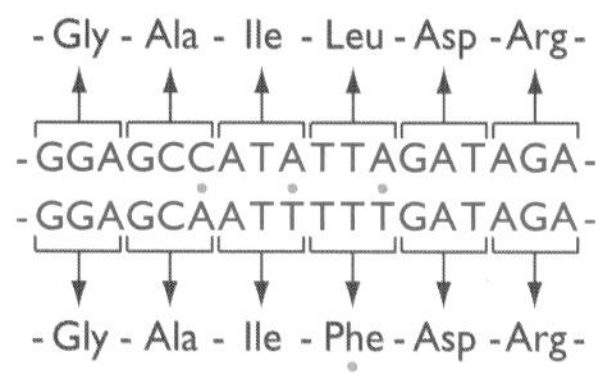

Figure 15.2 DNA yields more phylogenetic information than protein.

The two DNA sequences differ at three positions but the amino acid sequences differ at only one position.

which DNA samples for sequence analysis can be prepared by PCR (Technical Note 2.2, p. 20) is another key reason behind the predominance of DNA in modern molecular phylogenetics.

As well as DNA sequences, molecular phylogenetics also makes use of DNA markers such as RFLPs, SSLPs and SNPs (Section 2.2.2), especially for intraspecific studies such as those aimed at understanding migrations of prehistoric human populations (Section 15.3.2). Later in this chapter we will consider various examples of the use of both DNA sequences and DNA markers in molecular phylogenetics, but first we must make a more detailed study of the methodology used in this area of genome research.

15.2 THE RECONSTRUCTION OF DNA-BASED PHYLOGENETIC TREES

The objective of most phylogenetic studies is to reconstruct the tree-like pattern that describes the evolutionary relationships between the organisms being studied. Before examining the methodology for doing this we must take a closer look at a typical tree in order to familiarize ourselves with the basic terminology used in phylogenetic analysis.

15.2.1 The key features of DNA-based phylogenetic trees

A typical phylogenetic tree is shown in *Figure 15.3A*. This tree could have been reconstructed from any type of comparative data, but as we are interested in DNA sequences we will assume that the tree shows the relationships between four homologous genes, called *A*, *B*, *C* and *D*. The **topology** of this tree comprises four **external nodes**, each representing one of the four genes that we have compared, and two **internal nodes** representing ancestral genes, with the lengths of the **branches** indicating the degree of difference between the genes represented by the nodes. This degree of difference is calculated when the sequences are compared, as described in Section 15.2.2.

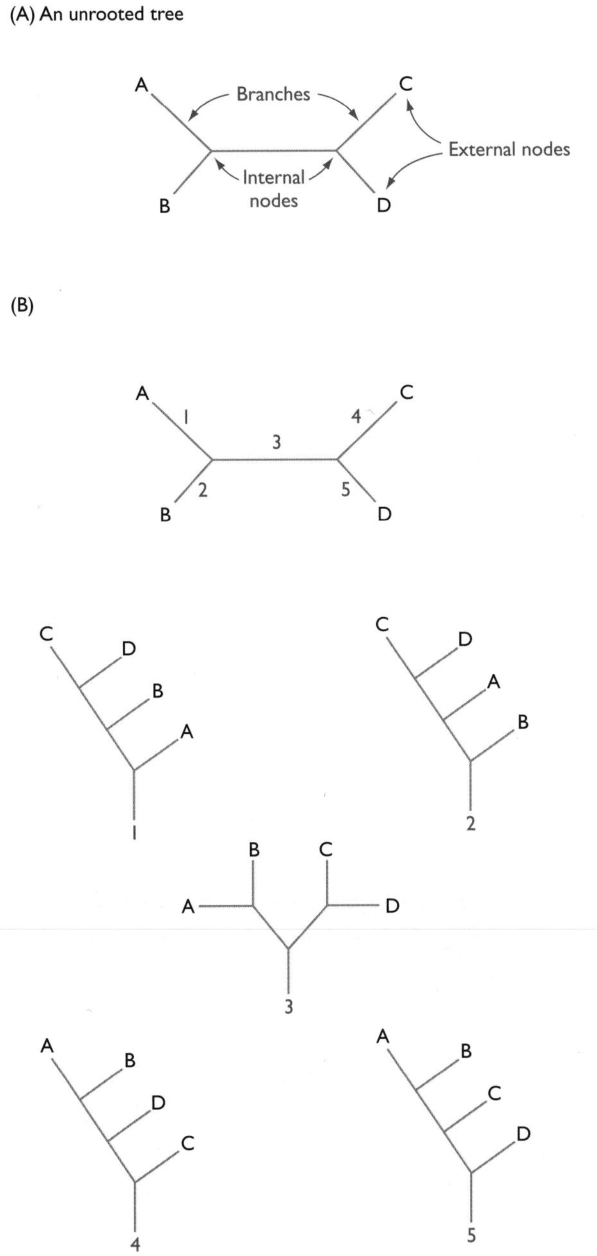

Figure 15.3 Phylogenetic trees.

(A) An unrooted tree with four external nodes. (B) The five rooted trees that can be drawn from the unrooted tree shown in part A. The positions of the roots are indicated by the numbers on the outline of the unrooted tree.

The tree in *Figure 15.3A* is **unrooted**, which means that it is only an illustration of the relationships between *A*, *B*, *C* and *D* and does not tell us anything about the series of evolutionary events that led to these genes. Five different evolutionary pathways are possible, each depicted by a different **rooted** tree, as shown in *Figure 15.3B*. To distinguish between them the phylogenetic analysis must include at least one **outgroup**, this being a homologous gene that we know is less closely related to *A*, *B*, *C* and *D* than these four genes are to each other. The outgroup enables the root of the tree to be located and the correct evolutionary pathway to be identified. The criteria used when choosing an outgroup depend very much on the type of analysis that is being carried out. As an example, let us say that the four homologous genes in our tree come from human, chimpanzee, gorilla and orangutan. We could then use as an outgroup the homologous gene from another primate, such as the baboon, which we know from palaeontological evidence branched away from the lineage leading to human, chimpanzee, gorilla and orangutan before the time of the common ancestor of those four species (*Figure 15.4*).

We refer to the rooted tree that we obtain by phylogenetic analysis as an **inferred tree**. This is to emphasize that it depicts the series of evolutionary events that are inferred from the data that were analyzed, and may not be the same as the **true tree**, the one that depicts the actual series of events that occurred. Sometimes we can be fairly confident that the inferred tree is the true tree, but most phylogenetic data analyses are prone to uncertainties which are likely to result in the inferred tree differing in some respects from the true tree. In Section 15.2.2 we will look at the various methods used to assign degrees of confidence to the branching pattern in an inferred tree, and later in the chapter we will examine some of the controversies that have arisen as a result of the imprecise nature of phylogenetic analysis.

Gene trees are not the same as species trees

The tree shown in *Figure 15.4* illustrates a common type of molecular phylogenetics project, where the objective is to

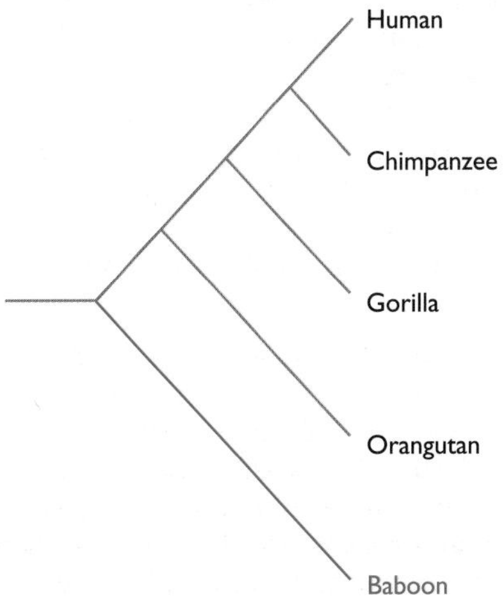

Figure 15.4 The use of an outgroup to root a phylogenetic tree.

The tree of human, chimpanzee, gorilla and orangutan genes is rooted with a baboon gene because we know from the fossil record that baboons split away from the primate lineage before the time of the common ancestor of the other four species. For more information on phylogenetic analysis of humans and other primates see Section 15.3.1.

use a **gene tree**, reconstructed from comparisons between the sequences of **orthologous** genes (those derived from the same ancestral sequence: see p. 93), to make inferences about the evolutionary history of the species from which the genes are obtained. The assumption is that the gene tree, based on molecular data with all its advantages, will be a more accurate and less ambiguous representation of the **species tree** than that obtainable by morphological comparisons. This assumption is often correct, but it does not mean that the gene tree is the *same* as the species tree. For that to be the case, the internal nodes in the gene and species trees would have to be precisely equivalent, and they are not (*Figure 15.5*):

- An internal node in a gene tree represents the divergence of an ancestral gene into two genes with different DNA sequences: this occurs by mutation.
- An internal node in a species tree represents a speciation event: this occurs by the population of the ancestral species splitting into two groups which are unable to interbreed, for example, because they are geographically isolated.

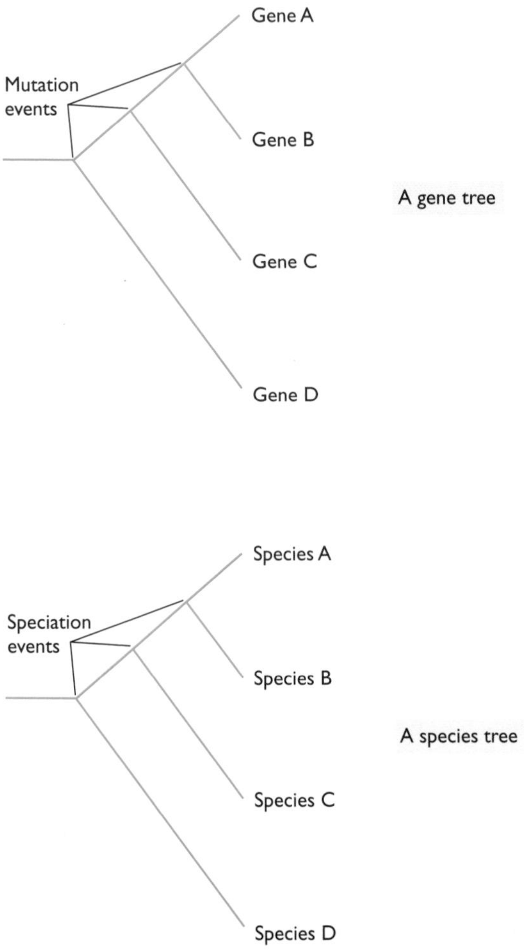

Figure 15.5 The difference between a gene tree and a species tree.

The important point is that these two events – mutation and speciation – are not expected to occur at the same time. For example, the mutation event could precede the speciation. This would mean that, to begin with, both alleles of the gene are present in the same, unsplit population of the ancestral species (*Figure 15.6*). When the population split occurs, it is likely that both alleles will still be present in each of the two resulting groups. After the split, the new populations evolve independently. One possibility is that the results of **random genetic drift** (see Box 15.5, p. 408) lead to one allele being lost from one population and the other being lost from the other population. This establishes the two separate genetic lineages that we infer from phylogenetic analysis of the gene sequences present in the modern species that result from the continued evolution of the two populations.

How do these considerations affect the coincidence between the gene and species trees? There are various implications, two of which are as follows:

- If a **molecular clock** (Section 15.2.2) is used to date the time at which the gene divergence took place, then it cannot be assumed that this is also the time of the speciation event. If the node being dated is ancient, 50 million or more years ago, for example, then the error may not be noticeable; but if the speciation event is recent, as when primates are being compared, then the date for the gene divergence might be significantly different to that for the speciation event.
- If the first speciation event is quickly followed by a second speciation in one of the two resulting populations, then the branching order of the gene tree might be different from that of the species tree. This can

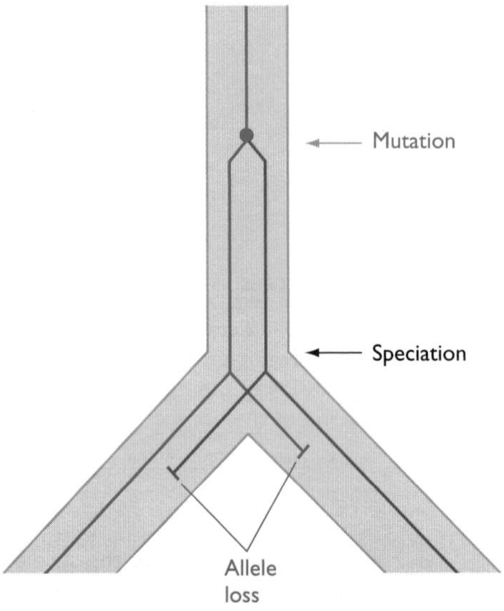

Figure 15.6 Mutation might precede speciation, giving an incorrect time for the latter if a molecular clock is used.

See the text for details. Based on Li (1997).

occur if the genes in the modern species are derived from alleles that had already appeared before the first of the two speciations, as illustrated in *Figure 15.7*.

15.2.2 Tree reconstruction

In this section we will look at how tree reconstruction is carried out with DNA sequences, concentrating on the four steps in the procedure:

- Aligning the DNA sequences and obtaining the comparative data that will be used to reconstruct the tree.
- The alternative methods that can be used to convert the comparative data into a reconstructed tree.
- The tests that are used to assign a degree of confidence to the tree as a whole and various segments within it.
- Using a **molecular clock** to assign dates to branch points within the tree.

Box 15.2: Terminology for molecular phylogenetics

The text includes definitions of most of the important terms used in molecular phylogenetics. A few additional definitions will complete the picture.

- **Operational taxonomic unit** or **OTU** is the term used to describe the organisms being compared when a tree is constructed from morphological data. 'OTU' is sometimes also used by molecular phylogeneticists as a synonym for 'gene or other nucleotide sequence', but it is perhaps best not used in this context because a gene is not, strictly speaking, a taxonomic unit. A taxonomic unit is a component of a classification scheme, such as 'species'. To describe a gene as a taxonomic unit runs the risk of confusing a gene tree with a species tree.
- **Monophyletic** refers to two or more DNA sequences that are derived from a single common ancestral DNA sequence.
- A **clade** is a group of monophyletic DNA sequences that comprises all the sequences included in the analysis that are descended from a particular common ancestral sequence.
- **Parsimony** was originally a philosophical term stating that, when deciding between competing hypotheses, preference should be given to the one that involves the fewest unconnected assumptions. In molecular phylogenetics, parsimony is an approach that decides between different tree topologies by identifying the one that involves the shortest evolutionary pathway, this being the pathway that requires the smallest number of nucleotide changes to go from the ancestral sequence, at the root of the tree, to all the present-day sequences that have been compared.

Sequence alignment is the essential preliminary to tree reconstruction

The data used in reconstruction of a DNA-based phylogenetic tree are obtained by comparing nucleotide sequences. These comparisons are made by aligning the

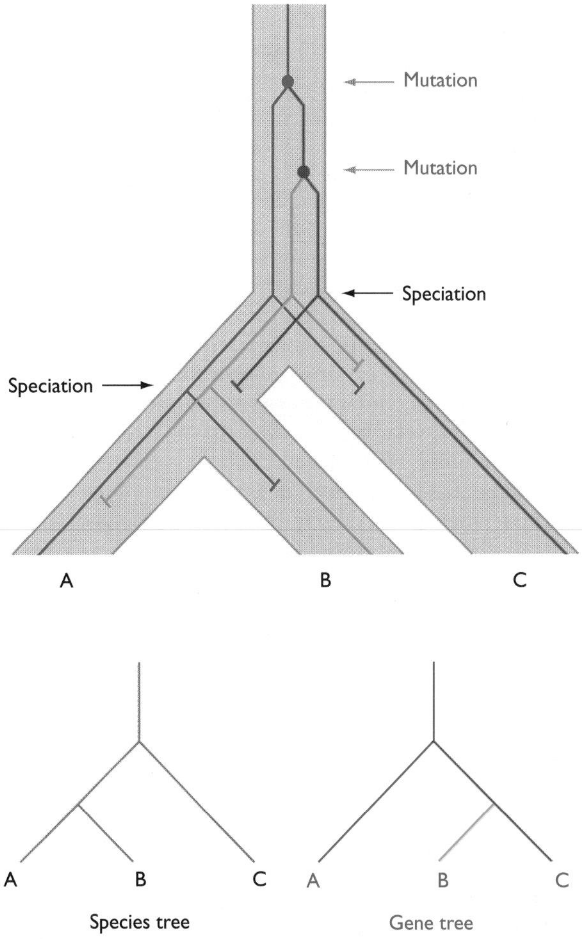

Figure 15.7 A gene tree can have a different branching order from a species tree.

In this example, the gene has undergone two mutations in the ancestral species, the first mutation giving rise to the 'blue' allele and the second to the 'green' allele. Random genetic drift in association with the two subsequent speciations result in the red allele lineage appearing in species A, the green allele lineage in species B and the blue allele lineage in species C. Molecular phylogenetics based on the gene sequences will reveal that the red-blue split occurred before the blue-green split, giving the gene tree shown on the right. But the actual species tree is different, as shown on the left. Based on Li (1997).

sequences so that nucleotide differences can be scored. This is the critical part of the entire enterprise because if the alignment is incorrect then the resulting tree will definitely not be the true tree.

The first issue to consider is whether the sequences being aligned are homologous. If they are homologous then they must, by definition, be derived from a common ancestral sequence (Section 5.2.1) and so there is a sound basis for the phylogenetic study. If they are not homologous then they do not share a common ancestor. The phylogenetic analysis will find one, because the methods used for tree reconstruction always produce a tree of some description even if the data are completely erroneous, but the resulting tree will have no biological relevance. With some DNA sequences – for example, the β-globin genes of different vertebrates – there is no difficulty in being sure that the sequences being compared are homologous. However, this is not always the case, as was described in Section 5.2.1 with reference to the use of homology analysis in assigning functions to newly-sequenced genes, and one of the commonest errors that arises during phylogenetic analysis is the inadvertent inclusion of a nonhomologous sequence.

Once it has been established that two DNA sequences are indeed homologous, the next step is to align the sequences so that homologous nucleotides can be compared. With some pairs of sequences this is a trivial exercise (*Figure 15.8A*), but it is not so easy if the sequences are relatively dissimilar and/ or they have diverged by the accumulation of insertions and deletions as well as point mutations. Insertions and deletions cannot be distinguished when pairs of sequences are compared so we refer to them as **indels**. Placing indels at their correct positions is often the most difficult part of sequence alignment (*Figure 15.8B*).

Some pairs of sequences can be aligned reliably by eye. For more complex pairs, alignment might be possible by the **dot matrix** method (*Figure 15.9*). The two sequences are written out on the x- and y-axes of a graph, and dots placed in the squares of the graph paper at positions corresponding to identical nucleotides in the two sequences. The alignment is indicated by a diagonal series of dots, broken by empty squares where the sequences have nucleotide differences, and shifting from one column to another at places where indels occur.

More rigorous mathematical approaches to sequence alignment have also been devised. The first of these is the **similarity approach** (Needleman and Wunsch, 1970), which aims to maximize the number of matched nucleotides, those that are identical in the two sequences. The complementary approach is the **distance method** (Waterman *et al.*, 1976), in which the objective is to minimize the number of mismatches. Often the two procedures will identify the same alignment as being the best one.

Usually the comparison involves many more than just two sequences, meaning that a **multiple alignment** is required. This can rarely be done effectively with pen and paper so, as in all steps in a phylogenetic analysis, a computer program is used. For multiple alignments, Clustal is often the most popular choice (Jeanmougin *et al.*, 1998). This and other software packages for phylogenetic analysis are described in Technical Note 15.1.

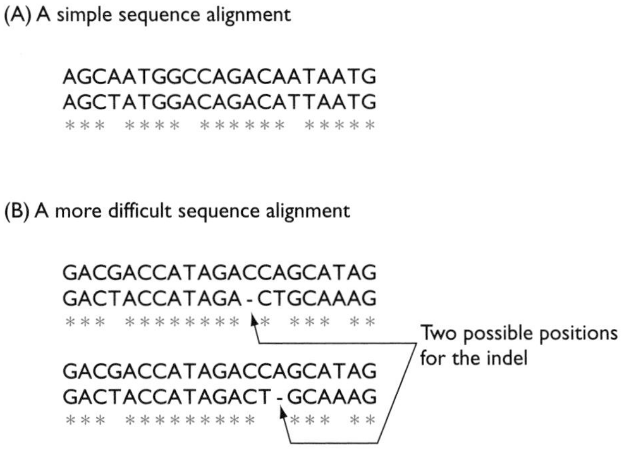

Figure 15.8 Sequence alignment.

(A) Two sequences which have not diverged to any great extent can be aligned easily by eye. (B) A more complicated alignment in which it is not possible to determine the correct position for an indel. If errors in indel placement are made in a multiple alignment then the tree reconstructed by phylogenetic analysis is likely to be incorrect. In this diagram, the red asterisks indicate nucleotides that are the same in both sequences.

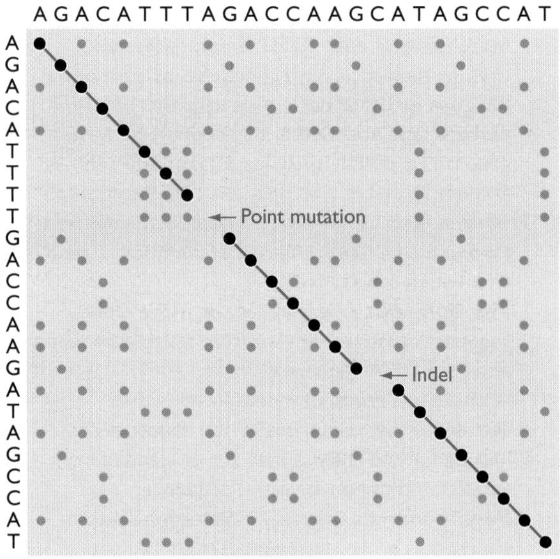

Figure 15.9 The dot matrix technique for sequence alignment.

The correct alignment stands out because it forms a diagonal of continuous dots, broken at point mutations and shifting to a different diagonal at indels.

TECHNICAL 15.1 NOTES

Phylogenetic analysis

Software packages for reconstruction of phylogenetic trees.

Few sets of DNA sequences are simple enough to be converted into phylogenetic trees entirely by hand and virtually all research in this area is carried out on the computer with the aid of any one of a variety of software packages designed specifically for tree reconstruction.

One of the easiest to use and most popular packages is Clustal, which was originally written in 1988 and has undergone several upgrades in the intervening years (Jeanmougin *et al.*, 1998). Clustal is primarily a program for carrying out multiple alignments of protein or DNA sequences, which it is able to do very effectively providing the sequences being compared do not contain extensive internal repeat motifs. Clustal is usually used in conjunction with NJplot, a simple program for tree reconstruction by the neighbor-joining method. One important advantage of Clustal and NJplot is that they do not require huge amounts of computer memory and so can be run on small PCs or Macintosh computers.

More comprehensive software packages enable the researcher to choose between a variety of different methods for tree reconstruction and to carry out more sophisticated types of phylogenetic analysis. The most widely used of these packages are PAUP (Swofford, 1993) and PHYLIP (Felsenstein, 1989). The tree-building programs in PAUP are often looked on as the most accurate ones currently available and are able to handle relatively large datasets (Eernisse, 1998). PHYLIP has the advantage of including a number of software tools not readily available from other sources. Other popular packages include MacClade and HENNIG86.

Box 15.3: Multiple substitutions

The evolutionary distance between a pair of nucleotide sequences corresponds with the number of nucleotide substitutions that have occurred in the two sequences since they diverged from their common ancestor. If the sequences do not display a great deal of divergence then the evolutionary distance can be calculated simply by counting the number of nucleotide differences, as in *Figure 15.10*. However, if the sequences are highly divergent, then the possibility of **multiple substitutions** or **multiple hits** must be taken into account. Multiple substitution occurs when a single site undergoes two or more changes, for example

Ancestral sequence ...ATGT...

Modern sequences ...AGGT... ...ACGT...

There is only one nucleotide difference between the two modern sequences, but two nucleotide substitutions have occurred. If this multiple hit is not recognized then the evolutionary distance between the two sequences will be significantly underestimated. Distance matrices for phylogenetic analysis are therefore usually constructed using mathematical methods that include statistical devices for estimating the amount of multiple substitution that has occurred.

Converting the alignment data into a phylogenetic tree

Once the sequences have been aligned accurately, an attempt can be made to reconstruct the phylogenetic tree. To date nobody has devised a perfect method for tree reconstruction and several different procedures are routinely used. Comparative tests have been run with artificial data, for which the true tree is known, but these have failed to identify any particular method as being better than any of the others (Felsenstein, 1988).

The main distinction between the different tree-building methods is the way in which the multiple sequence alignment is converted into numerical data that can be analyzed mathematically in order to reconstruct the tree. The simplest approach is to convert the sequence information into a **distance matrix**, which is simply a table showing the evolutionary distances between all pairs of sequences in the dataset (*Figure 15.10*). The evolutionary distance is calculated from the number of nucleotide differences between a pair of sequences and is used to establish the lengths of the branches connecting these two sequences in the reconstructed tree.

The **neighbor-joining method** (Saitou and Nei, 1987) is a popular tree-building procedure that uses the distance matrix approach. To begin the reconstruction, it is initially assumed that there is just one internal node from which branches leading to all the DNA sequences radiate in a star-like pattern (*Figure 15.11A*). This is virtually impossible in evolutionary terms but the pattern is just a starting point. Next, a pair of sequences is chosen at random, removed from the star, and attached to a second internal

Multiple alignment

```
1 AGGCCAAGCCATAGCTGTCC
2 AGGCAAAGACATACCTGACC
3 AGGCCAAGACATAGCTGTCC
4 AGGCAAAGACATACCTGTCC
```

Distance matrix

	1	2	3	4
1	–	0.20	0.05	0.15
2		–	0.15	0.05
3			–	0.10
4				–

Figure 15.10 A simple distance matrix.

The matrix shows the evolutionary distance between each pair of sequences in the alignment. In this example the evolutionary distance is expressed as the number of nucleotide differences per nucleotide site for each sequence pair. For example, sequences 1 and 2 are 20 nucleotides in length and have four differences, corresponding to an evolutionary difference of 4/20 = 0.2. See Box 15.3 for a description of the inaccuracy of this analysis.

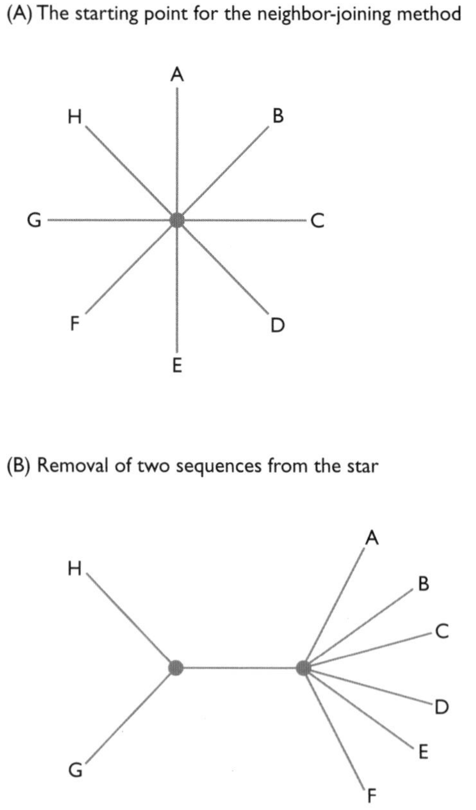

Figure 15.11 Manipulations carried out when using the neighbor-joining method for tree reconstruction.

See the text for details.

node, connected by a branch to the center of the star, as shown in *Figure 15.11B*. The total branch length in this new 'tree' is calculated. The sequences are then returned to their original positions and another pair attached to the second internal node, and again the total branch length is calculated. This operation is repeated until all the possible pairs have been examined, enabling the combination that gives the tree with the shortest total branch length to be identified. This pair of sequences will be neighbors in the final tree; in the interim, they are combined into a single unit, creating a new star with one less branch than the original one. The whole process of pair selection and tree length calculation is now repeated so that a second pair of neighboring sequences is identified, and then repeated again so a third pair is located, and so on. The result is a complete, reconstructed tree.

The advantage of the neighbor-joining method is that the data-handling is relatively easy to carry out, largely because the information content of the multiple alignment has been reduced to its simplest form. The disadvantage is that some of the information is lost, in particular that pertaining to the identities of the ancestral and derived nucleotides (equivalent to ancestral and derived character states as defined in Box 15.1, p. 393) at each position in the multiple alignment. The **maximum parsimony** method (Fitch, 1977) does take account of this information, utilizing it to recreate the series of nucleotide changes that resulted in the pattern of variation revealed by the multiple alignment. The assumption, possibly erroneous, is that evolution follows the shortest possible route and that the correct phylogenetic tree is therefore the one that requires the minimum number of nucleotide changes to produce the observed differences between the sequences. Trees are therefore constructed at random and the number of nucleotide changes that they involve calculated until all possible topologies have been examined and the one requiring the smallest number of steps identified. This is presented as the most likely inferred tree.

The maximum parsimony method is more rigorous in its approach compared with the neighbor-joining method, but this inevitably increases the amount of data handling that is involved. This is a significant problem because the number of possible trees that must be scrutinized increases rapidly as more sequences are added to the dataset. With just five sequences there are only 15 possible unrooted trees, but for ten sequences there are 2 027 025 unrooted trees and for 50 sequences the number exceeds the number of atoms in the universe (Eernisse, 1998). Even with a high speed computer it is not possible to check every one of these trees in a reasonable time, if at all, so often the maximum parsimony method is unable to carry out a comprehensive analysis. The same is true with many of the other more sophisticated methods for tree reconstruction.

Assessing the accuracy of a reconstructed tree

The limitations to the methods used in phylogenetic reconstruction lead inevitably to questions about the veracity of the resulting trees. Statistical tests of the accuracy of a reconstructed tree have been devised (Hillis, 1997) but these are necessarily complex because a tree is geometric rather than numerical, and the accuracy of one part of the topology may be greater or lesser than the accuracy of other parts.

In practice, the routine method for assigning confidence limits to different branch points within a tree is to carry out a **bootstrap analysis.** To do this we need a second multiple alignment that is different from, but equivalent to the real alignment. This new alignment is built up by taking columns, at random, from the real alignment, as illustrated in *Figure 15.12*. The new alignment comprises sequences that are different from the original, but it has a similar pattern of variability. This means that when we use the new alignment in tree reconstruction we do not simply reproduce the original analysis, but we should obtain the same tree.

In practice, 1000 new alignments are created from the real dataset so 1000 replicate trees are reconstructed. A **bootstrap value** can then be assigned to each internal node in the original tree, this value being the number of times that the branch pattern seen at that node in the original tree was reproduced in the replicate trees. If the bootstrap value is greater than 700/1000 then we can assign a reasonable degree of confidence to the topology at that particular internal node.

Molecular clocks enable the time of divergence of ancestral sequences to be estimated

When we carry out a phylogenetic analysis our primary objective is to infer the pattern of evolutionary relationships between the DNA sequences that are being compared. This relationship is revealed by the topology of the tree that is reconstructed. Often we also have a secondary objective: to discover when the ancestral sequences diverged to give the modern sequences. This information is interesting in the context of genome evolution, as we discovered when we looked at the evolutionary history of the human globin genes (see *Figure 14.8*, p. 377). The information is even more interesting on those occasions when we are able to equate a gene tree with a species tree, because now the times at which the ancestral sequences diverged approximate to the dates of speciation events.

To assign dates to branch points in a phylogenetic tree we must make use of a molecular clock. The molecular clock hypothesis, first proposed in the early 1960s, states that nucleotide substitutions (or amino acid substitutions if protein sequences are being compared) occur at a constant rate. This means that the degree of difference between two sequences can be used to assign a date to the time at which their ancestral sequence diverged. However, to be able to do this the molecular clock must be calibrated so that we know how many nucleotide substitutions to expect per million years. Calibration is usually achieved by reference to the fossil record. For example, fossils suggest that the most recent common ancestor of humans and orangutans lived 13 million years ago. To calibrate the human molecular clock we therefore compare human and orangutan DNA sequences to determine the amount of nucleotide substitution that has occurred, and then divide this figure by 13, followed by 2, to obtain a rate of substitution per million years (*Figure 15.13*).

At one time it was thought that there might be a universal molecular clock that applied to all genes in all organisms (Ochman and Wilson, 1987). Now we realize that molecular clocks are different in different organisms and are variable even within a single organism. The differences between organisms might be due to generation times, because a species with a short generation time is likely to accumulate DNA replication errors at a faster rate than a species with a longer generation time. This probably explains the observation that rodents have a faster molecular clock than primates (Gu and Li, 1992). Within an organism the variations are as follows:

- Synonymous substitutions occur at a faster rate than nonsynonymous ones. This is because a nonsynonymous mutation results in a change in the amino acid sequence of the protein coded by the gene. Many of these changes are deleterious to the organism, so their rate is reduced by the processes of natural selection (Box 15.5, p. 408).
- The molecular clock for mitochondrial genes is faster than for genes in the nuclear genome. This is probably because the mitochondrion lacks many of the DNA repair systems that operate on nuclear genes (Section 13.1.4; Gibbons, 1998).

Despite these complications, molecular clocks have become an immensely valuable adjunct to tree reconstruction, as we will see in the next section when we look at some typical molecular phylogenetics projects.

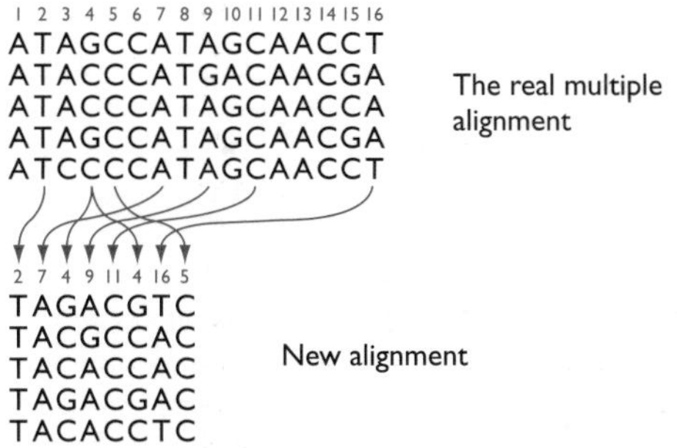

Figure 15.12 Constructing a new multiple alignment in order to bootstrap a phylogenetic tree.

The new alignment is built up by taking columns at random from the real alignment. Note that the same column can be sampled more than once.

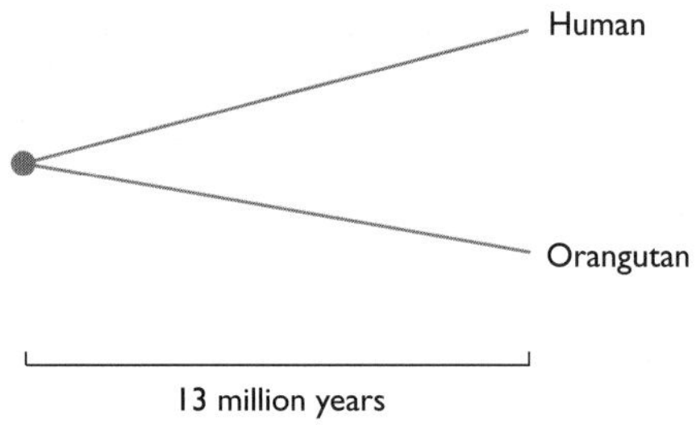

Number of substitutions $= x$

Number per lineage $= \frac{x}{2}$

Number per lineage per million years $= \frac{x}{2 \times 13}$

Figure 15.13 Calculating a human molecular clock.

The number of substitutions are determined for a pair of homologous genes from human and orangutan: call this number 'x'. The number of substitutions per lineage is therefore $\frac{x}{2}$, and the number per million years is $\frac{x}{2 \times 13}$.

15.3 THE APPLICATIONS OF MOLECULAR PHYLOGENETICS

Molecular phylogenetics has grown in stature since the start of the 1990s, largely because of the development of more rigorous methods for tree building, combined with the explosion of DNA sequence information obtained initially by PCR analysis and more recently by genome projects. The importance of molecular phylogenetics has also been enhanced by the successful application of tree reconstruction and other phylogenetic techniques to some of the more perplexing issues in biology. In this final section we will survey some of these successes.

15.3.1 Examples of the use of phylogenetic trees

First, we will consider four projects that illustrate the various ways in which conventional tree reconstruction is being used in modern molecular biology.

DNA phylogenetics has clarified the evolutionary relationships between humans and other primates

Darwin (1871) was the first biologist to speculate on the evolutionary relationships between humans and other primates. His view, that humans are closely related to the chimpanzee, gorilla and orangutan, was controversial when it was first proposed and fell out of favor, even among evolutionists, in the following decades. Indeed, biologists were among the most ardent advocates of an anthropocentric view of our place in the animal world (Goodman, 1962).

From studies of fossils, paleontologists had concluded prior to 1960 that chimpanzees and gorillas are our closest relatives but that the relationship was distant, the split, leading to humans on the one hand and chimpanzees and gorillas on the other, having occurred some 15 million years ago. The first molecular data, obtained by immunological studies in the 1960s (Goodman, 1962; Sarich and Wilson, 1967) confirmed that humans, chimpanzees and gorillas do indeed form a single clade (see Box 15.2, p. 397) but suggested that the relationship is much closer, a molecular clock indicating that this split occurred only 5 million years ago. This was one of the first attempts to apply a molecular clock to phylogenetic data and the result was, quite naturally, treated with some suspicion. In fact, an acrimonious debate opened up between paleontologists, who believed in the ancient split indicated by the fossil evidence, and biologists who had more confidence in the recent date suggested by the molecular data. This debate was eventually 'won' by the molecular biologists, whose view that the split occurred about 5 million years ago became generally accepted.

As more and more molecular data were obtained, the difficulties in establishing the exact pattern of the evolutionary events that led to humans, chimpanzees and gorillas became apparent. Comparisons of the mitochondrial genomes of the three species, by restriction mapping (Section 3.1) and DNA sequencing, suggested that the chimpanzee and gorilla are more closely related to each other than either are to humans (*Figure 15.14A*), whereas DNA–DNA hybridization data supported a closer relationship between humans and chimpanzees (*Figure 15.14B*). The reason for these conflicting results is the close similarities between DNA sequences in the three species, the differences being less than 3% for even the most divergent regions of the genomes (Section 14.4). This makes it difficult to establish relationships unambiguously.

The solution to the problem has been to make comparisons between as many different genes as possible and to target those loci that are expected to show the greatest amount of dissimilarity. By 1997, 14 different molecular datasets had been obtained, including sequences of variable loci such as pseudogenes and noncoding sequences (Ruvolo, 1997). These enable us to conclude with some conviction that the chimpanzee is the closest relative to humans, with our lineages diverging 4.6–5 million years ago, and that the gorilla is a slightly more distant cousin, its lineage having diverged from the human–chimp one between 0.3 and 2.8 million years earlier (*Figure 15.14C*).

The oldest life on Earth

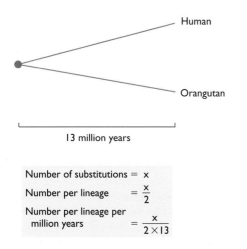

In Section 14.1 we reviewed the early history of the Earth and learnt that the planet was formed some 4.6 billion years ago, and that by 3.5 billion years ago cellular life forms had appeared (see *Figure 14.1*, p. 369). The existence of life at such an early date was surprising when it was first proposed, but the evidence – the presence of microfossils in rocks of this age – is now well established. Of course, it is difficult to be sure if such ancient microfossils are real or not, and some that were originally thought to be fossilized cells have now been reclassified as bubbles

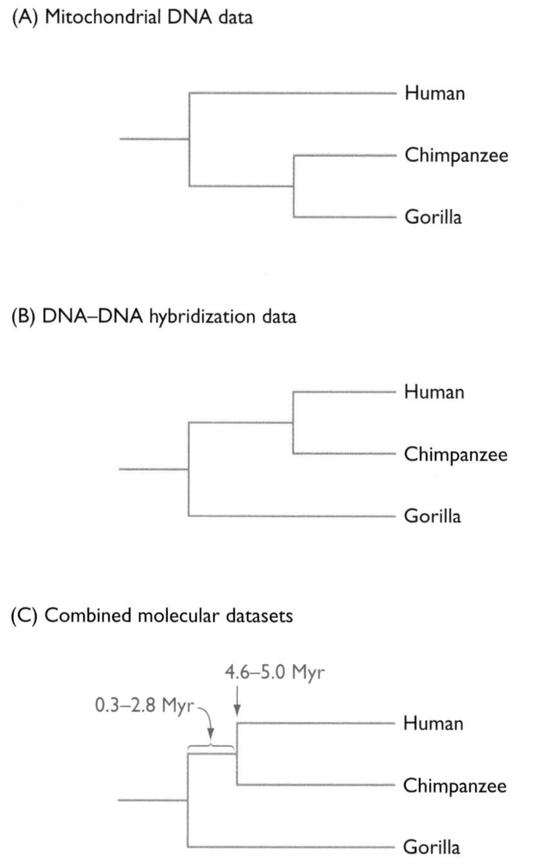

Figure 15.14 Different interpretations of the evolutionary relationships between humans, chimpanzees and gorillas.

See the text for details. Abbreviation: Myr, million years.

or other natural, nonbiological formations in the rock matrix. But microfossils from two sites have come through a stringent examination and been identified, with some confidence, as the remains of prokaryotes resembling modern cyanobacteria. As well as the structures that look like cells, the rocks also contain fossilized stromatolites, mats of cyanobacteria similar to those formed today at the edges of lakes (Mooers and Redfield, 1996).

If cells similar to modern cyanobacteria were living 3.5 billion years ago, then cellular evolution must have been fairly well advanced by that time. In particular, the first major branch in the evolutionary tree of life, separating the bacteria from the lineage that led to the archaea and the eukaryotic nucleus (*Figure 15.15*), must already have taken place. This branch can therefore be dated to no later than 3.5 billion years ago. Is it possible to use a molecular clock to confirm this date?

An attempt has been made to use molecular phylogenetics to date the split between bacteria and archaea/ eukaryotes but the result is not compatible with the existence of cyanobacteria microfossils from 3.5 billion years ago (Doolittle *et al.*, 1996). For this project, protein sequences were used rather than DNA, 531 sequences in total representing 57 important metabolic enzymes from living species taken from 15 different phylogenetic groups. With such a large amount of data included in the analysis we can be reasonably confident that the results will be accurate, at least with regard to the branching patterns that are revealed in the evolutionary histories of the species being compared. But applying a molecular clock to the resulting tree is less easy. There are plenty of well-dated branch points in the fossil record that can be used to calibrate the clock, but the most ancient of these, giving rise to the chordate and echinoderm lineages, leading to vertebrates and sea urchins respectively, occurred just 550 million years ago. This means that an accurate calibration of the clock can be achieved for the last half billion years, but extrapolation to the period before 550 million years ago is less easy.

The result of the phylogenetic analysis has been controversial. The archaea–eukaryote split is dated to 1.87 billion years ago and the split between bacteria and archaea/ eukaryotes to 2.16 billion years ago. This is only just over half the age of the cyanobacteria microfossils.

The balance of opinion within this particular controversy is probably against the molecular data and in favor of the microfossil evidence. This is because the fossils look so convincing and the extension of the molecular clock calibration to such an ancient period has never been attempted before. Many biologists are unconvinced that a consistent speed to the molecular clock can be assumed for periods so far back in the past. However, intuition suggests that if the molecular clock ran at a different speed in the earliest cells, then that speed would probably have been faster than at present, due perhaps to a more error-prone DNA replication process and less effective DNA repair systems in primitive cells (*Figure 15.16*). If this hypothesis is correct then dates for branch points assigned with the molecular clock should be earlier, not later, than the real dates. The molecular data can therefore be criticized, but these criticisms do not appear to tie in

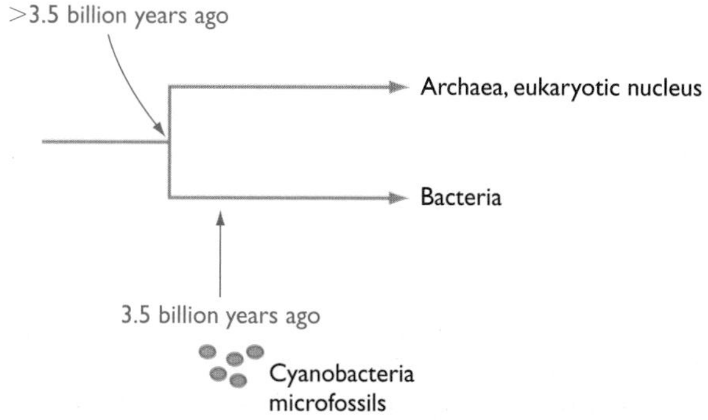

Figure 15.15 The presence of cyanobacteria microfossils in rocks that are 3.5 billion years old indicates that the split between the lineages leading to bacteria on the one hand and archaea and the eukaryotic nucleus on the other must have occurred before then.

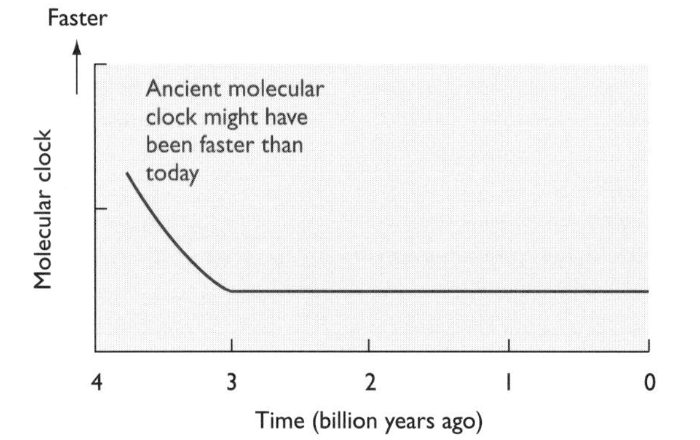

Figure 15.16 The molecular clock might have run faster in ancient times.

with the perceived problems with the results of the phylogenetic analysis.

The origins of AIDS

The global epidemic of AIDS has touched everyone's lives. AIDS is caused by human immunodeficiency virus 1 or HIV-1, a retrovirus (Section 6.3.2) that infects cells involved in the immune response. The demonstration in the early 1980s that HIV-1 is responsible for AIDS was quickly followed by speculations about the origins of the disease. These centered around the discovery that similar immunodeficiency viruses are present in primates such as the chimpanzee, sooty mangabey, mandrill and various monkeys. These simian immunodeficiency viruses (SIVs) are not pathogenic in their normal hosts but it was thought that if one had become transferred to humans then within this new species the virus might acquire new properties, such as the ability to cause disease and to spread rapidly through the population.

Retrovirus genomes accumulate mutations relatively quickly because reverse transcriptase, the enzyme that copies the RNA genome contained in the virus particle into the DNA version that integrates into the host genome (see Section 6.3.2), lacks an efficient proofreading activity (Section 12.3.2) and so tends to make errors when it carries out RNA-dependent DNA synthesis. This means that the molecular clock runs rapidly in retroviruses, and genomes that diverged quite recently display sufficient nucleotide dissimilarity for a phylogenetic analysis to be carried out. Even though the evolutionary period we are interested in is much less than 100 years, HIV and SIV genomes contain sufficient data for their relationships to be inferred by phylogenetic analysis.

The starting point for this phylogenetic analysis is RNA extracted from virus particles. RT-PCR (see Technical Note 5.1, p. 91) is therefore used to convert the RNA into a DNA copy and then to amplify the DNA so that sufficient amounts for nucleotide sequencing are obtained. Comparison between the DNA sequences resulted in the reconstructed tree shown in *Figure 15.17* (Leitner *et al.*, 1996; Wain-Hobson, 1998). This tree has a number of interesting features. First it shows that different samples of HIV-1 have slightly different sequences, the samples as a whole forming a tight cluster, almost a starlike pattern, that radiates from one end of the unrooted tree. The implication is that the global AIDS epidemic began with a very small number of viruses, perhaps just one, which have spread and diversified since entering the human population. The closest relative to HIV-1 among primates is the SIV of chimpanzees, the implication being that this virus jumped across the species barrier between chimps and humans and initiated the AIDS epidemic. However, this epidemic did not begin immediately because a relatively long, uninterrupted branch links the center of the HIV-1 radiation with the internal node leading to the relevant SIV sequence, suggesting that after transmission to humans, HIV-1 underwent a latent period when it remained restricted to a small part of the global human population, presumably in Africa, before beginning its rapid spread to other parts of the World. Other primate SIVs are less closely related to HIV-1, but one, the SIV from sooty mangabey, clusters in the tree with the second human immunodeficiency virus, HIV-2. It appears that HIV-2 was transferred to the human population independently of HIV-1, and from a different simian host. HIV-2 is also able to cause AIDS, but has not, as yet, become globally epidemic.

An intriguing addition to the HIV/SIV tree was made in 1998 when the sequence of an HIV-1 isolate from a blood sample taken from an African male in 1959 was sequenced (Zhu *et al.*, 1998). The RNA was very fragmented and only a short DNA sequence could be obtained, but this was sufficient for the sequence to be placed on the phylogenetic tree (see *Figure 15.17*). This sequence, called ZR59, attaches to the tree by a short branch that emerges from near the center of the HIV-1

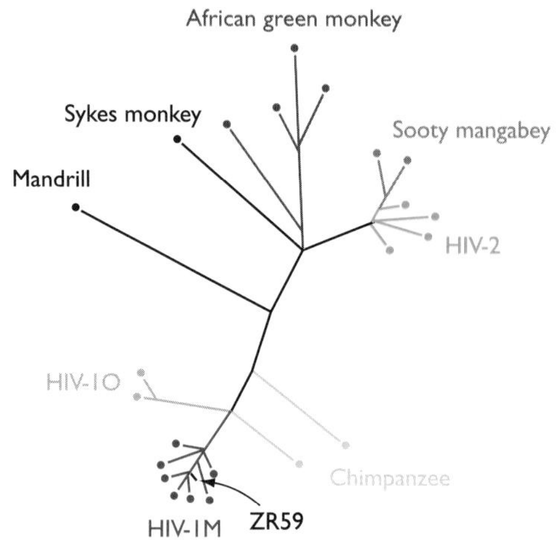

Figure 15.17 The phylogenetic tree reconstructed from HIV and SIV genome sequences.

The AIDS epidemic is due to the HIV-1M type of immunodeficiency virus. ZR59 is positioned near the root of the starlike pattern formed by genomes of this type. Based on Wain-Hobson (1998).

radiation. The positioning indicates that the ZR59 sequence represents one of the earliest versions of HIV-1 and dates the beginning of the global spread of HIV to the period around 1960. Pinning down the date in this way has enabled epidemiologists to begin an investigation of the historical and social conditions that might have been responsible for the start of the AIDS epidemic.

Problems with prions

The fourth example of molecular phylogenetics that we will consider concerns the odd 'proteinaceous infectious particles' called **prions**. What the name tells us is that a prion is a protein molecule that is infectious, meaning that when transferred to a susceptible host it is able to cause a disease, accompanied by production of more particles that spread the disease by moving on to new hosts. When first discovered, prions were interesting to biologists because they seemed to contravene the notion that infectious particles such as viruses must possess a nucleic acid genome in order to direct their replication. Prions lack nucleic acid and so appeared to be a 'protein life-form'. More recently, prions have come to the attention of a wider audience because they are the causative agents of neurological diseases such as bovine spongiform encephalopathy (BSE), whose possible transmission to humans has caused great concern in the UK and elsewhere.

It turns out that prions do not overturn our basic beliefs about genomes but are, nonetheless, extremely interesting. Prions are ordinary proteins, and most vertebrates have genes that code for them. The normal function of a prion protein is unknown, but knockout mice (Section 5.2.2) whose prion genes have been disrupted display abnormalities in the activity of the neurotransmitter γ-aminobutyric acid (GABA), suggesting that prions are involved in some aspect of signaling in the nervous system (Ng and Doig, 1997). The infectious properties arise because some prions can exist in two forms. The normal prion protein has a tertiary structure that is made up mainly of α-helices, but this can be converted into a very different tertiary structure in which β-sheets predominate. The cause of the structural change has not been identified and is certainly not a genetic mutation because the two forms of a prion protein have identical amino acid sequences. The critical feature, and the reason why the second form of the prion is infectious, is that one abnormal prion can trigger the conversion of normal prions into the β-sheeted form. Once one is present, more appear, resulting in insoluble 'plaques' of protein that are found in the brains of animals and humans suffering from a prion disease.

Prion genes have been sequenced from 33 different species and, as is now routine when a large number of homologous sequences are available, a phylogenetic tree has been reconstructed (*Figure 15.18*; Krakauer *et al.*, 1996). The topology of this tree has a few unusual features, suggesting that some pairs of species are more closely related than generally supposed (note that gorilla is placed next to humans, with chimpanzees more divergent), but these features can probably be explained by anomalies in this particular reconstruction. What is more exciting is an unexpected example of convergent evolution between the human and bovine proteins. Close inspection of the multiple alignment revealed that humans (as well as gorilla, chimpanzees and some other primates) have a serine at position 143 in the prion amino acid sequence and a histidine at position 155; most of the other prion sequences have asparagine and tyrosine, respectively, at these positions (*Figure 15.19*). The one other species to have these two substitutions is *Bos taurus*, the cow.

Could these shared features displayed by the prions of humans and cattle be linked with the susceptibility that humans display towards the infectious form of the cattle prion (Krakauer *et al.*, 1996)? The hypothesis is attractive because infection of humans by the sheep prion, which causes scrapie in sheep, has never been proven, so possibly we have a natural resistance to the sheep prion even though it is closely related to the cattle version, though lacking the two amino acid substitutions at positions 143 and 155. However, the hypothesis is far from proven and a strong counter-argument has been put forward on the grounds that mice and sheep can be infected with BSE even though their prion proteins lack the two substitutions seen in the cattle version (Goldman *et al.*, 1996).

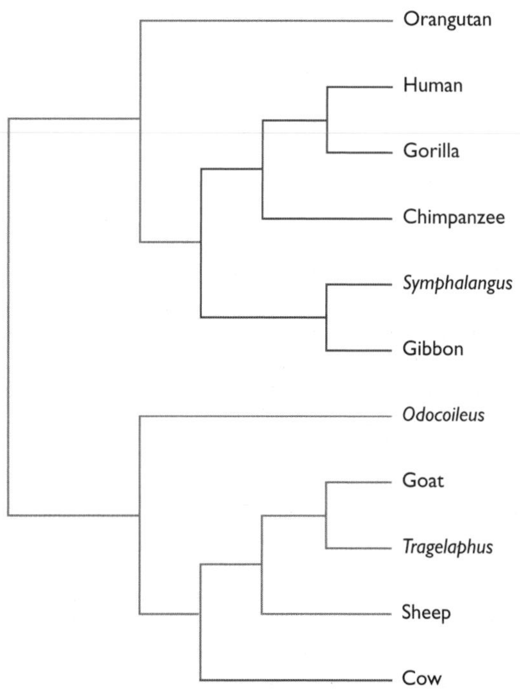

Figure 15.18 Evolutionary relationships between prion proteins.

The drawing shows part of a reconstructed tree of prions from 33 different species (Krakauer *et al.*, 1996). The highlighted branches are those leading to the primates and the cow prions which have the two amino acid substitutions shown in *Figure 15.19*.

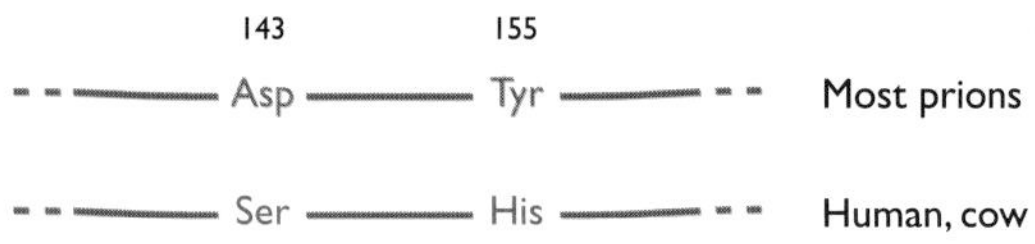

Figure 15.19 The two amino acid substitutions seen in the prion proteins of humans, cows and some other primates.

15.3.2 Molecular phylogenetics as a tool in the study of human prehistory

Now we will turn our attention to the use of molecular phylogenetics in intraspecific studies: the study of the evolutionary history of members of the same species. We could choose any one of several different organisms to illustrate the approaches and applications of intraspecific studies, but many people look on *Homo sapiens* as the most interesting organism so we will concentrate on humans and investigate how molecular phylogenetics is being used to deduce the origins of modern humans and the geographical patterns of their recent migrations in the Old and New Worlds.

Intraspecific studies require highly variable genetic loci

In any application of molecular phylogenetics, the genes chosen for analysis must display variability in the organisms being studied. If there is no variability then there is no phylogenetic information. This presents a problem in intraspecific studies because the organisms being compared are all members of the same species and so share a great deal of genetic similarity, even if the species has split into populations that interbreed only intermittently. This means that the DNA sequences that are used in the phylogenetic analysis must be the most variable ones that are available. In humans there are three possibilities.

- **Multiallelic genes**, such as members of the HLA family (Section 2.2.1), which exist in many different sequence forms.
- **Microsatellites**, which evolve not through mutation but by replication slippage (Section 13.1.1), which is difficult if not impossible to correct. Length changes therefore produce new microsatellite alleles relatively frequently.
- **Mitochondrial DNA** which, as mentioned in Section 15.2.2, accumulates nucleotide substitutions relatively rapidly because the mitochondrion lacks many of the repair systems that slow down the molecular clock in the human nucleus. The mitochondrial DNA variants present in a single species are called **haplotypes**.

It is important to note that it is not the potential for change that is critical to the application of these loci in phylogenetic analysis, it is the fact that different alleles or haplotypes of the locus coexist in the population as a whole. The loci are therefore **polymorphic** (see Box 15.5) and information pertaining to the relationships between different individuals can be obtained by comparing the combinations of alleles and/or haplotypes that those individuals possess.

The origins of modern humans – out of Africa or not?

It appears reasonably certain that the origins of humans lies in Africa because it is here that all of the oldest prehuman fossils have been found. The paleontological evidence reveals that hominids first moved outside of Africa over one million years ago, but these were not modern humans, they were an earlier species called *Homo erectus*. These were the first hominids to become geographically dispersed, eventually spreading to all parts of the Old World.

The events that followed the dispersal of *Homo erectus* are controversial. From comparisons using fossil skulls and bones, paleontologists have concluded that the *Homo erectus* populations that became located in different parts of the Old World gave rise to the modern human populations of those areas by a process called **multiregional evolution** (*Figure 15.20A*). There may have been a certain amount of interbreeding between humans from different geographical regions, but, to a large extent, these various populations remained separate throughout their evolutionary history.

Doubts about the multiregional hypothesis were first raised by reinterpretations of the fossil evidence and subsequently brought to a head by publication in 1987 of a phylogenetic tree reconstructed from mitochondrial

Box 15.4: Haplotypes

The terms 'allele' and 'haplotype' are sometimes used interchangeably. This is not correct. An allele is a sequence variation of a single gene. A haplotype is a collection of alleles that are often inherited together because the genes that they represent are closely linked in a DNA molecule and so are rarely separated by recombination during meiosis.

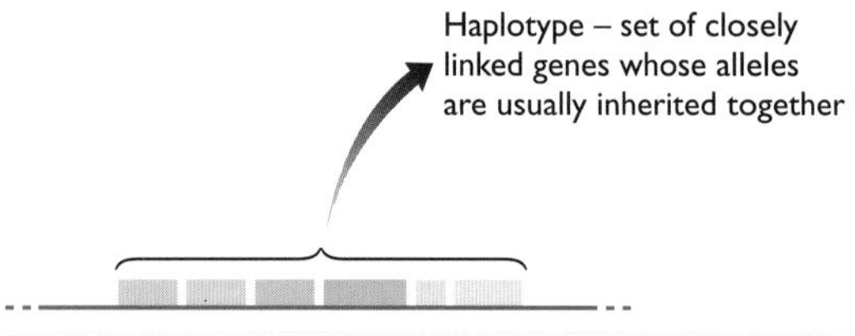

'Haplotype' is the correct term to use with mitochondrial DNA because the mitochondrial genome does not undergo recombination during meiosis or at any other stage of the cell cycle. All the genes and intergenic regions of a mitochondrial genome are inherited together, so variant sequences are haplotypes.

RFLP data obtained from 147 humans representing populations from all parts of the World (Cann *et al.*, 1987). The tree (*Figure 15.21*) confirmed that the ancestors of modern humans lived in Africa but suggested they were still there about 150 000 years ago. This inference was made by applying the mitochondrial molecular clock to the tree, which showed that the ancestral mitochondrial DNA, the one from which all modern mitochondrial DNAs are descended, existed between 140 000 and 290 000 years ago. Because this ancestral genome could be located in Africa, the person who possessed it, the so-called mitochondrial Eve (she had to be female because mitochondrial DNA is only inherited through the female line), must have been African.

The mitochondrial Eve discovery prompted a new scenario for the origins of modern humans. Rather than evolving in parallel throughout the world, as suggested by the multiregional hypothesis, **Out of Africa** states that

(A)

(B)

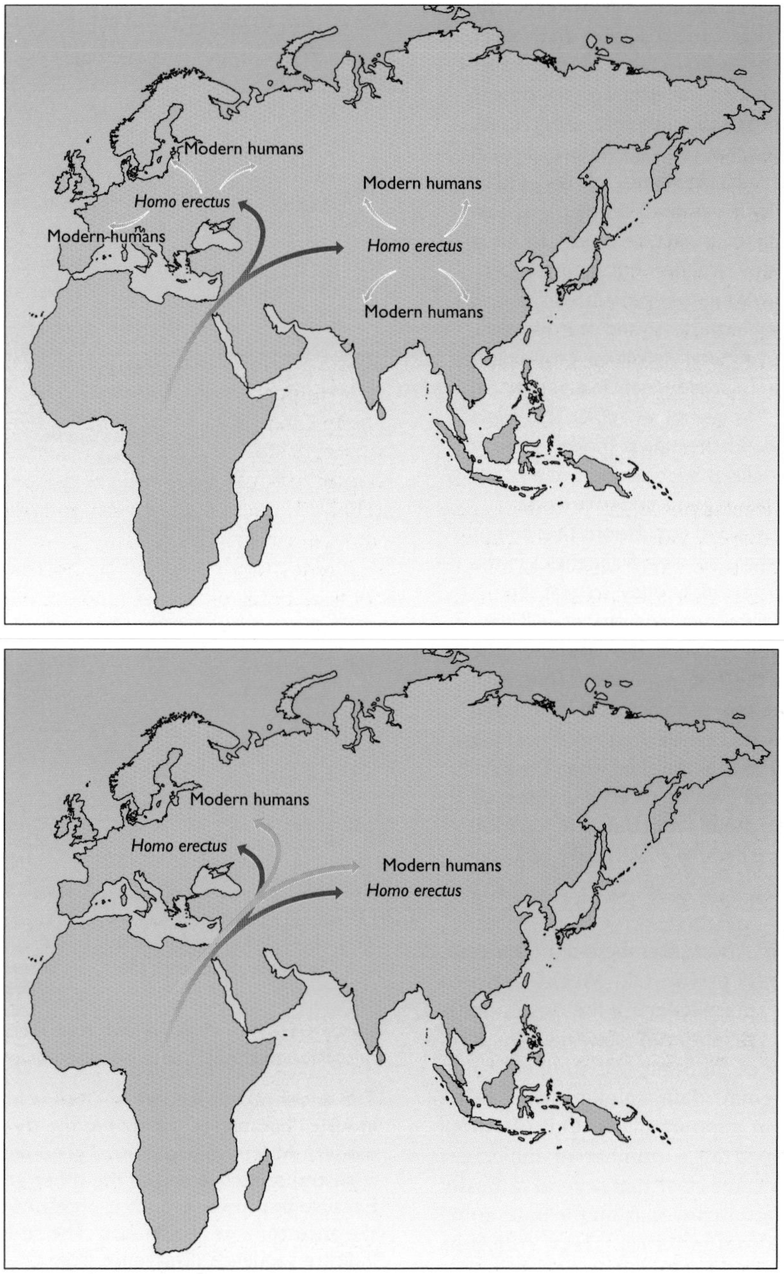

Figure 15.20 Two competing hypotheses for the origins of modern humans.

(A) The multiregional hypothesis states that *Homo erectus* left Africa over one million years ago and then evolved into modern humans in different parts of the Old World. (B) The Out of Africa hypothesis states that the populations of *Homo erectus* in the Old World were displaced by new populations of modern humans that followed them out of Africa.

Box 15.5: Genes in populations

New alleles and haplotypes appear in a population because of mutations that occur in the reproductive cells of individual organisms. This means that many genes are **polymorphic**, two or more alleles being present in the population as a whole, each with its own **allele frequency**. Allele frequencies change over time due to **natural selection** and **random genetic drift**. Natural selection occurs because of differences in **fitness**, the ability of an organism to survive and reproduce, that result in the 'preservation of favourable variations and the rejection of injurious variations' (Darwin, 1859). Natural selection, therefore, decreases the frequencies of alleles that reduce the fitness of an organism, and increases the frequency of alleles that improve fitness. In reality, few new alleles that arise in a population have a significant impact on fitness so most are not affected by natural selection, but their frequencies still change because of random genetic drift caused by the arbitrary nature of birth, death and reproduction.

Either because of natural selection or random drift, one allele can begin to predominate in a population and eventually achieve a frequency of 100%. This allele has now become **fixed**. Mathematical models predict that over time different alleles become fixed in a population, resulting in a series of **gene substitutions**.

If a species splits into two populations that do not interbreed extensively, then the allele frequencies in the two populations will change differently so that after a few tens of generations the two populations will have distinctive genetic features. Eventually, different gene substitutions occur in the two populations, but even before this happens we can distinguish between them because of the differences in allele frequencies. These differences can be used to date the time when the population split occurred and even to determine if one or both populations experienced a **bottleneck**, a period when the population size became substantially reduced.

Homo sapiens originated in Africa, members of this species then moving into the rest of the Old World between 100 000 and 50 000 years ago, displacing the descendents of *Homo erectus* that they encountered (see *Figure 15.20B*).

Such a radical change in thinking was unlikely to go unchallenged. The original data, along with DNA sequences put forward in support of the Out of Africa hypothesis, were examined by a number of molecular phylogeneticists and it became clear that several radically different trees could be reconstructed from the data, some of which did not have a root in Africa. These criticisms have been countered by more detailed mitochondrial DNA sequence datasets, most of which are compatible with a relatively recent African origin and so support the Out of Africa hypothesis rather than multiregional evolution, although the dates given to the ancestral sequence vary. However, further complications have arisen from the results of nuclear gene analyses. For example, detailed studies of β-globin sequences indicate that we have a relatively recent common ancestor but do not indicate an African origin for this ancestor (Harding *et al.*, 1997). More datasets, and hopefully some sort of Grand Synthesis, are eagerly awaited.

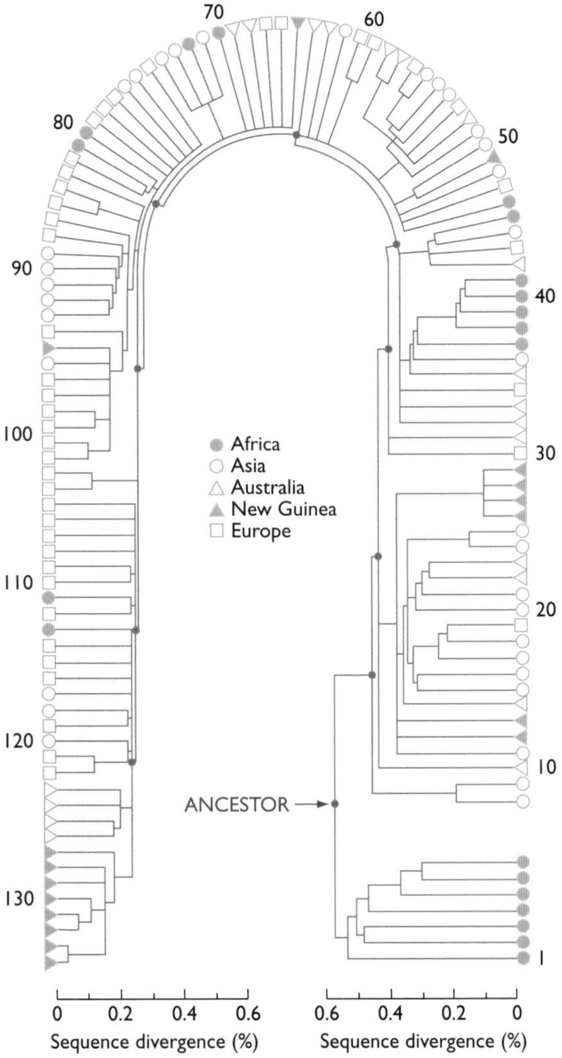

Figure 15.21 Phylogenetic tree reconstructed from mitochondrial RFLP data obtained from 147 modern humans.

The ancestral mitochondrial DNA is inferred to have existed in Africa because of the split in the tree between the seven modern African mitochondrial genomes placed below the ancestral sequence and all the other genomes above it. Because this lower branch is purely African it is deduced that the ancestor was also African. The scale bars at the bottom indicate sequence divergence from which, using the mitochondrial molecular clock, it is possible to assign dates to the branch points in the tree. The clock suggests that the ancestral sequence existed between 140 000 and 290 000 years ago. Reprinted with permission from Cann *et al.* (1987) *Nature*, **325**, 31–36. Copyright 1987 Macmillan Magazines Limited.

RESEARCH **15.1** BRIEFING

Neandertal DNA

Sequence analysis of 'ancient DNA' extracted from a fossil bone between 30 000 and 100 000 years old provides support for the Out of Africa hypothesis.

Neandertals are extinct hominids who lived in Europe between 300 000 and 30 000 years ago. They were descended from the *Homo erectus* populations who left Africa about one million years ago and, according to the Out of Africa hypothesis, were displaced when modern humans reached Europe about 50 000 years ago. Therefore, one prediction of the Out of Africa hypothesis is that there is no genetic continuity between Neandertals and the modern humans that live in Europe today. Bearing in mind that the last Neandertal died out 30 000 years ago, is there any way that we can test this hypothesis?

Ancient DNA could provide an answer. It has been known for some years that DNA molecules can survive the death of the organism in which they are contained, being recoverable centuries and possibly millennia later as short degraded molecules preserved in bones and other biological remains. There is never very much ancient DNA in a specimen, possibly no more than a few hundred genomes in a gram of bone, but that need not concern us because we can always use PCR to amplify these tiny amounts into larger quantities from which we can obtain DNA sequences.

The ancient DNA field has been plagued with controversies over the last 10 years. In the early 1990s there were many reports of ancient human DNA being detected in bones and other archaeological specimens, but often it turned out that what had been amplified by PCR was not ancient DNA at all, but contaminating modern DNA left on the specimen by the archaeologist who dug it up or the molecular biologist who carried out the DNA extraction. The worldwide success of *Jurassic Park* led to reports of DNA in insects preserved in amber and even in dinosaur bones, but all of these claims are now doubted. Many biologists started to wonder if ancient DNA existed at all, but gradually it became clear that if the work is carried out with extreme care it is sometimes possible to extract authentic ancient DNA from specimens up to about 50 000 years old. This is just old enough to include a few Neandertal bones.

DNA from a Neandertal bone

The Neandertal specimen selected for study is believed to be between 30 000 and 100 000 years old. DNA extraction was carried out with a fragment of bone weighing about 400 mg and a technique called **quantitative PCR** was used to determine if human DNA molecules were present and, if so, how many. The results indicated that the bone fragment contained about 1300 copies of the Neandertal genome, enough to suggest that sequence analysis was worth attempting.

PCRs were now directed at what was expected to be the most variable part of the Neandertal mitochondrial genome. Because it was anticipated that the DNA would be broken into very short pieces, a sequence was built up in sections by carrying out nine overlapping PCRs, none amplifying more than 170 bp of DNA but together giving a total length of 377 bp. This sequence provided the data used to test the Out of Africa hypothesis.

A phylogenetic tree was reconstructed using the sequence obtained from the Neandertal bone and the sequences of six mitochondrial DNA haplotypes from modern humans. The Neandertal sequence was positioned on a branch of its own connected to the root of the tree but not linked directly to any of the modern human sequences.

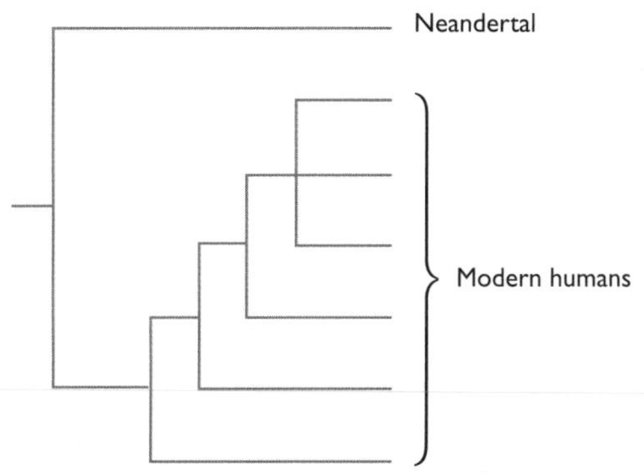

Next, a huge multiple alignment was made in order to compare the Neandertal sequence with 994 sequences from modern humans. The differences were striking. The Neandertal sequence differed from the modern sequences at an average of 27.2 ± 2.2 nucleotide positions whereas the modern sequences, which came from all over the world, not just Europe, differed from each other at only 8.0 ± 3.1 positions. This amount of variation is incompatible with the notion that modern Europeans are descended from Neandertals and strongly supports the Out of Africa hypothesis.

A few criticisms can be made, notably that just a single Neandertal specimen was examined. It could be that modern Europeans are descended from Neandertals, but not the population represented by this particular skeleton. The search is on for other Neandertal remains that contain ancient DNA.

Reference

Krings M, Stone A, Schmitz RW, Krainitzki H, Stoneking M and Pääbo S (1997) Neandertal DNA sequences and the origin of modern humans. *Cell*, **90**, 19–30.

The patterns of more recent migrations into Europe are also controversial

By whatever evolutionary pathway, modern humans were present throughout most of Europe by 40 000 years ago. This is clear from the fossil and archaeological records. The next controversial issue in human prehistory concerns whether or not these populations were displaced about 30 000 years later by other humans migrating into Europe from the Middle East.

The question centers on the process by which agriculture spread into Europe. The transition from hunting and gathering to farming occurred in the Middle East some 10 000 years ago, when early Neolithic villagers began to cultivate crops such as wheat and barley. After becoming established in the Middle East, farming spread into Asia, Europe and North Africa. By searching for evidence of agriculture at archaeological sites, for example, by looking for the remains of cultivated plants or for implements used in farming, it has been possible to trace the expansion of farming technology along two routes through Europe, one around the coast to Italy and Spain and the second through the Danube and Rhine valleys to northern Europe (*Figure 15.22*).

How did farming spread? The simplest explanation is that farmers migrated from one part of Europe to another, taking with them their implements, animals and crops, and displacing the indigenous, pre-agricultural human communities that were present in Europe at that time. This is the **wave of advance** model and support for it comes from a large scale phylogenetic analysis based on the frequencies of alleles for 95 nuclear genes in populations from across Europe (Cavalli-Sforza, 1998). Such a large and complex dataset cannot be analyzed in any meaningful way by conventional tree building and instead has to be examined by more advanced statistical methods, ones which are based more in population biology rather than phylogenetics. One such procedure is **principal component analysis**, which attempts to identify patterns in the dataset corresponding to the uneven geographical distribution of alleles, these uneven distributions possibly being indicative of past population migrations. The most striking pattern within the European dataset, accounting for about 28% of the total genetic variation, is a gradation of allele frequencies across Europe (*Figure 15.23*). This pattern implies that a migration of people occurred either from the Middle East to northeast Europe, or in the opposite direction. Because the former coincides with the expansion of farming, as revealed by the archaeological record, this first principal component is looked on as providing strong support for the wave of advance model.

The analysis looks convincing but two criticisms are possible. The first is that the data provide no indication of when the inferred migration took place, so the link between the first principal component and the spread of agriculture is based solely on the pattern of the allele gradation, not on any complementary evidence relating to the period when this gradation was set up. The second criticism arises because of the results of a second study of European human populations, one which does include a time dimension (Richards *et al.*, 1996). This study looked at mitochondrial DNA haplotypes in 821 individuals from various populations across Europe. It failed to confirm the gradation of allele frequencies detected in the nuclear DNA dataset, instead suggesting that European populations have remained relatively static over the last 20 000 years. The results do, however, leave room for a small scale migration of people from the Middle East into Europe at a time approximating with the spread of farming, leading to the suggestion that farming was introduced into Europe by 'pioneers' who interbred with the existing pre-farming communities rather than displacing them.

Prehistoric human migrations into the New World

Finally we will examine the completely different set of controversies surrounding the hypotheses regarding the patterns of human migration that led to the first entry of people into the New World. There is no evidence for the spread of *Homo erectus* into the Americas, so it is presumed that humans did not enter the New World until after modern *Homo sapiens* had evolved in, or migrated into Asia. The Bering Strait, between Asia and North America, are quite shallow and if the sea level dropped by 50 meters it would be possible to walk across from one continent to the other. It is believed that this was the route taken by the first humans to venture into the New World (*Figure 15.24*).

The sea was 50 meters or more below its current level for most of the last Ice Age, between about 60 000 and 11 000 years ago, but for most of this time the route would have been impassable because of the build-up of ice. Also, the northern parts of America would have been arctic during much of this period, providing few game animals for the migrants to hunt and very little wood with which they could make fires. These considerations, together with the absence of archaeological evidence of humans in North America before 11 500 years ago, led to the adoption of 'about 12 000 years ago' as the date for the first entry of humans into the New World. Recent discoveries of human occupation at sites dating to 20 000 years ago, both in North and South America, has prompted some rethinking, but it is still generally assumed that a substantial population migration into North America, possibly the one from which all modern Native Americans are descended, occurred about 12 000 years ago.

What information does molecular phylogenetics provide? The first relevant studies were carried out in the late 1980s using RFLP data. These indicated that Native Americans are descended from Asian ancestors and identified four distinct mitochondrial haplotypes among the population as a whole (Wallace *et al.*, 1985; Schurr *et al.*, 1990). Linguistic studies had already shown that American languages can be divided into three different groupings, suggesting that modern Native Americans are descended from three sets of people, each speaking a different language. The inference from the molecular data that there may in fact have been four ancestral

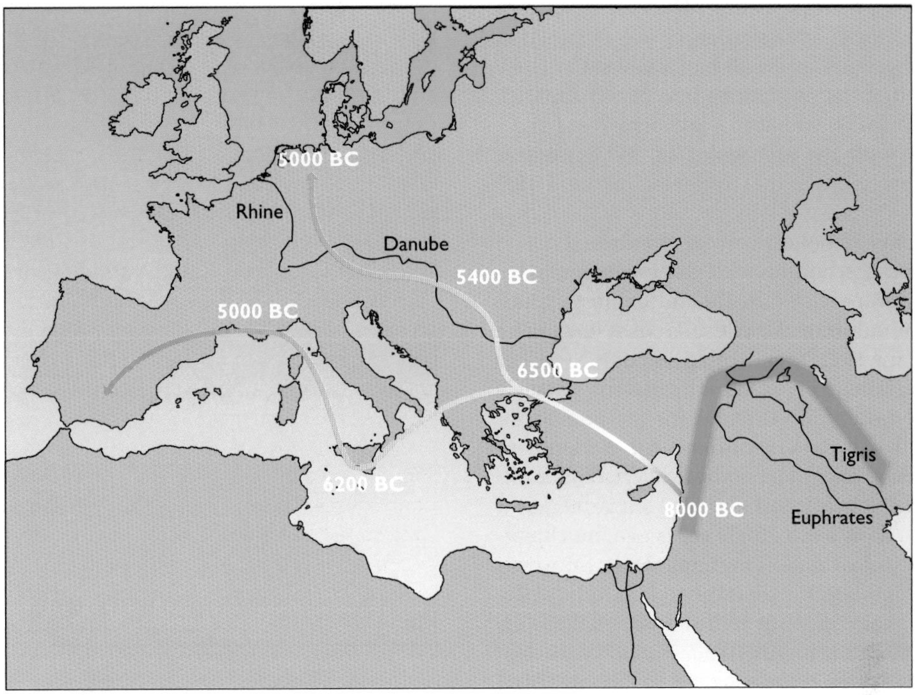

Figure 15.22 The spread of agriculture from the Middle East to Europe.

The green shaded area is the 'Fertile Crescent', the area of the Middle East where many of today's crops – wheat, barley, etc. – grow wild and where these plants are thought to have first been taken into cultivation.

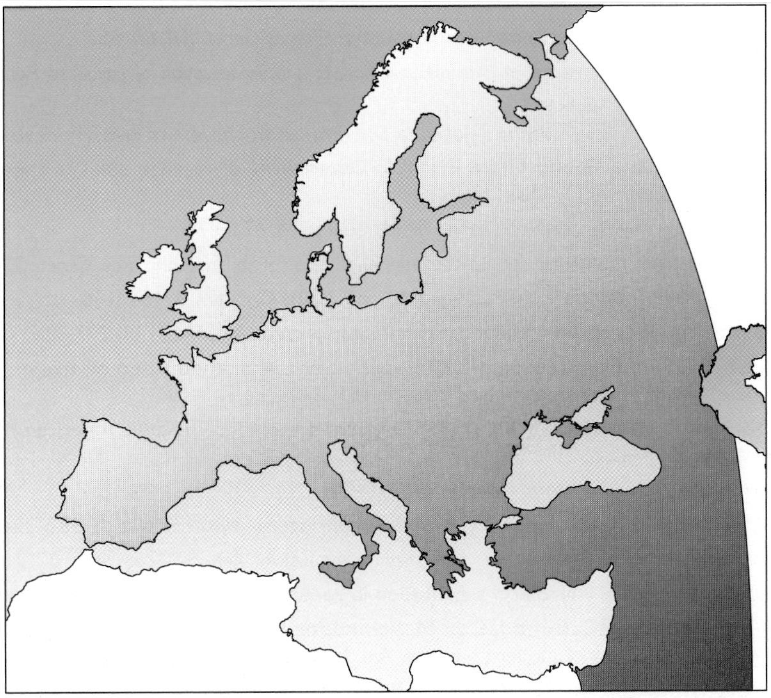

Figure 15.23 A genetic gradation across modern Europe.

See the text for details.

populations was not too disquieting. The first significant dataset of mitochondrial DNA sequences was obtained in 1991, enabling a rigorous application of a molecular clock which indicated that the migrations into North America occurred between 15 000 and 8000 years ago (Ward *et al.*, 1991), consistent with the archaeological evidence that humans were absent from the continent before 11 500 years ago.

These early phylogenetic analyses confirmed or at least were not too discordant with the complementary evidence provided by archaeological and linguistic studies. However, the additional molecular data that have been acquired since 1992 have tended to confuse rather than clarify the issue. For example, different datasets have provided a variety of estimates for the number of migrations into North America. The most comprehensive analysis, based on mitochondrial DNA (Forster *et al.*, 1996), puts this figure at just one, and suggests that it occurred between 25 000 and 20 000 years ago, much earlier than the traditional date. The implication from this study is that humans became established in North America about 20 000 years ago, but that the population suffered a major decline during the latter period of the last Ice Age, with numbers recovering again about 12 000 years ago. This hypothesis is still being evaluated by other molecular biologists and archaeologists.

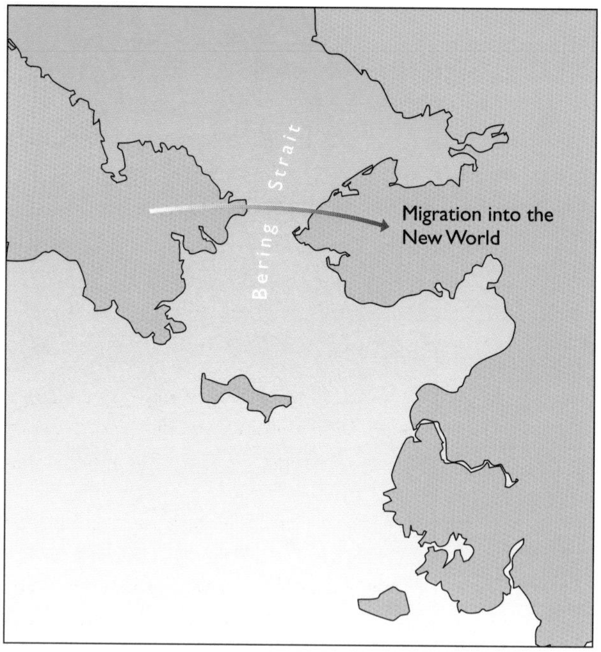

Figure 15.24 The route by which humans first entered the New World.

REFERENCES

Cann RL, Stoneking M and Wilson AC (1987) Mitochondrial DNA and human evolution. *Nature*, **325**, 31–36.

Cavalli-Sforza LL (1998) The DNA revolution in population genetics. *Trends Genet.*, **14**, 60–65.

Darwin C (1859) *The Origin of Species by Means of Natural Selection, or the Preservation of Favoured Races in the Struggle for Life.* Penguin Books, London.

Darwin C (1871) *The Descent of Man, and Selection in Relation to Sex.* Princeton University Press, Princeton, NJ.

Doolittle RF, Feng D-F, Tsang S, Cho G and Little E (1996) Determining divergence times of the major kingdoms of living organisms with a protein clock. *Science*, **271**, 470–477.

Eernisse DJ (1998) A brief guide to phylogenetic software. *Trends Genet.*, **14**, 473–475.

Felsenstein J (1988) Phylogenies from molecular sequences: inference and reliability. *Annu. Rev. Genet.*, **22**, 521–565.

Felsenstein J (1989) PHYLIP – Phylogeny Inference Package (Version 3.20). *Cladistics*, **5**, 164–166.

Fitch WM (1977) On the problem of discovering the most parsimonious tree. *Am. Nat.*, **111**, 223–257.

Fitch WM and Margoliash E (1967) Construction of phylogenetic trees. A method based on mutation distances as estimated from cytochrome *c* sequences is of general applicability. *Science*, **155**, 279–284.

Forster P, Harding R, Torroni A and Bandelt HJ (1996) Origin and evolution of native American mtDNA variation: a reappraisal. *Am. J. Hum. Genet.*, **59**, 935–945.

Gibbons A (1998) Calibrating the molecular clock. *Science*, **279**, 28–29.

Goldman W, Hunter N, Somerville R and Hope J (1996) Prion phylogeny revisited. *Nature*, **382**, 32–33.

Goodman M (1962) Immunochemistry of the primates and primate evolution. *Ann. NY Acad. Sci.*, **102**, 219–234.

Gu X and Li W-H (1992) Higher rates of amino acid substitution in rodents than in humans. *Mol. Phylogenet. Evol.*, **1**, 211–214.

Harding RM, Fullerton SM, Griffiths RC, Bond J, Cox M, Schneider JA, Moulin DS and Clegg JB (1997) Archaic African and Asian lineages in the genetic ancestry of modern humans. *Am. J. Hum. Genet.*, **60**, 772–789.

Hennig W (1966) *Phylogenetic Systematics.* University of Illinois Press, Urbana, IL.

Hillis DM (1997) Biology recapitulates phylogeny. *Science*, **276**, 218–219.

Jeanmougin F, Thompson JD, Gouy M, Higgins DG and Gibson TJ (1998) Multiple sequence alignment with Clustal X. *Trends Biochem. Sci.*, **23**, 403–405.

Krakauer DC, Pagel M, Southwood TRE and Zanotto M de A (1996) Phylogenesis of prion protein. *Nature*, **380**, 675.

Leitner T, Escanilla D, Franzen C, Uhlen M and Albert J (1996) Accurate reconstruction of a known HIV-1 transmission history by phylogenetic tree analysis. *Proc. Natl Acad. Sci. USA*, **93**, 10864–10869.

Li W-H (1997) *Molecular Evolution*. Sinauer, Sunderland, MA.

Michener CD and Sokal RR (1957) A quantitative approach to a problem in classification. *Evolution*, **11**, 130–162.

Mooers AØ and Redfield RJ (1996) Digging up the roots of life. *Nature*, **379**, 587–588.

Needleman SB and Wunsch CD (1970) A general method applicable to the search of similarities in the amino acid sequences of two proteins. *J. Mol. Biol.*, **48**, 443–453.

Ng SBL and Doig AJ (1997) Molecular and chemical basis of prion-related diseases. *Chem. Soc. Rev.*, **26**, 425–432.

Nuttall GHF (1904) *Blood Immunity and Blood Relationship*. Cambridge University Press, Cambridge.

Ochman H and Wilson AC (1987) Evolution in bacteria: evidence for a universal substitution rate in cellular genomes. *J. Mol. Evol.*, **26**, 74–86.

Richards M, Côrte-Real H, Forster P, Macaulay V, Wilkinson-Herbots H, Demaine A, Papiha S, Hedges R, Bandelt H-J and Sykes B (1996) Paleolithic and Neolithic lineages in the European mitochondrial gene pool. *Am. J. Hum. Genet.*, **59**, 185–203.

Ruvolo M (1997) Molecular phylogeny of the hominoids: inferences from multiple independent DNA sequence data sets. *Mol. Biol. Evol.*, **14**, 248–265.

Saitou N and Nei M (1987) The neighbor-joining method: a new method for reconstructing phylogenetic trees. *Mol. Biol. Evol.*, **4**, 406–425.

Sarich VM and Wilson AC (1967) Immunological time scale for hominid evolution. *Science*, **158**, 1200–1203.

Schurr TG, Ballinger SW, Gan YY, Hodge JA, Merriwether DA, Lawrence DN, Knowler WC, Weiss KM and Wallace DC (1990) Amerindian mitochondrial DNAs have rare Asian mutations at high frequencies suggesting they are derived from four primary maternal lineages. *Am. J. Hum. Genet.*, **46**, 613–623.

Swofford DL (1993) *PAUP: Phylogenetic Analysis Using Parsimony*. Illinois Natural History Survey, Champaign, Illinois.

Wain-Hobson S (1998) 1959 and all that. *Nature*, **391**, 531–532.

Wallace DC, Garrison K and Knowler WC (1985) Dramatic founder effects in Amerindian mitochondrial DNAs. *Am. J. Phys. Anthropol.*, **68**, 149–155.

Ward RH, Frazier BL, Dew-Jager K and Pääbo S (1991) Extensive mitochondrial diversity within a single Amerindian tribe. *Proc. Natl Acad. Sci. USA*, **88**, 8720–8724.

Waterman MS, Smith TF and Beyer WA (1976) Some biological sequence metrics. *Adv. Math.*, **20**, 367–387.

Zhu T, Korber BT, Nahmias AJ, Hooper E, Sharp PM and Ho DD (1998) An African HIV-1 sequence from 1959 and implications for the origin of the epidemic. *Nature*, **391**, 594–597.

FURTHER READING

Avise JC (1994) *Molecular Markers, Natural History and Evolution*. Chapman & Hall, New York. — *A detailed description of the use of molecular data in studies of evolution.*

Futuyama DJ (1998) *Evolutionary Biology*, 3rd edn. Sinauer, Sunderland, MA.

Hartl DL and Clark AG (1997) *Principles of Population Genetics*, 3rd edn. Sinauer, Sunderland, MA. — *A good introduction to population genetics that emphasizes the relevance of this subject to evolutionary biology.*

Hillis DM, Moritz C and Mable BK (eds) (1996) *Molecular Systematics*, 2nd edn. Sinauer, Sunderland, MA. — *Comprehensive coverage of techniques for phylogenetic tree reconstruction.*

Li WH and Graur D (1991) *Fundamentals of Molecular Evolution*. Sinauer, Sunderland, MA. — *Another excellent introduction.*

Nei M (1996) Phylogenetic analysis in molecular evolutionary genetics. *Annu. Rev. Genet.*, **30**, 371–403. — *Brief review of tree-building techniques.*

Appendix – Keeping up to Date

GENOME RESEARCH MOVES at a fast pace. If you check through the reading lists at the end of each chapter you will see that many of the papers cited in *Genomes* were published in 1998, the year in which the book was written. Important discoveries are likely to be made over the next 2–3 years and keeping up to date is a big challenge for students taking advanced courses in molecular biology.

In the past, the only way to learn about new advances was to read about them in research journals and other publications. This is still a good way to keep up to date, but today it is also possible to find out about the latest research, almost as it happens, through the Internet. This Appendix gives you some guidance about how to make the best use of these written and electronic resources.

KEEPING UP TO DATE BY READING THE LITERATURE

If you have made use of the reading lists at the end of each chapter, then by now you will be familiar with the best sources of up-to-date information on genome research. You will know that it is not necessary to tackle the research papers that describe the latest results – the 'primary' publications – because most of the important papers are summarized in more accessible reviews and news articles. These articles can be found in two types of journal (*Table A1*):

- **Review journals** are devoted to review articles and do not themselves contain any primary research publications. Most are published monthly and are therefore able to report new discoveries soon after they are announced. Others, such as the *Annual Reviews* series, come out once a year and aim to provide comprehensive surveys of the current status of various research areas.
- A few **research journals** publish not only primary papers, but also review articles that place these papers in context and explain their significance in a way that is easily understood by the general reader. They describe papers published in other journals as well as in their own pages, and so give a comprehensive

Table A1 Journals that publish review articles relevant to genome research

Journal	Frequency	Level[†]
Review journals		
Annual Review of Biochemistry	Yearly	3
Annual Review of Genetics	Yearly	3
Bioessays	Monthly	2
Current Biology	Twice monthly	2
Current Opinion in Cell Biology	Bimonthly	2
Current Opinion in Genetics and Development	Bimonthly	2
Current Opinion in Structural Biology	Bimonthly	2
New Scientist	Weekly	1
Scientific American	Monthly	1
Trends in Biochemical Sciences	Monthly	2
Trends in Biotechnology	Monthly	2
Trends in Genetics	Monthly	2
Research journals containing some review articles		
Cell	Biweekly	3
Nature	Weekly	2
Nature Biotechnology	Monthly	2
Nature Genetics	Monthly	2
Nature Structural Biology	Monthly	2
Science	Weekly	2

† Level 1 journals are aimed at a general readership and their articles are less detailed than the text of *Genomes*; the review articles in level 2 journals are roughly equivalent to *Genomes*; level 3 journals are more advanced.

coverage of the latest results. They are so useful and important that there are few scientists, in any area of research, who do not read *Science* and/or *Nature* every week.

KEEPING UP TO DATE USING THE INTERNET

Many research labs have their own Internet sites where they describe their ongoing research projects and latest discoveries. There are too many of these to list here, but they are easily found if you know the name of the university or research institute in which the group is located, which is always noted in their written publications. Simply find the home page for the university or institute (using a search engine) and usually you will see links to the research pages you are looking for.

Even more exciting is the possibility of examining new results by logging on to the various Internet sites that researchers use to keep themselves up to date. There is at least one site devoted to each organism whose genome is being or has been sequenced, as well as many sites specializing in more general areas of genome research (*Table A2*). These sites are designed for research use and the information they contain is sometimes difficult to understand if you are not an expert in that particular area, but many include at least some general information and all of them are interesting. You rarely need a password to gain access because most genome research operates on the basis that everyone should be able to see the latest results and data, so feel free to log on and see what there is.

Table A2 Internet addresses relevant to genome research

Description	Address
Sites relevant to the Human Genome Project	
CEPH (YAC mapping)	http://www.cephb.fr/bio/ceph-genethon-map.html
dBEST (EST sequences)	http://www.ncbi.nih.gov/dbEST/index.html
Généthon (SSLP maps)	http://www.genethon.fr/
Radiation hybrid database	http://www.ebi.ac.uk/RHdb
Whitehead Institute (physical mapping)	http://www.genome.mit.edu/
Sites relevant to other eukaryotic genome projects	
Microbial Eukaryotes	
Candida albicans	http://alces.med.umn.edu/Candida.html
Dictyostelium discoideum	http://glamdring.ucsd.edu/others/dsmith/dictydb.html
Plasmodium falciparum	http://ben.vub.ac.be/malaria/mad.html
	http://www.wehi.edu.au/biology/malaria/who.html
	http://parasite.arf.efl.edu/malaria.html
Saccharomyces cerevisiae	http://genome-www.stanford.edu/Saccharomyces/
	http://www.mips.biochem.mpg.de/
	http://www.sanger.ac.uk/Projects/
Schizosaccharomyces pombe	http://www.mips.biochem.mpg.de/
Invertebrates	
Bombyx mori	http://www.ab.a.u-tokyo.ac.jp/sericulture/shimada.html
Caenorhabditis elegans	http://www.sanger.ac.uk/Projects/C_elegans/
	http://moulon.inra.fr/acedb/acedb.html
	http://www.ddbj.nig.ac.jp/htmls/c-elegans/html/CE_INDEX.html
Drosophila melanogaster	http://flybase.bio.indiana.edu/
Mosquito	http://klab.agsci.colostate.edu/
Vertebrates	
Chicken	http://www.ri.bbsrc.ac.uk/chickmap/chickgbase/manager.html
Cow	http://locus.jouy.inra.fr/cgi-bin/bovmap/intro2.pl
Dog	http://mendel.berkeley.edu/dog.html
Mekada fish	http://niol1.bio.nagoya-u.ac.jp:8000/
Mouse	http://www.genome.wi.mit.edu/cgi-bin/mouse
	http://www.genethon.fr/
	http://www.informatics.jax.org/
Pig	http://www.ri.bbsrc.as.uk/pigmap/pigbase/pigbase.html
Pufferfish	http://fugu.hgmp.mrc.ac.uk
Rat	http://ratmap.gen.gu.se/
Sheep	http://dirk.invermay.cri.nz/
Zebrafish	http://zfish.uoregon.edu/

Plants

Arabidopsis thaliana	http://lenti.med.umn.edu/arabidopsis/arab_top_page.html
	http://cbil.humgen.upenn.edu/~atgc/ATGCUP.html
	http://www.arabidopsis.com/bb/b950928z.html
	http://nasc.nott.Ac.uk/JIC-contigs/JIC-contigs.html
	http://genome-www.stanford.edu/Arabidopsis/
	http://www.tigr.org/tdb/at/at.html
Beans	http://scaffold.biologie.uni-kl.de/Beanref
Cotton	http://algodon.tamu.edu/
Forest trees	http://s27w007.pswfs.gov/
Maize	http://teosinte.agron.missouri.edu/
	http://moulon.moulon.inra/imgd
Rice	http://www.staff.or.jp/
Soybean	http://mendel.agron.iastate.edu:8000/main.html

Sites relevant to prokaryotic genome projects

General	http://www.tigr.org/tdb/mdb/mdb.html
Archaeoglobus fulgidis	http://www.tigr.org/tdb/mdb/afdb/af_bg.html
Bacillus subtilis	http://www.pasteur.fr/Bio/SubtiList.html
	http://bacillus.tokyo-center.genome.ad.jp:8008/BSORF-DB.html
Borrelia burgdorferi	http://www.tigr.org/tdb/mdb/bbdb/bb_bg.html
Chlamydia trachomatis	http://chlamydia-www.berkeley.edu:4231/
Clostridium acetobutylicum	http://www.genomecorp.cpm/htdocs/sequences/clostridium/clospage.html
Escherichia coli	http://www.genetics.wisc.edu:80/index.html
	http://genome4.aist-nara.ac.jp/
	http://susi.bio.uni-giessen.de/usr/local/www/html/ecdc.html
	http://mol.genes.nig.ac.jp/ecoli/
	http://www.ai.sri.com/ecocyc/ecocyc.html
	http://www.mbl.edu/html/ecoli.html
Haemophilus influenzae	http://www.tigr.org/tdb/mdb/hidb/hidb.html
	http://susi.bio.uni-giessen.de/usr/local/www/html/hidc.html
Helicobacter pylori	http://www.tigr.org/tdb/mdb/hpdb/hpdb.html
Methanobacterium thermoautotrophicum	http://www.genomecorp.com/htdocs/methanobacter/abstract.html
Methanococcus jannaschii	http://www.tigr.org/tdb/mdb/mjdb/mjdb.html
Mycobacterium leprae	http://www.sanger.ac.uk/Projects/
	http://kiev.physchem.kth.se/MycDB.html
	http://www.cric.com/htdocs/leprae/index.html
Mycobacterium tuberculosis	http://www.sanger.ac.uk/Projects/
	http://kiev.physchem.kth.se/MycDB.html
Mycoplasma genitalium	http://www.tigr.org/tdb/mdb/mgdb/mgdb.html
Mycoplasma pneumoniae	http://www.zmbh.uni-heidelberg.de/M_pneumoniae/
Neisseria gonorrhoeae	http://dna1.chem.uoknor.edu/gono.html
Neisseria meningitidis	http://www.sanger.ac.uk/Projects/
Pseudomonas aeruginosa	http://www.cmcb.uq.edu.au/aeruginosa/
Streptococcus pyogenes	http://dna1.chem.uoknor.edu/strep.html
Sulfolobus solfataricus	http://www.imb.nrc.ca/imb/sulfolob/sulhom_e.html
Synechocystis sp.	http://www.kazusa.or.jp/cyano/cyano.html
Treponema palladium	http://utmmg.med.uth.tmc.edu/treponema/tpall.html

Other interesting sites

Sequence databases	
European Bioinformatics Institute	http://www.ebi.ac.uk
Genbank	http://www.ncbi.nlm.nih.gov
DNA Database of Japan	http://www.nig.oc.jp/home.html
Codon usage data	http://www.dna.affrc.go.jp/~nakamura/codon.html
Inteins	http://www.uk.neb.com/neb/inteins/
Restriction endonucleases	http://www.neb.com/rebase/rebase.html

For more information see Brown TA (1998) *Molecular Biology Labfax*, 2nd edition. Academic Press, London.

Glossary

Acceptor arm Part of the structure of a tRNA molecule.

Acceptor site The splice site at the 3′-end of an intron.

Acidic domain A type of activation domain.

Acridine dye A chemical compound that causes a frameshift mutation by intercalating between adjacent base pairs of the double helix.

Activation domain The part of a transcription factor that makes contact with the initiation complex.

Activator A DNA-binding protein that increases the rate of transcription initiation at a promoter.

Acylation The attachment of a lipid sidechain to a polypeptide.

Ada enzyme An *Escherichia coli* enzyme that is involved in the direct repair of alkylation mutations.

Adenine A purine base found in DNA and RNA.

Adenylate cyclase The enzyme that converts ATP to cyclic AMP.

A-DNA A structural configuration of the double helix, present but not common in cellular DNA.

Affinity chromatography A column chromatography method that makes use of a ligand that binds to the molecule being purified.

Agarose gel electrophoresis Electrophoresis carried out in an agarose gel and used to separate DNA molecules between 100 bp and 50 kb in length.

Alkaline phosphatase An enzyme that removes phosphate groups from the 5′-ends of DNA molecules.

Alkylating agent A mutagen that acts by adding alkyl groups to nucleotide bases.

Allele One of two or more alternative forms of a gene.

Allele frequency The frequency of an allele in a population.

Allele-specific oligonucleotide (ASO) hybridization The use of an oligonucleotide probe to determine which of two alternative nucleotide sequences is contained in a DNA molecule.

Allopolyploid A polyploid nucleus derived from fusion between gametes from different species.

α-helix One of the commonest secondary structural conformations taken up by segments of polypeptides.

Alphoid DNA The tandemly repeated nucleotide sequences located in the centromeric regions of human chromosomes.

Alu A type of SINE found in the genomes of humans and related mammals.

***Alu* PCR** A clone fingerprinting technique that uses PCR to detect the relative positions of *Alu* sequences in cloned DNA fragments.

Amino acid One of the monomeric units of a protein molecule.

Aminoacylation Attachment of an amino acid to the acceptor arm of a tRNA.

Aminoacyl- or A-site The site in the ribosome occupied by the aminoacyl-tRNA during translation.

Aminoacyl-tRNA synthetase An enzyme that catalyzes the aminoacylation of one or more tRNAs.

2-aminopurine A base analog that can cause mutations by replacing adenine in a DNA molecule.

Amino terminus The end of a polypeptide that has a free amino group.

Ancestral character state A character state possessed by a remote common ancestor of a group of organisms.

Ancient DNA DNA preserved in ancient biological material.

Annealing Attachment of an oligonucleotide primer to a DNA or RNA template.

Anticodon The triplet of nucleotides, at positions 34–36 in a tRNA molecule, that base-pair with a codon in an mRNA molecule.

Anticodon arm Part of the structure of a tRNA molecule.

Antigen A substance which when injected into the bloodstream elicits an immune response.

Antitermination A bacterial mechanism for regulating the termination of transcription.

Antiterminator protein A protein that attaches to bacterial DNA and mediates antitermination.

AP (apurinic/apyrimidinic) site A position in a DNA molecule where the base component of the nucleotide is missing.

AP endonuclease An enzyme involved in base excision repair.

Apomorphic character state A character state that evolved in a recent ancestor of a subset of organisms in a group being studied.

Apoptosis Programmed cell death.

Archaea One of the two main groups of prokaryotes, mostly found in extreme environments.

Ascospore One of the haploid products of meiosis in an ascomycete such as the yeast *Saccharomyces cerevisiae*.

Ascus The structure which contains the four ascospores produced by a single meiosis in the yeast *Saccharomyces cerevisiae*.

Attenuation A process used by some bacteria to regulate expression of an amino acid biosynthetic operon in accordance with the levels of the amino acid in the cell.

AU–AC intron A type of intron found in eukaryotic nuclear genes: the first two nucleotides in the intron are 5′-AU-3′ and the last two are 5′-AC-3′.

Autonomously replicating sequence (ARS) A DNA sequence, especially from yeast, that confers replicative ability on a nonreplicative plasmid.

Autopolyploid A polyploid nucleus derived from fusion between two gametes from the same species, neither of which are haploid.

Autoradiography The detection of radioactively labeled molecules by exposure of an X-ray-sensitive photographic film.

Autosome A chromosome that is not a sex chromosome.

Auxotroph A mutant microorganism that can grow only when supplied with a nutrient that is not needed by the wild type.

Bacteria One of the two main groups of prokaryotes.

Bacterial artificial chromosome (BAC) A high-capacity cloning vector based on the F plasmid of *Escherichia coli*.

Bacteriophage or phage A virus that infects a bacterium.

Barr body The highly condensed chromatin structure taken up by an inactivated X chromosome.

Basal promoter The position within a eukaryotic promoter where the initiation complex is assembled.

Basal rate of transcription The number of productive initiations of transcription occurring per unit time at a particular promoter.

Base analog A compound whose structural similarity to one of the bases in DNA enables it to act as a mutagen.

Base excision repair A DNA repair process that involves excision and replacement of an abnormal base.

Baseless site A position in a DNA molecule where the base component of the nucleotide is missing.

Base pair The hydrogen-bonded structure formed by two complementary nucleotides. When abbreviated to 'bp', the shortest unit of length for a double-stranded DNA molecule.

Base-pairing The attachment of one polynucleotide to another, or one part of a polynucleotide to another part of the same polynucleotide, by base pairs.

Base-stacking The hydrophobic interactions that occur between adjacent base pairs in a double-stranded DNA molecule.

Basic domain A type of DNA-binding domain.

B chromosome A chromosome possessed by some individuals in a population, but not all.

B-DNA The commonest structural conformation of the DNA double helix in living cells.

Beads-on-a-string An unpacked form of chromatin consisting of nucleosome beads on a string of DNA.

β-sheet One of the commonest secondary structural conformations taken up by segments of polypeptides.

β-turn A sequence of four amino acids, the second usually glycine, which causes a polypeptide to change direction.

Biological information The information contained in the genome of an organism and which directs the development and maintenance of that organism.

Biotechnology The use of living organisms, often, but not always microbes, in industrial processes.

Bivalent The structure formed when a pair of homologous chromosomes lines up during meiosis.

Blunt end An end of a double-stranded DNA molecule where both strands terminate at the same nucleotide position with no single-stranded extension.

Bootstrap analysis A method for inferring the degree of confidence that can be assigned to a branch point in a phylogenetic tree.

Bottleneck A temporary reduction in the size of a population.

–25 Box A component of the bacterial promoter.

Branch migration A step in the Holliday model for homologous recombination, involving exchange of polynucleotides between a pair of recombining double-stranded DNA molecules.

5-bromouracil A base analog that can cause mutations by replacing thymine in a DNA molecule.

Buoyant density The density possessed by a molecule or particle when suspended in an aqueous salt or sugar solution.

Cap The chemical modification at the 5′-end of most eukaryotic mRNA molecules.

Cap binding complex The complex, also called eIF-4F and comprising the initiation factors eIF-4A, eIF-4E and eIF-4G, which makes the initial attachment to the cap structure at the beginning of the scanning phase of eukaryotic translation.

Capping Attachment of a cap to the 5′-end of a eukaryotic mRNA.

Capsid The protein coat that surrounds the DNA or RNA genome of a virus.

CAP site A DNA-binding site for the catabolite activator protein.

Carboxyl terminus The end of a polypeptide that has a free carboxyl group.

CASPs (CTD-associated SR-like proteins) Proteins thought to play regulatory roles during splicing of GU-AG introns.

Catabolite activator protein A regulatory protein that binds to various sites in a bacterial genome and activates transcription initiation at downstream promoters.

Catabolite repression The means by which extracellular glucose levels dictate whether genes for sugar utilization are switched on or off in bacteria.

cDNA A double-stranded DNA copy of an mRNA molecule.

cDNA capture or cDNA selection Repeated hybridization probing of a pool of cDNA with the objective of obtaining a subpool enriched in certain sequences.

Cell cycle The series of events occurring in a cell between one division and the next.

Cell cycle checkpoint A period before entry into S or M phase of the cell cycle, a key point at which regulation is exerted.

Cell-free protein synthesizing system A cell extract containing all the components needed for protein synthesis and able to translate added mRNA molecules.

Centromere The constricted region of a chromosome that is the position at which the pair of chromatids are held together.

Chain termination method A DNA sequencing method that involves enzymatic synthesis of polynucleotide chains that terminate at specific nucleotide positions.

Chaperonin A multisubunit protein that forms a structure that aids the folding of other proteins.

Character state One of at least two alternative forms of a character used in phylogenetic analysis.

Chemical degradation sequencing A DNA sequencing method that involves the use of chemicals that cut DNA molecules at specific nucleotide positions.

Chemical shift The change in the rotation of a chemical nucleus, used as the basis of NMR.

Chi form An intermediate structure seen during recombination between DNA molecules.

Chimera An organism composed of two or more genetically different cell types.

Chi site A repeated nucleotide sequence in the *Escherichia coli* genome that is involved in the initiation of homologous recombination.

Chloroplast One of the photosynthetic organelles of a eukaryotic cell.

Chloroplast genome The genome present in the chloroplasts of a photosynthetic eukaryotic cell.

Chromatid The arm of a chromosome.

Chromatin The complex of DNA and histone proteins found in chromosomes.

30 nm chromatin fiber A relatively unpacked form of chromatin consisting of a possibly helical array of nucleosomes in a fiber approximately 30 nm in diameter.

Chromatin remodeling A process that results in repositioning of nucleosomes on a eukaryotic DNA molecule.

Chromatin remodeling machine (CRM) Enzymes that carry out chromatin remodeling.

Chromosome One of the DNA-protein structures that contains part of the nuclear genome of a eukaryote. Less accurately, the DNA molecule(s) that contains a prokaryotic genome.

Chromosome walking A technique that can be used to construct a clone contig by identifying overlapping fragments of cloned DNA.

Clade A group of monophyletic organisms or DNA sequences that include all of those in the analysis that are descended from a particular common ancestor.

Cladistics A phylogenetic approach that stresses the importance of understanding the evolutionary relevance of the characters that are studied.

Cleavage and polyadenylation specificity factor (CPSF) A protein that plays an ancillary role during polyadenylation of eukaryotic mRNAs.

Cleavage stimulation factor (CstF) A protein that plays an ancillary role during polyadenylation of eukaryotic mRNAs.

Clone A group of cells that contain the same recombinant DNA molecule.

Clone contig A collection of clones whose DNA fragments overlap.

Clone contig approach A genome sequencing strategy in which the molecules to be sequenced are broken into manageable segments, each a few hundred kb or few Mb in length, which are sequenced individually.

Clone fingerprinting Any one of several techniques that compares cloned DNA fragments in order to identify ones that overlap.

Cloning vector A DNA molecule that is able to replicate inside a host cell and therefore can be used to clone other fragments of DNA.

Closed promoter complex The structure formed during the initial step in assembly of the transcription initiation complex. The closed promoter complex consists of the RNA polymerase and/or accessory proteins attached to the promoter, before the DNA has been opened up by breakage of base pairs.

Cloverleaf A two-dimensional representation of the structure of a tRNA molecule.

Coding RNA An RNA molecule that codes for a protein: an mRNA.

Codominance The relationship between a pair of alleles which both contribute to the phenotype of a heterozygote.

Codon A triplet of nucleotides coding for a single amino acid.

Codon–anticodon recognition The interaction between a codon on an mRNA molecule and the corresponding anticodon on a tRNA.

Codon bias Refers to the fact that not all codons are used equally frequently in the genes of a particular organism.

Cohesive end An end of a double-stranded DNA molecule where there is a single-stranded extension.

Cointegrate An intermediate in the pathway resulting in replicative transposition.

Commitment complex The initial structure formed during splicing of a GU–AG intron.

Comparative genomics A research strategy that uses information obtained from the study of one genome to make inferences about the map positions and functions of genes in a second genome.

Competent Refers to a culture of bacteria that have been treated, for example, by soaking in calcium chloride, so that their ability to take up DNA molecules is enhanced.

Complementary Refers to two nucleotides or nucleotide sequences that are able to base-pair with one another.

Complementary DNA (cDNA) A double-stranded DNA copy of an mRNA molecule.

Composite transposon A DNA transposon comprising a pair of insertion sequences flanking a segment of DNA usually containing one or more genes.

Concatamer A DNA molecule made up of linear genomes linked head-to-tail.

Conditional-lethal mutation A mutation that results in a cell or organism able to survive only under permissive conditions.

Conjugation Transfer of DNA between two bacteria which come into physical contact with one another.

Conjugation mapping A technique for mapping bacterial genes by determining the time it takes for each gene to be transferred during conjugation.

Consensus sequence A nucleotide sequence that represents an 'average' of a number of related but nonidentical sequences.

Conservative replication A hypothetical mode of DNA replication in which one daughter double helix is made up of the two parental polynucleotides and the other is made up of two newly synthesized polynucleotides.

Conservative transposition Transposition that does not result in copying of the transposable element.

Constitutive heterochromatin Chromatin that is permanently in a compact organization.

Constitutive mutation A mutation that results in continuous expression of a gene or set of genes that is normally subject to regulatory control.

Contig A contiguous set of overlapping DNA sequences.

Contour clamped homogeneous electric fields (CHEF) An electrophoresis method used to separate large DNA molecules.

Conventional pseudogene A gene that has become inactive because of the accumulation of mutations.

Convergent evolution The situation that occurs when the same character state evolves independently in two lineages.

Core enzyme The version of *E. coli* RNA polymerase, subunit composition $\alpha_2\beta\beta'$, that carries out RNA synthesis but is unable to locate promoters efficiently.

Core octamer The central component of a nucleosome, made up of two subunits each of histones H2A, H2B, H3 and H4, around which DNA is wound.

Co-repressor A small molecule that must be bound to a repressor protein before the latter is able to attach to its operator site.

Core promoter The position within a eukaryotic promoter where the initiation complex is assembled.

Cosmid A high-capacity cloning vector consisting of the λ cos site inserted into a plasmid.

Cotransduction Transfer of two or more genes from one bacterium to another via a transducing phage.

Cotransformation Uptake of two or more genes on a single DNA molecule during transformation of a bacterium.

CpG island A GC-rich DNA region located upstream of approximately 56% of the genes in the human genome.

CREB An important transcription factor.

Crossing-over The exchange of DNA between chromosomes during meiosis.

Cryptic splice site A site whose sequence resembles an authentic splice site and which might be selected instead of the authentic site during aberrant splicing.

Cryptogene One of several genes in the trypanosome mitochondrial genome which specify abbreviated RNAs that must undergo pan-editing in order to become functional.

Cyanelle A photosynthetic organelle that resembles an ingested cyanobacterium.

Cyclic AMP A modified version of AMP in which an intramolecular phosphodiester bond links the 5′ and 3′ carbons.

Cyclin A regulatory protein whose abundance varies during the cell cycle and which regulates biochemical events in a cell cycle-specific manner.

Cyclobutyl dimer A dimer between two adjacent pyrimidine bases in a polynucleotide, formed by ultraviolet irradiation.

Cys_2His_2 finger A type of zinc finger DNA-binding domain.

Cytosine One of the pyrimidine bases found in DNA and RNA.

Dark repair A type of nucleotide excision repair process that corrects cyclobutyl dimers.

D arm Part of the structure of a tRNA molecule.

Deadenylation-dependent pathway A process for degradation of eukaryotic mRNAs that is initiated by removal of the poly(A) tail.

Deadenylation-independent pathway A process for degradation of eukaryotic mRNAs that is initiated by the presence of an internal termination codon.

Deaminating agent A mutagen that acts by removing amino groups from nucleotide bases.

Degenerate Refers to the fact that the genetic code has more than one codon for most amino acids.

Degradosome A multienzyme complex responsible for degradation of bacterial mRNAs.

Delayed-onset mutation A mutation whose effect is not apparent until a relatively late stage in the life of the mutant organism.

Deletion mutation A mutation resulting from deletion of one or more nucleotides from a DNA sequence.

Denaturation Breakdown by chemical or physical means of the noncovalent interactions, such as hydrogen bonding, that maintain the secondary and higher levels of structure of proteins and nucleic acids.

***De novo* methylation** Addition of methyl groups to new positions on a DNA molecule.

Density gradient centrifugation A technique in which a cell fraction is centrifuged through a dense solution, in the form of a gradient, so that individual components are separated.

Deoxyribonuclease An enzyme that cleaves phosphodiester bonds in a DNA molecule.

Derived character state A character state that evolved in a recent ancestor of a subset of organisms in a group being studied.

Development A coordinated series of transient and permanent changes that occurs during the life history of a cell or organism.

Diauxie The phenomenon whereby a bacterium, when provided with a mixture of sugars, uses up one sugar before beginning to metabolize the second sugar.

Dideoxynucleotide A modified nucleotide that lacks the 3′ hydroxyl group and so terminates strand synthesis when incorporated into a polynucleotide.

Differential centrifugation A technique that separates cell components by centrifuging an extract at different speeds.

Differentiation The adoption by a cell of a specialized biochemical and/or physiological role.

Dihybrid cross A sexual cross in which the inheritance of two pairs of alleles is followed.

Dimer A protein or other structure that comprises two subunits.

Diploid A nucleus that has two copies of each chromosome.

Directed evolution A set of experimental techniques that is used to obtain novel genes with improved products.

Directed shotgun approach A genome-sequencing strategy which combines random shotgun sequencing with a genome map, the latter used to aid assembly of the master sequence.

Direct readout The recognition of a DNA sequence by a binding protein that makes contacts with the outside of a double helix.

Direct repair A DNA repair system that acts directly on a damaged nucleotide.

Direct repeat A nucleotide sequence that is repeated twice or more frequently in a DNA molecule.

Discontinuous gene A gene that is split into exons and introns.

Dispersive replication A hypothetical mode of DNA replication in which both polynucleotides of each daughter double helix are made up partly of parental DNA and partly of newly synthesized DNA.

Displacement replication A mode of replication which involves continuous copying of one strand of the helix, the second strand being displaced and subsequently copied after synthesis of the first daughter strand has been completed.

Distance matrix A table showing the evolutionary distances between all pairs of nucleotide sequences in a dataset.

Distance method A rigorous mathematical approach to alignment of nucleotide sequences.

Disulfide bridge A covalent bond linking cysteine amino acids on different polypeptides or at different positions on the same polypeptide.

D-loop An intermediate structure formed during the Meselson–Radding model for homologous recombination. Also an intermediate formed during displacement replication.

DNA Deoxyribonucleic acid, one of the two forms of nucleic acid in living cells; the genetic material for all cellular lifeforms and many viruses.

DNA adenine methylase (Dam) An enzyme involved in methylation of *E. coli* DNA.

DNA bending A type of conformational change introduced into a DNA molecule by a binding protein.

DNA-binding motif The part of a DNA-binding protein that makes contact with the double helix.

DNA-binding protein A protein that attaches to a DNA molecule.

DNA chip A high-density array of DNA molecules used for parallel hybridization analyses.

DNA cytosine methylase (Dcm) An enzyme involved in methylation of *E. coli* DNA.

DNA-dependent DNA polymerase An enzyme that makes a DNA copy of a DNA template.

DNA-dependent RNA polymerase An enzyme that makes an RNA copy of a DNA template.

DNA glycolyase An enzyme that cleaves the β-*N*-glycosidic bond between a base and the sugar component of a nucleotide as part of the base excision and mismatch repair processes.

DNA gyrase A Type II topoisomerase of *E. coli*.

DNA ligase An enzyme that synthesizes phosphodiester bonds as part of DNA replication, repair and recombination processes.

DNA marker A DNA sequence that exists as two or more readily distinguished versions and which can therefore be used to mark a map position on a genetic, physical or integrated genome map.

DNA methylation Refers to the chemical modification of DNA by attachment of methyl groups, which has various regulatory functions in prokaryotes and eukaryotes.

DNA photolyase An bacterial enzyme involved in photoreactivation repair.

DNA polymerase An enzyme that synthesizes DNA on a DNA or RNA template.

DNA polymerase I The bacterial enzyme that completes synthesis of Okazaki fragments during DNA replication.

DNA polymerase II A bacterial DNA polymerase involved in DNA repair.

DNA polymerase III The main DNA replicating enzyme of bacteria.

DNA polymerase α The enzyme that primes DNA replication in eukaryotes.

DNA polymerase δ The main eukaryotic DNA replicating enzyme.

DNA polymerase γ The enzyme responsible for replicating the mitochondrial genome.

DNA repair The biochemical processes that correct mutations arising from replication errors and the effects of mutagenic agents.

DNA replication Synthesis of a new copy of the genome.

DNase I hypersensitive site A short region of eukaryotic DNA that is relatively easily cleaved with deoxyribonuclease I, possibly coinciding with positions where nucleosomes are absent.

DNA sequencing The technique for determining the order of nucleotides in a DNA molecule.

DNA shuffling A PCR-based procedure that results in directed evolution of a DNA sequence.

DNA topoisomerase An enzyme that introduces or removes turns from the double helix by breakage and reunion of one or both polynucleotides.

DNA transposon A transposon whose transposition mechanism does not involve an RNA intermediate.

DNA tumor virus A virus with a DNA genome, able to cause cancer after infection of an animal cell.

Domain duplication Duplication of a gene segment coding for a structural domain in the protein product.

Domain shuffling Rearrangement of segments of one or more genes, each segment coding for a structural domain in the gene product, to create a new gene.

Dominant The allele that is expressed in a heterozygote.

Donor site The splice site at the 5′-end of an intron.

Dot matrix A method for aligning nucleotide sequences.

Double helix The base-paired, double-stranded structure that is the natural form of DNA in the cell.

Double heterozygote A nucleus that is heterozygous for two genes.

Double homozygote A nucleus that is homozygous for two genes.

Double restriction Digestion of DNA with two restriction endonucleases at the same time.

Double-strand break repair A DNA repair process that mends double-stranded breaks.

Double-stranded Comprising two polynucleotides attached to one another by base-pairing.

Double-stranded RNA adenosine deaminase (dsRAD) An enzyme that edits various eukaryotic mRNAs by deaminating adenosine to inosine.

Double-stranded RNA-binding domain (dsRBD) A common type of RNA-binding domain.

Downstream Towards the 3′-end of a polynucleotide.

Dynamic allele-specific hybridization (DASH) A solution hybridization technique used to type SNPs.

Electrophoresis Separation of molecules on the basis of their net electrical charge.

Electrostatic interactions Ionic bonds that form between charged chemical groups.

Elongation factor A protein that plays an ancillary role in the elongation step of transcription or translation.

Embryonic stem (ES) cell A totipotent cell from the embryo of a mouse or other organism.

End-labeling The attachment of a radioactive or other label to one end of a DNA or RNA molecule.

End-modification The chemical alteration of the end of an RNA molecule.

Endogenous retrovirus (ERV) An active or inactive retroviral genome integrated into a host chromosome.

Endonuclease An enzyme that breaks phosphodiester bonds within a nucleic acid molecule.

Endosymbiont theory A theory that states that the mitochondria and chloroplasts of eukaryotic cells are derived from symbiotic prokaryotes.

Enhancer A regulatory sequence that increases the rate of transcription of a gene or genes located some distance away in either direction.

Episome A plasmid that is able to integrate into the host cell's chromosome.

Episome transfer Transfer of some or all of a bacterial chromosome by integration into a plasmid.

E site A position within a bacterial ribosome to which a tRNA moves immediately after deacylation.

Ethidium bromide A type of intercalating agent that causes mutations by inserting between adjacent base pairs in a double-stranded DNA molecule.

Ethylmethane sulfonate (EMS) A mutagen that acts by adding alkyl groups to nucleotide bases.

Euchromatin Regions of a eukaryotic chromosome that are relatively uncondensed, thought to contain active genes.

Eukaryote An organism whose cells contain membrane-bound nuclei.

Excision repair A DNA repair process that corrects various types of DNA damage by excising and resynthesizing a region of polynucleotide.

Exit site A position within a bacterial ribosome to which a tRNA moves immediately after deacylation.

Exon A coding region within a discontinuous gene.

Exon–intron boundary The nucleotide sequence at the junction between an exon and an intron. See **Acceptor site** and **Donor site**.

Exon skipping Aberrant splicing in which one or more or exons are omitted from the spliced RNA.

Exon theory of genes An 'introns early' hypothesis that holds that introns were formed when the first DNA genomes were constructed.

Exon trapping A method, based on cloning, for identifying the positions of exons in a DNA sequence.

Exonuclease An enzyme that removes nucleotides from the ends of a nucleic acid molecule.

Exportin A protein involved in transport of molecules out of the nucleus.

Expressed sequence tag (EST) A cDNA that is sequenced in order to gain rapid access to the genes in a genome.

External node The end of a branch in a phylogenetic tree, representing one of the organisms or DNA sequences being studied.

Extrachromosomal gene A gene in a mitochondrial or chloroplast genome.

Facultative heterochromatin Chromatin that has a compact organization in some, but not all cells, thought to contain genes that are inactive in some cells or at some periods of the cell cycle.

FEN1 The 'flap endonuclease' involved in replication of the lagging strand in eukaryotes.

Field inversion gel electrophoresis (FIGE) An electrophoresis method used to separate large DNA molecules.

Fitness The ability of an organism or allele to survive and reproduce.

Fixation Refers to the situation that occurs when a single allele reaches a frequency of 100% in a population.

Flow cytometry A method for the separation of chromosomes.

FLpter value The unit used in FISH to describe the position of a hybridization signal relative to the end of the short arm of the chromosome.

Fluorescent *in situ* hybridization (FISH) A technique for locating markers on chromosomes by observing the hybridization positions of fluorescent markers.

fMet *N*-formylmethionine, the modified amino acid carried by the tRNA that is used during the initiation of translation in bacteria.

Folding domain A segment of a polypeptide that folds independently of other segments.

Footprinting A range of techniques used for locating bound proteins on DNA molecules.

Fosmid A high-capacity vector carrying the F plasmid origin of replication and a λ *cos* site.

F plasmid A fertility plasmid that directs conjugal transfer of DNA between bacteria.

Fragile site A position in a chromosome that is prone to breakage because it contains an expanded trinucleotide repeat sequence.

Frameshifting The controlled movement of a ribosome from one reading frame to another at an internal position within a gene.

Frameshift mutation A mutation resulting from insertion or deletion of a group of nucleotides that is not a multiple of three and which therefore changes the frame in which translation occurs.

Functional analysis The area of genome research devoted to identifying the functions of unknown genes.

Functional domain A region of eukaryotic DNA around a gene or group of genes that can be delineated by treatment with deoxyribonuclease I.

Fusion protein A protein that consists of a fusion of two polypeptides, or parts of polypeptides, normally coded by separate genes.

Gain-of-function mutation A mutation that results in an organism acquiring a new function.

Gamete A reproductive cell, usually haploid, that fuses with a second gamete to produce a new cell during sexual reproduction.

Gap genes Developmental genes that play a role in establishing positional information within the *Drosophila* embryo.

Gap period One of two intermediate periods within the cell cycle.

GAPs (GTPase activating proteins) A set of proteins that are intermediates in the Ras signal transduction pathway.

GC content The percentage of nucleotides in a genome that are G or C.

Gel electrophoresis Electrophoresis performed in a gel so that molecules of similar electrical charge can be separated on the basis of size.

Gel retardation analysis A technique that identifies protein-binding sites on DNA molecules by virtue of the effect that a bound protein has on the mobility of the DNA fragments during gel electrophoresis.

Gel stretching A technique for preparing restricted DNA molecules for optical mapping.

Gene A DNA segment containing biological information and hence coding for an RNA and/or polypeptide molecule.

Gene cloning Insertion of a fragment of DNA, containing a gene, into a cloning vector, and subsequent propagation of the recombinant DNA molecule in a host organism.

Gene conversion A process that results in the four haploid products of meiosis displaying an unusual segregation pattern.

Gene expression The series of events by which the biological information carried by a gene is released and made available to the cell.

Gene fragment A gene relic consisting of short isolated regions from within a gene.

General recombination Recombination between two homologous double-stranded DNA molecules.

General transcription factor (GTF) A protein or protein complex that is a transient or permanent component of the initiation complex formed during eukaryotic transcription.

Gene substitution The replacement of an allele that at one time was fixed in the population by a second allele, this second allele arising by mutation and increasing in frequency until itself reaching fixation.

Gene superfamily A group of two or more evolutionarily related multigene families.

Genes-within-genes Refers to a gene whose intron contains a second gene.

Genetic code The rules that determine which triplet of nucleotides codes for which amino acid during protein synthesis.

Genetic footprinting A technique for the rapid functional analysis of many genes at once.

Genetic linkage The physical association between two genes that are on the same chromosome.

Genetic mapping The use of genetic techniques to construct a genome map.

Genetic marker A gene that exists as two or more readily distinguished alleles and whose inheritance can therefore be followed during a genetic cross, enabling the map position of the gene to be determined.

Genetic profile The banding pattern revealed after electrophoresis of the products of PCRs directed at a range of microsatellite loci.

Genetic redundancy The situation that occurs when two genes in the same genome perform the same function.

Genetics The branch of biology devoted to the study of genes.

Gene tree A phylogenetic tree that shows the evolutionary relationships between a group of genes or other DNA sequences.

Genome The entire genetic complement of a living organism.

Genome-wide repeat Sequences that recur at many dispersed positions within a genome.

Genomic imprinting Inactivation by methylation of gene on one of a pair of homologous chromosomes.

Genotype A description of the genetic composition of an organism.

Glutamine-rich domain A type of activation domain.

Glycolyase An enzyme that cleaves the β-*N*-glycosidic bond between a base and the sugar component of a nucleotide, as part of the base excision and mismatch repair processes.

β-*N*-glycosidic bond The linkage between the base and sugar of a nucleotide.

Glycosylation The attachment of sugar units to a polypeptide.

GNRPs (guanine nucleotide-releasing proteins) A set of proteins that are intermediates in the Ras signal transduction pathway.

G1 phase The first gap period of the cell cycle.

G2 phase The second gap period of the cell cycle.

Group I intron A type of intron found mainly in organelle genes.

Group II intron A type of intron found in organelle genes.

Group III intron A type of intron found in organelle genes.

GU–AG intron The commonest type of intron in eukaryotic nuclear genes. The first two nucleotides of the intron are 5′-GU-3′ and the last two are 5′-AG-3′.

Guanine One of the purine nucleotides found in DNA and RNA.

Guanine methyltransferase The enzyme that attaches a methyl group to the 5′-end of a eukaryotic mRNA during the capping reaction.

Guanylyl transferase The enzyme that attaches a GTP to the 5′-end of a eukaryotic mRNA at the start of the capping reaction.

Guide RNA A short RNA that specifies the positions at which one or more nucleotides are inserted into an abbreviated RNA by pan-editing.

Haploid A nucleus that has a single copy of each chromosome.

Haploinsufficiency The situation where inactivation of a gene on one of a pair of homologous chromosomes results in a change in the phenotype of the mutant organism.

Haplotype A collection of alleles that are usually inherited together.

Helicase An enzyme that breaks base pairs in a double-stranded DNA molecule.

Helix-loop-helix motif A dimerization domain commonly found in DNA-binding proteins.

Helix-turn-helix motif A common structural motif for attachment of a protein to a DNA molecule.

Heterochromatin Chromatin that is relatively condensed and is thought to contain DNA that is not being transcribed.

Heteroduplex analysis Transcript mapping by analysis of DNA–RNA hybrids with a single-strand-specific nuclease such as S1.

Heterogenous nuclear RNA (hnRNA) The nuclear RNA fraction that comprises unprocessed transcripts synthesized by RNA polymerase II.

Heteropolymer An artificial RNA comprising a mixture of different nucleotides.

Heterozygosity The probability that a person chosen at random from the population will be heterozygous for a particular marker.

Heterozygous A diploid nucleus that contains two different alleles for a particular gene.

High-performance liquid chromatography (HPLC) A column chromatography method with many applications in biochemistry.

Histone One of the basic proteins found in nucleosomes.

Holliday structure An intermediate structure formed during recombination between two DNA molecules.

Holoenzyme The version of the *E. coli* RNA polymerase, subunit composition $\alpha_2\beta\beta'\sigma$, that is able to recognize promoter sequences.

Homeodomain A DNA-binding motif found in many proteins involved in developmental regulation of gene expression.

Homeotic mutation A mutation that results in the transformation of one body part into another.

Homeotic selector gene A gene that establishes the identity of a body part such as a segment of the *Drosophila* embryo.

Homologous chromosomes Two or more identical chromosomes present in a single nucleus.

Homologous genes Genes that share a common evolutionary ancestor.

Homologous recombination Recombination between two homologous double-stranded DNA molecules, i.e. ones which share extensive nucleotide sequence similarity.

Homology searching A technique in which genes with sequences similar to that of an unknown gene are sought, the objective being to gain an insight into the function of the unknown gene.

Homoplasy The situation that occurs when the same character state evolves independently in two lineages.

Homopolymer An artificial RNA comprising just one nucleotide.

Homozygous A diploid nucleus that contains two identical alleles for a particular gene.

Horizontal gene transfer Transfer of a gene from one species to another.

Hormone response element A nucleotide sequence upstream of a gene that mediates the regulatory effect of a steroid hormone.

Hsp70 chaperone A family of proteins that bind to hydrophobic regions in other proteins in order to aid their folding.

Hybridization The attachment by base-pairing of two complementary polynucleotides.

Hybridization probing A technique that uses a labeled nucleic acid molecule as a probe to identify complementary or homologous molecules to which it base-pairs.

Hydrogen bond A weak electrostatic attraction between an electronegative atom such as oxygen or nitrogen and a hydrogen atom attached to a second electronegative atom.

Hydrophobic effects Chemical interactions that result in hydrophobic groups becoming buried inside a protein.

Hypermutation An increase in the mutation rate of a genome.

Immunocytochemistry A technique that uses antibody probing to locate the position of a protein in a tissue.

Immunoelectron microscopy An electron microscopy technique that uses antibody labeling to identify the positions of specific proteins on the surface of a structure such as a ribosome.

Immunoscreening The use of an antibody probe to detect a polypeptide synthesized by a cloned gene.

Importin A protein involved in transport of molecules into the nucleus.

Incomplete dominance Refers to a pair of alleles, neither of which displays dominance, the phenotype of a heterozygote being intermediate between the phenotypes of the two homozygotes.

Indel A position in an alignment between two DNA sequences where an insertion or deletion has occurred.

Inducer A molecule that induces expression of a gene or operon by binding to a repressor protein and preventing the repressor from attaching to the operator.

Inferred tree A rooted tree obtained by phylogenetic analysis.

Inhibitor-resistant mutant A mutant that is able to resist the toxic effects of an antibiotic or other type of inhibitor.

Initiation codon The codon, usually but not exclusively 5′-AUG-3′, found at the start of the coding region of a gene.

Initiation complex The complex of proteins that initiates transcription. Also the complex that initiates translation.

Initiation factor A protein that plays an ancillary role during initiation of translation.

Initiation of transcription The assembly upstream of a gene of the complex of proteins that will subsequently copy the gene into RNA.

Initiation region A region of eukaryotic chromosomal DNA within which replication initiates at positions that are not clearly defined.

Initiator (Inr) sequence A component of the RNA polymerase II core promoter.

Initiator tRNA The tRNA, aminoacylated with methionine in eukaryotes or *N*-formylmethionine in bacteria, that recognizes the initiation codon during protein synthesis.

Inosine A modified version of guanosine, sometimes found at the wobble position of an anticodon.

Insertional editing A less extensive form of pan-editing that occurs during processing of some viral RNAs.

Insertion mutation A mutation that arises by insertion of one or more nucleotides into a DNA sequence.

Insertion sequence A short transposable element found in bacteria.

Integrase A Type I topoisomerase that catalyzes insertion of the λ genome into *Escherichia coli* DNA.

Intein An internal segment of a polypeptide that is removed by a splicing process after translation.

Intein homing The conversion of a gene coding for a protein that lacks an intein into one coding for an intein-plus protein, catalyzed by the spliced component of the intein.

Intercalating agent A compound that can enter the space between adjacent base pairs of a double-stranded DNA molecule, often causing mutations.

Internal node A branch point within a phylogenetic tree, representing an organism or DNA sequence that is ancestral to those being studied.

Internal ribosome entry site (IRES) A nucleotide sequence that enables the ribosome to assemble at an internal position in some eukaryotic mRNAs.

Interphase The period between cell divisions.

Interspersed repeat element PCR (IRE-PCR) A clone fingerprinting technique that uses PCR to detect the relative positions of genome-wide repeats in cloned DNA fragments.

Intrinsic terminator A position in bacterial DNA where termination of transcription occurs without the involvement of Rho.

Intron A noncoding region within a discontinuous gene.

Intron homing The conversion of a gene lacking an intron into one that contains an intron, catalyzed by a protein coded by that intron.

Introns early The hypothesis that introns evolved relatively early and are gradually being lost from eukaryotic genomes.

Introns late The hypothesis that introns evolved relatively late and are gradually accumulating in eukaryotic genomes.

Inverted repeat Two identical nucleotide sequences repeated in opposite orientations in a DNA molecule.

***In vitro* mutagenesis** Techniques used to produce a specified mutation at a predetermined position in a DNA molecule.

***In vitro* packaging** Synthesis of infective λ particles from a preparation of λ proteins and a concatamer of λ DNA molecules.

Isoaccepting tRNAs Two or more tRNAs that are charged with the same amino acid.

Isotope One of two or more atoms that have the same atomic number but different atomic weights.

Janus kinase (JAK) A type of kinase that plays an intermediary role in some types of signal transduction involving STATs.

Karyogram The entire chromosome complement of a cell, with each chromosome described in terms of its appearance at metaphase.

K-homology domain A type of RNA-binding domain.

Kilobase pair (kb) 1000 base pairs.

Kinetochore The part of the centromere to which spindle microtubules attach.

Knockout mouse A mouse that has been engineered so that it carries an inactivated gene.

Kozak consensus The nucleotide sequence surrounding the initiation codon of a eukaryotic mRNA.

Lactose operon The cluster of three genes that code for enzymes involved in utilization of lactose by *E. coli*.

Lactose repressor The regulatory protein that controls transcription of the *lac* operon in response to the presence or absence of lactose in the environment.

Lagging strand The strand of the double helix which is copied in a discontinuous fashion during DNA replication.

Lariat Refers to the lariat-shaped intron RNA that results from splicing a GU-AG intron.

Leader segment The untranslated region of an mRNA upstream of the initiation codon.

Leading strand The strand of the double helix which is copied in a continuous fashion during DNA replication.

Leaky mutation A mutation that results in partial loss of a characteristic.

(6-4) lesion A dimer between two adjacent pyrimidine bases in a polynucleotide, formed by ultraviolet irradiation.

Lethal mutation A mutation that results in death of the cell or organism.

Leucine zipper A dimerization domain commonly found in DNA-binding proteins.

LINE (long interspersed nuclear element) A type of genome-wide repeat, often with transposable activity.

LINE-1 One type of human LINE.

Linkage The physical association between two genes that are on the same chromosome.

Linker DNA The DNA that links nucleosomes: the 'string' in the beads-on-a-string model for chromatin structure.

Linker histone A histone, such as H1, that is located outside of the nucleosome core octamer.

Locus The chromosomal location of a genetic or DNA marker.

Locus control region (LCR) A DNA sequence that maintains a functional domain in an open, active configuration.

Lod score A statistical measure of linkage as revealed by pedigree analysis.

Long patch repair A nucleotide excision repair process of *Escherichia coli* that results in excision and resynthesis of up to 2 kb of DNA.

Loss-of-function mutation A mutation that reduces or abolishes a protein's activity.

LTR element A type of genome-wide repeat typified by the presence of long terminal repeats (LTRs).

Lysogenic pathway The type of bacteriophage infection that involves integration of the phage genome into the host DNA molecule.

Lytic pathway The type of bacteriophage infection that involves lysis of the host cell immediately after the initial infection, with no integration of the phage DNA molecule into the host genome.

Macrochromosome One of the larger, gene-deficient chromosomes seen in the nuclei of chickens and various other species.

MADS box A DNA-binding domain found in several transcription factors involved in plant development.

Maintenance methylation Addition of methyl groups to positions on newly synthesized DNA strands that correspond with the positions of methylation on the parent strand.

Major groove The larger of the two grooves that spiral around the surface of the B-form of DNA.

Major histocompatibility complex A mammalian multigene family coding for cell surface proteins and including several multiallelic genes.

Map A chart showing the positions of genetic and/or physical markers in a genome.

MAP kinase A signal transduction pathway.

Mapping reagent A collection of DNA fragments spanning a chromosome or the entire genome and used in STS mapping.

Marker A distinctive feature on a genome map. Also a gene, carried by a cloning vector, that codes for a distinctive protein product and/or phenotype and so can be used to determine if a cell contains a copy of the cloning vector.

Maternal-effect gene A *Drosophila* gene that is expressed in the parent and whose mRNA is subsequently injected into the egg, after which it influences development of the embryo.

Mating type The equivalent of male and female for a eukaryotic microorganism.

Mating type switching The ability of yeast cells to change from a to α mating type, or *vice versa*, by gene conversion.

Matrix-associated region (MAR) An AT-rich segment of a eukaryotic genome that acts as an attachment point to the nuclear matrix.

Maturase A protein, coded by a gene in an intron, thought to be involved in splicing.

Maximum parsimony method A method for construction of phylogenetic trees.

Mediator A protein complex that forms a contact between various transcription factors and the C-terminal domain of the largest subunit of the yeast RNA polymerase II.

Megabase pair (Mb) 1000 kb, 1 000 000 bp.

Meiosis The series of events, involving two nuclear divisions, by which diploid nuclei are converted to haploid gametes.

Melting Denaturation of a double-stranded DNA molecule.

Melting temperature (T_m) The temperature at which the two strands of a double-stranded nucleic acid molecule or base-paired hybrid detach due to complete breakage of hydrogen bonding.

Messenger RNA (mRNA) The transcript of a protein-coding gene.

Metaphase chromosome A chromosome at the metaphase stage of cell division, when the chromatin takes on its most condensed structure and features such as the banding pattern can be visualized.

Methyl-CpG-binding protein (MeCP) A protein that binds to methylated CpG islands and may influence acetylation of nearby histones.

MGMT (O^6-methylguanine-DNA methyltransferase) An enzyme involved in the direct repair of alkylation mutations.

Microarray A low-density array of DNA molecules used for parallel hybridization analysis.

Microsatellite A type of simple sequence length polymorphism comprising tandem copies of, usually, di-, tri- or tetranucleotide repeat units. Also called a simple tandem repeat (STR).

Minichromosome One of the smaller, gene-rich chromosomes seen in the nucleus of chickens and various other species.

Minigene The name given to the pair of exons carried by a cloning vector used in the exon-trapping procedure.

Minimal medium A medium that provides only the minimum nutritional requirements for growth of a microorganism.

Minisatellite A type of simple sequence length polymorphism comprising tandem copies of repeats that are a few tens of nucleotides in length. Also called a variable number of tandem repeats (VNTR).

Minor groove The smaller of the two grooves that spiral around the surface of the B-form of DNA.

Mismatch A position in a double-stranded DNA molecule where base-pairing does not occur because the nucleotides are not complementary; in particular, a nonbase-paired position resulting from an error in replication.

Mismatch repair A DNA repair process that corrects mismatched nucleotide pairs by replacing the incorrect nucleotide in the daughter polynucleotide.

Missense mutation An alteration in a nucleotide sequence that converts a codon for one amino acid into a codon for a second amino acid.

Mitochondrial genome The genome present in the mitochondria of a eukaryotic cell.

Mitochondrion One of the energy-generating organelles of eukaryotic cells.

Mitosis The series of events that result in nuclear division.

Model organism An organism which is relatively easy to study and hence can be used to obtain information that is relevant to the biology of a second organism that is more difficult to study.

Modification assay A range of techniques used for locating bound proteins on DNA molecules.

Modification interference A technique used to identify nucleotides involved in interactions with a DNA-binding protein.

Molecular chaperone A protein that helps other proteins to fold.

Molecular clock A device based on the inferred mutation rate that enables times to be assigned to the branch points in a gene tree.

Molecular combing A technique for preparing restricted DNA molecules for optical mapping.

Molecular evolution The gradual changes that occur in genomes over time due to the accumulation of mutations and structural rearrangements resulting from recombination and transposition.

Molecular life sciences An area of research comprising molecular biology, biochemistry and cell biology, as well as some aspects of genetics and physiology.

Molecular phylogenetics A set of techniques that enable the evolutionary relationships between DNA sequences to be inferred by making comparisons between those sequences.

Monohybrid cross A sexual cross in which the inheritance of one pair of alleles is followed.

Monophyletic Refers to two or more organisms or DNA sequences that are derived from a single ancestral organism or DNA sequence.

M phase The stage of the cell cycle when mitosis or meiosis occurs.

Multicopy A gene, cloning vector or other genetic element that is present in multiple copies in a single cell.

Multicysteine zinc finger A type of zinc finger DNA-binding domain.

Multigene family A group of genes, clustered or dispersed, with related nucleotide sequences.

Multiple alignment An alignment of three or more nucleotide sequences.

Multiple alleles The different alternative forms of a gene that has more than two alleles.

Multiple hit or multiple substitution The situation that occurs when a single nucleotide in a DNA sequence undergoes two mutational changes, giving rise to two new alleles, both of which differ from each other and from the parent at that nucleotide position.

Multipoint cross A genetic cross in which the inheritance of three or more markers is followed.

Multiregional evolution A hypothesis that holds that modern humans in the Old World are descended from *Homo erectus* populations that left Africa over one million years ago.

Mutagen A chemical or physical agent that can cause a mutation in a DNA molecule.

Mutagenesis Treatment of a group of cells or organisms with a mutagen as a means of inducing mutations.

Mutant A cell or organism that possesses a mutation.

Mutasome A protein complex that is constructed during the SOS response of *Escherichia coli*.

Mutation An alteration in the nucleotide sequence of a DNA molecule.

Mutation scanning A set of techniques for detection of mutations in DNA molecules.

Mutation screening A set of techniques for determining if a DNA molecule contains a specific mutation.

Natural selection The preservation of favorable alleles and the rejection of injurious ones.

N-degron An N-terminal amino acid sequence that influences the degradation of a protein in which it is found.

Neighbor-joining method A method for construction of phylogenetic trees.

Nick A position in a double-stranded DNA molecule where one of the polynucleotides is broken due to the absence of a phosphodiester bond.

Nitrogenous base One of the purines or pyrimidines that form part of the molecular structure of a nucleotide.

N-linked glycosylation The attachment of sugar units to an asparagine in a polypeptide.

Noncoding RNA An RNA molecule that does not code for a protein.

Nonhomologous end-joining (NHEG) Another name for the double-strand break repair process.

Nonpenetrance The situation whereby the effect of mutation is never observed during the lifetime of a mutant organism.

Nonpolar A hydrophobic (water-hating) chemical group.

Nonsense mutation An alteration in a nucleotide sequence that changes a triplet coding for an amino acid into a termination codon.

Nonsynonymous mutation A mutation that converts a codon for one amino acid into a codon for a second amino acid.

Northern blotting The transfer of RNA from an electrophoresis gel to a membrane prior to northern hybridization.

Northern hybridization A technique used for detection of a specific RNA molecule against a background of many other RNA molecules.

Nuclear genome The DNA molecules present in the nucleus of a eukaryotic cell.

Nuclear magnetic resonance (NMR) spectroscopy A technique for determining the three-dimensional structure of large molecules.

Nuclear matrix A proteinaceous scaffold-like network that permeates the cell.

Nuclear pore complex The complex of proteins present at a nuclear pore.

Nuclear receptor superfamily A family of receptor proteins that bind hormones as an intermediate step in modulation of genome activity by these hormones.

Nuclease An enzyme that degrades a nucleic acid molecule.

Nuclease protection experiment A technique that uses nuclease digestion to determine the positions of proteins on DNA or RNA molecules.

Nucleic acid The term first used to describe the acidic chemical compound isolated from the nuclei of eukaryotic cells. Now used specifically to describe a polymeric molecule comprising nucleotide monomers, such as DNA and RNA.

Nucleic acid hybridization Formation of a double-stranded hybrid by base-pairing between complementary polynucleotides.

Nucleoid The DNA-containing region of a prokaryotic cell.

Nucleolus The region of the eukaryotic nucleus in which rRNA transcription occurs.

Nucleoside A purine or pyrimidine base attached to a five-carbon sugar.

Nucleosome The assembly of histones and DNA that is the basic structural unit in chromatin.

Nucleotide A purine or pyrimidine base attached to a five-carbon sugar, to which a mono-, di-, or triphosphate is also attached. The monomeric unit of DNA and RNA.

Nucleotide excision repair A repair process that corrects various types of DNA damage by excising and resynthesizing a region of a polynucleotide.

Nucleus The membrane-bound structure of a eukaryotic cell in which the chromosomes are contained.

3′-OH terminus The end of a polynucleotide that terminates with a hydroxyl group attached to the 3′-carbon of the sugar.

Okazaki fragment One of the short segments of RNA-primed DNA synthesized during replication of the lagging strand of the double helix.

Oligonucleotide-directed mutagenesis An *in vitro* mutagenesis technique in which a synthetic oligonucleotide is used to introduce a predetermined nucleotide alteration into the gene to be mutated.

O-linked glycosylation The attachment of sugar units to a serine or threonine in a polypeptide.

Open promoter complex A structure formed during assembly of the transcription initiation complex consisting of the RNA polymerase and/or accessory proteins attached to the promoter, after the DNA has been opened up by breakage of base pairs.

Open reading frame (ORF) A series of codons starting with an initiation codon and ending with a termination codon. The part of a protein-coding gene that is translated into protein.

Operational taxonomic unit (OTU) One of the organisms being compared in a phylogenetic analysis.

Operator The nucleotide sequence to which a repressor protein binds to prevent transcription of a gene or operon.

Optical mapping A technique for the direct visual examination of restricted DNA molecules.

Origin of replication A site on a DNA molecule where replication initiates.

Origin recognition complex (ORC) A set of proteins that binds to the origin recognition sequence.

Origin recognition sequence A component of a yeast origin of replication.

Orphan family A group of homologous genes whose functions are unknown.

Orthogonal field alternation gel electrophoresis (OFAGE) An electrophoresis system in which the field alternates between pairs of electrodes set at an angle of 45°, used to separate large DNA molecules.

Orthologous Refers to homologous genes located in the genomes of different organisms.

Outgroup An organism or DNA sequence that is used to root a phylogenetic tree.

Out of Africa A hypothesis that holds that modern humans evolved in Africa, moving to the rest of the Old World between 100 000 and 50 000 years ago, displacing the descendants of *Homo erectus* that they encountered.

Overlapping genes Two genes whose coding regions overlap.

Pair-rule genes Developmental genes that establish the basic segmentation pattern of the *Drosophila* embryo.

Pan-editing The extensive insertion of nucleotides into an abbreviated RNA, resulting in a functional molecule.

Paralogous Refers to two or more homologous genes located in the same genome.

Paranemic Refers to a helix whose strands can be separated without unwinding.

Parsimony An approach that decides between different phylogenetic tree topologies by identifying the one that involves the shortest evolutionary pathway.

Partial linkage The type of linkage usually displayed by a pair of genetic and/or physical markers on the same chromosome, the markers not always being inherited together because of the possibility of recombination between them.

Partial restriction Digestion of DNA with a restriction endonuclease under limiting conditions so that not all restriction sites are cut.

P1-derived artificial chromosome (PAC) A high-capacity vector that combines features of bacteriophage P1 vectors and PACs.

Pedigree A chart showing the genetic relationships between the members of a human family.

Pedigree analysis The use of pedigree charts to analyze the inheritance of a genetic or DNA marker in a human family.

P element A DNA transposon of *Drosophila*.

Pentose A sugar comprising 5 carbon atoms.

Peptide bond The chemical link between adjacent amino acids in a polypeptide.

Peptide nucleic acid (PNA) A polynucleotide analog in which the sugar–phosphate backbone is replaced by amide bonds.

Peptidyl- or P-site The site in the ribosome occupied by the tRNA attached to the growing polypeptide during translation.

Peptidyl transferase The enzyme activity that synthesizes peptide bonds during translation.

Permissive conditions Conditions under which a conditional-lethal mutant is able to survive.

PEST sequences Amino acid sequences that influence the degradation of proteins in which they are found.

Phage display A technique for identifying proteins that interact with one another.

Phenetics A classificatory approach based on the numerical typing of as many characters as possible.

Phenotype The observable characteristics displayed by a cell or organism.

Phosphodiesterase A type of nuclease.

Phosphodiester bond The chemical link between adjacent nucleotides in a polynucleotide.

Phosphorimaging An electronic method for determining the positions of radioactive markers in a microarray or on a hybridization membrane.

Photolyase An *Escherichia coli* enzyme involved in photoreactivation repair.

Photoproduct A modified nucleotide resulting from treatment of DNA with UV radiation.

(6-4) photoproduct photolyase An enzyme involved in photoreactivation repair.

Photoreactivation A DNA repair process in which cyclobutyl dimers and (6–4) photoproducts are corrected by a light-activated enzyme.

Phylogeny A classification scheme that indicates the evolutionary relationships between organisms.

Physical mapping The use of molecular biology techniques to construct a genome map.

Pilus A structure involved in bringing a pair of bacteria together during conjugation; possibly the tube through which DNA is transferred.

π-π interactions The hydrophobic interactions that occur between adjacent base pairs in a double-stranded DNA molecule.

Plaque A zone of clearing on a lawn of bacteria caused by lysis of the cells by infecting bacteriophages.

Plasmid A usually circular piece of DNA often found in bacteria and some other types of cell.

Plectonemic Refers to a helix whose strands cannot be separated without unwinding.

Plesiomorphic character state A character state possessed by a remote common ancestor of a group of organisms.

Point mutation A mutation that results from a single nucleotide change in a DNA molecule.

Polar A hydrophilic (water-loving) chemical group.

Polyacrylamide gel electrophoresis Electrophoresis carried out in a polyacrylamide gel and used to separate DNA molecules between 10 and 1500 bp in length.

Polyadenylate-binding protein A protein that aids poly(A) polymerase during polyadenylation of eukaryotic mRNAs, and which plays a role in maintenance of the tail after synthesis.

Polyadenylation The post-transcriptional addition of a series of As to the 3′-end of a eukaryotic mRNA.

Polyadenylation editing A form of editing that occurs with many animal mitochondrial RNAs, resulting in a termination codon being formed by adding a poly(A) tail to an mRNA that ends with the nucleotides U or UA.

Poly(A) polymerase The enzyme that attaches a poly(A) tail to the 3′-end of a eukaryotic mRNA.

Poly(A) tail A series of A nucleotides attached to the 3′-end of a eukaryotic mRNA.

Polymer A compound made up of a long chain of identical or similar units.

Polymerase chain reaction (PCR) A technique that results in exponential amplification of a selected region of a DNA molecule.

Polymorphic Refers to a locus that is represented by a number of different alleles or haplotypes in the population as a whole.

Polynucleotide A single-stranded DNA or RNA molecule.

Polynucleotide kinase An enzyme that adds phosphate groups to the 5′-ends of DNA molecules.

Polypeptide A polymer of amino acids.

Polyprotein A translation product consisting of a series of linked proteins which are processed by proteolytic cleavage to release the mature proteins.

Polypyrimidine tract A pyrimidine-rich region near the 3′-end of a GU-AG intron.

Polysome An mRNA molecule that is being translated by more than one ribosome at the same time.

Positional cloning A procedure that uses information on the map position of a gene to obtain a clone of that gene.

Post-replication complex (post-RC) A complex of proteins, derived from a pre-RC, that forms at a eukaryotic origin of replication during the replication process and ensures that the genome is copied just once per cell cycle.

POU domain A DNA-binding motif found in a variety of proteins.

Preinitiation complex The structure comprising the small subunit of the ribosome, the initiator tRNA plus ancillary factors that forms the initial association with the mRNA during protein synthesis.

Pre-mRNA The primary transcript of a protein-coding gene.

Prepriming complex A complex of proteins formed during initiation of replication in bacteria.

Prereplication complex (pre-RC) A protein complex that is constructed at a eukaryotic origin of replication and enables initiation of replication to occur.

Pre-RNA The initial product of transcription of a gene or group of genes, subsequently processed to give the mature transcript(s).

Pre-rRNA The primary transcript of a gene or group of genes specifying rRNA molecules.

Pre-tRNA The primary transcript of a gene or group of genes specifying tRNA molecules.

Pribnow box A component of the bacterial promoter.

Primary structure The sequence of amino acids in a polypeptide.

Primary transcript The initial product of transcription of a gene or group of genes, subsequently processed to give the mature transcript(s).

Primase The RNA polymerase enzyme that synthesizes RNA primers during bacterial DNA replication.

Primer A short oligonucleotide that is attached to a single-stranded DNA molecule in order to provide a start point for strand synthesis.

Primosome A protein complex involved in DNA replication.

Principal component analysis A procedure that attempts to identify patterns in a large dataset of variable character states.

Prion An unusual infectious agent that consists purely of protein.

Processed pseudogene A pseudogene that results from integration into the genome of a reverse-transcribed copy of an mRNA.

Processivity Refers to the length of polynucleotide that is synthesized by a DNA polymerase before it dissociates from the template.

Programmed mutation The possibility that under some circumstances an organism can increase the rate at which mutations occur in a specific gene.

Prokaryote An organism whose cells lack a distinct nucleus.

Proliferating cell nuclear antigen (PCNA) An accessory protein involved in DNA replication in eukaryotes.

Proline-rich domain A type of activation domain.

Promiscuous DNA DNA that has been transferred from one organelle genome to another.

Promoter The nucleotide sequence, upstream of a gene, to which RNA polymerase binds in order to initiate transcription.

Promoter clearance The completion of successful initiation of transcription that occurs when the RNA polymerase moves away from the promoter sequence.

Proofreading The $3'\rightarrow5'$ exonuclease activity possessed by some DNA polymerases which enables the enzyme to replace a misincorporated nucleotide.

Prophage The integrated form of the genome of a lysogenic bacteriophage.

Protease An enzyme that degrades protein.

Proteasome A multisubunit protein structure that is involved in the degradation of other proteins.

Protein The polymeric compound made up from amino acid monomers.

Protein engineering Various techniques for making directed alterations in protein molecules, often to improve the properties of enzymes used in industrial processes.

Protein folding The adoption of a folded structure by a polypeptide.

Protein–protein crosslinking A technique that links together adjacent proteins in order to identify proteins that are positioned close to one another in a structure such as a ribosome.

Proteome The complete protein content of a cell.

Protogenome An RNA genome that existed during the RNA world.

Prototroph An organism that has no nutritional requirements beyond those of the wild type and which can grow on minimal medium.

Pseudogene An inactivated and hence nonfunctional copy of a gene.

5′-P terminus The end of a polynucleotide that terminates with a mono-, di- or triphosphate attached to the 5′-carbon of the sugar.

Punctuation codon A codon that specifies either the start or the end of a gene.

Punnett square A tabular analysis for predicting the genotypes of the progeny resulting from a genetic cross.

Purine One of the two types of nitrogenous base found in nucleotides.

Pyrimidine One of the two types of nitrogenous base found in nucleotides.

Pyrosequencing A novel DNA sequencing method in which addition of a nucleotide to the end of a growing polynucleotide is directly detected by conversion of the released pyrophosphate into a flash of chemiluminescence.

Quantitative PCR A PCR method that enables the number of DNA molecules in a sample to be estimated.

Quaternary structure The structure resulting from the association of two or more polypeptides.

RACE (rapid amplification of cDNA ends) A PCR-based technique for mapping the end of an RNA molecule.

Radioactive marker A radioactive atom incorporated into a molecule and whose radioactive emissions are subsequently used to detect and follow that molecule during a biochemical reaction.

Radiolabeling The technique for attaching a radioactive atom to a molecule.

Random genetic drift The process that leads to alleles gradually changing their frequency in a population.

Ras A protein involved in signal transduction.

Reading frame One of the three overlapping sequences of triplet codons contained in a DNA sequence.

Readthrough mutation A mutation that changes a termination codon into a codon specifying an amino acid, and hence results in readthrough of the termination codon.

RecA An *E. coli* protein involved in homologous recombination.

RecBCD enzyme An enzyme complex involved in homologous recombination in *E. coli*.

Recessive The allele that is not expressed in a heterozygote.

Reciprocal strand exchange The exchange of DNA between two double-stranded molecules, occurring as a result of recombination, such that the end of one molecule is exchanged for the end of the other molecule.

Recognition helix An α-helix in a DNA-binding protein, one that is responsible for recognition of the target nucleotide sequence.

Recombinant A progeny member that possesses neither of the combinations of alleles displayed by the parents.

Recombinant DNA molecule A DNA molecule created in the test-tube by ligating pieces of DNA that are not normally joined together.

Recombinant DNA technology The techniques involved in the construction, study and use of recombinant DNA molecules.

Recombinant protein A protein synthesized in a recombinant cell due to expression of a cloned gene.

Recombinase A diverse family of enzymes that catalyze site-specific recombination events.

Recombination A large-scale rearrangement of a DNA molecule.

Recombination frequency The proportion of recombinant progeny arising from a genetic cross.

Recombination hotspot A region of a chromosome where crossovers occur at a higher frequency than the average for the chromosome as a whole.

Recombination repair A DNA repair process that mends double-stranded breaks.

Regulatory mutant A mutant that has a defect in a promoter or other regulatory sequence.

Release factor A protein that plays an ancillary role during termination of translation.

Renaturation The return of a denatured molecule to its natural state.

Repetitive DNA A DNA sequence that is repeated two or more times in a DNA molecule or genome.

Repetitive DNA fingerprinting A clone fingerprinting technique that involves determining the positions of genome-wide repeats in cloned DNA fragments.

Repetitive DNA PCR A clone fingerprinting technique that uses PCR to detect the relative positions of genome-wide repeats in cloned DNA fragments.

Replica plating A technique for transfer of colonies from one Petri dish to another, such that their relative positions on the surface of the agar medium are retained.

Replication factor C (RFC) A multisubunit accessory protein involved in eukaryotic DNA replication.

Replication factory A large structure attached to the nuclear matrix; the site of DNA replication.

Replication fork The region of a double-stranded DNA molecule that is being opened up to enable DNA replication to occur.

Replication licensing factors (RLFs) A set of proteins that regulate DNA replication, in particular by ensuring that only one round of DNA replication occurs per cell cycle.

Replication origin A site on a DNA molecule where replication initiates.

Replication protein A (RPA) The main single-strand binding protein involved in replication of eukaryotic DNA.

Replication slippage An error in replication that leads to an increase or decrease in the number of repeat units in a tandem repeat such as a microsatellite.

Replicative transposition Transposition that results in copying of the transposable element.

Replisome A complex of proteins involved in DNA replication.

Reporter gene A gene whose phenotype can be assayed and which can therefore be used to determine the function of a regulatory DNA sequence.

Resolution Separation of a pair of recombining double-stranded DNA molecules.

Restriction endonuclease An enzyme that cuts DNA molecules at a limited number of specific nucleotide sequences.

Restriction fragment length polymorphism (RFLP) A restriction fragment whose length is variable because of the presence of a polymorphic restriction site at one or both ends.

Restriction mapping Determination of the positions of restriction sites in a DNA molecule by analyzing the sizes of restriction fragments.

Restrictive conditions Conditions under which a conditional-lethal mutant is unable to survive.

Retroelement A genetic element that transposes via an RNA intermediate.

Retroposon A retroelement that does not have LTRs.

Retrotransposition Transposition via an RNA intermediate.

Retrotransposon A genome-wide repeat with a sequence similar to an integrated retroviral genome and possibly with retrotransposition activity.

Retroviral-like element (RTVL) A truncated retroviral genome integrated into a host chromosome.

Retrovirus A virus with an RNA genome that integrates into the genome of its host cell.

Reverse transcriptase A polymerase that synthesizes DNA on an RNA template.

Reverse transcriptase PCR (RT-PCR) PCR in which the first step is carried out by reverse transcriptase, so RNA can be used as the starting material.

Rho helicase A protein required for termination of some bacterial transcripts.

Rho-dependent terminator A position in bacterial DNA where termination of transcription occurs with the involvement of Rho.

Ribbon-helix-ribbon motif A type of DNA-binding domain.

Ribonuclease An enzyme that degrades RNA.

Ribonuclease D An enzyme involved in processing pre-tRNA in bacteria.

Ribonuclease MRP An enzyme involved in processing eukaryotic pre-rRNA.

Ribonuclease P An enzyme involved in processing pre-tRNA in bacteria.

Ribonucleoprotein (RNP) domain A common type of RNA-binding domain.

Ribosomal protein One of the protein components of a ribosome.

Ribosomal RNA (rRNA) The RNA molecules that are components of ribosomes.

Ribosome One of the protein–RNA assemblies on which translation occurs.

Ribosome binding site The nucleotide sequence that acts as the attachment site for the small subunit of the ribosome during initiation or translation in bacteria.

Ribozyme An RNA molecule that has catalytic activity.

RNA Ribonucleic acid, one of the two forms of nucleic acid in living cells; the genetic material for some viruses.

RNA-dependent DNA polymerase An enzyme that makes a DNA copy of an RNA template; a reverse transcriptase.

RNA-dependent RNA polymerase An enzyme that makes an RNA copy of an RNA template.

RNA editing A process by which nucleotides not coded by a gene are introduced at specific positions in an RNA molecule after transcription.

RNA polymerase An enzyme that synthesizes RNA on a DNA or RNA template.

RNA polymerase I The eukaryotic RNA polymerase that transcribes ribosomal RNA genes.

RNA polymerase II The eukaryotic RNA polymerase that transcribes protein-coding and snRNA genes.

RNA polymerase III The eukaryotic RNA polymerase that transcribes transfer RNA and other short genes.

RNA transcript An RNA copy of a gene.

RNA world The early period of evolution when all biological reactions were centered on RNA.

Rolling circle replication A replication process that involves continual synthesis of a polynucleotide which is 'rolled-off' of a circular template molecule.

Rooted Refers to a phylogenetic tree that provides information on the past evolutionary events that have led to the organisms or DNA sequences being studied.

SAP (stress activated protein) kinase A stress-activated signal transduction pathway.

Satellite DNA Repetitive DNA that forms a satellite band in a density gradient.

Scaffold attachment region (SAR) An AT-rich segment of a eukaryotic genome that acts as an attachment point to the nuclear matrix.

Scanning A system used during initiation of eukaryotic translation, in which the preinitiation complex attaches to the 5′-terminal cap structure of the mRNA and then scans along the molecule until it reaches an initiation codon.

Secondary structure The conformations, such as α-helix and β-sheet, taken up by a polypeptide.

Second messenger An intermediate in a certain type of signal transduction pathway.

Sedimentation analysis The centrifugal technique used to measure the sedimentation coefficient of a molecule or structure.

Sedimentation coefficient The value used to express the velocity at which a molecule or structure sediments when centrifuged in a dense solution.

Segment polarity genes Developmental genes that provide greater definition to the segmentation pattern of the *Drosophila* embryo established by the action of the pair-rule genes.

Segregation The separation of homologous chromosomes, or members of allele pairs, into different gametes during meiosis.

Selective medium A medium that supports the growth of only those cells that carry a particular genetic marker.

Selfish DNA DNA that appears to have no function and apparently contributes nothing to the cell in which it is found.

Semiconservative replication The mode of DNA replication in which each daughter double helix is made up of one polynucleotide from the parent and one newly synthesized polynucleotide.

Sequence skimming A method for rapid sequence acquisition in which a few random sequences are obtained from a cloned fragment, the rationale being that if the fragment contains any genes then there is a good chance that at least some of them will be revealed by these random sequences.

Sequence tagged site (STS) A DNA sequence that is unique in the genome.

Sex cell A reproductive cell; a cell that divides by meiosis.

Sex chromosome A chromosome which is involved in sex determination.

Shine–Dalgarno sequence The ribosome binding site upstream of an *E. coli* gene.

Short patch repair A nucleotide excision repair process of *E. coli* that results in excision and resynthesis of about 12 nucleotides of DNA.

Short tandem repeat (STR) A type of simple sequence length polymorphism comprising tandem copies of, usually, di-, tri- or tetranucleotide repeat units. Also called a microsatellite.

Shotgun approach A genome-sequencing strategy in which the molecules to be sequenced are randomly broken into fragments which are then individually sequenced.

Signal peptide A short sequence at the N-terminus of some proteins that directs the protein across a membrane.

Silencer A regulatory sequence that reduces the rate of transcription of a gene or genes located some distance away in either direction.

Silent mutation A change in a DNA sequence that has no effect on the expression or functioning of any gene or gene product.

Similarity approach A rigorous mathematical approach to alignment of nucleotide sequences.

Simple sequence length polymorphism (SSLP) An array of repeat sequences that display length variations.

SINE (short interspersed nuclear element) A type of genome-wide repeat, typified by the *Alu* sequences found in the human genome.

Single-copy DNA A DNA sequence that is not repeated elsewhere in the genome.

Single nucleotide polymorphism (SNP) A point mutation that is carried by some individuals of a population.

Single orphan A single gene, with no homolog, whose function is unknown.

Single-strand binding protein (SSB) One of the proteins that attach to single-stranded DNA in the region of the replication fork, preventing base pairs forming between the two parent strands before they have been copied.

Single-stranded A DNA or RNA molecule that comprises just a single polynucleotide.

Site-directed hydroxyl radical probing A technique for locating the position of a protein in a protein–RNA complex, such as a ribosome, by making use of the ability of Fe(II) ions to generate hydroxyl radicals which cleave nearby RNA phosphodiester bonds.

Site-directed mutagenesis Techniques used to produce a specified mutation at a predetermined position in a DNA molecule.

Site-specific recombination Recombination between two double-stranded DNA molecules which have only short regions of nucleotide sequence similarity.

Slippage The translocation of a ribosome along a short noncoding nucleotide sequence between the termination codon of one gene and the initiation codon of a second gene.

SMAD family A group of proteins involved in signal transduction.

Small cytoplasmic RNA (scRNA) A type of short eukaryotic RNA molecule with various roles in the cell.

Small nuclear ribonucleoprotein (snRNP) Structures involved in splicing GU-AG and AU-AC introns and other RNA processing events, comprising one or two snRNA molecules complexed with proteins.

Small nuclear RNA (snRNA) A type of short eukaryotic RNA molecule involved in splicing GU-AG and AU-AC introns and other RNA processing events.

Small nucleolar RNA (snoRNA) A type of short eukaryotic RNA molecule involved in chemical modification of rRNA.

S1 nuclease An enzyme that degrades single-stranded DNA or RNA molecules including single-stranded regions in predominantly double-stranded molecules.

Somatic cell A nonreproductive cell; a cell that divides by mitosis.

SOS response A series of biochemical changes that occur in *E. coli* in response to damage to the genome and other stimuli.

Southern hybridization A technique used for detection of a specific restriction fragment against a background of many other restriction fragments.

Species tree A phylogenetic tree that shows the evolutionary relationships between a group of species.

S phase The stage of the cell cycle when DNA synthesis occurs.

Spliceosome The protein–RNA complex involved in splicing GU-AG or AU-AC introns.

Splicing The removal of introns from the primary transcript of a discontinuous gene.

Splicing enhancer A nucleotide sequence that plays a regulatory role during splicing of GU-AG introns.

Spontaneous mutation A mutation that arises from an error in replication.

SR protein A protein that plays a role in splice-site selection during splicing of GU-AG introns.

STAT (signal transducer and activator of transcription) A type of protein that responds to binding of an extracellular signaling compound to a cell surface receptor by activating a transcription factor.

Stem-loop structure A structure made up of a base-paired stem and nonbase-paired loop, which can form in a single-stranded polynucleotide that contains an inverted repeat.

Steroid hormone A type of extracellular signaling compound.

Steroid receptor A protein that binds a steroid hormone after the latter has entered the cell, as an intermediate step in modulation of genome activity.

Sticky end An end of a double-stranded DNA molecule where there is a single-stranded extension.

Strong promoter A promoter that directs a relatively large number of productive initiations per unit time.

Structural domain A segment of a polypeptide that folds independently of other segments. Also, a loop of eukaryotic DNA, predominantly in the form of the 30 nm chromatin fiber, attached to the nuclear matrix.

STS mapping A physical mapping procedure that locates the positions of sequence tagged sites (STSs) in a genome.

Stuffer fragment A DNA fragment contained within a λ vector that is replaced by the DNA to be cloned.

Substitution mutation Commonly used as a synonym for a point mutation.

Supercoiling A conformational state in which a double helix is overwound or underwound so that superhelical coiling occurs.

Suppressor mutation A mutation in one gene that reverses the effect of a mutation in a second gene.

S value The unit of measurement for a sedimentation coefficient.

Syncytium A cell-like structure comprising a mass of cytoplasm and many nuclei.

Synonymous mutation A mutation that changes a codon into a second codon that specifies the same amino acid.

Synteny Refers to a pair of genomes in which at least some of the genes are located at similar map positions.

Tandem repeat Direct repeats that are adjacent to each other.

TATA-binding protein (TBP) A component of the general transcription factor TFIID, the part that recognizes the TATA box of the RNA polymerase II promoter.

TATA box A component of the RNA polymerase II core promoter.

Tautomeric shift The spontaneous change of a molecule from one structural isomer to another.

Tautomers Structural isomers that are in dynamic equilibrium.

TBP-associated factor (TAF) One of several components of the general transcription factor TFIID, playing ancillary roles in recognition of the TATA box.

TBP domain A type of DNA-binding domain.

Telomerase The enzyme that maintains the ends of eukaryotic chromosomes by synthesizing telomeric repeat sequences.

Telomere The end of a eukaryotic chromosome.

Telomere binding protein (TBP) A protein that binds to and regulates the length of a telomere.

Temperature-sensitive mutation A type of conditional-lethal mutation, one that is expressed only above a threshold temperature.

Template The polynucleotide that is copied during a strand synthesis reaction catalyzed by a DNA or RNA polymerase.

Template-dependent DNA synthesis Synthesis of a DNA molecule on a DNA or RNA template.

Template-dependent RNA synthesis Synthesis of an RNA molecule on a DNA or RNA template.

Template-independent RNA polymerase An enzyme that polymerizes RNA without the use of a template.

Template strand The polynucleotide that acts as the template for RNA synthesis during transcription of a gene.

Terminal deoxynucleotidyl transferase An enzyme that adds one or more nucleotides to the 3′-ends of DNA molecules.

Termination codon One of the three codons that mark the position where translation of an mRNA should stop.

Termination factor A protein that plays an ancillary role in termination of transcription.

Terminator sequence One of several sequences on a bacterial genome involved in termination of DNA replication.

Tertiary structure The structure resulting from folding the secondary structural units of a polypeptide.

Test cross A genetic cross between a heterozygote and the recessive homozygote.

Thermal cycle sequencing A DNA sequencing method that uses PCR to generate chain-terminated polynucleotides.

Thymine One of the pyrimidine bases found in DNA.

T_m Melting temperature.

Tn3-type transposon A type of DNA transposon that does not have flanking insertion sequences.

Topology The branching pattern of a phylogenetic tree.

Totipotent Refers to a cell that is not committed to a single developmental pathway and can hence give rise to all types of differentiated cell.

TΨC arm Part of the structure of a tRNA molecule.

Trailer segment The untranslated region of an mRNA downstream of the termination codon.

Transcript An RNA copy of a gene.

Transcription The synthesis of an RNA copy of a gene.

Transcription bubble The nonbase-paired region of the double helix, maintained by RNA polymerase, within which transcription occurs.

Transcription-coupled repair A nucleotide excision repair process that results in repair of the template strands of genes.

Transcription factor A protein that activates initiation of eukaryotic transcription.

Transcription factory A large structure attached to the nuclear matrix, the site of RNA synthesis.

Transcription initiation The assembly upstream of a gene of the complex of proteins that will subsequently copy the gene into RNA.

Transcriptome The entire mRNA content of a cell.

Transduction Transfer of bacterial genes from one cell to another by packaging in a phage particle.

Transduction mapping The use of transduction to map the relative positions of genes in a bacterial genome.

Transfer-messenger RNA (tmRNA) A bacterial RNA involved in protein degradation.

Transfer RNA (tRNA) A small RNA molecule that acts as an adaptor during translation and is responsible for decoding the genetic code.

Transformant A cell that has become transformed by the uptake of naked DNA.

Transformation The acquisition by a cell of new genes by the uptake of naked DNA.

Transformation mapping The use of transformation to map the relative positions of genes in a bacterial genome.

Transgenic mouse A mouse that carries a cloned gene.

Transition A point mutation that replaces a purine with another purine, or a pyrimidine with another pyrimidine.

Translation The synthesis of a polypeptide, the amino acid sequence of which is determined by the nucleotide sequence of an mRNA in accordance with the rules of the genetic code.

Translocation The movement of a ribosome along an mRNA molecule during translation.

Transposable element A genetic element that can move from one position to another in a DNA molecule.

Transposable phage A bacteriophage that transposes as part of its infection cycle.

Transposase An enzyme that catalyzes transposition of a transposable genetic element.

Transposition The movement of a genetic element from one site to another in a DNA molecule.

Transposon A genetic element that can move from one position to another in a DNA molecule.

Transposon tagging A gene isolation technique that involves inactivation of a gene by movement of a transposon into its coding sequence, followed by the use of a transposon-specific hybridization probe to isolate a copy of the tagged gene from a clone library.

Transversion A point mutation that involves a purine being replaced by a pyrimidine, or *vice versa*.

Trinucleotide repeat expansion disease A disease that results from the expansion of an array of trinucleotide repeats in or near to a gene.

Triplet binding assay A technique for determining the coding specificity of a triplet of nucleotides.

Triplex A DNA structure comprising three polynucleotides.

Trisomy The presence of three copies of a homologous chromosome in a nucleus that is otherwise diploid.

tRNA nucleotidyltransferase The enzyme responsible for the post-transcriptional attachment of the triplet 5′-CCA-3′ to the 3′-end of a tRNA molecule.

True tree A phylogenetic tree that depicts the actual series of evolutionary events that led to the group of organisms or DNA sequences being studied.

Truncated gene A gene relic that lacks a stretch from one end of the original, complete gene.

Tus The protein that binds to a bacterial terminator sequence and mediates termination of replication.

Twintron A composite structure made up of two or more Group II and/or Group III introns embedded in each other.

Type 0 cap The basic cap structure, consisting of 7-methylguanosine attached to the 5´-end of an mRNA.

Type 1 cap A cap structure comprising the basic 5´-terminal cap plus an additional methylation of the ribose of the second nucleotide.

Type 2 cap A cap structure comprising the basic 5´-terminal cap plus methylation of the riboses of the second and third nucleotides.

Ubiquitin A 76-amino-acid protein which, when attached to a second protein, acts as a tag directing that protein for degradation.

Unequal crossing-over A recombination event that results in duplication of a segment of DNA.

Unequal sister chromatid exchange A recombination event that results in duplication of a segment of DNA.

Unrooted Refers to a phylogenetic tree that merely illustrates relationships between the organisms or DNA sequences being studied, without providing information about the past evolutionary events that have occurred.

Upstream Towards the 5´-end of a polynucleotide.

Upstream control element A component of an RNA polymerase I promoter.

Upstream promoter element Components of a eukaryotic promoter that lie upstream of the position where the initiation complex is assembled.

Uracil One of the pyrimidine bases found in RNA.

U-RNA A uracil-rich nuclear RNA molecule including the snRNAs and snoRNAs.

UvrABC endonuclease A multienzyme complex involved in the short patch repair process of *E. coli*.

van der Waals forces A particular type of attractive or repulsive noncovalent bond.

Variable number of tandem repeats (VNTR) A type of simple sequence length polymorphism comprising tandem copies of repeats that are a few tens of nucleotides in length. Also called a minisatellite.

Vegetative cell A nonreproductive cell; a cell that divides by mitosis.

V loop Part of the structure of a tRNA molecule.

Wave of advance A hypothesis that holds that the spread of agriculture into Europe was accompanied by a large scale movement of human populations.

Weak promoter A promoter that directs relatively few productive initiations per unit time.

Wild type A gene, cell or organism that displays the typical phenotype and/or genotype for the species and is therefore adopted as a standard.

Winged helix-turn-helix A type of DNA-binding domain.

Wobble hypothesis The process by which a single tRNA can decode more than one codon.

X inactivation Inactivation by methylation of most of the genes on one copy of the X chromosome in a female nucleus.

X-ray crystallography A technique for determining the three-dimensional structure of a large molecule.

X-ray diffraction The diffraction of X-rays that occurs during passage through a crystal.

X-ray diffraction pattern The pattern obtained after diffraction of X-rays through a crystal.

Yeast artificial chromosome (YAC) A high-capacity cloning vector constructed from the components of a yeast chromosome.

Yeast two-hybrid system A technique for identifying proteins that interact with each other.

Z-DNA A conformation of DNA in which the two polynucleotides are wound into a left-handed helix.

Zinc finger A common structural motif for attachment of a protein to a DNA molecule.

Zoo blotting A technique that attempts to determine if a DNA fragment contains a gene by hybridizing that fragment to DNA preparations from related species, on the basis that genes have similar sequences in related species and so give positive hybridization signals, whereas the regions between genes have less similar sequences and so do not hybridize.

Zygote The cell resulting from fusion of gametes during meiosis.

Index

Note: Entries which are simply page numbers refer to the main text. Other entries have the following abbreviations immediately after the page number: B, Box; F, Figure; G, Glossary; RB, Research Briefing; T, Table; TN, Technical Note.

A

Acceptor arm, 233, 233F, 234F, 419G
Acceptor site, 213, 213F, 419G
Ac/Ds, 139
Acetylation of proteins, 175–176, 257T
Acidic domain, 190, 419G
Acquired immunodeficiency disease (AIDS), 138, 404–405
Acridine dye, 339, 419G
Acrylamide, 62TN
Activation domain, 190, 419G
Activator, 103, 105F, 189–190, 190T, 191, 419G
Acylation, 256, 257T, 419G
Ada enzyme, 348, 419G
Adaptive mutation, 346, 347RB
Adenine, 3F, 145, 419G
Adenosine 5′-triphosphate, 145
Adenylate cyclase, 270, 271F, 419G
A-DNA, 150, 150T, 151, 419G
Affinity chromatography, 157–158, 159F, 419G
Agarose gel electrophoresis, 18, 43TN, 44, 44F, 91TN, 419G
Aging, 288RB
Agriculture, 408–411, 411F
Agrobacterium tumefaciens, 131T
AIDS, 138, 404–405
Alanine, 154F, 155T
Alkaline phosphatase, 46TN, 64TN, 419G
Alkylating agents, 339, 348, 419G
Allele, 15, 419G
Allele frequency, 408B, 419G
Allele-specific oligonucleotide, 334TN, 419G
Allopolyploid, 383, 419G
Alphoid DNA, 118–119, 136, 419G
Alternative splicing, 198, 200F, 216, 218F
*Alu*I, 40TN
Alu-PCR, 79, 419G
Alu sequence, 79, 139, 419G
Amino acid, 150, 152F, 153–154, 154F, 419G
Aminoacylation, 232–235, 234F, 236F, 419G
Aminoacyl site, 246, 247F, 420G
Aminoacyl-tRNA synthetase, 234–235, 235T, 420G
γ-aminobutyric acid (GABA), 405
2-aminopurine, 338, 420G
Amino terminus, 150, 152F, 420G
Ammonium persulfate, 62TN
Amphioxus, 380
Amyotrophic lateral sclerosis, 106T
Anaphase, 25F, 26F
Ancestral character state, 393B, 420G
Ancient DNA, 409RB, 420G
Annealing, 61, 420G
Annual Review of Biochemistry, 415T
Annual Review of Genetics, 415T
ANT-C, 291–292, 291F
Antennapedia, 190T, 247B, 291
Antennapedia complex, 291–292, 291F
Anticodon, 233, 233F, 234F, 240F, 420G
Anticodon arm, 233, 233F, 420G
Antigen, 279, 420G
Antirrhinum, 292B
Antitermination, 203, 206F, 207, 420G
Antiterminator protein, 203, 420G
AP endonuclease, 349, 350F, 420G
Apolipoprotein-B, 226–227, 227F
Apomorphic character state, 393B, 420G
Apoptosis, 349B, 420G
AP site, 340, 341F, 348, 350F, 420G
Apurinic site, 340, 341F, 348, 350F, 420G
Apyrimidinic site, 340, 341F, 348, 350F, 420G
Aquifex aeolicus genome, 11T, 134
Arabidopsis thaliana genome, 9T, 11T, 292B, 417T
Arabidopsis thaliana mitochondrial genome, 127T
Archaea
  definition, 7, 420G
  DNA replication, 318B
  DNA topoisomerases, 306T
  evolution, 383B, 403, 403F
  genomes, 132, 134, 417T
  inteins, 257
  introns, 212T, 225T
  proteasome, 259
  RNA polymerase, 179
  transcription initiation, 181B
  translation, 251B
Archaeoglobus fulgidus genome, 11T, 417T
Arginine, 154F, 155T
ARS, 308–309, 310RB, 420G
Ascospore, 356, 358F, 420G
Ascus, 356, 358F, 420G
A site, 246, 247F, 420G
Asparagine, 154F, 155T
Aspartic acid, 154F, 155T
Aspergillus nidulans genome, 9T

Aspergillus nidulans mitochondrial genome, 127T
Ataxia telangiectasia, 106T, 348B
ATP, 145
Attenuation, 206, 208F, 420G
AU–AC intron, 212T, 216–219, 420G
Automated DNA sequencing, 68–69, 68F
Autonomously replicating sequence (ARS), 308–309, 310RB, 420G
Autopolyploid, 374–375, 374F, 375F, 420G
Autoradiography, 19TN, 46TN, 420G
Autosome, 4, 420G
Auxotroph, 34, 344, 344F, 420G

B

BAC, 75, 81, 420G
Bacillus megaterium genome, 9T
Bacillus subtilis genome, 11T, 134, 417T
Bacteria
  definition, 7, 420G
  DNA replication, 307–308, 314–317, 319–321
  gene mapping, 32–35
  genomes, 128–134, 417T
  promoters, 179, 180F, 180T, 185–186
  RNA polymerase, 179
  transcription elongation, 201–203
  transcription initiation, 181–182, 182F
  transcription regulation, 185–188
  transcription termination, 203–206
  translation, 243–248
  transposons, 139–140
Bacterial artificial chromosome (BAC), 75, 81, 420G
Bacteriophage
  definition, 421G
  infection pathways, 148RB, 357–358
  transduction, 32, 32F
Bacteriophage λ
  antitermination, 207B
  cloning vectors, 73–75, 74TN
  rolling circle replication, 303B
Bacteriophage φX174 genome, 124B
Bacteriophage M13 vector, 65, 65F, 100TN
Bacteriophage P1 vector, 75
Bacteriophage SP6, 91TN
Bacteriophage T1, 347RB
Bacteriophage T2, 148RB
Bacteriophage T3, 91TN
Bacteriophage T7, 91TN
*Bam*HI, 40TN
Barr body, 280, 421G
β-barrel dimer, 164T
Basal promoter, 179, 421G
Basal rate of transcription, 185, 187F, 421G
Base analogs, 337–338, 338F, 421G
Base excision repair, 348–349, 350F, 421G
Baseless site, 340, 341F, 421G
Base pair, 3, 421G
Base-pairing, 147, 147B, 149F, 421G
Base-stacking, 147, 421G
Basic domain, 164T, 167
B chromosome, 120B, 421G
B-DNA, 147–150, 149F, 150T, 151F, 421G
Beads-on-a-string structure, 117, 119F, 421G
Bean genome, 417T
*Bgl*I, 40TN
Bicoid, 190T, 289, 290F
Big Bang, 368, 369F
Bioessays, 415T
Biological information, 2, 421G
Biotechnology, 101TN, 421G
Bithorax complex, 291–292, 291F
Bivalent, 24, 26F, 421G
BLAST, 95
Blood groups, 16
Bloom's syndrome, 348B
Blunt end, 40TN, 421G
Bombyx mori genome, 9T, 416T
Bootstrap analysis, 401, 401F, 421G
Borrelia burgdorferi genome, 11T, 129, 131, 417T
Bottleneck, 408B, 421G
Bovine spongiform encephalopathy (BSE), 405
-25 box, 180, 180F, 421G
Branch migration, 353, 354F, 356, 421G
Branch, phylogenetic tree, 394, 395F
Brassica oleracea mitochondrial genome, 127T
Breast cancer, 348B
5-bromo-4-chloro-3-indolyl-β-D-galactopyranoside, 53TN
5-bromouracil, 310RB, 338, 338F, 421G
BSE, 405
*Bsr*BI, 40TN
*Bss*HII, 43
Buoyant density, 123TN, 136F, 421G
Buttonhead, 291F
BX-C, 291–292, 291F

C

C1b5p, 326
C1b6p, 326
Cabbage mitochondrial genome, 127T
Caenorhabditis elegans
  aging, 288RB
  functional analysis, 109TN
  genome size, 9T, 11T
  internet address, 11T, 416T
  micrograph, 296F
  model organism, 285–286
  zinc finger proteins, 166
Calcium, 275
Calmodulin, 275
cAMP, 270, 271F, 424G
Cancer genes, 106T, 107, 348B
Candida albicans genome, 416T
Cap, 207–209, 209F, 421G
Cap binding complex, 245, 246F, 422G
Capping, 207–209, 209F, 422G
Capsid, 148RB, 422G
CAP site, 271F, 422G
Carboxyl terminus, 150, 152F, 422G
Carcinogen, 337T

CASP, 216, 422G
Catabolite activator protein, 187–188, 270, 271F, 422G
Catabolite repression, 269–270, 271F, 422G
Caudal, 289, 290F
C-banding, 120T
Cdc6p, 325, 325F
Cdc7p-Dbf4p, 326
cDNA, 23TN, 49, 50F, 422G, 423G
cDNA capture, 90–91, 422G
cDNA selection, 90–91, 422G
Cell, 415T
Cell cycle, 324–327, 325F, 326F, 422G
Cell cycle checkpoint, 325, 326F, 422G
Cell-free protein synthesizing system, 238RB, 422G
CENP-A, 119
Centre d'Études du Polymorphisme Humaine (CEPH) family, 30–32, 33RB, 55, 416T
Centromere, 22, 422G
Cephalosporinase, 379RB
CEPH family, 30–32, 33RB, 55, 416T
Cesium chloride, 123TN
Chain termination DNA sequencing, 60–67, 422G
  outline, 61–64, 63F
  primer, 67
  template DNA, 64–67
Chaperonin, 252, 254F, 422G
Character state, 393, 422G
CHEF, 44, 44F, 424G
Chemical degradation DNA sequencing, 60, 61B, 422G
Chemical shift, 161, 163F, 422G
Chemiluminescence, 46TN
Chicken genome, 120B, 416T
Chi form, 353, 354F, 422G
Chimera, 98, 422G
Chimpanzee genome, 387–388, 388F, 402
Chi site, 355, 422G
Chlamydia trachomatis genome, 417T
Chlamydomonas reinhardtii chloroplast genome, 127T
Chlamydomonas reinhardtii mitochondrial genome, 127T
Chloroplast, 8, 422G
Chloroplast genome, 125–126, 127T, 128F, 422G
Chloroplast RNA polymerase, 178B
Chondrus crispus mitochondrial genome, 127T
Chromatid, 24, 422G
Chromatin, 117–118, 422G
Chromatin fiber, 30 nm, 118, 119F, 173, 174F, 422G
Chromatin remodeling, 176–178, 422G
Chromatin remodeling machine (CRM), 178, 422G
Chromomycin A3, 51
Chromosome
  bacterial, 129
  banding patterns, 120T, 121F
  definition, 4, 423G
  FISH, 46–48
  human maps, 33RB, 56RB
  mechanically stretched, 47
  metaphase, 47, 118–119
  nonmetaphase, 47
  numbers, 117T
  radiation hybrid, 50–51
  separation by flow cytometry, 51, 54F
  unusual types, 120B
Chromosome walking, 75, 77F, 423G
Chymotrypsin, 377–378
cis, 188B
Clade, 397B, 423G
Cladistics, 393, 393B, 423G
Clastogen, 337T
Cleavage and polyadenylation stimulation factor (CPSF), 210, 212F, 423G
Cleavage stimulation factor (CstF), 210, 212F, 423G
Clone, 52TN, 423G
Clone contig, 51–54, 423G
Clone contig approach for sequence assembly, 15, 16F, 73–79, 423G
Clone fingerprinting, 78, 79F, 423G
Clone library, 51–54
Cloning
  high-capacity vectors, 73–75, 74TN
  importance, 52TN
  methodology, 52TN
  vectors, 53TN, 65, 73–75, 74TN
Cloning vector
  BACs, 75, 81
  cosmids, 74TN, 75
  definition, 53TN, 423G
  fosmids, 75
  λ vectors, 73–75, 74TN
  M13 vectors, 65, 65F, 100TN
  P1 vectors, 75
  PACs, 75
  phagemids, 65
  plasmid vectors, 53TN, 65
  pUC8, 53TN
  pYAC3, 76F
  YACs, 75, 76F
Closed promoter complex, 181, 181F, 423G
Clostridium acetobutylicum genome, 417T
Cloverleaf, 233, 233F, 423G
Clustal, 399TN
Cockayne syndrome, 348B
Coding RNA, 196, 423G
Codominance, 17B, 423G
Codon, 86, 235, 240F, 423G
Codon–anticodon recognition, 235–241, 240F, 423G
Codon bias, 87, 88T, 417T, 423G
Codon Usage Database, 88T, 417T
Cohesive end, 40TN, 52TN, 423G
Cointegrate, 360, 361F, 423G
Colinearity, 235, 237F
Collagen gene, 213T, 381–382, 382F
Colon cancer, 106T
Col plasmid, 131T
Combinatorial screening, 77–78, 78F
Commitment complex, 215, 217F, 423G
Comparative genomics, 103–106, 423G
  gene mapping, 103–106
  human disease genes, 106
  Human Genome Project, 103, 105
Competent, 52TN, 423G
Complementary, 2, 423G
Complementary DNA (cDNA), 23TN, 49, 50F, 422G, 423G

Composite transposon, 139, 140F, 423G
Computer analysis
functional analysis, 93–96
locating open reading frames, 86–89
Concatamer, 303F, 423G
Concerted evolution, 380
Conditional-lethal mutation, 344, 424G
Confocal microscopy, 23TN, 106, 106F
Conjugation, 32, 32F, 424G
Conjugation mapping, 34F, 424G
Consensus sequence, 87, 179, 424G
Conservative DNA replication, 302, 302F, 304F, 424G
Conservative transposition, 139, 139F, 359–361, 360F, 361F, 424G
Constitutive heterochromatin, 120T, 173, 424G
Constitutive mutation, 345, 345F, 424G
Context-dependent codons, 237–238
Contig, 70, 424G
Contour clamped homogeneous electric fields (CHEF), 44, 44F, 424G
Conventional pseudogene, 124, 424G
Convergent evolution, 393B, 424G
COOH-terminus, 150, 152F
Core enzyme, 181, 424G
Core octamer, 176F, 183RB, 424G
Co-repressor, 187, 424G
Core promoter, 179, 424G
Cosmid, 74TN, 75, 424G
cos site, 74TN
Cotransduction, 34F, 424G
Cotransformation, 34F, 424G
Cotton genome, 417T
Cow genome, 416T
CpG island, 89, 122T, 424G
CPSF, 210, 212F, 423G
CREB, 275, 424G
CRM, 178, 422G
Crossing-over, 24, 27F, 424G
Crown gall disease, 131T
Cryptic splice site selection, 215, 216, 216F, 218F, 424G
Cryptogene, 228B
CstF, 210, 212F, 423G
CTD-associated SR-like protein (CASP), 216, 422G
C-terminus, 150, 152F
CTP, 145
Cucumis melo mitochondrial genome, 127T
Current Biology, 415T
Current Opinion in Cell Biology, 415T
Current Opinion in Genetics and Development, 415T
Current Opinion in Structural Biology, 415T
Cyanelle, 126–128, 424G
Cyanophora paradoxa, 126
Cyclic AMP (cAMP), 270, 271F, 424G
Cyclin, 326, 326F, 424G
Cyclobutyl dimer, 339, 340F, 348, 425G
Cysteine, 154F, 155T
Cys_2His_2 finger, 164T, 166, 166F, 425G
Cystic fibrosis, 106T
Cytidine 5´-triphosphate, 145
Cytosine, 3F, 145, 425G

D

DAG, 275, 277F
Dam, 351, 426G
DAPI, 44
Dark repair, 349, 425G
D arm, 233, 233F, 234F, 425G
DASH, 22, 427G
dATP, 145
Dcm, 351–352, 426G
dCTP, 145
Deadenylation-dependent pathway, 228, 425G
Deadenylation-independent pathway, 228, 425G
Deaminating agents, 338–339, 339F, 425G
Deformed, 281–282
Degeneracy, 235, 425G
Degradosome, 227–228, 425G
Deiodinase, 240T
Delayed-onset mutation, 344, 425G
Deletion cassette, 97, 97F
Deletion mutation, 330, 342, 342F, 425G
Denaturation, 251, 253F, 425G
De novo methylation, 177B, 425G
Density gradient centrifugation, 123TN, 136F, 425G
Dentatoribral-pallidoluysian atrophy, 337T
2′-deoxyadenosine 5´-triphosphate, 145
2′-deoxycytidine 5´-triphosphate, 145
2′-deoxyguanosine 5´-triphosphate, 145
Deoxyribonuclease, 64TN, 425G
Deoxyribonuclease I (DNase I) footprinting, 157, 158F
Deoxyribonuclease I (DNase I) hypersensitive site, 175, 427G
Deoxyribonuclease I (DNase I) sensitive site, 174–175
Deoxyribopyrimidine photolyase, 348
2′-deoxyribose, 145
2′-deoxythymidine 5′-triphosphate, 145
Derived character state, 393B, 425G
Development, 264, 425G
dGTP, 145
1,2–diacylglycerol (DAG), 275, 277F
4,6–diamino-2–phenylindole dihydrochloride (DAPI), 44
Diauxie, 269, 271F, 425G
Dictyostelium discoideum genome, 416T
Dideoxynucleotide, 62, 63F, 425G
Differential centrifugation, 123TN, 425G
Differentiation, 264, 425G
Dihybrid cross, 17B, 425G
Dihydrouridine, 224T
Dimer, 152–153, 425G
Dimethyl sulfate (DMS), 61B, 157, 159F
Dimethyl sulfate (DMS) modification interference, 157, 159F
Dioxetane, 46TN
Diploid, 5, 425G
Directed evolution, 379RB, 425G
Directed shotgun approach for sequence assembly, 15, 16F, 79–80, 80F, 425G
Direct readout, 163, 163F, 425G
Direct repair, 348, 348F, 426G
Direct repeats, 359, 360F, 426G
Discontinuous gene, 6F, 7, 9, 10F, 426G

Dispersive DNA replication, 302, 302F, 304F, 426G
Displacement replication, 303B, 426G
Distance matrix, 399, 400F, 426G
Distance method, 398, 426G
Disulfide bridge, 152, 426G
D-loop, 354, 355F, 356, 426G
DMS, 61B, 157, 159F
DMS modification interference, 157, 159F
DnaA, 308, 308F
DNA adenine methylase (Dam), 351, 426G
DNA bending, 164, 191F, 426G
DnaB helicase, 308, 308F, 315, 315F
DNA-binding motifs, 164–167, 426G
  β-barrel dimer, 164T
  basic domain, 164T, 167
  Cys_2His_2 finger, 164T, 166, 166F
  examples, 164T
  helix-turn-helix, 164–166, 164T, 165F
  high mobility group (HMG) domain, 164T
  histone fold, 164T
  homeodomain, 164T, 165, 165B, 165F
  HU/IHF motif, 164T
  multicysteine zinc finger, 164T, 166, 166F
  paired homeodomain, 164T
  polymerase cleft, 164T
  POU domain, 164T, 166
  Rel homology domain, 164T
  ribbon-helix-helix, 164T, 167, 167F
  TBP domain, 164T, 167
  winged helix-turn-helix, 164T, 166
  zinc binuclear cluster, 164T
  zinc finger, 164T, 166
DNA-binding proteins, 156–168, 426G
  contacts with DNA, 167–168
  examples, 162T
  functions, 162T
  locating binding sites, 156–157
  purification, 157–158, 159F
  structural studies, 160–162
  types, 162–167
DNA chip, 21, 23TN, 69, 70F, 109TN, 426G
DNA cloning
  high-capacity vectors, 73–75, 74TN
  importance, 52TN
  methodology, 52TN
  vectors, 53TN, 65, 73–75, 74TN
DNA cytosine methylase (Dcm), 351–352, 426G
DNA Database of Japan, 417T
DNA, definition, 2, 426G
DNA-dependent RNA polymerase, 178, 426G
DNA glycolyase, 349, 350F, 426G, 430G
DNA gyrase, 129, 306T, 426G
DNA helicase II, 352, 353F
DNA labeling
  fluorescent labels, 45, 46TN, 68–69
  heavy nitrogen, 305
  methods, 46TN
  radioactive labels, 45, 46TN
  uses, 46TN
DNA ligase
  definition, 64TN, 426G
  DNA repair, 348B, 348F, 350F, 352F, 353F
  DNA replication, 317, 317F, 320F
  double-strand repair, 357B
  recombinant DNA technology, 40TN, 52TN, 64TN
DNA ligase IV, 357B
DNA marker, 18–22, 426G
DNA methylation, 177B, 352, 352F, 426G
DNA microarray, 23TN, 106, 435G
DNA packaging, 117–118, 173–178
DNA photolyase, 348, 426G
DNA-PK_{CS}, 357B
DNA polymerase
  definition, 426G
  processivity, 66B
  recombinant DNA technology, 64TN, 66B
  types, 311, 311T
DNA polymerase I, 311, 311T, 317, 317F, 349, 352, 353F, 426G
DNA polymerase I, Klenow fragment, 66B
DNA polymerase II, 311, 313T, 427G
DNA polymerase III, 311, 313T, 314F, 315–317, 317F, 427G
DNA polymerase α, 311, 313T, 314F, 318, 427G
DNA polymerase β, 313T
DNA polymerase γ, 311, 313T, 427G
DNA polymerase δ, 311, 312F, 314F, 318–319, 427G
DNA polymerase ε, 313T, 326
DNA–protein interaction, 163–164, 163F, 167–168
DNA repair, 346–353, 427G
  base excision repair, 348–349, 350F
  dark repair, 349
  direct repair, 348, 348F
  double-strand break repair, 357B
  human diseases, 348B
  mismatch repair, 351–353, 353F
  nucleotide excision repair, 349–353, 353F
  recombination repair, 357B
  types, 346–347
DNA replication, 300–327, 427G
  archaea, 318B
  bacterial replication fork, 314–317
  DNA polymerases, 311, 313T
  DNA topoisomerases, 305–306, 306B, 306F, 315F
  elongation phase, 311–319
  eukaryotic replication fork, 317–319
  history of early research, 301–305
  initiation in *Escherichia coli*, 307–308, 308F
  initiation in higher eukaryotes, 309, 310RB
  initiation in *Saccharomyces cerevisiae*, 308–309, 309F
  Meselson-Stahl experiment, 302–305, 304F
  priming, 314, 314F
  regulation, 323–327
  relevant issues, 300–301
  telomeres, 322–323
  termination in *Escherichia coli*, 319–321
  termination in eukaryotes, 321–323
  topological problem, 301–306
DNase I footprinting, 157, 158F
DNase I hypersensitive site, 175, 427G
DNase I sensitive site, 174–175
DNA sequencing, 60–83, 427G
  chain termination sequencing, 60–67

DNA sequencing (*continued*)
  chemical degradation sequencing, 60, 61B
  databases, 417T
  human genome, 81–82
  methodology, 60–69
  new methods, 69
  sequence assembly, 70–80
  thermal cycle sequencing, 67–69
DNA shuffling, 379RB, 427G
DNA structure
  different forms, 147–150, 149F, 150T, 151F
  double helix, 145–150, 150F, 151F
  nucleotides, 145
  outline, 2, 3F
  polynucleotide, 145, 146F
  side-by-side model, 302
DNA template, 61, 178
DNA topoisomerase, 305–306, 306B, 306F, 315F, 427G
DNA topoisomerase I, 129
DNA transposon, 6F, 7, 139–140, 359–361, 361F, 427G
DNA tumor virus, 427G
Dog genome, 416T
Domain duplication, 380–382, 381F, 427G
Domain shuffling, 380–382, 381F, 427G
Dominant, 17B, 427G
Donor site, 213, 213F, 427G
Dot blot hybridization analysis, 107, 107F
Dot matrix, 398, 398F, 427G
Double helix, 3F, 145–150, 151F, 427G
  key features, 147
  structural flexibility, 147–150, 150F
  structure, 3F, 145–150, 151F
Double heterozygote, 28, 427G
Double homozygote, 28, 427G
Double restriction, 41, 427G
Double-strand break model for recombination, 356–357, 359F
Double-strand break repair, 357B, 427G
Double-stranded, 2, 427G
Double-stranded RNA adenosine deaminase (dsRAD), 227, 427G
Double-stranded RNA binding domain (dsRBD), 165B, 427G
Downstream, 7B, 427G
Drosophila melanogaster
  aging, 288RB
  chromosome number, 117T
  comparative genomics, 106
  development genes, 287–293
  genetic mapping, 15–16, 22–23, 24–26
  genome size, 9T, 117T
  HOM-C evolution, 379–381
  horizontal gene transfer, 383
  internet address, 11T, 416T
  mitochondrial genome, 127T, 239T
  P element, 288RB, 383
  segmentation pattern, 291F
  sex determination genes, 216, 218F
  telomeres, 322B
dsRAD, 227, 427G
dsRBD, 165B, 427G
dTTP, 145
Dynamic allele-specific hybridization (DASH), 22, 427G
Dystrophin gene, 213T

E

*Eco*RI, 18, 40TN
eEF-1, 246
eEF-2, 248
EF-G, 247F, 248
EF-Ts, 247, 247F
EF-Tu, 245, 247, 247F
eIF-2, 245, 248B
eIF-3, 245, 246F
eIF-4A, 245, 246F
eIF-4B, 245, 246F
eIF-4E, 245, 246F
eIF-4F, 245, 246F
eIF-4G, 245, 246F
Electron density map, 160, 161F
Electrophoresis, 43TN, 427G
Electrostatic interaction, 153B, 427G
Elongation factor, transcription, 210, 210T, 427G
Elongation factor, translation, 245–248, 427G
Embryonic stem (ES) cell, 98, 428G
Empty spiracles, 291F
EMS, 339, 428G
End-labeling, 64TN, 428G
End-modification, 198, 198F, 428G
Endogenous retrovirus (ERV), 138, 138F, 428G
Endonuclease, 64TN, 428G
Endosymbiont theory, 126–128, 428G
engrailed, 292
Enhanceosome, 189
Enhancer, 188, 428G
Eocyte, 383B
Epidermal growth factor, 382, 382F
Episome, 32, 428G
Episome transfer, 32, 32F, 428G
eRF, 248
ERV, 138, 138F, 428G
ES cell, 98, 428G
Escherichia coli
  adaptive mutation, 347RB
  DNA topoisomerases, 306T
  gene catalog, 135T
  genetic mapping, 32–35
  genome segment, 9–11, 10F
  internet address, 11T, 417T
  map, 132F
  physical organization of the genome, 128–129, 130RB, 131F
  plasmids, 131, 131T
  size, 9, 9T, 11T
  SOS response, 345
E site, 248, 428G
EST, 49, 55, 56RB, 81, 82F, 112T, 416T, 428G
Estradiol, 269F
Estrogen, 268
Ethidium bromide, 43TN, 339, 339F, 428G

Ethylmethane sulfonate (EMS), 339, 428G
Euchromatin, 173–174, 428G
Eukaryotes
  cell structure, 7, 8F
  definition, 428G
  evolution, 387B
  genome organization, 3–9, 6F, 10F
  genomes sizes, 9T
European Bioinformatics Institute, 417T
even-skipped, 292
Excinuclease, 349
Excision repair, 348–353, 428G
Exit site, 248, 428G
Exon
  definition, 7, 7B, 428G
  examples in genes, 6F
  locating in a DNA sequence, 87–88
Exon–intron boundary sequences, 87–88
Exon skipping, 215, 216F, 428G
Exon theory of genes, 385, 385F, 428G
Exon trapping, 92, 94F, 428G
Exonuclease, 64TN, 66B, 428G
Exportin, 199–201, 428G
Expressed sequence tag (EST), 49, 55, 56RB, 81, 82F, 112T, 416T, 428G
External node, 394, 395F, 428G
Extrachromosomal gene, 125, 428G

F

Factor VIII gene, 213T
Facultative heterochromatin, 173, 428G
Fanconi's anemia, 348B
Feedback loops, 281–282
FEN-1, 318, 319F, 320F, 429G
Ferritin, 248B
Fibre-fish, 48
Fibronectin, 382, 382F
Field inversion gel electrophoresis (FIGE), 44, 44F, 428G
FIGE, 44, 44F, 429G
Finger module, 382, 382F
Fish, 380
FISH, 45–48
Fitness, 408B, 428G
Fixation, 408B, 429G
Flap endonuclease (FEN-1), 318, 319F, 320F, 429G
Flow cytometry, 51, 54F, 429G
Flower development, 292B
FLpter value, 47, 429G
Fluorescent *in situ* hybridization (FISH), 45–48
  basis, 45–46
  definition, 39, 47F, 429G
  high resolution, 47–48
  low resolution, 46–47
fMet, 235, 236F, 244, 245F, 257T, 429G
Folding domain, 253, 429G
Footprinting, 157, 158F, 429G
Forest tree genome, 417T
Formate dehydrogenase, 240T
N-formylation, 257T
N-formylmethionine, 235, 236F, 244, 245F, 257T, 429G
Fosmid, 75, 429G
F plasmid, 75, 131T, 429G
Fragile site, 336, 337T, 429G
Fragile site expansions, 337T
Fragile XE mental retardation, 337T
Fragile X syndrome, 337T
Frameshifting, 250B, 429G
Frameshift mutation, 335–336, 342, 429G
Friedreich's ataxia, 337T
Fritillaria assyriaca genome, 9T
Fruit fly
  aging, 288RB
  chromosome number, 117T
  comparative genomics, 106
  development genes, 287–293
  genetic mapping, 15–16, 22–23, 24–26
  genome size, 9T, 117T
  HOM-C evolution, 379–381
  horizontal gene transfer, 383
  internet address, 11T, 416T
  mitochondrial genome, 127T, 239T
  P element, 288RB, 383
  segmentation pattern, 291F
  sex determination genes, 216, 218F
  telomeres, 322B
Fugu rubripes genome, 9T, 105, 416T
Functional analysis, 86–110, 429G
  comparative genomics, 103–106
  gene function analysis, 92–103
  locating genes, 86–92
Functional domain, 174
Fushi tarazu, 292
Fusion protein, 103, 104F, 429G

G

GABA, 405
Gain-of-function mutation, 344, 429G
α-galactosidase, 227T
β-galactosidase, 53TN, 102T
Gamete, 5, 429G
γ-complex, 315
GAP, 275, 277F, 429G
Gap 1 phase, 324, 325F, 430G
Gap 2 phase, 324, 325F, 431G
Gap genes, 289–290, 291F, 429G
G-banding, 120T
GC content, 41, 136F, 429G
Gel electrophoresis, 43TN, 44, 44F, 429G
Gel retardation analysis, 156–157, 156F, 429G
Gel stretching, 44, 45F, 429G
Genbank, 417T
Gene cloning
  definition, 430G
  high-capacity vectors, 73–75, 74TN
  importance, 52TN
  methodology, 52TN
  vectors, 53TN, 65, 73–75, 74TN
Gene conversion, 278, 356, 358F, 430G

Gene, definition, 2, 429G
Gene duplication, 373–374, 375–380
Gene expression, outline, 2, 4F, 430G
Gene fragment, 125, 125T, 430G
Gene inactivation, 97–98, 97F
Gene overexpression, 98–100, 100F
General recombination, 353–357, 430G
General transcription factor (GTF), 182, 184F, 184T, 430G
Gene rearrangement, 380–382
Gene relic, 124
Gene segments, 6, 6F
Gene substitution, 408B, 430G
Gene superfamily, 124, 430G
Genes-within-genes, 124B, 430G
Genetic code
  assumptions, 236, 237F
  codons, 237F
  context-dependent codons, 237–238
  definition, 7B, 430G
  elucidation, 238RB
  modified codes, 236–238, 239T, 240B
  origins and evolution, 240B
Genetic footprinting, 99RB, 430G
Genetic linkage, 17B, 430G
Genetic mapping, 14–35
  bacteria, 32–35
  comparison with a physical map, 39F
  definition, 15, 430G
  DNA markers, 18–22
  gene markers, 15–16
  humans, 29–32
  limitations, 38–39
  linkage analysis, 22–35
  mapping techniques, 22–35
  pedigree analysis, 29–32
Genetic markers, 15–16, 430G
Genetic profile, 137, 137F, 430G
Genetic redundancy, 378B, 430G
Genetics, 11, 430G
Gene tree, 395–397, 396F, 397F, 430G
Genome anatomy, 116–142
  eukaryotic nuclear genomes, 116–124
  organelle genomes, 125–128
  prokaryotic genomes, 128–134
  repetitive DNA, 134–141
Genome duplication, 374–375, 376RB
Genome evolution
  acquisition of new genes, 372–383
  definition, 436G
  early origins, 368–372
  first DNA genomes, 370–372
  gene duplication, 373–374, 375–380
  gene rearrangement, 380–382
  genome duplication, 374–375, 376RB
  horizontal gene transfer, 383
  human genome, 387–388
  introns, 385–387
  molecular basis, 300–364
  noncoding DNA, 384–387
  transposable elements, 384–385
Genome regulation, 264–293
  chromatin changes, 280–281
  examples, 266T
  feedback loops, 281–282
  genome rearrangements, 278–279
  permanent and semipermanent changes, 277–282
  signal transduction, 270–277
  transient changes, 266–267
Genomes, 1–413, 430G
Genome sizes, 8, 9T, 117, 117T
Genome-wide repeat
  definition, 7, 430G
  DNA transposons, 6F, 7
  LINEs, 6F, 7
  LTR elements, 6F, 7, 9, 10F
  positions in a genome, 6F, 9, 10F
  SINEs, 6F, 7
Genomic imprinting, 177B, 430G
Genotype, 17B, 430G
Giant, 291F
Giemsa, 120T
Globin genes, 112–124, 123F
  α, 122–124, 123F, 377, 377F, 384, 387, 387F
  β, 122–124, 123F, 174–175, 175F, 212F, 213T, 310RB, 377, 377F, 384, 387, 387F
Glucocorticoid hormone, 268
β-glucuronidase, 102T
Glutamate receptor, 227T
Glutamic acid, 154F, 155T
Glutamine, 154F, 155T
Glutamine-rich domain, 190, 430G
Glutathione peroxidase, 240T
Glycine, 154F, 155T
Glycine reductase, 240T
Glycolyase, 349, 350F, 430G
β-*N*-glycosidic bond, 145, 430G
Glycosylation, 255–256, 258F, 430G
GNRP, 275, 277F, 430G
G1 phase, 324, 325F, 431G
G2 phase, 324, 325F, 430G
G-protein-coupled receptor, 272T
Green fluorescent protein, 102T
GroEL/GroES, 252, 254F
Group I intron, 212T, 221–222, 222F, 223F, 431G
Group II intron, 212T, 225T, 431G
Group III intron, 212T, 225T, 431G
Growth factor domain, 382, 382F
GTF, 182, 184F, 184T, 430G
GTP, 145
GTPase activating protein (GAP), 275, 277F, 429G
GU–AG intron, 213–216, 217F, 431G
Guanine, 3F, 145, 431G
Guanine methyltransferase, 207, 209F, 431G
Guanine nucleotide releasing protein (GNRP), 275, 277F, 430G
Guanosine 5′-triphosphate, 145
Guanylyl transferase, 207, 209F, 431G
G-U base pair, 239, 241F
Guide RNA, 228B, 431G

H

Haemophilus influenzae genome
  gene catalog, 135T
  internet address, 11T, 417T
  sequencing project, 70–73, 71F, 72F
  size, 11T
Hammerhead, 222T
Haploid, 5, 431G
Haploinsufficiency, 343, 431G
Haplotype, 406, 406B, 431G
HAT medium, 50–51
Heat, 340, 341F
Heat shock promoter, 185
Heavy nitrogen, 305
Hedgehog, 257
Helicase, 308, 308F, 315, 315F, 352, 353F, 431G
Helicobacter pylori genome, 11T, 417T
α-helix, 152, 152F, 419G
Helix-loop-helix motif, 168, 431G
Helix-turn-helix motif, 164–166, 164T, 165F, 431G
Helper phage, 65
HENNIG86, 399TN
Hereditary nonpolyposis colorectal cancer (HNPCC), 348B
Hershey-Chase experiment, 148RB
HeT-A, 322B
Heterochromatin, 173, 431G
Heteroduplex, 353, 354F
Heteroduplex analysis, 92, 94F, 431G
Heterogenous nuclear RNA (hnRNA), 197F, 198, 431G
Heteropolymer, 238RB, 431G
Heterozygosity, 33RB, 431G
Heterozygous, 17B, 431G
High mobility group (HMG) domain, 164T
High-performance liquid chromatography (HPLC), 334TN, 431G
*Hin*fI, 40TN
Histidine, 154F, 155T
Histone, 117, 431G
Histone acetylation, 175–176, 255, 257F
Histone deacetylase, 176
Histone fold, 164T, 183RB
Histone methylation, 257F
Histone transacetylase, 176
HIV-1, 404–405, 404F
HIV-2, 404, 404F
HMG domain, 164T
HNPCC, 348B
hnRNA, 197F, 198, 431G
Hoechst, 33258, 51
Holliday model, 353–355, 354F, 355F
Holliday structure, 353, 354F, 431G
Holoenzyme, 179, 431G
HOM-C, 293, 293F
Homeodomain, 164T, 165, 165B, 165F, 431G
Homeotic mutant, 291, 431G
Homeotic selector genes, 291–292, 291F, 431G
Homo erectus, 387F, 406–408, 407F, 410
Homologous chromosomes, 24, 431G
Homologous genes, 89, 93–94
Homologous recombination
  definition, 432G
  gene inactivation, 97–98, 97F
  model, 353–357
Homology, 93–94
Homology analysis, 95–96, 376RB
Homology searching, 89, 183RB, 432G
Homoplasy, 393B, 432G
Homopolymer, 238RB, 432G
Homozygous, 17B, 432G
Horizontal gene transfer, 139, 383, 432G
Hormone response element, 268, 269F, 432G
Hox genes, 292–293, 293F, 380
HPLC, 334TN, 431G
HPRT, 50
Hsp70 chaperone, 252, 254F, 432G
Hucklebein, 291F
HU, 129, 308
HU/IHF motif, 164T
Human gene mapping, 29–32, 33RB
Human genetic map, 33RB
Human genome
  5′-CG-3′ deficiency, 41–42, 44F
  chromosome number, 117T
  content, 135B
  DNA repair defects, 348B
  evolution, 387–388, 402
  gene number, 122T
  genetic content, 6–7, 6F
  genetic map, 33RB, 55
  introns, 213T
  karyogram, 121F
  mitochondrial genome, 4, 5F, 126F, 127T, 228B, 240, 242B
  nuclear genome, 3–4, 5F
  physical map, 55–57, 56RB
  physical structure, 3–5
  population studies, 406–411
  size, 5, 5F, 9T, 11T, 117T
  telomerase, 323T
Human Genome Diversity Project, 82
Human Genome Project
  comparative genomics, 103, 105
  directed shotgun approach, 80
  ethics, 82
  goals, 6
  importance, 11–12
  internet addresses, 11T, 416T
  progress, 54–57
  sequencing phase, 81–82
  timetable, 6
Human immunodeficiency virus (HIV), 138
Human leukocyte antigen, 16
Human prehistory
  Neandertal DNA, 409RB
  New World, 410–412
  origins of agriculture, 408–410
  origins of modern humans, 406–408
Hunchback, 289, 290F
Huntington's disease, 337T
Hybridization, 19TN, 432G

Hybridization probing, 18, 19TN, 158, 159F, 432G
Hydrogen bond, 153B, 432G
Hydrophobic effect, 153B, 432G
Hydroxylation, 257T
Hypermutation, 345–346, 432G
Hypoxanthine, 339, 339F, 349
Hypoxanthine + aminopterin + thymidine (HAT medium), 50–51
Hypoxanthine phosphoribosyl transferase (HPRT), 50

I

IF-2, 245, 245F
IF-3, 244, 245F
Immunocytochemistry, 103, 432G
Immunoelectron microscopy, 242, 432G
Immunoglobulin genes, 247B, 278–279, 279F, 279T, 345–346, 346F
Immunological similarity, 394
Immunoscreening, 432G
Importin, 199, 432G
Incomplete dominance, 17B, 432G
Indel, 398, 398F, 432G
Inducer, 186, 187F, 432G
Inferred tree, 395, 432G
Inhibitor-resistant mutation, 344, 432G
Initiation codon, 7B, 86, 237F, 244, 245, 245F, 432G
Initiation complex, 246, 432G
Initiation factor, 244, 432G
Initiation of transcription, 172–193, 432G
  archaea, 181B
  assembly of the initiation complex, 181–184
  chromatin packaging, 173–178
  outline, 172F
  promoters, 179–181
  regulation, 185–192
  RNA polymerases, 178–179
Initiation region, 309, 310RB, 432G
Initiator (Inr) sequence, 180, 180F, 433G
Initiator-tRNA, 244, 245, 433G
Inosine, 224T, 240, 241F, 433G
Inositol-1,4,5–triphosphate (Ins(1,4,5)P_3), 275, 277F
Inr sequence, 180, 180F, 433G
Ins(1,4,5)P_3, 275, 277F
Insertional editing, 228B, 433G
Insertion mutation, 330, 342, 433G
Insertion sequence, 10F, 11, 139
Insulin gene, 213T
Integrase, 358, 433G
Intein, 256–258, 417T, 433G
Intein homing, 258, 259F, 433G
Intercalating agents, 339, 339F, 433G
Internal node, 394, 395F, 433G
Internal ribosome entry site (IRES), 247B, 433G
Internet addresses, 11T, 40TN, 88T, 416T-417T
Interphase, 25F, 26F, 433G
Interspersed repeat element PCR (IRE-PCR), 79, 79F, 433G
Intrinsic terminator, 203, 204F, 433G
Intron
  definition, 7, 7B, 433G
  evolution, 385–387
  examples in genes, 6F
  genes-within-genes, 124B
  locating in a DNA sequence, 87–88
Intron homing, 258, 433G
Introns early, 213, 385–387, 433G
Introns late, 213, 385–387, 433G
Intron splicing, 212–219
  AU–AC introns, 216–219
  group I introns, 221–222
  GU–AG introns, 213–216, 217F
  pre-rRNA introns, 221–222
  pre-tRNA introns, 222
Inverted repeat, 433G
In vitro mutagenesis, 101TN, 433G
In vitro packaging, 52TN, 74TN, 433G
Ion channel, 272T
Ionizing radiation, 339
IRE-PCR, 79, 79F, 433G
IRES, 247B, 433G
Iron-response element, 248B
Isoaccepting tRNA, 232, 433G
Isochore, 120
Isoleucine, 154F, 155T
Isotope, 46TN, 433G

J

Jacobsen syndrome, 337T
JAK, 270, 273F, 433G
Janus kinase (JAK), 270, 273F, 433G
Junk DNA, 386B

K

Karyogram, 118, 121F, 433G
K-homology domain, 165B, 433G
Kilobase pair, 4, 434G
Kinetochore, 119, 434G
Klenow polymerase, 66B
Knirps, 291F
Knockout mouse, 98, 434G
Kozak consensus, 246, 246F, 434G
Kringle structure, 382, 382F
Krüppel, 291F
Ku, 357B

L

Lactoferrin, 267
Lactose operator, 186, 187F
Lactose operon, 133–134, 133F, 180F, 180T, 186–187, 187F, 188B, 244T, 269–270, 271F, 347RB, 434G
Lactose repressor, 165, 186–187, 187F, 188B, 434G
lacZ', 53TN
Lagging strand, 311, 312F, 314–319, 316F, 317F, 434G
λ bacteriophage
  antitermination, 207B

cloning vectors, 73–75, 74TN
rolling circle replication, 303B
Lariat, 214, 215F, 434G
LCR, 174–175, 175F, 434G
Leader segment, 7B, 434G
Leading strand, 311, 312F, 314–319, 316F, 317F, 434G
Leaky mutation, 345, 434G
Leghemoglobin gene, 387
(6–4) lesion, 339, 340F, 348, 434G
Lethal mutation, 331, 343, 434G
Leucine, 154F, 155T
Leucine zipper, 167–168, 168F, 434G
LIN-1, 287, 287F
LINE, 6F, 7, 138F, 139, 141F, 434G
LINE-1, 139, 322B, 384, 434G
Linkage, 17B, 430G, 434G
Linkage analysis, 22–35
bacteria, 32–35
basis to, 22–26
eukaryotes, 27–29
humans, 29–32
pedigree analysis, 29–32
Linker DNA, 117, 434G
Linker histone, 117, 119F, 434G
Liverwort chloroplast genome, 127T
Locus, 15, 434G
Locus control region (LCR), 174–175, 175F, 434G
Locusta migratoria genome, 9T
Lod score, 30, 434G
Long interspersed nuclear element (LINE), 6F, 7, 138F, 139, 141F, 434G
Long patch repair, 351, 434G
Loss-of-function mutation, 343, 344F, 434G
LTR element, 6F, 7, 9, 10F, 138–139, 363–364, 434G
Luciferase, 102T
Lysine, 154F, 155T
Lysogenic pathway, 357, 434G
Lytic pathway, 148RB, 357, 434G

M

M13 bacteriophage vector, 65, 65F, 100TN
MacClade, 399TN
Machado-Joseph disease, 337T
Macrochromosome, 120B, 434G
MADS box, 292B, 434G
Maintenance methylation, 177B, 434G
Maize genome
chromosome number, 117T
genome organization, 9, 10F
internet address, 417T
size, 9T, 117T
Maize mitochondrial genome, 127T
Major groove, 149F, 150, 435G
Major histocompatibility complex, 16, 435G
Malaria parasite genome, 416T
MAP kinase, 275, 276F, 435G
Mapping reagent, 49, 435G
MAR, 174, 174F, 435G
Marchantia polymorpha chloroplast genome, 127T
mariner, 139, 383
Markers
biochemical, 17T, 32
definition, 435G
DNA, 18–22
gene cloning, 52TN, 53TN
genetic mapping, 15–22
Mass spectrometry, 109
Maternal-effect genes, 289, 290F, 435G
Mating type, 278, 278F, 435G
Mating type switching, 278, 278F, 435G
Matrix-associated region (MAR), 174, 174F, 435G
Maturase, 222, 435G
Maximum parsimony method, 400–401, 435G
Measles, 227
Mechanically stretched chromosome, 47
MeCP, 177B, 435G
Mediator, 190, 191F, 435G
Megabase pair, 4, 435G
Meiosis, 22, 26F, 435G
Mek, 275, 276F
Mekada fish genome, 416T
Melon mitochondrial genome, 127T
Melittin, 253, 256F
Melting, 22F, 435G
Melting temperature, 22F, 394, 435G
Mendelian genetics, 17B
Meselson-Radding model for recombination, 354–355, 355F
Meselson-Stahl experiment, 302–305, 304F
Messenger RNA
definition, 7B, 435G
synthesis, 201–210
Metaphase, 25F, 26F
Metaphase chromosome
definition, 435G
FISH, 46–47
structure, 118–119
Methanobacterium thermoautotrophicum genome, 11T, 417T
Methanococcus jannaschii genome, 11T, 134, 417T
Methionine, 154F, 155T
methuselah, 288RB
Methylation of proteins, 257T
Methyl-CpG-binding protein (MeCP), 177B, 435G
5-methylcytosine, 44F
N,N′-methylenebisacrylamide, 62TN
7-methylguanosine, 224T
Metridium senile mitochondrial genome, 127T
MGMT, 348, 435G
Microarray, 23TN, 106, 435G
Microprojectile, 52TN
Microsatellite
definition, 7, 136–137, 435G
genetic marker, 18–21, 31F, 33RB
genetic profiling, 137, 137F
origin, 384B
population studies, 406
positions in a genome, 6F
Minichromosome, 120B, 435G
Minigene, 92, 435G
Minimal medium, 344F, 435G

Minisatellite, 18–21, 136–137, 436G
Minor groove, 149F, 150, 436G
Mismatch, 331–332, 436G
Mismatch repair, 351–353, 353F, 436G
Missense mutation, 342, 436G
Mitochondrial genome, 4, 5F, 125–126, 126F, 127T, 242B, 406–407, 406B, 410F, 436G
Mitochondrial RNA polymerase, 178B
Mitochondrion, 4, 436G
Mitosis, 22, 25F, 324, 325F, 436G
Model organism, 97–98, 436G
Modification assay, 157, 436G
Modification interference assay, 157, 159F, 436G
Molecular chaperone, 155, 252–253, 436G
Molecular clock, 396, 401–402, 402F, 403, 404F, 436G
Molecular combing, 44, 48, 48F, 436G
Molecular evolution
  acquisition of new genes, 372–383
  definition, 436G
  early origins, 368–372
  first DNA genomes, 370–372
  gene duplication, 373–374, 375–380
  gene rearrangement, 380–382
  genome duplication, 374–375, 376RB
  horizontal gene transfer, 383
  human genome, 387–388
  introns, 385–387
  molecular basis, 300–364
  noncoding DNA, 384–387
  transposable elements, 384–385
Molecular life sciences, 11, 436G
Molecular phylogenetics, 392–412, 436G
  applications, 402–411
  features of trees, 394–397
  human populations, 406–411
  molecular clock, 401
  origins, 392–394
  sequence alignment, 397–398
  software, 399TN
  terminology, 397B
  tree reconstruction, 397–402
Monohybrid cross, 17B, 436G
Monophyletic, 397B, 436G
Mosquito genome, 416T
Mouse genome, 9T, 11T, 416T
Mouse mitochondrial genome, 127T
Moxalactam, 379RB
M phase, 324, 325F, 436G
mRNA
  definition, 7B, 435G
  synthesis, 201–210
Multicopy, 98–100, 436G
Multicysteine zinc finger, 164T, 166, 166F, 436G
Multigene families, 120–124, 436G
Multiple alignment, 398, 436G
Multiple alleles, 16
Multiple hit, 399B, 436G
Multiple substitution, 399B, 436G
Multipoint cross, 30B, 437G
Multiregional evolution, 406–408, 407F, 437G
Mus musculus genome, 9T, 11T, 416T
Mus musculus mitochondrial genome, 127T
Mut, 352–353
Mutagenesis, 101TN, 437G
Mutagens, 337–340, 437G
  alkylating agents, 339, 348
  base analogs, 337–338, 338F
  deaminating agents, 338–339, 339F
  heat, 340, 341F
  intercalating agents, 339, 339F
  ionizing radiation, 339
  UV radiation, 339
Mutant, 330, 437G
Mutasome, 345, 437G
Mutation, 331–346, 437G
  adaptive mutation, 346, 347RB
  causes, 331–340
  detection, 334B
  DNA repair, 346–353
  effects, 340–345
  hypermutation, 345–346
  monitoring damage, 349B
  mutagens, 337–340
  programmed mutation, 346, 347RB
  replication errors, 332–337
Mutation scanning, 334TN, 437G
Mutation screening, 334TN, 437G
Mycobacterium leprae genome, 417T
Mycobacterium tuberculosis genome, 11T, 417T
Mycoplasma genitalium genome, 9T, 11T, 73, 135T, 237, 239T, 417T
Mycoplasma pneumoniae genome, 11T, 237, 239T, 417T
MyoD, 281
Myoglobin, 377, 377F
Myotonic dystrophy, 106T, 337T
N-myristoylation, 257T

N

Nanos, 289, 290F
Natural selection, 408B, 437G
Nature, 415T
Nature Biotechnology, 415T
Nature Genetics, 415T
Nature Structural Biology, 415T
N-degron, 259, 437G
Neandertal DNA, 409RB
Neighbor-joining method, 399–400, 400F, 437G
Neisseria gonorrhoeae genome, 417T
Neisseria meningitidis genome, 417T
Neurofibromatosis, 106T, 227T
Neurofibromatosis gene, 124B
New Scientist, 415T
NF-κB, 273
NH_2–terminus, 150, 152F
NHEG, 357B, 437G
Nick, 348, 348F, 437G
Nick translation, 64TN
Nicotiana tabacum chloroplast genome, 127T
NiFeSe hydrogenase, 239T
Nitrogenous base, 145, 437G

N-linked glycosylation, 255–256, 257T, 258F, 437G
NMR spectroscopy, 160–162, 162F, 243, 437G
Noncoding DNA, 384–387, 386B
Noncoding RNA, 196, 437G
Nonhomologous end joining (NHEG), 357B, 437G
Nonpenetrance, 344, 437G
Nonpolar, 153, 154F, 437G
Nonsense mutation, 342, 342F, 437G
Nonsynonymous mutation, 341, 342F, 437G
Northern hybridization, 90, 90F, 91TN, 437G
*Not*I, 40TN, 43
N-terminus, 150, 152F
Nuclear genome, 116–124, 437G
Nuclear magnetic resonance (NMR) spectroscopy, 160–162, 162F, 243, 437G
Nuclear matrix, 174, 174F, 437G
Nuclear pore complex, 199, 201F, 437G
Nuclear receptor superfamily, 268, 437G
Nuclease, 64TN, 437G
Nuclease protection experiment, 117, 118F, 242, 437G
Nucleic acid, 2, 144–150, 438G
Nucleic acid hybridization, 45, 438G
Nucleoid, 8F, 128–132, 130RB, 438G
Nucleolus, 8F, 438G
Nucleoside, 145, 438G
Nucleosome, 117, 119F, 173, 173F, 438G
Nucleosome core octamer, 176F, 183RB, 424G
Nucleotide, 145, 438G
Nucleotide excision repair, 349–353, 353F, 438G
Nucleotide structure, 3F, 145
Nucleus, 8F, 438G

O

O^6-methylguanine-DNA methyltransferase (MGMT), 348, 435G
OFAGE, 44, 44F, 438G
3′-OH terminus, 3F, 145, 146F, 438G
Okazaki fragment, 314, 315–317, 317F, 318, 320F, 322–323, 323F, 325F, 438G
Oligonucleotide-directed mutagenesis, 100–102, 100TN, 438G
Oligonucleotide hybridization analysis, 21, 22F
Oligonucleotide synthesis, 23TN
O-linked glycosylation, 255–256, 257T, 258F, 438G
Oncogen, 337T
Open promoter complex, 181, 181F, 438G
Open reading frame (ORF), 7B, 86, 438G
Operational taxonomic unit (OTU), 397B, 438G
Operator, 186, 187F, 438G
Operon, 10F, 10–11, 133–134, 134F
Optical mapping, 44, 438G
ORC, 309, 309F, 325, 438G
ORF, 7B, 86, 438G
ORF scanning, 87–89, 88F
Organelle genomes, 125–128
  chloroplast genomes, 125–126, 127T, 128F
  gene contents, 125–128
  introns, 222, 225T
  mitochondrial genomes, 125, 126F, 127T
  origins, 126–128
  physical features, 125
oriC, 307–308, 308F
Origin firing, 326
Origin of replication, 307–309, 307F, 308F, 309F, 310RB, 438G
Origin recognition complex (ORC), 309, 309F, 325, 438G
Origin recognition sequence, 309, 309F, 438G
Orphan family, 96, 438G
Orthodenticle, 291F
Orthogonal field alternation gel electrophoresis (OFAGE), 44, 44F, 438G
Orthology, 93, 396, 438G
Oryza sativa chloroplast genome, 127T, 128F
Oryza sativa genome, 9T, 11T, 106, 417T
OTU, 397B, 438G
Outgroup, 395, 395F, 438G
Out of Africa hypothesis, 406–408, 407F, 438G
Ovarian cancer, 348B
Overlapping genes, 124B, 439G
Oxytricha telomeres, 323T

P

p300/CBP, 176, 275
PAC, 75, 439G
Paired homeodomain, 164T
Pair-rule genes, 290–291, 439G
Pan-editing, 228B, 439G
Paralogy, 94, 439G
Paranemic, 302, 439G
Parental genotype, 25
Parsimony, 397B, 439G
Partial linkage, 22, 24F, 25–26, 439G
Partial restriction, 41, 439G
PAUP, 399TN
P1 bacteriophage vector, 75
PCNA, 311, 441G
PCR, 18, 20TN, 67–69, 67F, 68F, 91TN, 92, 440G
P1-derived artificial chromosome (PAC), 75, 439G
Pea chloroplast genome, 127T
Pea genome, 9T
Pedigree, 29, 31F, 439G
Pedigree analysis, 29–32, 439G
P element, 288RB, 383, 439G
Pentose, 145, 439G
Peptide bond, 150, 152F, 439G
Peptide nucleic acid (PNA), 371B, 439G
Peptidyl site, 246, 247F, 439G
Peptidyl transferase, 247, 249RB, 439G
Permissive conditions, 344, 439G
PEST sequence, 259, 439G
Phage display, 103, 104F, 439G
Phagemid, 65
Phenetics, 393, 393B, 439G
Phenotype, 15, 98T, 439G
Phenylalanine, 154F, 155T
Phosphatidylinositol-4,5–biphosphate (PtdIns(4,5)P_2), 275, 277F
Phosphodiesterase, 349, 350F, 439G

Phosphodiester bond, 3F, 145, 439G
Phospholipase, 277F
Phosphorimaging, 23TN, 46TN, 439G
Phosphorylation, 257T
Photolyase, 348, 439G
Photoproduct, 339, 340F, 439G
(6–4) photoproduct photolyase, 348, 439G
Photoreactivation, 348, 439G
PHYLIP, 399TN
Phylogeny, 392, 440G
Physical gap, 72F
Physical mapping, 38–54
  comparison with a genetic map, 39F
  definition, 15, 440G
  fluorescent *in situ* hybridization (FISH), 44–48
  restriction mapping, 39–44
  sequence tagged site (STS) mapping, 48–54, 55, 56RB
Picornavirus, 247B
Pig genome, 416T
π-π interactions, 147, 440G
Pioneer model, 410
Piperidine, 61B
Pisum sativum chloroplast genome, 127T
Pisum sativum genome, 9T
Plaque, 74TN, 440G
Plasmid
  cloning vectors, 53TN, 65
  definition, 10, 10F, 440G
  types, 129–131, 131T
Plasminogen, 382, 382F
Plasmodium falciparum genome, 416T
Plectonemic, 302, 440G
Plesiomorphic character state, 393B, 440G
PNA, 371B, 439G
Point mutation, 21, 330, 332B, 440G
Polar, 153, 154F, 440G
Poliovirus, 247B
Polyacrylamide gel electrophoresis, 43TN, 61F, 62TN, 440G
Polyadenylate-binding protein, 210, 212F, 245, 440G
Polyadenylation, 210, 212F, 440G
Polyadenylation editing, 228B, 440G
Poly(A) polymerase, 178, 210, 212F, 440G
Poly(A) tail, 210, 440G
Polycomb, 280, 281F
Polyglutamine expansion, 337T
Polymer, 2, 440G
Polymerase chain reaction (PCR), 18, 20TN, 67–69., 67F, 68F, 91TN, 92, 440G
Polymerase cleft, 164T
Polymorphic, 406, 408B, 440G
Polynucleotide, 3F, 145, 440G
Polynucleotide kinase, 64TN, 440G
Polypeptide, 150, 440G
Polyprotein, 254–255, 257F, 440G
Polypyrimidine tract, 213, 213F, 440G
Polysome, 440G
Positional cloning, 78, 440G
Post-RC, 325, 440G
Post-replication complex (post-RC), 325, 440G
Post-translational modification, 248–258
  chemical modification, 255–257, 257T
  folding, 250–253
  intein splicing, 257–258
  proteolytic cleavage, 253–255
POU domain, 164T, 166, 441G
Precursor RNA, 197–198, 197F
Preinitiation complex, 245, 246F, 441G
Pre-mRNA, 197F, 198, 441G
Prepriming complex, 308, 441G
Pre-RC, 325–326, 441G
Pre-replication complex (pre-RC), 325–326, 441G
Pre-RNA, 197F, 198, 441G
Pre-rRNA, 197F, 198, 441G
Pre-rRNA chemical modification, 225–226
Pre-rRNA processing, 219–221
Pre-tRNA, 197F, 198, 441G
Pre-tRNA chemical modification, 222–225, 224F
Pre-tRNA processing, 219–221
Pribnow box, 179, 441G
Primary structure, 150, 441G
Primary transcript, 198, 441G
Primase, 314, 314F, 316, 441G
Primer, 20TN, 61, 67, 67F, 311, 312F, 314F, 362F, 363, 441G
Primosome, 315, 441G
Principal component analysis, 408, 441G
Prion, 405, 405F, 406F, 441G
Processed pseudogene, 124, 124F, 214B, 441G
Processivity, 66B, 441G
Progesterone, 268
Programmed mutation, 346, 347RB, 441G
Progressive myoclonus epilepsy, 337T
Prokaryotes
  cell structure, 7, 8F, 10F
  definition, 441G
  distinctiveness genes, 134
  gene catalogs, 135T
  genome organization, 9–11, 10F, 128–134
  genomes sizes, 9T
  minimal gene content, 134
Proliferating cell nuclear antigen (PCNA), 311, 441G
Proline, 154F, 155T
Proline-rich domain, 190, 441G
Promelittin, 253, 256F
Promiscuous DNA, 127–128, 441G
Promoter, 179–181, 185–186, 441G
  bacterial, 179, 180F, 180T, 185–186
  eukaryotic, 179–181, 180F
Promoter clearance, 181, 181F, 441G
Proofreading, 311, 332, 335F, 441G
Proopiomelanocortin, 255, 257F
Prophage, 358, 441G
Prophase, 25F, 26F
Protease, 248, 441G
Proteasome, 259–260, 441G
Protein, 150–156, 442G
Protein electrophoresis, 394
Protein engineering, 100, 442G
Protein folding, 155, 155F, 250–253, 442G
Protein function, 152T
Protein–protein crosslinking, 242, 442G

Protein structure, 150–156
  amino acid diversity, 152–154
  levels, 150–152, 154–156
Protein synthesis, 243–248, 249RB
  archaea, 251B
  elongation, 246–248, 247F
  initiation in bacteria, 243–245, 245F
  initiation in eukaryotes, 245, 246F
  regulation, 248B
  termination, 248, 251F
Protein turnover, 258–260
Proteome
  definition, 109, 442G
  synthesis of, 232–260
Protogenome, 370, 442G
Prototroph, 344, 442G
Pseudogene, 6F, 7, 123F, 124, 124F, 442G
Pseudomonas aeruginosa genome, 417T
Pseudomonas putida, 131T
Pseudouridine, 224T
P site, 246, 247F, 439G
PtdIns(4,5)P_2, 275, 277F
5′-P terminus, 3F, 145, 146F, 442G
pUC8, 53TN
Pufferfish genome, 9T, 105, 416T
Punctuation codon, 235, 442G
Punnett square, 17B, 442G
Purine, 145, 442G
pYAC3, 76F
Pyrimidine, 145, 442G
Pyrosequencing, 69, 69F, 442G

Q

Quantitative PCR, 409RB, 442G
Quaternary structure, 152–153, 442G
Q-banding, 120T
Queosine, 224T
Quinacrine, 120T

R

Rac, 275
RACE, 92, 93F, 442G
RAD51, 357
Radiation hybrids, 50F, 50–51, 416T
Radioactive marker, 46TN, 442G
Radiolabeling, 46TN, 442G
Raf, 275, 276F, 277F
Random genetic drift, 396, 408B, 442G
Rapid amplification of cDNA ends (RACE), 92, 93F, 442G
Ras, 275, 277F, 442G
Rat genome, 416T
R-banding, 120T
Rbk plasmid, 131T
Reading frame, 87, 87F, 442G
Readthrough mutation, 342, 342F, 442G
Reb1p, 219, 220F
REBASE, 417T
RecA protein, 355F, 356, 442G
RecBCD enzyme, 355–356, 355F, 359B, 442G
RecE pathway, 355, 359B
Recessive, 17B, 442G
RecF pathway, 355, 359B
RecG protein, 356
Reciprocal strand exchange, 354, 354F, 442G
RecJ protein, 359B
Reclinomonas americana mitochondrial genome, 127T
Recognition helix, 165, 442G
Recombinant, 25, 443G
Recombinant DNA molecule, 52TN, 443G
Recombinant genotype, 25
Recombinant protein, 52TN, 443G
Recombinase, 358, 443G
Recombination
  crossing-over, 24
  definition, 443G
  double-strand break model, 356–357, 359F
  general recombination, 353–357
  Holliday model, 353–355, 354F, 355F
  homologous recombination, 353–357
  Meselson-Radding model, 354–355, 355F
  proteins, 355–356, 355F, 356F, 359B
  retrotransposition, 363–364, 362F., 363F
  site-specific, 357–359, 360F
  transposition, 359–364, 361F, 362F, 363F
Recombination frequency, 26, 28F, 443G
Recombination hotspot, 26, 443G
Recombination repair, 357B, 443G
RecO protein, 359B
RecQ protein, 359B
Regulatory mutation, 343, 343F, 345, 345F, 443G
Rel, 273
Release factor, 248, 251F, 443G
Rel homology domain, 164T
Renaturation, 251, 443G
Repetitive DNA
  blocking in a probe, 46, 47F
  definition, 443G
  genome architecture, 140, 141F
  interspersed repeats, 137–140
  tandem repeats, 136–137
Repetitive DNA fingerprinting, 79, 79F, 443G
Repetitive DNA PCR, 79, 79F, 443G
Replica plating, 344F, 443G
Replication errors, 332–337
Replication factor C (RFC), 318, 327, 443G
Replication factory, 318, 320F, 443G
Replication fork, 301F, 443G
Replication licensing factor (RLF), 325, 443G
Replication origin, 307–309, 307F, 308F, 309F, 310RB, 443G
Replication protein A (RPA), 315F, 317, 326, 443G
Replication slippage, 336, 336F, 377, 443G
Replicative transposition, 139, 139F, 359–361, 360F, 361F, 443G
Replisome, 316, 443G
Reporter gene, 102, 102F, 102T, 443G
Repressor, 165, 186–187, 190–191
Resolution, 353, 354F, 443G

Restriction endonuclease, 18, 40TN, 41–42, 52TN, 53TN, 417T, 443G
Restriction fragment length polymorphism (RFLP), 18, 18F, 39F, 55, 406–407, 410, 410F, 444G
Restriction mapping, 39–44
basic methodology, 39–41
clone fingerprinting, 78, 79F
definition, 39, 444G
limitations, 41–44
worked example, 42F
Restrictive conditions, 344, 444G
Retroelement, 138–139, 138F, 363–364, 443G
Retroposon, 138F, 139, 444G
Retrotransposition, 137, 137F, 363–364, 362F, 363F, 444G
Retrotransposon, 138–139, 138F, 444G
Retroviral-like element (RTVL), 138, 444G
Retrovirus, 138, 138F, 444G
Reverse gyrase, 306T
Reverse transcriptase, 50F, 64TN, 91, 137, 362F, 363, 444G
Reverse transcriptase PCR (RT-PCR), 91TN, 92, 404, 444G
RF1, 248, 251F
RF2, 248, 251F
RF3, 248, 251F
RFC, 318, 327, 443G
RFLP, 18, 18F, 39F, 55, 406–407, 410, 410F, 444G
rho, 275
Rho-dependent terminator, 203, 204F, 444G
Rho helicase, 203, 204F, 444G
Ribbon-helix-helix motif, 164T, 167, 167F, 444G
Ribonuclease, 64TN, 444G
Ribonuclease III, 220F
Ribonuclease D, 221F, 444G
Ribonuclease E, 221F
Ribonuclease F, 220F, 221F
Ribonuclease H (RNase H), 50, 317F, 318, 320F, 362F, 363
Ribonuclease M5, 220F
Ribonuclease M16, 220F
Ribonuclease M23, 220F
Ribonuclease MRP, 221, 444G
Ribonuclease P, 220F, 221F, 222T, 444G
Ribonucleoprotein (RNP) domain, 165B, 444G
Ribonucleotide structure, 145, 147F
Ribosomal protein, 242, 243F, 248B, 444G
Ribosomal RNA
cell content, 197F
definition, 7B, 197, 444G
gene evolution, 380
gene families, 122
ribozyme, 249RB
structure, 242, 243F, 244F
Ribosome
definition, 444G
role in translation, 240, 243–248, 249RB
structure, 240–243, 243F
Ribosome binding site, 244, 244F, 244T, 444G
Ribosome slippage, 250B, 446G
Ribozyme, 221, 222T, 249RB, 370–371, 372F, 444G
Rice chloroplast genome, 127T, 128F
Rice genome, 9T, 11T, 106, 417T
Rickettsia genome, 126
RLF, 325, 443G
RNA
base pairs, 147B, 241F
cell content, 196–201, 197F
definition, 444G
PCR of, 91TN, 92
processing events, summary, 199T
pyranosyl version, 371B
structure, 145, 147F
RNA-binding proteins, 162T, 165B
RNA-dependent RNA polymerase, 178, 444G
RNA editing, 226–227, 228B, 444G
RNA labeling, 91TN
RNA polymerase, 178, 444G
RNA polymerase, archaeal, 179
RNA polymerase, bacterial, 179, 201–203, 203F, 205RB
RNA polymerase I
enzyme, 178, 179T, 444G
promoters, 180, 180F
regulation, 189B
transcription, 219
RNA polymerase II
enzyme, 178, 179T, 444G
promoters, 180, 180F
transcription, 207–212
RNA polymerase III
enzyme, 178, 179T, 445G
promoters, 180, 180F
transcription, 219
RNA polymerases, 178–179
RNase H, 50, 317F, 318, 320F, 362F, 363
RNA sequencing, 91TN
RNA transcript, 2, 445G
RNA transport, 199–201
RNA transposon, 138–139
RNA turnover, 227–229
RNA world, 222, 368–370, 445G
RNP domain, 165B, 444G
Rolling circle replication, 303B, 445G
Rooted tree, 395, 395F, 445G
RPA, 315F, 317, 326, 443G
rRNA
cell content, 197F
definition, 7B, 197
gene evolution, 380
gene families, 122
ribozyme, 249RB
structure, 242, 243F, 244F
Rsk, 275, 276F
RT-PCR, 91TN, 92, 404, 444G
RTVL, 138, 444G
Ruv, 356, 356F

S

Saccharomyces cerevisiae
biochemical markers, 17T
chromosome number, 117T
comparative genomics, 106
copper regulation, 268, 268F
DNA replication control, 325–326
genetic footprinting, 99RB

genetic mapping, 28
genetic *vs* physical map, 39F
genome duplication, 375, 376RB
genome organization, 8–9, 10F
genome size, 8, 9T, 11T, 117T
homology analysis, 95–96, 96F
intein, 256
internet address, 11T, 416T
mating type switching, 278, 278F
mitochondrial genome, 125, 126F, 127T, 239T
orphans, 96, 96F
phenotypes, 98T
protein turnover, 259–260
RNA turnover, 228–229
sporulation genes, 108RB
transcriptome, 107, 108RB
Salamander genome, 117T
*Sap*I, 41
SAP kinase, 275, 445G
SAR, 174, 174F, 445G
Satellite DNA, 136, 136F, 445G
*Sau*3AI, 40TN
Scaffold-associated region (SAR), 174, 174F, 445G
Scanning, 245, 246F, 445G
Schizosaccharomyces pombe genome, 416T
Science, 415T
Scientific American, 415T
Scintillation counting, 46TN
scRNA, 197, 197F, 446G
Sea urchin genome, 9T
Sea urchin mitochondrial genome, 127T
Secondary structure, 150–152, 445G
Second messengers, 275, 277F, 445G
Sedimentation analysis, 219, 445G
Sedimentation coefficient, 123TN, 445G
Segment polarity genes, 291, 445G
Segregation, 26F, 445G
Selective medium, 53TN, 445G
Selenocysteine, 153, 155F, 235, 236F, 237–238, 239T
Selenoprotein P, 239T
Selenoprotein W, 239T
Selfish DNA, 258, 386B, 445G
Semiconservative DNA replication, 302–305, 302F, 303B, 304F, 445G
Sequenase, 64TN, 66B
Sequence alignment, 397–398
Sequence contig, 70
Sequence databases, 417T
Sequence gap, 72F
Sequence skimming, 81, 445G
Sequence tagged site (STS), 48–49, 445G
Sequence tagged site (STS) mapping, 48–54
clone fingerprinting, 79, 79F
clone library, 51–54
definition, 39, 447G
human genome, 55, 56RB
mapping reagents, 49
radiation hybrids, 50–51
types of STS, 49
Serine, 154F, 155T
Serine-threonine kinase receptor, 272T
Serum albumin gene, 213T
Sex cell, 5, 445G
Sex chromosome, 4, 445G
*Sgf*I, 41
Sheep genome, 416T
β-sheet, 152, 152F, 421G
Shine-Dalgarno sequence, 243, 445G
Short patch repair, 349, 351F, 445G
Shotgun method for sequence assembly, 14, 14F, 15F, 70–73, 71F, 446G
Side-by-side model, 302
σ subunit
Bacillus sporulation, 284–285, 284F, 285F
heat shock, 186, 186F
RNA polymerase component, 179
Signal peptide, 254, 256F, 446G
Signal transducer and activator of transcription (STAT), 190T, 270–273, 273F, 447G
Signal transduction, 270–277
MAP kinase, 275, 276F
outline, 272F
Ras, 275, 277F
SAP kinase, 275
second messengers, 275, 277F
SMADs, 274RB
STATs, 270–273, 273F
types of receptor, 272T
Silencer, 188, 446G
Silent mutation, 341, 446G
Silkworm genome, 9T, 416T
Simian immunodeficiency virus (SIV), 404, 404F
Similarity approach, 398, 446G
Simple sequence length polymorphism (SSLP), 18–21, 21F, 49, 55, 416T, 446G
Simple tandem repeat (STR), 18–21
SINE, 6F, 7, 138F, 139, 446G
Single-copy DNA, 7, 446G
Single nucleotide polymorphism (SNP), 21–22, 21F, 22F, 446G
Single orphan, 96, 446G
Single-strand binding protein (SSB), 315, 316F, 446G
Single-stranded, 2, 446G
Site-directed hydroxyl radical probing, 242–243, 446G
Site-directed mutagenesis, 100–102, 100TN, 446G
Site-specific recombination, 357–359, 360F, 446G
SIV, 404, 404F
Sliding clamp, 311
Slime mold genome, 416T
Slippage, 250B, 446G
7SL RNA, 139
SMAD, 274RB, 446G
*Sma*I, 43
Small cytoplasmic RNA (scRNA), 197, 197F, 446G
Small nuclear RNA (snRNA), 178, 197, 197F, 208–209, 215–216, 446G
Small nucleolar RNA (snoRNA), 225–226, 226F, 197, 197F, 446G
SMA/MAD related family (SMAD), 274RB, 446G
Snapdragon, 292B
SNP, 21–22, 21F, 22F, 446G
snoRNA, 225–226, 226F, 197, 197F, 446G

snRNA, 178, 197, 197F, 208–209, 215–216, 446G
S1 nuclease, 92, 334TN, 446G
S1 nuclease mapping, 92, 94F
Somatic cell, 5, 446G
SOS response, 345, 446G
Southern hybridization, 18, 19TN, 446G
Soybean genome, 417T
SP6 Bacteriophage, 91TN
Species tree, 395–397, 396F, 397F, 446G
S phase, 324, 325F, 447G
Spinal and bulbar muscular atrophy, 337T
Spinocerebellar ataxia, 337T
Spliceosome, 216, 217, 447G
Splice site mutation, 343
Splicing, 212–219, 447G
  AU–AC introns, 216–219
  group I introns, 221–222
  GU–AG introns, 213–216, 217F
  pre-rRNA introns, 221–222
  pre-tRNA introns, 222
Splicing enhancer, 216, 447G
Spontaneous mutation, 332–337, 447G
SR protein, 216, 218F, 447G
SSB, 315, 316F, 446G
SSLP, 18–21, 21F, 49, 55, 416T, 446G
STAT, 190T, 270–273, 273F, 447G
Stem-loop structure, 203, 204F, 447G
Steroid hormone, 268, 269F, 447G
Steroid receptor, 164T, 166F, 268, 269F, 270F, 447G
Sticky end, 40TN, 52TN, 447G
STR, 18–21
Streptococcus pyogenes genome, 417T
Stress activated protein (SAP) kinase, 275, 445G
Stromatolite, 403
Strong promoter, 185, 447G
Strongylocentrotus pupuratus genome, 9T
Structural domain, 174, 447G
STS, 48–49, 445G
STS mapping, 48–54
  clone fingerprinting, 79, 79F
  clone library, 51–54
  definition, 39, 447G
  human genome, 55, 56RB
  mapping reagents, 49
  radiation hybrids, 50–51
  types of STS, 49
Stuffer fragment, 74TN, 447G
Substitution mutation, 21, 330, 332B, 447G
Sucrose, 123TN
Suillis grisellus mitochondrial genome, 127T
Sulfolobus solfataricus genome, 417T
Supercoiling, 129, 129F, 130RB, 306B, 447G
Superwobble, 240
Suppressor mutation, 238RB, 447G
S value, 123TN, 447G
Svedberg unit, 123TN
Syncytium, 289, 290F, 447G
Synechocystis genome, 11T, 417T
Synonymous mutation, 341, 342F, 447G
Synteny, 104, 447G
Synthesis phase, 324, 325F

T

T1 bacteriophage, 347RB
T2 bacteriophage, 148RB
T3 bacteriophage, 91TN
T4 polynucleotide kinase, 64TN, 440G
T7 bacteriophage, 91TN
TAF, 182, 183RB, 184F, 447G
Tandemly repeated DNA, 136–137, 141F, 447G
  microsatellites, 136–137
  minisatellites, 136–137
  satellite DNA, 136
Taq DNA polymerase, 20TN, 64TN
TART, 322B
TATA-binding protein (TBP), 167, 182, 184, 184F, 447G
TATA box, 180, 180F, 447G
Tautomeric shift, 332, 447G
Tautomers, 332, 335F, 338, 338F, 447G
TBP (TATA-binding protein), 167, 182, 184, 184F, 447G
TBP (telomere binding protein), 323, 448G
TBP-associated factor (TAF), 182, 183RB, 184F, 447G
TBP domain, 164T, 167, 447G
T cell receptor, 6, 6F, 278–279
β T-cell receptor locus, 6, 6F
Telomerase, 322, 323T, 324F, 448G
Telomere
  definition, 119, 122F, 448G
  DNA, 136, 141F
  replication of, 322–323
Telomere binding protein (TBP), 323, 448G
Telophase, 25F, 26F
TEMED, 62TN
Temperature-sensitive mutation, 344, 448G
Template-dependent DNA synthesis, 147, 312F, 448G
Template-dependent RNA synthesis, 147, 202F, 448G
Template DNA, 61, 178, 448G
Template-independent RNA synthesis, 178, 448G
Template strand, 178, 448G
Teratogen, 337T
Terminal deoxynucleotidyl transferase, 64TN, 448G
Termination codon, 7B, 86–87, 448G
Termination factor, 189B, 448G
Terminator sequence, 319–321, 321F, 448G
Tertiary structure, 152, 448G
Test cross, 28, 29F, 448G
Tetrahymena intron, 221–222, 222F, 223F
Tetrahymena telomere, 323T
Tetrahymena pyriformis genome, 9T
N,N,N′,N′-tetramethylethylenediamine (TEMED), 62TN
TFIIB, 182, 184F, 184T, 191
TFIID, 182, 183RB, 184F, 184T, 191
TFIIE, 182, 184F, 184T
TFIIF, 182, 184F, 184T
TFIIH, 182, 184F, 184T, 351
TFIIIB, 184
TGF-β, 274RB
Thermal cycle DNA sequencing, 67–69, 68F, 448G
4-thiouridine, 205RB, 224T
Threonine, 154F, 155T
Thymidine kinase (TK), 50
Thymine, 3F, 145, 448G

Ti plasmid, 131T
Tissue plasminogen activator (TPA), 382, 382F
TK, 50
T_m, 22F, 394, 448G
tmRNA, 197, 197F, 449G
Tn3–type transposon, 139, 140F, 360, 448G
Tobacco chloroplast genome, 127T
TOL plasmid, 131T
Topology, phylogenetic tree, 394, 395F, 448G
Totipotent, 264, 448G
TPA, 382, 382F
Trailer segment, 7B, 448G
trans, 188B
Transcript, 2, 448G
Transcription, 2, 172–192, 196–229, 448G
Transcription activation, 187–188
Transcription bubble, 201, 448G
Transcription-coupled repair, 348B, 351, 448G
Transcription elongation, 201–203, 209–210, 211RB
Transcription factor, 103, 105F, 189–190, 190T, 191, 448G
Transcription factory, 320F, 448G
Transcription initiation, 172–193, 449G
  archaea, 181B
  assembly of the initiation complex, 181–184
  chromatin packaging, 173–178
  outline, 172F
  promoters, 179–181
  regulation, 185–192
  RNA polymerases, 178–179
Transcription termination, 203–206
Transcriptome, 106, 196, 449G
Transduction, 32, 32F, 449G
Transduction mapping, 34F, 449G
Transfer-messenger RNA (tmRNA), 197, 197F, 449G
Transfer RNA
  aminoacylation, 232–235, 234F, 236F
  cell content, 197F
  codon-anticodon recognition, 235–241, 239F
  definition, 7B, 9, 197, 449G
  primer, 362F, 363
  role, 232, 233F
  structure, 233–234, 233F, 234F
  transcription, 178
  wobble, 238–240
Transferrin receptor, 248B
Transformant, 32, 449G
Transformation, 32, 32F, 52TN, 449G
Transformation mapping, 34F, 449G
Transforming growth factor-β (TGF-β), 274RB
Transgenic mouse, 100, 100F, 449G
Transition, 332B, 449G
Translation, 243–248, 249RB, 449G
  archaea, 251B
  elongation, 246–248, 247F
  initiation in bacteria, 243–245, 245F
  initiation in eukaryotes, 245, 246F
  regulation, 248B
  termination, 248, 251F
Translocation, 247, 247F, 449G
Transposable element, 11, 137, 177B, 384–385, 449G
Transposable phage, 140, 140F, 360, 449G
Transposase, 139, 449G
Transposition, 137–138, 359–364, 361F, 362F, 363F, 449G
Transposon, 11, 137, 177B, 384–385, 449G
Transposon tagging, 288RB, 449G
Transversion, 332B, 449G
Tree genome, 417T
Tree of life, 392, 393F
Trends in Biochemical Sciences, 415T
Trends in Biotechnology, 415T
Trends in Genetics, 415T
Treponema pallidum genome, 11T, 131–132, 417T
Trichomonad, 127
Trichothiodystrophy, 348B
Trimethylpsoralen, 130RB
Trinucleotide repeat expansion disease, 336, 337T, 449G
Triplet binding assay, 238RB, 449G
Triplex, 356, 449G
Trisomy, 374, 449G
Triticum aestivum genome, 9T, 106
tRNA
  aminoacylation, 232–235, 234F, 236F
  cell content, 197F
  codon-anticodon recognition, 235–241, 239F
  definition, 7B, 9, 197, 449G
  primer, 362F, 363
  role, 232, 233F
  structure, 233–234, 233F, 234F
  transcription, 178
  wobble, 238–240
tRNA nucleotidyltransferase, 221, 449G
True tree, 395, 449G
Truncated gene, 124–125, 125F, 449G
Trypsin, 377–378
Trypsinogen gene, 6, 6F
Tryptophan, 154F, 155T
Tryptophan operon, 180T, 187, 189F, 206, 250B
TTF-1, 219, 220F
β-turn, 165, 421G
Tus, 319, 321F, 449G
tudor domain, 95, 96F
Twintron, 212T, 225T, 450G
Two-dimensional polyacrylamide gel electrophoresis, 109
Two-step gene replacement, 102
Ty1/copia retroelement, 138
Ty1 mutagenesis, 99RB
Ty3/gypsy retroelement, 138, 138F, 141F
TΨC arm, 233, 233F, 234F, 448G
Type 0 cap, 208, 209F, 450G
Type 1 cap, 208, 209F, 450G
Type IA topoisomerase, 305, 306F, 306T
Type IB topoisomerase, 305, 306F, 306T
Type 2 cap, 208, 209F, 450G
Type 2 topoisomerase, 305, 306F, 306T, 356
Tyrosine, 154F, 155T
Tyrosine kinase associated receptor, 272T
Tyrosine kinase receptor, 272T

U

U1-snRNP, 215, 217F

U2AF, 215, 216, 217F
U2-snRNP, 215, 217F
U4atac/U6atac-snRNP, 217
U4/U6-snRNP, 215, 217F
U5-snRNP, 215, 217, 217F
U6-snRNP, 216, 217F
U11-snRNP, 217
U12-snRNP, 217
UBF, 184
Ubiquitin, 259–260, 450G
Ultracentrifugation, 123TN, 242
UmuD$'_2$C complex, 345
Unequal crossing-over, 375, 378F, 450G
Unequal sister chromatid exchange, 375, 378F, 450G
Universal primer, 67, 67F
Unrooted tree, 395, 395F, 450G
Upstream, 7B, 450G
Upstream control element, 179, 180, 450G
Upstream promoter element, 179–180, 180F, 450G
Uracil, 145, 450G
Urea, 251, 253F
Uridine 5′-triphosphate, 145
U-RNA, 215, 450G
UTP, 145
UvrABC endonuclease, 349–351, 351F, 450G
UV radiation, 339

V

Valine, 154F, 155T
van der Waals forces, 153B, 450G
Variable number of tandem repeats (VNTR), 18
Vegetative cell, 22, 450G
V loop, 233, 233F, 450G
VNTR, 18

W

Wave of advance model, 408, 411F, 450G
Weak promoter, 185, 450G
Werner's syndrome, 106, 106T
Wheat genome, 9T, 106
Wild type, 32, 450G
Wilm's tumor, 227T
Wilson's disease, 106T
Winged helix-turn-helix motif, 164T, 166, 450G
Wobble, 238–240, 241F, 450G

X

Xanthine, 339
Xeroderma pigmentosum, 348B
X inactivation, 177B, 280–281, 450G
Xist, 281
X-ray crystallography, 160, 161F, 257, 243, 450G
X-ray diffraction, 160, 161F, 450G
X-ray diffraction pattern, 160, 161F, 450G
XRCC4, 357B

Y

YAC, 55, 55F, 75, 76F, 451G
Yeast
  biochemical markers, 17T
  chromosome number, 117T
  comparative genomics, 106
  copper regulation, 268, 268F
  DNA replication control, 325–326
  genetic footprinting, 99RB
  genetic mapping, 28
  genetic *vs* physical map, 39F
  genome duplication, 375, 376RB
  genome organization, 8–9, 10F
  genome size, 8, 9T, 11T, 117T
  homology analysis, 95–96, 96F
  intein, 256
  internet address, 11T, 416T
  mating type switching, 278, 278F
  mitochondrial genome, 125, 126F, 127T, 239T
  orphans, 96, 96F
  phenotypes, 98T
  protein turnover, 259–260
  RNA turnover, 228–229
  sporulation genes, 108RB
  transcriptome, 107, 108RB
Yeast artificial chromosome (YAC), 55, 55F, 75, 76F, 451G
Yeast two-hybrid system, 103, 105F, 451G

Z

Z-DNA, 150, 150T, 151F, 451G
Zea mays genome, 9T
Zea mays mitochondrial genome, 127T
Zebrafish genome, 416T
Zinc binuclear cluster, 164T
Zinc finger, 164T, 166, 451G
Zoo-blotting, 90, 92F, 451G
Zygote, 356, 358F, 451G